SOME PHYSICAL PROPERTIES

Air (dry, at 20° C and 1 atm)
    Density   1.29 kg/m³
    Specific heat at constant pressure 1.00×10^3 J/kg·K
         0.240 cal/g·K
    Ratio of specific heats (γ)   1.40
    Speed of sound   343 m/s
         1130 ft/s

Water (20° C and 1 atm)
    Density 1.00×10^3 kg/m³
         1.00 g/cm³
    Speed of sound   1460 m/s
         4790 ft/s
    Index of refraction ($\lambda = 589$ nm)   1.33
    Specific heat at constant pressure   4180 J/kg·K
         1.00 cal/g·K
    Heat of fusion (0° C) 3.33×10^5 J/kg
         79.7 cal/g
    Heat of vaporization (100° C) $2.26 = 10^6$ J/kg
         539 cal/g

The Earth
    Mass 5.98×10^{24} kg
    Mean radius 6.37×10^6 m
         3960 mi
    Mean earth-sun distance 1.50×10^8 km
         9.30×10^7 mi
    Mean earth-moon distance 3.84×10^5 km
         2.39×10^5 mi
    Standard gravity   9.81 m/s²
         32.2 ft/s²
    Standard atmosphere 1.01×10^5 Pa
         14.7 lb/in²
         760 mm-Hg
         29.9 in-Hg

physics

PART
TWO
Extended Version

BOOKS BY HALLIDAY (D.) AND RESNICK (R.)

Physics, Parts I and II Combined, Third Edition

Physics, Part II, Third Edition

Fundamentals of Physics, 2nd Revised Printing

Fundamentals of Physics, 2nd Extended Revised Printing

BOOKS BY RESNICK (R.) AND HALLIDAY (D.)

Physics, Part I, Third Edition

Basic Concepts in Relativity and
Early Quantum Theory, 2nd
Available also in Paper Edition

BY R. RESNICK

Introduction to Special Relativity
Available in Paper Edition

BY ROBERT EISBERG AND ROBERT RESNICK

Quantum Physics of Atoms, Molecules,
Solids, Nuclei, and Particles, 2nd

physics

PART TWO
THIRD EDITION
EXTENDED VERSION

PROFESSOR OF PHYSICS **DAVID HALLIDAY**
UNIVERSITY OF PITTSBURGH

PROFESSOR OF PHYSICS **ROBERT RESNICK**
RENSSELAER POLYTECHNIC INSTITUTE

John Wiley & Sons,
New York   Chichester   Brisbane   Toronto   Singapore

SUPPLEMENTS

STUDENT STUDY GUIDE REVISED to accompany PHYSICS, 3RD and FUNDAMENTALS OF PHYSICS, 2ND and FUNDAMENTALS OF PHYSICS, 2ND EXTENDED

Stanley A. Williams, Iowa State University
Kenneth L. Brownstein, University of Maine
Robert L. Gray, University of Massachusetts

Developed for student use with this text and also FUNDAMENTALS OF PHYSICS, 2ND. Includes learning objectives carefully and thoroughly developed, plus exercises in problem solving for each chapter of the textbooks.

SELECTED SOLUTIONS FOR PHYSICS, 3RD REVISED

David Halliday, University of Pittsburgh
Robert Resnick, Rensselaer Polytechnic Institute
Edward Derringh, Wentworth Institute of Technology

Contains solutions to selected problems for student use with PHYSICS, 3RD.

PHYSICS PROBLEMS FOR PROGRAMMABLE CALCULATORS

J. Richard Christman, U.S. Coast Guard Academy
David Halliday, University of Pittsburgh
Robert Resnick, Rensselaer Polytechnic Institute

Designed for use with Halliday and Resnick PHYSICS, 3RD and FUNDAMENTALS OF PHYSICS, 2ND, as a supplement demonstrating to students the solution of physics problems using the calculator.

Coyright © 1960, 1962, 1966, 1978, 1986, by John Wiley & Sons, Inc.

All rights reserved. Published simultaneously in Canada.

Reproduction or translation of any part of this work beyond that permitted by Sections 107 and 108 of the 1976 United States Copyright Act without the permission of the copyright owner is unlawful. Requests for permission or further information should be addressed to the Permissions Department, John Wiley & Sons.

Library of Congress Cataloging in Publication Data:
Halliday, David, 1916–
Physics.

Includes index.
1. Physics.   I. Resnick, Robert, 1923–
II. Title.

QC21.2.H355   1986   530   86-5606
ISBN 0-471-83202-2 (v. 2)

Printed in the United States of America

10   9   8   7   6   5   4   3

preface to physics, part two, extended version

There has been increasing interest in including a significant treatment of modern physics in the introductory physics course. At many institutions the time required has been made available by trimming back on classical topics, such as fluids and thermodynamics. Usually a selection of modern material is made, to be treated not in a shallow encyclopaedic way but with a depth comparable to the treatment of the classical topics.

In response to this trend we have prepared an extended version of *Physics, Part II.* We have modified the last two chapters, Light and Quantum Physics, and The Wave Nature of Matter, and have added five new ones, namely, The Structure of Atoms, Atomic Physics: Three Selected Topics, Electrical Conduction in Solids, Nuclear Physics — An Introduction, and Energy from the Nucleus.

Users of our text *Fundamentals of Physics, Extended Version* will recognize the similarity in modern topics. As in the recent revised printing of that book, however, we have made a significant expansion in the problem sets to provide a better and more uniform coverage of the material and to tie them more closely to the text. We have also increased the number of worked examples, making the sets more interesting overall and improving their pedagogic impact. We have taken advantage of the four years of classroom experience with that material to make further improvements and to bring the material right up to date. For example, we have added new interesting applied topics, and we discuss more devices. We have put greater stress on the concepts unifying this material, namely, the wave theme and quantum ideas. Figures and illustrations have been improved and their number have been expanded. And, to better enable the experienced instrucvi

tor to select the material to cover and to skip, we have rearranged sections within chapters and have used a larger number of shorter sections.

We have been cognizant of the somewhat different audiences for the texts, leading to amplification of the chapter on electrical conductivity in solids and to a revised and somewhat more basic chapter on one-electron atoms. It is possible moreover to get a coherent treatment by following different paths through this material and even by omitting whole chapters where time requirements demand this or one's taste so dictates.

We are grateful to Dr. Philip Casabella (Rensselaer Polytechnic Institute) for his valuable and constructive comments on this material, to our editor Robert McConnin for his efficient management of the production process, and to Kay Guyette for her varied secretarial contributions.

We hope this extended version permits greater flexibility than heretofore in constructing the introductory physics course and adds to the students' interest in the subject.

Hanover, New Hampshire 03755   DAVID HALLIDAY
3 Clement Road

January 1986   ROBERT RESNICK
Troy, New York 12181   Department of Physics
Rensselaer Polytechnic Institute

preface to the third edition of part two

Physics is available in a single volume or in two separate parts; Part One includes mechanics, sound and heat, and Part Two includes electromagnetism, optics and quantum physics. The first edition was published in 1960 *(Physics for Students of Science and Engineering)* and the second in 1966 *(Physics).*

The text is intended for students studying calculus concurrently, such as students of science and engineering. The emphasis is on building a strong foundation in the principles of classical physics and on solving problems. Attention is given, however, to practical application, to the most modern theories, and to historical and philosophic issues throughout the book. This is accomplished by inclusion of special sections and thought questions, and by the entire manner of presentation of the material. There is a large set of worked-out examples, interspersed throughout the book, and an extensive collection of problems at the end of each chapter. Much care has been given to pedagogic devices that have proved effective for learning.

It has been eleven years since the publication of the second edition of *Physics.* During that time the book has continued to be well received throughout the world. We have had abundant correspondence with users over those years and concluded that a new edition is now appropriate.

In accordance with the increasing use of metric units in the United States and their general use throughout the world, we have greatly increased the emphasis on the metric system, using the Système Internationale (SI) units and nomenclature throughout. Where it seems to be sensible, in this transition period for the United States, we retain some features of the British Engineering system.

The entire book was carefully reviewed for pedagogic improvement, based chiefly on the experiences of users —, students and teachers,— and on the most recent scientific literature. As a result, we have rewritten selected areas significantly for improvements in presentation, accuracy, or physics. We have included new worked-out examples for topics or areas needing them. We have modernized all references, added new ones, and have improved many figures for greater clarity. The tables and the appendices have been expanded and updated to give newer data and more information than before. And we have added a supplementary topic on special relativity, in which the applications of this theory, scattered throughout Parts One and Two, are brought together as a cohesive whole.

Some subjects, not included in the second edition, are treated significantly in this third edition of Part Two. These include semiconductors, mutual inductance, earth magnetism, radio astronomy, virtual objects, and optical instruments. The long chapter on electromagnetic oscillations of the second edition has been divided into two chapters here, with extensive rewriting for greater clarity, and an entirely new chapter on alternating currents, for which there has been much demand, has been added.

As in Part One, we have made major improvements in the questions and the problems. In Parts One and Two combined, the number of questions has increased by 57%, from 778 in the second edition to 1219 in the present edition. For the problems the increase is 29%, from 1441 to 1864. Both problems and questions have been carefully edited, and most of the new ones have been classroom tested.

To assist students and teachers in organizing and evaluating the large number of problems, we have done several things. First, we have grouped problems within each chapter by section number, namely the first section needed to be covered in order to be able to work out the problem. Then, within each set of section problems, we have arranged the problems in the approximate order of increasing difficulty. Naturally, neither the assignment by section nor by difficulty is absolute, given different ways of solving some problems and different pedagogic values and tastes. We have coded the illustrations to the problems and have put the answers to the odd-numbered problems right at the end of these problems rather than at the end of the book. Finally, we have blended the supplementary problems, which appeared at the end of Part Two in the second edition, with the problems at the end of each chapter.

We have restyled the physical layout of the book to give it a less crowded appearance than formerly, making it easier now for the student to read the material, to make notations, and to differentiate between the various components of each chapter (text, figures, examples, tables, quotes, references, questions, problems, and so forth). We continue the practice of using somewhat reduced print for material which, in the context of a chapter, is of an advanced, specialized, or historical character.

We are grateful to John Wiley and Sons and to Donald Deneck, physics editor, for outstanding cooperation. We acknowledge the valuable assistance of Dr. Edward Derringh with the problem sets and of Mrs. Carolyn Clemente with the wide range of secretarial services required. We are indebted to the many teachers and students who have sent us constructive criticisms of the 1966 edition and particularly to Robert P. Bauman, Kenneth Brownstein, Robert Karplus, and Brian A. McInnes, who have advised or assisted us in many ways. We hope that this third edition of *Physics* will contribute to the improvement of physics education.

Hanover, New Hampshire 03755 DAVID HALLIDAY
             3 Clement Road

January 1978   ROBERT RESNICK
Troy, New York 12181   Department of Physics
             Rensselaer Polytechnic Institute

contents

26
charge and matter

The science of electricity has its roots in the observation, known to Thales of Miletus in 600 B.C., that a rubbed piece of amber will attract bits of straw. The study of magnetism goes back to the observation that naturally occurring "stones" (that is, magnetite) will attract iron. These two sciences developed quite separately until 1820, when Hans Christian Oersted (1777–1851) observed a connection between them, namely, that an *electric* current in a wire can affect a *magnetic* compass needle (Section 33-1).

The new science of electromagnetism was developed further by many workers, of whom one of the most important was Michael Faraday (1791–1867). It fell to James Clerk Maxwell (1831–1879) to put the laws of electromagnetism in essentially the form in which we know them today. These laws, called *Maxwell's equations,* are displayed in Table 40-2, which you may want to examine at this time. These laws play the same role in electromagnetism that Newton's laws of motion and of gravitation do in mechanics.

Although Maxwell's synthesis of electromagnetism rests heavily on the work of his predecessors, his own contribution was central and vital. Maxwell deduced that light is electromagnetic in nature and that its speed can be found by making purely electric and magnetic measurements. Thus the science of optics was intimately connected with those of electricity and of magnetism. The scope of Maxwell's equations is remarkable, including as it does the fundamental principles of all large-scale electromagnetic and optical devices such as motors, radio, television, microwave radar, microscopes, and telescopes.

The development of classical electromagnetism did not end with Maxwell. The English physicist Oliver Heaviside (1850–1925) and es-

26-1
ELECTROMAGNETISM— A PREVIEW

pecially the Dutch physicist H. A. Lorentz (1853–1928) contributed substantially to the clarification of Maxwell's theory. Heinrich Hertz (1857–1894)* took a great step forward when, more than twenty years after Maxwell set up his theory, he produced in the laboratory electromagnetic "Maxwellian waves" of a kind that we would now call short radio waves. It remained for Marconi and others to exploit the practical application of the electromagnetic waves of Maxwell and Hertz.

Present interest in electromagnetism takes two forms. At the level of engineering applications Maxwell's equations are used constantly and universally in the solution of a wide variety of practical problems. At the level of the foundations of the theory there is a continuing effort to extend its scope in such a way that electromagnetism is revealed as a special case of a more general theory. Such a theory would also include (say) the theories of gravitation and of quantum physics. This grand synthesis has not yet been achieved.

The rest of this chapter deals with electric charge and its relationship to matter. We can show that there are *two kinds* of charge by rubbing a glass rod with silk and hanging it from a long thread as in Fig. 26-1. If a second rod is rubbed with silk and held near the rubbed end of the first rod, the rods will repel each other. On the other hand, a rod of plastic (Lucite, say) rubbed with fur will *attract* the glass rod. Two plastic rods rubbed with fur will repel each other. We explain these facts by saying that rubbing a rod gives it an *electric charge* and that the charges on the two rods exert forces on each other. Clearly the charges on the glass and on the plastic must be different in nature.

Benjamin Franklin (1706–1790), who, among his other achievements, was the first American physicist‡, named the kind of electricity that appears on the glass *positive* and the kind that appears on the plastic (sealing wax or shell-lac in Franklin's day) *negative;* these names have remained to this day. We can sum up these experiments by saying that *like charges repel and unlike charges attract.*

Electric effects are not limited to glass rubbed with silk or to plastic rubbed with fur. Any substance rubbed with any other under suitable conditions will become charged to some extent; by comparing the unknown charge with a glass rod which had been rubbed with silk or a plastic rod which had been rubbed with fur, it can be labeled as either positive or negative.

The modern view of bulk matter is that, in its normal or neutral state, it contains equal amounts of positive and negative electricity. If two bodies like glass and silk are rubbed together, a small amount of charge is transferred from one to the other, upsetting the electric neutrality of each. In this case the glass would become positive, the silk negative.

26-2
ELECTRIC CHARGE†

Silk thread

F

−F

figure 26-1
Two positively charged glass rods repel each other.

* "Heinrich Hertz," by P. and E. Morrison, *Scientific American*, December 1957.

† To learn about practical applications of static electric charges, as in fly-ash precipitators, paint sprayers, electrostatic copying machines, etc., see "Modern Electrostatics" by A. W. Bright, *Physics Education*, **9**, 381 (1974), and "Electrostatics" by A. D. Moore, *Scientific American*, March 1972.

‡ The science historian I. Bernard Cohen of Harvard University says of Franklin in his book *Franklin and Newton:* "To say . . . that had Franklin 'Not been famous as a publisher and a statesman, he might never have been heard of as a scientist,' is absolutely wrong. Just the opposite is more nearly the case; his international fame and public reknown as a scientist was in no small measure responsible for his success in international statesmanship." See also "The Lightning Discharge" by Richard E. Orville, *The Physics Teacher*, January 1976 for a description of Franklin's famous kite experiment and a review of modern concepts about the nature of lightning.

A metal rod held in the hand and rubbed with fur will not seem to develop a charge. It is possible to charge such a rod, however, if it is furnished with a glass or plastic handle and if the metal is not touched with the hands while rubbing it. The explanation is that metals, the human body, and the earth are *conductors* of electricity and that glass, plastics, etc., are *insulators* (also called *dielectrics*).

In conductors electric charges are free to move through the material, whereas in insulators they are not. Although there are no perfect insulators, the insulating ability of fused quartz is about 10^{25} times as great as that of copper, so that for many practical purposes some materials behave as if they were perfect insulators.

In metals a fairly subtle experiment called the Hall effect (see Section 33-5) shows that only negative charge is free to move. Positive charge is as immobile as it is in glass or in any other dielectric. The actual charge carriers in metals are the *free electrons*. When isolated atoms are combined to form a metallic solid, the outer electrons of the atom do not remain attached to individual atoms but become free to move throughout the volume of the solid. For some conductors, such as electrolytes, both positive and negative charges can move.

A class of materials called *semiconductors* is intermediate between conductors and insulators in its ability to conduct electricity. Among the elements, silicon and germanium are well-known examples. Semiconductors have many practical applications, among which is their use in the construction of transistors. The way a semiconductor works cannot be described adequately without some understanding of the basic principles of quantum physics. Figure 26-2, however, suggests the principal features of the distinction between conductors, semiconductors, and insulators.

In solids, electrons have energies that are restricted to certain levels, the levels being confined to certain bands. The intervals between bands are forbidden, in the sense that electrons in the solid may not possess such energies. Electrons are assigned two to a level and they may not increase their energy (which means that they may not move freely through the solid) unless there are empty levels at higher energies into which they can readily move.

Figure 26-2a shows a conductor, such as copper. Band 1 is only partially filled so that electrons can easily move to the higher empty levels and thus travel through the solid. Figure 26-2b shows a (intrinsic) semiconductor such as silicon. Here band 1 is completely filled but band 2 is so close energetically that electrons can easily "jump" (absorbing energy from, say, thermal fluctuations) into the unfilled levels of that band. Figure 22-6c shows an insulator, such as sodium chloride. Here again band 1 is filled, but band 2 is too far above band 1 energetically to permit any appreciable number of the band-1 electrons to jump the energy gap.

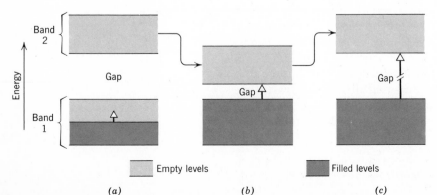

figure 26-2
Suggesting (*a*) a conductor, (*b*) an intrinsic semiconductor, and (*c*) an insulator. In (*b*) the gap is relatively small but in (*c*) it is relatively large. In intrinsic semiconductors the electrical conductivity can often be greatly increased by adding very small amounts of other elements such as arsenic or boron, a process called "doping."

Charles Augustin Coulomb (1736–1806) measured electrical attractions and repulsions quantitatively and deduced the law that governs them. His apparatus, shown in Fig. 26-3, resembles the hanging rod of Fig. 26-1, except that the charges in Fig. 26-3 are confined to small spheres a and b.

If a and b are charged, the electric force on a will tend to twist the suspension fiber. Coulomb canceled out this twisting effect by turning the suspension head through the angle θ needed to keep the two charges at the particular distance apart in which he was interested. The angle θ is then a relative measure of the electric force acting on charge a. The device of Fig. 26-3 is called a *torsion balance*; a similar arrangement was used later by Cavendish to measure gravitational attractions (Section 16-3).

Coulomb's first experimental results can be represented by

$$F \propto \frac{1}{r^2}.$$

Here F is the magnitude of the interaction force that acts on each of the two charges a and b; r is their distance apart. These forces, as Newton's third law requires, act along the line joining the charges but point in opposite directions. Note that the magnitude of the force on each charge is the same, even though the charges may be different.

The force between charges depends also on the magnitude of the charges. Specifically, it is proportional to their product. Although Coulomb did not prove this rigorously, he implied it and thus we arrive at

$$F \propto \frac{q_1 q_2}{r^2}, \tag{26-1}$$

where q_1 and q_2 are relative measures of the charges on spheres a and b. Equation 26-1, which is called *Coulomb's law*, holds only for charged objects whose sizes are much smaller than the distance between them. We often say that it holds only for *point charges*.

Coulomb's law resembles the inverse square law of gravitation which was already more than 100 years old at the time of Coulomb's experiments; q plays the role of m in that law. In gravity, however, the forces are always attractive; this corresponds to the fact that there are two kinds of electricity but (apparently) only one kind of mass.

Our belief in Coulomb's law does not rest quantitatively on Coulomb's experiments. Torsion balance measurements are difficult to make to an accuracy of better than a few percent. Such measurements could not, for example, convince us that the exponent in Eq. 26-1 is exactly 2 and not, say, 2.01. In Section 28-7 we show that Coulomb's law can also be deduced from an indirect experiment (1971) which shows that the exponent in Eq. 26-1 lies between the limits $2 \pm 3 \times 10^{-16}$.

Although we have established the physical concept of electric charge, we have not yet defined a unit in which it may be measured. It is possible to do so operationally by putting equal charges q on the spheres of a torsion balance and by measuring the magnitude F of the force that acts on each when the charges are a measured distance r apart. One could then define q to have a unit value if a unit force acts on each charge when the charges are separated by a unit distance and one can give a name to the unit of charge so defined.*

*This scheme is the basis for the definition of the unit of charge called the *statcoulomb*. However, in this book we do not use this unit or the systems of units of which it is a part; see Appendix L, however.

26-4
COULOMB'S LAW

Suspension head

Fiber

figure 26-3
Coulomb's torsion balance, from his 1785 memoir to the French Academy of Sciences.

For practical reasons having to do with the accuracy of measurements, the SI unit of charge is not defined using a torsion balance but is derived from the unit of electric current. If the ends of a long wire are connected to the terminals of a battery, it is common knowledge that an *electric current i* is set up in the wire. We visualize this current as a flow of charge. The SI unit of current is the *ampere* (abbr. A). In Section 34-4 we describe the operational procedures in terms of which the ampere is defined.

The SI unit of charge is the *coulomb* (abbr. C). *A coulomb is defined as the amount of charge that flows through any cross section of a wire in 1 second if there is a steady current of 1 ampere in the wire.* In symbols

$$q = it, \tag{26-2}$$

where q is in coulombs if i is in amperes and t is in seconds. Thus, if a wire is connected to an insulated metal sphere, a charge of 10^{-6} C can be put on the sphere if a current of 1.0 A exists in the wire for 10^{-6} s.

A copper penny has a mass of 3.1 g. Being electrically neutral, it contains equal amounts of positive and negative electricity. What is the magnitude q of these charges? A copper atom has a positive nuclear charge of 4.6×10^{-18} C and a negative electronic charge of equal magnitude.

The number N of copper atoms in a penny is found from the ratio

$$\frac{N}{N_0} = \frac{m}{M},$$

where N_0 is the Avogadro number, m the mass of the coin, and M the atomic weight of copper. This yields, solving for N,

$$N = \frac{(6.0 \times 10^{23} \text{ atoms/mole})(3.1 \text{ g})}{64 \text{ g/mole}} = 2.9 \times 10^{22} \text{ atoms}.$$

The charge q is

$$q = (4.6 \times 10^{-18} \text{ C/atom})(2.9 \times 10^{22} \text{ atoms}) = 1.3 \times 10^5 \text{ C}.$$

In a 100-watt, 110-volt light bulb the current is 0.91 ampere. Verify that it would take 40 h for a charge of this amount to pass through this bulb.

EXAMPLE 1

Equation 26-1 can be written as an equality by inserting a constant of proportionality. Instead of writing this simply as, say, k, it is usually written in a more complex way as $1/4\pi\epsilon_0$ or

$$F = \frac{1}{4\pi\epsilon_0} \frac{q_1 q_2}{r^2}. \tag{26-3}$$

Certain equations that are derived from Eq. 26-3, but are used more often than it is, will be simpler in form if we do this.

In SI units we can measure q_1, q_2, r, and F in Eq. 26-3 in ways that do not depend on Coulomb's law. Numbers with units can be assigned to them. There is no choice about the so-called *permittivity constant* ϵ_0; it must have that value which makes the right-hand side of Eq. 26-3 equal to the left-hand side. This (measured) value turns out to be*

$$\epsilon_0 = 8.854187818 \times 10^{-12} \text{ C}^2/\text{N} \cdot \text{m}^2.$$

---

* For practical reasons this value is not actually measured by direct application of Eq. 26-3 but in an equivalent although more circuitous way.

In this book the value 8.9×10^{-12} C²/N · m² will be accurate enough for most problems. For direct application of Coulomb's law or in any problem in which the quantity $1/4\pi\epsilon_0$ occurs we may use, with sufficient accuracy for this book,

$$1/4\pi\epsilon_0 = 9.0 \times 10^9 \text{ N·m}^2/\text{C}^2.$$

EXAMPLE 2

Let the total positive and the total negative charges in a copper penny be separated to a distance such that their force of attraction is 1.0 lb (= 4.5 N). How far apart must they be?

We have (Eq. 26-3)

$$F = \frac{1}{4\pi\epsilon_0} \frac{q_1 q_2}{r^2}.$$

Putting $q_1 q_2 = q^2$ (see Example 1) and solving for r yields

$$r = q \sqrt{\frac{1/4\pi\epsilon_0}{F}} = 1.3 \times 10^5 \text{ C} \sqrt{\frac{9.0 \times 10^9 \text{ N · m}^2/\text{C}^2}{4.5 \text{ N}}}$$

$$= 5.8 \times 10^9 \text{ m} = 3.6 \times 10^6 \text{ mi.}$$

This is 910 earth radii and it suggests that it is not possible to upset the electrical neutrality of gross objects by any very large amount. What would be the force between the two charges if they were placed 1.0 m apart?

If more than two charges are present, Eq. 26-3 holds for every pair of charges. Let the charges be q_1, q_2, q_3, etc.; we calculate the force exerted on any one (say q_1) by all the others from the vector equation

$$\mathbf{F}_1 = \mathbf{F}_{12} + \mathbf{F}_{13} + \mathbf{F}_{14} + \cdots, \tag{26-4}$$

where $\mathbf{F}_{12}$, for example, is the force exerted on q_1 by q_2.

EXAMPLE 3

Figure 26-4 shows three fixed charges q_1, q_2, and q_3. What force acts on q_1? Assume that $q_1 = -1.0 \times 10^{-6}$ C, $q_2 = +3.0 \times 10^{-6}$ C, $q_3 = -2.0 \times 10^{-6}$ C, $r_{12} = 15$ cm, $r_{13} = 10$ cm, and $\theta = 30°$.

From Eq. 26-3, ignoring the signs of the charges, since we are interested only in the magnitudes of the forces,

$$F_{12} = \frac{1}{4\pi\epsilon_0} \frac{q_1 q_2}{r_{12}^2}$$

$$= \frac{(9.0 \times 10^9 \text{ N · m}^2/\text{C}^2)(1.0 \times 10^{-6} \text{ C})(3.0 \times 10^{-6} \text{ C})}{(1.5 \times 10^{-1} \text{ m})^2}$$

$$= 1.2 \text{ N}$$

and

$$F_{13} = \frac{(9.0 \times 10^9 \text{ N · m}^2/\text{C}^2)(1.0 \times 10^{-6} \text{ C})(2.0 \times 10^{-6} \text{ C})}{(1.0 \times 10^{-1} \text{ m})^2}$$

$$= 1.8 \text{ N}.$$

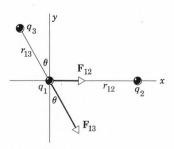

figure 26-4
Example 3. Showing the forces exerted on q_1 by q_2 and q_3.

The directions of $\mathbf{F}_{12}$ and $\mathbf{F}_{13}$ are as shown in the figure.

The components of the resultant force $\mathbf{F}_1$ acting on q_1 (see Eq. 26-4) are

$$F_{1x} = F_{12x} + F_{13x} = F_{12} + F_{13} \sin \theta$$

$$= 1.2 \text{ N} + (1.8 \text{ N})(\sin 30°) = 2.1 \text{ N}$$

and

$$F_{1y} = F_{12y} + F_{13y} = 0 - F_{13} \cos \theta$$

$$= -(1.8 \text{ N})(\cos 30°) = -1.6 \text{ N}.$$

Find the magnitude of $\mathbf{F}_1$ and the angle it makes with the x-axis.

In Franklin's day electric charge was thought of as a continuous fluid, an idea that was useful for many purposes. The atomic theory of matter, however, has shown that fluids themselves, such as water and air, are not continuous but are made up of atoms. Experiment shows that the "electric fluid" is not continuous either but that it is made up of integral multiples of a certain minimum electric charge. This fundamental charge, to which we give the symbol e, has the magnitude 1.6021892 $\times 10^{-19}$C. Any physically existing charge q, no matter what its origin, can be written as ne where n is a positive or a negative integer.

When a physical property such as charge exists in discrete "packets" rather than in continuous amounts, the property is said to be *quantized*. Quantization is basic to modern quantum physics. The existence of atoms and of particles such as the electron and the proton indicates that *mass* is quantized also. Later you will learn that several other properties prove to be quantized when suitably examined on the atomic scale; among them are energy and angular momentum.

The quantum of charge e is so small that the "graininess" of electricity does not show up in large-scale experiments, just as we do not realize that the air we breathe is made up of atoms. In an ordinary 110-volt, 100-watt light bulb, for example, 6×10^{18} elementary charges enter and leave the bulb every second.

There exists today no theory that predicts the quantization of charge (or the quantization of mass, that is, the existence of fundamental particles such as protons, electrons, muons, pions, etc. Even assuming quantization, the classical theory of electromagnetism and Newtonian mechanics are incomplete in that they do not correctly describe the behavior of charge and matter on the atomic scale. The classical theory of electromagnetism, for example, describes correctly what happens when a bar magnet is thrust through a closed copper loop; it fails, however, if we wish to explain the magnetism of the bar in terms of the atoms that make it up. The more detailed theories of quantum physics are needed for this and similar problems.

Matter as we ordinarily experience it can be regarded as composed of three kinds of particles, the proton, the neutron, and the electron. Table 26-1 shows their masses and charges. Note that the masses of the neutron and the proton are approximately equal but that the electron is less massive by a factor of about 1840.

26-5
CHARGE IS QUANTIZED

26-6
CHARGE AND MATTER

Table 26-1
Some properties of three particles

Particle	Symbol	Charge	Mass
Proton	p	$+e$	$1.6726485 \times 10^{-27}$ kg
Neutron	n	0	$1.6749543 \times 10^{-27}$ kg
Electron	e^-	$-e$	9.109534×10^{-31} kg

Atoms are made up of a dense, positively charged *nucleus*, surrounded by a cloud of electrons; see Fig. 26-5. The nucleus varies in radius from about 1×10^{-15} m for hydrogen to about 7×10^{-15} m for the heaviest atoms. The outer diameter of the electron cloud, that is, the diameter of the atom, lies in the range $1-3 \times 10^{-10}$ m, about 10^5 times larger than the nuclear diameter.

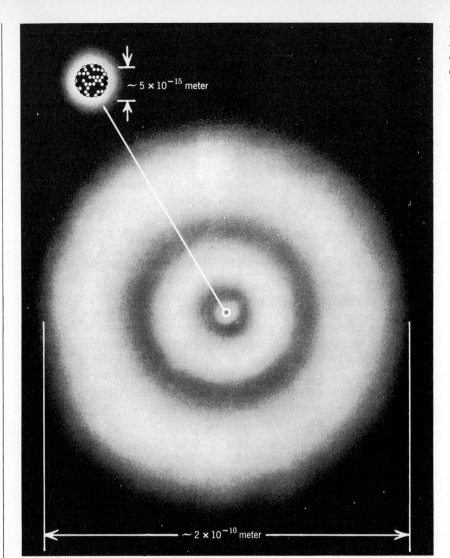

figure 26-5
An atom, suggesting the electron cloud and, above, an enlarged view of the nucleus.

EXAMPLE 4

The distance r between the electron and the proton in the hydrogen atom is about 5.3×10^{-11} m. What are the magnitudes of *(a)* the electrical force and *(b)* the gravitational force between these two particles?

From Coulomb's law,

$$F_e = \frac{1}{4\pi\epsilon_0} \frac{q_1 q_2}{r^2}$$

$$= \frac{(9.0 \times 10^9 \text{ N} \cdot \text{m}^2/\text{C}^2)(1.6 \times 10^{-19} \text{ C})^2}{(5.3 \times 10^{-11} \text{ m})^2}$$

$$= 8.1 \times 10^{-8} \text{ N}.$$

The gravitational force is given by Eq. 16-1, or

$$F_g = G \frac{m_1 m_2}{r^2}$$

$$= \frac{(6.7 \times 10^{-11} \text{ N} \cdot \text{m}^2/\text{kg}^2)(9.1 \times 10^{-31} \text{ kg})(1.7 \times 10^{-27} \text{ kg})}{(5.3 \times 10^{-11} \text{ m})^2}$$

$$= 3.7 \times 10^{-47} \text{ N}.$$

Thus the electrical force is about 10^{39} times stronger than the gravitational force.

The significance of Coulomb's law goes far beyond the description of the forces acting between charged spheres. This law, when incorporated into the structure of quantum physics, correctly describes (a) the electric forces that bind the electrons of an atom to its nucleus, (b) the forces that bind atoms together to form molecules, and (c) the forces that bind atoms or molecules together to form solids or liquids. Thus most of the forces of our daily experience that are not gravitational in nature are electrical. A force transmitted by a steel cable is basically an electrical force because, if we pass an imaginary plane through the cable at right angles to it, it is only the attractive electrical interatomic forces acting between atoms on opposite sides of the plane that keep the cable from parting. We ourselves are an assembly of nuclei and electrons bound together in a stable configuration by Coulomb forces.

In the atomic *nucleus* we encounter a new force which is neither gravitational nor electrical in nature. This strong attractive force, which binds together the protons and neutrons that make up the nucleus, is called simply *the nuclear force* or *the strong interaction*. If this force were not present, the nucleus would fly apart at once because of the strong Coulomb repulsion force that acts between its protons. The nature of the nuclear force is only partially understood today and forms the central problem of present day researches in nuclear physics.

EXAMPLE 5

What repulsive Coulomb force exists between two protons in a nucleus of iron? Assume a separation of 4.0×10^{-15} m.

From Coulomb's law,

$$F = \frac{1}{4\pi\epsilon_0} \frac{q_1 q_2}{r^2}$$

$$= \frac{(9.0 \times 10^9 \text{ N} \cdot \text{m}^2/\text{C}^2)(1.6 \times 10^{-19} \text{ C})^2}{(4.0 \times 10^{-15} \text{ m})^2}$$

$$= 14 \text{ N}.$$

This enormous repulsive force (3.2 lb acting on a single proton) must be more than compensated for by the strong attractive nuclear forces. This example, combined with Example 4, shows that nuclear binding forces are much stronger than atomic binding forces. Atomic binding forces are, in turn, much stronger than gravitational forces for the same particles separated by the same distance.

The repulsive Coulomb forces acting between the protons in a nucleus make the nucleus less stable than it otherwise would be. The spontaneous emission of alpha particles from heavy nuclei and the phenomenon of nuclear fission are evidences of this instability.

The fact that heavy nuclei contain significantly more neutrons than protons is still another effect of the Coulomb forces. Consider Fig. 26-6 in which a particular atomic species is represented by a circle, the coordinates being Z, the number of protons in the nucleus (that is, the *atomic number*), and N, the number of neutrons in the nucleus (that is, the *neutron number*). Stable nuclei are represented by filled circles and radioactive nuclei, that is, nuclei which disintegrate spontaneously, emitting electrons or α-particles, by open circles. Note that all elements (xenon, for example, for which $Z = 54$; see arrow) exist in a number of different forms, called *isotopes*.

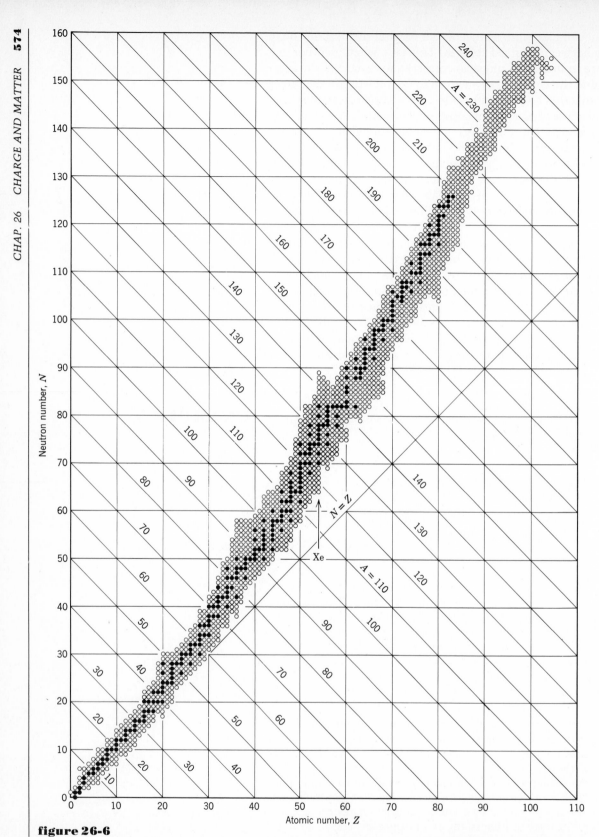

figure 26-6

The filled circles represent stable nuclei and the open ones represent radioactive nuclei. Note, for example, that xenon (Z = 54) has 26 isotopes, 9 of them stable and 17 radioactive. Each xenon isotope has 54 protons (and 54 extranuclear electrons for neutral atoms). The number of neutrons ranges from N = 64 to N = 89 and the mass number A (= N + Z) ranges from 118 to 143. No other element has as many isotopes.

Figure 26-6 shows that light nuclei, for which the Coulomb forces are relatively unimportant,* lie on or close to the line labeled "$N = Z$" and thus have about equal numbers of neutrons and protons. The heavier nuclei have a pronounced neutron excess, U^{238} having 92 protons and $238 - 92$ or 146 neutrons.† In the absence of Coulomb forces we would assume, extending the $N = Z$ rule, that the most stable nucleus with 238 particles would have 119 protons and 119 neutrons. However, such a nucleus, if assembled, would fly apart at once because of Coulomb repulsion. Relative stability is found only if 27 of the protons are replaced by neutrons, thus diluting the total Coulomb repulsion. Even in U^{238} Coulomb repulsion is still very important because (a) this nucleus is radioactive and emits α-particles and (b) it may break up into two large fragments (fission) if it absorbs a neutron; both processes result in separation of the nuclear charge and are Coulomb repulsion effects. Figure 26-6 shows that all nuclei with $Z \gtrsim 83$ are unstable.

We have pointed out that matter, as we ordinarily experience it, is made up of electrons, neutrons, and protons. Nature exhibits much more variety than this, however. There are very many more particles than these. Appendix F, which lists some properties of some of these particles, shows that, like the more familiar particles of Table 26-1, their charges are quantized, the quantum of charge again being e. An understanding of the nature of these particles and of their relationships to each other is one of the most significant research goals of modern physics.

26-7
CHARGE IS CONSERVED

When a glass rod is rubbed with silk, a positive charge appears on the rod. Measurement shows that a negative charge of equal magnitude appears on the silk. This suggests that rubbing does not create charge but merely transfers it from one object to another, disturbing slightly the electrical neutrality of each. This hypothesis of the *conservation of charge* has stood up under close experimental scrutiny both for large-scale events and at the atomic and nuclear level; no exceptions have ever been found.

An interesting example of charge conservation comes about when an electron (charge $= -e$) and a positron (charge $= +e$) are brought close to each other. The two particles may simply disappear, converting all their rest mass into energy according to the well-known $E = mc^2$ relationship; this annihilation process was described in Section 8-9. The energy appears in the form of two oppositely directed gamma rays, which are similar in character to X-rays; thus:

$$e^- + e^+ \rightarrow \gamma + \gamma. \qquad (26\text{-}5)$$

The net charge is zero both before and after the event so that charge is conserved. Rest mass is *not* conserved, being turned completely into energy.

Another example of charge conservation is found in radioactive decay, of which the following process is typical:

$$U^{238} \rightarrow Th^{234} + He^4. \qquad (26\text{-}6)$$

The radioactive "parent" nucleus, U^{238}, contains 92 protons (that is, its

---

* Coulomb forces are important in relation to the strong nuclear attractive forces only for large nuclei, because Coulomb repulsion occurs between *every pair* of protons in the nucleus but the attractive nuclear force does not. In U^{238}, for example, every proton exerts a force of repulsion on each of the other 91 protons. However, each proton (and neutron) exerts a nuclear attraction on only a small number of other neutrons and protons that happen to be near it. As we proceed to larger nuclei, the amount of energy associated with the repulsive Coulomb forces increases much faster than that associated with the attractive nuclear forces.

† The superscript in this notation is the *mass number A* $(= N + Z)$. This is the total number of particles in the nucleus. See the sloping lines in Figure 26-6.

atomic number $Z = 92$). It disintegrates spontaneously by emitting an α-particle (He^4; $Z = 2$) transmuting itself into the nucleus Th^{234}, with $Z = 90$. Thus the amount of charge present before disintegration $(+92e)$ is the same as that present after the disintegration.

An additional example of charge conservation is found in nuclear reactions, of which the bombardment of Ca^{44} with cyclotron-accelerated protons is typical. In a particular collision a neutron may emerge from the nucleus, leaving Sc^{44} as a "residual" nucleus:

$$Ca^{44} + p \rightarrow Sc^{44} + n.$$

The sum of the atomic numbers before the reaction $(20 + 1)$ is exactly equal to the sum of the atomic numbers after the reaction $(21 + 0)$. Again charge is conserved.

A final example of charge conservation is the decay of the K-meson (see Appendix F) which, in one mode, goes as

$$K^0 \rightarrow \pi^+ + \pi^-.$$

The resultant charge is zero both before and after this decay process.

questions

1. You are given two metal spheres mounted on portable insulating supports. Find a way to give them equal and opposite charges. You may use a glass rod rubbed with silk but may not touch it to the spheres. Do the spheres have to be of equal size for your method to work?

2. In Question 1, find a way to give the spheres equal charges of the same sign. Again, do the spheres need to be of equal size for your method to work?

3. A charge rod attracts bits of dry cork dust which, after touching the rod, often jump violently away from it. Explain.

4. In Section 26-2 can there not be four kinds of charge, that is, on glass, silk, plastic, and fur? What is the argument against this?

5. If you rub a coin briskly between your fingers, it will not seem to become charged by friction. Why?

6. If you walk briskly down the carpeted corridor of a hotel, you often experience a spark upon touching a door knob. (a) What causes this? (b) How might it be prevented?

7. Why do electrostatic experiments not work well on humid days?

8. An insulated rod is said to carry an electric charge. How could you verify this and determine the sign of the charge?

9. If a charged glass rod is held near one end of an insulated uncharged metal rod as in Fig. 26-7, electrons are drawn to one end, as shown. Why does the flow of electrons cease? There is an almost inexhaustible supply of them in the metal rod.

10. In Fig. 26-7 does any net electrical force act on the metal rod? Explain.

11. A person standing on an insulated stool touches a charged, insulated conductor. Is the conductor discharged completely?

12. (a) A positively charged glass rod attracts a suspended object. Can we conclude that the object is negatively charged? (b) A positively charged glass rod repels a suspended object. Can we conclude that the object is positively charged?

13. Is the Coulomb force that one charge exerts on another changed if other charges are brought nearby?

14. You are given a collection of small charged spheres, the sign and magnitude of the charge and the mass of the sphere being at your disposal. Do you think

Metal

Glass rod

Insulating support

figure 26-7
Questions 9, 10

that a *stable* equilibrium position is possible, involving only electrostatic forces? Test several arrangements. A rigorous answer is not required.

15. Suppose that someone told you that in Eq. 26-3 the product of the charges $(q_1 q_2)$ should be replaced by their algebraic sum $(q_1 + q_2)$. What experimental facts refute this statement? What if the square root of their product $\sqrt{q_1 q_2}$ were proposed?

16. The quantum of charge is 1.60×10^{-19} C. Is there a corresponding single quantum of mass?

17. A nucleus U^{238} splits into two identical parts. Are the nuclei so produced likely to be stable or radioactive?

18. In the decay mode

$$\Xi^0 \rightarrow \Lambda + \pi^0$$

what is the charge of the Λ particle? See Appendix F.

19. Verify the fact that the decay schemes for the elementary particles in Appendix F are consistent with charge conservation.

20. What does it mean to say that a physical quantity is (a) quantized or (b) conserved? Give some examples.

problems

SECTION 26-4

1. The electrostatic force between two like ions that are separated by a distance of 5.0×10^{-10} m is 3.7×10^{-9} N. (a) What is the charge on each ion? (b) How many electrons are missing from each ion?
   Answer: (a) 3.2×10^{-19} C (b) Two.

2. Two fixed charges, $+1.0 \times 10^{-6}$ C and -3.0×10^{-6}, are 10 cm apart. (a) Where may a third charge be located so that no force acts on it? (b) Is the equilibrium of this third charge stable or unstable?

3. Each of two small spheres is charged positively, the combined charge being 5.0×10^{-5} C. If each sphere is repelled from the other by a force of 1.0 N when the spheres are 2.0 m apart, how is the total charge distributed between the spheres? *Answer:* 1.2×10^{-5} C and 3.8×10^{-5} C.

4. Two equal positive point charges are held a fixed distance $2a$ apart. A point test charge is located in a plane which is normal to the line joining these charges and midway between them. (a) Calculate the radius r of the circle of symmetry in this plane for which the force on the test charge has a maximum value. (b) What is the direction of this force, assuming a positive test charge.

5. A certain charge Q is to be divided into two parts, q and $Q - q$. What is the relationship of Q to q if the two parts, placed a given distance apart, are to have a maximum Coulomb repulsion? *Answer:* $q = \frac{1}{2}Q$.

6. How far apart must two protons be if the electrical repulsive force acting on either one is equal to its weight at the earth's surface. The mass of a proton is 1.7×10^{-27} kg.

7. Two equal positive charges, Q, are fixed at a distance $2a$ apart. The force on a small positive test charge q midway between them is zero. If the test charge is displaced a short distance either (a) toward one of the charges or (b) at right angles to the line joining the charges, find the direction of the force on it. Is the equilibrium stable or unstable in each case?
   Answer: (a) Toward the original position; stable (b) Away from the original position; unstable.

8. Two *free* point charges $+q$ and $+4q$ are a distance l apart. A third charge is so placed that the entire system is in equilibrium. Find the location, magnitude, and sign of the third charge. Is the equilibrium stable?

9. Two similar conducting balls of mass m are hung from silk threads of length l and carry similar charges q as in Fig. 26-8. Assume that θ is so small that $\tan \theta$ can be replaced by its approximate equal, $\sin \theta$. To this approximation (a) show that

$$x = \left(\frac{q^2 l}{2\pi\epsilon_0 mg}\right)^{1/3}$$

where x is the separation between the balls. (b) If $l = 120$ cm, $m = 10$ g, and $x = 5.0$ cm, what is q?   *Answer:* (b) $\pm 2.4 \times 10^{-8}$ C.

10. Assume that each ball in Problem 9 is losing charge at the rate of 1.0×10^{-9} C/s. At what instantaneous relative speed ($= dx/dt$) do the balls approach each other initially?

11. If the balls of Fig. 26-8 are conducting, (a) what happens to them after one is discharged? (b) Find the new equilibrium separation.
Answer: (a) They touch and repel. (b) 3.1 cm.

figure 26-8
Problems 9, 10, 11

12. The charges and coordinates of two charged particles held fixed in the x-y plane are: $q_1 = +3.0 \times 10^{-6}$ C; $x = 3.5$ cm, $y = 0.50$ cm, and $q_2 = -4.0 \times 10^{-6}$ C; $x = -2.0$ cm, $y = 1.5$ cm. (a) Find the magnitude and direction of the force on q_2. (b) Where you locate a third charge $q_3 = +4.0 \times 10^{-6}$ C such that the total force on q_2 is zero?

13. Two identical conducting spheres, having charges of opposite sign, attract each other with a force of 0.108 N when separated by 0.500 m. The spheres are connected by a conducting wire, which is then removed, and thereafter repel each other with a force of 0.0360 N. What were the initial charges on the spheres?
Answer: $\pm 1.0 \times 10^{-6}$ C; $\mp 3.0 \times 10^{-6}$ C.

14. Two engineering students (John at 200 lb and Mary at 100 lb) are 100 ft apart. Let each have an 0.01% imbalance in their amount of positive and negative charge, one student being positive and the other negative. Estimate roughly the electrostatic force of attraction between them. (Hint: Replace the students by equivalent spheres of water.)

15. Two equally charged particles, 3.2×10^{-3} m apart, are released from rest. The acceleration of the first particle is observed to be 7.0 m/s² and that of the second to be 9.0 m/s². If the mass of the first particle is 6.3×10^{-7} kg, what are (a) the mass of the second particle and (b) the common charge? *Answer:* (a) 4.9×10^{-7} kg. (b) 7.1×10^{-11} C.

16. (a) How many electrons would have to be removed from a penny to leave it with a charge of $+1.0 \times 10^{-7}$ C (b) What fraction of the electrons in the penny does this correspond to?

17. (a) What equal positive charges would have to be placed on the earth and on the moon to neutralize their gravitational attraction? (b) Do you need to know the lunar distance to solve this problem? (c) How many tons of hydrogen would be needed to provide the positive charge calculated in a? *Answer:* (a) 5.7×10^{13} C. (b) No. (c) 630 tons.

18. Estimate *roughly* the number of coulombs of positive charge in a glass of water. Assume the volume of the water to be 250 cm³.

19. Protons in the cosmic rays strike the earth's upper atmosphere at a rate, averaged over the earth's surface, of 0.15 protons/cm²·s. What total current does the earth receive from beyond its atmosphere in the form of incident cosmic ray protons? The earth's radius is 6.4×10^6 m.   *Answer:* 0.12 A.

20. Three charged particles lie on a straight line and are separated by a distance d as shown in Fig. 26-9. Charges q_1 and q_2 are held fixed. If q_3 is free to move but in fact remains stationary, then how are q_1 and q_2 related?

figure 26-9
Problem 20

21. Three small balls, each of mass 10 g, are suspended separately from a common point by silk threads, each 1.0 m. long. The balls are identically charged and hang at the corners of an equilateral triangle 0.1 m long on a side. What is the charge on each ball?   *Answer:* 6.0×10^{-8} C.

22. Three point charges of $+4.0 \times 10^{-6}$ C are fixed at the corners of an equilateral triangle of side 10 cm. What force (magnitude and direction) acts on a typical charge?

23. A charge Q is fixed at each of two opposite corners of a square. A charge q is placed at each of the other two corners. (a) If the resultant electrical force on Q is zero, how are Q and q related? (b) Could q be chosen to make the resultant force on *every* charge zero?
    Answer: (a) $Q = -2\sqrt{2}\, q$. (b) No.

24. In Fig. 26-10 what is the resultant force on the charge in the lower left corner of the square? Assume that $q = 1.0 \times 10^{-7}$ C and $a = 5.0$ cm. The changes are fixed in position.

25. A cube of edge a carries a point charge q at each corner. (a) Show that the magnitude of the resultant force on any one of the charges is

$$F = \frac{0.262 q^2}{\epsilon_0 a^2}.$$

(b) What is the direction of **F** relative to the cube edges?
    Answer: (b) Along a body diagonal, directed away from the cube.

26. Figure 26-11 shows a long insulating, massless rod of length l, pivoted at its center and balanced with a weight W at a distance x from the left end. At the left and right ends of the rod are attached positive charges q and $2q$, respectively. A distance h directly beneath each of these charges is a fixed positive charge Q. (a) Find the distance x for the position of the weight when the rod is balanced. (b) What value should h have so that the rod exerts no vertical force on the bearing when balanced? Neglect the interaction between charges at the opposite ends of the rod.

figure 26-10
Problem 24

figure 26-11
Problem 26

SECTION 26-7

27. An electron is projected with an initial speed of 3.24×10^5 m/s directly toward a proton that is essentially at rest. If the electron is initially a great distance from the proton, at what distance from the proton is its speed instantaneously equal to twice its initial value? (Hint: Use the work-energy theorem.)
    Answer: 1.6×10^{-9} m.

28. In the radioactive decay of U^{238} (see Eq. 26-6) the center of the emerging α-particle is, at a certain instant, 9.0×10^{-15} m from the center of the residual nucleus Th^{234}. At this instant (a) what is the force on the α-particle and (b) what is its acceleration?

27
the electric field

With every point in space near the earth we can associate a *gravitational field* vector **g** (see Eq. 16-12). This is the gravitational acceleration that a test body, placed at that point and released, would experience. If m is the mass of the body and **F** the gravitational force acting on it, **g** is given by

$$\mathbf{g} = \mathbf{F}/m. \qquad (27\text{-}1)$$

This is an example of a *vector field*. For points near the surface of the earth the field is often taken as *uniform*; that is, **g** is the same for all points.

The flow of water in a river provides another example of a vector field, called a *flow field* (see Section 18-7). Every point in the water has associated with it a vector quantity, the velocity **v** with which the water flows past the point. If **g** and **v** do not change with time, the corresponding fields are described as *stationary*. In the case of the river note that even though the water is moving, the vector **v** at any point does not change with time for steady-flow conditions.

If we place a test charge in the space near a charged rod, an electrostatic force will act on the charge. We speak of an *electric field* in this space. In a similar way we speak of a *magnetic field* in the space around a bar magnet. In the classical theory of electromagnetism the electric and magnetic fields are central concepts. In this chapter we deal with electric fields associated with charges viewed from a reference frame in which they are at rest, that is, with *electrostatics*.

Before Faraday's time, the force acting between charged particles was thought of as a direct and instantaneous interaction between the two particles. This *action-at-a-distance* view was also held for magnetic and for gravitational forces. Today we prefer to think in terms of electric fields as follows:

1. Charge q_1 in Fig. 27-1 sets up an electric field in the space around itself. This field is suggested by the shading in the figure; later we shall show how to represent electric fields more concretely.
2. The field acts on charge q_2; this shows up in the force **F** that q_2 experiences.

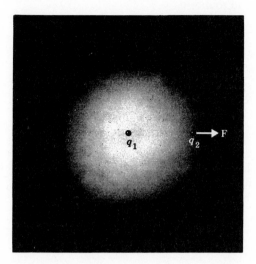

figure 27-1
Charge q_1 sets up a field that exerts a force **F** on charge q_2.

The field plays an intermediary role in the forces between charges. There are two separate problems: (*a*) calculating the fields that are set up by given distributions of charge and (*b*) calculating the forces that given fields will exert on charges placed in them. We think in terms of

$$\text{charge} \rightleftharpoons \text{field} \rightleftharpoons \text{charge}$$

and not, as in the action-at-a-distance point of view, in terms of

$$\text{charge} \rightleftharpoons \text{charge}.$$

In Fig. 27-1 we can also imagine that q_2 sets up a field and that this field acts on q_1, producing a force $-\mathbf{F}$ on it in accord with Newton's third law. The situation is completely symmetrical, each charge being immersed in a field associated with the other charge.

If the only problem in electromagnetism was that of the forces between stationary charges, the field and the action-at-a-distance points of view would be perfectly equivalent. Suppose, however, that q_1 in Fig. 27-1 suddenly accelerates to the right. How quickly does the charge q_2 learn that q_1 has moved and that the force which it (q_2) experiences must increase? Electromagnetic theory predicts that q_2 learns about q_1's motion by a *field disturbance* that emanates from q_1, traveling with the speed of light. The action-at-a-distance point of view requires that information about q_1's acceleration be communicated *instantaneously* to q_2; this is not in accord with experiment.* Accelerating electrons in the antenna of a radio transmitter influence electrons in a distant receiving antenna only after a time l/c where l is the separation of the antennas and c is the speed of light.

---

*By introducing other considerations it *is* possible to develop a consistent program of electromagnetism from the action-at-a-distance point of view. This is not commonly done however and we will not do so in this book.

To define the electric field operationally, we place a small test charge q_0 (assumed positive for convenience) at the point in space that is to be examined, and we measure the electrical force **F** (if any) that acts on this body. The *electric field* **E** at the point is defined as*

$$\mathbf{E} = \mathbf{F}/q_0. \qquad (27\text{-}2)$$

Here **E** is a vector because **F** is one, q_0 being a scalar. The direction of **E** is the direction of **F**, that is, it is the direction in which a resting positive charge placed at the point would tend to move.

The definition of gravitational field **g** is much like that of electric field, except that the mass of the test body rather than its charge is the property of interest. Although the units of **g** are usually written as m/s², they could also be written as N/kg (Eq. 27-1); those for **E** are N/C (Eq. 27-2). Thus both **g** and **E** are expressed as a force divided by a property (mass or charge) of the test body.

27-2
THE ELECTRIC FIELD **E**

EXAMPLE 1

What is the magnitude of the electric field **E** such that an electron, placed in the field, would experience an electrical force equal to its weight?

From Eq. 27-2, replacing q_0 by e and F by mg, where m is the electron mass, we have

$$E = \frac{F}{q_0} = \frac{mg}{e}$$

$$= \frac{(9.1 \times 10^{-31} \text{ kg})(9.8 \text{ m/s}^2)}{1.6 \times 10^{-19} \text{ C}}$$

$$= 5.6 \times 10^{-11} \text{ N/C}.$$

This is a very weak electric field. Which way will **E** have to point if the electric force is to cancel the gravitational force?

In applying Eq. 27-2 we must use a test charge as small as possible. A large test charge might disturb the primary charges that are responsible for the field, thus changing the very quantity that we are trying to measure. Equation 27-2 should, strictly, be replaced by

$$\mathbf{E} = \lim_{q_0 \to 0} \frac{\mathbf{F}}{q_0}. \qquad (27\text{-}3)$$

This equation instructs us to use a smaller and smaller test charge q_0, evaluating the ratio $\mathbf{F}/q_0$ at every step. The electric field **E** is then the limit of this ratio as the size of the test charge approaches zero.†

The concept of the electric field as a vector was not appreciated by Michael Faraday, who always thought in terms of *lines of force*. The lines of force still form a convenient way of visualizing electric-field patterns. We shall use them for this purpose but we shall not employ them quantitatively.

The relationship between the (imaginary) lines of force and the electric field vector is this:

27-3
LINES OF FORCE

*This definition of **E**, though conceptually sound and quite appropriate to our present purpose, is rarely carried out in practice because of experimental difficulties. **E** is normally found by calculation from more readily measurable quantities such as the electric potential; see Section 29-7.
† Of course, q_0 can never be less than the electronic charge e

1. The tangent to a line of force at any point gives the *direction* of **E** at that point.
2. The lines of force are drawn so that the number of lines per unit cross-sectional area (perpendicular to the lines) is proportional to the *magnitude* of **E**. Where the lines are close together *E* is large and where they are far apart *E* is small.

It is not obvious that it is possible to draw a continuous set of lines to meet these requirements. Indeed, it turns out that if Coulomb's law were not true, it would *not* be possible to do so; see Problem 7.

Figure 27-2 shows the lines of force for a uniform sheet of positive charge. We assume that the sheet is infinitely large, which, for a sheet of finite dimensions, is equivalent to considering only those points whose distance from the sheet is small compared to the distance to the nearest edge of the sheet. A positive test charge, released in front of such a sheet, would move away from the sheet along a perpendicular line. Thus the electric field vector at any point near the sheet must be at right angles to the sheet. The lines of force are uniformly spaced, which means that **E** has the same magnitude for all points near the sheet.

Figure 27-3 shows the lines of force for a negatively charged sphere. From symmetry, the lines must lie along radii. They point inward because a free positive charge would be accelerated in this direction. The electric field *E* is not constant but decreases with increasing distance from the charge. This is evident in the lines of force, which are farther apart at greater distances. From symmetry, *E* is the same for all points that lie a given distance from the center of the charge.

figure 27-2
Lines of force for a section of an infinitely large sheet of positive charge.

figure 27-3
Lines of force for a negatively charged sphere.

In Fig. 27-3 how does *E* vary with the distance *r* from the center of the charged sphere?

Suppose that *N* lines terminate on the sphere. Draw an imaginary concentric sphere of radius *r*; the number of lines per unit cross-sectional area at every point on the sphere is $N/4\pi r^2$. Since *E* is proportional to this, we can write that

$$E \propto 1/r^2.$$

We derive an exact relationship in Section 27-4. How does *E* vary with distance from an infinitely long uniform cylinder of charge?

EXAMPLE 2

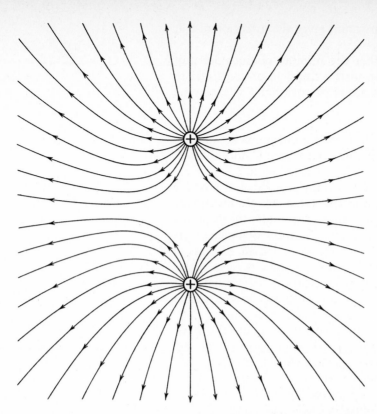

figure 27-4
Lines of force for two equal
positive charges.

Figures 27-4 and 27-5 show the lines of force for two equal like charges and for two equal unlike charges, respectively. Michael Faraday, as we have said, used lines of force a great deal in his thinking. They were more real for him than they are for most scientists and engineers today. It is possible to sympathize with Faraday's point of view. Can

figure 27-5
Lines of force for equal but opposite charges.

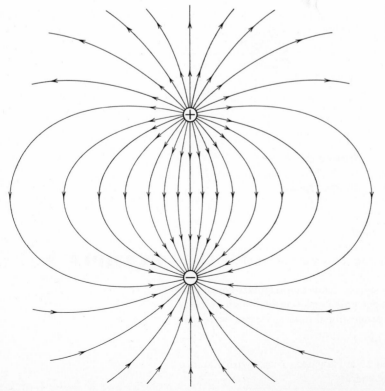

we not almost "see" the charges being pushed apart in Fig. 27-4 and pulled together in Fig 27-5 by the lines of force? Compare Fig. 27-5 with Fig. 18-14, which represents a flow field. Figure 27-6 shows a representation of lines of force around charged conductors, using grass seeds suspended in an insulating liquid.

Lines of force give a vivid picture of the way **E** varies through a given region of space. However, the equations of electromagnetism (see Table 40-2) are written in terms of the electric field **E** and other field vectors and not in terms of the lines of force. The electric field **E** varies in a perfectly continuous way as any path in the field is traversed; see Fig. 27-7.

figure 27-6
Photographs of the patterns of electric lines of force around (a) a charged plate (compare Fig. 27-2), and (b) two rods with equal and opposite charges (compare Fig. 27-5). The patterns were made by suspending grass seed in an insulating liquid. (Courtesy Educational Services Incorporated, Watertown, Mass.)

figure 27-7
E varies continuously as we move along any path AB in the field set up by point charge $+Q$. In general, the path AB and the field vectors **E** will not lie in the plane of the figure.

In this section we deal with the charge-field interaction by showing how we may calculate **E** for various points near given charge distributions, starting with the simple case of a point charge q.

Let a test charge q_0 be placed a distance r from a point charge q. The magnitude of the force acting on q_0 is given by Coulomb's law, or

$$F = \frac{1}{4\pi\epsilon_0}\frac{qq_0}{r^2}.$$

The electric field at the site of the test charge is given by Eq. 27-2, or

$$E = \frac{F}{q_0} = \frac{1}{4\pi\epsilon_0}\frac{q}{r^2}. \tag{27-4}$$

The direction of **E** is on a radial line from q, pointing outward if q is positive and inward if q is negative.

To find **E** for a group of point charges: (a) Calculate $\mathbf{E}_n$ due to each charge at the given point *as if it were the only charge present*. (b) Add these separately calculated fields vectorially to find the resultant field **E** at the point. In equation form,

$$\mathbf{E} = \mathbf{E}_1 + \mathbf{E}_2 + \mathbf{E}_3 + \cdots = \Sigma\mathbf{E}_n \qquad n = 1, 2, 3, \ldots . \tag{27-5}$$

The sum is a vector sum, taken over all the charges. Equation 27-5 (like Eq. 26-4) is an example of the *principle of superposition* which states, in this context, that at a given point the electric fields due to separate charge distributions simply add up (vectorially) or superimpose independently. The principle of superposition is very important in physics. It applies, for example, to gravitational and magnetic situations as well.*

If the charge distribution is a continuous one, the field it sets up at any point P can be computed by dividing the charge into infinitesimal elements dq. The field $d\mathbf{E}$ due to each element at the point in question is then calculated, treating the elements as point charges. The magnitude of $d\mathbf{E}$ (see Eq. 27-4) is given by

$$dE = \frac{1}{4\pi\epsilon_0}\frac{dq}{r^2}, \tag{27-6}$$

where r is the distance from the charge element dq to the point P. The resultant field at P is then found from the superposition principle by adding (that is, integrating) the field contributions due to all the charge elements, or,

$$\mathbf{E} = \int d\mathbf{E}. \tag{27-7}$$

The integration, like the sum in Eq. 27-5, is a vector operation; in Example 5 we will see how such an integral is handled in a simple case.

An electric dipole. Figure 27-8 shows a positive and a negative charge of equal magnitude q placed a distance $2a$ apart, a configuration called an electric dipole. The pattern of lines of force is that of Fig. 27-5, which also shows an electric dipole. What is the field **E** due to these charges at point P, a distance r along the perpendicular bisector of the line joining the charges? Assume $r \gg a$.

EXAMPLE 3

* Formally, the principle of superposition holds in physics only to the extent that the differential equation defining the situation is linear. To the extent that the amplitudes of mechanical or electromagnetic oscillations become relatively large the principle tends to fail. We do not discuss such cases in this book. In particular, the principle holds absolutely in electrostatics.

Equation 27-5 gives the vector equation

$$\mathbf{E} = \mathbf{E}_1 + \mathbf{E}_2,$$

where, from Eq. 27-4,*

$$E_1 = E_2 = \frac{1}{4\pi\epsilon_0}\frac{q}{a^2 + r^2}.$$

The vector sum of $\mathbf{E}_1$ and $\mathbf{E}_2$ points vertically downward and has the magnitude

$$E = 2E_1 \cos\theta.$$

From the figure we see that

$$\cos\theta = \frac{a}{\sqrt{a^2 + r^2}}.$$

Substituting the expressions for E_1 and for $\cos\theta$ into that for E yields

$$E = \frac{2}{4\pi\epsilon_0}\frac{q}{(a^2 + r^2)}\frac{a}{\sqrt{a^2 + r^2}} = \frac{1}{4\pi\epsilon_0}\frac{2aq}{(a^2 + r^2)^{3/2}}.$$

If $r \gg a$, we can neglect a in the denominator; this equation then reduces to

$$E \cong \frac{1}{4\pi\epsilon_0}\frac{(2a)(q)}{r^3}. \tag{27-8a}$$

The essential properties of the charge distribution in Fig. 27-8, the magnitude of the charge q and the separation $2a$ between the charges, enter Eq. 27-8a only as a product. This means that, if we measure $\mathbf{E}$ at various distances from the electric dipole (assuming $r \gg a$), we can never deduce q and $2a$ separately but only the product $2aq$; if q were doubled and a simultaneously cut in half, the electric field *at large distances from the dipole* would not change.

The product $2aq$ is called the *electric dipole moment p*. Thus we can rewrite this equation for E, *for distant points along the perpendicular bisector*, as

$$E = \frac{1}{4\pi\epsilon_0}\frac{p}{r^3}. \tag{27-8b}$$

The result for distant points *along the dipole axis* (see Problem 25) and the general result for any distant point (see Problem 28) also contain the quantities $2a$ and q only as the product $2aq \ (= p)$. The variation of E with r in the general result for distant points is also as $1/r^3$, as in Eq. 27-8b.

The dipole of Fig. 27-8 is two equal and opposite charges placed close to each other so that their separate fields at distant points almost, but not quite, cancel. On this point of view it is easy to understand that $E(r)$ for a dipole varies at large distances as $1/r^3$ (Eq. 27-8b), whereas for a point charge $E(r)$ drops off more slowly, namely as $1/r^2$ (Eq. 27-4).

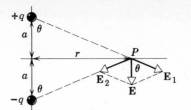

figure 27-8
Example 3.

EXAMPLE 4

Figure 27-9 shows a charge $q_1 \ (= +1.0 \times 10^{-6}$ C) 10 cm from a charge q_2 $(= +2.0 \times 10^{-6}$ C). At what point on the line joining the two charges is the electric field zero?

The point must lie between the charges because only here do the forces exerted by q_1 and q_2 on a test charge (no matter whether it is positive or negative) oppose each other. If $\mathbf{E}_1$ is the electric field due to q_1 and $\mathbf{E}_2$ that due to q_2, we must have

$$E_1 = E_2$$

or (see Eq. 27-4)

$$\frac{1}{4\pi\epsilon_0}\frac{q_1}{x^2} = \frac{1}{4\pi\epsilon_0}\frac{q_2}{(l - x)^2}.$$

figure 27-9
Example 4.

* Note that the r's in Eq. 27-4 and in this equation have different meanings.

where x is the distance from q_1 and l equals 10 cm. Solving for x yields

$$x = \frac{l}{1 + \sqrt{q_2/q_1}} = \frac{10 \text{ cm}}{1 + \sqrt{2}} = 4.1 \text{ cm.}$$

Supply the missing steps. On what basis was the second root of the resulting quadratic equation discarded?

***

EXAMPLE 5

Ring of charge. Figure 27-10 shows a ring of charge q and radius a. Calculate $\mathbf{E}$ for points on the axis of the ring a distance x from its center.

Consider a differential element of the ring of length ds, located at the top of the ring in Fig. 27-10. It contains an element of charge given by

$$dq = q \, \frac{ds}{2\pi a}$$

where $2\pi a$ is the circumference of the ring. This element sets up a differential electric field $d\mathbf{E}$ at point P.

The resultant field $\mathbf{E}$ at P is found by integrating the effects of all the elements that make up the ring. From symmetry this resultant field must lie along the ring axis. Thus only the component of $d\mathbf{E}$ parallel to this axis contributes to the final result. The component perpendicular to the axis is canceled out by an equal but opposite component established by the charge element on the opposite side of the ring.

Thus the general vector integral (Eq. 27-7)

$$\mathbf{E} = \int d\mathbf{E}$$

becomes a scalar integral $E = \int dE \cos \theta.$

The quantity dE follows from Eq. 27-6, or

$$dE = \frac{1}{4\pi\epsilon_0} \frac{dq}{r^2} = \frac{1}{4\pi\epsilon_0} \left(\frac{q \, ds}{2\pi a} \right) \frac{1}{a^2 + x^2}.$$

From Fig. 27-10 we have $\cos \theta = \dfrac{x}{\sqrt{a^2 + x^2}}.$

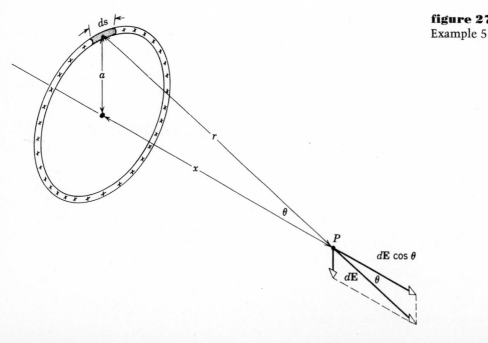

figure 27-10
Example 5

Noting that, for a given point P, x has the same value for all charge elements and is not a variable and that s is the variable of integration, we obtain

$$E = \int dE \cos \theta = \int \frac{1}{4\pi\epsilon_0} \frac{q \, ds}{(2\pi a)(a^2 + x^2)} \frac{x}{\sqrt{a^2 + x^2}}$$

$$= \frac{1}{4\pi\epsilon_0} \frac{qx}{(2\pi a)(a^2 + x^2)^{3/2}} \int ds.$$

The integral is simply the circumference of the ring $(= 2\pi a)$, so that

$$E = \frac{1}{4\pi\epsilon_0} \frac{qx}{(a^2 + x^2)^{3/2}}.$$

Does this expression for E reduce to an expected result for $x = 0$? For $x \gg a$ we can neglect a in the denominator of this equation, yielding

$$E \cong \frac{1}{4\pi\epsilon_0} \frac{q}{x^2}.$$

This is an expected result (compare Eq. 27-4) because at great enough distances the ring behaves like a point charge q.

Infinite line of charge. Figure 27-11 shows a section of an infinite line of charge whose linear charge density (that is, the charge per unit length, measured in C/m) has the constant value λ. Calculate the field $\mathbf{E}$ a distance y from the line.

The magnitude of the field contribution dE due to charge element $dq \, (= \lambda \, dx)$ is given, using Eq. 27-6, by

$$dE = \frac{1}{4\pi\epsilon_0} \frac{dq}{r^2} = \frac{1}{4\pi\epsilon_0} \frac{\lambda \, dx}{y^2 + x^2}.$$

EXAMPLE 6

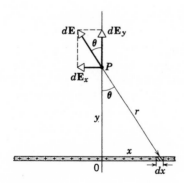

figure 27-11
Example 6. A section of an infinite line of charge.

The vector $d\mathbf{E}$, as Fig. 27-11 shows, has the components

$$dE_x = -dE \sin \theta \quad \text{and} \quad dE_y = dE \cos \theta.$$

The minus sign in front of dE_x indicates that $d\mathbf{E}_x$ points in the negative x direction. The x and y components of the resultant vector $\mathbf{E}$ at point P are given by

$$E_x = \int dE_x = -\int_{x=-\infty}^{x=+\infty} \sin \theta \, dE \quad \text{and} \quad E_y = \int dE_y = \int_{x=-\infty}^{x=+\infty} \cos \theta \, dE.$$

E_x must be zero because every charge element on the right has a corresponding element on the left such that their field contributions in the x direction cancel. Thus $\mathbf{E}$ points entirely in the y direction. Because the contributions to E_y from the right- and left-hand halves of the rod are equal, we can write

$$E = E_y = 2\int_{x=0}^{x=+\infty} \cos \theta \, dE.$$

Note that we have changed the lower limit of integration and have introduced a compensating factor of two.

Substituting the expression for dE into this equation gives

$$E = \frac{\lambda}{2\pi\epsilon_0} \int_{x=0}^{x=\infty} \cos\theta \, \frac{dx}{y^2 + x^2}.$$

From Fig. 27-11 we see that the quantities θ and x are not independent. We must eliminate one of them, say x. The relation between x and θ is (see figure)

$$x = y \tan \theta.$$

Differentiating, we obtain $\quad dx = y \sec^2 \theta \, d\theta.$

Substituting these two expressions leads finally to

$$E = \frac{\lambda}{2\pi\epsilon_0 y} \int_{\theta=0}^{\theta=\pi/2} \cos\theta \, d\theta.$$

You should check this step carefully, noting that the limits must now be on θ and not on x. For example, as $x \to +\infty$, $\theta \to \pi/2$, as Fig. 27-11 shows. This equation integrates readily to

$$E = \frac{\lambda}{2\pi\epsilon_0 y} (\sin\theta) \Big|_0^{\pi/2} = \frac{\lambda}{2\pi\epsilon_0 y}.$$

You may wonder about the usefulness of solving a problem involving an infinite rod of charge when any actual rod must have a finite length (see Problem 23). However, for points close enough to finite rods and not near their ends, the equation that we have just derived yields results that are so close to the correct values that the difference can be ignored in many practical situations. It is usually unnecessary to solve exactly every geometry encountered in practical problems. Indeed, if idealizations or approximations are not made, the vast majority of significant problems of all kinds in physics and engineering cannot be solved at all.

27-5 A POINT CHARGE IN AN ELECTRIC FIELD

Here and in the following section, in contrast with Section 27-4, we investigate the other half of the charge-field interaction, namely, if we are given a field **E**, what forces and torques will act on a charge configuration placed in it? We start with the simple case of a point charge in a uniform electric field.

An electric field will exert a force on a charged particle given by (Eq. 27-2)

$$\mathbf{F} = \mathbf{E}q.$$

This force will produce an acceleration

$$\mathbf{a} = \mathbf{F}/m,$$

where m is the mass of the particle. We will consider two examples of the acceleration of a charged particle in a uniform electric field. Such a field can be produced by connecting the terminals of a battery to two parallel metal plates which are otherwise insulated from each other. If the spacing between the plates is small compared with the dimensions of the plates, the field between them will be fairly uniform except near the edges. Note that in calculating the motion of a particle in a field set up by external charges the field due to the particle itself (that is, its *self-field*) is ignored. In the same way, the earth's gravitational field can exert no net force on the earth itself but only on a second object, say a stone, placed in that field.

A particle of mass m and charge q is placed at rest in a uniform electric field (Fig. 27-12) and released. Describe its motion.

EXAMPLE 7

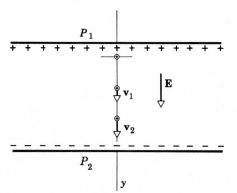

The motion resembles that of a body falling in the earth's gravitational field. The (constant) acceleration is given by

$$a = \frac{F}{m} = \frac{qE}{m}.$$

The equations for uniformly accelerated motion (Table 3-1) then apply. With $v_0 = 0$, they are

$$v = at = \frac{qEt}{m},$$

$$y = \tfrac{1}{2}at^2 = \frac{qEt^2}{2m},$$

and

$$v^2 = 2ay = \frac{2qEy}{m}.$$

The kinetic energy attained after moving a distance y is found from

$$K = \tfrac{1}{2}mv^2 = \tfrac{1}{2}m\left(\frac{2qEy}{m}\right) = qEy.$$

This result also follows directly from the work-energy theorem because a constant force qE acts over a distance y.

Deflecting an electron beam. Figure 27-13 shows an electron of mass m and charge e projected with speed v_0 at right angles to a uniform field $\mathbf{E}$. Describe its motion.

EXAMPLE 8

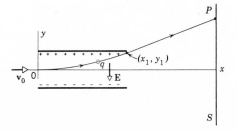

figure 27-13
Example 8. An electron is projected into a uniform electric field.

The motion is like that of a projectile fired horizontally in the earth's gravitational field. The considerations of Section 4-3 apply, the horizontal (x) and vertical (y) motions being given by

$$x = v_0 t$$

and

$$y = \tfrac{1}{2}at^2 = \frac{eE}{2m}\,t^2.$$

Eliminating t yields

$$y = \frac{eE}{2mv_0^2}\,x^2 \qquad\qquad (27\text{-}9)$$

for the equation of the trajectory.

When the electron emerges from the plates in Fig. 27-13, it travels (neglecting gravity) in a straight line tangent to the parabola of Eq. 27-9 at the exit point. We can let it fall on a fluorescent screen S placed some distance beyond the plates. Together with other electrons following the same path, it will then make itself visible as a small luminous spot; this is the principle of the electrostatic *cathode-ray oscilloscope*.

The electric field between the plates of a cathode-ray oscilloscope is 1.2×10^4 N/C. What deflection will an electron experience if it enters at right angles to the field with a kinetic energy of 2000 eV $(= 3.2 \times 10^{-16}$ J), a typical value? The deflecting assembly is 1.5 cm long.

Recalling that $K_0 = \tfrac{1}{2}mv_0^2$, we can rewrite Eq. 27-9 as

$$y = \frac{eEx^2}{4K_0}.$$

If x_1 is the horizontal position of the far edge of the plate, y_1 will be the corresponding deflection (see Fig. 27-13), or

$$y_1 = \frac{eEx_1^2}{4K_0}$$

$$= \frac{(1.6 \times 10^{-19}\ \text{C})(1.2 \times 10^4\ \text{N/C})(1.5 \times 10^{-2}\ \text{m})^2}{(4)(3.2 \times 10^{-16}\ \text{J})}$$

$$= 3.4 \times 10^{-4}\ \text{m} = 0.34\ \text{mm}.$$

The deflection measured, not at the deflecting plates but at the fluorescent screen, is much larger.

EXAMPLE 9

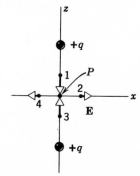

figure 27-14
Example 10. The electric field at four points near a point P which is centered between two equal positive charges.

A positive point test charge q_0 is placed halfway between two equal fixed positive charges q. What force acts on it at or near this point P?

From symmetry the force *at* the point is zero so that the particle is in equilibrium; the nature of the equilibrium remains to be found. Figure 27-14 (compare Fig. 27-4) shows the **E** vectors for four points near P. If the test charge is moved along the z axis, a *restoring* force is brought into play; however, the equilibrium is unstable for motion along the x (and y) axes. Thus we have the three-dimensional equivalent of *saddle point equilibrium*; see Fig. 14-8. What is the nature of the equilibrium for a negative test charge?

EXAMPLE 10

An electric dipole moment can be regarded as a vector **p** whose magnitude p, for a dipole like that described in Example 3, is the product $2aq$ of the magnitude of either charge q and the distance $2a$ between the charges. The *direction* of **p** for such a dipole is from the *negative* to the *positive* charge. The vector nature of the electric dipole moment permits us to cast many expressions involving electric dipoles into concise form, as we shall see.

27-6
A DIPOLE IN AN ELECTRIC FIELD

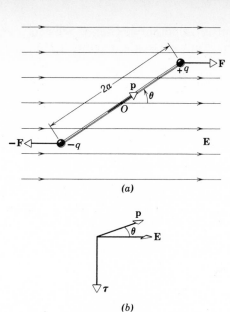

figure 27-15
(a) An electric dipole in a uniform external field. (b) Illustrating $\tau = \mathbf{p} \times \mathbf{E}$.

Figure 27-15a shows an electric dipole formed by placing two charges $+q$ and $-q$ a fixed distance $2a$ apart. The arrangement is placed in a uniform *external* electric field $\mathbf{E}$, its dipole moment $\mathbf{p}$ making an angle θ with this field. Two equal and opposite forces $\mathbf{F}$ and $-\mathbf{F}$ act as shown, where

$$F = qE.$$

The net force is clearly zero, but there is a net torque about an axis through O (see Eq. 12-2) given by

$$\tau = 2F(a \sin \theta) = 2aF \sin \theta.$$

Combining these two equations and recalling that $p = (2a)(q)$, we obtain

$$\tau = 2aqE \sin \theta = pE \sin \theta. \qquad (27\text{-}10)$$

Thus an electric dipole placed in an external electric field $\mathbf{E}$ experiences a torque tending to align it with the field. Equation 27-10 can be written in vector form as

$$\tau = \mathbf{p} \times \mathbf{E}, \qquad (27\text{-}11)$$

the appropriate vectors being shown in Fig. 27-15b.

Work (positive or negative) must be done by an external agent to change the orientation of an electric dipole in an external field. This work is stored as potential energy U in the system consisting of the dipole and the arrangement used to set up the external field. If θ in Fig. 27-15a has the initial value θ_0, the work required to turn the dipole axis to an angle θ is given (see Table 12-2) from

$$W = \int dW = \int_{\theta_0}^{\theta} \tau \, d\theta = U,$$

where τ is the torque exerted by the agent that does the work. Combining this equation with Eq. 27-10 yields

$$U = \int_{\theta_0}^{\theta} pE \sin \theta \, d\theta = pE \int_{\theta_0}^{\theta} \sin \theta \, d\theta$$

$$= pE \left(- \cos \theta\right)\Big|_{\theta_0}^{\theta}.$$

Since we are interested only in *changes* in potential energy, we can choose the reference orientation θ_0 to have any convenient value, in this case 90°. This gives

$$U = -pE \cos \theta \qquad (27\text{-}12)$$

or, in vector terms,

$$U = -\mathbf{p} \cdot \mathbf{E}. \qquad (27\text{-}13)$$

EXAMPLE 11

An electric dipole consists of two opposite charges of magnitude $q = 1.0 \times 10^{-6}$ C separated by $d = 2.0$ cm. The dipole is placed in an external field of 1.0×10^5 N/C.

(*a*) What maximum torque does the field exert on the dipole? The maximum torque is found by putting $\theta = 90°$ in Eq. 27-10 or

$$\tau = pE \sin \theta = qdE \sin \theta$$

$$= (1.0 \times 10^{-6} \text{ C})(0.020 \text{ m})(1.0 \times 10^5 \text{ N/C})(\sin 90°)$$

$$= 2.0 \times 10^{-3} \text{ N} \cdot \text{m}.$$

(*b*) How much work must an external agent do to turn the dipole end for end, starting from a position of alignment $(\theta = 0)$? The work is the difference in potential energy U between the positions $\theta = 180°$ and $\theta = 0$. From Eq. 27-12,

$$W = U_{180°} - U_{0°} = (-pE \cos 180°) - (-pE \cos 0)$$

$$= 2pE = 2qdE$$

$$= (2)(1.0 \times 10^{-6} \text{ C})(0.020 \text{ m})(1.0 \times 10^5 \text{ N/C})$$

$$= 4.0 \times 10^{-3} \text{ J}.$$

---

questions

1. Name as many scalar fields and vector fields as you can.

2. (*a*) In the gravitational attraction between the earth and a stone, can we say that the earth lies in the gravitational field of the stone? (*b*) How is the gravitational field due to the stone related to that due to the earth?

3. A positively charged ball hangs from a long silk thread. We wish to measure **E** at a point in the same horizontal plane as that of the hanging charge. To do so, we put a positive test charge q_0 at the point and measure $\mathbf{F}/q_0$. Will F/q_0 be less than, equal to, or greater than E at the point in question?

4. Taking into account the quantization of electric charge (the single electron providing the basic charge unit), how can we justify the procedure suggested by Eq. 27-3?

5. In exploring electric fields with a test charge we have often assumed, for convenience, that the test charge was positive. Does this really make any difference in determining the field? Illustrate in a simple case of your own devising.

6. Electric lines of force never cross. Why?

7. In Fig. 27-4 why do the lines of force around the edge of the figure appear, when extended backward, to radiate uniformly from the center of the figure?

8. Figure 27-2 shows that **E** has the same value for all points in front of an infinite uniformly charged sheet. Is this reasonable? One might think that the field should be stronger near the sheet because the charges are so much closer.

9. If a point charge q of mass m is released from rest in a nonuniform field, will it follow a line of force?

10. A point charge is moving in an electric field at right angles to the lines of force. Does any force **F** act on it?

11. In Fig. 27-7 path AB is not a line of force. How can you tell?

12. In Fig. 27-6, why should grass seeds line up with electric lines of force? Grass seeds normally carry no electric charge. See "Demonstration of the Electric Fields of Current-Carrying Conductors" by O. Jefimenko, *American Journal of Physics.* January 1962.

13. Two point charges of unknown magnitude and sign are a distance d apart. The electric field is zero at one point between them, on the line joining them. What can you conclude about the charges?

14. Compare the way E varies with r for (a) a point charge (Eq. 27-4), (b) a dipole (Eq. 27-8a), and (c) a quadrupole (Problem 33).

15. Charges $+Q$ and $-Q$ are fixed a distance L apart and a long straight line is drawn through them. What is the direction of **E** on this line for points (a) between the charges, (b) outside the charges in the direction of $+Q$, and (c) outside the charges in the direction of $-Q$?

16. Two point charges of unknown sign and magnitude are fixed a distance L apart. Can we have **E** = 0 for off-axis points (excluding ∞)? Explain.

17. In what way does Eq. 27-8b fail to represent the lines of force of Fig. 27-5 if we relax the requirement that $r \gg a$?

18. If two dipoles of moments $\mathbf{p}_1$ and $\mathbf{p}_2$ are superimposed, is the dipole moment of the resulting configuration given by $\mathbf{p}_1 + \mathbf{p}_2$?

19. In Fig. 27-5 the force on the lower charge points up and is finite. The crowding of the lines of force, however, suggests that E is infinitely great at the site of this (point) charge. A charge immersed in an infinitely great field should have an infinitely great force acting on it. What is the solution to this dilemma?

20. An electric dipole is placed in a *nonuniform* electric field. Is there a net force on it?

21. An electric dipole is placed at rest in a uniform external electric field, as in Fig. 27-15a, and released. Discuss its motion.

22. An electric dipole has its dipole moment **p** aligned with a uniform external electric field **E**. (a) Is the equilibrium stable or unstable? (b) Discuss the nature of the equilibrium if **p** and **E** point in opposite directions.

problems

SECTION 27-2

1. What is the magnitude of a point charge chosen so that the electric field 50 cm away has the magnitude 2.0 N/C? *Answer:* 5.6×10^{-11} C.

2. What is the magnitude and direction of an electric field that will balance the weight of (a) an electron and (b) an alpha particle?

3. An electric field **E** with an average magnitude of about 150 N/C points downward in the earth's atmosphere. We wish to "float" a sulfur sphere weighing 1.0 lb in this field by charging it. (a) What charge (sign and magnitude) must be used? (b) Why is the experiment not practical? Give a qualitative reason supported by a very rough numerical calculation to prove your point.
   Answer: (a) -0.030 C. (b) Sphere would blow up because of mutual Coulomb repulsion.

4. At some instant the velocity components of an electron moving between two charged parallel plates are $v_x = 1.5 \times 10^5$ m/s and $v_y = 0.30 \times 10^4$ m/s. If the electric field between the plates is given by $\mathbf{E} = \mathbf{j}\ 1.2 \times 10^4$ N/C, (a) what is the acceleration of the electron? (b) When the x-coordinate of the electron has changed by 2.0 cm what will be the velocity of the electron?

5. A particle having a charge of -2.0×10^{-9} C is acted on by a downward electric force of 3.0×10^{-6} N in a uniform electric field. (a) What is the strength of the electric field? (b) What is the magnitude and direction of the electric force exerted on a proton placed in this field? (c) What is the gravitational force on the proton? (d) What is the ratio of the electric to the gravitational forces in this case?

  Answer: (a) 1.5×10^3 N/C. (b) 2.4×10^{-16} N, up. (c) 1.6×10^{-26} N. (d) 1.5×10^{10}.

6. A uniform vertical field **E** is established in the space between two large parallel plates. In this field one suspends a small conducting sphere of mass m from a string of length l. Find the period of this pendulum when the sphere is given a charge $+q$ if the lower plate (a) is charged positively; (b) is charged negatively.

SECTION 27-3

7. Assume that the exponent in Coulomb's law is not "two" but n. Show that for $n \neq 2$ it is impossible to construct lines that will have the properties listed for lines of force in Section 27-3. For simplicity, treat an isolated point charge.

8. Sketch qualitatively the lines of force associated with a thin, circular, uniformly charged disk of radius R. (Hint: Consider as limiting cases points very close to the surface and points very far from it.) Show the lines only in a plane containing the axis of the disk.

9. Sketch qualitatively the lines of force between two concentric conducting spherical shells, charge $+q_1$ being placed on the inner sphere and $-q_2$ on the outer. Consider the cases $q_1 > q_2$, $q_1 = q_2$, $q_1 < q_2$.

10. (a) Sketch qualitatively the lines of force associated with three long parallel lines of charge, in a perpendicular plane. Assume that the intersections of the lines of charge with such a plane form an equilateral triangle and that each line of charge has the same linear charge density λ (C/m). (b) Discuss the nature of the equilibrium of a test charge placed on the central axis of the charge assembly.

11. In Fig. 27-4 consider two neighboring lines of force leaving the upper charge at small angles with a straight line connecting the charges. If the angle between their tangents for points near the charge is θ, it becomes $\theta/\sqrt{2}$ at great distances. Verify this statement and explain it. (Hint: Consider how the lines must behave both close to either charge and far from the charges.)

SECTION 27-4

12. Three charges are arranged in an equilateral triangle as in Fig. 27-16. What is the direction of the force on $+q$?

13. Two equal and opposite charges of magnitude 2.0×10^{-7} C are 15 cm apart. (a) What are the magnitude and direction of **E** at a point midway between the charges? (b) What force (magnitude and direction) would act on an electron placed there?

  Answer: (a) 6.4×10^5 N/C, toward the negative charge. (b) 1.0×10^{-13} N, toward the positive charge.

14. Two point charges of magnitude $+2.0 \times 10^{-7}$ C and $+8.5 \times 10^{-8}$ C are 12 cm apart. (a) What electric field does each produce at the site of the other? (b) What force acts on each?

15. Two point charges of unknown magnitude and sign are placed a distance d apart. (a) If it is possible to have **E** $= 0$ at any point *not* between the charges but on the line joining them, what are the necessary conditions and where is the point located? (b) Is it possible, for any arrangement of two point charges, to find *two* points (neither at infinity) at which **E** $= 0$; if so, under what conditions?

  Answer: (a) Charges must be opposite in sign, the nearer charge being smaller in magnitude than the farther charge. (b) No.

16. (a) In Fig. 27-17 locate the point (or points) at which the electric field is zero. (b) Sketch qualitatively the lines of force. Take $a = 50$ cm.

figure 27-16
Problem 12

figure 27-17
Problem 16

figure 27-18
Problem 17

17. Two point charges are a distance d apart (Fig. 27-18). Plot $E(x)$, assuming $x = 0$ at the left-hand charge. Consider both positive and negative values of x. Plot E as positive if $\mathbf{E}$ points to the right and negative if $\mathbf{E}$ points to the left. Assume $q_1 = +1.0 \times 10^{-6}$ C, $q_2 = +3.0 \times 10^{-6}$ C, and $d = 10$ cm.

18. What is $\mathbf{E}$ in magnitude and direction at the center of the square of Fig. 27-19? Assume that $q = 1.0 \times 10^{-8}$ C and $a = 5.0$ cm.

19. In Fig. 27-8 assume that both charges are positive. (a) Show that E at point P in that figure, assuming $r \gg a$, is given by

$$E = \frac{1}{4\pi\epsilon_0} \frac{2q}{r^2}.$$

(b) What is the direction of $\mathbf{E}$? (c) Is it reasonable that E should vary as r^{-2} here and as r^{-3} for the dipole of Fig. 27-8?
Answer: (b) At right angles to the axis and away from it

20. Charges $+q$ and $-2q$ are fixed a distance d apart as in Fig. 27-20. (a) Find $\mathbf{E}$ at points A, B, and C. (b) Sketch roughly the lines of force.

figure 27-19
Problem 18

figure 27-20
Problem 20

21. Calculate $\mathbf{E}$ (direction and magnitude) at point P in Fig. 27-21.
Answer: $E = q/\pi\epsilon_0 a^2$, along bisector, away from triangle.

22. A thin glass rod is bent into a semicircle of radius R. A charge $+Q$ is uniformly distributed along the upper half and a charge $-Q$ is uniformly distributed along the lower half, as shown in Fig. 27-22. Find the electric field $\mathbf{E}$ at P, the center of the semicircle.

23. A thin nonconducting rod of finite length l carries a total charge q, spread uniformly along it. Show that E at point P on the perpendicular bisector in Fig. 27-23 is given by

$$E = \frac{q}{2\pi\epsilon_0 y} \frac{1}{\sqrt{l^2 + 4y^2}}.$$

Show that as $l \to \infty$ this result approaches that of Example 6.

24. An electron is constrained to move along the axis of the ring of charge in Example 5. Show that the electron can perform oscillations whose frequency is given by

$$\omega = \sqrt{\frac{eq}{4\pi\epsilon_0 ma^3}}.$$

This formula holds only for small oscillations, that is, for $x \ll a$ in Fig. 27-10. (Hint: Show that the motion is simple harmonic and use Eq. 15-11.)

25. *Axial field due to an electric dipole.* In Fig. 27-8, consider a point a distance r from the center of the dipole along its axis. (a) Show that, at large values of r, the electric field is

$$E = \frac{1}{2\pi\epsilon_0} \frac{p}{r^3},$$

which is twice the value given for the conditions of Example 3. (b) What is the direction of $\mathbf{E}$?
Answer: (b) Parallel to $\mathbf{p}$.

figure 27-21
Problem 21

figure 27-22
Problem 22

figure 27-23
Problem 23

26. For the ring of charge in Example 5, show that the maximum value of E occurs at $x = a/\sqrt{2}$.

27. Consider the ring of charge of Example 5. Suppose that the charge q is not distributed uniformly over the ring but that charge q_1 is distributed uniformly over half the circumference and charge q_2 is distributed uniformly over the other half. Let $q_1 + q_2 = q$. (a) Find the component of the electric field at any point on the axis directed *along* the axis and compare with the uniform case of Example 5. (b) Find the component of the electric field at any point on the axis *perpendicular* to the axis and compare with the uniform case of Example 5.

   Answer: (a) $E = \dfrac{1}{4\pi\epsilon_0} \dfrac{qx}{(a^2 + x^2)^{3/2}}$; (b) $E = \dfrac{1}{2\pi^2\epsilon_0} \dfrac{(q_1 - q_2)a}{(a^2 + x^2)^{3/2}}$.

28. *Field due to an electric dipole.* Show that the components of **E** due to a dipole are given, at distant points, by

$$E_x = \frac{1}{4\pi\epsilon_0} \frac{3pxy}{(x^2 + y^2)^{5/2}}$$

$$E_y = \frac{1}{4\pi\epsilon_0} \frac{p(2y^2 - x^2)}{(x^2 + y^2)^{5/2}},$$

where x and y are coordinates of a point in Fig. 27-24. Show that this general result includes the special results of Eq. 27-8b and of Problem 25.

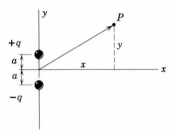

figure 27-24
Problem 28.

29. A "semi-infinite" insulating rod (Fig. 27-25) carries a constant charge per unit length of λ. Show that the electric field at the point P makes an angle of 45° with the rod and that this result is independent of the distance R.

figure 27-25
Problem 29.

30. A nonconducting hemispherical cup of inner radius a has a total charge q spread uniformly over its inner surface. Find the electric field at the center of curvature.

31. A thin nonconducting rod is bent to form the arc of a circle of radius a and subtends an angle θ_0 at the center of the circle. A total charge q is spread uniformly along its length. Find the electric field at the center of the circle in terms of a, q, and θ_0.

   Answer: $E = \dfrac{q}{2\pi\epsilon_0\theta_0 a^2} \sin{(\theta_0/2)}$.

32. A thin circular disk of radius a is charged uniformly so as to have a charge per unit area of σ. Find the electric field on the axis of the disk at a distance r from the disk.

33. *Electric quadrupole.* Figure 27-26 shows a typical electric quadrupole. It consists of two dipoles whose effects at external points do not quite cancel. Show that the value of E on the axis of the quadrupole for points distant r from its center (assume $r \gg a$) is given by

$$E = \frac{3Q}{4\pi\epsilon_0 r^4}$$

where $Q \; (= 2qa^2)$ is called the *quadrupole moment* of the charge distribution.

figure 27-26
Problem 33.

34. One type of "electric quadrupole" is formed by four charges located at the vertices of a square of side $2a$. Point P lies a distance R from the center of the quadrupole on a line parallel to two sides of the square as shown in

Fig. 27-27. For $R \gg a$, show that the electric field at P is approximately given by

$$E = \frac{3(2qa^2)}{2\pi\epsilon_0 R^4}.$$

(Hint: Treat the quadrupole as two dipoles.)

SECTION 27-5

35. A uniform electric field exists in a region between two oppositely charged plates. An electron is released from rest at the surface of the negatively charged plate and strikes the surface of the opposite plate, 2.0 cm away, in a time 1.5×10^{-8} s. (a) What is the speed of the electron as it strikes the second plate? (b) What is the magnitude of the electric field $\mathbf{E}$?
Answer: (a) 2.7×10^6 m/s. (b) 1.0×10^3 N/C.

36. An electron moving with a speed of 5.0×10^8 cm/s is shot parallel to an electric field of strength 1.0×10^3 N/C arranged so as to retard its motion. (a) How far will the electron travel in the field before coming (momentarily) to rest and (b) how much time will elapse? (c) If the electric field ends abruptly after 0.8 cm, what fraction of its initial kinetic energy will the electron lose in traversing it?

37. (a) What is the acceleration of an electron in a uniform electric field of 1.0×10^6 N/C? (b) How long would it take for the electron, starting from rest, to attain one-tenth the speed of light? Assume that Newtonian mechanics holds.
Answer: (a) 1.8×10^{17} m/s². (b) 1.7×10^{-10} s.

38. An electron is projected as in Fig. 27-28 at a speed of 6.0×10^6 m/s and at an angle θ of 45°; $E = 2.0 \times 10^3$ N/C (directed upward), $d = 2.0$ cm, and $l = 10.0$ cm. (a) Will the electron strike either of the plates? (b) If it strikes a plate, where does it do so?

39. *Millikan's oil drop experiment.* R. A. Millikan set up an apparatus (Fig. 27-29) in which a tiny, charged oil drop, placed in an electric field $\mathbf{E}$, could be

figure 27-27
Problem 34.

figure 27-28
Problem 38.

figure 27-29, Problem 39
Millikan's oil drop apparatus. Charged oil drops from the atomizer fall through the hole in the central plate.

"balanced" by adjusting E until the electric force on the drop was equal and opposite to its weight. If the radius of the drop is 1.64×10^{-4} cm and E at balance is 1.92×10^5 N/C, (a) what charge is on the drop in terms of the electronic charge e? (b) Why did Millikan not try to balance electrons in his apparatus instead of oil drops? The density of the oil is 0.851 g/cm³. (Millikan first measured the electronic charge in this way. He measured the drop radius by observing the limiting speed that the drops attained when they fell in air with the electric field turned off. He charged the oil drops by irradiating them with bursts of X-rays.) See *The Electron* by Robert Millikan, 2d ed., University of Chicago Press, 1924.

Answer: (a) 5.0 e. (b) Cannot see electrons; also E at balance would be inconveniently small.

40. In a particular early run (1911), Millikan observed that the following measured charges, among others, appeared at different times on a single drop:

6.563×10^{-19} C	13.13×10^{-19} C	19.71×10^{-19} C
8.204×10^{-19} C	16.48×10^{-19} C	22.89×10^{-19} C
$11.50\ \times 10^{-19}$ C	18.08×10^{-19} C	26.13×10^{-19} C

What value for the elementary charge e can be deduced from these data?

SECTION 27-6

41. *Dipole in a nonuniform field.* (a) Derive an expression for dE/dz at a point midway between two equal positive charges, where z is the distance from one of the charges, measured along the line joining them. (b) Would there be a force on a small dipole placed at this point, its axis being aligned with the z axis? Recall that $\mathbf{E} = 0$ at this point.

Answer: (a) $dE/dz = -8q/\pi\epsilon_0 d^3$, where d is the distance between the charges. (b) Yes.

42. Find the frequency of oscillation of an electric dipole, of moment p and rotational inertia I, for small amplitudes of oscillation about its equilibrium position in a uniform electric field E.

43. A charge $q = 3.0 \times 10^{-6}$ C is 30 cm from a small dipole along its perpendicular bisector. The magnitude of the force on the charge is 5.0×10^{-6} N. Show on a diagram (a) the direction of the force on the charge, (b) the direction of the force on the dipole, and (c) determine the magnitude of the force on the dipole.

Answer: (a) Opposite to **p**. (b, c) 5.0×10^{-6} N, parallel to **p**.

28
gauss's law

In the preceding chapter we saw how we could use Coulomb's law to calculate **E** at various points if we knew enough about the distribution of charges that set up the field. This method always works. It is straightforward but, except in the simplest cases, laborious. Given a versatile enough computor, however, we can always find the answer to any such problem, no matter how complicated.

We can express Coulomb's law in another form, called *Gauss's law*. If we use this formulation, the calculations are not laborious. However, the number of problems that we can solve by the Gauss law formulation is small. Those that we can solve we solve with grace and elegance but, by and large, the Gauss law formulation is more useful for the insights that it gives rather than for practical problem solving.

Before we discuss Gauss's law we must develop a new concept, that of the flux of a vector field.

Flux (symbol Φ) is a property of all vector fields. We are concerned in this chapter with the flux Φ_E of the electric field **E**. By way of introduction however, let us discuss semiquantitatively the more familiar flux of fluid flow Φ_v (see Chapter 18). The word *flux* is derived from the Latin word *fluere* (to flow).

Figure 28-1 shows a stationary, uniform field of fluid flow (water, say) characterized by a constant flow vector **v,** the constant velocity of the fluid at any given point. Figure 28-1*a* suggests, in cross section, a hypothetical plane surface, a circle of radius R and area $\mathbf{A}_{a}$, immersed in the flow field at right angles to **v.** The mass flux $\Phi_{v,a}$ (kg/s) through

figure 28-1

figure 28-1
Showing four hypothetical surfaces immersed in a stationary, uniform flow field of an incompressible fluid (water, say) characterized by a constant field vector **v**, the velocity of the fluid at any given point. The horizontal lines are stream lines. R, in all four cases, is the radius of a circle at right angles to the stream lines.

this surface is given by

$$\Phi_{v,a} = \rho \, v \, A_a \qquad (28\text{-}1a)$$

in which ρ is the fluid density (kg/m³). Check that the dimensions are correct. We may also write this equation in vector notation as

$$\Phi_{v,a} = \rho \mathbf{v} \cdot \mathbf{A}_a. \qquad (28\text{-}1b)$$

Note that flux is a scalar.

Figure 28-1b suggests a plane surface whose projected area $(A_b \cos \theta)$ is equal to A_a. It seems clear that the mass flux Φ_v (kg/s) through surface b must be the same as that through surface a. To gain some insight we can write

$$\Phi_{v,b} = \Phi_{v,a} = \rho v A_a = \rho v (A_b \cos \theta)$$
$$= \rho \mathbf{v} \cdot \mathbf{A}_b. \qquad (28\text{-}2)$$

Figure 28-1c suggests a curved hypothetical surface whose projected area is said, without proof, to equal A_a. Once more it seems clear that $\Phi_{v,c} = \Phi_{v,a}$.

Figure 28-1d suggests a *closed* surface, the preceding three having been open. We assert that the flux $\Phi_{v,d}$ for this closed surface in this flow field is zero and justify it by noting that the amount of fluid (kg/s) that enters the left portion of the surface per unit time also leaves through the right portion. In this case the fluid (assumed incompressible) neither builds up nor disappears within the surface. In the language of Chapter 18 we say that there happen to be no sources or sinks of fluid within the surface. Every stream line that enters on the left leaves on the right.

After these preliminaries we now turn our attention from Φ_v to Φ_E, the flux of the electric field. It may seem that in the latter case nothing is flowing. In a formal sense, however, Eqs. 28-1b and 28-2 do not concern themselves with flow either but deal with the (constant in this case) field vector **v**. If, in Fig. 28-1 we replace **v** by **E** and regard the stream lines as lines of force, all the discussion of this section remains true.

Finally, in what follows we deal only with closed surfaces immersed in the **E** field. This is because we are concerned here with Gauss's law, which is expressed only in terms of closed surfaces.

In the flow of incompressible fluids it is not true in general that, as in the special case of Fig. 28-1d, $\Phi_v = 0$ for *all* closed surfaces. There may be sources or sinks of fluid within the surface, as suggested in Fig. 18-14. In such cases $\Phi_v \neq 0$.

In the same way it is not true that $\Phi_E = 0$ for every closed surface. There *are* sources (positive charges; here $\Phi_E > 0$) and sinks (negative charges; here $\Phi_E < 0$) of **E** that may be located within the hypothetical closed surface immersed in the **E** field.

For closed surfaces in an electric field we shall see below that Φ_E is positive if the lines of force point outward everywhere and negative if they point inward. Figure 28-2 shows two equal and opposite charges and their lines of force. Curves S_1, S_2, S_3, and S_4 are the intersections with the plane of the figure of four hypothetical closed surfaces. From the statement just given, Φ_E is positive for surface S_1 and negative for S_2. Φ_E for surface S_3 (compare Fig. 28-1d) is zero. We shall discuss Φ_E for surface S_4 in Section 28-4. The flux of the electric field is important because Gauss's law, one of the four basic equations of electromagnetism (see Table 40-2), is expressed in terms of it.

To define Φ_E precisely, consider Fig. 28-3, which shows an arbitrary closed surface immersed in a nonuniform electric field. Let the surface be divided into elementary squares ΔS, each of which is small enough so that it may be considered to be plane. Such an element of area can be represented as a vector $\Delta \mathbf{S}$, whose magnitude is the area ΔS; the direction of $\Delta \mathbf{S}$ is taken as the *outward-drawn normal* to the surface.

At every square in Fig. 28-3 we can also construct an electric field vector **E**. Since the squares have been taken to be arbitrarily small, **E** may be taken as constant for all points in a given square.

The vectors **E** and $\Delta \mathbf{S}$ that characterize each square make an angle θ with each other. Figure 28-3b shows an enlarged view of the three

28-3
FLUX OF THE ELECTRIC FIELD

figure 28-2
Two equal and opposite charges. The dashed lines represent the intersection with the plane of the figure of hypothetical closed surfaces.

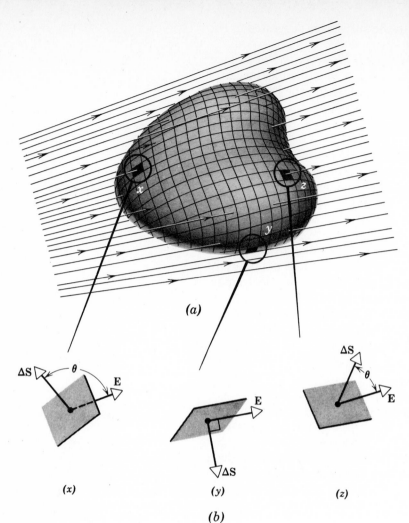

figure 28-3
(*a*) A hypothetical surface immersed
in a nonuniform electric field.
(*b*) Three elements of area on this
surface, shown enlarged.

(*a*)

ΔS θ E
(*x*)

E ΔS
(*y*)

ΔS θ E
(*z*)

(*b*)

squares on the surface of Fig. 28-3*a* marked *x*, *y*, and *z*. Note that at
x, $\theta > 90°$ (**E** points in); at *y*, $\theta = 90°$ (**E** is parallel to the surface); and
at *z*, $\theta < 90°$ (**E** points out).

A semiquantitative definition of flux is

$$\Phi_E \cong \Sigma \mathbf{E} \cdot \Delta \mathbf{S}, \qquad (28\text{-}3)$$

which instructs us to add up the scalar quantity $\mathbf{E} \cdot \Delta\mathbf{S}$ for all elements
of area into which the surface has been divided. For points such as *x* in
Fig. 28-3 the contribution to the flux is negative; at *y* it is zero and at *z*
it is positive. Thus if **E** is everywhere outward, $\theta < 90°$, $\mathbf{E} \cdot \Delta\mathbf{S}$ will be
positive, and Φ_E for the entire surface will be positive; see Fig. 28-2,
surface S_1. If **E** is everywhere inward, $\theta > 90°$, $\mathbf{E} \cdot \Delta\mathbf{S}$ will be negative,
and Φ_E for the surface will be negative; see Fig. 28-2, surface S_2. From
Eq. 28-3 we see that the appropriate SI unit for Φ_E is the newton meter²/
coulomb (N · m²/C).

The exact definition of electric flux is found in the differential limit
of Eq. 28-3. Replacing the sum over the surface by an integral over the
surface yields

$$\Phi_E = \oint \mathbf{E} \cdot d\mathbf{S}. \qquad (28\text{-}4)$$

This *surface integral* indicates that the surface in question is to be divided into infinitesimal elements of area $d\mathbf{S}$ and that the scalar quantity $\mathbf{E} \cdot d\mathbf{S}$ is to be evaluated for each element and the sum taken for the entire surface. The circle on the integral sign indicates that the surface of integration is a closed surface.*

Figure 28-4 shows a hypothetical closed cylinder of radius R immersed in a uniform electric field $\mathbf{E}$, the cylinder axis being parallel to the field. What is Φ_E for this closed surface?

EXAMPLE 1

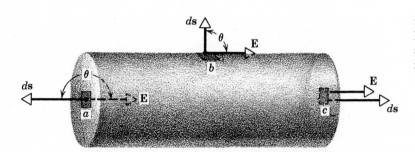

figure 28-4
Example 1. A closed cylindrical surface immersed in a uniform field $\mathbf{E}$ parallel to its axis.

The flux Φ_E can be written as the sum of three terms, an integral over (a) the left cylinder cap, (b) the cylindrical surface, and (c) the right cap. Thus

$$\Phi_E = \oint \mathbf{E} \cdot d\mathbf{S}$$
$$= \int_{(a)} \mathbf{E} \cdot d\mathbf{S} + \int_{(b)} \mathbf{E} \cdot d\mathbf{S} + \int_{(c)} \mathbf{E} \cdot d\mathbf{S}.$$

For the left cap, the angle θ for all points is 180°, $\mathbf{E}$ has a constant value, and the vectors $d\mathbf{S}$ are all parallel. Thus

$$\int_{(a)} \mathbf{E} \cdot d\mathbf{S} = \int E \cos 180° \, dS$$
$$= -E \int dS = -ES,$$

where S $(= \pi R^2)$ is the cap area. Similarly, for the right cap,

$$\int_{(c)} \mathbf{E} \cdot d\mathbf{S} = +ES,$$

the angle θ for all points being zero here. Finally, for the cylinder wall,

$$\int_{(b)} \mathbf{E} \cdot d\mathbf{S} = 0,$$

because $\theta = 90°$, hence $\mathbf{E} \cdot d\mathbf{S} = 0$ for all points on the cylindrical surface. Thus

$$\Phi_E = -ES + 0 + ES = 0.$$

As we shall see in Section 28-4 we expect this because there are no sources or sinks of $\mathbf{E}$, that is, charges, within the closed surface of Fig. 28-4. Lines of (constant) $\mathbf{E}$ enter at the left and emerge at the right, just as in Fig. 28-1d.

---

* Similarly, a circle on a *line* integral sign indicates a closed *path*. It will be clear from the context and from the differential element ($d\mathbf{S}$ in this case) whether we are dealing with a surface integral or a line integral.

Gauss's law, which applies to any closed hypothetical surface (called a *Gaussian surface*), gives a connection between Φ_E for the surface and the net charge q enclosed by the surface. It is

$$\epsilon_0\Phi_E = q \qquad (28\text{-}5)$$

or, using Eq. 28-4,

$$\epsilon_0\oint \mathbf{E} \cdot d\mathbf{S} = q. \qquad (28\text{-}6)$$

The fact that Φ_E proves to be zero in Example 1 is predicted by Gauss's law because no charge is enclosed by the Gaussian surface in Fig. 28-4 $(q = 0)$.

Note that q in Eq. 28-5 (or in Eq. 28-6) is the *net* charge, taking its algebraic sign into account. If a surface encloses equal and opposite charges, the flux Φ_E is zero. Charge *outside* the surface makes no contribution to the value of q, nor does the exact location of the inside charges affect this value.

Gauss's law can be used to evaluate $\mathbf{E}$ if the charge distribution is so symmetric that by proper choice of a Gaussian surface we can easily evaluate the integral in Eq. 28-6. Conversely, if $\mathbf{E}$ is known for all points on a given closed surface, Gauss's law can be used to compute the charge inside. If $\mathbf{E}$ has an outward component for every point on a closed surface, Φ_E, as Eq. 28-4 shows, will be positive and, from Eq. 28-6, there must be a net positive charge within the surface (see Fig. 28-2, surface S_1). If $\mathbf{E}$ has an inward component for every point on a closed surface, there must be a net negative charge within the surface (see Fig. 28-2, surface S_2). Surface S_3 in Fig. 28-2 encloses no charge, so that Gauss's law predicts that $\Phi_E = 0$. This is consistent with the fact that lines of $\mathbf{E}$ pass directly through surface S_3, the contribution to the integral on one side canceling that on the other. For surface S_4 in Fig. 28-2, $\Phi_E = 0$ because the algebraic sum of the charges within the surface is zero. Put another way, as for surface S_3, as many lines of force leave the surface as enter it.

Coulomb's law can be deduced from Gauss's law and symmetry considerations. To do so, let us apply Gauss's law to an isolated point charge q as in Fig. 28-5. Although Gauss's law holds for any surface whatever, information can most readily be extracted for a spherical surface of radius r centered on the charge. The advantage of this surface is that, from symmetry, $\mathbf{E}$ must be normal to it and must have the same (as yet unknown) magnitude for all points on the surface.

In Fig. 28-5 both $\mathbf{E}$ and $d\mathbf{S}$ at any point on the Gaussian surface are directed radially outward. The angle between them is zero and the quantity $\mathbf{E} \cdot d\mathbf{S}$ becomes simply $E\, dS$. Gauss's law (Eq. 28-6) thus reduces to

$$\epsilon_0\oint \mathbf{E} \cdot d\mathbf{S} = \epsilon_0\oint E\, dS = q.$$

Because E is constant for all points on the sphere, it can be factored from inside the integral sign, leaving

$$\epsilon_0 E\oint dS = q,$$

where the integral is simply the area of the sphere.* This equation gives

---

28-4
GAUSS'S LAW†

28-5
GAUSS'S LAW AND COULOMB'S LAW

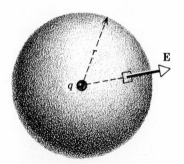

figure 28-5
A spherical Gaussian surface of radius r surrounding a point charge q.

---

* The usefulness of Gauss's law depends on our ability to find a surface over which, from symmetry, both E and θ (see Fig. 28-3) have constant values. Then $E \cos \theta$ can be factored out of the integral and E can be found simply, as in this example.

† See "Gauss," Ian Stewart, *Scientific American*, July 1977 for a fascinating account of the life of this remarkable man.

$$\epsilon_0 E(4\pi r^2) = q$$

or $$E = \frac{1}{4\pi\epsilon_0}\frac{q}{r^2}. \qquad (28\text{-}7)$$

Equation 28-7 gives the magnitude of the electric field **E** at any point a distance r from an isolated point charge q; compare Eq. 27-4. The direction of **E** is already known from symmetry.

Let us put a second point charge q_0 at the point at which **E** is calculated. The magnitude of the force that acts on it (see Eq. 27-2) is

$$F = Eq_0.$$

Combining with Eq. 28-7 gives

$$F = \frac{1}{4\pi\epsilon_0}\frac{qq_0}{r^2},$$

which is precisely Coulomb's law. Thus we have deduced Coulomb's law from Gauss's law and considerations of symmetry.

Gauss's law is one of the fundamental equations of electromagnetic theory and is displayed in Table 40-2 as one of Maxwell's equations. Coulomb's law is not listed in that table because, as we have just proved, it can be deduced from Gauss's law and from simple assumptions about the symmetry of **E** due to a point charge.

It is interesting to note that writing the proportionality constant in Coulomb's law as $1/4\pi\epsilon_0$ (see Eq. 26-3) permits a particularly simple form for Gauss's law (Eq. 28-5). If we had written the Coulomb law constant simply as k, Gauss's law would have to be written as $(1/4\pi k)\Phi_E = q$. We prefer to leave the factor 4π in Coulomb's law so that it will not appear in Gauss's law or in other much used relations that will be derived later.

Gauss's law can be used to make an important prediction, namely: *An excess charge, placed on an insulated conductor, resides entirely on its outer surface.* This hypothesis was shown to be true by experiment (see Section 28-7) before either Gauss's law or Coulomb's law was advanced. Indeed, the experimental proof of the hypothesis is the experimental foundation upon which both laws rest: We have already pointed out that Coulomb's torsion balance experiments, although direct and convincing, are not capable of great accuracy. In showing that the italicized hypothesis is predicted by Gauss's law, we are simply reversing the historical situation.

Figure 28-6 is a cross section of an insulated conductor of arbitrary shape carrying an excess charge q. The dashed lines show a Gaussian surface that lies a small distance below the actual surface of the conductor. Although the Gaussian surface can be as close to the actual surface as we wish, it is important to keep in mind that the Gaussian surface is *inside* the conductor.

When an excess charge is placed at random on an insulated conductor, it will set up electric fields inside the conductor. These fields act on the charge carriers of the conductor (electrons) and cause them to move, that is, they set up internal currents. These currents redistribute the excess charge in such a way that the internal electric fields are automatically reduced in magnitude. Eventually the electric fields inside the conductor become zero everywhere, the currents automatically stop, and electrostatic conditions prevail. This redistribution of charge normally takes place in a time that is negligible for most purposes. What

28-6
AN INSULATED CONDUCTOR

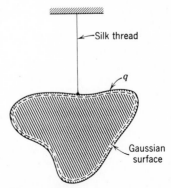

figure 28-6
An insulated metallic conductor.

can be said about the distribution of the excess charge when such electrostatic conditions have been achieved?

If, at electrostatic equilibrium, **E** is zero everywhere inside the conductor, it must be zero for every point on the Gaussian surface because this surface lies inside the conductor. This means that the flux Φ_E for this surface must be zero. Gauss's law then predicts (see Eq. 28-5) that there must be no net charge inside the Gaussian surface. If the excess charge q is not *inside* this surface, it can only be *outside* it, that is, *it must be on the actual surface of the conductor.*

Let us turn to the experiments that prove that the hypothesis of Section 28-6 is true. For a simple test, charge a metal ball and lower it with a silk thread deep into a metal can as in Fig. 28-7. Touch the ball to the inside of the can; when the ball is removed from the can, all its charge will have vanished. When the metal ball touches the can, the ball and can together form an "insulated conductor" to which the hypothesis of Section 28-6 applies. That the charge moves entirely to the outside surface of the can can be shown by touching a small insulated metal object to the can; only on the *outside* of the can will it be possible to pick up a charge.

Benjamin Franklin seems to have been the first to notice that there can be no charge inside an insulated metal can. In 1755 he wrote to a friend:

> I electrified a silver pint cann, on an electric stand, and then lowered into it a cork-ball, of about an inch diameter, hanging by a silk string, till the cork touched the bottom of the cann. The cork was not attracted to the inside of the cann as it would have been to the outside, and though it touched the bottom, yet when drawn out, it was not found to be electrified by that touch, as it would have been by touching the outside. The fact is singular. You require the reason; I do not know it. . . .

About ten years later Franklin recommended this "singular fact" to the attention of his friend Joseph Priestley (1733–1804). In 1767 (about twenty years before Coulomb's experiments) Priestley checked Franklin's observation and, with remarkable insight, realized that the inverse square law of force followed from it. Thus the indirect approach is not only more accurate than the direct approach of Section 26-4 but was carried out earlier.

Priestley, reasoning by analogy with gravitation, said that the fact that no electric force acted on Franklin's cork ball when it was surrounded by a deep metal can is similar to the fact (see Section 16-6) that no gravitational force acts on a mass inside a spherical shell of matter; if gravitation obeys an inverse square law, perhaps the electrical force does also. In Priestley's words:

> May we not infer from this [that is, Franklin's experiment] that the attraction of electricity is subject to the same laws with that of gravitation and is therefore according to the squares of the distances; since it is easily demonstrated that were the earth in the form of a shell, a body in the inside of it would not be attracted to one side more than another?

Michael Faraday also carried out experiments designed to show that excess charge resides on the outside surface of a conductor. In particular, he built a large metal-covered box which he mounted on insulating supports and charged with a powerful electrostatic generator. In Faraday's words:

> I went into the cube and lived in it, and using lighted candles, electrometers, and all other tests of electrical states, I could not find the least influence upon them . . . though all the time the outside of the cube was very powerfully charged, and large sparks and brushes were darting off from every part of its outer surface.

28-7
EXPERIMENTAL PROOF OF GAUSS'S AND COULOMB'S LAWS

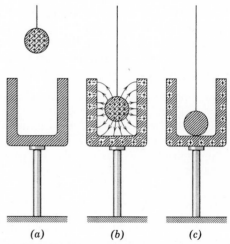

(a) $\quad$ (b) $\quad$ (c)

figure 28-7
The *entire* charge on the ball is transferred to the *outside* of the can. This statement and the discussion of the first paragraph of Section 28-7 are strictly correct only if the can is provided with a conducting lid which can be closed after the ball is inserted; otherwise we cannot define "outside surface."

table 28-1
Test of Coulomb's inverse square law[a]

Experimenters	Date	n
Benjamin Franklin[c]	1755	—
Joseph Priestley[c]	1767	". . . according to the squares of the distance . . ."
John Robison[b]	1769	$\leqslant 0.06$
Henry Cavendish[b,c]	1773	$\leqslant 0.02$
Charles A. Coulomb	1785	a few percent at most
James Clerk Maxwell[c]	1873	$\leqslant 5 \times 10^{-5}$
Samuel J. Plimpton and Willard E. Lawton[c,d]	1936	$\leqslant 2 \times 10^{-9}$
Edwin R. Williams, James E. Faller, and Henry A. Hill[c,e]	1971	$\leqslant 2 \times 10^{-16}$

[a] Values of n (see Eq. 28-8) are subject to a probable error, not shown. All results are consistent with $n = 0$.
[b] Robison's and Cavendish's results were not made public until after Coulomb had published his results.
[c] These are "Gauss's law" experiments, in the spirit of Fig. 28-7. The others are direct tests of Coulomb's law.
[d] Work done at Worcester Polytechnic Institute.
[e] Work done at Wesleyan University.

For many reasons it is important to know whether or not the exponent in Coulomb's law is exactly "2"; experiments based on Gauss's law can help to determine this. Let us write Coulomb's law as

$$F = \frac{1}{4\pi\epsilon_0} \frac{q_1 q_2}{r^{2+n}} \qquad (28\text{-}8)$$

in which $n = 0$ yields an exact inverse square law.* Table 28-1 shows the progress made in determining how closely n in Eq. 28-8 approaches zero.

Figure 28-8 is an idealized sketch of the apparatus of Plimpton and

figure 28-8
A representation of the apparatus of Plimpton and Lawton.

---

*If $n \neq 0$ Maxwell's equations (Table 40-2), including Gauss's law, must be modified and the photon, the particle aspect of light, must have a nonzero mass, contrary to the usual expectation. See "The Mass of the Photon" by Alfred Scharff Goldhaber and Michael Martin Nieto, *Scientific American*, May 1976 (p. 86) for a readable account, including magnetic as well as electric field considerations.

Lawton; see Table 28-1. It consists in principle of two concentric metal shells, A and B, the former being 5 ft in diameter. The inner shell contains a sensitive electrometer E connected so that it will indicate whether any charge moves between shells A and B. If the shells are connected electrically, any charge placed on the shell assembly should reside entirely on shell A (see Section 28-6) if Gauss's law—and thus Coulomb's law—are correct as stated.

By throwing switch S to the left, a substantial charge can be placed on the sphere assembly. If any of this charge moves to shell B, it will have to pass through the electrometer and will cause a deflection, which can be observed optically using telescope T, mirror M, and windows W.

However, when the switch S is thrown alternately from left to right, thus connecting the shell assembly either to the battery or to the ground, no effect is observed on the galvanometer. Knowing the sensitivity of their electrometer, Plimpton and Lawton calculated that n in Eq. 28-8 has the value shown in Table 28-1.

28-8
GAUSS'S LAW—SOME APPLICATIONS

Gauss's law can be used to calculate **E** if the symmetry of the charge distribution is high. One example of this, the calculation of **E** for a point charge, has already been discussed (Eq. 28-7). Here we present other examples.

EXAMPLE 2

Spherically symmetric charge distribution. Figure 28-9 shows a spherical distribution of charge of radius R. The *charge density* ρ (that is, the charge per unit volume, measured in C/m³) at any point depends only on the distance of the point from the center and not on the direction, a condition called *spherical symmetry*. Find an expression for E for points (a) outside and (b) inside the charge distribution. Note that the object in Fig. 28-9 cannot be a conductor or, as we have seen, the excess charge would reside on its surface.

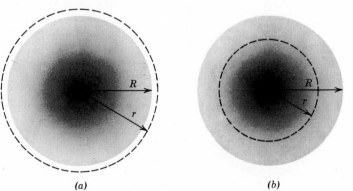

(a)   *(b)*

figure 28-9
Example 2. A spherically symmetric charge distribution, showing two Gaussian surfaces. The density of charge, as the shading suggests, varies with distance from the center but not with direction.

Applying Gauss's law to a spherical Gaussian surface of radius r in Fig. 28-9a (see Section 28-5) leads exactly to Eq. 28-7, or

$$E = \frac{1}{4\pi\epsilon_0} \frac{q}{r^2}, \qquad (28\text{-}7)$$

where q is the total charge. Thus for points outside a spherically symmetric distribution of charge, the electric field has the value that it would have if the charge were concentrated at its center. This reminds us that a spherically symmetric distribution of mass m behaves gravitationally, for outside points, as if the mass were concentrated at its center. At the root of this similarity lies the fact that both Coulomb's law and the law of gravitation are inverse square laws. The gravitational case was proved in detail in Section 16-6; the proof using Gauss's law in the electrostatic case is certainly much simpler.

Figure 28-9b shows a spherical Gaussian surface of radius r drawn *inside* the charge distribution. Gauss's law (Eq. 28-6) gives

$$\epsilon_0 \oint \mathbf{E} \cdot d\mathbf{S} = \epsilon_0 E(4\pi r^2) = q'$$

or

$$E = \frac{1}{4\pi\epsilon_0} \frac{q'}{r^2},$$

in which q' is that part of q contained within the sphere of radius r. The part of q that lies outside this sphere makes no contribution to $\mathbf{E}$ at radius r. This corresponds, in the gravitational case (Section 16-6), to the fact that a spherical shell of matter exerts no gravitational force on a body inside it.

A special case of a spherically symmetric charge distribution is a uniform sphere of charge. For such a sphere, which would be suggested by uniform shading in Fig. 28-9, the charge density ρ would have a constant value for all points within a sphere of radius R and would be zero for all points outside this sphere. For points inside such a uniform sphere of charge we can put

$$q' = q \frac{\frac{4}{3}\pi r^3}{\frac{4}{3}\pi R^3}$$

or

$$q' = q\left(\frac{r}{R}\right)^3$$

where $\frac{4}{3}\pi R^3$ is the volume of the spherical charge distribution. The expression for E then becomes

$$E = \frac{1}{4\pi\epsilon_0} \frac{qr}{R^3}. \qquad (28\text{-}9)$$

This equation becomes zero, as it should, for $r = 0$. Note that Eqs. 28-7 and 28-9 give the same result, as they must, for points on the surface of the charge distribution (that is, if $r = R$). Note that Eq. 28-9 does not apply to the charge distribution of Fig. 28-9b because the charge density, suggested by the shading, is *not* constant in that case.

EXAMPLE 3

The Thomson atom model. At one time the positive charge in the atom was thought to be distributed uniformly throughout a sphere with a radius of about 1.0×10^{-10} m, that is, throughout the entire atom. Calculate the electric field at the surface of a gold atom ($Z = 79$) on this (erroneous) assumption. Neglect the effect of the electrons.

The positive charge of the atom is Ze or $(79)(1.6 \times 10^{-19}$ C). Equation 28-7 yields, for E at the surface,

$$E = \frac{1}{4\pi\epsilon_0} \frac{q}{r^2}$$

$$= \frac{(9.0 \times 10^9 \text{ N·m}^2/\text{C}^2)(79)(1.6 \times 10^{-19} \text{ C})}{(1.0 \times 10^{-10} \text{ m})^2}$$

$$= 1.1 \times 10^{13} \text{ N/C}.$$

Figure 28-10 is a plot of E as a function of distance from the center of the atom,

figure 28-10
Example 3. The electric field due to the positive charge in a gold atom, according to the (erroneous) Thomson model.

using Eqs. 28-7 and 28-9. We see that E has its maximum value on the surface and decreases linearly to zero at the center (see Eq. 28-9). Outside the sphere E decreases as the inverse square of the distance (see Eq. 28-7).

The Rutherford, or nuclear, atom model. We shall see in Section 28-9 that the positive charge of the atom is *not* spread uniformly throughout the atom (see Example 3) but is concentrated in a small region (the *nucleus*) at the center of the atom. For gold the radius of the nucleus is about 6.9×10^{-15} m. What is the electric field at the nuclear surface? Again neglect effects associated with the atomic electrons.

EXAMPLE 4

The problem is the same as that of Example 3, except that the radius is much smaller. This will make the electric field at the surface larger, in proportion to the ratio of the squares of the radii. Thus

$$E = (1.1 \times 10^{13} \text{ N/C}) \frac{(1.0 \times 10^{-10} \text{ m})^2}{(6.9 \times 10^{-15} \text{ m})^2}$$

$$= 2.3 \times 10^{21} \text{ N/C}.$$

This is an enormous electric field, much stronger than could be produced and maintained in the laboratory. It is about 10^8 times as large as the field calculated in Example 3.

---

*An infinite line of charge.** Figure 28-11 shows a section of an infinite line of charge, the *linear charge density* λ (that is, the charge per unit length, measured in C/m) being constant for all points on the line. Find an expression for E at a distance r from the line.

EXAMPLE 5

figure 28-11
Example 5. An infinite line of charge, showing a closed, coaxial cylindrical Gaussian surface.

---

* There are in practice, of course, no such things as infinite lines or sheets of charge. It is a convenient abstraction however, as, for example, was the ideal gas concept developed in Chapter 23. In all cases discussed in this book the assumption of infinite lines (or sheets) of charge allows us to ignore end (or edge) effects.

From symmetry, **E** due to a uniform linear charge can only be radially directed. As a Gaussian surface we choose a circular cylinder of radius r and length h, closed at each end by plane caps normal to the axis. E is constant over the cylindrical surface and the flux of **E** through this surface is $E(2\pi rh)$ where $2\pi rh$ is the area of the surface. There is no flux through the circular caps because **E** here lies in the surface at every point.

The charge enclosed by the Gaussian surface of Fig. 28-11 is λh. Gauss's law (Eq. 28-6),

$$\epsilon_0 \oint \mathbf{E} \cdot d\mathbf{S} = q,$$

then becomes

$$\epsilon_0 E(2\pi rh) = \lambda h,$$

whence

$$E = \frac{\lambda}{2\pi\epsilon_0 r}. \qquad (28\text{-}10)$$

The direction of **E** is radially outward for a line of positive charge.

Note how much simpler is the solution using Gauss's law than that using integration methods, as in Example 6, Chapter 27. Note too that the solution using Gauss's law is possible only if we choose our Gaussian surface to take full advantage of the cylindrical symmetry of the electric field set up by a long line of charge. We are free to choose any surface, such as a cube or a sphere, for a Gaussian surface. Even though Gauss's law holds for all such surfaces, they are not all useful for the problem at hand; only the cylindrical surface of Fig. 28-11 is appropriate in this case.

Gauss's law has the property that it provides a useful technique for calculation only in problems that have a certain degree of symmetry, but in these problems the solutions are strikingly simple.

An infinite sheet of charge. Figure 28-12 shows a portion of a thin, *nonconducting*, infinite sheet of charge, the *surface charge density* σ (that is, the charge per unit area, measured in C/m^2) being constant. What is **E** at a distance r in front of the sheet?

EXAMPLE 6

figure 28-12
Example 6. An infinite nonconducting sheet of charge pierced by a cylindrical Gaussian surface.

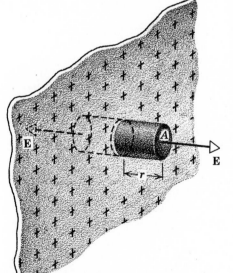

A convenient Gaussian surface is a closed cylinder of cross-sectional area A and height $2r$, arranged to pierce the plane as shown. From symmetry, **E** points at right angles to the end caps and away from the plane. Since **E** does not pierce the cylindrical surface, there is no contribution to the flux from this source. Thus Gauss's law,

$$\epsilon_0 \oint \mathbf{E} \cdot d\mathbf{S} = q$$

becomes

$$\epsilon_0(EA + EA) = \sigma A$$

where σA is the enclosed charge. This gives

$$E = \frac{\sigma}{2\epsilon_0}. \qquad (28\text{-}11)$$

Note that E is the same for all points on each side of the plane; compare Fig. 27-2. Although an infinite sheet of charge cannot exist physically, this derivation is still useful in that Eq. 28-11 yields substantially correct results for real (not infinite) charge sheets if we consider only points not near the edges whose distance from the sheet is small compared to the dimensions of the sheet.

EXAMPLE 7

A charged conductor. Figure 28-13 shows a *conductor* carrying on its surface a charge whose surface charge density at any point is σ; in general σ will vary from point to point. What is $\mathbf{E}$ for points a short distance above the surface?

The direction of $\mathbf{E}$ for points close to the surface is at right angles to the surface, pointing away from the surface if the charge is positive. If $\mathbf{E}$ were *not* normal to the surface, it would have a component lying in the surface. Such a component would act on the charge carriers in the conductor and set up surface currents. Since there are no such currents under the assumed electrostatic conditions, $\mathbf{E}$ must be normal to the surface.

The magnitude of $\mathbf{E}$ can be found from Gauss's law using a small flat closed cylinder of cross section A as a Gaussian surface. Since $\mathbf{E}$ equals zero everywhere inside the conductor (see Section 28-6), the only contribution to Φ_E is through the plane cap of area A that lies outside the conductor. Gauss's law

$$\epsilon_0 \oint \mathbf{E} \cdot d\mathbf{S} = q$$

becomes

$$\epsilon_0(EA) = \sigma A$$

where σA is the net charge within the Gaussian surface. This yields

$$E = \frac{\sigma}{\epsilon_0}. \qquad (28\text{-}12)$$

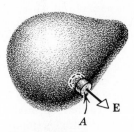

figure 28-13
Example 7. A charged insulated conductor, showing a Gaussian surface.

Comparison with Eq. 28-11 shows that the electric field is twice as great near a *conductor* carrying a charge whose surface charge density is σ as that near a *nonconducting sheet* with the same surface charge density. Compare the Gaussian surfaces in Figs. 28-12 and 28-13 carefully. In Fig. 28-12 lines of force leave the surface through *each* end cap, an electric field existing on *both* sides of the sheet. In Fig. 28-13 the lines of force leave only through the *outside* end cap, the inner end cap being inside the conductor where no electrical field exists. If we assume the same surface charge density and cross-sectional area A for the two Gaussian surfaces, the enclosed charge $(= \sigma A)$ will be the same. Since, from Gauss's law, the flux Φ_E must then be the same in each case, it follows that $E = (\Phi_E/A)$ must be twice as large in Fig. 28-13 as in Fig. 28-12. It is helpful to note that in Fig. 28-12 half the flux emerges from one side of the surface and half from the other, whereas in Fig. 28-13 all the flux emerges from the outside surface.

28-9
THE NUCLEAR MODEL OF THE ATOM

Ernest Rutherford (1871–1937) was first led, in 1911, to assume that the atomic nucleus existed when he tried to interpret some experiments carried out at the University of Manchester by his collaborators H. Geiger and E. Marsden.* The results of Examples 3 and 4 played an important part in Rutherford's analysis of these experiments.

* See "The Birth of the Nuclear Atom," E. N. da C. Andrade, *Scientific American*, November 1956. See also Example 6, Chapter 10.

These workers, at Rutherford's suggestion, allowed a beam of α-particles* to strike and be deflected by a thin film of a heavy element such as gold. They counted the number of particles deflected through various angles ϕ. Figure 28-14 shows the experimental setup schematically. Figure 28-15 shows the paths taken by typical α-particles as they scatter from atoms in a gold foil; the angles ϕ through which the α-particles are deflected range from 0 to 180° as the character of the collision varies from "grazing" to "head-on."

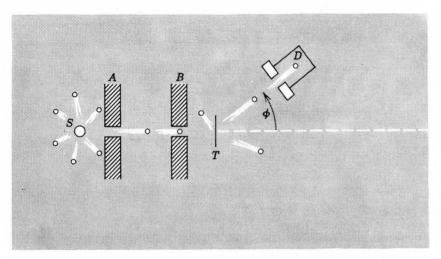

figure 28-14
Experimental arrangement for studying the scattering of α-particles. Particles from radioactive source S are allowed to fall on a thin metal "target" T; α-particles scattered by the target through an (adjustable) angle ϕ are counted by detector D.

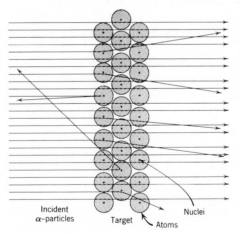

figure 28-15
The deflection of the incident α-particles depends on the nature of the nuclear collision. (From "The Birth of the Nuclear Atom," E. N. Andrade, *Scientific American*, November 1956.)

The electrons in the gold atom, being so light, have almost no effect on the motion of an oncoming α-particle; the electrons are themselves strongly deflected, just as a swarm of insects would be by a stone hurled through them. Any deflection of the α-particle must be caused by the repulsive action of the positive charge of the gold atom, which was known to possess most of the mass of the atom.

At the time of these experiments most physicists believed in the so-called "plum pudding" model of the atom that had been suggested by J. J. Thomson (1856–1940). In this view (see Example 3) the positive

---

* α-particles are helium nuclei that are emitted spontaneously by some radioactive materials such as radium. They move with speeds of the order of one-thirtieth that of light when so emitted.

charge of the atom was thought to be spread out through the whole atom, that is, through a spherical volume of radius about 10^{-10} m. The electrons were thought to vibrate about fixed centers inside this sphere.

Rutherford showed that this model of the atom was not consistent with the α-scattering experiments and proposed instead the nuclear model of the atom that we now accept. Here the positive charge is confined to a very much smaller sphere whose radius is about 10^{-14} m (the *nucleus*). The electrons move around this nucleus and occupy a roughly spherical volume of radius about 10^{-10} m. This brilliant deduction by Rutherford laid the foundation for modern atomic and nuclear physics.

The feature of the α-scattering experiments that attracted Rutherford's attention at once was that a few α-particles are deflected through very large angles, up to 180°. To scientists accustomed to thinking in terms of the "plum pudding" model, this was a very surprising result. In Rutherford's words: "It was quite the most incredible event that ever happened to me in my life. It was almost as incredible as if you had fired a 15-inch shell at a piece of tissue paper and it came back and hit you."

The α-particle must pass through a region in which the electric field is very high indeed in order to be deflected so strongly.* Example 3 shows that, in Thomson's model, the maximum electric field is 1.1×10^{13} N/C. Compare this with the value calculated in Example 4 for a point on the surface of a gold nucleus (2.3×10^{21} N/C). Thus the deflecting force acting on an α-particle can be up to 10^8 times as great if the positive charge of the atom is compressed into a small enough region (the nucleus) at the center of the atoms. Rutherford made his hypothesis about the existence of nuclei only after a much more detailed mathematical analysis than that given here.

questions

1. By analogy with Φ_E, how would you define the flux Φ_g of a gravitational field? What is the flux of the earth's gravitational field through the boundaries of a room, assumed to contain no matter?

2. A point charge is placed at the center of a spherical Gaussian surface. Is Φ_E changed (a) if the surface is replaced by a cube of the same volume, (b) if the sphere is replaced by a cube of one-tenth the volume, (c) if the charge is moved off-center in the original sphere, still remaining inside, (d) if the charge is moved just outside the original sphere, (e) if a second charge is placed near, and outside, the original sphere, and (f) if a second charge is placed inside the Gaussian surface?

3. In Gauss's law,

$$\epsilon_0 \oint \mathbf{E} \cdot d\mathbf{S} = q,$$

is $\mathbf{E}$ the electric field attributable to the charge q?

4. Show that Eq. 18-3 illustrates what might be called *Gauss's law for incompressible fluids*, or

$$\Phi_v = \oint \mathbf{v} \cdot d\mathbf{S} = 0.$$

5. A surface encloses an electric dipole. What can you say about Φ_E for this surface?

6. Suppose that a Gaussian surface encloses no net charge. Does Gauss's law require that $\mathbf{E}$ equal zero for all points on the surface? Is the converse of this statement true, that is, if $\mathbf{E}$ equals zero everywhere on the surface, does Gauss's law require that there be no net charge inside?

---

* The chance that a big deflection can result from the combined effects of many small deflections can be shown to be very small.

7. Is Gauss's law useful in calculating the field due to three equal charges located at the corners of an equilateral triangle? Explain.

8. The use of line, surface, and volume densities of charge to calculate the charge contained in an element of a charged object implies a continuous distribution of charge, whereas, in fact, charge on the microscopic scale is discontinuous. How, then, is this procedure justified?

9. Is **E** necessarily zero inside a charged rubber balloon if the balloon is (a) spherical or (b) sausage-shaped? For each shape assume the charge to be distributed uniformly over the surface.

10. A spherical rubber balloon carries a charge that is uniformly distributed over its surface. How does E vary for points (a) inside the balloon, (b) at the surface of the balloon, and (c) outside the balloon, as the balloon is blown up?

11. We have seen that there are sources (positive charges) and sinks (negative charges) for the **E** field. Are there sources and/or sinks for (a) the **v** field in fluid flow and (b) for the gravitational **g** field?

12. Is it precisely true that Gauss's law states that the total number of lines of force crossing any closed surface in the outward direction is proportional to the net positive charge enclosed within the surface?

13. In Section 28-5 we have seen that Coulomb's law can be derived from Gauss's law. Does this *necessarily* mean that Gauss's law can be derived from Coulomb's law?

14. A large, insulated, hollow conductor carries a charge $+q$. A small metal ball carrying a charge $-q$ is lowered by a thread through a small opening in the top of the conductor, allowed to touch the inner surface, and then withdrawn. What is then the charge on (a) the conductor and (b) the ball?

15. Can we deduce from the argument of Section 28-6 that the electrons in the wires of a house wiring system move along the surfaces of those wires. If not, why not?

16. Does Gauss's law, as applied in Section 28-6, require that all the conduction electrons in an insulated conductor reside on the surface?

17. In Section 28-6, we assumed that **E** equals zero everywhere inside a conductor. However, there are certainly very large electric fields inside the conductor, at points close to the electrons or to the nuclei. Does this invalidate the proof of Section 28-6?

18. It is sometimes said that excess charge resides entirely on the outer surface of a conductor because like charges repel and try to get as far away as possible from one another. Comment on this plausibility argument.

19. Would Gauss's law hold if the exponent in Coulomb's law were not exactly two?

20. As you penetrate a uniform sphere of charge, E should decrease because less charge lies inside a sphere drawn through the observation point. On the other hand, E should increase because you are closer to the center of this charge. Which effect predominates and why?

21. Given a spherically symmetric charge distribution (not of uniform density of charge), is E necessarily a maximum at the surface? Comment on various possibilities.

22. Explain in your own words the factor of 2 that distinguishes Eq. 28-11 from Eq. 28-12.

23. Does Eq. 28-7 hold true for Fig. 28-9a if (a) there is a concentric spherical cavity in the body, (b) a point charge Q is at the center of this cavity, and (c) the charge Q is inside the cavity but not at its center?

24. An atom is normally *electrically neutral*. Why then should an α-particle be deflected by the atom under any circumstances?

25. If an α-particle, fired at a gold nucleus, is deflected through 135°, can you conclude (a) that any force has acted on the α-particle or (b) that any net work has been done on it?

26. Explain in your own words why the α-scattering experiments of Rutherford and his colleagues (see Example 4) render the Thomson atom model (see Example 3) untenable.

problems

SECTION 28-1

1. Calculate Φ_E through a hemisphere of radius R. The field of $\mathbf{E}$ is uniform and is parallel to the axis of the hemisphere. *Answer:* $\pi R^2 E$.

2. A butterfly net is in a uniform electric field as shown in Fig. 28-16. The rim, a circle of radius a, is aligned perpendicular to the field. Find the electric flux through the netting.

figure 28-16
Problem 2

SECTION 28-3

3. In Example 1 compute Φ_E for the cylinder if it is turned so that its axis is perpendicular to the electric field. Calculate the flux directly without using Gauss's law. *Answer:* Zero.

4. A point charge of 1.0×10^{-6} C is at the center of a cubical Gaussian surface 0.50 m on edge. What is Φ_E for the surface?

SECTION 28-4

5. Charge on an originally uncharged insulated conductor is separated by holding a positively charged rod nearby, as in Fig. 28-17. What can you learn from Gauss's law about the flux for the five Gaussian surfaces shown? The induced negative charge on the conductor is equal to the positive charge on the rod.
    Answer: $+q$ = charge on rod. $\Phi_{S_1} = q/\epsilon_0$. $\Phi_{S_2} = -q/\epsilon_0$. $\Phi_{S_3} = q/\epsilon_0$. $\Phi_{S_4} = 0$.
    $\Phi_{S_5} = q/\epsilon_0$.

figure 28-17
Problem 5

6. A uniformly charged conducting sphere of 1.0 m diameter has a surface charge density of 8.0 C/m^2. What is the total electric flux leaving the surface of the sphere?

7. The intensity of the earth's electric field near its surface is ~ 130 N/C, pointing down. What is the earth's charge, assuming that this field is caused by it? *Answer:* -6×10^5 C.

8. A point charge q is placed at one corner of a cube of edge a. What is the flux through each of the cube faces? (Hint: Use Gauss's law and symmetry arguments.)

9. "Gauss's law for gravitation" is

$$\frac{1}{4\pi G}\, \Phi_g = \frac{1}{4\pi G} \oint \mathbf{g} \cdot d\mathbf{S} = -m,$$

    where m is the enclosed mass and G is the universal gravitation constant (Section 16-3). Derive Newton's law of gravitation from this.

10. The electric field components in Fig. 28-18 are $E_x = bx^{1/2}$, $E_y = E_z = 0$, in which $b = 800$ N/C $\cdot$ m$^{1/2}$. Calculate (a) the flux Φ_E through the cube and (b) the charge within the cube. Assume that $a = 10$ cm.

figure 28-18
Problem 10

SECTION 28-6

11. Equation 28-12 ($E = \sigma/\epsilon_0$) gives the electric field at points near a charged conducting surface. Show that this equation leads to a believable result when applied to a conducting sphere of radius r, carrying a charge q.

    Answer: It leads to $E = q/4\pi\epsilon_0 r^2$.

12. An insulated conductor carries a net charge of $+10 \times 10^{-6}$ C. Inside the conductor is a hollow cavity inside of which is a point charge Q of $+3.0 \times 10^{-6}$ C. What is the charge (a) on the cavity wall and (b) on the outer surface of the conductor?

13. Figure 28-19 shows a point charge of 1.0×10^{-7} C at the center of a spherical cavity of radius 3.0 cm in a piece of metal. Use Gauss's law to find the electric field (a) at point a, halfway from the center to the surface and (b) at point b. *Answer:* (a) 4.0×10^6 N/C. (b) Zero.

14. Figure 28-20 shows a spherical nonconducting shell of charge of uniform density ρ (C/m³). Plot E for distances r from the center of the shell ranging from zero to 30 cm. Assume that $\rho = 1.0 \times 10^{-6}$ C/m³, $a = 10$ cm, and $b = 20$ cm.

15. An uncharged, spherical, thin, metallic shell has a point charge q at its center. Give expressions for the electric field (a) inside the shell and (b) outside the shell, using Gauss's law. (c) Has the shell any effect on the field due to q? (d) Has the presence of q any effect on the shell? (e) If a second point charge is held outside the shell, does this outside charge experience a force? (f) Does the inside charge experience a force? (g) Is there a contradiction with Newton's third law here?

    Answer: (a) $E = q/4\pi\epsilon_0 r^2$, radially outward. (b) Same as (a). (c) No. (d) Yes, charges are induced on the surfaces. (e) Yes. (f) No. (g) No.

16. A thin, metallic, spherical shell of radius a carries a charge q_a. Concentric with it is another thin, metallic, spherical shell of radius b ($b > a$) carrying a charge q_b. Use Gauss's law to find the electric field at radial points r where (a) $r < a$; (b) $a < r < b$; (c) $r > b$. (d) Discuss the criterion one would use to determine how the charges are distributed on the inner and outer surfaces of each shell.

17. A *nonconducting* sphere of radius a is placed at the center of a spherical *conducting* shell of inner radius b and outer radius c, as in Fig. 28-21. A charge $+Q$ is distributed uniformly through the inner sphere (charge density ρ, C/m³). The outer shell carries $-Q$. Find $E(r)$, (a) within the sphere ($r < a$), (b) between the sphere and the shell ($a < r < b$), (c) inside the shell ($b < r < c$), and (d) outside the shell ($r > c$). (e) What charges appear on the inner and outer surfaces of the shell?

    Answer: (a) $E = (Q/4\pi\epsilon_0 a^3)r$. (b) $E = Q/4\pi\epsilon_0 r^2$. (c) Zero. (d) Zero. (e) Inner, $-Q$; Outer, zero.

18. An irregularly shaped conductor has an irregularly shaped cavity inside. A charge $+Q$ is placed on the conductor but there is no charge inside the cavity. Show that (a) $E = 0$ inside the cavity and (b) there is no charge on the cavity walls.

19. Two concentric conducting spherical shells have radii $R_1 = 0.145$ m and $R_2 = 0.207$ m. The inner sphere bears a charge -6.00×10^{-8} C. An electron escapes from the inner sphere with negligible speed. Assuming that the region between the spheres is a vacuum, compute the speed with which the electron strikes the outer sphere. *Answer:* 2.0×10^7 m/s.

20. The spherical region $a < r < b$ carries a charge per unit volume of $\rho = A/r$, where A is constant. At the center ($r = 0$) of the enclosed cavity is a point charge Q. What should the value of A be so that the electric field in the region $a < r < b$ has constant magnitude?

21. A solid nonconducting sphere carries a uniform charge per unit volume ρ. Let $\mathbf{r}$ be the vector from the center of the sphere to a general point P within the sphere. (a) Show that the electric field at P is given by $\mathbf{E} = \rho\mathbf{r}/3\epsilon_0$. (b) A spherical cavity is created in the above sphere, as shown in Fig. 28-22.

figure 28-19
Problem 13

figure 28-20
Problem 14

figure 28-21
Problem 17

figure 28-22
Problem 21

Using superposition concepts show that the electric field at all points within the cavity is $\mathbf{E} = \rho\mathbf{a}/3\epsilon_0$ (uniform field), where $\mathbf{a}$ is the vector connecting the center of the sphere with the center of the cavity. Note that both these results are independent of the radii of the sphere and the cavity.

22. Figure 28-23 shows a section through a long, thin-walled metal tube of radius R, carrying a charge per unit length λ on its surface. Derive expressions for E for various distances r from the tube axis, considering both (a) $r > R$ and (b) $r < R$. Plot your results for the range $r = 0$ to $r = 5$ cm, assuming that $\lambda = 2.0 \times 10^{-8}$ C/m and $R = 3.0$ cm.

23. A long conducting cylinder (length l) carrying a total charge $+q$ is surrounded by a conducting cylindrical shell of total charge $-2q$, as shown in cross section in Fig. 28-24. Use Gauss's law to find (a) the electric field at points outside the conducting shell, (b) the distribution of the charge on the conducting shell, and (c) the electric field in the region between the cylinders.

Answer: (a) $E = q/2\pi\epsilon_0 lr$, radially inward. (b) $-q$ on both inner and outer surfaces. (c) $E = q/2\pi\epsilon_0 lr$, radially outward.

24. Two charged concentric cylinders have radii of 3.0 cm and 6.0 cm. The charge per unit length on the inner cylinder is 5.0×10^{-6} C/m and that on the outer cylinder is -7.0×10^{-6} C/m. Find the electric field at (a) $r = 4.0$ cm and (b) at $r = 8.0$ cm.

25. Figure 28-25 shows a section through two long concentric cylinders of radii a and b. The cylinders carry equal and opposite charges per unit length λ. Using Gauss's law prove (a) that $E = 0$ for $r > b$ and for $r < a$ and (b) that between the cylinders E is given by

$$E = \frac{1}{2\pi\epsilon_0}\frac{\lambda}{r}.$$

26. In Problem 25 a positive electron revolves in a circular path of radius r, between and concentric with the cylinders. What must be its kinetic energy K? Assume $a = 2.0$ cm, $b = 3.0$ cm, and $\lambda = 3.0 \times 10^{-8}$ C/m.

27. Charge is distributed uniformly throughout an infinitely long cylinder of radius R. (a) Show that E at a distance r from the cylinder axis ($r < R$) is given by

$$E = \frac{\rho r}{2\epsilon_0},$$

where ρ is the density of charge (C/m³). (b) What result do you expect for $r > R$? Answer: (b) $\rho R^2/2\epsilon_0 r$.

28. A metal plate 8.0 cm on a side carries a total charge of 6.0×10^{-6} C. (a) Estimate the electric field 0.50 cm above the surface of the plate near the plate's center. (b) Estimate the field at a distance of 3.0 m.

29. Two large metal plates face each other as in Fig. 28-26 and carry charges with surface charge density $+\sigma$ and $-\sigma$, respectively, on their inner surfaces. What is $\mathbf{E}$ at points (a) to the left of the sheets, (b) between them, and (c) to the right of the sheets. Consider only points not near the edges whose distance from the sheets is small compared to the dimensions of the sheet. Answer: (a) Zero. (b) $E = \sigma/\epsilon_0$, to the left. (c) Zero.

30. Two large metal plates of area 1.0 m² face each other. They are 5.0 cm apart and carry equal and opposite charges on their inner surfaces. If E between the plates is 55 N/C, what is the charge on the plates? Neglect edge effects.

31. Two large nonconducting sheets of positive charge face each other as in Fig. 28-27. What is $\mathbf{E}$ at points (a) to the left of the sheets, (b) between them, and (c) to the right of the sheets? Assume the same surface charge density σ for each sheet. Consider only points not near the edges whose distance from the sheets is small compared to the dimensions of the sheet. (Hint: $\mathbf{E}$ at any point is the vector sum of the separate electric fields set up by each sheet.) Answer: (a) $E = \sigma/\epsilon_0$, to the left. (b) $E = 0$. (c) $E = \sigma/\epsilon_0$, to the right.

figure 28-23
Problem 22

figure 28-24
Problem 23

figure 28-25
Problem 25

figure 28-26
Problem 29

figure 28-27
Problem 31

32. A nonconducting plane slab of thickness d has a uniform volume charge density ρ. Find the magnitude of the electric field at all points in space both (*a*) inside and (*b*) outside the slab.

33. A 100-eV electron is fired directly toward a large metal plate that has a surface charge density of -2.0×10^{-6} C/m². From what distance must the electron be fired if it is to just fail to strike the plate?   *Answer:* 0.44 mm.

34. A small sphere whose mass m is 1.0×10^{-3} g carries a charge q of 2.0×10^{-8} C. It hangs from a silk thread which makes an angle of 30° with a large, charged nonconducting sheet as in Fig. 28-28. Calculate the surface charge density σ for the sheet.

35. Show that stable equilibrium under the action of electrostatic forces alone is impossible. (Hint: Assume that at a certain point P in an **E** field a charge $+q$ would be in stable equilibrium if it were placed there — which it is not. Draw a spherical Gaussian surface about P, imagine how **E** must point on this surface, and apply Gauss's law.)

figure 28-28
Problem 34

SECTION 28-9

36. A gold foil used in a Rutherford scattering experiment is 3×10^{-5} cm thick. (*a*) What fraction of its surface area is "blocked out" by gold nuclei, assuming a nuclear radius of 6.9×10^{-15} m? Assume that no nucleus is screened by any other. (*b*) What fraction of the volume of the foil is occupied by the nuclei? (*c*) What fills all the rest of the space in the foil?

37. An α-particle, approaching the surface of a nucleus of gold, is a distance equal to one nuclear radius $(6.9 \times 10^{-15}$ m) away from that surface. What are (*a*) the force on the α-particle and (*b*) its acceleration at that point? The mass of the α-particle, which may be treated as a point, is 6.7×10^{-27} kg.
Answer: (*a*) 190 N. (*b*) 2.9×10^{28} m/s².

29
electric potential

The electric field around a charged rod can be described not only by a (vector) electric field **E** but also by a scalar quantity, the *electric potential V*. These quantities are intimately related, and often it is only a matter of convenience which is used in a given problem.

To find the *electric potential difference* between two points A and B in an electric field, we move a test charge q_0 from A to B, always keeping it in equilibrium, and we measure the work W_{AB} that must be done by the agent moving the charge. The electric potential difference* is defined from

$$V_B - V_A = \frac{W_{AB}}{q_0}. \qquad (29\text{-}1)$$

The work W_{AB} may be (*a*) positive, (*b*) negative, or (*c*) zero. In these cases the electric potential at B will be (*a*) higher, (*b*) lower, or (*c*) the same as the electric potential at A.

The SI unit of potential difference that follows from Eq. 29-1 is the joule/coulomb. This combination occurs so often that a special unit, the *volt* (abbr. V), is used to represent it; that is,

$$1 \text{ volt} = 1 \text{ joule/coulomb}.$$

Usually point A is chosen to be at a large (strictly an infinite) distance from all charges, and the electric potential V_A at this infinite distance is arbitrarily taken as zero. This allows us to define the *electric potential*

---

* This definition of potential difference, though conceptually sound and suitable for our present purpose, is rarely carried out in practice because of technical difficulties. Equivalent and more technically feasible methods are usually adopted.

at a point. Putting $V_A = 0$ in Eq. 29-1 and dropping the subscripts leads to

$$V = \frac{W}{q_0},\qquad(29\text{-}2)$$

where W is the work that an external agent must do to move the test charge q_0 from infinity to the point in question. Keep in mind that *potential differences* are of fundamental concern and that Eq. 29-2 depends on the arbitrary assignment of the value zero to the potential V_A at the reference position (infinity); this reference potential could equally well have been chosen as any other value, say -100 V. Similarly, any other agreed-upon point could be chosen as a reference position. In many problems the earth is taken as a reference of potential and assigned the value zero.

Bearing in mind the assumptions made about the reference position, we see from Eq. 29-2 that V near an isolated positive charge is positive because positive work must be done by an outside agent to push a (positive) test charge in from infinity. Similarly, the potential near an isolated negative charge is negative because an outside agent must exert a restraining force on (that is, must do negative work on) a (positive) test charge as it comes in from infinity. Electric potential as defined in Eq. 29-2 is a scalar because W and q_0 in that equation are scalars.

Both W_{AB} and $V_B - V_A$ in Eq. 29-1 are independent of the path followed in moving the test charge from point A to point B. If this were not so, point B would not have a unique electric potential (with respect to point A as a defined reference position) and the concept of potential would not be useful.

Let us now show that the potential difference is path independent for the field due to the point charge of Fig. 29-1. Then we shall show that path independence holds in *all* electrostatic situations. Figure 29-1 shows two points A and B set up in the field of a point charge q, assumed positive.

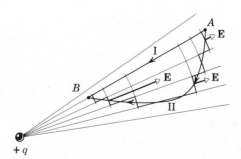

figure 29-1
A test charge q_0 is moved from A to B in the field of charge q along either of two paths. The open arrows show **E** at three points on path II.

Let us imagine a positive test charge q_0 being moved by an external agent from point A to point B by two different paths. Path I is a simple radial line. Path II is a completely arbitrary path between the two points.

We may approximate path II by a broken path made up of alternating elements of circular arc and of radius. Since these elements can be arbitrarily small, the broken path can be made arbitrarily close to the actual path. On path II the external agent does work *only along the radial segments* because along the arcs the force **F** and the displacement $d\mathbf{l}$ are at right angles, $\mathbf{F} \cdot d\mathbf{l}$ being zero in such cases. The sum of the work done on the radial segments that make up path II is the same as the work done on path I because each path has the same array of radial segments. Since path II is arbitrary, we have proved that the work done is the same for *all* paths connecting *A* and *B*.

This proof holds only for the field due to a point charge. However, any charge distribution (discrete or continuous) can be considered as made up of an assembly of point charges or differential charge elements. Hence, from the principle of superposition, we conclude that path independence for the electrostatic potential holds for *all* electrostatic charge configurations.

The locus of points, all of which have the same electric potential, is called an *equipotential surface*. A family of equipotential surfaces, each surface corresponding to a different value of the potential, can be used to give a general description of the electric field in a certain region of space. We have seen earlier (Section 27-3) that electric lines of force can also be used for this purpose; in later sections (see, for example, Fig. 29-15) we explore the intimate connection between these two ways of describing the electric field.

No work is required to move a test charge between any two points on an equipotential surface. This follows from Eq. 29-1,

$$V_B - V_A = \frac{W_{AB}}{q_0},$$

because W_{AB} must be zero if $V_A = V_B$. This is true, because of the path independence of potential difference, even if the path connecting *A* and *B* does not lie entirely in the equipotential surface.

Figure 29-2 shows an arbitrary family of equipotential surfaces. The work to move a charge along paths I and II is zero because both these paths begin and end on the same equipotential surface. The work to move a charge along paths I′ and II′ is not zero but is the same for each path because the initial and the final potentials are identical; paths I′ and II′ connect the same pair of equipotential surfaces.

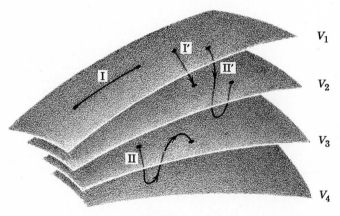

figure 29-2
Portions of four equipotential surfaces. The heavy lines show four paths along which a test charge is moved.

From symmetry, the equipotential surfaces for a spherical charge are a family of concentric spheres. For a uniform field they are a family of planes at right angles to the field. In all cases (including these two examples) the equipotential surfaces are at right angles to the lines of force and thus to **E** (see Fig. 29-15). If **E** were *not* at right angles to the equipotential surface, it would have a component lying in that surface. Then work would have to be done in moving a test charge about on the surface. Work cannot be done if the surface is an equipotential, so **E** must be at right angles to the surface.

There is a strong analogy between electrostatic forces and gravitational forces, based on the fact that their fundamental laws are inverse square laws (see Eqs. 26-3 and 16-1):

$$F_E = \frac{1}{4\pi\epsilon_0} \frac{q_1 q_2}{r^2} \quad \text{and} \quad F_g = G\frac{m_1 m_2}{r^2},$$

and also on the fact that the forces are proportional to the magnitude of the test body (charge in one case, mass in the other). Thus we can define the *gravitational potential* V_g (compare Eq. 29-2) from

$$V_g = \frac{W}{m},$$

where W is the work required to move a test body of mass m from infinity to the point in question. Gravitational equipotential surfaces can also be constructed; they prove to be everywhere at right angles to the gravitational field vector **g**. For a uniform gravitational field, such as that near the surface of the earth, these surfaces are horizontal planes. This correlates with the facts that (*a*) no net work is required to move a stone of mass m between two points with the same elevation and (*b*) the same net work is required to move a stone along any path starting on a given horizontal surface and ending on another. Gravitational forces, like coulombic forces, are conservative; see Section 8-2.

29-2
POTENTIAL AND THE ELECTRIC FIELD

Let A and B in Fig. 29-3 be two points in a uniform electric field **E**, set up by an arrangement of charges not shown, and let A be a distance d in the field direction from B. Assume that a positive test charge q_0 is moved, by an external agent and without acceleration, from A to B along the straight line connecting them.

The electric force on the charge is q_0**E** and points down. To move the charge in the way we have described we must counteract this force by applying an external force **F** of the same magnitude but directed up. The work W done by the agent that supplies this force is

$$W_{AB} = Fd = q_0 Ed. \tag{29-3}$$

Substituting this into Eq. 29-1 yields

$$V_B - V_A = \frac{W_{AB}}{q_0} = Ed. \tag{29-4}$$

This equation shows the connection between potential difference and field strength for a simple special case. Note from this equation that another SI unit for **E** is the volt/meter (V/m). You may wish to prove that a volt/meter is identical with a newton/coulomb (N/C); this latter unit was the one first presented for **E** in Section 27-2.

In Fig. 29-3 B has a higher potential than A. This is reasonable because an external agent would have to do positive work to push a positive test charge from A to B. Figure 29-3 could be used as it stands to illustrate the act of lifting a stone from A to B in the uniform gravitational field near the earth's surface.

What is the connection between V and **E** in the more general case in which the field is *not* uniform and in which the test body is moved along a path that is *not* straight, as in Fig. 29-4? The electric field exerts a force q_0**E** on the test charge, as shown. To keep the test charge from accelerating, an external agent must apply a force **F** chosen to be exactly equal to $-q_0$**E** for all positions of the test body.

If the external agent causes the test body to move through a displacement d**l** along the path from A to B, the element of work done by the

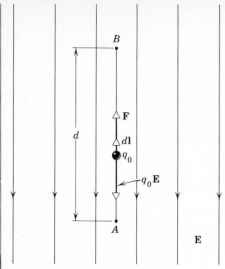

figure 29-3
A positive test charge q_0 is moved from A to B in a uniform electric field $\mathbf{E}$ by an external agent that exerts a force $\mathbf{F}$ on it.

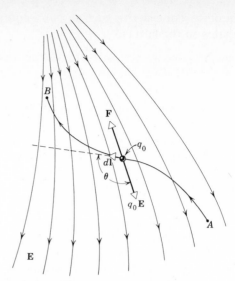

figure 29-4
A positive test charge q_0 is moved from A to B in a nonuniform electric field by an external agent that exerts a force $\mathbf{F}$ on it.

external agent is $\mathbf{F} \cdot d\mathbf{l}$. To find the total work W_{AB} done by the external agent in moving the test charge from A to B, we add up (that is, integrate) the work contributions for all the infinitesimal segments into which the path is divided. This leads to

$$W_{AB} = \int_A^B \mathbf{F} \cdot d\mathbf{l} = -q_0 \int_A^B \mathbf{E} \cdot d\mathbf{l}.$$

Such an integral is called a *line integral.* Note that we have substituted $-q_0\mathbf{E}$ for its equal, $\mathbf{F}$.

Substituting this expression for W_{AB} into Eq. 29-1 leads to

$$V_B - V_A = \frac{W_{AB}}{q_0} = -\int_A^B \mathbf{E} \cdot d\mathbf{l}. \qquad (29\text{-}5)$$

If point A is taken to be infinitely distant and the potential V_A at infinity is taken to be zero, this equation gives the potential V at point B, or, dropping the subscript B,

$$V = -\int_\infty^B \mathbf{E} \cdot d\mathbf{l}. \qquad (29\text{-}6)$$

These two equations allow us to calculate the potential difference between any two points (or the potential at any point) if $\mathbf{E}$ is known at various points in the field.

In Fig. 29-3 calculate $V_B - V_A$ using Eq. 29-5. Compare the result with that obtained by direct analysis of this special case (Eq. 29-4).

In moving the test charge the element of path $d\mathbf{l}$ always points in the direction of motion; this is upward in Fig. 29-3. The electric field $\mathbf{E}$ in this figure points down so that the angle θ between $\mathbf{E}$ and $d\mathbf{l}$ is 180°.

Equation 29-5 then becomes

$$V_B - V_A = -\int_A^B \mathbf{E} \cdot d\mathbf{l} = -\int_A^B E \cos 180° \, dl = \int_A^B E \, dl.$$

EXAMPLE 1

E is constant for all parts of the path in this problem and can thus be taken outside the integral sign, giving

$$V_B - V_A = E \int_A^B dl = Ed,$$

which agrees with Eq. 29-4, as it must.

EXAMPLE 2

In Fig. 29-5 let a test charge q_0 be moved without acceleration from A to B over the path shown. Compute the potential difference between A and B.

For the path AC we have $\theta = 135°$ and, from Eq. 29-5,

$$V_C - V_A = -\int_A^C \mathbf{E} \cdot d\mathbf{l} = -\int_A^C E \cos 135° \, dl = \frac{E}{\sqrt{2}} \int_A^C dl.$$

The integral is the length of the line AC which is $\sqrt{2}d$. Thus

$$V_C - V_A = \frac{E}{\sqrt{2}} (\sqrt{2}d) = Ed.$$

Points B and C have the same potential because no work is done in moving a charge between them, $\mathbf{E}$ and $d\mathbf{l}$ being at right angles for all points on the line CB. In other words, B and C lie on the same equipotential surface at right angles to the lines of force. Thus

$$V_B - V_A = V_C - V_A = Ed.$$

This is the same value derived for a direct path connecting A and B, a result to be expected because the potential difference between two points is path independent.

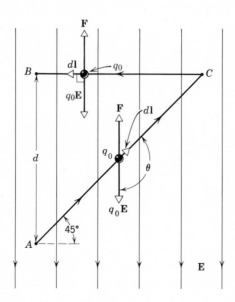

figure 29-5
Example 2. A test charge q_0 is moved along path ACB in a uniform electric field by an external agent.

29-3
POTENTIAL DUE TO A POINT CHARGE

Figure 29-6 shows two points A and B near an isolated positive point charge q. For simplicity we assume that A, B, and q lie on a straight line. Let us compute the potential difference between points A and B, assuming that a test charge q_0 is moved without acceleration along a radial line from A to B.

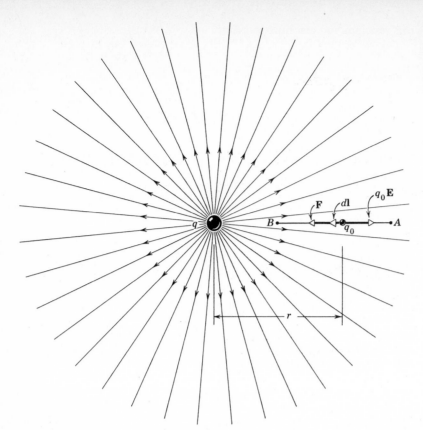

figure 29-6
A test charge q_0 is moved by an external agent from A to B in the field set up by a positive point charge q.

In Fig. 29-6 **E** points to the right and $d\mathbf{l}$, which is always in the direction of motion, points to the left. Therefore, in Eq. 29-5,

$$\mathbf{E} \cdot d\mathbf{l} = E \cos 180° \, dl = -E \, dl.$$

However, as we move a distance dl to the left, we are moving in the direction of decreasing r because r is measured from q as an origin. Thus

$$dl = -dr.$$

Combining yields $\qquad \mathbf{E} \cdot d\mathbf{l} = E \, dr.$

Substituting this into Eq. 29-5 gives

$$V_B - V_A = -\int_A^B \mathbf{E} \cdot d\mathbf{l} = -\int_{r_A}^{r_B} E \, dr.$$

Combining with Eq. 27-4,

$$E = \frac{1}{4\pi\epsilon_0} \frac{q}{r^2}$$

leads to $\qquad V_B - V_A = -\frac{q}{4\pi\epsilon_0} \int_{r_A}^{r_B} \frac{dr}{r^2} = \frac{q}{4\pi\epsilon_0} \left(\frac{1}{r_B} - \frac{1}{r_A} \right).$ (29-7)

Choosing reference position A to be at infinity (that is, letting $r_A \to \infty$), choosing $V_A = 0$ at this position, and dropping the subscript B leads to

$$V = \frac{1}{4\pi\epsilon_0} \frac{q}{r}.$$ (29-8)

This equation shows clearly that equipotential surfaces for an isolated point charge are spheres concentric with the point charge (see Fig. 29-15a). A study of the derivation will show that this relation also holds for points external to spherically symmetric charge distributions.

EXAMPLE 3

What must the magnitude of an isolated positive point charge be for the electric potential at 10 cm from the charge to be $+100$ V?

Solving Eq. 29-8 for q yields

$$q = V4\pi\epsilon_0 r = (100 \text{ V})(4\pi)(8.9 \times 10^{-12} \text{ C}^2/\text{N}\cdot\text{m}^2)(0.10 \text{ m})$$
$$= 1.1 \times 10^{-9} \text{ C}.$$

This charge is comparable to charges that can be produced by friction.

EXAMPLE 4

What is the electric potential at the surface of a gold nucleus? The radius is 6.6×10^{-15} m and the atomic number $Z = 79$.

The nucleus, assumed spherically symmetrical, behaves electrically for external points as if it were a point charge. Thus we can use Eq. 29-8, or, recalling that the proton charge is 1.6×10^{-19} C,

$$V = \frac{1}{4\pi\epsilon_0}\frac{q}{r} = \frac{(9.0 \times 10^9 \text{ N}\cdot\text{m}^2/\text{C}^2)(79)(1.6 \times 10^{-19} \text{ C})}{6.6 \times 10^{-15} \text{ m}}$$

$$= 1.7 \times 10^7 \text{ V}.$$

29-4
A GROUP OF POINT CHARGES

The potential at any point due to a group of point charges is found by (a) calculating the potential V_n due to each charge, as if the other charges were not present and (b) adding the quantities so obtained, or (see Eq. 29-8)

$$V = \sum_n V_n = \frac{1}{4\pi\epsilon_0} \sum_n \frac{q_n}{r_n}, \qquad (29\text{-}9)$$

where q_n is the value of the nth charge and r_n is the distance of this charge from the point in question. The sum used to calculate V is an *algebraic sum* and not a vector sum like the one used to calculate $\mathbf{E}$ for a group of point charges (see Eq. 27-5). Herein lies an important computational advantage of potential over electric field.

If the charge distribution is continuous, rather than being a collection of points, the sum in Eq. 29-9 must be replaced by an integral, or

$$V = \int dV = \frac{1}{4\pi\epsilon_0} \int \frac{dq}{r}, \qquad (29\text{-}10)$$

where dq is a differential element of the charge distribution, r is its distance from the point at which V is to be calculated, and dV is the potential it establishes at that point.

EXAMPLE 5

What is the potential at the center of the square of Fig. 29-7? Assume that $q_1 = +1.0 \times 10^{-8}$ C, $q_2 = -2.0 \times 10^{-8}$ C, $q_3 = +3.0 \times 10^{-8}$ C, $q_4 = +2.0 \times 10^{-8}$ C, and $a = 1.0$ m.

The distance r of each charge from P is $a/\sqrt{2}$ or 0.71 m. From Eq. 29-9

$$V = \sum_n V_n = \frac{1}{4\pi\epsilon_0}\frac{q_1 + q_2 + q_3 + q_4}{r}$$

$$= \frac{(9.0 \times 10^9 \text{ N}\cdot\text{m}^2/\text{C}^2)(1.0 - 2.0 + 3.0 + 2.0) \times 10^{-8} \text{ C}}{0.71 \text{ m}}$$

$$= 500 \text{ V}.$$

Is the potential constant within the square? Does any point inside have a negative potential? Sketch roughly the intersection of the plane of Fig. 29-7 with the equipotential surface corresponding to zero volts.

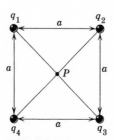

figure 29-7
Example 5

A charged disk. Find the electric potential for points on the axis of a uniformly charged circular disk whose surface charge density is σ (see Fig. 29-8).

Consider a charge element dq consisting of a flat circular strip of radius y and width dy. We have

$$dq = \sigma(2\pi y)(dy),$$

where $(2\pi y)(dy)$ is the area of the strip. All parts of this charge element are the same distance $r'\ (= \sqrt{y^2 + r^2})$ from axial point P so that their contribution dV to the electric potential at P is given by Eq. 29-8, or

$$dV = \frac{1}{4\pi\epsilon_0}\frac{dq}{r'} = \frac{1}{4\pi\epsilon_0}\frac{\sigma 2\pi y\, dy}{\sqrt{y^2 + r^2}}.$$

The potential V is found by integrating over all the strips into which the disk can be divided (Eq. 29-10), or

$$V = \int dV = \frac{\sigma}{2\epsilon_0}\int_0^a (y^2 + r^2)^{-1/2} y\, dy$$

$$= \frac{\sigma}{2\epsilon_0}\left(\sqrt{a^2 + r^2} - r\right).$$

This general result is valid for all values of r. In the special case of $r \gg a$ the quantity $\sqrt{a^2 + r^2}$ can be approximated as

$$\sqrt{a^2 + r^2} = r\left(1 + \frac{a^2}{r^2}\right)^{1/2} = r\left(1 + \frac{1}{2}\frac{a^2}{r^2} + \cdots\right) \cong r + \frac{a^2}{2r},$$

in which the quantity in parentheses in the second member of this equation has been expanded by the binomial theorem (see Appendix I). This equation means that V becomes

$$V \cong \frac{\sigma}{2\epsilon_0}\left(r + \frac{a^2}{2r} - r\right) = \frac{\sigma\pi a^2}{4\pi\epsilon_0 r} = \frac{1}{4\pi\epsilon_0}\frac{q}{r},$$

where $q\ (= \sigma\pi a^2)$ is the total charge on the disk. This limiting result is expected because the disk behaves like a point charge for $r \gg a$.

EXAMPLE 6

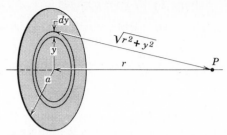

figure 29-8

Example 6. A point P on the axis of a uniformly charged circular disk of radius a.

Two equal charges of opposite sign, $\pm q$, separated by a distance $2a$, constitute an electric dipole; see Example 3, Chapter 27. The electric dipole moment $\mathbf{p}$ has the magnitude $2aq$ and points from the negative charge to the positive charge. Here we derive an expression for the electric potential V at any point of space due to a dipole, provided only that the point is not too close to the dipole $(r \gg a)$.

A point P is specified by giving the quantities r and θ in Fig. 29-9. From symmetry, it is clear that the potential will not change as point P rotates about the z axis, r and θ being fixed. Thus we need only find $V(r,\theta)$ for any plane containing this axis; the plane of Fig. 29-9 is such a plane. Applying Eq. 29-9 gives,

$$V = \sum_n V_n = V_1 + V_2 = \frac{1}{4\pi\epsilon_0}\left(\frac{q}{r_1} - \frac{q}{r_2}\right) = \frac{q}{4\pi\epsilon_0}\frac{r_2 - r_1}{r_1 r_2},$$

which is an exact relationship.

We now limit consideration to points such that $r \gg 2a$. These approximate relations then follow from Fig. 29-9:

$$r_2 - r_1 \cong 2a\cos\theta \quad\text{and}\quad r_1 r_2 \cong r^2,$$

and the potential reduces to

$$V = \frac{q}{4\pi\epsilon_0}\frac{2a\cos\theta}{r^2} = \frac{1}{4\pi\epsilon_0}\frac{p\cos\theta}{r^2}, \tag{29-11}$$

29-5
POTENTIAL DUE TO A DIPOLE

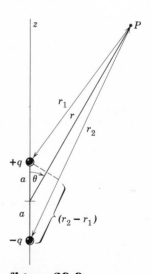

figure 29-9

A point P in the field of an electric dipole.

in which p $(= 2aq)$ is the dipole moment. Note that V vanishes everywhere in the equatorial plane $(\theta = 90°)$. This reflects the fact that it takes no work to bring a test charge in from infinity along the perpendicular bisector of the dipole. For a given radius, V has its greatest positive value for $\theta = 0$ and its greatest negative value for $\theta = 180°$. Note that the potential does not depend separately on q and $2a$ but only on their product p.

It is convenient to call *any* assembly of charges, for which V at distant points is given by Eq. 29-11, an *electric dipole*. Two point charges separated by a small distance behave this way, as we have just proved. However, other charge configurations also obey Eq. 29-11. Suppose that by measurement at points outside an imaginary box (Fig. 29-10) we find a pattern of lines of force that can be described quantitatively by Eq. 29-11. We then declare that the object inside the box is an *electric dipole*, that its axis is the line zz', and that its dipole moment **p** points vertically upward.

Many molecules have electric dipole moments. That for H_2O in its vapor state is 6.1×10^{-30} C·m. Figure 29-11 is a representation of this molecule, showing the three nuclei and the surrounding electron cloud. The dipole moment **p** is represented by the arrow on the axis of symmetry of the molecule. In this molecule the effective center of positive charge does not coincide with the effective center of negative charge. It is precisely because of this separation that the dipole moment exists.

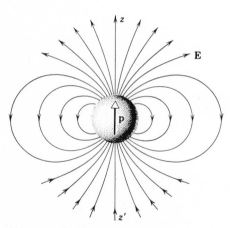

figure 29-10
If an object inside the spherical box sets up the electric field shown (described quantitatively by Eq. 29-11), it is an electric dipole.

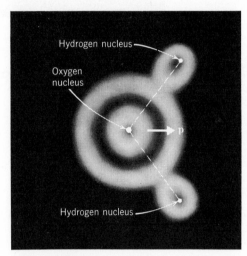

figure 29-11
A schematic representation of a water molecule, showing the three nuclei, the electron cloud, and the orientation of the dipole moment.

Atoms, and many molecules, do not have permanent dipole moments. However, dipole moments may be induced by placing any atom or molecule in an external electric field. The action of the field (Fig. 29-12) is to separate the centers of positive and of negative charge. We say that the atom becomes *polarized* and acquires an *induced electric dipole moment*. Induced dipole moments disappear when the electric field is removed.

Electric dipoles are important in situations other than atomic and molecular ones. Radio and radar antennas are often in the form of a metal wire or rod in which electrons surge back and forth periodically. At a certain time one end of the wire or rod will be negative and the other end positive. Half a cycle later the

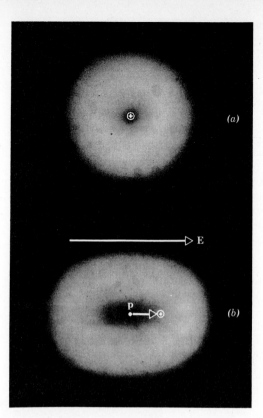

figure 29-12
(*a*) An atom, showing the nucleus and the electron cloud. The center of negative charge coincides with the center of positive charge, that is, with the nucleus. (*b*) If an external field **E** is applied, the electron cloud is distorted so that the center of negative charge, marked by the dot, and the center of positive charge no longer coincide. An electric dipole appears. The distortion is greatly exaggerated.

polarity of the ends is exactly reversed. This is an *oscillating* electric dipole. It is so named because its dipole moment changes in a periodic way with time.

EXAMPLE 7

An electric quadrupole. An electric quadrupole, of which Fig. 27-26 is an example, consists of two electric dipoles so arranged that they almost, but not quite, cancel each other in their electric effects at distant points. Calculate $V(r)$ for points on the axis of this quadrupole.

Applying Eq. 29-9 to Fig. 27-26 yields,

$$V = \sum_n V_n = \frac{1}{4\pi\epsilon_0}\left(\frac{q}{r-a} - \frac{2q}{r} + \frac{q}{r+a}\right)$$

$$= \frac{q}{4\pi\epsilon_0}\frac{2a^2}{(r-a)(r)(r+a)}.$$

Assuming that $r \gg a$ allows us to put $a = 0$ in the sum and difference terms in the denominator, yielding

$$V = \frac{1}{4\pi\epsilon_0}\frac{Q}{r^3},$$

where $Q\,(= 2qa^2)$ is the *electric quadrupole moment* of the charge assembly of Fig. 27-26. Note that V varies (*a*) as $1/r$ for a point charge (see Eq. 29-8), (*b*) as $1/r^2$ for a dipole (see Eq. 29-11), and (*c*) as $1/r^3$ for a quadrupole.

Note too that (*a*) a dipole is two equal and opposite charges that do not quite coincide in space so that their electric effects at distant points do not quite cancel, and (*b*) a quadrupole is two equal and opposite dipoles that do not quite coincide in space so that their electric effects at distant points again do not quite cancel. This pattern can be extended to define higher orders of charge distribution such as *octupoles*.

The potential at points at distances from an arbitrary charge distribution (continuous or discrete) that are large compared with the size of the distribution can always be written as the sum of separate potential distributions due to (a) a single charge—sometimes, in this context, called a *monopole*—(b) a dipole, (c) a quadrupole, etc. This process is called an *expansion in multipoles* and is a very useful technique in problem solving.

29-6 ELECTRIC POTENTIAL ENERGY*

figure 29-13

If we raise a stone from the earth's surface, the work that we do against the earth's gravitational attraction is stored as *potential energy* in the system earth + stone. If we release the stone, the stored potential energy changes steadily into kinetic energy as the stone drops. After the stone comes to rest on the earth, this kinetic energy, equal in magnitude just before the time of contact to the originally stored potential energy, is transformed into thermal energy in the system earth + stone.

A similar situation exists in electrostatics. Consider two charges q_1 and q_2 a distance r apart, as in Fig. 29-13. If we increase the separation between them, an external agent must do work that will be positive if the charges are opposite in sign and negative otherwise. The energy represented by this work can be thought of as stored in the system $q_1 + q_2$ as *electric potential energy*. This energy, like all varieties of potential energy, can be transformed into other forms. If q_1 and q_2, for example, are charges of opposite sign and we release them, they will accelerate toward each other, transforming the stored potential energy into kinetic energy of the accelerating masses. The analogy to the earth + stone system is exact, save for the fact that electric forces may be either attractive or repulsive whereas gravitational forces are always attractive.

We define the electric potential energy of a system of point charges as the work required to assemble this system of charges by bringing them in from an infinite distance. We assume that the charges are all at rest when they are infinitely separated, that is, they have no initial kinetic energy.

In Fig. 29-13 let us imagine q_2 removed to infinity and at rest. The *electric potential* at the original site of q_2, caused by q_1, is given by Eq. 29-8, or

$$V = \frac{1}{4\pi\epsilon_0}\frac{q_1}{r}.$$

If q_2 is moved in from infinity to the original distance r, the work required is, from the definition of electric potential, that is, from Eq. 29-2,

$$W = Vq_2. \tag{29-12}$$

Combining these two equations and recalling that this work W is pre-

*In mechanics the concept of *potential energy* (of compressed springs, bodies in the earth's gravitational field, etc.) is perhaps more commonly used than that of *potential* (gravitational, say). In electricity and magnetism, on the other hand, *potential* is perhaps more commonly used than *potential energy*. A shorthand distinction is that potential is potential energy *per unit charge*. With potential in units of volts and potential energy in units of joules, the volt is equivalent, therefore, to a joule per coulomb.

cisely the *electric potential energy* U of the system $q_1 + q_2$ yields

$$U \;(= W) = \frac{1}{4\pi\epsilon_0}\frac{q_1 q_2}{r_{12}}. \qquad (29\text{-}13)$$

The subscript of r emphasizes that the distance involved is that between the point charges q_1 and q_2.

For systems containing more than two charges the procedure is to compute the potential energy for every pair of charges separately and to add the results algebraically. This procedure rests on a physical picture in which (a) charge q_1 is brought into position, (b) q_2 is brought from infinity to its position near q_1, (c) q_3 is brought from infinity to its position near q_1 and q_2, etc.

The potential energy of continuous charge distributions (an ellipsoid of charge, for example) can be found by dividing the distribution into differential elements dq, treating each such element as a point charge, and using the procedures of the preceding paragraph, with the summation process replaced by an integration. We have not considered such problems in this text.

EXAMPLE 8

Two protons in a nucleus of U^{238} are 6.0×10^{-15} m apart. What is their mutual electric potential energy?

From Eq. 29-13

$$U = \frac{1}{4\pi\epsilon_0}\frac{q_1 q_2}{r} = \frac{(9.0 \times 10^9 \text{ N}\cdot\text{m}^2/\text{C}^2)(1.6 \times 10^{-19} \text{ C})^2}{6.0 \times 10^{-15} \text{ m}}$$

$$= 3.8 \times 10^{-14} \text{ J} = 2.4 \times 10^5 \text{ eV}.$$

EXAMPLE 9

Three charges are held fixed as in Fig. 29-14. What is their mutual electric potential energy? Assume that $q = 1.0 \times 10^{-7}$ C and $a = 10$ cm.

The total energy of the configuration is the sum of the energies of each pair of particles. From Eq. 29-13,

$$U = U_{12} + U_{13} + U_{23}$$

$$= \frac{1}{4\pi\epsilon_0}\left[\frac{(+q)(-4q)}{a} + \frac{(+q)(+2q)}{a} + \frac{(-4q)(+2q)}{a}\right]$$

$$= -\frac{10}{4\pi\epsilon_0}\frac{q^2}{a}$$

$$= -\frac{(9.0 + 10^9 \text{ N}\cdot\text{m}^2/\text{C}^2)(10)(1.0 \times 10^{-7} \text{ C})^2}{0.10 \text{ m}} = -9.0 \times 10^{-3} \text{ J}.$$

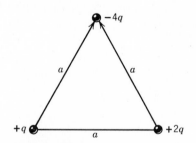

figure 29-14
Example 9. Three charges are fixed, as shown, by external forces.

The fact that the total energy is negative means that negative work would have to be done to assemble this structure, starting with the three charges separated and at rest at infinity. Expressed otherwise, 9.0×10^{-3} J of work must be done to dismantle this structure, removing the charges to an infinite separation from one another.

When, as is common practice, infinity is taken as the zero of electric potential, a positive potential energy, in simple problems like these, (as in Example 8) corresponds to repulsive electric forces and a negative potential energy (as in this example) to attractive electric forces. If the protons in Example 8 were not held in place by attractive (nonelectrical) nuclear forces, they would move away from each other. If the three particles in this example were released from their fixed positions, in which they are held by external forces, they would move toward each other.

We have stated that V and $\mathbf{E}$ are equivalent descriptions of electric fields, and have determined (Eq. 29-6) how to calculate V from $\mathbf{E}$. Let us now consider how to calculate $\mathbf{E}$ if we know V throughout a certain region.

This problem has already been solved graphically. If $\mathbf{E}$ is known at every point in space, the lines of force can be drawn; then a family of equipotentials can be sketched in by drawing surfaces at right angles. These equipotentials describe the behavior of V. Conversely, if V is given as a function of position, a set of equipotential surfaces can be drawn. The lines of force can then be found by drawing lines at right angles, thus describing the behavior of $\mathbf{E}$. It is the mathematical equivalent of this second graphical process that we seek here. Figure 29-15 shows three examples of lines of force and of the corresponding equipotential surfaces.

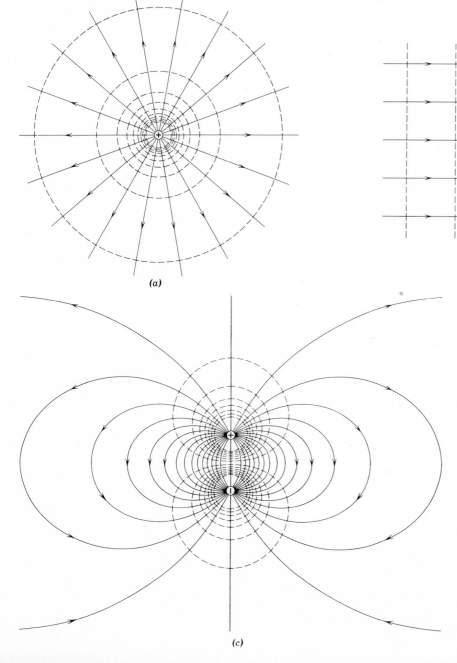

(a)

(b)

(c)

figure 29-15
Equipotential surfaces (dashed lines) and lines of force (solid lines) for (a) a point charge, (b) a uniform electric field set up by charges not shown, and (c) an electric dipole. In (a) and (c) the dashed lines represent intersections with the plane of the figure of closed surfaces; in (b) the dashed lines represent infinite sheets. In all figures there is a constant difference of potential ΔV between adjacent equipotential surfaces. Thus from Eq. 29-14, written for the case of $\theta = 180°$ as $\Delta l = -\Delta V/E$, the surfaces will be relatively close together where E is relatively large and relatively far apart where E is small. Similarly (see Section 27-3) the lines of force are relatively close together where E is large and far apart where E is small. See discussion and figures of Section 18-7 for other examples.

Figure 29-16 shows the intersection with the plane of the figure of a family of equipotential surfaces. The figure shows that **E** at a typical point P is at right angles to the equipotential surface through P, as it must be.

Let us move a test charge q_0 from P along the path marked $\Delta \mathbf{l}$ to the equipotential surface marked $V + \Delta V$. The work that must be done by the agent exerting the force **F** (see Eq. 29-1) is $q_0 \Delta V$.

From another point of view we can calculate the work from*

$$\Delta W = \mathbf{F} \cdot \Delta \mathbf{l},$$

where **F** is the force that must be exerted on the charge to overcome exactly the electrical force $q_0 \mathbf{E}$. Since **F** and $q_0 \mathbf{E}$ have opposite signs and are equal in magnitude,

$$\Delta W = -q_0 \mathbf{E} \cdot \Delta \mathbf{l} = -q_0 E \cos (\pi - \theta) \, \Delta l = q_0 E \cos \theta \, \Delta l.$$

These two expressions for the work must be equal, which gives

$$q_0 \, \Delta V = q_0 E \cos \theta \, \Delta l$$

or

$$E \cos \theta = \frac{\Delta V}{\Delta l}. \tag{29-14}$$

Now $E \cos \theta$ is the component of **E** in the direction $-\mathbf{l}$ in Fig. 29-16; the quantity $-E \cos \theta$, which we call E_l, would then be the component of **E** in the $+\mathbf{l}$ direction. In the differential limit Eq. 29-14 can then be written as

$$E_l = -\frac{dV}{dl}. \tag{29-15}$$

In words, this equation says: If we travel through an electric field along a straight line and measure V as we go, the rate of change of V with distance that we observe, when changed in sign, is the component of **E** in that direction. The minus sign implies that **E** points in the direction of decreasing V, as in Fig. 29-16. It is clear from Eq. 29-15 that appropriate units for **E** are volt/meter (V/m).

There will be one direction $\mathbf{l}$ for which the quantity $-dV/dl$ is a maximum. From Eq. 29-15, E_l will also be a maximum for this direction and will in fact be E itself. Thus

$$E = -\left(\frac{dV}{dl}\right)_{\max}. \tag{29-16}$$

The maximum value of dV/dl at a given point is called the *potential gradient* at that point. The direction $\mathbf{l}$ for which dV/dl has its maximum value is always at right angles to the equipotential surface, corresponding to $\theta = 0$ in Fig. 29-16.

If we take the direction $\mathbf{l}$ to be, in turn, the directions of the x, y, and z axes, we can find the three components of **E** at any point, from Eq. 29-15.

$$E_x = -\frac{\partial V}{\partial x}; \qquad E_y = -\frac{\partial V}{\partial y}; \qquad E_z = -\frac{\partial V}{\partial z}. \tag{29-17}$$

Thus if V is known for all points of space, that is, if the function $V(x, y, z)$ is known, the components of **E**, and thus **E** itself, can be found by taking derivatives.†

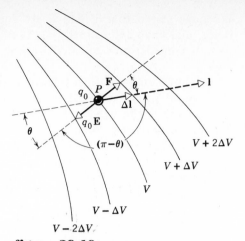

figure 29-16
A test charge q_0 is moved from one equipotential surface to another along an arbitrarily selected direction marked **l**.

---

* We assume that the equipotentials are so close together that **F** is constant for all parts of the path $\Delta \mathbf{l}$. In the limit of a differential path ($d\mathbf{l}$) there will be no difficulty.

† The symbol $\partial V/\partial x$ denotes a *partial derivative*. In taking this derivative of the function $V(x, y, z)$ the quantity x is to be viewed as a variable and y and z are to be regarded as constants. Similar considerations hold for $\partial V/\partial y$ and $\partial V/\partial z$.

Calculate $E(r)$ for a point charge q, using Eq. 29-16 and assuming that $V(r)$ is given as (see Eq. 29-8)

EXAMPLE 10

$$V = \frac{1}{4\pi\epsilon_0} \frac{q}{r}.$$

From symmetry, **E** must be directed radially outward for a (positive) point charge. Consider a point P in the field a distance r from the charge. It is clear that $-dV/dl$ at P has its greatest value if the direction l is identified with that of r. Thus, from Eq. 29-16,

$$E = -\frac{dV}{dr} = -\frac{d}{dr}\left(\frac{1}{4\pi\epsilon_0}\frac{q}{r}\right)$$

$$= -\frac{q}{4\pi\epsilon_0}\frac{d}{dr}\left(\frac{1}{r}\right) = \frac{1}{4\pi\epsilon_0}\frac{q}{r^2}.$$

This result agrees exactly with Eq. 27-4, as it must.

E *for a dipole.* Figure 29-17 shows a (distant) point P in the field of a dipole located at the origin of an xy-axis system. V is given by Eq. 29-11, or

EXAMPLE 11

$$V = \frac{1}{4\pi\epsilon_0}\frac{p\cos\theta}{r^2}.$$

Calculate **E** as a function of position.

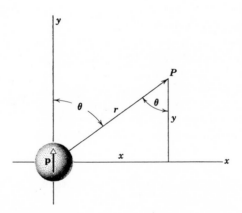

figure 29-17
Showing a point P in the field of an electric dipole **p**.

From symmetry, **E**, for points in the plane of Fig. 29-17, lies in this plane. Thus it can be expressed in terms of its components E_x and E_y, E_z being zero. Let us first express the potential function in rectangular coordinates rather than polar coordinates, making use of

$$r = (x^2 + y^2)^{1/2} \quad \text{and} \quad \cos\theta = \frac{y}{(x^2+y^2)^{1/2}}.$$

The result is

$$V = \frac{p}{4\pi\epsilon_0}\frac{y}{(x^2+y^2)^{3/2}}.$$

We find E_y from Eq. 29-17, recalling that x is to be treated as a constant in this calculation:

$$E_y = -\frac{\partial V}{\partial y} = -\frac{p}{4\pi\epsilon_0}\frac{(x^2+y^2)^{3/2} - y\frac{3}{2}(x^2+y^2)^{1/2}(2y)}{(x^2+y^2)^3}$$

$$= -\frac{p}{4\pi\epsilon_0}\frac{x^2 - 2y^2}{(x^2+y^2)^{5/2}}.$$

Note that putting $x = 0$ describes distant points along the dipole axis (that is, the

y axis), and the expression for E_y reduces to

$$E_y = \frac{2p}{4\pi\epsilon_0}\frac{1}{y^3}.$$

This result agrees exactly with that found in Chapter 27 (see Problem 25), for, from symmetry, E_x equals zero on the dipole axis.

Putting $y = 0$ in the expression for E_y describes distant points in the median plane of the dipole and yields

$$E_y = -\frac{p}{4\pi\epsilon_0}\frac{1}{x^3},$$

which agrees exactly with the result found in Chapter 27 (see Example 3), for, again from symmetry, E_x equals zero in the median plane. The minus sign in this equation indicates that $\mathbf{E}$ points in the negative y direction (see Fig. 29-10).

The component E_x is also found from Eq. 29-17, recalling that y is to be taken as a constant during this calculation:

$$E_x = -\frac{\partial V}{\partial x} = -\frac{py}{4\pi\epsilon_0}\left(-\tfrac{3}{2}\right)(x^2 + y^2)^{-5/2}(2x)$$

$$= \frac{3p}{4\pi\epsilon_0}\frac{xy}{(x^2 + y^2)^{5/2}}.$$

As expected, E_x vanishes both on the dipole axis ($x = 0$) and in the median plane ($y = 0$); see Fig. 29-10.

29-8 AN INSULATED CONDUCTOR

We proved in Section 28-6, using Gauss's law, that after a steady state is reached an excess charge q placed on an insulated conductor will move to its outer surface. We now assert that this charge q will distribute itself on this surface so that all points of the conductor, including those on the surface and those inside, have the same potential.

Consider any two points A and B in or on the conductor. If they were not at the same potential, the charge carriers in the conductor near the point of lower potential would tend to move toward the point of higher potential. We have assumed, however, that a steady-state situation, in which such currents do not exist, has been reached; thus all points, both on the surface and inside it, must have the same potential. Since the surface of the conductor is an equipotential surface, $\mathbf{E}$ for points on the surface must be at right angles to the surface.

We saw in Section 28-6 that a charge placed on an insulated conductor will spread over the surface until $\mathbf{E}$ equals zero for all points inside. We now have an alternative way of saying the same thing; the charge will move until all points of the conductor (surface points and interior points) are brought to the same potential, for if V is constant in the conductor, then $\mathbf{E}$ is zero everywhere in the conductor ($E_l = -dV/dl$).

Figure 29-18a is a plot of potential against radial distance for an isolated spherical conducting shell of 1.0-m radius carrying a positive charge of 1.0×10^{-6} C. For points outside the shell $V(r)$ can be calculated from Eq. 29-8 because the charge q behaves, for such points, as if it were concentrated at the center of the sphere. Equation 29-8 is correct right up to the surface of the shell. Now let us push the test charge through the surface, assuming that there is a small hole, and into the interior. No extra work is needed because no electrical forces act on the test charge once it is inside the shell. Thus the potential everywhere inside is the same as that on the surface, as Fig. 29-18a shows.

Figure 29-18b shows the electric field for this same sphere. Note that E equals zero inside. The lower of these curves can be derived from the

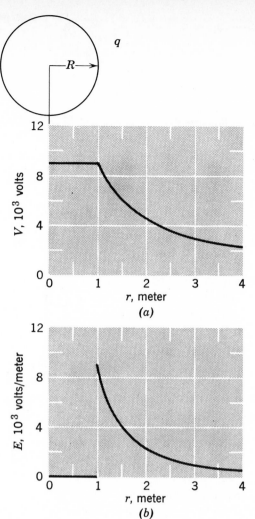

upper by differentiation with respect to r, using Eq. 29-16; the upper can be derived from the lower by integration with respect to r, using Eq. 29-6.

Figure 29-18 holds without change if the conductor is a solid sphere rather than a spherical shell as we have assumed. It is constructive to compare Fig. 29-18b (conducting shell or sphere) with Fig. 28-10, which holds for a *nonconducting* sphere. Try to understand the difference between these two figures, bearing in mind that in the first the charge lies on the surface, whereas in the second it was assumed to have been spread uniformly throughout the volume of the sphere.

Finally we note that, as a general rule, the charge density tends to be high on isolated conducting surfaces whose radii of curvature are small, and conversely. For example, the charge density tends to be relatively high on sharp points and relatively low on plane regions on a conducting surface. The electric field E at points immediately above a charged surface is proportional to the charge density σ so that E may also reach very high values near sharp points. Glow discharges from sharp points during thunderstorms are a familiar example. The lightning rod acts in this way to neutralize charged clouds and thus prevent lightning strokes.

We can examine the qualitative relationship between σ and the curvature of the surface in a particular case by considering two spheres of different radii con-

figure 29-19
Two spheres connected by a long
fine wire.

nected by a very long fine wire; see Fig. 29-19. Suppose that this entire assembly is raised to some arbitrary potential V. The (equal) potentials of the two spheres are, from Eq. 29-8,*

$$V = \frac{1}{4\pi\epsilon_0}\frac{q_1}{R_1} = \frac{1}{4\pi\epsilon_0}\frac{q_2}{R_2},$$

which yields

$$\frac{q_1}{q_2} = \frac{R_1}{R_2}, \qquad (29\text{-}18)$$

where q_1 is the charge on the sphere of radius R_1 and q_2 is the charge on the sphere of radius R_2.

The *surface charge densities* for each sphere are given by

$$\sigma_1 = \frac{q_1}{4\pi R_1{}^2} \qquad \text{and} \qquad \sigma_2 = \frac{q_2}{4\pi R_2{}^2}.$$

Dividing gives

$$\frac{\sigma_1}{\sigma_2} = \frac{q_1}{q_2}\frac{R_2{}^2}{R_1{}^2}.$$

Combining with Eq. 29-18 yields

$$\frac{\sigma_1}{\sigma_2} = \frac{R_2}{R_1},$$

which is consistent with our qualitative statement above. Note that the larger sphere has the larger total charge but the smaller charge density.

The fact that σ, and thus **E,** can become very large near sharp points is important in the design of high-voltage equipment. *Corona discharge* can result from such points if the conducting object is raised to high potential and surrounded by air. Normally air is thought of as a nonconductor. However, it contains a small number of ions produced, for example, by the cosmic rays. A positively charged conductor will attract negative ions from the surrounding air and thus will slowly neutralize itself.

If the charged conductor has sharp points, the value of E in the air near the points can be very high. If the value is high enough, the ions, as they are drawn toward the conductor, will receive such large accelerations that, by collision with air molecules, they will produce vast additional numbers of ions. The air is thus made much more conducting, and the discharge of the conductor by this corona discharge may be very rapid indeed. The air surrounding sharp conducting points may even glow visibly because of light emitted from the air molecules during these collisions.

The electrostatic generator was conceived by Lord Kelvin in 1890 and put into useful practice in essentially its modern form by R. J. Van de Graaff in 1931. It is a device for producing electric potential differences of the order of several millions of volts. Its chief application in physics is the use of this potential difference to accelerate charged particles to

29-9
THE ELECTROSTATIC GENERATOR

---

*Equation 29-8 holds only for an isolated point charge or spherically symmetric charge distribution. The spheres must be assumed to be so far apart that the charge on either one has a negligible effect on the distribution of charge on the other.

high energies. Beams of energetic particles made in this way can be used in many different "atom-smashing" experiments. The technique is to let a charged particle "fall" through a potential difference V, gaining kinetic energy as it does so.

Let a particle of (positive) charge q move in a vacuum under the influence of an electric field from one position A to another position B whose electric potential is lower by V. The electric potential energy of the system is reduced by qV because this is the work that an external agent would have to do to restore the system to its original condition. This decrease in potential energy appears as kinetic energy of the particle, or

$$K = qV. \tag{29-19}$$

K is in joules if q is in coulombs and V in volts. If the particle is an electron or a proton, q will be the quantum of charge e.

If we adopt the quantum of charge e as a unit in place of the coulomb, we arrive at another unit for energy, the *electron volt* (abbr. eV), which we have used before and is used extensively in atomic and nuclear physics. By substituting into Eq. 29-19,

$$1 \text{ electron volt} = (1 \text{ quantum of charge})(1 \text{ volt})$$
$$= (1.60 \times 10^{-19} \text{ coulomb})(1.00 \text{ volt})$$
$$= 1.60 \times 10^{-19} \text{ joule}.$$

The electron volt can be used interchangeably with any other energy unit. Thus a 10-g object moving at 1000 cm/s can be said to have a kinetic energy of 3.1×10^{18} eV. Most physicists would prefer to express this result as 0.50 J, the electron volt being inconveniently small. In atomic, nuclear, and elementary particle physics, however, the electron volt (eV) and its multiples the keV (= 10^3 eV), MeV (= 10^6 eV), and the GeV (= 10^9 eV) are the usual units of choice.

figure 29-20
Example 12. A small charged sphere of radius r is suspended inside a charged spherical shell whose outer surface has radius R.

EXAMPLE 12

The electrostatic or Van de Graaff generator. Figure 29-20, which illustrates the basic operating principle of the electrostatic generator, shows a small sphere of radius r placed inside a large spherical shell of radius R. The two spheres carry charges q and Q, respectively. Calculate their potential difference.

The potential of the large sphere is caused in part by its own charge and in part because it lies in the field set up by the charge q on the small sphere. From Eq. 29-8,

$$V_R = \frac{1}{4\pi\epsilon_0}\left(\frac{Q}{R} + \frac{q}{R}\right).$$

The potential of the small sphere is caused in part by its own charge and in part because it is inside the large sphere; see Fig. 29-18a. From Eq. 29-8,

$$V_r = \frac{1}{4\pi\epsilon_0}\left(\frac{q}{r} + \frac{Q}{R}\right).$$

The potential difference is

$$V_r - V_R = \frac{q}{4\pi\epsilon_0}\left(\frac{1}{r} - \frac{1}{R}\right).$$

Thus, assuming q is positive, the inner sphere will always be higher in potential than the outer sphere. If the spheres are connected by a fine wire, the charge q will flow *entirely* to the outer sphere, regardless of the charge Q that may already be present.

Removable tank cover

Electronic circuits

Built-in 2 kW power supply

Charge collector

Equipotential planes

Insulating column 18′ long

Windows

2–50 hp 1,800 rpm motors

Belt tension adjustment

Lead shielding

Dry ice trap

Pumping system

Equipotential shield, 68″ I.D.

High voltage terminal, 38″ diam

Positive ion source

400 lb/in.² steel pressure tank

Field control rods

Insulating belt

Positive ion accelerating tube

Manhole

Movable platform

Main valve

Flexible coupling

Analyzing magnet

Beam axis

Adjustable magnet base

0   2   4   6   8
Feet

figure 29-21

An electrostatic generator at MIT capable of producing 9-MeV protons. The proton beam is accelerated vertically downward, being deflected into a horizontal plane by the analyzing magnet shown at the bottom. (Courtesy of J. G. Trump.)

From another point of view, we note that since the spheres when electrically connected form a single conductor at electrostatic equilibrium there can be only a single potential. This means that $V_r - V_R = 0$, which can occur only if $q = 0$.

In actual electrostatic generators charge is carried into the shell on rapidly moving belts made of insulating material. Charge is "sprayed" onto the belts outside the shell by corona discharge from a series of sharp metallic points connected to a source of moderately high potential difference. Charge is removed from the belts inside the shell by a similar series of points connected to the shell. Electrostatic generators can be built commercially to accelerate protons to energies up to 10 MeV, using a single acceleration. Figure 29-21 shows a schematic diagram of an electrostatic generator at MIT that can produce 9-MeV protons.

Generators can be built in which the accelerated particles are subject to two or three successive accelerations.

questions

1. Are we free to call the potential of the earth +100 volts instead of zero? What effect would such an assumption have on measured values of (*a*) potentials and (*b*) potential differences?

2. What would happen to a person on an insulated stand if his potential was increased by 10,000 volts?

3. Do electrons tend to go to regions of high potential or of low potential?

4. Suppose that the earth has a net charge that is *not* zero. Is it still possible to adopt the earth as a standard reference point of potential and to assign the potential $V=0$ to it?

5. Does the potential of a positively charged insulated conductor have to be positive? Give an example to prove your point.

6. Can two different equipotential surfaces intersect?

7. An electrical worker is accidentally electrocuted and a newspaper account reported: "He accidentally touched a high voltage cable and 20,000 volts of electricity surged through his body." Criticize this statement.

8. Does the amount of work per unit charge required to transfer electric charge from one point to another in an electrostatic field depend on the amount of charge transferred?

9. Advice to mountaineers caught in lightning and thunderstorms is (*a*) get rapidly off peaks and ridges and (*b*) put both feet together and crouch in the open, only the feet touching the ground. What is the basis for this good advice?

10. If **E** equals zero at a given point, must *V* equal zero for that point? Give some examples to prove your answer.

11. If you know **E** at a given point, can you calculate *V* at that point? If not, what further information do you need?

12. In Fig. 29-2 is the electric field **E** greater at the left or the right side of the figure?

13. In Fig. 29-6 is it necessary to assume that *A, B,* and *q* lie on a straight line in order to prove that Eq. 29-8 is true?

14. Is the uniformly charged, nonconducting disk of Example 6 a surface of constant potential? Explain.

15. Why can an isolated atom not have a permanent electric dipole moment?

16. If *V* equals a constant throughout a given region of space, what can you say about **E** in that region?

17. In Section 16-6 we saw that the gravitational field strength is zero inside a spherical shell of matter. The electrical field strength is zero not only inside an isolated charged spherical conductor but inside an isolated conductor of *any* shape. Is the gravitational field strength inside, say, a cubical shell of matter zero? If not, in what respect is the analogy not complete?

18. How can you insure that the electric potential in a given region of space will have a constant value?

19. A charge is placed on an insulated conductor in the form of a perfect cube. What will be the relative charge density at various points on the cube (surfaces, edges, corners); what will happen to the charge if the cube is in air?

20. We have seen (Fig. 29-18*a*) that the potential inside a thin conducting spherical shell is the same as that on its surface. (*a*) What if the sphere is solid? (*b*) What if it is solid and irregularly shaped? (*c*) What if it is solid, is irregularly shaped, and has an irregularly shaped cavity inside? In particular, what is *V* within the hollow cavity? (*d*) Same as (*c*) but suppose a point charge is suspended within the cavity?

21. A closed metal box in the form of a pyramid is placed on an insulating support and charged to a potential $+V$. Is the average potential inside the pyramid (a) greater than, (b) equal to, or (c) less than V?

22. An isolated conducting spherical shell carries a negative charge. What will happen if a positively charged metal object is placed in contact with the shell interior? Assume that the positive charge is (a) less than, (b) equal to, and (c) greater than the negative charge in magnitude.

23. An uncharged metal sphere suspended by a silk thread is placed in a uniform external electric field **E.** What is the magnitude of the electric field for points inside the sphere? Is your answer changed if the sphere carries a charge?

problems

SECTION 29-1

1. In a typical lightning flash the potential difference between discharge points is about 10^9 V and the quantity of charge transferred is about 30 C. How much ice would it melt at 0°C if all the energy released could be used for this purpose?   *Answer:* 99 tons.

2. A charge q is distributed uniformly throughout a nonconducting spherical volume of radius R. (a) Show that the potential a distance r from the center, where $r < R$, is given by

$$V = \frac{q(3R^2 - r^2)}{8\pi\epsilon_0 R^3}.$$

(b) Is it reasonable that, according to this expression, V is not zero at the center of the sphere?

3. A gold nucleus contains a positive charge equal to that of 79 protons. An α-particle $(Z = 2)$ has a kinetic energy K at points far from this charge and is traveling directly toward the charge. The particle just touches the surface of the charge (assumed spherical) and is reversed in direction. (a) Calculate K, assuming a nuclear radius of 5.0×10^{-15} m. (b) The actual α-particle energy used in the experiments of Rutherford and his collaborators (see Section 28-9) was 5.0 MeV. What do you conclude?
   Answer: (a) 45 MeV. (b) In these important experiments the α-particles approached, but did not "touch" the gold nuclei.

4. (a) Through what potential difference must an electron fall, according to Newtonian mechanics, to acquire a speed v equal to the speed c of light? (b) Newtonian mechanics fails as $v \rightarrow c$. Therefore, using the correct relativistic expression for the kinetic energy

$$K = mc^2 \left[\frac{1}{\sqrt{1 - (v/c)^2}} - 1 \right]$$

in place of the Newtonian expression $K = \frac{1}{2}mv^2$, determine the actual electron speed acquired in falling through the potential difference computed in (a). Express this speed as an appropriate fraction of the speed of light.

SECTION 29-2

5. An infinite charged sheet has a surface charge density σ of 1.0×10^{-7} C/m². How far apart are the equipotential surfaces whose potentials differ by 5.0 V? *Answer:* 0.89 mm.

6. Two large parallel conducting plates are 10 cm apart and carry equal but opposite charges on their facing surfaces. An electron placed midway between the two plates experiences a force of 1.6×10^{-15} N. What is the potential difference between the plates?

7. In the Millikan oil drop experiment (see Fig. 27-29) an electric field of 1.92 $\times 10^5$ N/C is maintained at balance across two plates separated by 1.50 cm. Find the potential difference between the plates.   *Answer:* 2900 V.

SECTION 29-3

8. Consider a point charge with $q = 1.5 \times 10^{-8}$ C. (a) What is the radius of an equipotential surface having a potential of 30 V? (b) Are surfaces whose potentials differ by a constant amount (1.0 V, say) evenly spaced in radius?

SECTION 29-4

9. A point charge has $q = +1.0 \times 10^{-6}$ C. Consider point A which is 2.0 m distant and point B which is 1.0 m distant in a direction diametrically opposite, as in Fig. 29-22a. (a) What is the potential difference $V_A - V_B$? (b) Repeat if points A and B are located as in Fig. 29-22b.
Answer: (a) -4500 V. *(b)* Same as (a) because potential is a scalar quantity.

10. In Fig. 29-23, locate the points (a) where $V = 0$ and (b) where $\mathbf{E} = 0$. Consider only points on the axis and choose $d = 1.0$ m.

11. In Fig. 29-23 (see Problem 10) sketch qualitatively (a) the lines of force and (b) the intersections of the equipotential surfaces with the plane of the figure. (Hint: Consider the behavior close to each point charge and at considerable distances from the pair of charges.)

12. (a) In Fig. 29-24 derive an expression for $V_A - V_B$. (b) Does your result reduce to the expected answer when $d = 0$? When $q = 0$?

13. For the charge configuration of Fig. 29-25, show that $V(r)$ for points on the vertical axis, assuming $r \gg a$, is given by

$$V = \frac{1}{4\pi\epsilon_0}\left(\frac{q}{r} + \frac{2qa}{r^2}\right).$$

Is this an expected result? (Hint: The charge configuration can be viewed as the sum of an isolated charge and a dipole.)

SECTION 29-5

14. Calculate the dipole moment of a water molecule under the assumption that all ten electrons in the molecule circulate symmetrically about the oxygen atom, that the OH distance is 0.96×10^{-8} cm, and that the angle between the two OH bonds is 104°. Compare with the value quoted on p. 631; see Fig. 29-11.

SECTION 29-6

15. A particle of mass m, charge $q > 0$, and initial kinetic energy K is projected (from "infinity") toward a heavy nucleus of charge Q, assumed to have a fixed position in our reference frame. (a) If the aim is "perfect," how close to the center of the nucleus is the particle when it comes instantaneously to rest? (b) With a particular imperfect aim the particle's closest approach to the nucleus is twice the distance determined in part (a). Determine the speed of the particle at this closest distance of approach.
Answer: (a) $qQ/4\pi\epsilon_0 K$. *(b)* $\sqrt{K/m}$.

16. The charges and coordinates of two charges located on the x-y plane are: $q_1 = +3.0 \times 10^{-6}$ C; $x = +3.5$ cm, $y = +0.50$ cm, and $q_2 = -4.0 \times 10^{-6}$ C; $x = -2.0$ cm, $y = +1.5$ cm. (a) Find the electric potential at the origin. (b) How much work must be done to locate these charges at their given positions, starting from infinity?

17. Derive an expression for the work required to put the four charges together as indicated in Fig. 29-26. *Answer:* $-0.21q^2/\epsilon_0 a$.

18. A particle of charge Q is kept in a fixed position at a point P and a second particle of mass m, having the same charge Q, is initially held at rest a distance r_1 from P. The second particle is then released and is repelled from the first one. Determine its speed at the instant it is a distance r_2 from P. Let $Q = 3.1 \times 10^{-6}$ C, $m = 2.0 \times 10^{-5}$ kg, $r_1 = 9.0 \times 10^{-4}$ m, and $r_2 = 25 \times 10^{-4}$ m.

19. What is the electric potential energy of the charge configuration of Fig. 29-6? Use the numerical values of Example 5. *Answer:* -6.4×10^{-7} J.

20. In the rectangle shown in Fig. 29-27, the sides have lengths 5.0 cm and 15.0 cm, $q_1 = -5.0 \times 10^{-6}$ C and $q_2 = +2.0 \times 10^{-6}$ C. (a) What is the electric potential at corner B? At corner A? (b) How much work is involved in moving

figure 29-22
Problem 9

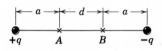
figure 29-23
Problems 10, 11

figure 29-24
Problem 12

figure 29-25
Problem 13

figure 29-26
Problem 17

figure 29-27
Problem 20

a third charge $q_3 = +3.0 \times 10^{-6}$ C *from B to A* along a diagonal of the rectangle? (c) In this process, is external work converted into electrostatic potential energy or vice versa? Explain.

21. Two charges q ($= +2.0 \times 10^{-6}$ C) are fixed in space a distance d (2.0 cm) apart, as shown in Fig. 29-28. (a) What is the electric potential at point C? (b) You bring a third charge q ($+2.0 \times 10^{-6}$ C) very slowly from infinity to C. How much work must you do? (c) What is the potential energy U of the configuration when the third charge is in place?
    Answer: (a) 2.5×10^6 V. (b) 5.1 J. (c) 6.9 J.

figure 29-28
Problem 21

22. Three charges of +0.1 C each are placed on the corner of an equilateral triangle, 1.0 m on a side. If energy is supplied at the rate of 1.0 kW, how many days would be required to move one of the charges onto the midpoint of the line joining the other two?

23. Two electrons are 2.0 m apart. Another electron is shot from infinity and comes to rest midway between the two. What must its initial velocity be?
    Answer: 32 m/s.

24. Calculate (a) the electric potential established by the nucleus of a hydrogen atom at the mean distance of the circulating electron ($r = 5.3 \times 10^{-11}$ m), (b) the electric potential energy of the atom when the electron is at this radius, and (c) the kinetic energy of the electron, assuming it to be moving in a circular orbit of this radius centered on the nucleus. (d) How much energy is required to ionize the hydrogen atom? Express all energies in electron volts.

25. A particle of (positive) charge Q is assumed to have a fixed position at P. A second particle of mass m and (negative) charge $-q$ moves at constant speed in a circle of radius r_1, centered at P. Derive an expression for the work W that must be done by an external agent on the second particle in order to increase the radius of the circle of motion, centered at P, to r_2.

    Answer: $W = \dfrac{qQ}{8\pi\epsilon_0} \left[\dfrac{1}{r_1} - \dfrac{1}{r_2} \right].$

26. Devise an arrangement of three point charges, separated by finite distances, that has zero electric potential energy.

27. Figure 29-29 shows an idealized representation of a U^{238} nucleus ($Z=92$) on the verge of fission. Calculate (a) the repulsive force acting on each fragment and (b) the mutual electric potential energy of the two fragments. Assume that the fragments are equal in size and charge, spherical, and just touching. The radius of the initially spherical U^{238} nucleus is 8.0×10^{-15} m. Assume that the material out of which nuclei are made has a constant density.   *Answer:* (a) 3.0×10^3 N. (b) 3.8×10^{-11} J, or 240 MeV.

figure 29-29
Problem 27

SECTION 29-7

28. The electric potential varies along the x-axis as shown in the graph of Fig. 29-30. For each of the intervals shown (ignore the behavior at the end points of the intervals) determine the x-component of the electric field and plot E_x vs. x.

figure 29-30
Problem 28

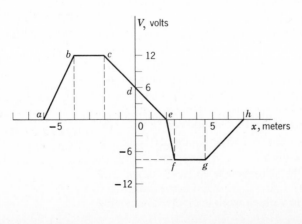

29. (a) Show that the electric potential at a point on the axis of a ring of charge of radius a, computed directly from Eq. 29-10, is given by

$$V = \frac{1}{4\pi\epsilon_0} \frac{q}{\sqrt{x^2 + a^2}}.$$

(b) From this result derive an expression for E at axial points; compare with the direct calculation of E in Example 5, Chapter 27.

30. What is the potential gradient, in V/m, at the surface of a gold nucleus? See Problem 3.

31. In Example 6 the potential at an axial point for a charged disk was shown to be

$$V = \frac{\sigma}{2\epsilon_0} (\sqrt{a^2 + r^2} - r).$$

(a) From this result show that E for axial points is given by

$$E = \frac{\sigma}{2\epsilon_0} \left(1 - \frac{r}{\sqrt{a^2 + r^2}}\right).$$

(b) Does this expression for E reduce to an expected result for $r \gg a$ and for $r = 0$?

32. (a) Starting from Eq. 29-11, find the magnitude E_r of the radial component of the electric field due to a dipole. (b) For what value of θ is E_r zero?

33. A charge per unit length λ is distributed uniformly along a straight-line segment of length L. (a) Determine the electrostatic potential (chosen to be zero at infinity) at a point P a distance y from one end of the charged segment and in line with it (see Fig. 29-31). (b) Use the result of (a) to compute the component of the electric field at P in the y-direction (along the line). (c) Determine the component of the electric field at P in a direction perpendicular to the straight line.

Answer: (a) $\dfrac{\lambda}{4\pi\epsilon_0} \ln \dfrac{L+y}{y}.$

(b) $\dfrac{\lambda}{4\pi\epsilon_0} \dfrac{L}{y(L+y)}.$

(c) Zero.

figure 29-31
Problem 33

34. On a thin rod of length L lying along the x-axis with one end at the origin $(x = 0)$, as in Fig. 29-32, there is distributed a charge per unit length given by $\lambda = kx$, where k is a constant. (a) Taking the electrostatic potential at infinity to be zero, find V at the point P on the y-axis. (b) Determine the vertical component, E_y, of the electric field intensity at P from the result of part (a) and also by direct calculation. (c) Why cannot E_x, the horizontal component of the electric field at P, be found using the result of part (a)?

figure 29-32
Problem 34

SECTION 29-8

35. What is the charge density σ on the surface of a conducting sphere of radius 0.15 m whose potential is 200 V? *Answer:* 1.2×10^{-8} C/m².

36. A charged sphere of radius 1.5 m contains a total charge of 3.0×10^{-6} C. (a) What is the electric field at the sphere's surface? (b) What is the electric potential at the sphere's surface? (c) At what distance from the sphere's surface has the electric potential decreased by 5000 V?

figure 29-33
Problem 37

37. The metal object in Fig. 29-33 is a figure of revolution about the horizontal axis. If it is charged negatively, sketch roughly a few equipotentials and lines of force. Use physical reasoning rather than mathematical analysis.

38. If the earth had a net charge equivalent to 1 electron/m² of surface area (a very artificial assumption), (a) what would the earth's potential be? (b) What would the electric field due to the earth be just outside its surface?

39. A charge of 10^{-8} C can be produced by simple rubbing. To what potential would such a charge raise an insulated conducting sphere of 10-cm radius? *Answer: 900 V.*

40. For the spheres of Fig. 29-19, what is the ratio of electric fields at the surface?

41. Consider a thin, isolated, conducting, spherical shell which is uniformly charged to a constant charge density σ (C/m²). How much work does it take to move a small positive test charge q_0 (a) from the surface of the shell to the interior, through a small hole, (b) from one point on the surface to another, regardless of path, (c) from point to point inside the shell, and (d) from any point P outside the shell over any path, whether or not it pierces the shell, back to P? (e) For the conditions given, does it matter whether or not the shell is conducting?   *Answer: (a) Zero. (b) Zero. (c) Zero. (d) Zero. (e) No.*

42. A thin, spherical, conducting shell of radius R is mounted on an insulated support and charged to a potential $-V$. An electron is fired from point P a distance r from the center of the shell $(r \gg R)$ with an initial speed v_0, directed radially inward. What is the value of v_0 chosen so that the electron will just reach the shell?

43. Can a conducting sphere 10 cm in radius hold a charge of 4×10^{-6} C in air without breakdown? The dielectric strength (minimum field required to produce breakdown) of air at 1 atm is 3×10^6 V/m.   *Answer: No.*

44. Two thin, insulated, concentric conducting spheres of radii R_1 and R_2 carry charges q_1 and q_2. Derive expressions for $E(r)$ and $V(r)$, where r is the distance from the center of the spheres. Plot $E(r)$ and $V(r)$ from $r = 0$ to $r = 4.0$ m for $R_1 = 0.50$ m, $R_2 = 1.0$ m, $q_1 = +2.0 \times 10^{-6}$ C, and $q_2 = +1.0 \times 10^{-6}$ C. Compare with Fig. 29-18.

45. The space between two concentric spheres of radii r_1 and r_2 is filled with a nonconducting material having a uniform charge density ρ. Find the electric potential V as a function of the distance r from the center of the spheres, considering the regions (a) $r > r_2$, (b) $r_2 > r > r_1$, and (c) $r < r_1$. (d) Do these solutions agree at $r = r_2$ and at $r = r_1$?

Answer: (a) $\dfrac{\rho}{3\epsilon_0}\dfrac{(r_2^3 - r_1^3)}{r}$. *(b)* $\dfrac{\rho}{3\epsilon_0}\left(\dfrac{3}{2}r_2^2 - \dfrac{r^2}{2} - \dfrac{r_1^3}{r}\right)$. *(c)* $\dfrac{\rho}{2\epsilon_0}(r_2^2 - r_1^2)$. *(d) Yes.*

46. Two metal spheres are 3.0 cm in radius and carry charges of $+1.0 \times 10^{-8}$ C and -3.0×10^{-8} C, respectively, assumed to be uniformly distributed. If their centers are 2.0 m apart, calculate (a) the potential of the point halfway between their centers and (b) the potential of each sphere.

47. Two identical conducting spheres of radius $r = 0.15$ m are separated by a distance $a = 10.0$ m. What is the charge on each sphere if the potential of one is $+1500$ V and if the other is -1500 V?   *Answer: $\pm 2.5 \times 10^{-8}$ C.*

48. Two conducting spheres, one of radius 6.0 cm and the other of radius 12.0 cm, each have a charge of 3×10^{-8} C and are very far apart. If the spheres are connected by a conducting wire, find (a) the direction of motion and the magnitude of the charge transferred and (b) the final charge on and potential of each sphere.

49. In Fig. 29-19 let $R_1 = 1.0$ cm and $R_2 = 2.0$ cm. Before the spheres are connected by the fine wire, a charge of 2.0×10^{-7} C is placed on the smaller sphere, the larger sphere being uncharged. Calculate (a) the charge, (b) the surface charge density, and (c) the potential for each sphere after they are connected.
Answer: (a) $q_1 = 0.67 \times 10^{-7}$ C, $q_2 = 1.33 \times 10^{-7}$ C. (b) $\sigma_1 = 5.33 \times 10^{-5}$ C/m², $\sigma_2 = 2.65 \times 10^{-5}$ C/m². (c) $V_1 = V_2 = 6.0 \times 10^4$ V.

50. A spherical drop of water carrying a charge of 3×10^{-11} C has a potential of 500 V at its surface. (a) What is the radius of the drop? (b) If two such drops of the same charge and radius combine to form a single spherical drop, what is the potential at the surface of the new drop so formed?

SECTION 29-9

51. (a) How much charge is required to raise an isolated metallic sphere of 1.0-m radius to a potential of 1.0×10^6 V? Repeat for a sphere of 1.0-cm radius. (b) Why use a large sphere in an electrostatic generator since the same potential can be achieved for a smaller charge with a small sphere?
    Answer: (a) 1.1×10^{-4} C; 1.1×10^{-6} C. (b) Because of the larger E field at the surface of the smaller sphere (verify), charge will leak off more rapidly.

52. An alpha particle is accelerated through a potential difference of one million volts in an electrostatic generator. (a) What kinetic energy does it acquire? (b) What kinetic energy would a proton acquire under these same circumstances? (c) Which particle would acquire the greater speed, starting from rest?

53. Let the potential difference between the shell of an electrostatic generator and the point at which charges are sprayed onto the moving belt be 3.0×10^6 V. If the belt transfers charge to the shell at the rate of 3.0×10^{-3} C/s, what power must be provided to drive the belt, considering only electrical forces?   *Answer:* 9.0 kW.

54. The high-voltage electrode of an electrostatic generator is a charged spherical metal shell having a potential V ($+9.0 \times 10^6$ V). (a) Electrical breakdown occurs in the gas in this machine at a field E (1.0×10^8 V/m). In order to prevent such breakdown, what restriction must be made on the radius r of the shell? (b) A long moving rubber belt transfers charge to the shell at 3.0×10^{-4} C/s, the potential of the shell remaining constant because of leakage. What minimum power is required to transfer the charge? (c) The belt is of width w (0.50 m) and travels at speed v (30 m/s). What is the surface charge density on the belt?

30
capacitors and dielectrics

Figure 30-1 shows a generalized *capacitor*, consisting of two insulated conductors, *a* and *b*, of arbitrary shape (later, regardless of their geometry, we will call them *plates*). We assume that they are totally isolated from objects in their surroundings and carry equal and opposite charges $+q$ and $-q$, respectively. Every line of force that originates on *a* terminates on *b*. We further assume, for the time being, that the conductors *a* and *b* exist in a vacuum.

The capacitor of Fig. 30-1 is characterized by q, the magnitude of the *charge on either conductor*, and by V, the *potential difference between the conductors*. Note (*a*) that q is *not* the net charge on the capacitor, which is zero, and (*b*) that V is *not* the potential of either conductor, referred perhaps to $V \rightarrow 0$ at ∞, but the potential difference between them.

It is not hard to put equal and opposite charges on conductors such as those of Fig. 30-1. We need not charge them separately. All that we have to do is to connect the conductors momentarily to opposite poles of a battery. Equal and opposite charges $(\pm q)$ will automatically appear.

For the moment we state without proof that q and V for a capacitor are proportional to each other, or

$$q = CV \tag{30-1}$$

in which C, the constant of proportionality, is called the *capacitance* of the capacitor. We state, again for the moment without proof, that C depends on the shapes and relative positions of the conductors. We will show in Section 30-2, for three important special cases, that C does indeed depend on these variables. C also depends on the medium in

650

figure 30-1
Two insulated conductors, totally
isolated from their surroundings and
carrying equal and opposite charges,
form a capacitor.

651   CAPACITANCE   SEC. 30-1

which the conductors are immersed but for the present (see Section 30-4
however) we assume this to be a vacuum.

The SI unit of capacitance that follows from Eq. 30-1 is the coulomb/
volt. A special unit, the *farad* (abbr. F), is used to represent it. It is
named in honor of Michael Faraday who, among other contributions,
developed the concept of capacitance. Thus

$$1 \text{ farad} = 1 \text{ coulomb/volt.}$$

The submultiples of the farad, the *microfarad* $(1 \ \mu\text{F} = 10^{-6} \text{ F})$ and the
picofarad $(1 \text{ pF} = 10^{-12} \text{ F})$, are more convenient units in practice.

An analogy can be made between a capacitor carrying a charge q and a rigid container of volume V containing n moles of an ideal gas. The gas pressure p is
directly proportional to n for a fixed temperature, according to the ideal gas
law (Eq. 23-2)

$$n = \left(\frac{V}{RT}\right) p.$$

For the capacitor (Eq. 30-1)

$$q = (C)V.$$

Comparison shows that the capacitance C of the capacitor, assuming a fixed
temperature for the gas, is analogous to the volume V of the container.

Note that any amount of charge can be put on the capacitor, and any mass of
gas can be put in the container, up to certain limits. These correspond to electrical breakdown ("arcing over") for the capacitor and to rupture of the walls for
the container.

Capacitors are very useful devices, of great interest to engineers and
physicists. For example:

1. In this book we stress the importance of *fields* to the understanding
   of natural phenomena. A capacitor can be used to establish desired
   electric field configurations for various purposes. In Section 27-5 we
   described the deflection of an electron beam in a uniform field set up
   by a capacitor, although we did not use this term in that section. In
   later sections we discuss the behavior of dielectric materials when
   placed in an electric field (provided conveniently by a capacitor) and

we shall see how the laws of electromagnetism can be generalized to take the presence of dielectric bodies into account.

2. A second concept stressed in this book is *energy*. By analyzing a charged capacitor we show that electric energy may be considered to be stored in the electric field between the plates and indeed in any electric field, however generated. Because capacitors can confine strong electric fields to small volumes, they can serve as useful devices for storing energy. In electron synchrotrons, which are cyclotron-like devices for accelerating electrons, energy accumulated and stored in a large bank of capacitors over a relatively long period of time is made available intermittently to accelerate the electrons by discharging the capacitor in a much shorter time. Many researches and devices in plasma physics also make use of bursts of energy stored in this way.

3. The electronic age could not exist without capacitors. They are used, in conjunction with other devices, to reduce voltage fluctuations in electronic power supplies, to transmit pulsed signals, to generate or detect electromagnetic oscillations at radio frequencies, to provide electronic time delays, and in many other ways. In most of these applications the potential difference between the plates will not be constant, as we assume in this chapter, but will vary with time, often in a sinusoidal or a pulsed fashion. In later chapters we consider some aspects of the capacitor used as a circuit element.

30-2 CALCULATING CAPACITANCE

Figure 30-2 shows a *parallel-plate* capacitor in which the conductors of Fig. 30-1 take the form of two parallel plates of area A separated by a distance d. If we connect each plate momentarily to the terminals of a battery, a charge $+q$ will automatically appear on one plate and a charge $-q$ on the other. If d is small compared with the plate dimensions, the electric field $\mathbf{E}$ between the plates will be uniform, which means that the lines of force will be parallel and evenly spaced. The laws of electromagnetism (see Problem 14, Chapter 35) require that there be some "fringing" of the lines at the edges of the plates, but for small enough d we can neglect it for our present purpose.

We can calculate the capacitance of this device using Gauss's law, another illustration of the usefulness of this law in situations of simple geometry. Figure 30-2 shows (dashed lines) a Gaussian surface of height h closed by plane caps of area A that are the shape and size of the capacitor plates. The flux of $\mathbf{E}$ is zero for the part of the Gaussian surface that lies inside the top capacitor plate because the electric field inside a conductor carrying a static charge is zero. The flux of $\mathbf{E}$ through the vertical wall of the Gaussian surface is zero because, to the extent that we can neglect the fringing of the lines of force, $\mathbf{E}$ lies in the wall.

This leaves only the face of the Gaussian surface that lies between

figure 30-2
A parallel-plate capacitor with conductors (plates) of area A. The dashed line represents a Gaussian surface whose height is h and whose top and bottom caps are the same shape and size as the capacitor plates.

the plates. Here **E** is constant and the flux Φ_E is simply EA. Gauss's law (Eq. 28-5) gives

$$\epsilon_0 \Phi_E = \epsilon_0 EA = q. \qquad (30\text{-}2)$$

The work required to carry a test charge q_0 from one plate to the other can be expressed either as $q_0 V$ (see Eq. 29-1) or as the product of a force $q_0 E$ times a distance d or $q_0 Ed$. These expressions must be equal, or

$$V = Ed. \qquad (30\text{-}3)$$

Formally, Eq. 30-3 is a special case of the general relation (Eq. 29-5; see also Example 1, Chapter 29)

$$V = -\int \mathbf{E} \cdot d\mathbf{l},$$

where V is the difference in potential between the plates. The integral may be taken over any path that starts on one plate and ends on the other because each plate is an equipotential surface and the electrostatic force is path independent. Although the simplest path between the plates is a perpendicular straight line, Eq. 30-3 follows no matter what path of integration we choose.

If we substitute Eqs. 30-2 and 30-3 into the relation $C = q/V$, we obtain

$$C = \frac{q}{V} = \frac{\epsilon_0 EA}{Ed} = \frac{\epsilon_0 A}{d}. \qquad (30\text{-}4)$$

Equation 30-4 holds only for capacitors of the parallel-plate type; different formulas hold for capacitors of different geometry. This equation shows, for a particular case, that C does indeed depend on the geometry of the conductors (plates) as we pointed out in Section 30-1. Both A and d are geometrical factors.

In Section 26-4 we stated that ϵ_0, which we first met in connection with Coulomb's law, was not measured in terms of that law because of experimental difficulties. Equation 30-4 suggests that we might determine ϵ_0 by building a capacitor of accurately known plate area and plate spacing and measuring its capacitance experimentally by measuring q and V in the relation $C = q/V$. Thus we can solve Eq. 30-4 for ϵ_0 and find a numerical value in terms of the measured quantities A, d, and C; ϵ_0 has indeed been measured in this way.

EXAMPLE 1

The parallel plates of an air-filled capacitor are everywhere 1.0 mm apart. What must the plate area be if the capacitance is to be 1.0 F?

From Eq. 30-4

$$A = \frac{dC}{\epsilon_0} = \frac{(1.0 \times 10^{-3} \text{ m})(1.0 \text{ F})}{8.9 \times 10^{-12} \text{ C}^2/\text{N}\cdot\text{m}^2} = 1.1 \times 10^8 \text{ m}^2.$$

This is the area of a square sheet more than 6 miles on edge; the farad is indeed a large unit.

EXAMPLE 2

A cylindrical capacitor. A cylindrical capacitor consists of two coaxial cylinders (Fig. 30-3) of radii a and b and length l. What is the capacitance of this device? Assume that the capacitor is very long (that is, that $l \gg b$) so that we can ignore the fringing of the lines of force at the ends for the purpose of calculating the capacitance.

As a Gaussian surface construct a coaxial cylinder of radius r and length l, closed by plane caps. Gauss's law (Eq. 28-6)

$$\epsilon_0 \oint \mathbf{E} \cdot d\mathbf{S} = q$$

gives

$$\epsilon_0 E(2\pi r)(l) = q,$$

the flux being entirely through the cylindrical surface and not through the end caps. Solving for E yields

$$E = \frac{q}{2\pi\epsilon_0 rl}.$$

The potential difference between the plates is given by Eq. 29-5 [note that $\mathbf{E}$ and $d\mathbf{l}$ ($= d\mathbf{r}$) point in opposite directions] or

$$V = -\int_a^b \mathbf{E} \cdot d\mathbf{l} = \int_a^b E\, dr = \int_a^b \frac{q}{2\pi\epsilon_0 l}\frac{dr}{r} = \frac{q}{2\pi\epsilon_0 l}\ln\frac{b}{a}.$$

Finally, the capacitance is given by

$$C = \frac{q}{V} = \frac{2\pi\epsilon_0 l}{\ln(b/a)}.$$

Like the relation for the parallel-plate capacitor (Eq. 30-4), this relation also depends only on geometrical factors, l, b, and a.

figure 30-3
Example 2. A cross section of a cylindrical capacitor. The dashed circle is a cross section of a closed cylindrical Gaussian surface of radius r and length l.

EXAMPLE 3

The Capacitance of an isolated sphere. In Section 29-8 we showed that the potential of an isolated conducting sphere of radius R carrying a charge q is given by

$$V = \frac{1}{4\pi\epsilon_0}\frac{q}{R}. \qquad (30\text{-}5)$$

We can regard this sphere as one plate of a capacitor, the other plate being a conducting sphere of infinite radius, with V chosen to be zero on the sphere at infinity.

The capacitance of the sphere of radius R is then given, from Eq. 30-5, by

$$C = \frac{q}{V} = 4\pi\epsilon_0 R. \qquad (30\text{-}6)$$

Again, the only relevant geometric factor, the sphere radius R, appears.

EXAMPLE 4

Capacitors in parallel. Figure 30-4 shows three capacitors connected in parallel. What single capacitance C is equivalent to this combination? "Equivalent" means that if the parallel combination and the single capacitor were each in a box with wires a and b connected to terminals, it would not be possible to distinguish the two by electrical measurements external to the box.

The potential difference across each capacitor in a parallel arrangement (Fig. 30-4) *is the same.* This follows because all of the upper plates are connected together and to terminal a, whereas all of the lower plates are connected together and to terminal b. Applying the relation $q = CV$ to each capacitor yields

$$q_1 = C_1 V; \qquad q_2 = C_2 V; \qquad \text{and} \qquad q_3 = C_3 V.$$

The total charge q on the combination is

$$q = q_1 + q_2 + q_3$$
$$= (C_1 + C_2 + C_3)V.$$

The equivalent capacitance C is

figure 30-4
Example 4. Three capacitors in parallel. The symbol for a capacitor (—||—) is chosen for its simplicity. Although it suggests a parallel-plate capacitor, it is meant to suggest a capacitor of *any* geometry.

$$C = \frac{q}{V} = C_1 + C_2 + C_3.$$

We can easily extend this result to any number of parallel-connected capacitors.

Capacitors in series. Figure 30-5 shows three capacitors connected in series. What single capacitance C is "equivalent" (see Example 4) to this combination?

For capacitors connected in a series arrangement (Fig. 30-5) *the magnitude q of the charge on each plate must be the same.* This is true because the net charge on the part of the circuit enclosed by the dashed line in Fig. 30-5 must be zero; that is, the charge present on these plates initially is zero and connecting a battery between a and b will only produce a charge separation, the *net* charge on these plates still being zero. Assuming that neither C_1 nor C_2 "sparks over," there is no way for charge to enter or leave the volume suggested by the dashed line.

Applying the relation $q = CV$ to each capacitor yields

$$V_1 = q/C_1; \qquad V_2 = q/C_2; \qquad \text{and} \qquad V_3 = q/C_3.$$

The potential difference for the series combination is

$$V = V_1 + V_2 + V_3$$

$$= q \left(\frac{1}{C_1} + \frac{1}{C_2} + \frac{1}{C_3} \right).$$

The equivalent capacitance

$$C = \frac{q}{V} = \frac{1}{\dfrac{1}{C_1} + \dfrac{1}{C_2} + \dfrac{1}{C_3}},$$

or

$$\frac{1}{C} = \frac{1}{C_1} + \frac{1}{C_2} + \frac{1}{C_3}.$$

The equivalent series capacitance is always less than the smallest capacitance in the chain.

EXAMPLE 5

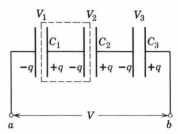

figure 30-5
Example 5. Three capacitors in series.

30-3
ENERGY STORAGE IN AN ELECTRIC FIELD

In Section 29-6 we saw that all charge configurations have a certain *electric potential energy U*, equal to the work W (which may be positive or negative) that must be done to assemble them from their individual components, originally assumed to be infinitely far apart and at rest. This potential energy reminds us of the potential energy stored in a compressed spring or the gravitational potential energy stored in, say, the earth-moon system.

For a simple example, work must be done to separate two equal and opposite charges. This energy is stored in the system and can be recovered if the charges are allowed to come together again. Similarly, a charged capacitor has stored in it an electrical potential energy U equal to the work W required to charge it. This energy can be recovered if the capacitor is allowed to discharge. We can visualize the work of charging by imagining that an external agent pulls electrons from the positive plate and pushes them onto the negative plate, thus bringing about the charge separation; normally the work of charging is done by a battery, at the expense of its store of chemical energy.

Suppose that at a time t a charge $q'(t)$ has been transferred from one plate to the other. The potential difference $V(t)$ between the plates at that moment will be $q'(t)/C$. If an extra increment of charge dq' is transferred, the small amount of additional work needed will be

$$dW = V\,dq = \left(\frac{q'}{C}\right)dq'.$$

If this process is continued until a total charge q has been transferred, the total work will be found from

$$W = \int dW = \int_0^q \frac{q'}{C}\,dq' = \frac{1}{2}\frac{q^2}{C}. \qquad (30\text{-}7)$$

From the relation $q = CV$ we can also write this as

$$W\ (= U) = \tfrac{1}{2}CV^2. \qquad (30\text{-}8)$$

It is reasonable to suppose that the energy stored in a capacitor resides in the electric field. As q or V in Eqs. 30-7 and 30-8 increase, for example, so does the electric field E; when q and V are zero, so is E.

In a parallel-plate capacitor, neglecting fringing, the electric field has the same value for all points between the plates. Thus the *energy density u*, which is the stored energy per unit volume, should also be uniform; u (see Eq. 30-8) is given by

$$u = \frac{U}{Ad} = \frac{\tfrac{1}{2}CV^2}{Ad},$$

where Ad is the volume between the plates. Substituting the relation $C = \epsilon_0 A/d$ (Eq. 30-4) leads to

$$u = \frac{\epsilon_0}{2}\left(\frac{V}{d}\right)^2.$$

However, V/d is the electric field E, so that

$$u = \tfrac{1}{2}\epsilon_0 E^2. \qquad (30\text{-}9)$$

Although we derived this equation for the special case of a parallel-plate capacitor, it is true in general. *If an electric field* **E** *exists at any point in space (a vacuum), we can think of that point as the site of stored energy in amount, per unit volume, of* $\tfrac{1}{2}\epsilon_0 E^2$.

A capacitor C_1 is charged to a potential difference V_0. This charging battery is then removed and the capacitor is connected as in Fig. 30-6 to an uncharged capacitor C_2.

(a) What is the final potential difference V across the combination?

The original charge q_0 is now shared by the two capacitors. Thus

$$q_0 = q_1 + q_2.$$

Applying the relation $q = CV$ to each of these terms yields

$$C_1 V_0 = C_1 V + C_2 V$$

or

$$V = V_0 \frac{C_1}{C_1 + C_2}.$$

This suggests a way to measure an unknown capacitance (C_2, say) in terms of a known one.

(b) What is the stored energy before and after the switch in Fig. 30-6 is thrown?

The initial stored energy is

$$U_0 = \tfrac{1}{2}C_1 V_0^2.$$

The final stored energy is

EXAMPLE 6

figure 30-6
Example 6. C_1 is charged to a potential difference V_0 and then connected to C_2 by closing switch S.

$$U = \tfrac{1}{2}C_1V^2 + \tfrac{1}{2}C_2V^2 = \tfrac{1}{2}(C_1 + C_2)\left(\frac{V_0C_1}{C_1 + C_2}\right)^2 = \left(\frac{C_1}{C_1 + C_2}\right)U_0.$$

Thus U is less than U_0! For $C_1 = C_2$, in fact, $U = \tfrac{1}{2}U_0$.

This is *not* a violation of the principle of the conservation of energy. The example tacitly assumes an ideal (rather than an actual laboratory) circuit in that the resistance (Chapter 31) and the inductance (Chapter 36) of the connecting wires have both been assumed to be zero. In an actual laboratory circuit the "missing" energy would appear as thermal energy in the wires and/or as energy radiated away from the circuit as electromagnetic radiation (Chapter 41). For a good discussion see "On Conservation of Energy in Electric Circuits" by Camillo Cuvaj, *American Journal of Physics*, 1968.

An isolated conducting sphere of radius R, in a vacuum, carries a charge q. (a) Compute the total electrostatic energy stored in the surrounding space. At any radius r from the center of the sphere (assuming $r > R$) E is given by

EXAMPLE 7

$$E = \frac{1}{4\pi\epsilon_0}\frac{q}{r^2}.$$

The energy density at any radius r is found from Eq. 30-9, or

$$u = \tfrac{1}{2}\epsilon_0 E^2 = \frac{q^2}{32\pi^2\epsilon_0 r^4}.$$

The energy dU that lies in a spherical shell between the radii r and $r + dr$ is

$$dU = (4\pi r^2)(dr)u = \frac{q^2}{8\pi\epsilon_0}\frac{dr}{r^2},$$

where $(4\pi r^2)(dr)$ is the volume of the spherical shell. The total energy U is found by integration, or

$$U = \int dU = \frac{q^2}{8\pi\epsilon_0}\int_R^\infty \frac{dr}{r^2} = \frac{q^2}{8\pi\epsilon_0 R}.$$

Note that this relation follows at once from Eq. 30-7 $(U = q^2/2C)$, where C (see Example 3) is the capacitance $(= 4\pi\epsilon_0 R)$ of an isolated sphere of radius R.

(b) What is the radius R_0 of a spherical surface such that half the stored energy lies within it?

In the equation just given we put

$$\tfrac{1}{2}U = \frac{q^2}{8\pi\epsilon_0}\int_R^{R_0}\frac{dr}{r^2}$$

or

$$\frac{q^2}{16\pi\epsilon_0 R} = \frac{q^2}{8\pi\epsilon_0}\left(\frac{1}{R} - \frac{1}{R_0}\right),$$

which yields, after some rearrangement,

$$R_0 = 2R.$$

Thus, most of the energy is stored in space rather close to the sphere.

Equation 30-4 holds only for a parallel-plate capacitor with its plates in a vacuum. Michael Faraday, in 1837, first investigated the effect of filling the space between the plates with a dielectric (see Table 30-1 for a sampling of dielectrics used today). In Faraday's words:

30-4
PARALLEL-PLATE CAPACITOR WITH DIELECTRIC

The question may be stated thus: suppose A an electrified plate of metal suspended in air, and B and C two exactly similar plates, placed parallel to and on each side of A at equal distances and insulated; A will then induce equally to-

(b)

figure 30-7
(*a*) Battery *B* supplies the same potential difference to each capacitor; the one on the right has the higher charge. (*b*) Both capacitors carry the same charge; the one on the right has the lower potential difference, as indicated by the meter readings.

ward *B* and *C* [that is, equal charges will appear on these plates]. If in this position of the plates some other dielectric than air, as shell-lac, be introduced between *A* and *C*, will the induction between them remain the same? Will the relation of *C* and *B* to *A* be unaltered, notwithstanding the difference of the dielectrics interposed between them?

Faraday answered this question by constructing two identical capacitors, in one of which he placed a dielectric, the other containing air at normal pressure. When both capacitors were charged to the *same potential difference*, Faraday found by experiment that *the charge on the one containing the dielectric was greater than that on the other*; see Fig. 30-7a.

Since *q* is larger, for the same *V*, if a dielectric is present, it follows from the relation $C = q/V$ that the capacitance of a capacitor increases if a dielectric is placed between the plates. The ratio of the capacitance with the dielectric* to that without is called the *dielectric constant* κ of the material; see Table 30-1.

Instead of maintaining the two capacitors at the same potential difference, we can place the *same charge* on them, as in Fig. 30-7b. Experiment then shows that the potential difference V_d between the plates of the right-hand capacitor is smaller than that for the left-hand capacitor by the factor $1/\kappa$, or

$$V_d = V_0/\kappa.$$

We are led once again to conclude, from the relation $C = q/V$, that the effect of the dielectric is to increase the capacitance by a factor κ.

For a parallel-plate capacitor we can write, as an experimental result,

$$C = \frac{\kappa \epsilon_0 A}{d}. \qquad (30\text{-}10)$$

Equation 30-4 is a special case of this relation found by putting $\kappa = 1$, corresponding to a vacuum between the plates. Experiment shows that the capacitance of *all* types of capacitor is increased by the factor κ if the space between the plates is filled with a dielectric. Thus the capacitance of any capacitor can be written as

$$C = \kappa \epsilon_0 L,$$

where *L* depends on the geometry and has the dimensions of a length. For a parallel-plate capacitor (see Eq. 30-4) *L* is A/d; for a cylindrical capacitor (see Example 2) it is $2\pi l/\ln(b/a)$.

---

* Assumed to fill completely the space between the plates.

Table 30-1
Properties of some dielectrics*

Material	Dielectric Constant	Dielectric Strength** (kV/mm)
Vacuum	1.00000	∞
Air	1.00054	0.8
Water	78	—
Paper	3.5	14
Ruby mica	5.4	160
Porcelain	6.5	4
Fused quartz	3.8	8
Pyrex glass	4.5	13
Bakelite	4.8	12
Polyethylene	2.3	50
Amber	2.7	90
Polystyrene	2.6	25
Teflon	2.1	60
Neoprene	6.9	12
Transformer oil	4.5	12
Titanium dioxide	100	6

* These properties are at approximately room temperature and for conditions such that the electric field **E** in the dielectric does not vary with time.
** This is the maximum potential gradient that may exist in the dielectric without the occurrence of electrical breakdown. Dielectrics are often placed between conducting plates to permit a higher potential difference to be applied between them than would be possible with air as the dielectric.

EXAMPLE 8

A parallel-plate capacitor has plates with area A and separation d. A battery charges the plates to a potential difference V_0. The battery is then disconnected, and a dielectric slab of thickness d is introduced. Calculate the stored energy both before and after the slab is introduced and account for any difference.

The energy U_0 before introducing the slab is

$$U_0 = \tfrac{1}{2} C_0 V_0{}^2.$$

After the slab is in place, we have

$$C = \kappa C_0 \quad \text{and} \quad V = V_0/\kappa$$

and thus

$$U = \tfrac{1}{2} C V^2 = \tfrac{1}{2} \kappa C_0 \left(\frac{V_0}{\kappa}\right)^2 = \frac{1}{\kappa} U_0.$$

The energy after the slab is introduced is *less* by a factor $1/\kappa$. The "missing" energy would be apparent to the person who inserted the slab. He would feel a "tug" on the slab and would have to restrain it if he wished to insert the slab without acceleration. This means that he would have to do negative work on it, or, alternatively, that the capacitor + slab system would do positive work on him. This positive work is

$$W = U_0 - U = \tfrac{1}{2} C_0 V_0{}^2 \left(1 - \frac{1}{\kappa}\right).$$

As expected, $W = 0$ for the case of $\kappa = 1$.

The following section will give some detailed insight into how the "tug" referred to above arises, in terms of the attraction between what we will call "free" charges on the capacitor plates and "induced" charges on the dielectric.

Note from the relation $U = \frac{1}{2}CV^2$ (see Eq. 30-8) that we can derive the *energy density u*, for a parallel-plate capacitor in which a dielectric slab is present, from

$$u = \frac{U}{(Ad)} = \left(\frac{1}{Ad}\right)\left(\frac{1}{2}CV^2\right) = \left(\frac{1}{Ad}\right)\left(\frac{1}{2}\right)\left(\frac{\epsilon_0\kappa A}{d}\right)(V^2).$$

But $E = V/d$ so that we have

$$u = \frac{1}{2}\epsilon_0\kappa E^2.$$

As for Eq. 30-9, this relation, although derived for a parallel-plate capacitor, holds in general; that is, at any point P in a dielectric of constant κ. As we expect, for $\kappa = 1$, this new relation reduces to Eq. 30-9.

We now seek to understand, in atomic terms, what happens when we place a dielectric in an electric field. There are two possibilities. The molecules of some dielectrics, like water (see Fig. 29-11), have permanent electric dipole moments. In such materials (called *polar*) the electric dipole moments **p** tend to align themselves with an external electric field, as in Fig. 30-8b; see also section 27-6. Because the molecules are in constant thermal agitation, the degree of alignment will not be complete but will increase as the applied electric field is increased or as the temperature is decreased.

30-5
DIELECTRICS— AN ATOMIC VIEW

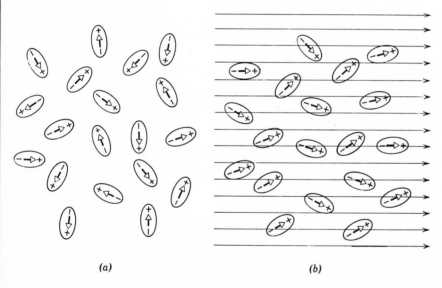

(a)   (b)

figure 30-8
(a) Molecules with a permanent electric dipole moment, showing their random orientation in the absence of an external electric field. (b) An electric field is applied, producing partial alignment of the dipoles. Thermal agitation prevents complete alignment.

Whether or not the molecules have permanent electric dipole moments, they acquire them by *induction* when placed in an electric field. In Section 29-5 we saw that the external electric field tends to separate the negative and the positive charge in the atom or molecule. This *induced electric dipole moment* is present only when the electric field is present. It is proportional to the electric field (for normal field strengths) and is created already lined up with the electric field as Fig. 29-12 suggests.

Let us use a parallel-plate capacitor, carrying a fixed charge q and not connected to a battery (see Fig. 30-7b), to provide a uniform external electric field $\mathbf{E}_0$ into which we place a dielectric slab. The over-all effect of alignment and induction is to separate the center of positive charge of the entire slab slightly from the center of negative charge. The slab, as

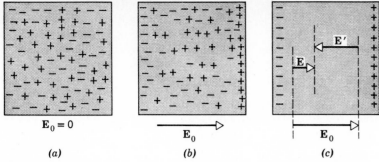

figure 30-9
(*a*) A dielectric slab, showing the random distribution of plus and minus charges. (*b*) An external field $\mathbf{E}_0$, established by putting the slab between the plates of a parallel-plate capacitor (not shown), separates the center of plus charge in the slab slightly from the center of minus charge, resulting in the appearance of surface charges. No *net* charge exists in any volume element located in the *interior* of the slab. (*c*) The surface charges set up a field $\mathbf{E}'$ which opposes the external field $\mathbf{E}_0$ associated with the charges on the capacitor plates. The resultant field $\mathbf{E}$ $(= \mathbf{E}_0 + \mathbf{E}')$ in the dielectric is thus less than $\mathbf{E}_0$.

a whole, although remaining electrically neutral, becomes *polarized*, as Fig. 30-9*b* suggests. The net effect is a pile-up of positive charge on the right face of the slab and of negative charge on the left face; within the slab no excess charge appears in any given volume element. Since the slab as a whole remains neutral, the positive *induced surface charge* must be equal in magnitude to the negative induced surface charge. Note that in this process electrons in the dielectric are displaced from their equilibrium positions by distances that are considerably less than an atomic diameter. There is no transfer of charge over macroscopic distances such as occurs when a current is set up in a conductor.

Figure 30-9*c* shows that the induced surface charges will always appear in such a way that the electric field set up *by them* ($\mathbf{E}'$) opposes the external electric field $\mathbf{E}_0$. The *resultant* field in the dielectric $\mathbf{E}$ is the vector sum of $\mathbf{E}_0$ and $\mathbf{E}'$. It points in the same direction as $\mathbf{E}_0$ but is smaller. *If we place a dielectric in an electric field, induced surface charges appear which tend to weaken the original field within the dielectric.*

This weakening of the electric field reveals itself in Fig. 30-7*b* as a reduction in potential difference between the plates of a charged isolated capacitor when a dielectric is introduced between the plates. The relation $V = Ed$ for a parallel-plate capacitor (see Eq. 30-3) holds whether or not dielectric is present and shows that the reduction in V described in Fig. 30-7*b* is directly connected to the reduction in E described in Fig. 30-9. More specifically, if a dielectric slab is introduced into an isolated charged parallel-plate capacitor, then

$$\frac{E_0}{E} = \frac{V_0}{V_d} = \kappa \qquad (30\text{-}11)$$

where the symbols on the left refer to Fig. 30-9 and the symbols V_0 and V_d refer to Fig. 30-7*b*.*

---

* Equation 30-11 does not hold if the battery remains connected while the dielectric slab is introduced. In this case V (hence E) could not change. Instead, the charge q on the capacitor plates would increase by a factor κ, as Fig. 30-7*a* suggests.

figure 30-10
A charged rod attracts an uncharged
piece of paper because unbalanced
forces act on the induced surface
charges.

Induced surface charge is the explanation of the most elementary fact
of static electricity, namely, that a charged rod will attract uncharged
bits of paper, etc. Figure 30-10 shows a bit of paper in the field of a
charged rod. Surface charges appear on the paper as shown. The negatively charged end of the paper will be pulled toward the rod and the
positively charged end will be repelled. These two forces do not have the
same magnitude because the negative end, being closer to the rod, is in
a stronger field and experiences a stronger force. The net effect is an
attraction. A dielectric body in a *uniform* electric field will not experience a net force.

In Example 8 we pointed out that, if we insert a dielectric slab into a
parallel-plate capacitor carrying a fixed charge q, a force will act on the
slab drawing it into the capacitor. This force is provided by the electrostatic attraction between the charges $\pm q$ on the capacitor plates and the
induced surfaces charges $\mp q'$ on the dielectric slab. When the slab is
only part way into the capacitor neither q nor q' will be uniformly distributed. Sketch qualitatively a possible distribution for q and q' when
the slab is, say, halfway into the capacitor.

So far our use of Gauss's law has been confined to situations in which
no dielectric was present. Now let us apply this law to a parallel-plate
capacitor filled with a dielectric of dielectric constant κ.

30-6 DIELECTRICS AND GAUSS'S LAW

Figure 30-11 shows the capacitor both with and without the dielectric. We assume that the charge q on the plates is the same in each case.
Gaussian surfaces have been drawn after the fashion of Fig. 30-2.

If no dielectric is present (Fig. 30-11a), Gauss's law (see Eq. 30-2)
gives

$$\epsilon_0 \oint \mathbf{E} \cdot d\mathbf{S} = \epsilon_0 E_0 A = q$$

or

$$E_0 = \frac{q}{\epsilon_0 A}. \tag{30-12}$$

If the dielectric is present (Fig. 30-11b), Gauss's law gives

$$\epsilon_0 \oint \mathbf{E} \cdot d\mathbf{S} = \epsilon_0 E A = q - q'$$

or

$$E = \frac{q}{\epsilon_0 A} - \frac{q'}{\epsilon_0 A}, \tag{30-13}$$

in which $-q'$, the *induced surface charge*, must be distinguished from
q, the so-called *free charge* on the plates. These two charges, both of
which lie within the Gaussian surface, are opposite in sign; $q - q'$ is
the *net* charge within the Gaussian surface.

Equation 30-11 shows that in Fig. 30-11

$$E = \frac{E_0}{\kappa}.$$

Combining this with Eq. 30-12, we have

$$E = \frac{E_0}{\kappa} = \frac{q}{\kappa \epsilon_0 A}. \tag{30-14}$$

(a)

(b)

figure 30-11
A parallel-plate capacitor (a) without and (b) with a dielectric. The charge q on the plates is assumed to be the same in each case.

663

DIELECTRICS AND GAUSS'S LAW   SEC. 30-6

Inserting this in Eq. 30-13 yields

$$\frac{q}{\kappa\epsilon_0 A} = \frac{q}{\epsilon_0 A} - \frac{q'}{\epsilon_0 A} \qquad (30\text{-}15a)$$

or

$$q' = q\left(1 - \frac{1}{\kappa}\right). \qquad (30\text{-}15b)$$

This shows correctly that the induced surface charge q' is always less in magnitude than the free charge q and is equal to zero if no dielectric is present, that is, if $\kappa = 1$.

Now we write Gauss's law for the case of Fig. 30-11b in the form

$$\epsilon_0 \oint \mathbf{E} \cdot d\mathbf{S} = q - q', \qquad (30\text{-}16)$$

$q - q'$ again being the net charge within the Gaussian surface. Substituting from Eq. 30-15b for q' leads, after some rearrangement, to

$$\epsilon_0 \oint \kappa \mathbf{E} \cdot d\mathbf{S} = q. \qquad (30\text{-}17)$$

This important relation, although derived for a parallel-plate capacitor, is true generally and is the form in which Gauss's law is usually written when dielectrics are present. Note the following:

1. The flux integral now contains a factor κ.

2. The charge q contained within the Gaussian surface is taken to be the *free charge only*. Induced surface charge is deliberately ignored on the right side of this equation, having been taken into account by the introduction of κ on the left side. Equations 30-16 and 30-17 are completely equivalent formulations.

EXAMPLE 9

Figure 30-12 shows a dielectric slab of thickness b and dielectric constant κ placed between the plates of a parallel-plate capacitor of plate area A and separation d. A potential difference V_0 is applied with no dielectric present. The battery is then disconnected and the dielectric slab inserted. Assume that $A = 100 \text{ cm}^2$, $d = 1.0 \text{ cm}$, $b = 0.50 \text{ cm}$, $\kappa = 7.0$, and $V_0 = 100 \text{ V}$ and (a) calculate the capacitance C_0 before the slab is inserted.

From Eq. 30-4, C_0 is:

$$C_0 = \frac{\epsilon_0 A}{d} = \frac{(8.9 \times 10^{-12} \text{ C}^2/\text{N}\cdot\text{m}^2)(10^{-2} \text{ m}^2)}{10^{-2} \text{ m}} = 8.9 \times 10^{-12} \text{ F} = 8.9 \text{ pF}.$$

(*b*) Calculate the free charge *q*.
From Eq. 30-1,

$$q = C_0 V_0 = (8.9 \times 10^{-12} \text{ F})(100 \text{ V}) = 8.9 \times 10^{-10} \text{ C}.$$

Because of the technique used to charge the capacitor, the free charge remains unchanged as the slab is introduced. If the charging battery had *not* been disconnected, this would not be the case.

(*c*) Calculate the electric field in the gap.

Applying Gauss's law in the form given in Eq. 30-17 to the Gaussian surface of Fig. 30-12 (upper plate) yields

$$\epsilon_0 \oint \kappa \mathbf{E} \cdot d\mathbf{S} = \epsilon_0 E_0 A = q,$$

or $\qquad E_0 = \dfrac{q}{\epsilon_0 A} = \dfrac{8.9 \times 10^{-10} \text{ C}}{(8.9 \times 10^{-12} \text{ C}^2/\text{N·m}^2)(10^{-2} \text{ m}^2)} = 1.0 \times 10^4 \text{ V/m}.$

Note that we put $\kappa = 1$ here because the surface over which we evaluate the flux integral does not pass through any dielectric. Note too that E_0 remains unchanged when the slab is introduced; this derivation takes no specific account of the presence of the dielectric.

(*d*) Calculate the electric field in the dielectric.

figure 30-12
Example 9. A parallel-plate capacitor containing a dielectric slab.

Applying Eq. 30-17 to the Gaussian surface of Fig. 30-12 (lower plate) yields

$$\epsilon_0 \oint \kappa \mathbf{E} \cdot d\mathbf{S} = \epsilon_0 \kappa E A = q.$$

Note that κ appears here because the surface cuts through the dielectric and that only the free charge *q* appears on the right. Thus we have

$$E = \frac{q}{\kappa \epsilon_0 A} = \frac{E_0}{\kappa} = \frac{1.0 \times 10^4 \text{ V/m}}{7.0} = 0.14 \times 10^4 \text{ V/m}.$$

(*e*) Calculate the potential difference between the plates.

Applying Eq. 29-5 to a straight perpendicular path from the lower plate (*L*) to the upper one (*U*) yields

$$V = -\int_L^U \mathbf{E} \cdot d\mathbf{l} = -\int_L^U E \cos 180° \, dl = \int_L^U E \, dl = E_0(d - b) + Eb.$$

Numerically

$$V = (1.0 \times 10^4 \text{ V/m})(5 \times 10^{-3} \text{ m}) + (0.14 \times 10^4 \text{ V/m})(5 \times 10^{-3} \text{ m}) = 57 \text{ V}.$$

This contrasts with the original applied potential difference of 100 V; compare Fig. 30-7*b*.

(*f*) Calculate the capacitance with the slab in place.
From Eq. 30-1,

$$C = \frac{q}{V} = \frac{8.9 \times 10^{-10} \text{ C}}{57 \text{ V}} = 16 \text{ pF}.$$

When the dielectric slab is introduced, the potential difference drops from 100 to 57 V and the capacitance rises from 8.9 to 16 pF, a factor of 1.8. If the dielectric slab had filled the capacitor, the capacitance would have risen by a factor of $\kappa \, (= 7.0)$ to 62 pF.

For all situations that we encounter in this book our discussion of the behavior of dielectrics in an electric field is adequate. However, the problems that we treat are simple ones, such as that of a rectangular slab placed at right angles to a uniform external electric field. For more difficult problems, such as that of finding **E** at the center of a dielectric ellipsoid placed in a (possibly nonuniform) external electric field, it greatly simplifies the labor and leads to deeper insight if we introduce a new formalism. We do so largely so that students who take a second course in electromagnetism will have some familiarity with the concepts.

Let us rewrite Eq. 30-15a, which applies to a parallel-plate capacitor containing a dielectric, as

$$\frac{q}{A} = \epsilon_0 \left(\frac{q}{\kappa \epsilon_0 A} \right) + \frac{q'}{A}. \qquad (30\text{-}18)$$

The quantity in parentheses (see Eq. 30-14) is simply the electric field E in the dielectric. The last term in Eq. 30-18 is the *induced surface charge per unit area.* We call it the *electric polarization P,* or

$$P = \frac{q'}{A}. \qquad (30\text{-}19)$$

The name is suitable because the induced surface charge q' (also called the polarization charge) appears when the dielectric is polarized.

The electric polarization P can be defined in an equivalent way by multiplying the numerator and denominator in Eq. 30-19 by d, the thickness of the dielectric slab in Fig. 30-11,

$$P = \frac{q'd}{Ad}. \qquad (30\text{-}20)$$

The numerator is the product $q'd$ of the magnitude of the (equal and opposite) polarization charges by their separation. It is thus the induced electric dipole moment of the dielectric slab. Since the denominator Ad is the volume of the slab, we see that the electric polarization can also be defined as the induced electric dipole moment per unit volume in the dielectric. This definition suggests that since the electric dipole moment is a vector the electric polarization is also a vector, its magnitude being P. The direction of **P** is from the negative induced charge to the positive induced charge, as for any dipole. In Fig. 30-13, which shows a capacitor with a dielectric slab filling half the space between the plates, **P** points down.

We can now rewrite Eq. 30-18 as

$$\frac{q}{A} = \epsilon_0 E + P. \qquad (30\text{-}21)$$

The quantity on the right occurs so often in electrostatic problems that we give it the special name *electric displacement D,* or

$$D = \epsilon_0 E + P \qquad (30\text{-}22a)$$

in which

$$D = \frac{q}{A}. \qquad (30\text{-}22b)$$

The name has historical significance only.

Since **E** and **P** are vectors, **D** must also be one, so that in the more general case we have

$$\mathbf{D} = \epsilon_0 \mathbf{E} + \mathbf{P}. \qquad (30\text{-}23)$$

In Fig. 30-13 all three vectors point down and each has a constant magnitude for every point in the dielectric (and also at every point in the air gap) so that the vector nature of Eq. 30-23 is not very important in this case. In more compli-

figure 30-13
(a) Showing **D**, ϵ_0**E**, and **P** in the dielectric (*upper right*) and in the gap (*upper left*) for a parallel-plate capacitor. (b) Showing samples of the lines associated with **D** (free charge), ϵ_0**E** (all charges), and **P** (polarization charge).

cated problems, however, **E**, **P**, and **D** may vary in magnitude and direction from point to point.

From their definitions we see the following:

1. **D** (see Eq. 30-22b) is connected with the *free charge* only. We can represent the vector field of **D** by *lines of* **D**, just as we represent the field of **E** by lines of force. Figure 30-13b shows that the lines of **D** begin and end on the free charges.
2. **P** (see Eq. 30-19) is connected with the *polarization charge* only. It is also possible to represent this vector field by lines. Figure 30-13b shows that the lines of **P** begin and end on the polarization charges.
3. **E** is connected with *all* charges that are actually present, whether free or polarization. The lines of **E** reflect the presence of both kinds of charge, as Fig. 30-13b shows. Note (Eqs. 30-19 and 30-22b) that the units for **P** and **D** (C/m²) differ from those of **E** (N/C).

The electric field vector **E**, which is what determines the force that acts on a suitably placed test charge, remains of fundamental interest. **D** and **P** are auxiliary vectors useful as aids in the solution of problems more complex than that of Fig. 30-13.

The vectors **D** and **P** can both be expressed in terms of **E** alone. A convenient starting point is the identity

$$\frac{q}{A} = \kappa\epsilon_0 \left(\frac{q}{\kappa\epsilon_0 A}\right).$$

Comparison with Eqs. 30-14 and 30-22b shows that this, extended to vector form, can be written as

$$\mathbf{D} = \kappa\epsilon_0\mathbf{E}. \qquad (30\text{-}24)$$

We can also write the polarization (see Eqs. 30-19 and 30-15b) as

$$P = \frac{q'}{A} = \frac{q}{A}\left(1 - \frac{1}{\kappa}\right).$$

Since $q/A = D$, we can rewrite this, using Eq. 30-24 and casting the result into vector form, as

$$\mathbf{P} = \epsilon_0(\kappa - 1)\mathbf{E}. \qquad (30\text{-}25)$$

This shows clearly that in a vacuum ($\kappa = 1$) the polarization vector $\mathbf{P}$ is zero.* Equations 30-24 and 30-25 show that for isotropic materials, to which a single dielectric constant κ can be assigned, $\mathbf{D}$ and $\mathbf{P}$ both point in the direction of $\mathbf{E}$ at any given point.

The definition of $\mathbf{D}$ given by Eq. 30-24 allows us to write Eq. 30-17, that is, Gauss's law in the presence of a dielectric, simply as

$$\oint \mathbf{D} \cdot d\mathbf{S} = q, \qquad (30\text{-}26)$$

where, as before, q represents the free charge only, the induced surface charges being excluded.

In Figure 30-13, using data from Example 9, calculate E, D, and P: (a) in the dielectric and (b) in the air gap.

EXAMPLE 10

(a) The electric field in the dielectric is calculated in Example 9 to be 1.43×10^3 V/m. From Eq. 30-24,

$$D = \kappa\epsilon_0 E$$

$$= (7.0)(8.9 \times 10^{-12} \text{ C}^2/\text{N}\cdot\text{m}^2)(1.43 \times 10^3 \text{ V/m})$$

$$= 8.9 \times 10^{-8} \text{ C/m}^2$$

and, from Eq. 30-25,

$$P = \epsilon_0(\kappa - 1)E$$

$$= (8.9 \times 10^{-12} \text{ C}^2/\text{N}\cdot\text{m}^2)(7.0 - 1)(1.43 \times 10^3 \text{ V/m})$$

$$= 7.5 \times 10^{-8} \text{ C/m}^2.$$

(b) The electric field E_0 in the air gap is calculated in Example 9 to be 1.00×10^4 V/m. From Eq. 30-24,

$$D_0 = \kappa\epsilon_0 E_0$$

$$= (1)(8.9 \times 10^{-12} \text{ C}^2/\text{N}\cdot\text{m}^2)(1.00 \times 10^4 \text{ V/m})$$

$$= 8.9 \times 10^{-8} \text{ C/m}^2$$

and, from Eq. 30-25, recalling, as above, that $\kappa = 1$ in the air gap,

$$P_0 = \epsilon_0(\kappa - 1)E_0 = 0.$$

Note that $\mathbf{P}$ vanishes outside the dielectric, $\mathbf{D}$ has the same value in the dielectric and in the gap, and $\mathbf{E}$ has different values in the dielectric and in the gap. Verify that Eq. 30-23 ($\mathbf{D} = \epsilon_0\mathbf{E} + \mathbf{P}$) is correct both in the gap and in the dielectric.

It can be shown from Maxwell's equations that no matter how complex the problem the component of $\mathbf{D}$ *normal* to the surface of the dielectric has the same value on each side of the surface. In this problem $\mathbf{D}$ itself is normal to the surface, there being no component but the normal one. It can also be shown that the component of $\mathbf{E}$ *tangential* to the dielectric surface has the same value on each side of the surface. This *boundary condition*, like the one for $\mathbf{D}$, is trivial

---

* Certain waxes, when polarized in their molten state, retain a permanent polarization after solidifying, even though the external polarizing field is removed. *Electrets*, manufactured in this way, are the electrostatic analog of permanent magnets in that they possess a gross permanent electric dipole moment. Materials from which electrets can be constructed are called *ferroelectric*. Electrets do *not* obey Eq. 30-25 because they have a nonvanishing value of $\mathbf{P}$ even though $\mathbf{E} = 0$.

in this problem, both tangential components being zero. In more complex problems these boundary conditions on **D** and **E** are very important. Table 30-2 summarizes the properties of the electric vectors, **E**, **D**, and **P**.

Table 30-2
Three electric vectors

Name	Symbol	Associated with	Boundary Condition
Electric field	**E**	All charges	Tangential component continuous
Electric displacement	**D**	Free charges only	Normal component continuous
Polarization (electric dipole moment per unit volume)	**P**	Polarization charges only	Vanishes in a vacuum

Defining equation for **E**	$\mathbf{F} = q\mathbf{E}$	Eq. 27-2
General relation among the three vectors	$\mathbf{D} = \epsilon_0\mathbf{E} + \mathbf{P}$	Eq. 30-23
Gauss's law when dielectric media are present	$\oint \mathbf{D} \cdot d\mathbf{S} = q$	Eq. 30-26
	(q = free charge only)	
Empirical relations for certain dielectric materials*	$\mathbf{D} = \kappa\epsilon_0\mathbf{E}$ $\mathbf{P} = (\kappa - 1)\epsilon_0\mathbf{E}$	Eq. 30-24 Eq. 30-25

* Generally true, with κ independent of **E**, except for certain materials called *ferroelectrics*; see footnote on page 667.

questions

1. A capacitor is connected across a battery. (a) Why does each plate receive a charge of exactly the same magnitude? (b) Is this true even if the plates are of different sizes?

2. Can there be a potential difference between two adjacent conductors that carry the same positive charge?

3. The relation $\sigma \propto 1/R$, in which R is the radius of curvature (see Section 29-8), suggests that the charge placed on an isolated conductor concentrates on points and avoids flat surfaces, where $R = \infty$. How do we reconcile this with Fig. 30-2 in which the charge is definitely on the flat surface of either plate?

4. A sheet of aluminum foil of negligible thickness is placed between the plates of a capacitor as in Fig. 30-14. What effect has it on the capacitance if (a) the foil is electrically insulated and (b) the foil is connected to the upper plate?

figure 30-14
Question 4.

5. You are given two capacitors, C_1 and C_2, in which $C_1 \gg C_2$. Can C_1 always hold more charge than C_2? Explain.

6. In Fig. 30-1 suppose that a and b are nonconductors, the charge being distributed arbitrarily over their surfaces. (a) Would Eq. 30-1 ($q = CV$) hold, with C independent of the charge arrangements? (b) How would you define V in this case?

7. In connection with Eq. 30-1 ($q = CV$) we said that C is a constant. Yet we pointed out (see Eq. 30-4) that it depends on the geometry (and also, as we

saw later, on the medium). If C is indeed a constant, with respect to what variables does it remain constant?

8. For a finite A, does Eq. 30-4 $(C = \epsilon_0 A/d)$ hold as $d \to \infty$? If not, why not?

9. Suppose that in Example 4 the three capacitors shown are identical parallel-plate capacitors with the same (square) plates of area A and the same plate separation d. Develop an argument, based only on Eq. 30-4 $(C = \epsilon_0 A/d)$ that the equivalent capacitance is three times the individual capacitance, as Example 4 predicts.

10. You are given a parallel-plate capacitor with square plates of area A and separation d, in a vacuum. What is the qualitative effect of each of the following on its capacitance? (a) Reduce d. (b) Put a slab of copper between the plates, touching neither plate. (c) Double the area of both plates. (d) Double the area of one plate only. (e) Slide the plates parallel to each other so that the area of overlap is, say, 50%. (f) Double the potential difference between the plates. (g) Tilt one plate so that the separation remains d at one end but is $\frac{1}{2}d$ at the other.

11. Discuss similarities and differences when (a) a dielectric slab and (b) a conducting slab are inserted between the plates of a parallel-plate capacitor. Assume the slab thicknesses to be one-half the plate separation.

12. An oil-filled parallel-plate capacitor has been designed to have a capacitance C and to operate safely at or below a certain maximum potential difference V_m without arcing over. However, the designer did not do a good job and the capacitor occasionally arcs over. What can be done to redesign the capacitor, keeping C and V_m unchanged and using the same dielectric?

13. Would you expect the dielectric constant, for substances containing permanent molecular electric dipoles, to vary with temperature?

14. An isolated conducting sphere is given a positive charge. Does its mass increase, decrease, or remain the same?

15. A dielectric slab is inserted in one end of a charged parallel-plate capacitor (the plates being horizontal and the charging battery having been disconnected) and then released. Describe what happens. Neglect friction.

16. A capacitor is charged by using a battery, which is then disconnected. A dielectric slab is then slipped between the plates. Describe qualitatively what happens to the charge, the capacitance, the potential difference, the electric field, and the stored energy.

17. While a capacitor remains connected to a battery, a dielectric slab is slipped between the plates. Describe qualitatively what happens to the charge, the capacitance, the potential difference, the electric field, and the stored energy. Is work required to insert the slab?

18. Two identical capacitors are connected as shown in Fig. 30-15. A dielectric slab is slipped between the plates of one capacitor, the battery remaining connected. Describe qualitatively what happens to the charge, the capacitance, the potential difference, the electric field, and the stored energy for each capacitor.

19. Show that the dielectric constant of a conductor can be taken to be infinitely great.

20. For a given potential difference does a capacitor store more or less charge with a dielectric than it does without a dielectric (vacuum)? Explain in terms of the microscopic picture of the situation.

21. In this chapter we have assumed electrostatic conditions, that is, the potential difference V between the capacitor plates remains constant. Suppose however that, as it often does in practice, V varies sinusoidally with time with an angular frequency ω. Would you expect the dielectric constant κ to vary with ω?

22. In connection with Section 30-7 describe in your own words the differences between **D**, **E**, and **P** in Eq. 30-23.

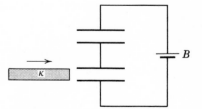

figure 30-15
Question 18.

SECTION 30-2

problems

1. A 100-pF capacitor is charged to a potential difference of 50 V, the charging battery then being disconnected. The capacitor is then connected in parallel with a second (initially uncharged) capacitor. If the measured potential difference drops to 35 V, what is the capacitance of this second capacitor? *Answer:* 43 pF.

2. A potential difference of 300 V is applied to a 2.0-μF capacitor and an 8.0-μF capacitor connected in series. (*a*) What are the charge and the potential difference for each capacitor? (*b*) The charged capacitors are reconnected with their positive plates together and their negative plates together, no external voltage being applied. What are the charge and the potential difference for each? (*c*) The charged capacitors in (*a*) are reconnected with plates of *opposite* sign together. What are the charge and the potential difference for each?

3. If we solve Eq. 30-4 for ϵ_0, we see that its SI units are farad/meter. Show that these units are equivalent to those obtained earlier for ϵ_0, namely coulomb^2/newton·meter^2.

4. A parallel-plate capacitor has circular plates of 8.0-cm radius and 1.0-mm separation. What charge will appear on the plates if a potential difference of 100 V is applied?

5. Figure 30-16 shows two capacitors in series, the rigid center section of length *b* being movable vertically. Show that the equivalent capacitance of the series combination is independent of the position of the center section and is given by

$$C = \frac{\epsilon_0 A}{a - b}.$$

figure 30-16
Problem 5

6. In Fig. 30-17 a variable air capacitor of the type used in tuning radios is shown. Alternate plates are connected together, one group being fixed in position, the other group being capable of rotation. Consider a pile of *n* plates of alternate polarity, each having an area *A* and separated from adjacent plates by a distance *d*. Show that this capacitor has a maximum capacitance of

$$C = \frac{(n - 1)\epsilon_0 A}{d}.$$

figure 30-17
Problem 6

7. A 6.0-μF capacitor is connected in series with a 4.0-μF capacitor and a potential difference of 200 V is applied across the pair. (*a*) What is the charge on each capacitor? (*b*) What is the potential difference across each capacitor? *Answer:* (*a*) 4.8×10^{-4} C. (*b*) $V_4 = 120$ V; $V_6 = 80$ V.

8. Repeat the previous problem for the same two capacitors connected in parallel.

9. How many 1.0-μF capacitors would need to be connected in parallel in order to store a charge of 1.0 C with a potential of 300 V across the capacitors? *Answer:* 3300.

figure 30-18
Problem 10

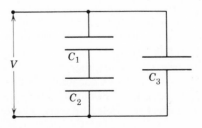

10. In Fig. 30-18 find the equivalent capacitance of the combination. Assume that $C_1 = 10$ μF, $C_2 = 5$ μF, and $C_3 = 4$ μF.

11. In Fig. 30-19 find the equivalent capacitance of the combination. Assume that $C_1 = 10$ μF, $C_2 = 5$ μF, and $C_3 = 4$ μF. *Answer:* 3.2 μF.

12. In Fig. 30-19 suppose that capacitor C_3 breaks down electrically, becoming equivalent to a conducting path. What *changes* in (a) the charge and (b) the potential difference occur for capacitor C_1? Assume $V = 100$ V.

13. A slab of copper of thickness b is thrust into a parallel-plate capacitor as shown in Fig. 30-20; it is exactly halfway between the plates. What is the capacitance (a) before and (b) after the slab is introduced?
Answer: (a) $\epsilon_0 A/d$. (b) $\epsilon_0 A/(d-b)$.

14. When switch S is thrown to the left in Fig. 30-21, the plates of the capacitor C_1 acquire a potential difference V_0. C_2 and C_3 are initially uncharged. The switch is now thrown to the right. What are the final charges q_1, q_2, q_3 on the corresponding capacitors?

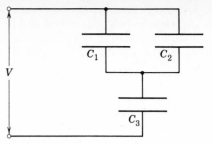

figure 30-19
Problems 11, 12, 31

figure 30-20
Problem 13

figure 30-21
Problem 14

15. Charges q_1, q_2, q_3 are placed on capacitors of capacitance C_1, C_2, C_3, respectively, arranged in series as shown in Fig. 30-22. Switch S is then closed. What are the final charges q_1', q_2', q_3' on the capacitors?

Answer: (a) $q_1' = \dfrac{-(C_1C_2 + C_1C_3)q_1 + C_1C_3q_2 + C_1C_2q_3}{C_1C_2 + C_1C_3 + C_2C_3}$.

(b) $q_2' = \dfrac{(C_1C_2 + C_2C_3)q_2 - C_1C_2q_3 - C_2C_3q_1}{C_1C_2 + C_1C_3 + C_2C_3}$.

(c) $q_3' = \dfrac{(C_1C_3 + C_2C_3)q_3 - C_1C_3q_2 - C_2C_3q_1}{C_1C_2 + C_1C_3 + C_2C_3}$.

figure 30-22
Problem 15

16. You have material available to construct two parallel-plate capacitors having a combined plate area A. How would you distribute this area between the two capacitors to obtain maximum total capacitance if you intend to connect them (a) in parallel and (b) in series?

17. Capacitors C_1 (1.0 μF) and C_2 (3.0 μF) are each charged to a potential V (100 V) but with opposite polarity, so that points a and c are on the side of the respective positive plates of C_1 and C_2, and points b and d are on the side of the respective negative plates (see Fig. 30-23). Switches S_1 and S_2 are now closed. (a) What is the potential difference between points e and f? (b) What is the charge on C_1? (c) What is the charge on C_2?
Answer: (a) 50 V. (b) 0.50×10^{-4} C. (c) 1.5×10^{-4} C.

18. Find the equivalent capacitance between points x and y in Fig. 30-24. Assume that $C_2 = 10$ μF and that the other capacitors are all 4.0 μF. (Hint: Apply a potential difference V between x and y and write down all the relationships that involve the charges and potential differences for the separate capacitors.)

figure 30-23
Problem 17

figure 30-24
Problem 18

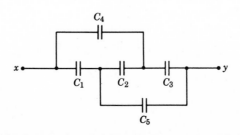

19. In Fig. 30-25 the battery B supplies 12 V. (a) Find the charge on each capacitor when switch S_1 is closed and (b) when (later) switch S_2 is also closed. Take $C_1 = 1$ μF, $C_2 = 2$ μF, $C_3 = 3$ μF, and $C_4 = 4$ μF.
   Answer: (a) $q_1 = 9.0\mu$C; $q_2 = 16\mu$C;
   $\quad q_3 = 9.0\mu$C; $q_4 = 16\mu$C.
   (b) $q_1 = 8.4\mu$C; $q_2 = 17\mu$C;
   $\quad q_3 = 11\mu$C; $q_4 = 14\mu$C.

figure 30-25
Problem 19

20. If you have available several 2.0-μF capacitors, each capable of withstanding 200 V without breakdown, how would you assemble a combination having an equivalent capacitance of (a) 0.40 μF or of (b) 1.2 μF, each capable of withstanding 1000 V?

21. Calculate the capacitance of the earth, viewed as a spherical conductor of radius 6400 km. *Answer:* 710 μF.

22. A spherical capacitor consists of two concentric spherical shells of radii a and b, with $b > a$. (a) Show that its capacitance is

$$C = 4\pi\epsilon_0 \frac{ab}{b - a}.$$

(b) Does this reduce (with $a = R$) to the result of Example 3 as $b \to \infty$?

23. Suppose that the two spherical shells of a spherical capacitor have their radii approximately equal. Under these conditions the device approximates a parallel-plate capacitor with $b - a = d$. Show that the formula in Problem 22 does indeed reduce to Eq. 30-4 in this case.

24. Two metallic spheres, radii a and b, are connected by a thin wire. Their separation is large compared with their dimensions. A charge Q is put onto this system and then the wire is disconnected. (a) How much charge resides on each sphere? (b) Apply the definition of capacitance to show that the capacitance of this system is $C = 4\pi\epsilon_0 (a + b)$.

25. A capacitor has square plates, each of side a, making an angle of θ with each other as shown in Fig. 30-26. Show that for small θ the capacitance is given by

$$C = \frac{\epsilon_0 a^2}{d} \left(1 - \frac{a\theta}{2d}\right).$$

(Hint: The capacitor may be divided into differential strips which are effectively in parallel.)

figure 30-26
Problem 25

SECTION 30-3
26. A parallel-plate air capacitor having area A (40 cm²) and spacing d (1.0 mm) is charged to a potential V (600 V). Find (a) the capacitance, (b) the magnitude of the charge on each plate, (c) the stored energy, (d) the electric field between the plates and (e) the energy density between the plates.

27. What would be the capacitance required to store energy U (10 kW·h) at a potential difference V (1000 V)? *Answer:* 72F.

28. Two capacitors (2.0 μF and 4.0 μF) are connected in parallel across a 300-V potential difference. Calculate the total stored energy in the system.

29. A parallel-connected bank of 2000 5.0-μF capacitors is used to store electric energy. What does it cost to charge this bank to 50,000 V, assuming a rate of 10¢/kW·h? *Answer:* 35¢.

30. A parallel-plate air capacitor has a capacitance of 100 pF. (a) What is the stored energy if the applied potential difference is 50 V? (b) Can you calculate the energy density for points between the plates?

31. In Fig. 30-19 find (a) the charge, (b) the potential difference, and (c) the stored energy for each capacitor. Assume the numerical values of Problem 11, with $V = 100$ V.
   Answer: (a) $q_1 = 2.1 \times 10^{-4}$ C; $q_2 = 1.1 \times 10^{-4}$ C; $q_3 = 3.2 \times 10^{-4}$ C.
   (b) $V_1 = V_2 = 21$ V; $V_3 = 79$ V.
   (c) $U_1 = 2.2 \times 10^{-3}$ J; $U_2 = 1.1 \times 10^{-3}$ J; $U_3 = 1.3 \times 10^{-2}$ J.

32. For the capacitors of Problem 2, compute the energy stored for the three different connections of parts (a), (b), and (c). Compare your answers and explain any differences.

33. A parallel-plate capacitor has plates of area A and separation d, and is charged to a potential difference V. The charging battery is then disconnected and the plates are pulled apart until their separation is $2d$. Derive expressions in terms of A, d, and V for (a) the new potential difference, (b) the initial and the final stored energy, and (c) the work required to separate the plates.

    Answer: (a) $V_f = 2V$. (b) $U_i = \frac{1}{2}\frac{\epsilon_0 A V^2}{d}$; $U_f = 2U_i$. (c) $W = \frac{1}{2}\frac{\epsilon_0 A V^2}{d}$.

34. In terms of the original capacitance C, find the work done in inserting a copper slab of thickness $d/2$ in Problem 13 if (a) the potential difference is held constant and (b) if the charge is held constant.

35. An isolated metal sphere whose diameter is 10 cm has a potential of 8000 V. What is the energy density at the surface of the sphere?   Answer: 0.11 J/m³.

36. (a) If the potential difference across a cylindrical capacitor is doubled, the energy stored in the capacitor is changed by what factor? (b) If the radii of the inner and outer cylinders are each doubled, keeping the charge constant, how does the stored energy change?

37. A cylindrical capacitor has radii a and b as in Fig. 30-3. Show that half the stored electric potential energy lies within a cylinder whose radius is

$$r = \sqrt{ab}.$$

38. Show that the plates of a parallel-plate capacitor attract each other with a force given by

$$F = \frac{q^2}{2\epsilon_0 A}.$$

    Prove this by calculating the work necessary to increase the plate separation from x to $x + dx$.

39. Using the result of Problem 38 show that the force per unit area (the electrostatic stress) acting on either capacitor plate is given by $\frac{1}{2}\epsilon_0 E^2$. Actually, this result is true in general, for a conductor of *any* shape with an electric field $\mathbf{E}$ at its surface.

40. A soap bubble of radius R_0 is slowly given a charge q. Because of mutual repulsion of the surface charges, the radius increases slightly to R. The air pressure inside the bubble drops, because of the expansion, to $p(V_0/V)$ where p is the atmospheric pressure, V_0 is the initial volume, and V the final volume. Show that

$$q^2 = 32 \ \pi^2 \epsilon_0 \ p \ R \ (R^3 - R_0^3).$$

    (Hint: Imagine the bubble to expand a further amount dR. Consider the energy changes associated with (a) the decrease in the stored electric field energy, (b) work ($= p \ dV$) done in pushing back the atmosphere, and (c) work done by the gas in the bubble. Apply conservation of energy. Neglect surface tension.)

SECTION 30- 4

41. A certain substance has a dielectric constant of 2.8 and a dielectric strength of 18×10^6 V/m. If it is used as the dielectric material in a parallel-plate capacitor, what minimum area may the plates of the capacitor have in order that the capacitance be $7.0 \times 10^{-2} \ \mu$F and that the capacitor be able to withstand a potential difference of 4000 V?   Answer: 0.63 m².

42. A parallel-plate capacitor is filled with two dielectrics as in Fig. 30-27. Show that the capacitance is given by

$$C = \frac{\epsilon_0 A}{d}\left(\frac{\kappa_1 + \kappa_2}{2}\right).$$

figure 30-27
Problem 42

Check this formula for all the limiting cases that you can think of. (Hint: Can you justify regarding this arrangement as two capacitors in parallel?)

43. A parallel-plate capacitor is filled with two dielectrics of equal thickness, as in Fig. 30-28. Show that the capacitance is given by

$$C = \frac{2\epsilon_0 A}{d} \left(\frac{\kappa_1 \kappa_2}{\kappa_1 + \kappa_2} \right).$$

Check this formula for all the limiting cases that you can think of. (Hint: Can you justify regarding this arrangement as two capacitors in series?)

44. What is capacitance of the capacitor in Fig. 30-29? The plate area is A.

45. For making a capacitor you have available two plates of copper, a sheet of mica (thickness = 0.10 mm, $\kappa = 6$), a sheet of glass (thickness = 2.0 mm, $\kappa = 7$), and a slab of paraffin (thickness = 1.0 cm, $\kappa = 2$). To obtain the largest capacitance, which sheet (or sheets) should you place between the copper plates?   *Answer:* The mica sheet.

SECTION 30-6

46. A parallel-plate capacitor has a capacitance of 100 pF, a plate area of 100 cm², and a mica dielectric ($\kappa = 5.4$). At 50 V potential difference, calculate (a) E in the mica, (b) the magnitude of the free charge on the plates, and (c) the magnitude of the induced surface charge.

47. Two parallel plates of area 100 cm² are each given equal but opposite charges of 8.9×10^{-7} C. Within the dielectric material filling the space between the plates the electric field is 1.4×10^6 V/m. (a) Find the dielectric constant of the material. (b) Determine the magnitude of the charge induced on each dielectric surface.   *Answer:* (a) 7.1. (b) 7.7×10^{-7} C.

48. In Example 9, suppose that the 100-V battery remains connected during the time that the dielectric slab is being introduced. Calculate (a) the charge on the capacitor plates, (b) the electric field in the gap, (c) the electric field in the slab, and (d) the capacitance. For all of these quantities give the numerical values before and after the slab is introduced. Contrast your results with those of Example 9 by constructing a tabular listing.

49. A parallel-plate capacitor has plates of area 0.12 m² and a separation of 1.2 cm. A battery charges the plates to a potential difference of 120 V and is then disconnected. A dielectric slab of thickness 0.4 cm and dielectric constant 4.8 is then placed symmetrically between the plates. (a) Find the capacitance before the slab is inserted. (b) What is the capacitance with the slab in place? (c) What is the free charge q before and after the slab is inserted? (d) Determine the electric field in the space between the plates and dielectric. (e) What is the electric field in the dielectric? (f) With the slab in place what is the potential difference across the plates? (g) How much external work is involved in the process of inserting the slab?
Answer: (a) 89 pF. (b) 120 pF. (c) 1.1×10^{-8} C; 1.1×10^{-8} C. (d) 10^4 V/m. (e) 2.1×10^3 V/m. (f) 88 V. (g) 1.7×10^{-7} J.

50. In the capacitor of Example 9 the dielectric slab fills half the space between the plates. (a) What percent of the energy is stored in the air gaps? (b) What percent is stored in the slab?

51. A dielectric slab of thickness b is inserted between the plates of a parallel-plate capacitor of plate separation d. Show that the capacitance is given by

$$C = \frac{\kappa \epsilon_0 A}{\kappa d - b(\kappa - 1)}.$$

(Hint: Derive the formula following the pattern of Example 9.) Does this formula predict the correct numerical result of Example 9? Does the formula seem reasonable for the special cases of $b = 0$, $\kappa = 1$, and $b = d$?

52. In Example 8, how does the *energy density u* between the plates compare, before and after the dielectric slab is introduced?

figure 30-28
Problem 43

figure 30-29
Problem 44

31
current
and resistance

The free electrons in an isolated metallic conductor, such as a length of copper wire, are in random motion like the molecules of a gas confined to a container. They have no net directed motion along the wire. If we pass a hypothetical plane through the wire, the rate at which electrons pass through it from right to left is the same as the rate at which they pass through from left to right; the *net* rate is zero.*

If the ends of the wire are connected to a battery, an electric field will be set up at every point within the wire. If the potential difference maintained by the battery is 10 V and if the wire (assumed uniform) is 5 m long, the strength of this field at every point will be 2 V/m. This field **E** will act on the electrons and will give them a resultant motion in the direction of $-\mathbf{E}$. We say that an *electric current i* is established; if a net charge q passes through any cross section of the conductor in time t, the current (assumed constant) is

$$i = q/t. \tag{31-1}$$

The appropriate SI units are amperes (abbr. A) for i, coulombs for q, and seconds for t. Recall (Section 26-4) that Eq. 31-1 is the defining equation for the coulomb and that we have not yet given an operational definition of the ampere; we do so in Section 34-4.

31-1
CURRENT AND
CURRENT DENSITY

---

*Actually, because the number of electrons is finite, there will be small statistical fluctuations in these rates and a conductor will contain a small, rapidly fluctuating current, even though, averaged over a long enough period of time, the net current i is zero. This is one aspect of the readily measurable *electrical noise* which is so familiar to those who know something about electronics.

If the rate of flow of charge with time is not constant, the current varies with time and is given by the differential limit of Eq. 31-1, or

$$i = dq/dt. \qquad (31\text{-}2)$$

In the rest of this chapter we consider only constant currents.*

The current i is the same for all cross sections of a conductor, even though the cross-sectional area may be different at different points. In the same way the rate at which water (assumed incompressible) flows past any cross section of a pipe is the same even if the cross section varies. The water flows faster where the pipe is smaller and slower where it is larger, so that the volume rate, measured perhaps in liters/ second, remains unchanged. This constancy of the electric current follows because charge must be conserved; it does not pile up steadily or drain away steadily from any point in the conductor under the assumed steady-state conditions. In the language of Section 18-3 there are no "sources" or "sinks" of charge.

The existence of an electric field inside a conductor does not contradict Section 28-6, in which we asserted that **E** equals zero inside a conductor. In that section, which dealt with a state in which all net motion of charge had stopped (electrostatics), we assumed that the conductor was insulated and that no potential difference was deliberately maintained between any two points on it, as by a battery. In this chapter, which deals with charges in motion, we relax this restriction.

The electric field exerts a force ($= -e\mathbf{E}$) on the electrons in a conductor but this force does not produce a *net* acceleration because the electrons keep colliding with the atoms (strictly, ions, Cu^+ in copper) that make up the conductor. This array of ions, coupled together by strong spring-like forces of electromagnetic origin, is called the *lattice* (see Fig. 21-5). The over-all effect of these collisions is to transfer kinetic energy from the accelerating electrons into vibrational energy of the lattice. The electrons acquire a constant average *drift speed* v_d in the direction $-\mathbf{E}$. There is a close analogy to a ball bearing falling in a uniform gravitational field **g** at a constant terminal speed through a viscous oil. The gravitational force (mg) acting on the ball as it falls does not go into increasing the ball's kinetic energy (which is constant) but is transferred to the fluid by molecular collisions, producing a small rise in temperature.

Although in metals the charge carriers are electrons, in electrolytes or in gaseous conductors (plasmas) they may also be positive or negative ions, or both. We need a convention for labeling the directions of currents because charges of opposite sign move in opposite directions in a given field. A positive charge moving in one direction is equivalent in nearly all external effects to a negative charge moving in the opposite direction. Hence, for simplicity and algebraic consistency, *we assume that all charge carriers are positive and we draw the current arrows in the direction that such charges would move.* If the charge carriers are negative, they simply move opposite to the direction of the current arrow (see Fig. 31-1). When we encounter a case (as in the *Hall effect*; see Section 33-5) in which the sign of the charge carriers makes a difference in the external effects, we will disregard the convention and take the actual situation into account.

Current i is a characteristic of a particular conductor. It is a macroscopic quantity, like the mass of an object, the volume of an object, or the length of a rod. A related microscopic quantity is the current density

* A much less formal example of current flow is lightning. See "The Lightning Discharge" by Richard E. Orville, *The Physics Teacher*, January 1976.

figure 31-1
Electrons drift in a direction opposite
to the electric field in a conductor.

677

CURRENT AND CURRENT DENSITY   SEC. 31-1

j. It is a vector and is characteristic of a point inside a conductor rather than of the conductor as a whole. If the current is distributed uniformly across a conductor of cross-sectional area A, as in Fig. 31-1, the magnitude of the current density for all points on that cross section is

$$j = i/A. \tag{31-3}$$

The vector **j** at any point is oriented in the direction that a positive charge carrier would move at that point. An electron at that point moves in the direction $-$**j**. In Fig. 31-1 **j** is a constant vector and points to the left; the electrons drift to the right.

The general relationship between **j** and i is that, for a particular surface (which need not be plane) in a conductor, i is the flux of the vector **j** over that surface, or

$$i = \int \mathbf{j} \cdot d\mathbf{S}, \tag{31-4}$$

where $d\mathbf{S}$ is an element of surface area and the integral is taken over the surface in question. Equation 31-3 (written as $i = jA$) is a special case of this relationship in which the surface of integration is a plane cross section of the conductor and in which **j** is constant over this surface and at right angles to it. However, we may apply Eq. 31-4 to any surface through which we wish to know the current. Equation 31-4 shows clearly that i is a scalar because the integrand $\mathbf{j} \cdot d\mathbf{S}$ is a scalar.

The arrow often associated with the current in a wire does not indicate that current is a vector but merely shows the *sense* of charge flow. Positive charge carriers either move in a certain direction along the wire or in the opposite direction, these two possibilities being represented by $+$ or $-$ in algebraic equations. Note that (a) the current in a wire remains unchanged if the wire is bent, tied into a knot, or otherwise distorted, and (b) the arrows representing the sense of currents do not in any way obey the laws of vector addition. Thus currents can not be vectors.

We can compute the drift speed v_d of charge carriers in a conductor from the current density j. Figure 31-1 shows the conduction electrons in a wire moving to the right at an assumed constant drift speed v_d. The number of conduction electrons in the wire is nAl where n is the number of conduction electrons per unit volume and Al is the volume of the wire. A charge of magnitude

$$q = (nAl)e$$

passes out of the wire, through its right end, in a time t given by

$$t = \frac{l}{v_d}.$$

The current i is given by

$$i = \frac{q}{t} = \frac{nAle}{l/v_d} = nAev_d.$$

Solving for v_d and recalling that $j = i/A$ (Eq. 31-3) yields

$$v_d = \frac{i}{nAe} = \frac{j}{ne}. \tag{31-5}$$

An aluminum wire whose diameter is 0.10 in. is welded end to end to a copper wire with a diameter of 0.064 in. The composite wire carries a steady current of 10 A. What is the current density in each wire?

EXAMPLE 1

The current is distributed uniformly over the cross section of each conductor, except near the junction, which means that the current density may be taken as constant for all points within each wire. The cross-sectional area of the aluminum wire is 0.0079 in.². Thus, from Eq. 31-3,

$$j_{Al} = \frac{i}{A} = \frac{10 \text{ A}}{0.0079 \text{ in.}^2} = 1300 \text{ A/in.}^2.$$

The cross-sectional area of the copper wire is 0.0032 in.². Thus

$$j_{Cu} = \frac{i}{A} = \frac{10 \text{ A}}{0.0032 \text{ in.}^2} = 3100 \text{ A/in.}^2.$$

The fact that the wires are of different materials does not enter into consideration here.

What is v_d for the copper wire in Example 1?

EXAMPLE 2

We can write the current density for the copper wire (3100 A/in.²) as 480 A/cm². To compute n we start from the fact that there is one free electron per atom in copper. The number of atoms per unit volume is dN_0/M where d is the density, N_0 is the Avogadro number, and M is the atomic weight. The number of free electrons per unit volume is then

$$n = \frac{dN_0}{M} = \frac{(9.0 \text{ g/cm}^3)(6.0 \times 10^{23} \text{ atoms/mol})(1 \text{ electron/atom})}{64 \text{ g/mol}}$$

$$= 8.4 \times 10^{22} \text{ electrons/cm}^3.$$

Finally, v_d is, from Eq. 31-5,

$$v_d = \frac{j}{ne} = \frac{480 \text{ A/cm}^2}{(8.4 \times 10^{22} \text{ electrons/cm}^3)(1.6 \times 10^{-19} \text{ C/electron})}$$

$$= 3.6 \times 10^{-2} \text{ cm/s}.$$

It takes 28 seconds for the electrons in this wire to drift 1.0 cm. Would you have guessed that v_d was so low? The drift speed of electrons must not be confused with the speed at which changes in the electric field configuration travel along wires, a speed which approaches that of light. When we apply a pressure to one end of a long water-filled garden hose, a *pressure wave* travels rapidly along the hose. The speed at which *water* moves through the hose is much lower, however.

If we apply the same potential difference between the ends of geometrically similar rods of copper and of wood, very different currents result. The characteristic of the conductor that enters here is its *resistance*. We define the resistance of a conductor (often called a *resistor*; symbol ‑www‑) between two points by applying a potential difference V between those points, measuring the current i, and dividing:

31-2
*RESISTANCE,
RESISTIVITY, AND
CONDUCTIVITY*

$$R = V/i. \qquad (31\text{-}6)$$

If V is in volts and i in amperes, the resistance R will be in *ohms* (abbr. Ω).

The flow of charge through a conductor is often compared with the

Table 31-1
Properties of metals as conductors

Metal	Resistivity (at 20°C) 10^{-8} $\Omega \cdot$m	Temperature Coefficient of Resistivity, α, per C° ($\times 10^{-5}$)†
Silver	1.6	380
Copper	1.7	390
Aluminum	2.8	390
Tungsten	5.6	450
Nickel	6.8	600
Iron	10	500
Steel	18	300
Manganin	44	1.0
Carbon*	3500	−50

* Carbon, not strictly a metal, is included for comparison.
† This quantity, defined from

$$\alpha = \frac{1}{\rho}\frac{d\rho}{dT} \qquad (31\text{-}7)$$

is the fractional change in resistivity $(d\rho/\rho)$ per unit change in temperature. It varies with temperature, the values here referring to 20°C. For copper $(\alpha = 3.9 \times 10^{-3}/\text{C}°)$ the resistivity increases by 0.39 percent for a temperature increase of 1°C near 20°C. Note that α for carbon is negative, which means that the resistivity *decreases* with increasing temperature.

flow of water through a pipe, which occurs because there is a difference in pressure between the ends of the pipe, established perhaps by a pump. This pressure difference can be compared with the potential difference established between the ends of a resistor by a battery. The flow of water (liters/second, say) is compared with the current (coulombs/second, or amperes). The rate of flow of water for a given pressure difference is determined by the nature of the pipe. Is it long or short? Is it narrow or wide? Is it empty or filled, perhaps with gravel? These characteristics of the pipe are analogous to the resistance of a conductor.

Primary standards of resistance, kept at the National Bureau of Standards, are spools of wire whose resistances have been accurately measured. Because resistance varies with temperature, these standards, when used, are placed in an oil bath at a controlled temperature. They are made of a special alloy, called *manganin*, for which the change of resistance with temperature is very small. They are carefully annealed to eliminate strains, which also affect the resistance. These primary standard resistors are used chiefly to calibrate secondary standards for other laboratories.

Operationally, the primary resistance standards are not measured by using Eq. 31-6 but are measured in an indirect way which involves magnetic fields. Equation 31-6 is, in fact, used to measure V by setting up an accurately known current i (using a *current balance*; see Section 34-4) in an accurately known resistance R. This operational procedure for potential difference is the one normally used in place of the conceptual definition introduced in Section 29-1, in which one measures the work per unit charge required to move a test charge between two points.

Related to resistance is the *resistivity* ρ, which is a characteristic of a material rather than of a particular specimen of a material; it is de-

$$\rho = \frac{E}{j}. \qquad (31\text{-}8a)$$

The resistivity of copper is $1.7 \times 10^{-8}\ \Omega \cdot \text{m}$; that of fused quartz is about $10^{16}\ \Omega \cdot \text{m}$. Few physical properties are measurable over such a range of values; Table 31-1 lists some electrical properties for common metals.

Often we prefer to speak of the *conductivity* (σ) of a material rather than its resistivity. These are reciprocal quantities, related by

$$\sigma = 1/\rho. \qquad (31\text{-}8b)$$

The SI units of σ are $(\Omega \cdot \text{m})^{-1}$.

Consider a cylindrical conductor, of cross-sectional area A and length l, carrying a steady current i. Let us apply a potential difference V between its ends. If the cylinder cross sections at each end are equipotential surfaces, the electric field and the current density will be constant for all points in the cylinder and will have the values

$$E = \frac{V}{l} \quad \text{and} \quad j = \frac{i}{A}.$$

We may then write the resistivity ρ as

$$\rho = \frac{E}{j} = \frac{V/l}{i/A}.$$

But V/i is the resistance R which leads to

$$R = \rho\,\frac{l}{A}. \qquad (31\text{-}9)$$

V, i, and R are *macroscopic* quantities, applying to a particular body or extended region. The corresponding *microscopic* quantities are $\mathbf{E}$, $\mathbf{j}$, and ρ; they have values at every point in a body. The macroscopic quantities are related to each other by Eq. 31-6 $(V = iR)$ and the microscopic quantities by Eq. 31-8a, which can be written in vector form as $\mathbf{E} = \mathbf{j}\rho$.

The macroscopic quantities can be found by integrating over the microscopic quantities, using relations already given, namely

$$i = \int \mathbf{j} \cdot d\mathbf{S} \qquad (31\text{-}4)$$

and

$$V_{ab} = -\int_a^b \mathbf{E} \cdot d\mathbf{l}. \qquad (29\text{-}5)$$

The integral in Eq. 31-4 is a surface integral, carried out over any cross section of the conductor. The integral in Eq. 29-5 is a line integral carried out along an arbitrary line drawn along the conductor, connecting any two equipotential surfaces, identified by a and b. For a long wire connected to a battery equipotential surface a might be chosen as a cross section of the wire near the positive battery terminal and b might be a cross section near the negative terminal.

We can express the resistance of a conductor between a and b in microscopic terms by dividing the two equations, or

$$R = \frac{V_{ab}}{i} = \frac{-\int_a^b \mathbf{E} \cdot d\mathbf{l}}{\int \mathbf{j} \cdot d\mathbf{S}}.$$

* These are materials whose properties (electrical in this case) do not vary with direction in the material.

If the conductor is a long cylinder of cross section A and length l, and if points a and b are its ends, the foregoing equation for R (see Eq. 31-8a) reduces to

$$R = \frac{El}{jA} = \rho \frac{l}{A},$$

which is Eq. 31-9.

The macroscopic quantities V, i, and R are of primary interest when we are making electrical measurements on real conducting objects. They are the quantities that one reads on meters. The microscopic quantities $\mathbf{E}$, $\mathbf{j}$, and ρ are of primary importance when we are concerned with the fundamental behavior of matter (rather than of specimens of matter), as we usually are in the research area of *solid state physics*. Thus Section 31-4 deals appropriately with an atomic view of the *resistivity* of a metal and not of the *resistance* of a metallic specimen. The microscopic quantities are also important when we are interested in the interior behavior of irregularly shaped conducting objects.

EXAMPLE 3

A rectangular carbon block has dimensions 1.0 cm × 1.0 cm × 50 cm. (*a*) What is the resistance measured between the two square ends and (*b*) between two opposing rectangular faces? The resistivity of carbon at 20°C is 3.5×10^{-5} $\Omega \cdot$ m.

(*a*) The area of a square end is 1.0 cm² or 1.0×10^{-4} m². Equation 31-9 gives for the resistance between the square ends

$$R = \rho \frac{l}{A} = \frac{(3.5 \times 10^{-5} \ \Omega \cdot \text{m})(0.50 \ \text{m})}{1.0 \times 10^{-4} \ \text{m}^2} = 0.18 \ \Omega.$$

(*b*) For the resistance between opposing rectangular faces (area = 5.0×10^{-3} m²), we have

$$R = \rho \frac{l}{A} = \frac{(3.5 \times 10^{-5} \ \Omega \cdot \text{m})(10^{-2} \ \text{m})}{5.0 \times 10^{-3} \ \text{m}^2} = 7.0 \times 10^{-5} \ \Omega.$$

Thus a given conductor can have any number of resistances, depending on how the potential difference is applied to it. The ratio of resistances for these two cases is 2600. We assume in each that the potential difference is applied to the block in such a way that the surfaces between which the resistance is desired are equipotential. Otherwise Eq. 31-9 would not be valid.

Figure 31-2 shows (solid curve) how the resistivity of copper varies with temperature. Sometimes, for practical use, such data are expressed in equation form. If we are interested in only a limited range of temperatures extending, say, from 0 to 500°C, we can fit a straight line to the curve of Fig. 31-2, making it pass through two arbitrarily selected points; see the dashed line. We choose the point labeled T_0, ρ_0 in the figure as a reference point, T_0 being 0° C in this case and ρ_0 being 1.56×10^{-8} $\Omega \cdot$ m. The resistivity ρ at any temperature T can be found from the empirical equation of the dashed straight line in Fig. 31-2, which is

$$\rho = \rho_0 \left[1 + \bar{\alpha}(T - T_0) \right]. \tag{31-10}$$

This relation shows correctly that $\rho \rightarrow \rho_0$ as $T \rightarrow T_0$.

If we solve Eq. 31-10 for $\bar{\alpha}$, we obtain

$$\bar{\alpha} = \frac{1}{\rho_0} \frac{\rho - \rho_0}{T - T_0}.$$

Comparison with Eq. 31-7 shows that $\bar{\alpha}$ is a *mean temperature coefficient of resistivity* for a selected pair of temperatures rather than the temperature coefficient of resistivity at a particular temperature, which is the definition of α. For most practical purposes Eq. 31-10 gives results that are within the acceptable range of accuracy.

figure 31-2
The resistivity of copper as a function of temperature. The dashed line is an approximation chosen to fit the curve at the two circled points. The point marked T_0, ρ_0 is chosen as a reference point.

The curve of Fig. 31-2 does not go to zero at the absolute zero of temperature, even though it appears to do so, the residual resistivity at this temperature being $0.02 \times 10^{-8}\ \Omega \cdot m$. For many substances the resistance *does* become zero at some low temperature. Figure 31-3 shows the resistance of a specimen of mercury for temperatures below 6 K. In the space of about 0.05 K the resistance drops abruptly to an immeasurably low value. This phenomenon, called *superconductivity*,* was discovered by Kamerlingh Onnes in the Netherlands in 1911. The resistance of materials in the superconducting state seems to be truly zero; currents, once established in closed superconducting circuits, persist for weeks without diminution, even though there is no battery in the circuit. If the temperature is raised slightly above the superconducting point, or if a large enough magnetic field is applied, such currents drop rapidly to zero.

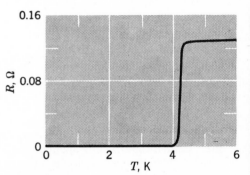

figure 31-3
The resistivity of mercury disappears below about 4 K.

Let us apply a variable potential difference V between the ends of a 100-foot coil of #18 copper wire. For each applied potential difference, let us measure the current i and plot it against V as in Fig. 31-4. The straight line that results means that *the resistance of this conductor is the same no matter what applied voltage we use to measure it.* This important result, which holds for metallic conductors, is known as *Ohm's law.* We assume that the temperature of the conductor is essentially constant throughout the measurements.

Many conductors do not obey Ohm's law. Figure 31-5, for example, shows a V-i plot for a type 2A3 vacuum tube. The plot is not straight and the resistance depends on the voltage used to measure it. Also, the current for this device is almost vanishingly small if the polarity of the applied potential difference is reversed. For metallic conductors the

31-3
OHM'S LAW

* See (a) "Superconductivity" by B. T. Matthias, *Scientific American*, November 1957; (b) "The Search for High-Temperature Superconductors" by B. T. Matthias, *Physics Today*, August 1971; (c) "Large-Scale Applications of Superconductivity" by Brian B. Schwartz and Simon Foner, *Physics Today*, July 1977.

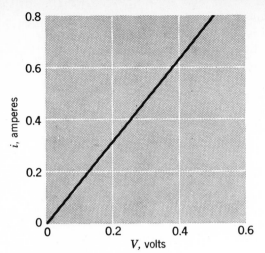

figure 31-4
The current in a particular copper conductor as a function of potential difference. This conductor obeys Ohm's law.

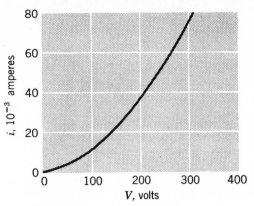

figure 31-5
The current in a type 2A3 vacuum tube as a function of potential difference. This conductor does *not* obey Ohm's law.

current reverses direction when the potential difference is reversed, but its magnitude does not change.

Figure 31-6 shows a typical *V-i* plot for another nonohmic device, a *thermistor*. This is a semiconductor (see Section 26-3) with a large and negative temperature coefficient of resistivity α (see Table 31-1) that varies greatly with temperature. We note that two different currents through the thermistor can correspond to the same potential difference between its ends. Thermistors are often used to measure the rate of energy flow in microwave beams by allowing the microwave beam to fall on the thermistor and heat it. The relatively small temperature rise so produced results in a relatively large change in resistance, which serves as a measure of the microwave power. Modern electronics, and therefore much of the character of our present technological civilization, depends in a fundamental way on the fact that many conductors, such as transistors and vacuum tubes, do *not* obey Ohm's law.

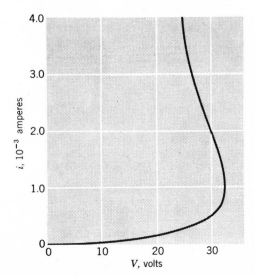

figure 31-6
A plot of current as a function of potential difference in a Western Electric 1-B thermistor. Again, this conductor does not obey Ohm's law. The shape of the curve can be accounted for in terms of the large negative temperature coefficient of resistivity of the material of which the device is made.

We stress that the relationship $V = iR$ is *not* a statement of Ohm's law. A conductor obeys this law only if its *V-i* curve is linear, that is, if R is independent of V and i. The relationship $R = V/i$ remains as the general definition of the resistance of a conductor whether or not the conductor obeys Ohm's law.

The microscopic equivalent of the relationship $V = iR$ is Eq. 31-8*a*, or $\mathbf{E} = \mathbf{j}\rho$. A conducting *material* is said to obey Ohm's law if a plot of *E* versus *j* is linear, that is, if the resistivity ρ is independent of *E* and *j*. Ohm's law is a specific property of certain materials and is not a general law of electromagnetism, for example, like Gauss's law.

A close analogy exists between the flow of charge because of a potential difference and the flow of heat because of a temperature difference. Consider a thin electrically conducting slab of thickness Δx and area A. Let a potential difference ΔV be maintained between opposing faces. The current i is given by Eqs. 31-6 ($i = V/R$) and 31-9 ($R = \rho l/A$), or

$$i = \frac{V_a - V_b}{R} = \frac{(V_a - V_b)A}{\rho l} = -\frac{(V_b - V_a)A}{\rho \Delta x}.$$

In the limiting case of a slab of thickness dx this becomes

$$i = -\frac{1}{\rho} A \frac{dV}{dx}$$

or

$$\frac{dq}{dt} = -\sigma A \frac{dV}{dx}, \qquad (31\text{-}11)$$

where $\sigma \ (= 1/\rho)$, as we have seen (Eq. 31-8*b*), is the *conductivity* of the material. Since positive charge flows in the direction of decreasing V, we introduce a minus sign into Eq. 31-11, that is, dq/dt is positive when dV/dx is negative.

The analogous heat flow equation (see Section 22-4) is

$$\frac{dQ}{dt} = -kA \frac{dT}{dx}, \qquad (31\text{-}12)$$

which shows that k, the thermal conductivity, corresponds to σ and dT/dx, the temperature gradient, corresponds to dV/dx, the potential gradient. For pure metals there is more than a formal mathematical analogy between Eqs. 31-11 and 31-12. Both heat energy and charge are carried by the free electrons in such metals; empirically, a good electrical conductor (silver, say) is also a good heat conductor.

31-4
OHM'S LAW—A MICROSCOPIC VIEW

As we have said earlier, Ohm's law is not a fundamental law of electromagnetism because it depends on the properties of the conducting medium. The law is very simple in form, and it is curious that many conductors obey it so well, whereas other conductors do not obey it at all (see Figs. 31-4, 31-5, and 31-6). Let us see if we can understand why metals obey Ohm's law, which we shall write (see Eq. 31-8*a*) in the microscopic form $\mathbf{E} = \rho \mathbf{j}$.

In a metal the valence electrons are not attached to individual atoms but are free to move about within the lattice and are called *conduction electrons*. In copper there is one such electron per atom, the other 28 remaining bound to the copper nuclei to form ionic cores.

Although the speed distribution of conduction electrons can be described correctly only in terms of quantum physics, the classical *free-electron model* will suit our purpose. It suffices to consider only a suitably defined average speed $\bar{v}$; for copper $\bar{v} = 1.6 \times 10^8$ cm/s. In the absence of an electric field, the directions in which the free or conduction) electrons move are completely random, like those of the molecules of a gas confined to a container.

The electrons collide constantly with the ionic cores of the conductor, that is, they interact with the lattice, often suffering sudden changes in speed and direction. These collisions remind us of the collisions of gas molecules confined to a container. As in the case of molecular collisions, we can describe electron-lattice collisions by a *mean free path* λ, where λ is the average distance that an electron travels between collisions.*

In an ideal metallic crystal at 0 K electron-lattice collisions would not occur, according to the predictions of quantum physics, that is, $\lambda \to \infty$ as $T \to 0$ K for ideal crystals. Collisions take place in actual crystals because (a) the ionic cores at any temperature T are vibrating about their equilibrium positions in a random way, (b) impurities, that is, foreign atoms, may be present, and (c) the crystal may contain lattice imperfections, such as missing atoms and displaced atoms. On this view it is not surprising that the resistivity of a metal can be increased by (a) raising its temperature, (b) adding small amounts of impurities, and (c) straining it severely, as by drawing it through a die, to increase the number of lattice imperfections.

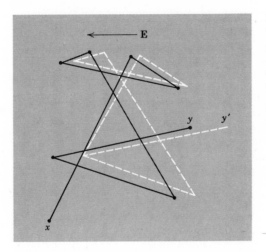

figure 31-7
The solid lines show an electron moving from x to y, making six collisions. The dashed curves show what the electron path *might* have been in the presence of an electric field **E**. Note the steady drift in the direction of $-\mathbf{E}$.

When we apply an electric field to a metal, the electrons modify their random motion in such a way that they drift slowly, in the opposite direction to that of the field, with an average drift speed v_d. This drift speed is very much less (by a factor of something like 10^{10}; see Example 2) than the effective average speed $\bar{v}$ mentioned above. Figure 31-7 suggests the relationship between these two speeds. The solid lines suggest a possible random path followed by an electron in the absence of an applied field; the electron proceeds from x to y, making six collisions on the way. The dashed curves show how this same event *might* have occurred if an electric field **E** had been applied. Note that the electron drifts steadily to the right, ending at y' rather than at y. In preparing Fig. 31-7, it has been assumed that the drift speed v_d is $0.02\bar{v}$; actually, it is more like $10^{-10}\bar{v}$, so that the "drift" exhibited in the figure is greatly exaggerated.

We can calculate the drift speed v_d in terms of the applied electric field E and of $\bar{v}$ and λ. When a field is applied to an electron in the metal

---

* It can be shown that collisions between electrons occur only rarely and have little effect on the resistivity.

it will experience a force eE which will impart to it an acceleration a given by Newton's second law,

$$a = \frac{eE}{m}.$$

Consider an electron that has just collided with an ion core. The collision, in general, will momentarily destroy the tendency to drift and the electron will have a truly random direction after the collision. During the time interval to the next collision the electron's velocity will have changed, on the average, by $a(\lambda/\bar{v})$ or $a\tau$ where τ is the mean time between collisions. We call this the drift speed v_d, or

$$v_d = a\tau = \frac{eE\tau}{m} \qquad (31\text{-}13)$$

The electron's motion through the conductor is analogous to the constant terminal rate of fall of a stone in water. The gravitational force F_g on the stone is opposed by a viscous resisting force that is proportional to the velocity, or

$$F_g = mg = bv,$$

where b is a viscous coefficient (see Section 15-9). Thus the constant terminal speed of the stone is

$$v = \left(\frac{1}{b}\right) F_g.$$

We can rewrite Eq. 31-13 as

$$v_d = \left(\frac{\tau}{m}\right) F_E,$$

where $F_E (= eE)$ is the electrical force. Comparison of these equations shows that the equivalent "viscous coefficient" for the motion of an electron in a particular conductor is m/τ. If τ is short, the conductor exhibits a greater "viscous effect" on the electron motion, and the drift speed v_d is proportionally lower.

We may express v_d in terms of the current density (Eq. 31-5) and combine with Eq. 31-13 to obtain

$$v_d = \frac{j}{ne} = \frac{eE\tau}{m}.$$

Combining this with Eq. 31-8a $(\rho = E/j)$ leads finally to*

$$\rho = \frac{m}{ne^2\tau}. \qquad (31\text{-}14)$$

Equation 31-14 can be taken as a statement that metals obey Ohm's law *if* we can show that τ does not depend on the applied electric field E. In this case ρ will not depend on E, which (see Section 31-3) is the criterion that a material obey Ohm's law. The quantity τ depends on the speed distribution of the conduction electrons. We have seen that this distribution is affected only very slightly by the application of even a relatively large electric field, since $\bar{v}$ is of the order of 10^8 cm/s and v_d (see Example 2) only of the order of 10^{-2} cm/s, a ratio of 10^{10}. We may be sure that whatever the value of τ is (for copper at 20°C, say) in

* See "Drift Speed and Collision Time" by Donald E. Tilley, *American Journal of Physics*, June 1976, for a full discussion.

the absence of a field it remains essentially unchanged when the field is applied. Thus the right side of Eq. 31-14 is independent of E (which means that ρ is independent of E) and the material obeys Ohm's law.

EXAMPLE 4

What are (a) the mean time τ between collisions and (b) the mean free path for free electrons in copper?

(a) From Eq. 31-14 (see also Example 2), we have

$$\tau = \frac{m}{ne^2\rho} = \frac{(9.1 \times 10^{-31} \text{ kg})}{(8.4 \times 10^{28}/\text{m}^3)(1.6 \times 10^{-19} \text{ C})^2(1.7 \times 10^{-8} \ \Omega \cdot \text{m})}$$

$$= 2.5 \times 10^{-14} \text{ s}.$$

(b) The mean free path is

$$\lambda = \tau \bar{v} = (2.5 \times 10^{-14} \text{ s})(1.6 \times 10^8 \text{ cm/s}) = 4.0 \times 10^{-6} \text{ cm}.$$

This is about 220 ionic diameters.*

Figure 31-8 shows a circuit consisting of a battery B connected to a "black box". A steady current i exists in the connecting wires and a steady potential difference V_{ab} exists between the terminals a and b. The box might contain a resistor, a motor, or a storage battery, among other things.

Terminal a, connected to the positive battery terminal, is at a higher potential than terminal b. If a charge dq moves through the box from a to b, this charge will decrease its electric potential energy by $dq \, V_{ab}$ (see Section 29-6). The conservation-of-energy principle tells us that this energy is transferred in the box from electric potential energy to some other form. What that other form will be depends on what is in the box. In a time dt the energy dU transferred inside the box is then

$$dU = dq \, V_{ab} = i \, dt \, V_{ab}.$$

We find the *rate* of energy transfer P by dividing by the time, or

$$P = \frac{dU}{dt} = iV_{ab}. \tag{31-15}$$

If the device in the box is a motor, the energy appears largely as mechanical work done by the motor; if the device is a storage battery that is being charged, the energy appears largely as stored chemical energy in this second battery.

If the device is a resistor, we assert that the energy appears as thermal energy in the resistor. To see this, consider a stone of mass m that falls through a height h. It decreases its gravitational potential energy by mgh. If the stone falls in a vacuum or—for practical purposes—in air, this energy is transformed into kinetic energy of the stone. If the stone falls into the depths of the ocean, however, its speed eventually becomes constant, which means that the kinetic energy no longer increases. The potential energy that is steadily being made available as the stone falls then appears as thermal energy in the stone and the surrounding water. It is the viscous, friction-like drag of the water on the surface of the stone that stops the stone from accelerating, and it is at this surface that thermal energy appears.

31-5
ENERGY TRANSFERS IN AN ELECTRIC CIRCUIT

figure 31-8
A battery B sets up a current in a circuit containing a "black box", that is, a box whose contents are not known to us.

* See footnote on p. 686.

The course of the electrons through the resistor is much like that of the stone through water. The electrons travel with a constant drift speed v_d and thus do not gain kinetic energy. The electric potential energy that they lose is transferred to the resistor as thermal energy. On a microscopic scale we can understand this in that collisions between the electrons and the lattice (see Fig. 21-5) increase the amplitude of the thermal vibrations of the lattice; on a macroscopic scale this corresponds to a temperature increase. There can be a flow of heat out of the resistor subsequently, if the environment is at a lower temperature than the resistor.

For a resistor we can combine Eqs. 31-6 ($R = V/i$) and 31-15 and obtain either

$$P = i^2R \qquad (31\text{-}16)$$

or

$$P = \frac{V^2}{R}. \qquad (31\text{-}17)$$

Note that Eq. 31-15 applies to electrical energy transfer of *all* kinds; Eqs. 31-16 and 31-17 apply only to the transfer of electrical energy to thermal energy in a resistor. Equations 31-16 and 31-17 are known as *Joule's law*. This law is a particular way of writing the conservation-of-energy principle for the special case in which electrical energy is transferred into thermal energy (Joule energy).

The unit of power that follows from Eq. 31-15 is the volt-ampere. We can write it as

$$1 \text{ volt·ampere} = 1 \text{ volt·ampere} \left(\frac{1 \text{ joule}}{1 \text{ volt} \times 1 \text{ coulomb}}\right)$$
$$\left(\frac{1 \text{ coulomb}}{1 \text{ ampere} \times 1 \text{ second}}\right)$$
$$= 1 \text{ joule/second.}$$

The first conversion factor in parentheses comes from the definition of the volt (Eq. 29-1); the second comes from the definition of the coulomb. The joule/second is such a common unit that it is given a special name of its own, the *watt* (abbr. W); see Section 7-7. Power is not an exclusively electrical concept, of course, and we can express in watts the power ($= \mathbf{F} \cdot \mathbf{v}$) expended by an agent that exerts a force $\mathbf{F}$ while it moves with a velocity $\mathbf{v}$.

EXAMPLE 5

You are given a 20-ft length of heating wire made of the special alloy Nichrome; it has a resistance of 24 Ω. Can you obtain more heat by winding one coil or by cutting the wire in two and winding two separate coils? In each case the coils are to be connected individually across a 110-V line.

The power P for the single coil is given by Eq. 31-17:

$$P = \frac{V^2}{R} = \frac{(110 \text{ V})^2}{24 \text{ Ω}} = 500 \text{ W.}$$

The power for a coil of half the length is given by

$$P' = \frac{(110 \text{ V})^2}{12 \text{ Ω}} = 1000 \text{ W.}$$

There are two "half-coils," so that the total power obtained by cutting the wire in half is 2000 W, or four times that for the single coil. This would seem to suggest that we could buy a 500-W heating coil, cut it in half, and rewind it to obtain 2000 W. Why is this not a practical idea?

---

1. Name other physical quantities that, like current, are scalars having a sense represented by an arrow in a diagram.

2. What conclusions can you draw by applying Eq. 31-4 to a closed surface through which a number of wires pass in random directions, carrying steady currents of different magnitudes? Does Gauss's law hold?

3. A potential difference V is applied to a copper wire of diameter d and length l. What is the effect on the electron drift speed of (a) doubling V, (b) doubling l, and (c) doubling d?

4. If the drift speeds of the electrons in a conductor under ordinary circumstances are so slow (see Example 2), why do the lights in a room turn on so quickly after the switch is closed?

5. Can you think of a way to measure the drift speed for electrons by timing their travel along a conductor?

6. Let a battery be connected to a copper cube at two corners defining a body diagonal. Pass a hypothetical plane completely through the cube, tilted at any angle. (a) Is the current i through the plane independent of the position and orientation of the plane? (b) Is there any position and orientation of the plane for which $\mathbf{j}$ is a constant in magnitude, direction, or both? (c) Does Eq. 31-4 hold for all orientations of the plane? (d) Does Eq. 31-4 hold for a closed surface of arbitrary shape, which may or may not lie entirely within the cube? If not, why; if so, what does it predict?

7. In our convention for the direction of current arrows (a) would it have been more convenient, or even possible, to have assumed all charge carriers to be negative? (b) Would it have been more convenient, or even possible, to have labeled the electron as positive, the proton as negative, etc.?

8. Explain in your own words why we can have $\mathbf{E} \neq 0$ inside a conductor in this chapter but that we took $\mathbf{E} = 0$ for granted in Chapter 28 (see, for example, Section 28-6).

9. A potential difference V is applied to a circular cylinder of carbon by clamping it between circular copper electrodes, as in Fig. 31-9. Discuss the difficulty of calculating the resistance of the carbon cylinder, using the relation $R = \rho L/A$.

10. How would you measure the resistance of a pretzel-shaped conductor? Give specific details to clarify the concept.

11. Discuss the difficulties of testing whether the filament of a light bulb obeys Ohm's law.

12. You are given a circular cylinder of aluminum, 1.00 cm in radius and 2.00 cm high. What practical laboratory arrangements would you make if you wanted to measure its resistance between parallel faces (assumed equipotentials) using Eq. 31-9 ($R = \rho l/A$).

13. You are given a cube of aluminum and access to two battery terminals. How would you connect the terminals to the cube to insure (a) a maximum and (b) a minimum resistance?

14. Does the relation $V = iR$ apply to resistors that do not obey Ohm's law?

15. The temperature coefficient of resistance of a thermistor is negative and varies greatly with temperature. Account qualitatively for the shape of the curve of i versus V for the thermistor of Fig. 31-6.

16. Why are the dashed white lines in Fig. 31-7 curved slightly?

17. A current i enters the top of a copper sphere of radius R and leaves at a diametrically opposite point. Are all parts of the sphere equally effective in dissipating thermal energy?

18. What special characteristics must (a) heating wire and (b) fuse wire have?

19. Equation 31-16 ($P = i^2R$) seems to suggest that the rate of increase of thermal energy in a resistor is reduced if the resistance is made less; Eq. 31-17 ($P = V^2/R$) seems to suggest just the opposite. How do you reconcile this apparent paradox?

Copper

Carbon

Copper

figure 31-9
Question 9

20. Is the filament resistance lower or higher in a 500-W light bulb than in a 100-W bulb? Both bulbs are designed to operate on 100 V.

21. Five wires of the same length and diameter are connected in turn between two points maintained at constant potential difference. Will Joule energy be developed at the faster rate in the wire of (a) the smallest or (b) the largest resistance?

22. The windings of a motor (connected to a load) have a resistance of 1.0 Ω. If we apply a potential difference of 100 V to the motor, does it follow that the current through the motor will be 110 V/1.0 Ω = 110 A?

23. A cow and a man are standing in a meadow when lightning strikes the ground nearby. Why is the cow more likely to be killed than the man? The responsible phenomenon is called "step voltage".

problems

SECTION 31-1

1. A current of 5 A exists in a 10-Ω resistor for 4 min. (a) How many coulombs and (b) how many electrons pass through any cross section of the resistor in this time?   *Answer:* (a) 1.2×10^3 C. (b) 7.5×10^{21} electrons.

2. A current is established in a gas discharge tube when a sufficiently high potential difference is applied across the two electrodes in the tube. The gas ionizes; electrons move toward the positive terminal and positive ions toward the negative terminal. What are the magnitude and sense of the current in a hydrogen discharge tube in which 3.1×10^{18} electrons and 1.1×10^{18} protons move past a cross-sectional area of the tube each second?

3. A steady beam of alpha particles ($q = 2e$) traveling with constant kinetic energy 20 MeV carries a current 0.25×10^{-6} A. (a) If the beam is directed perpendicular to a plane surface, how many alpha particles strike the surface in 3.0 s? (b) At any instant, how many alpha particles are there in a given 20-cm length of the beam? (c) Through what potential difference was it necessary to accelerate each alpha particle from rest to bring it to an energy of 20 MeV?   *Answer:* (a) 2.3×10^{12}. (b) 5.0×10^3. (c) 10^7 V.

4. We have 2.0×10^8 doubly charged positive ions per cubic centimeter, all moving north with a speed of 1.0×10^7 cm/s. (a) What is the current density **j**, in magnitude and direction? (b) Can you calculate the total current i in this ionic beam? If not, why?

5. A small but measurable current of 1.0×10^{-10} A exists in a copper wire whose diameter is 0.10 in. Calculate the electron drift speed.
Answer: 1.5×10^{-15} m/s.

6. The belt of an electrostatic generator is 50 cm wide and travels at 30 m/s. The belt carries charge into the sphere at a rate corresponding to 1.0×10^{-4} A. Compute the surface charge density on the belt.

7. A current i enters one corner of a square sheet of copper and leaves at the opposite corner. Sketch arrows for various points within the square to represent the relative values of **j.** Intuitive guesses rather than detailed mathematical analyses are called for.

8. You are given a conducting sphere of 10-cm radius. One wire carries a current of 1.0000020 A into it. Another wire carries a current of 1.0000000 A out of it. How long would it take for the sphere to increase in potential by 1000 V?

SECTION 31-2

9. A steel trolley-car rail has a cross-sectional area of 7.1 in.² What is the resistance of 10 miles of single track? The resistivity of the steel is 6.0×10^{-7} Ω·m.   *Answer:* 2.1 Ω.

10. A square aluminum rod is 1.0 m long and 5.0 mm on edge. (a) What is the resistance between its ends? (b) What must be the diameter of a circular 1.0 m copper rod if its resistance is to be the same?

11. A copper wire and an iron wire of the same length have the same potential

difference applied to them. (a) What must be the ratio of their radii if the current is to be the same? (b) Can the current density be made the same by suitable choices of the radii? *Answer:* (a) 2.4, iron being larger. (b) No.

12. A wire with a resistance of 6.0 Ω is drawn out through a die so that its new length is three times its original length. Find the resistance of the longer wire, assuming that the resistivity and density of the material are not changed during the drawing process.

13. A wire of Nichrome (a nickel-chromium alloy commonly used in heating elements) is 1.0 m long and 1.0 mm² in cross-sectional area. It carries a current of 4.0 A when a 2.0-V potential difference is applied between its ends. What is the conductivity σ of Nichrome?   *Answer:* 2.0 × 10⁶ S.

14. A rod of a certain metal is 1.00 m long and 0.550 cm in diameter. The resistance between its ends (at 20°C) is 2.87 × 10⁻³ Ω. A round disk is formed of this same material, 2.00 cm in diameter and 1.00 mm thick. (a) What is the resistance between the opposing round faces, assuming equipotential surfaces? (b) What is the material?

15. Two conductors are made of the same material and have the same length. Conductor A is a solid wire of diameter 1.0 mm. Conductor B is a hollow tube of outside diameter 2.0 mm and inside diameter 1.0 mm. What is the resistance ratio, R_A/R_B, measured between their ends?   *Answer:* 3.0

16. Nine copper wires of length l and diameter d are connected in parallel to form a single composite conductor of resistance R. What must be the diameter D of a single copper wire of length l if it is to have the same resistance?

17. A copper wire and an iron wire of equal length l and diameter d are joined and a potential difference V is applied between the ends of the composite wire. Calculate (a) the potential difference across each wire. Assume that $l = 10$ m, $d = 2.0$ mm, and $V = 100$ V. (b) Also calculate the current density in each wire, and (c) the electric field in each wire.
Answer: (a) 15 V (copper); 85 V (iron). (b) 8.5 × 10⁷ A/m². (c) 1.5 V/m (copper); 8.5 V/m (iron).

18. The resistance of an iron wire is 5.9 times that of a copper wire of the same dimensions. What must be the diameter of an iron wire if it is to have the same resistance as a copper wire 0.12 cm in diameter, both wires being the same length?

19. Circularly cylindrical aluminum and copper rods have the same length and are designed to have the same resistance. The resistivity of copper is 0.61 times that of aluminum but its density is 3.3 times that of aluminum. What is the ratio of the mass of the aluminum rod to that of the copper rod?
Answer: 0.50.

20. Conductors A and B, having equal lengths of 40 m and a cross-sectional area of 0.10 m², are connected in series. A potential of 60 V is applied across the terminal points of the connected wires. The resistances of the wires are 40 and 20 Ω respectively. Determine: (a) the resistivities of the two wires; (b) the magnitude of the electric field in each wire; (c) the current density in each wire; (d) the potential difference applied to each conductor.

21. A resistor is in the shape of a truncated right circular cone (Fig. 31-10). The end radii are a and b, the altitude is l. If the taper is small, we may assume that the current density is uniform across any cross section. (a) Calculate the resistance of this object. (b) Show that your answer reduces to $\rho(l/A)$ for the special case of zero taper ($a = b$).

Answer: (a) $R = \rho \dfrac{l}{\pi ab}$.

22. (a) At what temperature would the resistance of a copper conductor be double its resistance at 0°C? (b) Does this same temperature hold for all copper conductors, regardless of shape or size?

figure 31-10
Problem 21

23. The copper windings of a motor have a resistance of 50 Ω at 20°C when the motor is idle. After running for several hours the resistance rises to 58 Ω. What is the temperature of the windings?   *Answer:* 61°C.

24. When a metal rod is heated, not only its resistance but also its length and its cross-sectional area change. The relation $R = \rho l/A$ suggests that all three factors should be taken into account in measuring ρ at various temperatures. (*a*) If the temperature changes by 1.0°C, what percent changes in R, l, and A occur for a copper conductor? (*b*) What conclusion do you draw? The coefficient of linear expansion is $1.7 \times 10^{-5}/\text{C}°$.

SECTION 31-3

25. List in tabular form similarities and differences between the flow of charge along a conductor, the flow of water through a horizontal pipe, and the conduction of heat through a slab. Consider such quantities as what causes the flow, what opposes it, what particles (if any) participate, and the units in which the flow may be measured.

26. (*a*) Using data from Fig. 31-5, plot the resistance of the vacuum tube as a function of applied potential difference. (*b*) Repeat for the thermistor of Fig. 31-6.

SECTION 31-4

27. Explain why the momentum which conduction electrons transfer to the ions in a metal conductor does not give rise to a resultant force on the conductor.

    Answer: Because of Newton's third law there is no resultant force on the conductor (electrons + ions).

SECTION 31-5

28. Thermal energy is developed in a resistor at a rate of 100 W when the current is 3.0 A. What is the resistance in ohms?

29. An x-ray tube takes a current of 7.0 mA and operates at a potential difference of 80 kV. What power in watts is dissipated?   *Answer:* 560 W.

30. A potential difference of 1.0 V is applied to a 100-ft length of #18 copper wire (diameter = 0.040 in.). Calculate (*a*) the current, (*b*) the current density, (*c*) the electric field, and (*d*) the rate at which thermal energy is developed in the wire.

31. The National Board of Fire Underwriters has fixed safe current-carrying capacities for various sizes and types of wire. For #10 rubber-coated copper wire (wire diameter = 0.10 in.) the maximum safe current is 25 A. At this current, find (*a*) the current density, (*b*) the electric field, (*c*) the potential difference for 1000 ft of wire, and (*d*) the rate at which thermal energy ($= i^2R$) is developed for 1000 ft of wire.

    Answer: (*a*) 4.9×10^6 A/m². (*b*) 8.3×10^{-2} V/m. (*c*) 25 V. (*d*) 630 W.

32. A beam of 16-MeV deuterons from a cyclotron falls on a copper block. The beam is equivalent to a current of 15×10^{-6} A. (*a*) At what rate do deuterons strike the block? (*b*) At what rate is thermal energy produced in the block?

33. A 500-W immersion heater is placed in a pot containing 2.0 liters of water at 20°C. (*a*) How long will it take to bring the water to boiling temperature, assuming that 80% of the available energy is absorbed by the water? (*b*) How much longer will it take to boil half the water away?

    Answer: (*a*) 28 min. (*b*) 1.6 h.

34. A 500-W heating unit is designed to operate from a 115-V line. (*a*) By what percentage will its heat output drop if the line voltage drops to 110 V? Assume no change in resistance. (*b*) Taking the variation of resistance with temperature into account, would the actual heat output drop be larger or smaller than that calculated in (*a*)?

35. A 1250-W radiant heater is constructed to operate at 115 V. (*a*) What will be the current in the heater? (*b*) What is the resistance of the heating coil? (*c*) How many kilocalories are generated in one hour by the heater?

    Answer: (*a*) 11 A. (*b*) 11 Ω. (*c*) 1100 kcal.

36. A Nichrome heater dissipates 500 W when the applied potential difference is 110 V and the wire temperature is 800°C. How much power would it dissipate if the wire temperature were held to 200°C by immersion in a bath of cooling oil? The applied potential difference remains the same; $\bar{\alpha}$ for Nichrome is about $4 \times 10^{-4}/\text{C}°$.

37. (a) Derive the formulas $P = j^2\rho$ and $P = E^2/\rho$ where P = power per unit volume in a resistor. (b) A cylindrical resistor of radius 0.50 cm and length 2.0 cm has a resistivity of 3.5×10^{-5} $\Omega \cdot$m. What are the current density and potential difference when the power dissipation is 1.0 W?
    Answer: (b) $j = 1.3 \times 10^5$ A/m²; $V = 0.094$ V.

38. It is desired to make a long cylindrical conductor whose temperature coefficient of resistivity at 20°C will be close to zero. (a) If such a conductor is made by assembling alternate disks of iron and carbon, what is the ratio of the thickness of a carbon disk to that of an iron disk? Assume that the temperature remains essentially the same in each disk. (b) What is the ratio of thermal energy generation in a carbon disk to that in an iron disk? See Table 31-1.

39. An electron linear accelerator produces a pulsed beam of electrons. The pulse current is 0.50 A and the pulse duration 0.10 μs. (a) How many electrons are accelerated per pulse? (b) What is the average current for a machine operating at 500 pulses/s? (c) If the electrons are accelerated to an energy of 50 MeV, what are the average and peak power outputs of the accelerator?
    Answer: (a) 3.1×10^{11}. (b) 25 μA. (c) 1200 W (average); 2.5×10^7 W (peak).

32
electromotive
force and circuits

There exist in nature certain devices such as batteries and electric generators which are able to maintain a potential difference between two points to which they are attached. We call such devices seats of *electromotive force* (symbol ε; abbr. emf). In this chapter we do not discuss their internal construction or detailed mode of action but confine ourselves to describing their gross electrical characteristics and to exploring their usefulness in electric circuits.

Figure 32-1a shows a seat of emf B, represented by a battery*, connected to a resistor R. The seat of emf maintains its upper terminal positive and its lower terminal negative, as shown by the + and − signs. In the circuit external to B positive charge carriers would be driven in the direction shown by the arrows marked *i*. In other words, a clockwise current would be set up.

An emf is represented by an arrow which is placed next to the seat and points in the direction in which the seat, acting alone, would cause a positive charge carrier to move in the external circuit. We draw a small circle on the tail of an emf arrow so that we will not confuse it with a current arrow.

A seat of emf must be able to do work on charge carriers that enter it. In the circuit of Fig. 32-1a, for example, the seat acts to move positive charges from a point of low potential (the negative terminal) through the seat to a point of high potential (the positive terminal). This reminds us of a pump, which can cause water to move from a place of low gravitational potential to a place of high potential.

In Fig. 32-1a a charge dq passes through *any* cross section of the circuit in time dt. In particular, this charge enters the seat of emf ε at its low-potential end and leaves at its high-potential end. The seat must do

---

* Batteries are far from being the only source of emf. Among others are generators (see Chapter 39); devices activated by temperature differences (thermocouples, etc.); devices activated by light (see "The Photovoltaic Generation of Electricity" by Bruce Chalmers, *Scientific American*, October 1976); the human heart (see "The Electrocardiograph—Teaching Physics to Premeds" by Pierre Lafrance, *The Physics Teacher*, November 1972); certain fish (see "Seeing the World Through a New Sense: Electroreception in Fish" by T. H. Bullock, *American Scientist*, May–June 1973).

an amount of work dW on the (positive) charge carriers to force them to go to the point of higher potential. The emf ε of the seat is defined from

$$\varepsilon = dW/dq. \tag{32-1}$$

The unit of emf is the joule/coulomb (see Eq. 29-1) which is the *volt*. We might be inclined to say that a battery has an emf of 1 volt if it maintains a difference of potential of 1 volt between its terminals. This is true only under certain conditions, which we describe in Section 32-4. Note also from Eq. 32-1 that the electromotive force is not actually a force, that is, we cannot measure it in newtons. The name is involved with the early history of the subject.

If a seat of emf does work on a charge carrier, energy must be transferred within the seat. In a battery, for example, chemical energy is transferred into electrical energy. Thus we can describe a seat of emf as a device in which chemical, mechanical, or some other form of energy is changed (reversibly) into electrical energy. The chemical energy provided by the battery in Fig. 32-1a is stored in the electric and the magnetic* fields that surround the circuit. This stored energy does not increase because it is being drained away, by transfer to Joule (thermal) energy in the resistor, at the same rate at which it is supplied. The electric and magnetic fields play an intermediary role in the energy transfer process, acting as a storage reservoir.

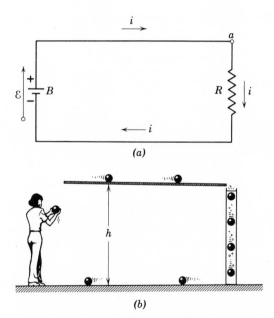

(a)

(b)

figure 32-1
(*a*) A simple electric circuit and (*b*) its gravitational analog.

Figure 32-1b shows a gravitational analog of Fig. 32-1a. In the top figure the seat of emf B does work on the charge carriers. This energy, stored in passage as electromagnetic field energy, appears eventually as thermal energy in resistor R. In the lower figure the person, in lifting the bowling balls from the floor to the shelf, does work on them. This energy is stored, in passage, as gravitational field energy. The balls roll slowly and uniformly along the shelf, dropping from the right end into a cylinder full of viscous oil. They sink to the bottom at an essentially

---

* A current in a wire is surrounded by a magnetic field, and this field, like the electric field, can also be viewed as a site of stored energy (see Section 36-4).

constant speed, are removed by a trapdoor mechanism not shown, and roll back along the floor to the left. The energy put into the system by the person appears eventually as thermal energy in the viscous fluid, resulting in a temperature rise. The energy supplied by the person comes from internal (chemical) energy. The circulation of charges in Fig. 32-1a will stop eventually if battery B is not recharged; the circulation of bowling balls in Fig. 32-1b will stop eventually if the person does not replenish the store of internal energy by eating.

Figure 32-2 shows a circuit containing two (ideal) batteries, A and B, a resistor R, and an (ideal) electric motor employed in lifting a weight. The batteries are connected so that they tend to send charges around the circuit in opposite directions; the actual direction of the current is determined by B, which supplies the larger potential difference. The energy transfers in this circuit are

figure 32-2
Two batteries, a resistor, and a motor, connected in a single-loop circuit. It is given that $\mathcal{E}_B > \mathcal{E}_A$.

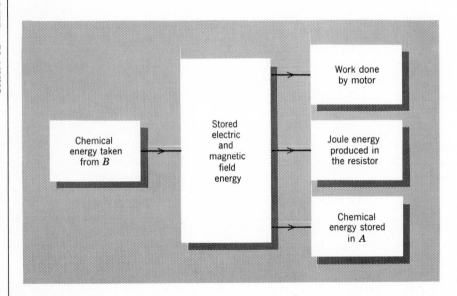

The chemical energy in B is steadily depleted, the energy appearing in the three forms shown on the right. Battery A is being charged while battery B is being discharged. Again, the electric and magnetic fields that surround the circuit act as an intermediary.

It is part of the definition of an emf that the energy transfer process be *reversible*, at least in principle. Recall that a reversible process is one that passes through equilibrium states; its course can be reversed by making an infinitesimal change in the environment of the system; see Section 25-2. A battery, for example, can either be on charge or discharge; a generator can be driven mechanically, producing electrical energy, or it can be operated backward as a motor. The (reversible) energy transfers here are

$$\text{electrical} \rightleftharpoons \text{chemical}$$

and

$$\text{electrical} \rightleftharpoons \text{mechanical}.$$

The energy transfer from electrical energy to thermal energy is not reversible. We can easily raise the temperature of a conductor by supplying electric energy to it, but it is *not* possible to set up a current in a closed copper loop by raising its temperature uniformly. Because of this lack of reversibility, we do not associate an emf with the Joule effect, that is, with energy transfers associated with Eqs. 31-16 and 31-17.

In a time dt an amount of energy given by $i^2R\,dt$ will appear in the resistor of Fig. 32-1a as Joule thermal energy. During this same time a charge $dq\ (= i\,dt)$ will have moved through the seat of emf, and the seat will have done work on this charge (see Eq. 32-1) given by

$$dW = \varepsilon\,dq = \varepsilon i\,dt.$$

From the conservation of energy principle, the work done by the seat must equal the thermal energy, or

$$\varepsilon i\,dt = i^2R\,dt.$$

Solving for i, we obtain

$$i = \varepsilon/R. \qquad (32\text{-}2)$$

We can also derive Eq. 32-2 by considering that, if electric potential is to have any meaning, a given point can have only one value of potential at any given time. If we start at any point in the circuit of Fig. 32-1a and, in imagination, go around the circuit in either direction, adding up algebraically the changes in potential that we encounter, we must arrive at the same potential when we return to our starting point. In other words, *the algebraic sum of the changes in potential encountered in a complete traversal of the circuit must be zero.*

In Fig. 32-1a let us start at point a, whose potential is V_a,* and traverse the circuit clockwise. In going through the resistor, there is a change in potential of $-iR$. The minus sign shows that the top of the resistor is higher in potential than the bottom, which must be true, because positive charge carriers move of their own accord from high to low potential. As we traverse the battery from bottom to top, there is an *increase* of potential $+\varepsilon$ because the battery does (positive) work on the charge carriers, that is, it moves them from a point of low potential to one of high potential. Adding the algebraic sum of the changes in potential to the initial potential V_a must yield the identical value V_a, or

$$V_a - iR + \varepsilon = V_a.$$

We write this as

$$-iR + \varepsilon = 0,$$

which is independent of the value of V_a and which asserts explicitly that the algebraic sum of the potential changes for a complete circuit traversal is zero. This relation leads directly to Eq. 32-2.

These two ways to find the current in single-loop circuits, based on the conservation of energy and on the concept of potential, are completely equivalent because potential differences are defined in terms of work and energy (see Section 29-1). The statement that the sum of the changes in potential encountered in making a complete loop is zero is called *Kirchhoff's second rule;* for brevity we call it the *loop theorem.* Always bear in mind that this theorem is simply a particular way of stating the law of conservation of energy for electric circuits.

To prepare for the study of more complex circuits, let us examine the rules for finding potential differences; these rules follow from the previous discussion. They are not meant to be memorized but rather to be

* The actual value of V_a depends on assumptions made in the definition of potential (as described in Section 29-1). The numerical value of V_a is not important because, as in most electric circuit situations, we are concerned here with *differences* of potential. Point a in Fig. 32-1a (or any other single point in that figure) could be connected to ground (symbol $\div$) and assigned the potential $V_a = 0$, following a common practice.

so thoroughly understood that it becomes trivial to re-derive them on each application.

1. If a resistor is traversed in the direction of the current, the change in potential is $-iR$; in the opposite direction it is $+iR$.
2. If a seat of emf is traversed in the direction of the emf, the change in potential is $+\varepsilon$; in the opposite direction it is $-\varepsilon$.

Figure 32-3a shows a circuit which emphasizes that all seats of emf have an intrinsic internal resistance r. This resistance cannot be removed—although we would usually like to do so—because it is an inherent part of the device. The figure shows the internal resistance r and the emf separately, although, actually, they occupy the same region of space.

If we apply the loop theorem, starting at b and going around clockwise, we obtain

$$V_b + \varepsilon - ir - iR = V_b$$

or

$$+\varepsilon - ir - iR = 0.$$

Compare these equations with Fig. 32-3b, which shows the changes in potential graphically. In writing these equations, note that we traversed r and R in the direction of the current and ε in the direction of the emf. The same equation follows if we start at any other point in the circuit or if we traverse the circuit in a counterclockwise direction. Solving for i gives

$$i = \frac{\varepsilon}{R + r}. \qquad (32\text{-}3)$$

32-3
OTHER SINGLE-LOOP CIRCUITS

figure 32-3
(a) A single-loop circuit. The rectangular block is a seat of emf with internal resistance r. (b) The same circuit is drawn for convenience as a straight line. Directly below are shown the changes in potential that one encounters in traversing the circuit clockwise from point b. Note that the two points marked "b" at the top of this figure are the same point.

(a)

(b)

Resistors in series. Resistors in series are connected so that there is only one conducting path through them, as in Fig. 32-4. What is the equivalent resistance R of this series combination? The equivalent resistance is the single resistance R which, substituted for the series combination between the terminals ab, will leave the current i unchanged.

Applying the loop theorem (going clockwise from a) yields

$$-iR_1 - iR_2 - iR_3 + \varepsilon = 0$$

or

$$i = \frac{\varepsilon}{R_1 + R_2 + R_3}.$$

For the equivalent resistance R

$$i = \frac{\varepsilon}{R}$$

or

$$R = R_1 + R_2 + R_3. \qquad (32\text{-}4)$$

The extension to more than three resistors is clear.

EXAMPLE 1

figure 32-4
Example 1. Three resistors are connected in series between terminals a and b.

32-4
POTENTIAL DIFFERENCES

We often want to compute the potential difference between two points in a circuit. In Fig. 32-3a, for example, what is the relationship between the potential difference V_{ab} $(= V_a - V_b)$ between points b and a and the fixed circuit parameters ε, r, and R? To find this relationship, let us start from point b and traverse the circuit to point a, passing through resistor R against the current. If V_b and V_a are the potentials at b and a, respectively, we have

$$V_b + iR = V_a$$

because we experience an increase in potential in traversing a resistor against the current arrow. We rewrite this relation as

$$V_{ab} = V_a - V_b = +iR,$$

which tells us that V_{ab}, the potential difference between points a and b, has the magnitude iR and that point a is more positive than point b. Combining this last equation with Eq. 32-3 yields

$$V_{ab} = \varepsilon \frac{R}{R + r}. \qquad (32\text{-}5)$$

To sum up: To find the potential difference between any two points in a circuit start at one point and traverse the circuit to the other, following any path,* and add up algebraically the potential changes encountered. This algebraic sum will be the potential difference. This procedure is similar to that for finding the current in a closed loop, except that here the potential differences are added up over part of a loop and not over the whole loop.

The potential difference between any two points can have only one value; thus we must obtain the same answer for all paths that connect these points. If we consider two points on the side of a hill, the measured difference in gravitational potential (that is, in altitude) between them is the same no matter what path is followed in going from one to the other. In Fig. 32-3a let us also calculate V_{ab}, using a path passing through the seat of emf. We have

---

*Recall (see Section 29-1) that path independence is a central feature of the potential concept.

$$V_b + \varepsilon - ir = V_a$$

or (see also Fig. 32-3b)

$$V_{ab} = V_a - V_b = +\varepsilon - ir.$$

Again, combining with Eq. 32-3 leads to Eq. 32-5.

The terminal potential difference of the battery V_{ab}, as Eq. 32-5 shows, is less than ε unless the battery has no internal resistance $(r = 0)$ or if it is on open circuit $(R = \infty)$; then V_{ab} is equal to ε. Thus the emf of a device is equal to its terminal potential difference *when on open circuit.*

EXAMPLE 2

In Fig. 32-5a let ε_1 and ε_2 be 2.0 V and 4.0 V, respectively; let the resistances r_1, r_2, and R be 1.0 Ω, 2.0 Ω, and 5.0 Ω, respectively. What is the current?

Emfs ε_1 and ε_2 oppose each other, but because ε_2 is larger it controls the direction of the current. Thus i will be counterclockwise. The loop theorem, going clockwise from a, yields

$$-\varepsilon_2 + ir_2 + iR + ir_1 + \varepsilon_1 = 0.$$

Check that the same result is obtained by going around counterclockwise. Also, compare this equation carefully with Fig. 32-5b, which shows the potential changes graphically.

Solving for i yields

$$i = \frac{\varepsilon_2 - \varepsilon_1}{R + r_1 + r_2} = \frac{4.0 \text{ V} - 2.0 \text{ V}}{5.0 \text{ Ω} + 1.0 \text{ Ω} + 2.0 \text{ Ω}}$$

$$= 0.25 \text{ A}.$$

It is not necessary to know in advance what the actual direction of the current is. To show this, let us assume that the current in Fig. 32-5a is clockwise, an assumption that we know is incorrect. The loop theorem then yields (going clockwise from a)

$$-\varepsilon_2 - ir_2 - iR - ir_1 + \varepsilon_1 = 0$$

or

$$i = \frac{\varepsilon_1 - \varepsilon_2}{R + r_1 + r_2}.$$

Substituting numerical values (see above) yields −0.25 A for the current. The minus sign tells us that the current is in the opposite direction from the one we have assumed.

In more complex circuit problems involving many loops and branches it is often impossible to know in advance the correct directions for the currents in all parts of the circuit. We can assume directions for the currents at random. Those currents for which positive numerical values are obtained will have the correct directions; those for which negative values are obtained will be exactly opposite to the assumed directions. In all cases the numerical values will be correct.

EXAMPLE 3

What is the potential difference (a) between points b and a in Fig. 32-5a, and (b) between points a and c?

(a) For points a and b we start at b and traverse the circuit to a, obtaining

$$V_{ab} \, (= V_a - V_b) = -ir_2 + \varepsilon_2 = -(0.25 \text{ A})(2.0 \text{ Ω}) + 4.0 \text{ V}$$

$$= +3.5 \text{ V}.$$

Thus a is more positive than b and the potential difference (3.5 V) is *less than* the emf (4.0 V); see Fig. 32-5b.

(b) For points c and a, we start at c and traverse the circuit to a, obtaining

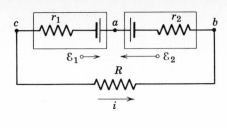

(a)

figure 32-5
Examples 2, 3. *(a)* A single-loop circuit. *(b)* The same circuit is shown schematically as a straight line, the potential differences encountered in traversing the circuit clockwise from point *a* being displayed directly below. In the lower figure the potential of point *a* was assumed to be zero for convenience. Note that the two points marked "*a*" in this figure are the same point.

(b)

$$V_{ac} \ (= V_a - V_c) = +\mathcal{E}_1 + ir_1 = +2.0 \text{ V} + (0.25 \text{ A})(1.0 \ \Omega)$$

$$= +2.25 \text{ V}.$$

This tells us that *a* is at a higher potential than *c*. The terminal potential difference of $\mathcal{E}_1$ (2.25 V) is *larger than* the emf (2.0 V); see Fig. 32-5*b*. Charge is being forced through $\mathcal{E}_1$ in a direction opposite to the one in which it would send charge if it were acting by itself; if $\mathcal{E}_1$ is a storage battery, it is being charged at the expense of $\mathcal{E}_2$.

Let us test the first result by proceeding from *b* to *a* along a different path, namely, through *R*, r_1, and $\mathcal{E}_1$. We have

$$V_{ab} = iR + ir_1 + \mathcal{E}_1 = (0.25 \text{ A})(5.0 \ \Omega) + (0.25 \text{ A})(1.0 \ \Omega) + 2.0 \text{ V} = +3.5 \text{ V},$$

which is the same as the earlier result.

32-5
MULTILOOP CIRCUITS

Figure 32-6 shows a circuit containing two loops. For simplicity, we have neglected the internal resistances of the batteries. There are two *junctions*, *b* and *d*, and three *branches* connecting these junctions. The branches are the left branch *bad*, the right branch *bcd*, and the central branch *bd*. If the emfs and the resistances are given, what are the currents in the various branches?

We label the currents in the branches as i_1, i_2, and i_3, as shown. Current i_1 has the same value for any cross section of the left branch from *b* to *d*. Similarly, i_2 has the same value everywhere in the right branch and i_3 in the central branch. The directions of the currents have been chosen arbitrarily. The careful reader will note that i_3 must point in a direction opposite to the one we have shown. We have deliberately drawn it in wrong to show how the formal mathematical procedures will always indicate this to us.

The three currents i_1, i_2, and i_3 carry charge either toward junction *d* or away from it. Charge does not accumulate at junction *d*, nor does it

figure 32-6
A multiloop circuit.

drain away from this junction because the circuit is in a steady-state condition. Thus charge must be removed from the junction by the currents at the same rate that it is brought into it. If we arbitrarily call a current approaching the junction positive and one leaving the junction negative, then

$$i_1 + i_3 - i_2 = 0.$$

This equation suggests a general principle for the solution of multiloop circuits: *At any junction the algebraic sum of the currents must be zero.* This *junction theorem* is also known as *Kirchhoff's first rule.* Note that it is simply a statement of the conservation of charge. Thus our basic tools for solving circuits are (a) the conservation of energy (see p. 697) and (b) the conservation of charge.

For the circuit of Fig. 32-6, the junction theorem yields only one relationship among the three unknowns. Applying the theorem at junction b leads to exactly the same equation, as you can easily verify. To solve for the three unknowns, we need two more independent equations; they can be found from the loop theorem.

In single-loop circuits there is only one conducting loop around which to apply the loop theorem, and the current is the same in all parts of this loop. In multiloop circuits there is more than one loop, and the current in general will not be the same in all parts of any given loop.

If we traverse the left loop of Fig. 32-6 in a counterclockwise direction, the loop theorem gives

$$\mathcal{E}_1 - i_1 R_1 + i_3 R_3 = 0. \tag{32-6}$$

The right loop gives

$$-i_3 R_3 - i_2 R_2 - \mathcal{E}_2 = 0. \tag{32-7}$$

These two equations, together with the relation derived earlier with the junction theorem, are the three simultaneous equations needed to solve for the unknowns i_1, i_2, and i_3. Doing so yields

$$i_1 = \frac{\mathcal{E}_1(R_2 + R_3) - \mathcal{E}_2 R_3}{R_1 R_2 + R_2 R_3 + R_1 R_3}, \tag{32-8a}$$

$$i_2 = \frac{\mathcal{E}_1 R_3 - \mathcal{E}_2(R_1 + R_3)}{R_1 R_2 + R_2 R_3 + R_1 R_3}, \tag{32-8b}$$

and

$$i_3 = \frac{-\mathcal{E}_1 R_2 - \mathcal{E}_2 R_1}{R_1 R_2 + R_2 R_3 + R_1 R_3}. \tag{32-8c}$$

Supply the missing steps. Equation 32-8c shows that no matter what numerical values are given to the emfs and to the resistances, the current i_3 will always have a negative value. This means that it will always point up in Fig. 32-6 rather than down, as we deliberately assumed. The currents i_1 and i_2 may be in either direction, depending on the particular numerical values given.

Verify that Eqs. 32-8 reduce to sensible conclusions in special cases. For $R_3 = \infty$, for example, we find

$$i_1 = i_2 = \frac{\varepsilon_1 - \varepsilon_2}{R_1 + R_2} \quad \text{and} \quad i_3 = 0.$$

What do these equations reduce to for $R_2 = \infty$?

The loop theorem can be applied to a large loop consisting of the entire circuit *abcda* of Fig. 32-6. This fact might suggest that there are more equations than we need, for there are only three unknowns and we already have three equations written in terms of them. However, the loop theorem yields for this loop

$$-i_1 R_1 - i_2 R_2 - \varepsilon_2 + \varepsilon_1 = 0,$$

which is nothing more than the sum of Eqs. 32-6 and 32-7. Thus this large loop does not yield another *independent* equation. It will never be found in solving multiloop circuits that there are more independent equations than variables.

Resistors in parallel. Figure 32-7 shows three resistors connected across the same seat of emf. Resistances across which the identical potential difference is applied are said to be in parallel. What is the equivalent resistance R of this parallel combination? The equivalent resistance is that single resistance which, substituted for the parallel combination between terminals *ab*, would leave the current i unchanged.

EXAMPLE 4

figure 32-7
Example 4. Three resistors are connected in parallel between terminals *a* and *b*.

The currents in the three branches are

$$i_1 = \frac{V}{R_1}, \quad i_2 = \frac{V}{R_2}, \quad \text{and} \quad i_3 = \frac{V}{R_3},$$

where V is the potential difference that appears between points *a* and *b*. The total current i is found by applying the junction theorem to junction *a*, or

$$i = i_1 + i_2 + i_3 = V \left(\frac{1}{R_1} + \frac{1}{R_2} + \frac{1}{R_3} \right).$$

If the equivalent resistance is used instead of the parallel combination, we have

$$i = \frac{V}{R}.$$

Combining these two equations gives

$$\frac{1}{R} = \frac{1}{R_1} + \frac{1}{R_2} + \frac{1}{R_3}. \tag{32-9}$$

This formula can easily be extended to more than three resistances. Note that the equivalent resistance of a parallel combination is less than any of the resistances that make it up.

A meter to measure currents is called an *ammeter* (or a *milliammeter, micro-ammeter*, etc., depending on the size of the current to be measured). To determine the current in a wire, it is necessary to break or cut the wire and to insert the ammeter, so that the current to be measured passes through the meter (see Fig. 32-8).*

It is essential that the resistance R_A of the ammeter be *small* compared to other resistances in the circuit. Otherwise the presence of the meter will in itself change the current to be measured. An ideal ammeter would have zero resistance. In the circuit of Fig. 32-8 the required condition, assuming that the voltmeter is not connected, is

$$R_A \ll r + R_1 + R_2.$$

A meter to measure potential differences is called a *voltmeter* (or a *milli-voltmeter* or *microvoltmeter*, etc.). To find the potential difference between two points in a circuit, it is necessary to connect one of the voltmeter terminals to each of the circuit points, without breaking the circuit (see Fig. 32-8).†

It is essential that the resistance of the voltmeter R_V be *large* compared to any circuit resistance across which the voltmeter is connected. Otherwise the meter will itself constitute an important circuit element and will alter the circuit current and the potential difference to be measured. An ideal voltmeter would have an infinite resistance. In Fig. 32-8 the required condition is

$$R_V \gg R_1.$$

In measuring potential difference in electronic circuits, where the effective circuit resistance may be of the order of 10^6 ohms or higher, it becomes necessary to use an electronic voltmeter, which is an electron tube or transistor device designed specifically to have an extremely high effective resistance between its input terminals.

Figure 32-9 shows the rudiments of a *potentiometer*, which is a device for measuring an unknown emf ε_x. The currents and emfs are marked as shown. Applying the loop theorem to loop *abcd* yields

$$-\varepsilon_x - ir + (i_0 - i)R = 0,$$

where $i_0 - i$, by application of the junction theorem at *a*, is the current in resistor *R*. Solving for *i* yields

$$i = \frac{i_0 R - \varepsilon_x}{R + r},$$

in which *R* is a variable resistor. This relation shows that if *R* is adjusted to have the value R_x where

$$i_0 R_x = \varepsilon_x, \tag{32-10}$$

the current *i* in the branch *abcd* becomes zero. To *balance* the potentiometer in this way, *R* must be adjusted manually until the sensitive current meter *G* reads zero.

The emf can be obtained from Eq. 32-10 if the current i_0 is known. However, it is standard practice to replace ε_x by a known standard emf ε_s, and once again to adjust *R* to the zero-current condition. This yields, assuming the current i_0 remains unchanged,

$$i_0 R_s = \varepsilon_s.$$

* The meter must be connected so that the direction of current through it (assuming positive charge carriers) is *into* the meter terminal marked +. Otherwise the meter will deflect in a direction opposite to that intended.

† The voltmeter terminal marked + must be connected to the point of higher potential. Otherwise the meter will deflect in a direction opposite to that intended.

32-6
MEASURING CURRENTS AND POTENTIAL DIFFERENCES

figure 32-8
An ammeter (*A*) is connected to read the current in a circuit, and a voltmeter (*V*) is connected to read the potential difference across resistor R_1.

32-7
THE POTENTIOMETER

figure 32-9
Elements of a potentiometer.
Point *d* represents a sliding contact.

Combining the last two equations yields

$$\mathcal{E}_x = \mathcal{E}_s \frac{R_x}{R_s},$$ (32-11)

which allows us to compare emfs with precision. Note that the internal resistance r of the emf plays no role. In practice, potentiometers are conveniently packaged units, containing a *standard cell* which, after calibration at the National Bureau of Standards or elsewhere, serves as a convenient known standard seat of emf $\mathcal{E}_s$. Switching arrangements for replacing the unknown emf by the standard and arrangements for ascertaining that the current i_0 remains constant are also incorporated.

32-8
RC *CIRCUITS*

The preceding sections dealt with circuits in which the circuit elements were resistors and in which the currents did not vary with time. Here we introduce the capacitor as a circuit element, which will lead us to the concept of time-varying currents. In Fig. 32-10 let switch S be thrown to position a. What current is set up in the single-loop circuit so formed? Let us apply conservation of energy principles.

In time dt a charge $dq \; (= i \; dt)$ moves through any cross section of the circuit. The work done by the seat of emf $(= \mathcal{E} \; dq;$ see Eq. 32-1) must equal the energy that appears as thermal energy in the resistor during time $dt \; (= i^2R \; dt)$ plus the increase in the amount of energy U that is stored in the capacitor $[= dU = d(q^2/2C);$ see Eq. 30-7]. In equation form

$$\mathcal{E} \; dq = i^2R \; dt + d\left(\frac{q^2}{2C}\right)$$

or

$$\mathcal{E} \; dq = i^2R \; dt + \frac{q}{C} \; dq.$$

Dividing by dt yields

$$\mathcal{E} \frac{dq}{dt} = i^2R + \frac{q}{C}\frac{dq}{dt}.$$

But dq/dt is simply i, so that this equation becomes

$$\mathcal{E} = iR + \frac{q}{C}.$$ (32-12)

This equation also follows from the loop theorem, as it must, since the loop theorem was derived from the conservation of energy principle. Starting from point x and traversing the circuit clockwise, we experience an increase in potential in going through the seat of emf and decreases in potential in traversing the resistor and the capacitor, or

$$\mathcal{E} - iR - \frac{q}{C} = 0,$$

which is identical to Eq. 32-12.

figure 32-10
An *RC* circuit.

We cannot immediately solve Eq. 32-12 because it contains two variables, q and i, which, however, are related by

$$i = \frac{dq}{dt}. \qquad (32\text{-}13)$$

Substituting for i into Eq. 32-12 gives

$$\varepsilon = R\,\frac{dq}{dt} + \frac{q}{C}. \qquad (32\text{-}14)$$

Our task now is to find the function $q(t)$ that satisfies this *differential equation*. Although this particular equation is not hard to solve, we choose to avoid mathematical complexity by simply presenting the solution, which is

$$q = C\varepsilon(1 - e^{-t/RC}). \qquad (32\text{-}15)$$

We can easily test whether this function $q(t)$ is really a solution of Eq. 32-14 by substituting it into that equation and seeing whether an identity results. Differentiating Eq. 32-15 with respect to time yields

$$\frac{dq}{dt}\,(= i) = \frac{\varepsilon}{R}\,e^{-t/RC}. \qquad (32\text{-}16)$$

Substituting q (Eq. 32-15) and dq/dt (Eq. 32-16) into Eq. 32-14 yields an identity, as you should verify. Thus Eq. 32-15 is a solution of Eq. 32-14.

Figure 32-11 shows some plots of Eqs. 32-15 and 32-16 for a particular case. Study of these plots and of the corresponding equations shows that (a) at $t = 0$, $q = 0$, and $i = \varepsilon/R$, and (b) as $t \to \infty$, $q \to C\varepsilon$, and $i \to 0$; that is, the current is initially ε/R and finally zero, and the charge on the capacitor plates is initially zero and finally $C\varepsilon$.

The quantity RC in Eqs. 32-15 and 32-16 has the dimensions of time (since the exponent must be dimensionless) and is called the *capacitive time constant* of the circuit. It is the time at which the charge on the capacitor has increased to within a factor of $(1 - e^{-1})\,(\cong 63\%)$ of its equilibrium value. To show this, we put $t = RC$ in Eq. 32-15 to obtain

$$q = C\varepsilon(1 - e^{-1}) = 0.63C\varepsilon.$$

Since $C\varepsilon$ is the equilibrium charge on the capacitor, corresponding to $t \to \infty$, the foregoing statement follows.

EXAMPLE 5

After how many time constants will the energy stored in the capacitor in Fig. 32-10 reach one-half its equilibrium value?

The energy is given by Eq. 30-7, or

$$U = \frac{1}{2C}\,q^2,$$

the equilibrium energy U_∞ being $(1/2C)(C\varepsilon)^2$. From Eq. 32-15, we can write the energy as

$$U = \frac{1}{2C}\,(C\varepsilon)^2(1 - e^{-t/RC})^2$$

or

$$U = U_\infty(1 - e^{-t/RC})^2.$$

Putting $U = \frac{1}{2}U_\infty$ yields

$$\tfrac{1}{2} = (1 - e^{-t/RC})^2$$

and solving this relation for t yields finally

$$t = 1.22\,RC = 1.22 \text{ time constants.}$$

(a)

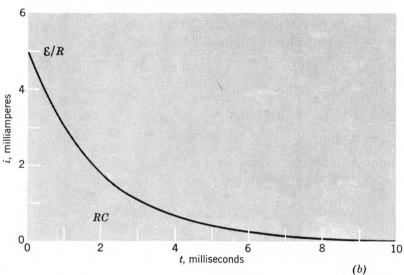

(b)

figure 32-11
If, in Fig. 32-10, we assume that $R = 2000\ \Omega$, $C = 1.0\ \mu F$, and $\varepsilon = 10$ V, then (a) shows the variation of q with t during the charging process and (b) the variations of i with t. The time constant is $RC = 2.0 \times 10^{-3}$ s.

figure 32-11
If, in Fig. 32-10, we assume that $R = 2000\ \Omega$, $C = 1.0\ \mu F$, and $\varepsilon = 10$ V, then (a) shows the variation of q with t during the charging process and (b) the variations of i with t. The time constant is $RC = 2.0 \times 10^{-3}$ s.

Figure 32-11 shows that if a resistance is included in the circuit, the rate of increase of the charge of a capacitor toward its final equilibrium value is *delayed* in a way measured by the time constant RC. With no resistor present $(RC = 0)$, the charge would rise immediately to its equilibrium value. Although we have shown that this time delay follows from an application of the loop theorem to RC circuits, it is important to develop a physical understanding of the causes of the delay.

When switch S in Fig. 32-10 is closed on a, the resistor experiences instantaneously an applied potential difference of ε, and an initial current of ε/R is set up. Initially, the capacitor experiences no potential difference because its initial charge is zero, the potential difference always being given by q/C. The flow of charge through the resistor starts to charge the capacitor, which has several effects. First, the existence of a capacitor charge means that there must now be a potential difference $(= q/C)$ across the capacitor; this, in turn, means that the potential difference across the resistor must decrease by this amount, since the sum of the two potential differences must always equal ε. This decrease in the potential difference across R means that the charging current is reduced. Thus the charge of the capacitor builds up and the charging current decreases until the capacitor is fully charged. At this point the full emf ε is applied to the capacitor, there being no potential drop $(iR = 0)$ across the resistor. This is precisely the reverse of the initial situation. Review the derivations of Eqs. 32-15 and 32-16 and study Fig. 32-11 with the qualitative arguments of this paragraph in mind.

Assume now that the switch S in Fig. 32-10 has been in position a for a time t such that $t \gg RC$. The capacitor is then fully charged for all practical purposes. The switch S is then thrown to position b. How do the charge of the capacitor and the current vary with time?

With the switch S closed on b, there is no emf in the circuit and Eq. 32-12 for the circuit, with $\varepsilon = 0$, becomes simply

$$iR + \frac{q}{C} = 0. \tag{32-17}$$

Putting $i = dq/dt$ allows us to write, as the differential equation of the circuit (compare Eq. 32-14),

$$R\frac{dq}{dt} + \frac{q}{C} = 0. \tag{32-18a}$$

The solution is

$$q = q_0 e^{-t/RC}, \tag{32-18b}$$

as you may readily verify by substitution, q_0 being the initial charge on the capacitor. The capacitive time constant RC appears in this expression for capacitor discharge as well as in that for the charging process (Eq. 32-15). We see that at a time such that $t = RC$ the capacitor charge is reduced to $q_0 e^{-1}$, which is about 37% of the initial charge q_0.

The current during discharge follows from differentiating Eq. 32-18b, or

$$i = \frac{dq}{dt} = -\frac{q_0}{RC} e^{-t/RC}. \tag{32-19}$$

The negative sign shows that the current is in the direction opposite to that shown in Fig. 32-10. This is as it should be, since the capacitor is discharging rather than charging. Since $q_0 = C\varepsilon$, we can write Eq. 32-19 as

$$i = -\frac{\varepsilon}{R} e^{-t/RC},$$

in which ε/R appears as the initial current, corresponding to $t = 0$. This is reasonable because the initial potential difference for the fully charged capacitor is ε.

The behavior of the RC circuit in Fig. 32-10 during charge and discharge can be studied with a cathode-ray oscilloscope. This familiar laboratory device can display on its fluorescent screen plots of the variation of potential with time. Figure 32-12 shows the circuit of Fig. 32-10 with connections made to display (a) the potential difference V_C across the capacitor and (b) the potential difference V_R across the resistor as functions of time. V_C and V_R are given by

$$V_C = \left(\frac{1}{C}\right) q$$

and

$$V_R = (R)i,$$

the former being proportional to the charge and the latter to the current.

figure 32-12
The circuit of Fig. 32-10 with connections made to display the potential variations across the resistor and the capacitor on a cathode-ray oscilloscope.

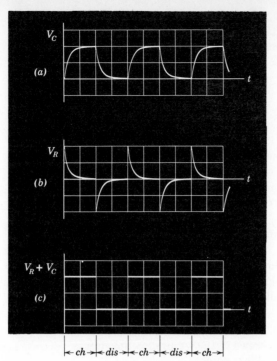

Figure 32-13 shows oscillograph plots of V_C and V_R that result when, in effect, switch S in Fig. 32-10 is thrown regularly back and forth between positions a and b, being left in each position for a time equal to several time constants. Intervals during which the charge is building up are labeled ch and those during which it is decaying are labeled dis.

The "charge" intervals in plot a (see Eq. 32-15) are represented by

$$V = \left(\frac{1}{C}\right) q = \varepsilon(1 - e^{-t/RC})$$

and the "discharge" intervals (see Eq. 32-18b) by

$$V = \left(\frac{1}{C}\right) q = \varepsilon e^{-t/RC}.$$

Note that the current, as indicated by plot b, is in opposite directions during the charge and discharge intervals, in agreement with Eqs. 32-16 and 32-19.

In plot c in Fig. 32-13 the oscillograph has been connected to show the algebraic sum of plots a and b. According to the loop theorem this sum should equal ε during the charge intervals and should be zero during the discharge intervals, when the battery is no longer in the circuit; that is,

$$V_R + V_C = \varepsilon \text{ during charge (see Eq. 32-12)},$$

$$V_R + V_C = 0 \text{ during discharge (see Eq. 32-17)}.$$

Plot c is in exact agreement with this expectation.

1. Does the direction of the emf provided by a battery depend on the direction of current flow through the battery?

2. In Fig. 32-2 discuss what changes would occur if we increased the mass m by such an amount that the "motor" reversed direction and became a "generator", that is, a seat of emf.

3. Discuss in detail the statement that the energy method and the loop theorem method for solving circuits are perfectly equivalent.

questions

4. It is possible to generate a 10,000-volt potential difference by rubbing a pocket comb with wool. Why is this large voltage not dangerous when the much lower voltage provided by an ordinary electric outlet is very dangerous?

5. Devise a method for measuring the emf and the internal resistance of a battery.

6. The current passing through a battery of emf ε and internal resistance r is decreased by some external means. Does the potential difference between the terminals of the battery necessarily decrease or increase? Explain.

7. In calculating V_{ab} in Fig. 32-3a, is it permissible to follow a path from a to b that does not lie in the conducting circuit?

8. A 25-watt, 110-volt bulb glows at normal brightness when connected across a bank of batteries. A 500-watt, 110-volt bulb glows only dimly when connected across the same bank. Explain.

9. Under what circumstances can the terminal potential difference of a battery exceed its emf?

10. What is the difference between an emf and a potential difference?

11. Compare and contrast the formulae for the equivalent values of (a) capacitors and (b) resistors, in series and in parallel.

12. On what physical laws do (a) the loop theorem and (b) the junction theorem depend?

13. In Fig. 32-6 convince yourself that i_3 is drawn in the wrong direction. It was drawn so deliberately to make a point. What point?

14. Explain in your own words why the resistance of an ammeter should be very small whereas that of a voltmeter should be very large.

15. In a potentiometer, the internal resistance r of the emf plays no role. Why? See Fig. 32-9.

16. Does the time required for the charge on a capacitor in an RC circuit to build up to a given fraction of its equilibrium value depend on the value of the applied emf?

17. Devise a method whereby an RC circuit can be used to measure very high resistances.

18. In Fig. 32-10 suppose that switch S is closed on a. Explain why, in view of the fact that the negative terminal of the battery is not connected to resistance R, the initial current in R, at $t = 0$, should be ε/R, as Eq. 32-16 predicts.

19. In Fig. 32-10 suppose that switch S is closed on a. Why does the charge on capacitor C not rise instantaneously to $q = C\varepsilon$? After all, the positive battery terminal is connected to one plate of C and the negative terminal to the other.

20. Show that the product RC in Eqs. 32-15 and 32-16 has the dimensions of time, that is, that 1 second = 1 ohm × 1 farad.

21. Can you construct a water-flow analogy to capacitor discharge, using, for example, two burettes to simulate the capacitor plates and a long capillary tube to simulate the resistor? See "A Water Flow Analogy to Capacitor Discharge" by Ernest L. Madsen, *The Physics Teacher*, April 1976, for a full discussion.

problems

SECTION 32-1

1. A 5.0-A current is set up in an external circuit by a 6.0-V storage battery for 6.0 min. By how much is the chemical energy of the battery reduced?
Answer: 1.1×10^4 J.

2. A certain car battery (12 V) carries an initial charge Q (120 A·h). Assuming that the potential across the terminals stays constant until the battery is completely discharged, for how many hours can it deliver power P (100 W)?

SECTION 32-2

3. The current in a simple series circuit is 5.0 A. When an additional resistance of 2.0 Ω is inserted, the current drops to 4.0 A. What was the resistance of the original circuit? *Answer:* 8.0 Ω.

4. A battery of emf ε (2.0 V) and internal resistance r (1.0 Ω) is driving a motor. The motor is lifting a weight W (2.0 N) at constant speed v (0.50 m/s). Assuming no power losses, find (a) the current i in the circuit and (b) the potential difference V across the terminals of the motor.

SECTION 32-3

5. Thermal energy is to be generated in a 0.10-Ω resistor at the rate of 10 W by connecting it to a battery whose emf is 1.5 V. (a) What is the internal resistance of the battery? (b) What potential difference exists across the resistor? *Answer:* (a) 0.050 Ω. (b) 1.0 V.

6. A wire of resistance 5.0 Ω is connected to a battery whose emf ε is 2.0 V and whose internal resistance is 1.0 Ω. In 2.0 min (a) how much energy is transferred from chemical to electric form? (b) How much energy appears in the wire as thermal energy? (c) Account for the difference between (a) and (b).

7. (a) In the circuit of Fig. 32-3a show that the power delivered to R as thermal energy is a maximum when R is equal to the internal resistance r of the battery. (b) Show that this maximum power is $P = \varepsilon^2/4r$.

8. (a) In Fig. 32-14 what value must R have if the current in the circuit is to be 0.0010 A? Take $\varepsilon_1 = 2.0$ V, $\varepsilon_2 = 3.0$ V, and $r_1 = r_2 = 3.0$ Ω. (b) What is the rate of thermal energy transfer in R?

figure 32-14
Problem 8

9. In Fig. 32-3a put $\varepsilon = 2.0$ V and $r = 100$ Ω. Plot (a) the current and (b) the potential difference across R, as functions of R over the range 0 to 500 Ω. Make both plots on the same graph. (c) Make a third plot by multiplying together, for each value of R, the two curves plotted. What is the physical significance of this plot?

10. You are given a number of 10-Ω resistors, each capable of dissipating only 1.0 W. What is the minimum number of such resistors that you need to combine in series or parallel combinations to make a 10-Ω resistor capable of dissipating 5.0 W?

SECTION 32-4

11. In Fig. 32-5(a) calculate the potential difference between a and c by considering a path that contains R and ε_2.
Answer: 2.25 V, as expected. See Example 3.

12. In Example 2 an ammeter whose resistance is 0.050 Ω is inserted in the circuit. What percent change in the current results because of the presence of the meter?

13. The section of circuit AB (see Fig. 32-15) absorbs power $P = 50.0$ W and a current $i = 1.0$ A passes through it in the indicated direction. (a) What is the potential difference between A and B? (b) If the element C does not have internal resistance, what is its emf? (c) What is its polarity?
Answer: (a) 50 V. (b) 48 V. (c) B is the negative terminal.

figure 32-15
Problem 13

14. Two batteries having the same emf ε but different internal resistances r_1 and r_2 are connected in series to an external resistance R. Find the value of R that makes the potential difference zero between the terminals of the first battery.

figure 32-16

Problem 19

SECTION 32-5

15. Two light bulbs, one of resistance R_1 and the other of resistance R_2 ($<R_1$) are connected (a) in parallel and (b) in series. Which bulb is brighter?
    Answer: (a) R_2. (b) R_1.

16. Two batteries of emf ε and internal resistance r are connected in parallel across a resistor R, as in Fig. 32-18b. (a) For what value of R is the thermal energy delivered to the resistor a maximum? (b) What is the maximum energy dissipation rate?

17. In Fig. 32-6 calculate the potential difference between points c and d by as many paths as possible. Assume that $\varepsilon_1 = 4.0$ V, $\varepsilon_2 = 1.0$ V, $R_1 = R_2 = 10\ \Omega$, and $R_3 = 5\ \Omega$.   *Answer: $V_d - V_c = +0.25$ V, by all paths.*

18. Two resistors, R_1 and R_2, may be connected either in series or parallel across a (resistanceless) battery with emf ε. We desire the thermal energy transfer rate for the parallel combination to be five times that for the series combination. If $R_1 = 100\ \Omega$, what is R_2?

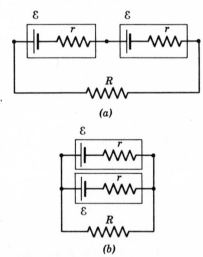

figure 32-17

Problem 23

19. (a) In Fig. 32-16 what is the equivalent resistance of the network shown? (b) What are the currents in each resistor? Put $R_1 = 100\ \Omega$, $R_2 = R_3 = 50\ \Omega$, $R_4 = 75\ \Omega$, and $\varepsilon = 6.0$ V.
    Answer: (a) 120 Ω; note that R_2, R_3, and R_4 are in parallel. (b) $i_1 = 50$ mA; $i_2 = i_3 = 20$ mA; $i_4 = 13$ mA.

20. A copper wire of radius a (0.25 mm) has an aluminum jacket of outside radius b (0.38 mm). (a) If there is a current i (2.0 A) in the wire, find the current in each material. (b) What is the wire length if potential difference V (12 V) maintains the current?

21. By using only two resistance coils—singly, in series, or in parallel—you are able to obtain resistances of 3, 4, 12, and 16 Ω. What are the separate resistances of the coils?   *Answer: 4.0 Ω and 12 Ω.*

22. Four 100-W heating coils are to be connected in all possible series-parallel combinations and plugged into a 100-V line. What different rates of Joule (thermal) energy dissipation are possible?

23. (a) In Fig. 32-17 what power appears as thermal energy in R_1? In R_2? In R_3? (b) What power is supplied by ε_1? By ε_2? (c) Discuss the energy balance in this circuit. Assume that $\varepsilon_1 = 3.0$ V, $\varepsilon_2 = 1.0$ V, $R_1 = 5.0\ \Omega$, $R_2 = 2.0\ \Omega$, and $R_3 = 4.0\ \Omega$.   *Answer: (a) 0.35 W; 0.050 W; 0.71 W. (b) 1.3 W; -0.16 W.*

figure 32-18

Problems 16, 24

24. You are given two batteries of emf ε and internal resistance r. They may be connected either in series or in parallel and are used to establish a current in a resistor R, as in Fig. 32-18. (a) Derive expressions for the current in R for both methods of connection. (b) Which connection yields the larger current if $R > r$ and if $R < r$?

25. In Fig. 32-19 find the current in each resistor and the potential difference between a and b. Put $\varepsilon_1 = 6.0$ V, $\varepsilon_2 = 5.0$ V, $\varepsilon_3 = 4.0$ V, $R_1 = 100\ \Omega$, and $R_2 = 50\ \Omega$.   *Answer: $i_1 = 50$ mA, to the right; $i_2 = 60$ mA, down; $V_{ab} = 9.0$ V.*

26. What is the equivalent resistance between the terminal points x and y of the circuits shown in (a) Fig. 32-20a, (b) Fig. 32-20b, and (c) Fig. 32-20c? Assume that the resistance of each resistor is 10 Ω. Do you detect any similarities between Figs. 32-20a and c?

27. (a) Find the three currents in Fig. 32-21. (b) Find V_{ab}. Assume that $R_1 = 1.0\ \Omega$, $R_2 = 2.0\ \Omega$, $\varepsilon_1 = 2.0$ V, and $\varepsilon_2 = \varepsilon_3 = 4.0$ V.
    Answer: (a) Left branch, 0.67 A, down; center branch, 0.33 A, up; right branch, 0.33 A, up. (b) 3.3 V.

28. What current, in terms of ε and R, does the meter M in Fig. 32-22 read? Assume that M has zero resistance.

figure 32-19

Problem 25

(a)　　　　　(b)　　　　　(c)

figure 32-20
Problem 26

713 PROBLEMS CHAP. 32

figure 32-21
Problem 27

figure 32-22
Problem 28

29. (a) Two batteries of emf and internal resistance ε_1, r_1 and ε_2, r_2, are connected in parallel. Show that the effective emf of this parallel combination is

$$\varepsilon = r\left(\frac{\varepsilon_1}{r_1} + \frac{\varepsilon_2}{r_2}\right)$$

where r is defined from

$$\frac{1}{r} = \frac{1}{r_1} + \frac{1}{r_2}.$$

See "Batteries Connected in Parallel" by J. S. Wallingford and H. W. Jones, *American Journal of Physics*, **36**, 639, (1968) for an extension of this problem to an indefinite number of batteries.

30. N identical batteries of emf ε and internal resistance r may be connected all in series or all in parallel. Show that each arrangement will give the same current in an external resistor R, if $R = r$.

31. *The Wheatstone bridge.* In Fig. 32-23 R_s is to be adjusted in value until points a and b are brought to exactly the same potential. (One tests for this condition by momentarily connecting a sensitive meter between a and b; if these points are at the same potential, the meter will not deflect.) Show that when this adjustment is made, the following relation holds:

$$R_x = R_s \frac{R_2}{R_1}.$$

Unknown resistors (R_x) can be measured in terms of standards (R_s) using this device, which is called a Wheatstone bridge.

32. If points a and b in Fig. 32-23 are connected by a wire of resistance r, show that the current in the wire is

$$i = \frac{\varepsilon(R_s - R_x)}{(R + 2r)(R_s + R_x) + 2R_sR_x},$$

where ε is the emf of the battery. Assume that R_1 and R_2 are equal ($R_1 = R_2 = R$) and that R_0 equals zero. Is this formula consistent with the result of Problem 31?

figure 32-23
Problem 31

figure 32-24
Problem 33

33. For manual control of the current in a circuit, you can use a parallel combination of variable resistors of the sliding contact type, as in Fig. 32-24,
with $R_1 = 20R_2$. (a) What procedure is used to adjust the current to the desired value? (b) Why is the parallel combination better than a single-variable
resistor? (c) Can you extend this technique to three resistors in parallel?
Answer: (a) Put R_1 roughly in the middle of its range; adjust current roughly
with R_2; make fine adjustment with R_1. (b) Relatively large percentage changes in R_1 cause only small percentage changes in the
resistance of the parallel combination, thus permitting fine adjustment. The ratio is 1:21.

figure 32-25
Problem 34

34. In Fig. 32-25 imagine an ammeter inserted in the branch containing R_3.
(a) What will it read, assuming $\varepsilon = 5.0$ V, $R_1 = 2.0$ Ω, $R_2 = 4.0$ Ω, and $R_3 =
6.0$ Ω? (b) The ammeter and the source of emf are now physically inter-
changed. Show that the ammeter reading remains unchanged.

35. Twelve resistors, each of resistance R ohms, form a cube (see Fig. 32-26).
(a) Find R_{AB}, the equivalent resistance of an edge. (b) Find R_{BC}, the equivalent resistance of a face diagonal. (c) Find R_{AC}, the equivalent resistance of
a body diagonal.   *Answer:* (a) $\frac{7}{12}R$. (b) $\frac{3}{4}R$. (c) $\frac{5}{6}R$.

SECTION 32-6

36. A voltmeter and an ammeter are used to determine two unknown resistances R_1 and R_2, one determination by each of the two methods shown in
Fig. 32-27. The voltmeter resistance is 307 Ω and the ammeter resistance
is 3.62 Ω; in method (a) the ammeter reads 0.317 A and the voltmeter reads
28.1 V, whereas in method (b) the ammeter reads 0.356 A and the voltmeter
23.7 V. Compute R_1 and R_2.

figure 32-26
Problem 35

figure 32-27
Problem 36

(a)   (b)

37. *Resistance measurement.* A voltmeter (resistance R_V) and an ammeter (resistance R_A) are connected to measure a resistance R, as in Fig. 32-28a. The
resistance is given by $R = V/i$, where V is the voltmeter reading and i is the
current *in the resistor R*. Some of the current registered by the ammeter
(i') goes through the voltmeter so that the ratio of the meter readings ($= V/i'$)

figure 32-28
Problems 37, 38

715 PROBLEMS CHAP. 32

(a)   (b)

gives only an *apparent* resistance reading R'. Show that R and R' are related by

$$\frac{1}{R} = \frac{1}{R'} - \frac{1}{R_V}.$$

Note that if $R_V \gg R$, then $R \cong R'$.

38. *Resistance measurement.* If meters are used to measure resistance, they may also be connected as they are in Fig. 32-28b. Again the ratio of the meter readings gives only an apparent resistance R'. Show that R' is related to R by

$$R = R' - R_A,$$

in which R_A is the ammeter resistance. Note that if $R_A \ll R$, then $R \cong R'$.

39. In Fig. 32-8 assume that $\varepsilon = 5.0$ V, $r = 2.0\ \Omega$, $R_1 = 5.0\ \Omega$, and $R_2 = 4.0\ \Omega$. If $R_A = 0.10\ \Omega$, what percent error is made in reading the current? Assume that the voltmeter is not present.   *Answer: 0.9%.*

40. In Fig. 32-8 assume that $\varepsilon = 5.0$ V, $r = 20\ \Omega$, $R_1 = 50\ \Omega$, and $R_2 = 40\ \Omega$. If $R_V = 1000\ \Omega$, what percent error is made in reading the potential differences across R_1? Ignore the presence of the ammeter.

SECTION 32-8

41. How many time constants must elapse before a capacitor in an RC circuit is charged to within 1.0 percent of its equilibrium charge?   *Answer: 4.6.*

42. A 10,000-Ω resistor and a capacitor are connected in series and a 10-V potential is suddenly applied. If the potential across the capacitor rises to 5.0 V in 1.0 μs, what is the capacitance of the capacitor?

43. The potential difference between the plates of a leaky capacitor C $(2.0\ \mu F)$ drops from V_0 to V $(\frac{1}{4}V_0)$ in time t $(2.0$ s$)$. What is the equivalent resistance between the capacitor plates?   *Answer: $7.2 \times 10^5\ \Omega$.*

44. An RC circuit is discharged by closing a switch at time $t = 0$. The initial potential difference across the capacitor is 100 V. If the potential difference has decreased to 1.0 V after 10 s, (a) what will the potential difference be 20 s after $t = 0$? (b) What is the time constant of the circuit?

45. Prove that when switch S in Fig. 32-10 is thrown from a to b all the energy stored in the capacitor is transformed into thermal energy in the resistor. Assume that the capacitor is fully charged before the switch is thrown.

46. A capacitor with capacitance $C = 1.0\ \mu$F and initial stored energy $U_0 = 0.50$ J is discharged through a resistance $R = 1.0 \times 10^6\ \Omega$. (a) What is the initial charge on the capacitor? (b) What is the current through the resistor when the discharge starts? (c) Determine V_C, the voltage across the capacitor, and V_R, the voltage across the resistor, as a function of time. (d) Express the rate of generation of thermal energy in the resistor as a function of time.

47. A 3.0×10^6-Ω resistor and a 1.0-μF capacitor are connected in a single-loop circuit with a seat of emf with $\varepsilon = 4.0$ V. At 1.0 s after the connection is made, what are the rates at which (a) the charge of the capacitor is increasing, (b) energy is being stored in the capacitor, (c) thermal energy is appearing in the resistor, and (d) energy is being delivered by the seat of emf? *Answer: (a) 9.6×10^{-7} C/s. (b) 1.1×10^{-6} W. (c) 2.8×10^{-6} W. (d) 3.8×10^{-6} W.*

figure 32-29
Problem 48

48. In the circuit of Fig. 32-29 let i_1, i_2, and i_3 be the currents through resistors R_1, R_2, and R_3, respectively, and let V_1, V_2, V_3, and V_C be the corresponding potential differences across the resistors and across the capacitor C. (a) Plot qualitatively, as a function of time after switch S is closed, the currents and voltages listed above. (b) After being closed for a large number of time constants, the switch S is now opened. Plot qualitatively, as a function of time after the switch is opened, the currents and voltages listed above.

49. In the circuit of Fig. 32-30, $\varepsilon = 1200$ V, $C = 6.50$ μF, $R_1 = R_2 = R_3 = 7.30 \times 10^5$ Ω. With C completely uncharged, the switch S is suddenly closed $(t = 0)$. (a) Determine the currents through each resistor for $t = 0$ and $t = \infty$. (b) Draw qualitatively a graph of the potential drop V_2 across R_2 from $t = 0$ to $t = \infty$. (c) What are the numerical values of V_2 at $t = 0$ and $t = \infty$? (d) Give the physical meaning of "$t = \infty$" and state a rough, but significant, numerical lower bound, in seconds, for "$t = \infty$" in this case.

Answer: (a) At $t = 0$, $i_1 = 1.1$ mA, $i_2 = i_3 = 0.55$ mA; at $t = \infty$, $i_1 = i_2 = 0.82$ mA, $i_3 = 0$. (c) At $t = 0$, $V_2 = 400$ V; at $t = \infty$, $V_2 = 600$ V. (d) The time constant is 7.1 s.

figure 32-30
Problem 49

33
the magnetic field

The science of magnetism grew from the observation that certain "stones" (magnetite) would attract bits of iron. The word *magnetism* comes from the district of Magnesia in Asia Minor, which is one of the places at which the stones were found. Figure 33-1 shows a modern permanent magnet, the lineal descendant of these natural magnets. Another "natural magnet" is the earth itself, whose orienting action on a magnetic compass needle has been known since ancient times.

In 1820 Oersted discovered that a current in a wire can also produce magnetic effects, namely, that it can change the orientation of a compass needle.* We pointed out in Section 26-1 how this important discovery linked the then separate sciences of magnetism and electricity. We can intensify the magnetic effect of a current in a wire by forming the wire into a coil of many turns and by providing an iron core. Figure 33-2 shows how this is done in a large electromagnet of a type commonly used for research involving magnetism.

We define the space around a magnet or a current-carrying conductor as the site of a *magnetic field,* just as we defined the space near a charged rod as the site of an electric field. The basic magnetic field vector **B,** which we define in the following section, is called the *magnetic induc-*

33-1
THE MAGNETIC FIELD

---

*In 1878 H. A. Rowland, at the Johns Hopkins University, discovered that a moving charged object (in his case, a rapidly rotating charged disk) also causes magnetic effects. Actually, it is far from obvious that a moving charge is equivalent to a current in a wire. See. "Rowland's Physics" by John D. Miller, *Physics Today,* July 1976 for a scientific biography of this outstanding American physicist, including a detailed discussion of this experiment.

figure 33-1
A permanent magnet. Lines of
magnetic induction leave the north
pole face, marked N, and enter the
south pole face on the other side
of the air gap.

figure 33-2
A research-type electromagnet
showing iron frame F, pole faces P,
and coils C. The pole faces are 12 in.
in diameter. (Courtesy Varian
Associates.)

tion.* We can represent it by *lines of induction,* just as we represent
the electric field by lines of force. As for the electric field (see Section
27-3), the magnetic field vector is related to its lines of induction in this
way:

1. The tangent to a line of induction at any point gives the *direction* of
   B at that point.
2. The lines of induction are drawn so that the number of lines per unit
   cross-sectional area (perpendicular to the lines) is proportional to the
   magnitude of **B.** Where the lines are close together B is large and
   where they are far apart B is small.

As for the electric field, the magnetic field vector **B** is of fundamental
importance, the lines of induction simply giving a graphic representation of the way **B** varies throughout a certain region of space. The *flux*
Φ_B for a magnetic field can be defined in exact analogy with the flux Φ_E

for the electric field, namely $\Phi_B = \int \mathbf{B} \cdot d\mathbf{S}$, in which the integral is taken

over the surface (closed or open) for which Φ_B is defined.

33-2
THE DEFINITION† OF **B**

In this chapter we are not concerned with the *causes* of the magnetic
field; we will explore them in the next chapter. Our related concerns
here are (a) to *define* operationally the magnitude and direction of **B** at
any point P near, say, a magnet, a current-carrying conductor, or a
moving charge and (b) to determine the *effect* of a magnetic field on
objects, such as moving charges, subject to its influence.

As we did for the electric field, let us consider a charge q_0 as a test
body. Let us place the test body *at rest* at a point P near, say, a perma-

---

* *Magnetic field strength* would be a more suitable name for **B**, but it has been usurped for
historical reasons by another vector connected with the magnetic field (see Section 37-8).
Often in this book we shall call **B**, for brevity, the "magnetic field" in analogy to calling **E**
the "electric field".
† The definition and method of measurement of **B** given in this section, though conceptually sound and suitable for our present purpose, are not carried out in practice because
of experimental difficulties. The following section will show how **B** may be more conveniently measured in the laboratory.

nent magnet such as that of Fig. 33-1. We would then find by experiment that *no force* (attributable *only* to the presence or absence of the magnet) acts on q_0. However, if we fire test body q_0 through point P with a velocity **v**, we find that a *sideways force* **F** acts on it if the magnet is in place; by a sideways force we mean a force at right angles to **v**. We shall define **B** at point P in terms of **F**, **v**, and q_0.

If we vary the direction of **v** through point P, keeping the magnitude of **v** unchanged, we find that although **F** will always remain at right angles to **v**, its magnitude F will change. For a particular orientation of **v** (and also for the opposite orientation −**v**) the force **F** becomes zero. We define this orientation as the *direction* of **B**, the specification of the *sense* of **B** (that is, the way it points along this line) being left to the more complete definition of **B** that we give below.

Having found the direction of **B**, we are now able to orient **v** so that the test charge moves at right angles to **B**. We will find that the force **F** is now a maximum, and we define the *magnitude* of **B** from the measured magnitude of this maximum force $F_\perp$, or

$$B = \frac{F_\perp}{q_0 v}. \tag{33-1}$$

Let us regard this definition of **B** (in which we have specified its magnitude and direction, but not its sense) as preliminary to the complete vector definition that we now give: *If we fire a positive test charge q_0 with velocity **v** through a point P and if a force **F** acts on the moving charge, a magnetic field **B** is present at point P, where **B** is the vector that satisfies the relation*

$$\mathbf{F} = q_0 \mathbf{v} \times \mathbf{B}, \tag{33-2}$$

v, q_0, and **F** being measured quantities. The magnitude of the magnetic deflecting force **F**, according to the rules for vector products, is given by *

$$F = q_0 v B \sin \theta, \tag{33-3}$$

in which θ is the angle between **v** and **B**.

Figure 33-3 shows the relations among the vectors. We see from Eq. 33-2 that **F**, being at right angles to the plane formed by **v** and **B**, will always be at right angles to **v** (and also to **B**) and thus will always be a sideways force. Equation 33-2 is consistent with the observed facts that (*a*) the magnetic force vanishes as $v \rightarrow 0$, (*b*) the magnetic force vanishes if **v** is either parallel or antiparallel to the direction of **B** (in these cases $\theta = 0$ or 180° and $\mathbf{v} \times \mathbf{B} = 0$), and (*c*) if **v** is at right angles to **B** ($\theta = 90°$), the deflecting force has its maximum value, given by Eq. 33-1, that is, $q_0 v B$.

This definition of **B** is similar in spirit, although more complex, than the definition of the electric field **E**, which we can cast into this form: *If we place a positive test charge q_0 at point P and if an (electric) force **F** acts on the stationary charge, an electric field **E** is present at P, where **E** is the vector satisfying the relation*

$$\mathbf{F} = q_0 \mathbf{E},$$

q_0 and **F** being measured quantities. In defining **E**, the only characteristic direction to appear is that of the electric force $\mathbf{F}_E$ which acts on the positive test body; the direction of **E** is taken to be that of $\mathbf{F}_E$. In defining **B**, *two* characteristic directions appear, those of **v** and of the magnetic force $\mathbf{F}_B$; they prove always to be at right angles. Although we can easily solve the above equation for **E**, we cannot solve Eq. 33-2 for **B**. Why not?

* You may wish to review Section 2-4, which deals with vector products.

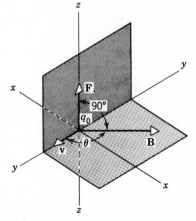

figure 33-3
Illustrating $\mathbf{F} = q_0 \mathbf{v} \times \mathbf{B}$ (Eq. 33-2). Test charge q_0 is fired through the origin with velocity **v**.

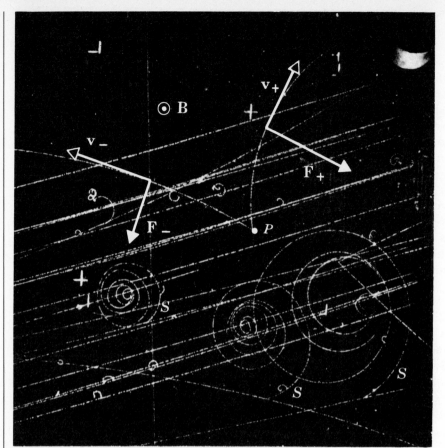

figure 33-4
A *bubble chamber* is a device for rendering visible, by means of small bubbles, the tracks of charged particles that pass through the chamber. The figure is a photograph taken with such a chamber immersed in a magnetic field **B** and exposed to radiations from a large cyclotron-like accelerator. The curved *V* at point *P* is formed by a positive and a negative electron, which deflect in opposite directions in the magnetic field. The spirals *S* are the tracks of three low-energy electrons. (Courtesy E. O. Lawrence Radiation Laboratory, University of California.)

In Fig. 33-4 a positive and a negative electron are created at point *P* in a bubble chamber. A magnetic field is perpendicular to the chamber, pointing out of the plane of the figure (symbol ⊙).* The relation $\mathbf{F} = q_0\mathbf{v} \times \mathbf{B}$ (Eq. 33-2) shows that the deflecting forces acting on the two particles are as indicated in the figure. These deflecting forces would make the tracks deflect as shown.

The unit of **B** that follows from Eq. 33-3 is the newton/(coulomb) (meter/second). This is given the SI name *tesla* (abbr. T) or weber/meter² (abbr. Wb/m²).** Recalling that a coulomb/second is an ampere, we have

$$1 \text{ tesla} = 1 \text{ weber/meter}^2 = 1 \text{ newton/(ampere} \cdot \text{meter).}$$

An earlier unit for **B**, still in common use, is the *gauss;* the relationship is

$$1 \text{ tesla} = 1 \text{ weber/meter}^2 = 10^4 \text{ gauss.}†$$

The fact that the magnetic force is always at right angles to the direction of motion means that (for steady magnetic fields at least) the work done by this force on the particle is zero. For an element of the path of the particle of length $d\mathbf{l}$, this work dW is $\mathbf{F}_B \cdot d\mathbf{l}$; dW is zero because $\mathbf{F}_B$ and $d\mathbf{l}$ are always at right angles. Thus a static magnetic field cannot

---

* The symbol ⊗ indicates a vector into the page, the × being thought of as the tail of an arrow; the symbol ⊙ indicates a vector out of the page, the dot being thought of as the tip of an arrow.

** The weber is used to measure the magnetic flux, a concept we discuss in succeeding chapters. See also Sec. 33-1.

† See "Megagauss Physics" by C. M. Fowler, *Science,* April 1973 for a fascinating account of the properties of such enormous fields.

change the kinetic energy of a moving charge; it can only deflect it sideways.*

If a charged particle moves through a region in which both an electric field and a magnetic field are present, the resultant force is found by combining Eqs. 27-2 and 33-2, or

$$\mathbf{F} = q_0\mathbf{E} + q_0\mathbf{v} \times \mathbf{B}. \qquad (33\text{-}4)$$

This is sometimes called the *Lorentz equation* in tribute to H. A. Lorentz who did so much to develop and clarify the concepts of the electric and magnetic fields.

EXAMPLE 1

A uniform magnetic field **B** points horizontally from south to north; its magnitude is 1.5 T. If a 5.0-MeV proton moves vertically downward through this field, what force will act on it?

The kinetic energy of the proton is

$$K = (5.0 \times 10^6 \text{ eV})(1.6 \times 10^{-19} \text{ J/eV}) = 8.0 \times 10^{-13} \text{ J}.$$

We can find its speed from the relation $K = \frac{1}{2}mv^2$, or

$$v = \sqrt{\frac{2K}{m}} = \sqrt{\frac{(2)(8.0 \times 10^{-13} \text{ J})}{1.7 \times 10^{-27} \text{ kg}}} = 3.1 \times 10^7 \text{ m/s}.$$

Equation 33-3 gives

$$F = qvB \sin\theta = (1.6 \times 10^{-19} \text{ C})(3.1 \times 10^7 \text{ m/s})(1.5 \text{ T})(\sin 90°)$$

$$= 7.4 \times 10^{-12} \text{ N}.$$

You can show that this force is about 4×10^{14} times greater than the weight of the proton.

The relation $\mathbf{F} = q\mathbf{v} \times \mathbf{B}$ shows that the *direction* of the deflecting force is to the east. If the particle had been negatively charged, the deflection would have been to the west. This is predicted automatically by Eq. 33-2 if we substitute $-e$ for q_0.

A current is an assembly of moving charges. Because a magnetic field exerts a sideways force on a moving charge, we expect that it will also exert a sideways force on a wire carrying a current. Figure 33-5 shows a length *l* of wire carrying a current *i* and placed in a magnetic field **B**. For simplicity we have oriented the wire so that the current density vector **j** is at right angles to **B**.

33-3
MAGNETIC FORCE ON A CURRENT

figure 33-5
A wire carrying a current *i* is placed at right angles to a magnetic field **B**. Only the drift velocity of the electrons, not their random motion, is suggested.

---

* Actually, $\mathbf{F}_B$ is at right angles to **v**, *even if* **B** varies with time. In this case, however, the theory of special relativity predicts that an *electric* field **E** will appear and act on the charged particle in such a way as to do work on it. See *Introduction to Special Relativity* by Robert Resnick, John Wiley & Sons, 1968 (Section 4.2) for details.

The current i in a metal wire is carried by the free (or conduction) electrons, n being the number of such electrons per unit volume of the wire. The magnitude of the average force on one such electron is given by Eq. 33-3, or, since $\theta = 90°$,

$$F' = q_0 v B \sin \theta = e v_d B$$

where v_d is the drift speed. From the relation $v_d = j/ne$ (Eq. 31-5),

$$F' = e\left(\frac{j}{ne}\right) B = \frac{jB}{n}.$$

The length l of the wire contains nAl free electrons, Al being the volume of the section of wire of cross section A that we are considering. The total force on the free electrons in the wire, and thus on the wire itself, is

$$F = (nAl)F' = nAl\,\frac{jB}{n}.$$

Since jA is the current i in the wire, we have

$$F = ilB. \tag{33-5}$$

The negative charges, which move to the right in the wire of Fig. 33-5, are equivalent to positive charges moving to the left, that is, in the direction of the current arrow. For such a positive charge the velocity $\mathbf{v}$ would point to the left and the force on the wire, given by Eq. 33-2 ($\mathbf{F} = q_0\mathbf{v} \times \mathbf{B}$) points up, out of the page. This same conclusion follows if we consider the actual negative charge carriers for which $\mathbf{v}$ points to the right but q_0 has a negative sign. Thus by measuring the sideways magnetic force on a wire carrying a current and placed in a magnetic field, we cannot tell whether the current carriers are negative charges moving in a given direction or positive charges moving in the opposite direction.

Equation 33-5 holds only if the wire is at right angles to $\mathbf{B}$. We can express the more general situation in vector form as

$$\mathbf{F} = i\mathbf{l} \times \mathbf{B}, \tag{33-6a}$$

where $\mathbf{l}$ is a vector whose magnitude is the length of the wire and which points along the (straight) wire in the direction of the current. Equation 33-6a is equivalent to the relation $\mathbf{F} = q_0\mathbf{v} \times \mathbf{B}$ (Eq. 33-2); either can be taken as a defining equation for $\mathbf{B}$. Note that the vector $\mathbf{l}$ in Fig. 33-5 points to the left and that the magnetic force $\mathbf{F}$ ($= i\mathbf{l} \times \mathbf{B}$) points up, out of the page. This agrees with the conclusion obtained by analyzing the forces that act on the individual charge carriers.

If we consider a differential element of a conductor of length $d\mathbf{l}$, we can find the force $d\mathbf{F}$ acting on it, by analogy with Eq. 33-6a, from

$$d\mathbf{F} = i\,d\mathbf{l} \times \mathbf{B}. \tag{33-6b}$$

By integrating this formula in an appropriate way we can find the force $\mathbf{F}$ on a conductor which is not straight.

EXAMPLE 2

A wire bent as shown in Fig. 33-6 carries a current i and is placed in a uniform magnetic field $\mathbf{B}$ that emerges from the plane of the figure. Calculate the force acting on the wire. The magnetic field is represented by lines of induction, shown emerging from the page. The dots show that the sense of $\mathbf{B}$ is up, out of the page.

figure 33-6
Example 2

The force on each straight section, from Eq. 33-6a, has the magnitude

$$F_1 = F_3 = ilB$$

and points down as shown by the arrows in the figure.

A segment of wire of length dl on the arc has a force $d\mathbf{F}$ on it whose magnitude is

$$dF = iB \; dl = iB(R \; d\theta)$$

and whose direction is radially toward O, the center of the arc. Only the downward component of this force is effective, the horizontal component being canceled by an oppositely directed component associated with the corresponding arc segment on the other side of O. Thus the total force on the semicircle of wire about O points down and is

$$F_2 = \int_0^\pi dF \sin \theta = \int_0^\pi (iBR \; d\theta) \sin \theta = iBR \int_0^\pi \sin \theta \; d\theta = 2iBR.$$

The resultant force on the whole wire is

$$F = F_1 + F_2 + F_3 = 2ilB + 2iBR = 2iB(l + R).$$

Notice that this force is the same as that acting on a straight wire of length $2l + 2R$.

Figure 33-7 shows the arrangement used by Thomas, Driscoll, and Hipple in 1949 at the National Bureau of Standards to measure the magnetic field provided by a laboratory magnet such as that of Fig. 33-2. The rectangle is a coil of nine turns whose width a and length b are about 10 cm and 70 cm, respectively. The lower end of the coil is placed in the magnetic field $\mathbf{B}$ and the upper end is hung from the arm of a balance; $\mathbf{B}$ enters the plane of the figure at right angles.

An accurately known current i of about 0.10 A is set up in the coil in the direction shown, and weights are placed in the right-hand balance pan until the system is balanced. The magnetic force $\mathbf{F}$ ($= i\mathbf{l} \times \mathbf{B}$; see Eq. 33-6a) on the bottom leg of the coil points upward, as shown in the figure. Equation 33-5 also shows that the force on each wire at the bottom of the coil is iaB. Since there are nine wires, the total force on the bottom leg of the coil is $9iaB$. The forces on the vertical sides of the coil ($= i\mathbf{l} \times \mathbf{B}$) are sideways; because they are equal and opposite, they cancel and produce no effect.

After balancing the system, the experimenters reversed the direction of the current, which changed the sign of all the magnetic forces acting on the coil. In particular, $\mathbf{F}$ then pointed downward, which caused the rest point of the bal-

figure 33-7
An apparatus used to measure **B**.
The zero point of the balance is
observed by means of a light beam
reflected from the mirror attached
to the balance beam.

ance to move. A mass m of about 8.78 g had to be added to the left balance pan
to restore the original rest point. The *change* in force when the current is reversed is $2F$, and this must equal the weight added to the left balance pan, or

$$mg = 2(9iaB) = 18iaB.$$

This gives

$$B = \frac{mg}{18ai} = \frac{(8.78 \times 10^{-3}\ \text{kg})(9.80\ \text{m/s}^2)}{(18)(0.10\ \text{m})(0.10\ \text{A})} = 0.48\ \text{T} = 4800\ \text{gauss}.$$

The Bureau of Standards' workers made this measurement with much more
precision than these approximate numbers suggest. In one series of measurements, for example, they found a magnetic field of 4697.55 gauss; even greater
precision is possible today, by this and other methods.

Figure 33-8 shows a rectangular loop of wire of length a and width b placed in a uniform magnetic field **B**, with sides 1 and 3 always normal to the field direction. The normal nn' to the plane of the loop makes an angle θ with the direction of **B**.

Assume the current to be as shown in the figure. Wires must be provided to lead the current into the loop and out of it. If these wires are twisted tightly together, there will be no net magnetic force on the twisted pair because the currents in the two wires are in opposite directions. Thus the lead wires may be ignored. Also, some way of supporting the loop must be provided. Let us imagine it to be suspended from a long string attached to the loop at its center of mass. In this way the loop will be free to turn, through a small angle at least, about any axis through the center of mass. Alternatively, we can imagine the experiment to be carried out in an orbiting earth satellite (Skylab, say) in which the effective value of **g** is zero.

The net force on the loop is the resultant of the forces on the four sides of the loop. On side 2 the vector **l** points in the direction of the current and has the magnitude b. The angle between **l** and **B** for side 2 (see Fig. 33-8b) is $90° - \theta$. Thus the magnitude of the force on this side is

$$F_2 = ibB \sin(90° - \theta) = ibB \cos\theta.$$

From the relation $\mathbf{F} = i\mathbf{l} \times \mathbf{B}$ (Eq. 33-6a), we find the direction of $\mathbf{F}_2$ to be out of the plane of Fig. 33-8b. You can show that the force $\mathbf{F}_4$ on side 4 has the same magnitude as $\mathbf{F}_2$ but points in the opposite direction. Thus $\mathbf{F}_2$ and $\mathbf{F}_4$, taken together, have no effect on the motion of the loop. The net force they provide is zero, and, since they have the same line of action, the net torque due to these forces is also zero.

The common magnitude of $\mathbf{F}_1$ and $\mathbf{F}_3$ is iaB. These forces, too, are oppositely directed so that they do not tend to move the coil bodily. As Fig. 33-8b shows, however, they do *not* have the same line of action if the coil is in the position shown; there is a net torque, which tends to rotate the coil clockwise about the line xx'. The coil can be supported on a rigid axis that lies along xx', with no loss of its freedom of motion. This torque can be represented in Fig. 33-8b by a vector pointing into the figure at point x' or in Fig. 33-8a by a vector pointing along the xx' axis from right to left.

The magnitude of the torque τ' is found by calculating the torque caused by $\mathbf{F}_1$ about axis xx' and doubling it, for $\mathbf{F}_3$ exerts the same torque about this axis that $\mathbf{F}_1$ does. Thus

(a)

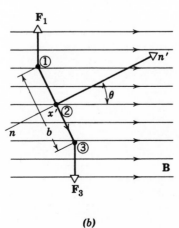

(b)

figure 33-8
A rectangular coil carrying a current i is placed in a uniform external magnetic field.

$$\tau' = 2(iaB)\left(\frac{b}{2}\right)(\sin\theta) = iabB\sin\theta.$$

This torque acts on every turn of the coil. If there are N turns, the torque on the entire coil is

$$\tau = N\tau' = NiabB\sin\theta = NiAB\sin\theta, \tag{33-7}$$

in which A, the area of the coil, is substituted for ab.

This equation can be shown to hold for *all plane loops of area A, whether they are rectangular or not.* A torque on a current loop is the basic operating principle of the electric motor and of most electric meters used for measuring current or potential difference.

EXAMPLE 3

A galvanometer. Figure 33-9 shows the rudiments of a galvanometer, which is the basic operating mechanism of ammeters and voltmeters. The coil is 2.0 cm high and 1.0 cm wide; it has 250 turns and is mounted so that it can rotate about a vertical axis in a uniform *radial* magnetic field with $B = 2000$ gauss. A spring *Sp* provides a countertorque that cancels out the magnetic torque, resulting in a steady angular deflection ϕ corresponding to a given current i in the coil. If a current of 1.0×10^{-4} A produces an angular deflection of 30°, what is the *torsional constant* κ of the spring (see Eq. 15-21)?

Equating the magnetic torque to the torque caused by the spring (see Eq. 33-7) yields

$$\tau = NiAB\sin\theta = \kappa\phi$$

or

$$\kappa = \frac{NiAB\sin\theta}{\phi}$$

$$= \frac{(250)(1.0 \times 10^{-4}\text{ A})(2.0 \times 10^{-4}\text{ m}^2)(0.20\text{ T})(\sin 90°)}{30°}$$

$$= 3.3 \times 10^{-8}\text{ N}\cdot\text{m/degree.}$$

Note that the normal to the plane of the coil (that is, the pointer P) is always at right angles to the (radial) magnetic field so that $\theta = 90°$.

figure 33-9
Example 3. The elements of a galvanometer, showing the coil, the helical spring *Sp*, and pointer *P*.

A current loop orienting itself in an external magnetic field reminds us of the action of a compass needle in such a field. One face of the loop behaves like the north pole of the needle;* the other face behaves like the south pole. Compass needles, bar magnets, and current loops can all be regarded as *magnetic dipoles.* We show this here for the current loop, reasoning entirely by analogy with *electric* dipoles.

A structure is called an electric dipole if (a) when placed in an *external* electric field it experiences a torque given by Eq. 27-11,

$$\tau = \mathbf{p} \times \mathbf{E}, \tag{33-8}$$

where $\mathbf{p}$ is the electric dipole moment, and (b) it sets up a field of its own at distant points, described qualitatively by the lines of force of Fig. 29-10 and quantitatively by Eq. 29-11. These two requirements are not independent; if one is fulfilled, the other follows automatically.

The magnitude of the torque described by Eq. 33-8 is

$$\tau = pE \sin \theta, \tag{33-9}$$

where θ is the angle between $\mathbf{p}$ and $\mathbf{E}$. Let us compare this with Eq. 33-7, the expression for the torque on a current loop:

$$\tau = (NiA)B \sin \theta. \tag{33-7}$$

In each case the appropriate field (E or B) appears, as does a term $\sin \theta$. Comparison suggests that NiA in Eq. 33-7 can be taken as the *magnetic dipole moment* μ, corresponding to p in Eq. 33-9, or

$$\mu = NiA. \tag{33-10}$$

Equation 33-7 suggests that we write the torque on a current loop in vector form, in analogy with Eq. 33-8, or

$$\tau = \mu \times \mathbf{B}. \tag{33-11}$$

The magnetic dipole moment of the loop μ must be taken to lie along an axis perpendicular to the plane of the loop; its direction is given by the following rule: Let the fingers of the right hand curl around the loop in the direction of the current; the extended right thumb will then point in the direction of μ. If μ is defined by this rule and Eq. 33-10, check carefully that Eq. 33-11 correctly describes in every detail the torque acting on a current loop in an external field (see Fig. 33-8).

Since a torque acts on a current loop, or other magnetic dipole, when it is placed in an external magnetic field, it follows that work (positive or negative) must be done by an external agent to change the orientation of such a dipole. Thus a magnetic dipole has *potential energy* associated with its orientation in an external magnetic field. This energy may be taken to be zero for any arbitrary position of the dipole. By analogy with the assumption made for electric dipoles in Section 27-6, we assume that the magnetic energy U is zero when μ and $\mathbf{B}$ are at right angles, that is, when $\theta = 90°$. This choice of a zero-energy configuration for U is arbitrary because we are interested only in the *changes* in energy that occur when the dipole is rotated.

The magnetic potential energy in any position θ is defined as the work that an external agent must do to turn the dipole from its zero-energy position ($\theta = 90°$) to the given position θ. Thus

* The north pole of a compass needle is the end that points toward the geographic north.

$$U = \int_{90°}^{\theta} \tau \, d\theta = \int_{90°}^{\theta} NiAB \sin \theta \, d\theta = \mu B \int_{90°}^{\theta} \sin \theta \, d\theta = -\mu B \cos \theta,$$

in which Eq. 33-7 is used to substitute for τ. In vector symbolism we can write this relation as

$$U = -\boldsymbol{\mu} \cdot \mathbf{B}, \tag{33-12}$$

which is in perfect correspondence with Eq. 27-13, the expression for the energy of an *electric* dipole in an external *electric* field,

$$U = -\mathbf{p} \cdot \mathbf{E}.$$

EXAMPLE 4

A circular coil of N turns has an effective radius a and carries a current i. How much work is required to turn it in an external magnetic field $\mathbf{B}$ from a position in which θ equals zero to one in which θ equals 180°? Assume that $N = 100$, $a = 5.0$ cm, $i = 0.10$ A, and $B = 1.5$ T.

The work required is the difference in energy between the two positions, or, from Eq. 33-12,

$$W = U_{\theta=180°} - U_{\theta=0} = (-\mu B \cos 180°) - (-\mu B \cos 0) = 2\mu B.$$

But $\mu = NiA$, so that

$$W = 2NiAB = 2Ni(\pi a^2)B$$

$$= (2)(100)(0.10 \text{ A})(\pi)(5 \times 10^{-2} \text{ m})^2(1.5 \text{ T}) = 0.24 \text{ J}.$$

33-5
THE HALL EFFECT

In 1879 E. H. Hall, at Harvard University, reported an experiment that gives the sign of the charge carriers in a conductor; see p. 676. Figure 33-10 shows a flat strip of copper carrying a current i in the direction shown. As usual, the direction of the current arrow, labeled i, is the direction in which the charge carriers would move *if* they were positive. The current arrow can represent either positive charges moving down (as in Fig. 33-10a) or negative charges moving up (as in Fig. 33-10b). The Hall effect can be used to decide between these two possibilities.*

A magnetic field $\mathbf{B}$ is set up at right angles to the strip by placing the strip between the polefaces of an electromagnet. This field exerts a deflecting force $\mathbf{F}$ on the strip (given by $i\mathbf{l} \times \mathbf{B}$), which points to the right in the figure. Since the sideways force on the strip is due to the sideways forces on the charge carriers (given by $q\mathbf{v} \times \mathbf{B}$), it follows that these carriers, whether they are positive or negative, will tend to drift toward the right in Fig. 33-10 as they drift along the strip, producing a *transverse Hall potential difference* V_{xy} between points such as x and y. The sign of the charge carriers is determined by the sign of this Hall poten-

---

* The connection between the eminent but self-effacing physicist H. A. Rowland and the Hall effect has often been glossed over. From the reference cited in the footnote on p. 717 we learn that Hall (Rowland's student at the Johns Hopkins University) used ". . . an experimental configuration devised by Rowland," and also, in Rowland's words (in 1894), ". . . I had already obtained the Hall effect on a small scale. . . ." Rowland's colleague, Joseph Ames, wrote ". . . There have been several striking cases where it might have seemed to an impartial observer that Rowland's name should have appeared on the title page.

The question of priority of discovery is complex. Before drawing conclusions, you should read carefully (a) the reference in the footnote on p. 717 and (b) Hall's paper, which can be found in "Source Book in Physics" by William Francis Magie, ed., McGraw-Hill Book Co., 1935, p. 541.

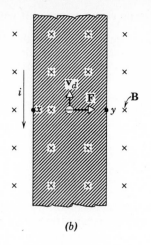

figure 33-10
A current i is set up in a copper strip placed in a magnetic field **B**, assuming (a) positive carriers and (b) negative carriers.

tial difference. If the carriers are positive, y will be at a higher potential than x; if they are negative, y will be at a lower potential than x. Experiment shows that in metals the charge carriers are negative.*

To analyze the Hall effect quantitatively, let us use the free-electron model of a metal, the same model used in Section 31-4 to help us understand why metals obey Ohm's law. The charge carriers can be assumed to move along the conductor with a certain constant drift speed v_d. The magnetic deflecting force that causes the moving charge carriers to drift toward the right edge of the strip is given by $q\mathbf{v}_d \times \mathbf{B}$ (see Eq. 33-2).

The charge carriers do not build up without limit on the right edge of the strip because the displacement of charge gives rise to a transverse *Hall electric field* $\mathbf{E}_H$, which acts, inside the conductor, to oppose the sideways drift of the carriers. This Hall electric field is another manifestation of the Hall potential difference and is related to it by

$$E_H = V_{xy}/d.$$

Eventually an equilibrium is reached in which the sideways magnetic deflecting force on the charge carriers is just canceled by the oppositely directed electric force $q\mathbf{E}_H$ caused by the Hall electric field, or

$$q\mathbf{E}_H + q\mathbf{v}_d \times \mathbf{B} = 0,$$

which we can write as

$$\mathbf{E}_H = -\mathbf{v}_d \times \mathbf{B}. \tag{33-13}$$

This equation shows explicitly that if we measure $\mathbf{E}_H$ and $\mathbf{B}$, we can find $\mathbf{v}_d$ both in magnitude and direction; given the direction of $\mathbf{v}_d$, the sign of the charge carriers follows at once, as Fig. 33-10 shows.

The number of charge carriers per unit volume (n) can also be found from Hall effect measurements. If we write Eq. 33-13 in terms of magnitudes, for the case in which $\mathbf{v}_d$ and $\mathbf{B}$ are at right angles, we obtain $E_H = v_d B$. Combining this with Eq. 31-5 ($v_d = j/ne$) leads to

$$E_H = \frac{j}{ne} B \quad \text{or} \quad n = \frac{jB}{eE_H}. \tag{33-14}$$

The agreement between experiment and Eq. 33-14 is rather good for monovalent metals, as Table 33-1 shows.

---

* At the time of the Hall-Rowland experiments the electron was not yet discovered (see Section 33-8). Hall's analysis was based on a fluid model of electricity but the general conclusions remain unchanged.

Table 33-1
Number of conduction electrons per unit volume

Metal	Based on Hall Effect Data, $10^{22}/cm^3$	Calculated, Assuming One Electron/Atom, $10^{22}/cm^3$
Li	3.7	4.8
Na	2.5	2.6
K	1.5	1.3
Cs	0.80	0.85
Cu	11	8.4
Ag	7.4	6.0
Au	8.7	5.9

For nonmonovalent metals, for iron and similar magnetic materials, and for semiconductors such as germanium, the simple interpretation of the Hall effect in terms of the free-electron model is not valid. A theoretical interpretation of the Hall effect based on quantum physics gives a reasonable agreement with experiment in all cases.

EXAMPLE 5

A copper strip 2.0 cm wide and 1.0 mm thick is placed in a magnetic field with $B = 1.5$ T, as in Fig. 33-10. If a current of 200 A is set up in the strip, what Hall potential difference appears across the strip?

From Eq. 33-14,

$$E_H = \frac{jB}{ne};$$

but

$$E_H = \frac{V_{xy}}{d} \quad \text{and} \quad j = \frac{i}{A} = \frac{i}{dh},$$

where h is the thickness of the strip. Combining these equations gives

$$V_{xy} = \frac{iB}{neh} = \frac{(200 \text{ A})(1.5 \text{ T})}{(8.4 \times 10^{28}/m^3)(1.6 \times 10^{-19} \text{ C})(1.0 \times 10^{-3} \text{ m})}$$

$$= 2.2 \times 10^{-5} \text{ V} = 22 \ \mu\text{V}.$$

These potential differences, though quite measurable, are not large. See p. 678 for the calculation of n.

33-6
CIRCULATING CHARGES

Figure 33-11 shows a negatively charged particle introduced with velocity **v** into a uniform magnetic field **B**. We assume that **v** is at right angles to **B** and thus lies entirely in the plane of the figure. The relation **F** = $q\mathbf{v} \times \mathbf{B}$ (Eq. 33-2) shows that the particle will experience a sideways deflecting force of magnitude qvB. This force will lie in the plane of the figure, which means that the particle cannot leave this plane.

This reminds us of a stone held by a rope and whirled in a horizontal circle on a smooth surface. Here, too, a force of constant magnitude, the tension in the rope, acts in a plane and at right angles to the velocity. The charged particle, like the stone, also moves with constant speed in a circular path. From Newton's second law we have

$$qvB = \frac{mv^2}{r} \quad \text{or} \quad r = \frac{mv}{qB}, \tag{33-15}$$

which gives the radius of the path. The three spirals in Fig. 33-4 show relatively low-energy electrons in a bubble chamber. The paths are not

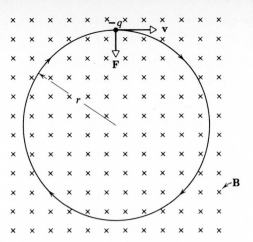

figure 33-11
A charge $-q$ circulates at right angles to a uniform magnetic field.

731 CIRCULATING CHARGES SEC. 33-6

circles because the electrons lose energy by collisions in the chamber as they move.

The angular velocity ω is given by v/r or, from Eq. 33-15,

$$\omega = \frac{v}{r} = \frac{qB}{m}.$$

The frequency ν is given by

$$\nu = \frac{\omega}{2\pi} = \frac{qB}{2\pi m}. \qquad (33\text{-}16)$$

Note that ν does not depend on the speed of the particle. Fast particles move in large circles (Eq. 33-15) and slow ones in small circles, but all require the same time T (the *period*) to complete one revolution in the field.

The frequency ν is a characteristic frequency for the charged particle in the field and may be compared to the characteristic frequency of a swinging pendulum in the earth's gravitational field or to the characteristic frequency of an oscillating mass-spring system. It is sometimes called the *cyclotron frequency* of the particle in the field because particles circulate at this frequency in a cyclotron.

A 10-eV electron is circulating in a plane at right angles to a uniform magnetic field of 1.0×10^{-4} T ($= 1.0$ gauss).

(*a*) What is its orbit radius?

The velocity of an electron whose kinetic energy is K can be found from

$$v = \sqrt{\frac{2K}{m}} \; .$$

Verify that this yields 1.9×10^6 m/s for v. Then, from Eq. 33-15,

$$r = \frac{mv}{qB} = \frac{(9.1 \times 10^{-31} \text{ kg})(1.9 \times 10^6 \text{ m/s})}{(1.6 \times 10^{-19} \text{ C})(1.0 \times 10^{-4} \text{ T})} = 0.11 \text{ m} = 11 \text{ cm}.$$

(*b*) What is the cyclotron frequency? From Eq. 33-16,

$$\nu = \frac{qB}{2\pi m} = \frac{(1.6 \times 10^{-19} \text{ C})(1.0 \times 10^{-4} \text{ T})}{(2\pi)(9.1 \times 10^{-31} \text{ kg})} = 2.8 \times 10^6 \text{ Hz}.$$

(*c*) What is the period of revolution T?

$$T = \frac{1}{\nu} = \frac{1}{2.8 \times 10^6 \text{ Hz}} = 3.6 \times 10^{-7} \text{ s}.$$

EXAMPLE 6

Thus an electron requires 0.36 μs to make one revolution in a 1.0-gauss field. (d) What is the direction of circulation as viewed by an observer sighting along the field?

In Fig. 33-11 the magnetic force $q\mathbf{v} \times \mathbf{B}$ must point radially inward, since it provides the centripetal force. Since $\mathbf{B}$ points into the plane of the paper, $\mathbf{v}$ would have to point to the left at the position shown in the figure if the charge q were positive. However, the charge is an electron, with $q = -e$, which means that $\mathbf{v}$ must point to the right. Thus the charge circulates clockwise as viewed by an observer sighting in the direction of $\mathbf{B}$.

The cyclotron, first made operational in 1932 by Ernest O. Lawrence at the University of California at Berkeley, accelerates charged particles such as hydrogen nuclei (protons) and heavy hydrogen nuclei (deuterons) to high energies so that they can be used in atom-smashing experiments.* Figure 33-12 shows a cyclotron formerly operated at the University of Pittsburgh. Although conventional cyclotrons of this type are no longer used, we discuss them for two reasons: (a) they provide an excellent framework within which to discuss the action of magnetic and electric fields on charged particles and (b) the cyclotron has led to the production of several generations of improved accelerators, the proton synchrotron being one. These later devices provide even more opportunities for studying the interactions of charged particles with magnetic and electric fields although that, of course, is not their primary purpose.

33-7
CYCLOTRONS AND SYNCHROTRONS

figure 33-12
The former University of Pittsburgh cyclotron. Note vacuum chamber V, magnet frame F, magnetic pole faces P, magnet coils C, and the deuteron beam emerging into the air of the laboratory through an aluminum foil "window." The rule is 6 ft. long. (Courtesy A. J. Allen.)

In an *ion source* at the center of the cyclotron molecules of deuterium (heavy hydrogen) are bombarded with electrons whose energy is high enough (say 100 eV) so that many positive ions are formed during the collisions. Many of these ions are free deuterons, which enter the cyclo-

* Lawrence received a Nobel prize in 1939 for this work.

tron proper through a small hole in the wall of the ion source and are available to be accelerated.

The cyclotron uses a modest potential difference for accelerating (say 10^5 V), but it requires the ion to pass through this potential difference a number of times. To reach 10 MeV with 10^5 V accelerating potential requires 100 passages. A magnetic field is used to bend the ions around so that they may pass again and again through the same accelerating potential.

Figure 33-13 is a top view of the part of the cyclotron that is inside the vacuum tank marked V in Fig. 33-12. The two D-shaped objects, called *dees*, are made of copper sheet and form part of an electric oscillator which establishes an accelerating potential difference across the gap between the dees. The direction of this potential difference is made to change sign some millions of times per second.

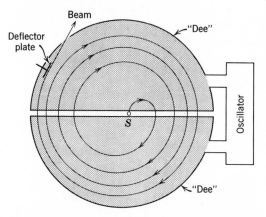

figure 33-13
The elements of a cyclotron showing the ion source S and the dees. The deflector plate, held at a suitable negative potential, deflects the particles out of the dee system.

The dees are immersed in a magnetic field ($B \cong 1.6$ T) whose direction is out of the plane of Fig. 33-13. The field is set up by a large electromagnet, part of which is shown by F in Fig. 33-12. Finally, the space in which the ions move is evacuated to a pressure of about 10^{-6} mm-Hg. If this were not done, the ions would continually collide with air molecules.

Suppose that a deuteron, emerging from the ion source, finds the dee that it is facing to be negative; it will accelerate toward this dee and will enter it. Once inside, it is screened from electric fields by the metal walls of the dees. The magnetic field is not screened by the dees so that the ion bends in a circular path whose radius, which depends on the velocity, is given by Eq. 33-15, or

$$r = \frac{mv}{qB}.$$

After a time t_0 the ion emerges from the dee on the other side of the ion source. Let us assume that the accelerating potential has now changed sign. Thus the ion *again* faces a negative dee, is further accelerated, and again describes a semicircle, of somewhat larger radius (see Eq. 33-15), in the dee. *The time of passage through this dee, however, is still t_0.* This follows because the period of revolution T of an ion circulating in a magnetic field does not depend on the speed of the ion; see Eq. 33-16. This process goes on until the ion reaches the outer edge of one dee where it is pulled out of the system by a negatively charged deflector plate.

The key to the operation of the cyclotron is that the characteristic frequency ν at which the ion circulates in the field must be equal to the fixed frequency ν_0 of the electric oscillator, or

$$\nu = \nu_0.$$

The *resonance condition* says that if the energy of the circulating ion is to increase, energy must be fed to it at a frequency ν_0 that is equal to the natural frequency ν at which the ion circulates in the field. In the same way we feed energy to a swing by pushing it at a frequency equal to the natural frequency of oscillation of the swing.

From Eq. 33-16 ($\nu = qB/2\pi m$), we can rewrite the resonance equation as

$$\frac{qB}{2\pi m} = \nu_0. \qquad (33\text{-}17)$$

Once we have selected an ion to be accelerated, q/m is fixed; usually the oscillator is designed to work at a single frequency ν_0. We then "tune" the cyclotron by varying B until Eq. 33-17 is satisfied and an accelerated beam appears.

The *energy* of the particles produced in the cyclotron depends on the radius R of the dees. From Eq. 33-15 ($r = mv/qB$), the velocity of a particle circulating at this radius is given by

$$v = \frac{qBR}{m}.$$

The kinetic energy is then

$$K = \tfrac{1}{2}mv^2 = \frac{q^2 B^2 R^2}{2m}. \qquad (33\text{-}18)$$

EXAMPLE 7

The University of Pittsburgh cyclotron had an oscillator frequency of 12×10^6 Hz and a dee radius of 53 cm ($= 21$ in.). (*a*) What value of B is needed to accelerate deuterons?

From Eq. 33-17, $\nu_0 = qB/2\pi m$, so that

$$B = \frac{2\pi \nu_0 m}{q} = \frac{(2\pi)(12 \times 10^6 \text{ Hz})(3.3 \times 10^{-27} \text{ kg})}{1.6 \times 10^{-19} \text{ C}} = 1.6 \text{ T}.$$

Note that the deuteron has the same charge as the proton but (very closely) twice the mass. (*b*) What deuteron energy results?

From Eq. 33-18,

$$K = \frac{q^2 B^2 R^2}{2m} = \frac{(1.6 \times 10^{-19} \text{ C})^2 (1.6 \text{ T})^2 (0.53 \text{ m})^2}{(2)(3.3 \times 10^{-27} \text{ kg})}$$

$$= (2.8 \times 10^{-12} \text{ J}) \left(\frac{1 \text{ eV}}{1.6 \times 10^{-19} \text{ J}} \right) = 17 \text{ MeV}.$$

There are two reasons why the conventional cyclotron that we have described runs into difficulties at high energies. One deals with physics, the other with cost. Let us discuss each in turn.

(*a*) The cyclotron fails to operate at high energies because one of its assumptions, that the frequency of rotation of an ion circulating in a magnetic field is independent of its speed, is true only for speeds much less than that of light. As the particle speed increases, we must use the *relativistic mass m* in Eq. 33-16.

The relativistic mass increases with velocity (Eq. 8-20) so that at high enough speeds ν decreases with velocity. Thus the ions get out of step with the electric oscillator, and eventually the energy of the circulating ion stops increasing.

(b) The second difficulty associated with the acceleration of particles to high energies is that the size of the magnet that would be required to guide such particles in a circular path is very large. For a 30-GeV proton, for example, in a field of 1.5 T (= 15,000 gauss) the radius of curvature is 65 m. A magnet of the cyclotron type of this size (about 430 ft in diameter) would be prohibitively expensive. Incidentally, a 30-GeV proton has a speed equal to 0.99998 that of light.

Both the relativistic and the economic limitations have been removed by techniques that can be understood in terms of Eq. 33-17 ($2\pi\nu_0 m = qB$) in which m is now taken to be the relativistic mass, given by Eq. 8-20, or

$$m = \frac{m_0}{\sqrt{1 - (v/c)^2}},$$

v being the speed of the particle and c being that of light.

As the particle speed increases, the relativistic mass m also increases. To maintain the equality in Eq. 33-17, and thus insure resonance, one may decrease the oscillator frequency ν_0 as the particle (assumed to be a proton) accelerates in such a way that the product $\nu_0 m$ remains constant. Accelerators that use this technique have been called *synchrocyclotrons*.

To ameliorate the magnet cost limitation one can vary *both* B and ν_0 in a cyclic fashion in such a way that not only is Eq. 33-17 satisfied at all times but the particle orbit radius remains essentially constant during the acceleration process. This permits the use of an annular (or ring-shaped) magnet, rather than the conventional cyclotron type, at great saving in cost. With the *two* variables B and ν_0 at our disposal, it is possible to preserve *two* equalities during the acceleration process, one being Eq. 33-17 and the other being the relation

$$v = \omega_0 R_0 = (2\pi\nu_0)R_0$$

in which R_0 is the desired (fixed) orbit radius. Accelerators that use this technique are called *synchrotrons*. Table 33-2 shows some characteristics of the accelerator built at the Brookhaven National Laboratory, embodying these principles.

Table 33-2
The Brookhaven Proton Synchrotron

Maximum proton energy	33 GeV
Mean orbit radius	128 m
Maximum orbit field	1.3 T
Injection energy	50 MeV
Pulse repetition rate	2.4 Hz
Beam aperture	18 cm × 8 cm
Total weight of magnets	4000 tons

Note that even the energy at which the protons are *injected* into this accelerator (50 MeV) far exceeds the capabilities of a conventional cyclotron.*

As of this writing (1977) the protron synchrotron producing the most energetic protons (500 GeV with 1000 GeV as a goal) is located in Batavia, Illinois.† The ring, of 954 separate magnets, is 6.3 km (= 4.1 miles) in circumference!

The system for injecting protons into this ring is impressive in itself. The

* See, for example, "Introduction to Nuclear Physics" by Harald Enge, John Wiley & Sons, 1966, for a readable account with much more information about particle accelerators than we can present here.
† See "The Batavia Accelerator" by R. R. Wilson, *Scientific American*, February 1974.

protons are first accelerated to 750 keV by a transformer-rectifier arrangement. They are then directed into a 145 m-long linear accelerator from which they emerge with an energy of 200 MeV. The protons are then led into a "medium size" proton synchrotron (the "booster") from which they emerge with an energy of 80 GeV. Only then are they injected into the main accelerator ring.

In all of these processes magnetic and electric fields not only accelerate the protons, they direct them in desired directions so that they may be used for experiments; above all, they focus them so that a well-defined proton beam emerges after as much as 10^6 miles of travel through the accelerator complex. Although this is not its purpose, no better "laboratory" for demonstrating the action of magnetic and electric fields on charged particles has yet been devised. As an indication of its size and scope, we point out that its yearly budget for electric energy alone is several million (1976) dollars.

33-8
THE DISCOVERY OF THE ELECTRON

This crucial experiment, performed in 1897 by J. J. Thomson in the Cavendish Laboratory in Cambridge, England, was a measurement of the ratio of charge e to mass m of the electron by observing its deflection in combined magnetic and electric fields. It amounted to the discovery of the electron as a fundamental particle and we discuss it here as another practical example of the action of magnetic and electric fields on charged particles.

In Fig. 33-14, which is a modernized version of Thomson's apparatus, electrons are emitted from hot filament F and accelerated by an applied potential difference V. They then enter a region in which they move at right angles to an electric field $\mathbf{E}$ and a magnetic field $\mathbf{B}$; $\mathbf{E}$ and $\mathbf{B}$ are right angles to each other. The beam is made visible as a spot of light when it strikes fluorescent screen S. The entire region in which the electrons move is highly evacuated so that collisions with air molecules will not occur.

The resultant force on a charged particle moving through an electric and a magnetic field is given by Eq. 33-4, or

$$\mathbf{F} = q_0\mathbf{E} + q_0\mathbf{v} \times \mathbf{B}.$$

Study of Fig. 33-14 shows that the electric field deflects the particle upward and the magnetic field deflects it downward. If these deflecting forces are to cancel (that is, if $\mathbf{F} = 0$), this equation, for this problem, reduces to

$$eE = evB$$

or
$$E = vB. \tag{33-19}$$

Thus for a given electron speed v the condition for zero deflection can be satisfied by adjusting E or B.

Thomson's procedure was (a) to note the position of the undeflected beam spot, with $\mathbf{E}$ and $\mathbf{B}$ both equal to zero; (b) to apply a fixed electric field $\mathbf{E}$, measuring on the fluorescent screen the deflection so caused; and (c) to apply a magnetic field and adjust its value until the beam deflection is restored to zero.

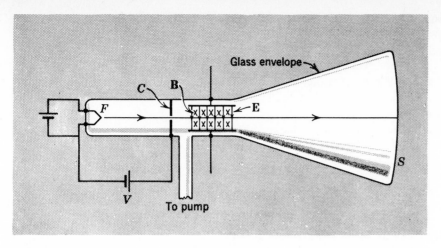

figure 33-14
Electrons from the heated filament *F* are accelerated by a potential difference *V* and pass through a hole in the screen *C*. After passing through a region in which perpendicular electric and magnetic fields are present, they strike the fluorescent screen *S*.

In Section 27-5 we saw that the deflection *y* of an electron in a purely electric field (step *b*), measured at the far edge of the deflecting plates, is given by Eq. 27-9, or, with small changes in notation,

$$y = \frac{eEl^2}{2mv^2},$$

where *v* is the electron speed and *l* is the length of the deflecting plates; *y* is not measurable directly, but it may be calculated from the measured displacement of the spot on the screen if the geometry of the apparatus is known. Thus *y*, *E*, and *l* are known; the ratio *e/m* and the velocity *v* are unknown. We cannot calculate *e/m* until we have found the velocity, which is the purpose of step *c* above.

If (step *c*) the electric force is set equal and opposite to the magnetic force, the net force is zero and we can write (Eq. 33-19)

$$v = \frac{E}{B}.$$

Substituting this equation into the equation for *y* and solving for the ratio *e/m* leads to

$$\frac{e}{m} = \frac{2yE}{B^2l^2}, \qquad (33\text{-}20)$$

in which all the quantities on the right can be measured. Thomson's value for *e/m* was 1.7×10^{11} C/kg, in full agreement with the 1977 value of 1.758805×10^{11} C/kg.

1. In a letter to the editor of *Sky and Telescope*, August 1976, Cicely M. Botley says, in part, "The [geomagnetic] pole in northern Canada is more correctly the *dip pole*, where a freely suspended magnetic needle is vertical. The *geomagnetic pole*, from which geomagnetic latitudes are measured, is in northwest Greenland." Discuss these two concepts.

2. Of the three vectors in the equation $\mathbf{F} = q\mathbf{v} \times \mathbf{B}$, which pairs are always at right angles? Which may have any angle between them?

3. Why do we not simply define the magnetic field **B** to point in the direction of the magnetic force that acts on the moving charge?

4. Imagine that you are sitting in a room with your back to one wall and that an electron beam, traveling horizontally from the back wall toward the front wall, is deflected to your right. What is the direction of the magnetic field that exists in the room?

questions

5. If an electron is not deflected in passing through a certain region of space, can we be sure that there is no magnetic field in that region?

6. If a moving electron is deflected sideways in passing through a certain region of space, can we be sure that a magnetic field exists in that region?

7. A beam of protons is deflected sideways. Could this deflection be caused (a) by an electric field? (b) By a magnetic field? (c) If either could be responsible, how would you be able to tell which was present?

8. A conductor, even though it is carrying a current, has zero net charge. Why, then, does a magnetic field exert a force on it?

9. In Example 2 (see Fig. 33-6) we saw that the magnetic force was the same as if the semicircular arc had been replaced by a straight wire of length $2R$. Would this same conclusion hold if we replaced the semicircular arc by a curve of irregular shape? Give a specific example to prove your point, one way or the other.

10. Does Eq. 33-6a ($\mathbf{F} = i\mathbf{l} \times \mathbf{B}$) hold for a straight wire whose cross section varies irregularly along its length (a "lumpy" wire)?

11. A straight copper wire carrying a current i is immersed in a magnetic field $\mathbf{B}$, at right angles to it. We know that $\mathbf{B}$ exerts a sideways force on the free (or conduction) electrons. Does it do so on the bound electrons? After all, they are not at rest. Discuss.

12. In Section 33-3 we state that a magnetic field $\mathbf{B}$ exerts a sideways force on the conduction electrons in, say, a copper wire carrying a current i. We have tacitly assumed that this same force acts on the conductor itself. Are there some missing steps in this argument?

13. Equation 33-11 ($\boldsymbol{\tau} = \boldsymbol{\mu} \times \mathbf{B}$) shows that there is no torque on a current loop in an external magnetic field if the angle between the axis of the loop and the field is (a) 0° or (b) 180°. Discuss the nature of the equilibrium (that is, is it stable, neutral, or unstable?) for these two positions.

14. In Example 4 we showed that the work required to turn a current loop end-for-end in an external magnetic field is $2\mu B$. Does this hold no matter what the original orientation of the loop was?

15. Imagine that the room in which you are seated is filled with a uniform magnetic field with $\mathbf{B}$ pointing vertically upward. A circular loop of wire has its plane horizontal. For what direction of current in the loop, as viewed from above, will the loop be in stable equilibrium with respect to forces and torques of magnetic origin?

16. A rectangular current loop is in an arbitrary orientation in an external magnetic field. Is any work required to rotate the loop about an axis perpendicular to its plane?

17. You wish to modify a galvanometer (see Example 3) to make it into (a) an ammeter and (b) a voltmeter. What do you need to do in each case?

18. (a) In measuring Hall potential differences, why must we be careful that points x and y in Fig. 33-10 are exactly opposite each other? (b) If one of the contacts is movable, what procedure might we follow in adjusting it to make sure that the two points are properly located?

19. A uniform magnetic field fills a certain cubical region of space. Can an electron be fired into this cube from the outside in such a way that it will travel in a closed circular path inside the cube?

20. Imagine the room in which you are seated to be filled with a uniform magnetic field with $\mathbf{B}$ pointing vertically downward. At the center of the room two electrons are suddenly projected horizontally with the same speed but in opposite directions. (a) Discuss their motions. (b) Discuss their motions if one particle is an electron and one a positron, that is, a positively charged electron.

21. In Fig. 33-4 why are the low-energy electron tracks spirals? That is, why does the radius of curvature change in the constant magnetic field in which the chamber is immersed?

22. What are the primary functions of (a) the electric field and (b) the magnetic field in the cyclotron?

23. What central fact makes the operation of a conventional cyclotron possible? Ignore relativistic considerations.

24. For Thomson's *e/m* experiment to work properly (Section 33-8), is it essential that the electrons have a fairly constant speed?

25. The arrangement of crossed electric and magnetic fields shown in the central part of Fig. 33-14 is sometimes called a *velocity filter*. How can this name be justified?

problems

SECTION 33-2

1. Particles 1, 2, and 3 follow the paths shown in Fig. 33-15 as they pass through the magnetic field there. What can one conclude about each particle?
   Answer: Particle 1 is positive, particle 2 is neutral, particle 3 is negative.

2. The electrons in the beam of a television tube have an energy of 12 keV. The tube is oriented so that the electrons move horizontally from south to north. The earth's magnetic field points down and has $B = 5.5 \times 10^{-5}$ T. (a) In what direction will the beam deflect? (b) What is the acceleration of a given electron? (c) How far will the beam deflect in moving 20 cm through the television tube?

3. An electron has a velocity given in m/s by $\mathbf{v} = 2.0 \times 10^6 \mathbf{i} + 3.0 \times 10^6 \mathbf{j}$. It enters a magnetic field given in T by $\mathbf{B} = 0.03\mathbf{i} - 0.15\mathbf{j}$. (a) Find the magnitude and direction of the force on the electron. (b) Repeat your calculation for a deuteron having the same velocity.
   Answer: (a) 6.2×10^{-14} **k**, N. (b) -6.2×10^{-14} **k**, N.

4. A beam of electrons whose kinetic energy is K emerges from a "window" at the end of an accelerator tube. There is a metal plate a distance d from this window and at right angles to the direction of the emerging beam. Show that we can stop the beam from hitting the plate if we apply a magnetic field B such that

$$B \geq \left(\frac{2mK}{e^2 d^2}\right)^{1/2},$$

in which m and e are the electron mass and charge. How should **B** be oriented?

figure 33-15
Problem 1

(a)   (b)   (c)   (d)

figure 33-16
Problem 5

SECTION 33-3

5. Figure 33-16 shows a magnet and a straight wire in which a cluster of electrons is flowing out of the page at right angles to it. In which case will there be a force on the wire that points toward the top of the page?   *Answer:* (b).

6. A metal wire of mass m slides without friction on two horizontal rails spaced a distance d apart, as in Fig. 33-17. The track lies in a vertical uniform magnetic field **B**. A *constant current i* flows from generator G along one rail, across the wire, and back down the other rail. Find the velocity (speed and direction) of the wire as a function of time, assuming it to be at rest at $t = 0$.

7. A wire 1.0 m long carries a current of 10 A and makes an angle of 30° with a uniform magnetic field with $B = 1.5$ T. Calculate the magnitude and direction of the force on the wire.
   Answer: 7.5 N, perpendicular to both the wire and to **B**.

figure 33-17
Problem 6

8. A wire 50 cm long lying along the x-axis carries a current of 0.50 A in the positive x-direction. A magnetic field is present that is given in T by $\mathbf{B} = 0.0030\mathbf{j} + 0.010\mathbf{k}$. Find the components of the force on the wire.

9. A wire of 60 cm length and mass 10 g is suspended by a pair of flexible leads in a magnetic field of 0.40 T. What are the magnitude and direction of the current required to remove the tension in the supporting leads? See Fig. 33-18.   *Answer:* 0.41 A, from left to right.

10. An electron in a uniform magnetic field $\mathbf{B}$ has a velocity $\mathbf{v} = 4.0 \times 10^5\mathbf{i} + 7.1 \times 10^5\mathbf{j}$, in m/s. It experiences a force $\mathbf{F} = -2.7 \times 10^{-13}\mathbf{i} + 1.5 \times 10^{-13}\mathbf{j}$, in N. If $B_x = 0$, find the magnetic field.

11. Consider the possibility of a new design for an electric train. The engine is driven by the force due to the vertical component of the earth's magnetic field on a conducting axle. Current is passed down one rail, into a conducting wheel, through the axle, through another conducting wheel, and then back to the source via the other rail. (a) What current is needed to provide a modest 10,000 N force? Take the vertical component of $\mathbf{B}$ to be 10^{-5} T and the length of the axle to be 3.0 m. (b) How much power would be lost for each ohm of resistance in the rails? (c) Is such a train totally unrealistic or marginally unrealistic?
Answer: (a) 3.3×10^9 A. (b) 1.0×10^{17} W. (c) Totally unrealistic.

12. A U-shaped wire of mass m and length l is immersed with its two ends in mercury (Fig. 33-19). The wire is in a homogeneous magnetic field $\mathbf{B}$. If a charge, that is, a current pulse $q = \int i\,dt$, is sent through the wire, the wire will jump up. Calculate, from the height h that the wire reaches, the size of the charge or current pulse, assuming that the time of the current pulse is very small in comparison with the time of flight. Make use of the fact that impulse of force equals $\int F\,dt$, which equals mv. (Hint: Try to relate $\int i\,dt$ to $\int F\,dt$.) Evaluate q for $B = 0.10$ T, $m = 10$ g, $l = 20$ cm, and $h = 3$ m.

13. Figure 33-20 shows a wire of arbitrary shape carrying a current i between points a and b. The wire lies in a plane at right angles to a uniform magnetic field $\mathbf{B}$. Prove that the force on the wire is the same as that on a straight wire carrying a current i directly from a to b. (Hint: Replace the wire by a series of "steps" parallel and perpendicular to the straight line joining a and b.)

14. Figure 33-21 shows a wire ring of radius a at right angles to the general direction of a radially symmetric diverging magnetic field. The magnetic field at the ring is everywhere of the same magnitude B, and its direction at the ring is everywhere at an angle θ with a normal to the plane of the ring. The twisted lead wires have no effect on the problem. Find the magnitude and direction of the force the field exerts on the ring if the ring carries a current i as shown in the figure.

SECTION 33-4

15. An N-turn circular coil of radius R is suspended in a uniform magnetic field $\mathbf{B}$ that points vertically downward. The coil can rotate about a horizontal axis through its center. A mass m hangs by a string from the bottom of the coil. When a current i is put through the coil it eventually assumes an equilibrium position, in which the perpendicular to the plane of the coil makes an angle ϕ with the direction of $\mathbf{B}$. Find ϕ and draw a sketch of this equilibrium position. Take $B = 0.50$ T, $R = 10$ cm, $N = 10$, $m = 500$ g, and $i = 1.0$ A.   *Answer:* $\phi = 72°$.

16. Figure 33-22 shows a wooden cylinder with a mass m of 0.25 kg, a radius R, and a length l of 0.1 m with N equal to ten turns of wire wrapped around it longitudinally, so that the plane of the wire loop contains the axis of the cylinder. What is the least current through the loop that will prevent the cylinder from rolling down an inclined plane whose surface is inclined at

figure 33-18
Problem 9

figure 33-19
Problem 12

figure 33-20
Problem 13

figure 33-21
Problem 14

figure 33-22
Problem 16

an angle θ to the horizontal, in the presence of a vertical field of magnetic induction 0.5 T, if the plane of the windings is parallel to the inclined plane?

17. A certain galvanometer has a resistance of 75.3 Ω; its needle experiences a full-scale deflection when a current 1.62×10^{-3} A passes through its coil. (a) Determine the value of the auxiliary resistance required to convert the galvanometer into a voltmeter that reads 1.000 V at full-scale deflection. How is it to be connected? (b) Determine the value of the auxiliary resistance required to convert the galvanometer into an ammeter that reads 0.0500 A at full-scale deflection. How is it to be connected?
Answer: (a) 540 Ω, connected in series. (b) 2.52 Ω, connected in parallel.

18. A circular loop of wire having a radius of 8.0 cm carries a current of 0.20 A. A unit vector parallel to the dipole moment μ of the loop is given in A · m² by $0.60\mathbf{i} - 0.80\mathbf{j}$. If the loop is located in a magnetic field given in T by $\mathbf{B} = 0.25\mathbf{i} + 0.30\mathbf{k}$, find (a) the magnitude and direction of the torque on the loop and (b) the magnetic potential energy of the loop. Assume the same zero-energy configuration that we assumed in Section 33-4.

19. Figure 33-23 shows a rectangular twenty-turn loop of wire, 10 cm by 5.0 cm. It carries a current of 0.10 A and is hinged at one side. What torque (direction and magnitude) acts on the loop if it is mounted with its plane at an angle of 30° to the direction of a uniform magnetic field of 0.50 T?
Answer: 4.3×10^{-3} N·m. The torque vector is parallel to the long side of the coil and points down.

figure 33-23
Problem 19

20. Prove that the relation $\tau = NAiB \sin \theta$ holds for closed loops of arbitrary shape and not only for rectangular loops as in Fig. 33-8. (Hint: Replace the loop of arbitrary shape by an assembly of adjacent long, thin — approximately rectangular — loops which are equivalent to it as far as the distribution of current is concerned.)

21. A length l of wire carries a current i. Show that if the wire is formed into a circular coil, the maximum torque in a given magnetic field is developed when the coil has *one* turn only and the maximum torque has the value

$$\tau = \frac{1}{4\pi} l^2 iB.$$

SECTION 33-5

22. In a Hall effect experiment a current of 3.0 A lengthwise in a conductor 1.0 cm wide, 4.0 cm long, and 10^{-3} cm thick produced a transverse Hall voltage (across the width) of 1.0×10^{-5} V when a magnetic field of 1.5 T passed perpendicularly through the thin conductor. From these data, find (a) the drift velocity of the charge carriers and (b) the number of carriers per cubic centimeter. (c) Show on a diagram the polarity of the Hall voltage with a given current and magnetic field direction, assuming the charge carriers are (negative) electrons.

23. A current i, indicated by the crosses in Fig. 33-24, is established in a strip of copper of height h and width w. A uniform magnetic field **B** is applied at right angles to the strip. (a) Calculate the drift speed v_d for the electrons. (b) What are the magnitude and direction of the magnetic force **F** acting on the electrons? (c) What would the magnitude and direction of a homogeneous electric field **E** have to be in order to counter-balance the effect of the magnetic field? (d) What is the voltage V necessary between two sides of the conductor in order to create this field **E**? Between which sides of the conductor would this voltage have to be applied? (e) If no electric field is applied *from the outside*, the electrons will be pushed somewhat to one side and therefore will give rise to a uniform Hall electric field E_H across the conductor until the forces of this electrostatic field E_H balance the magnetic forces encountered in part (b). What will be the magnitude and direction of the field E_H? Assume that n, the number of conduction electrons per unit volume, is $1.1 \times 10^{29}/\text{m}^3$ and that $h = 0.020$ m, $w = 0.10$ cm, $i = 50$ A, and $B = 2.0$ T.

Answer: (a) 1.4×10^{-4} m/s. (b) 4.5×10^{-23} N; down. (c) 2.8×10^{-4} V/m; down. (d) 5.7×10^{-6} V; top +, bottom −. (e) Same as (c).

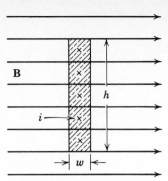

figure 33-24
Problem 23

24. (a) Show that the ratio of the Hall electric field E_H to the electric field E responsible for the current is

$$\frac{E_H}{E} = \frac{B}{ne\rho},$$

where ρ is the resistivity of the material.

(b) What is the angle between E_H and **E**? (c) Evaluate this ratio for the conditions of Example 5.

SECTION 33-6

25. A proton, a deuteron, and an α-particle, accelerated through the same potential difference, enter a region of uniform magnetic field, moving at right angles to **B**. (a) Compare their kinetic energies. (b) If the radius of the proton's circular path is 10 cm, what are the radii of the deuteron and the α-particle paths?   *Answer:* (a) $K_p = K_d = \frac{1}{2} K_\alpha$. (b) $R_d = R_\alpha = 14$ cm.

26. A proton, a deuteron, and an α-particle with the same kinetic energies enter a region of uniform magnetic field, moving at right angles to **B**. Compare the radii of their circular paths.

27. An electron is accelerated through 15,000 V and is then allowed to circulate at right angles to a uniform magnetic field with $B = 250$ gauss; 10^4 gauss = 1 T. What is its path radius?   *Answer:* 1.7 cm.

28. In a nuclear experiment a 1.0-MeV proton moves in a uniform magnetic field in a circular path. What energy must (a) an alpha particle and (b) a deuteron have if they are to circulate in the same orbit?

29. (a) In a magnetic field with $B = 0.50$ T, for what path radius will an electron circulate at 0.10 the speed of light? (b) What will its kinetic energy be?
Answer: (a) 0.34 mm. (b) 2.6 keV.

30. (a) What speed would a proton need to circle the earth at the equator, if the earth's magnetic field is everywhere horizontal there and directed along longitudinal lines? Take the magnitude of the earth's magnetic field to be 0.41×10^{-4} T at the equator. (b) Draw the velocity and magnetic field vectors corresponding to this situation.

31. What uniform magnetic field can be set up in space to permit a proton of speed 1.0×10^7 m/s to move in a circle the size of the earth's equator?
Answer: 1.6×10^{-8} T.

32. An α-particle travels in a circular path of radius 0.45 m in a magnetic field with $B = 1.2$ T. Calculate (a) its speed, (b) its period of revolution, (c) its kinetic energy, and (d) the potential difference through which it would have to be accelerated to achieve this energy.

33. A neutral particle, viewed in a reference frame in which it is at rest, lies in a homogeneous magnetic field of magnitude B. At time $t = 0$ it decays into two charged particles each of mass m. (a) If the charge of one of the particles is $+q$, what is the charge of the other? The two particles move off in separate paths both of which lie in the plane perpendicular to **B**. (b) At a later time the particles collide. Express the time from decay until collision, t, in terms of m, B, and q.   *Answer:* (a) $-q$. (b) $t = \pi m/qB$.

34. Singly ionized chlorine atoms of 35 u (= unified atomic mass units) and 37 u, traveling with speed 2.0×10^5 m/s, enter perpendicularly a uniform magnetic field of 0.50 T. After bending through $180°$ the atoms strike a photographic film. What is the separation distance between the two spots on the film? $(1.00 \text{ u} = 1.67 \times 10^{-27}$ kg.)

35. Show that the radius of curvature of a charged particle moving at right angles to a magnetic field is proportional to its momentum.

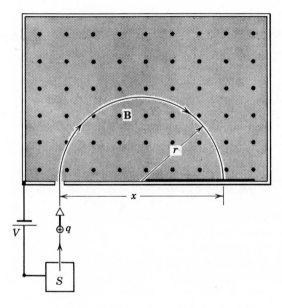

figure 33-25
Problem 36.

36. *Mass spectrometer.* Figure 33-25 shows an arrangement used by Dempster to measure the masses of ions. An ion of mass M and charge $+q$ is produced essentially at rest in source S, a chamber in which a gas discharge is taking place. The ion is accelerated by potential difference V and allowed to enter a magnetic field **B.** In the field it moves in a semicircle, striking a photographic plate at distance x from the entry slit is recorded. Show that the mass M is given by

$$M = \frac{B^2 q}{8V}\, x^2.$$

37. Two types of singly ionized atoms having the same charge q, and mass differing by a small amount ΔM are introduced into the mass spectrometer described in Problem 36. (a) Calculate the difference in mass in terms of V, q, M (of either), B, and the distance Δx between the spots on the photographic plate. (b) Calculate Δx for a beam of singly ionized chlorine atoms of masses 35 and 37 u if $V = 7.3 \times 10^3$ V and $B = 0.50$ T $(1.00 \text{ u} = 1.67 \times 10^{-27}$ kg).

Answer: (a) $B\left(\dfrac{mq}{2V}\right)^{1/2}\Delta x$. (b) 8.2 mm.

figure 33-26
Problem 38.

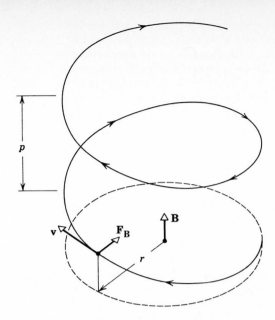

38. A 2-keV positron is projected into a uniform magnetic field **B** of 0.10 T with its velocity vector making an angle of 89° with **B**. (a) Convince yourself that the path will be a helix, its axis being the direction of **B**. Find (b) the period, (c) the pitch p, and (d) the radius r of the helix; see Fig. 33-26.

39. *Time-of-flight spectrometer.* S. A. Goudsmit has devised a method for measuring accurately the masses of heavy ions by timing their period of circulation in a known magnetic field. A singly charged ion of iodine makes 7.00 rev in a field of 4.50×10^{-2} T in about 1.29×10^{-3} s. What (approximately) is its mass in kilograms? Actually, the mass measurements are carried out to much greater accuracy than these approximate data suggest.
   Answer: 2.11×10^{-25} kg, or about 127 proton masses.

40. *Zeeman effect.* In Bohr's theory of the hydrogen atom the electron can be thought of as moving in a circular orbit of radius r about the proton. Suppose that such an atom is placed in a magnetic field, with the plane of the orbit at right angles to **B**. (a) If the electron is circulating clockwise, as viewed by an observer sighting along **B**, will the angular frequency increase or decrease? (b) What if the electron is circulating counterclockwise? Assume that the orbit radius does not change. [Hint: The centripetal force is now partially electric (**F**$_E$) and partially magnetic (**F**$_B$) in origin.]

41. In Problem 40 show that the change in frequency of rotation caused by the magnetic field is given approximately by

$$\Delta\nu = \pm\frac{Be}{4\pi m}.$$

Such frequency shifts were observed by Zeeman in 1896. (Hint: Calculate the frequency of rotation without the magnetic field and also with it. Subtract, bearing in mind that because the effect of the magnetic field is very small, some—but not all—terms containing B can be set equal to zero with little error.)

42. (a) What is the cyclotron frequency of an electron with an energy of 100 eV in the earth's magnetic field of 1.0×10^{-4} T? (b) What is the radius of curvature of the path of this electron if its velocity is perpendicular to the magnetic field?

43. A 10-KeV electron is circulating in a plane at right angles to a uniform magnetic field. The orbit radius is 25 cm. Find (a) the magnetic field, (b) the cyclotron frequency, and (c) the period of the motion.
   Answer: (a) 1.4×10^{-3} T. (b) 3.8×10^7 Hz. (c) 2.6×10^{-8} s.

SECTION 33-7

44. The cyclotron shown in Fig. 33-12 was normally adjusted to accelerate deuterons. (a) What energy of protons could it produce, using the same oscillator frequency as that used for deuterons? (b) What magnetic field would be required? (c) What energy of protons could be produced if the magnetic field was left at the value used for deuterons? (d) What oscillator frequency would then be required? (e) Answer the same questions for α-particles.

45. A deuteron in a large cyclotron is moving in a magnetic field with $B = 1.5$ T and an orbit radius of 2.0 m. Because of a grazing collision with a target, the deuteron breaks up, with a negligible loss of kinetic energy, into a proton and a neutron. Discuss the subsequent motions of each. Assume that the deuteron energy is shared equally by the proton and neutron at breakup.
Answer: Neutron moves tangent to the original path. Proton moves in a circular orbit of 1.0-m radius.

46. In a certain cyclotron a proton moves in a circle of radius $r = 0.50$ m. The magnitude of the **B** field is 1.2 T. (a) What is the cyclotron frequency? (b) What is the kinetic energy of the proton?

47. Estimate the total path length traversed by a deuteron in the cyclotron shown in Fig. 33-12 during the acceleration process. Assume an accelerating potential between the dees of 80,000 V. *Answer:* About 240 m.

SECTION 33-8

48. An electric field of 1500 V/m and a magnetic field of 0.40 T act on a moving electron to produce no force. (a) Calculate the minimum electron speed v. (b) Draw the vectors **E**, **B**, and **v**.

49. An electron is accelerated through a potential difference of 1000 V and directed into a region between two parallel plates separated by 0.02 m with a potential difference of 100 V between them. If the electron enters moving perpendicular to the electric field between the plates, what magnetic field is necessary perpendicular to both the electron path and the electric field so that the electron travels in a straight line?
Answer: 2.7×10^{-4} T.

50. A 10-KeV electron moving horizontally enters a region of space in which there is a downward-directed electric field of magnitude 100 V/cm. (a) What are the magnitude and direction of the (smallest) magnetic field that will allow the electron to continue to move horizontally? Ignore the gravitational force, which is rather small. (b) Is it possible for a proton to pass through this combination of fields undeflected? If so, under what circumstances?

51. A positive point charge Q travels in a straight line with constant speed through an evacuated region in which there is a uniform electric field **E** and a uniform magnetic field **B**. (a) If **E** is directed vertically up and the charge travels horizontally from north to south with speed v, determine the least value of the magnitude of **B** and the corresponding direction of **B**. (b) Explain why **B** is not uniquely determined when **E** and **v** alone are given. (c) Suppose the charge is a proton which enters the region after having been accelerated through a potential difference of 3.10×10^5 V. If $E = 1.90 \times 10^5$ V/m, compute the value of B in part (a). (d) If in part (c) the electric field **E** is turned off, determine the radius r of the circle in which the proton now moves.
Answer: (a) $B = E/v$; from east to west. (c) 2.47×10^{-2} T. (d) 3.26 m.

34
ampère's law

One class of problems involving magnetic fields, dealt with in Chapter 33, concerns the forces *exerted by* a magnetic field on a moving charge or on a current-carrying conductor and the torque exerted on a magnetic dipole (a bar magnet or a current loop, say). A second class of problems concerns the *production* of a magnetic field by a magnet, a current-carrying conductor, or by moving charges. This chapter deals with problems of this second class.

The discovery that currents produce magnetic effects was made by Hans Christian Oersted in 1820. Oersted made his discovery in connection with a classroom demonstration. In his paper entitled *The Action of Currents on Magnets,* Oersted wrote, translated from the Latin,

The first experiments on the subject which I undertook to illustrate were set on foot in the classes for electricity, galvanism, and magnetism, which were held by me in the winter just past.

Because of the importance of Oersted's discovery (a fundamental connection between electricity and magnetism) and especially because of its context in a teaching situation, the medal awarded annually by the American Association of Physics Teachers to a physics teacher especially noted for his or her impact on the teaching of physics is named after Oersted.

If we may deal for the moment with current-carrying wires as typical sources of magnetic fields and as typical objects on which magnetic fields may act, we may write in analogy with the argument of Section 27-1 for electric fields.

$$\text{current} \rightleftharpoons \text{field } (\mathbf{B}) \rightleftharpoons \text{current,}$$

34-1
AMPÈRE'S LAW

which suggests (a) that currents generate magnetic fields and (b) that magnetic fields exert forces on currents. We dealt with (b) in Section 33-3; we deal with (a) in this chapter.

Figure 34-1, which shows a wire surrounded by a number of small magnets, shows a modification of Oersted's experiment. If there is no current in the wire, all the magnets are aligned with the horizontal component of the earth's magnetic field. When a strong current is present, the magnets point so as to suggest that the magnetic field lines form closed circles around the wire. This view is strengthened by the experiment in Fig. 34-2, which shows iron filings on a horizontal glass plate, through the center of which a current-carrying conductor passes.

Today we write the quantitative relationship between current i and the magnetic field **B** as

$$\oint \mathbf{B} \cdot d\mathbf{l} = \mu_0 i, \qquad (34\text{-}1)$$

which is known as *Ampère's law*. Ampère, being an advocate of the action-at-a-distance point of view, did not formulate his results in terms of fields; this was first done by Maxwell. Ampère's law, including an important extension of it made later by Maxwell, is one of the basic equations of electromagnetism (see Table 40-2).

We can gain an appreciation of the way Ampère's law developed historically by considering a hypothetical experiment which has, in fact, much in common with experiments that were actually carried out. The experiment consists of measuring **B** at various distances r from a long

figure 34-1
An array of compass needles near a central wire carrying a strong current. The black ends of the compass needles are their north poles. The central dot shows the current emerging from the page. As usual, the direction of a current is taken as the direction of flow of positive charge.

figure 34-2
Iron filings around a wire carrying a strong current. (Courtesy Physical Science Study Committee.)

straight wire of circular cross section and carrying a current i. This can be done by making quantitative the qualitative observation of Fig. 34-1.

Let us put a small compass needle a distance r from the wire. Such a needle, a small magnetic dipole, tends to line up with an external magnetic field, with its north pole pointing in the direction of **B**. Figure 34-1 makes it clear that **B** at the site of the dipole is tangent to a circle of radius r centered on the wire.

If the current in the wire of Fig. 34-1 is reversed in direction, all the compass needles would reverse end-for-end. This experimental result leads to the "right-hand rule" for finding the direction of **B** near a wire carrying a current i: *Grasp the wire with the right hand, the thumb pointing in the direction of the current. The fingers will curl around the wire in the direction of* **B.**

Let us now turn the dipole through an angle θ from its equilibrium position. To do this, we must exert an external torque just large enough to overcome the restoring torque τ that will act on the dipole. τ, θ, and **B** are related by Eq. 33-11 $(\tau = \mu \times \mathbf{B})$, which we can write in terms of magnitude as

$$\tau = \mu B \sin \theta \tag{34-2}$$

in which μ is the magnitude of the magnetic moment of the dipole and θ is the angle between the vectors μ and **B**. Even though we may not know the value of μ for the compass needle, we may take it to be a constant, independent of the position or orientation of the needle. Thus by measuring τ and θ in Eq. 34-2 we can obtain a *relative* measure of B for various distances r and for various currents i in the wire. We can describe the experimental results by the proportionality

$$B \propto \frac{i}{r}. \tag{34-3}$$

We can convert this proportionality into an equality by inserting a proportionality constant. As in the case of Coulomb's law, and for similar reasons (see Section 26-4), we do not write this constant simply as, say, k but in a more complex form, namely $\mu_0/2\pi$, in which μ_0 is called the *permeability constant.** Equation 34-3 then becomes

$$B = \frac{\mu_0 i}{2\pi r}, \tag{34-4}$$

which we choose to write in the form

$$(B)(2\pi r) = \mu_0 i. \tag{34-5}$$

The left side of Eq. 34-5 is $\oint \mathbf{B} \cdot d\mathbf{l}$ for a path consisting of a circle of radius r centered on the wire. For all points on this circle **B** has the same (constant) magnitude B and $d\mathbf{l}$, which is always tangent to the path of integration, points in the same direction as **B**, as Fig. 34-3 shows. Thus

$$\oint \mathbf{B} \cdot d\mathbf{l} = \oint B \, dl = B \oint dl = (B)(2\pi r),$$

$\oint dl$ being simply the circumference of the circle. In this special case, therefore, we can write the experimentally observed connection between the field and the current as

$$\oint \mathbf{B} \cdot d\mathbf{l} = \mu_0 i, \tag{34-1}$$

* μ_0 has no connection with the dipole moment μ that appears in Eq. 34-2.

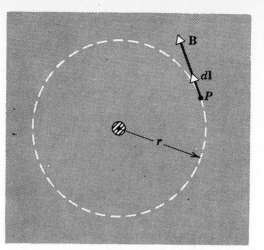

figure 34-3
A circular path of integration surrounding a wire. The central dot suggests a current i in the wire emerging from the page. Note that the angle between **B** and $d\mathbf{l}$ is zero so that $\mathbf{B} \cdot d\mathbf{l} = B \, dl$.

which is Ampère's law. A host of other experiments suggests that Eq. 34-1 is true in general for *any* magnetic field configuration, for *any* assembly of currents, and for *any* path of integration.*

In applying Ampère's law in the general case, we construct a *closed linear path* in the magnetic field as shown in Fig. 34-4. This path is divided into elements of length $d\mathbf{l}$, and for each element the quantity $\mathbf{B} \cdot d\mathbf{l}$ is evaluated. Recall that $\mathbf{B} \cdot d\mathbf{l}$ has the magnitude $B \, dl \cos \theta$ and can be interpreted as the product of dl and the component of $\mathbf{B} \, (= B \cos \theta)$ parallel to $d\mathbf{l}$. The integral is the sum of the quantities $\mathbf{B} \cdot d\mathbf{l}$ for all path elements in the complete loop; it is a *line integral* around a closed path. The term i on the right of Eq. 34-1 is the *net* current that passes through the area bounded by the closed path.

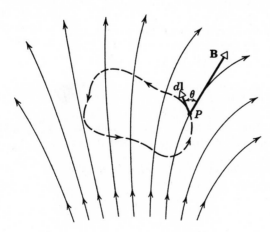

figure 34-4
A path of integration in a magnetic field.

The permeability constant in Ampère's law has an assigned value of

$$\mu_0 = 4\pi \times 10^{-7} \text{ tesla} \cdot \text{meter/ampere}.$$

Both this and the permittivity constant (ϵ_0) occur in electromagnetic formulas when SI units are used.

---

*We must modify Eq. 34-1 if a time-varying electric field or magnetic material are present within the path of integration. In this chapter we assume that if electric fields are present, they are constant in magnitude and direction and that no magnetic material is present.

You may wonder why ϵ_0 in Coulomb's law is a measured quantity, whereas μ_0 in Ampère's law is an assigned quantity. The answer is that the ampere, which is the SI unit for the current i in Ampère's law, is defined by a laboratory technique (the *current balance*) that involves forces exerted by magnetic fields and in which this same constant μ_0 appears. In effect, as we show in detail in Section 34-4, the size of the current that we agree to define as one ampere is adjusted so that μ_0 may have exactly the value assigned to it above. In Coulomb's law, on the other hand, the quantities $\mathbf{F}$, q, and r are measured in ways in which the constant ϵ_0 plays no role. This constant must then take on the particular value that makes the left side of Coulomb's law equal to the right side; no arbitrary assignment is possible.

34-2
B NEAR A LONG WIRE

We have seen that the lines of $\mathbf{B}$ for a long straight cylindrical wire carrying a current i are concentric circles centered on the axis of the wire and that B at a distance r from this axis is given by Eq. 34-4:

$$B = \frac{\mu_0 i}{2\pi r}. \tag{34-4}$$

We may regard this as an experimental result consistent with, and readily derivable from, Ampère's law.

It is interesting to compare Eq. 34-4 with the expression for the electric field near a long line of charge, or

$$E = \frac{1}{2\pi\epsilon_0}\frac{\lambda}{r}. \tag{28-10}$$

In each case there are multiplying constants, namely $\mu_0/2\pi$ and $1/2\pi\epsilon_0$, and factors describing the device responsible for the field, namely i and λ. Finally, each field varies as $1/r$.

Equation 28-10 may be derived from Gauss's law by relating the electric field at a Gaussian surface to the net charge within this surface. The (surface) integral in Gauss's law is evaluated for a closed cylindrical surface to which the lines of $\mathbf{E}$ are everywhere perpendicular.

Equation 34-4 may be derived from Ampère's law by relating the magnetic field at a path of integration to the net current that pierces this path. The (line) integral in Ampère's law is evaluated for a closed circular path to which the lines of $\mathbf{B}$ are everywhere tangent.

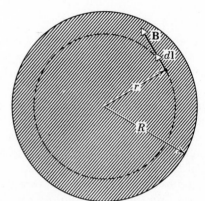

figure 34-5
Example 1. A circular path of integration inside a wire. A current i_0, distributed uniformly over the cross section of the wire, emerges from the page.

EXAMPLE 1

Derive an expression for $\mathbf{B}$ at a distance r from the center of a long cylindrical wire of radius R, where $r < R$. The wire carries a current i_0, distributed uniformly over the cross section of the wire.

Figure 34-5 shows a circular path of integration inside the wire. Symmetry suggests that $\mathbf{B}$ is tangent to the path as shown. Ampère's law,

$$\oint \mathbf{B} \cdot d\mathbf{l} = \mu_0 i,$$

gives

$$(B)(2\pi r) = \mu_0 i_0 \frac{\pi r^2}{\pi R^2},$$

since only the fraction of the current that passes through the path of integration is included in the factor i on the right. Solving for B and dropping the subscript on the current yields

$$B = \frac{\mu_0 i r}{2\pi R^2}.$$

At the surface of the wire $(r = R)$ this equation reduces to the same expression as that found by putting $r = R$ in Eq. 34-4 $(B = \mu_0 i/2\pi R)$.

Figure 34-6 shows a flat strip of copper of width a and negligible thickness carrying a current i. Find the magnetic field **B** at point P, at a distance R from the center of the strip, at right angles to the strip.

EXAMPLE 2

Let us subdivide the strip into long infinitesimal filaments of width dx, each of which may be treated as a wire carrying a current di given by $i(dx/a)$. The field contribution dB at point P in Fig. 34-6 is given, for the element shown, by the differential form of Eq. 34-4, or

$$dB = \frac{\mu_0}{2\pi} \frac{di}{r} = \frac{\mu_0}{2\pi} \frac{i(dx/a)}{R \sec \theta}.$$

Note that the vector $d\mathbf{B}$ is at right angles to the line marked r.

Only the horizontal component of $d\mathbf{B}$, namely $dB \cos \theta$, is effective, the vertical component being canceled by the contribution associated with a symmetrically located filament on the other side of the origin. Thus B at point P is given by the (scalar) integral

$$B = \int dB \cos \theta = \int \frac{\mu_0 i(dx/a)}{2\pi R \sec \theta} \cos \theta$$

$$= \frac{\mu_0 i}{2\pi a R} \int \frac{dx}{\sec^2 \theta}.$$

The variables x and θ are not independent, being related by

$$x = R \tan \theta$$

or

$$dx = R \sec^2 \theta \, d\theta.$$

Bearing in mind that the limits on θ are $\pm\tan^{-1}(a/2R)$ and eliminating dx from this expression for B, we find

$$B = \frac{\mu_0 i}{2\pi aR} \int \frac{R \sec^2 \theta \, d\theta}{\sec^2 \theta}$$

$$= \frac{\mu_0 i}{2\pi a} \int_{-\tan^{-1} a/2R}^{+\tan^{-1} a/2R} d\theta = \frac{\mu_0 i}{\pi a} \tan^{-1} \frac{a}{2R}.$$

At points far from the strip, $a/2R$ is a small angle, for which $\tan^{-1} \alpha \cong \alpha$. Thus we have, as an approximate result,

$$B \cong \frac{\mu_0 i}{\pi a}\left(\frac{a}{2R}\right) = \frac{\mu_0}{2\pi}\frac{i}{R}.$$

This result is expected because at distant points the strip cannot be distinguished from a cylindrical wire (see Eq. 34-4).

Figure 34-7 shows the lines of **B** representing the field of **B** near a long straight wire. Note the increase in the spacing of the lines with increasing distance from the wire. This represents the $1/r$ decrease in B predicted by Eq. 34-4.

Figure 34-8 shows the resultant magnetic lines associated with a current in a wire that is oriented at right angles to a uniform *external* field $\mathbf{B}_e$ directed to the right. At any point the resultant magnetic field **B** will be the vector sum of $\mathbf{B}_e$ and $\mathbf{B}_i$, where $\mathbf{B}_i$ is the magnetic field set up by the current in the wire. The fields $\mathbf{B}_e$ and $\mathbf{B}_i$ tend to cancel above the wire and to reenforce each other below the wire. At point P in Fig. 34-8 $\mathbf{B}_e$ and $\mathbf{B}_i$ cancel exactly. Very near the wire the field is represented by circular lines and is essentially $\mathbf{B}_i$.

Michael Faraday, who originated the concept of magnetic lines, endowed them with more reality than they are currently given. He imagined that, like stretched rubber bands, they represent the site of mechanical forces. On this picture can we not visualize that the wire in Fig. 34-8 will be pushed up? Today we use lines of **B** largely for purposes of visualization. For quantitative calculations we use the field vectors, describing the force on the wire in Fig. 34-8, for example, from the relation $\mathbf{F} = i\mathbf{l} \times \mathbf{B}$.

In applying this relation to Fig. 34-8, we recall that **B** is always the *external field* in which the wire is immersed; that is, it is $\mathbf{B}_e$ and thus points to the right. Since **l** points out of the page, the magnetic force on the wire ($= i\mathbf{l} \times \mathbf{B}_e$) does indeed point up. It is necessary to use only the external field in such calculations because the field set up by the current in the wire cannot exert a force on the wire, just as the gravitational

34-3
LINES OF **B**

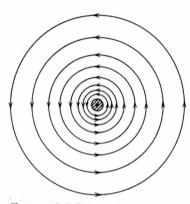

figure 34-7
Lines of **B** near a long, circularly cylindrical wire. A current i, suggested by the central dot, emerges from the page.

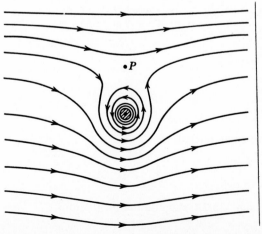

figure 34-8
Lines of **B** near a long current-carrying wire immersed in a uniform external field $\mathbf{B}_e$ that points to the right. The current i is emerging from the page. At point P, $\mathbf{B} = 0$.

field of the earth cannot exert a force on the earth itself but only on another body. In Fig. 34-7, for example, there is no magnetic force on the wire because no external magnetic field is present.

Faraday's idea of lines of **B** was instrumental in overthrowing the older action-at-a-distance theory of magnetic (and electric) attraction. Like many new ideas, it was not immediately accepted. In 1851, for example, Faraday wrote:

> I cannot refrain from again expressing my conviction of the truthfulness of the representation, which the idea of lines of force affords in regard to magnetic action. All the points that are experimentally established in regard to that action—i.e., all that is not hypothetical—appear to be well and truly represented by it.

On the other hand, four years later another well-known British scientist, Sir George Airy, wrote:

> I declare that I can hardly imagine anyone who practically and numerically knows this agreement [with the action-at-a-distance theory] to hesitate an instant in the choice between this simple and precise action, on the one hand, and anything so vague as lines of force, on the other hand.

Students who imagine that scientific pronouncements are absolute can do well to compare these two statements, each made by a distinguished contemporary. Modern examples are not lacking.

Figure 34-9 shows two long parallel wires separated by a distance d and carrying currents i_a and i_b. It is an experimental fact, noted by Ampère only one week after word of Oersted's experiments reached Paris, that two such conductors attract each other.

Some of Ampère's colleagues thought that in view of Oersted's experiment this attraction between two conductors was an obvious result and did not need to be proved. They reasoned that if wire a and wire b each exert forces on a compass needle they should exert forces on each other. This conclusion is wrong. When he heard it, Arago, a contemporary of Ampère, drew two iron keys from his pocket and replied, "Each of these keys attracts a magnet. Do you believe that they therefore also attract each other?"

Wire a in Fig. 34-9 will produce a magnetic field $\mathbf{B}_a$ at all nearby points. The magnitude of $\mathbf{B}_a$, due to the current i_a, at the site of the second wire is, from Eq. 34-4,

$$B_a = \frac{\mu_0 i_a}{2\pi d}.$$

The right-hand rule shows that the direction of $\mathbf{B}_a$ at wire b is down, as shown in the figure.

Wire b, which carries a current i_b, finds itself immersed in an *external* magnetic field $\mathbf{B}_a$. A length l of this wire will experience a sideways magnetic force (= $i\mathbf{l} \times \mathbf{B}$) whose magnitude is

$$F_b = i_b l B_a = \frac{\mu_0 l i_b i_a}{2\pi d}. \qquad (34\text{-}6)$$

The vector rule of signs tells us that $\mathbf{F}_b$ lies in the plane of the wires and points to the left in Fig. 34-9.

We could have started with wire b, computed the magnetic field which it produces at the site of wire a, and then computed the force on wire a. The force on wire a would, for parallel currents, point to the right. The forces that the two wires exert on each other are equal and

34-4
TWO PARALLEL CONDUCTORS

figure 34-9
Two parallel wires that carry parallel currents attract each other.

opposite, as they must be according to Newton's law of action and reaction. For antiparallel currents the two wires repel each other.

This discussion reminds us of our discussion of the electric field between two point charges in Section 27-1. There we saw that the charges act on each other through the intermediary of the electric field. The conductors in Fig. 34-9 act on each other, as we have said earlier, through the intermediary of the magnetic field **B.** We think in terms of

$$\text{current} \rightleftharpoons \text{field } (\mathbf{B}) \rightleftharpoons \text{current}$$

and not, as in the action-at-a-distance point of view, in terms of

$$\text{current} \rightleftharpoons \text{current.}$$

The attraction between long parallel wires is used to define the ampere. Suppose that the wires are one meter apart $(d = 1.0 \text{ m, ex-}$ actly) and that the two currents are equal $(i_a = i_b = i)$. If this common current is adjusted until, by measurement, the force of attraction per unit length between the wires is 2×10^{-7} N/m exactly, the current is defined to be one ampere. From Eq. 34-6,

$$\frac{F}{l} = \frac{\mu_0 i^2}{2\pi d} = \frac{(4\pi \times 10^{-7} \text{ T} \cdot \text{m/A})(1 \text{ A})^2}{(2\pi)(1 \text{ m})}$$

$$= 2 \times 10^{-7} \text{ N/m}$$

as expected.*

At the National Bureau of Standards primary measurements of current are made with a *current balance.* This consists of a carefully wound coil placed between two other coils, as in Fig. 34-10. The outer pair of coils is fastened to

figure 34-10
A current balance.

---

*Note that μ_0 appears in this relation used to define the ampere. As stated on page 749, μ_0 is assigned the (arbitrary) value of $4\pi \times 10^{-7}$ tesla·meter/ampere, and the size of the current that we define as one ampere is adjusted to give the required force of attraction per unit length.

the table, and the inner one is hung from the arm of a balance. The coils are so connected that the current to be measured exists, as a common current, in all three of them.

The coils exert forces on one another—just as the parallel wires of Fig. 34-9 do—which can be measured by loading weights on the balance pan. The current is defined in terms of this measured force and the dimensions of the coils. The current balance is perfectly equivalent to the long parallel wires of Fig. 34-9 but is a much more practical arrangement. Current balance measurements are used primarily to standardize other, more convenient, secondary methods of measuring currents.

EXAMPLE 3

A long horizontal rigidly supported wire carries a current i_a of 100 A. Directly above it and parallel to it is a fine wire that carries a current i_b of 20 A and weighs 0.0050 lb/ft (= 0.073 N/m). How far above the lower wire should this second wire be strung if we hope to support it by magnetic repulsion?

To provide a repulsion, the two currents must point in opposite directions. For equilibrium, the magnetic force per unit length must equal the weight per unit length and must be oppositely directed. Solving Eq. 34-6 for d yields

$$d = \frac{\mu_0 i_a i_b}{2\pi(F/l)} = \frac{(4\pi \times 10^{-7} \text{ T} \cdot \text{m/A})(100 \text{ A})(20 \text{ A})}{(2\pi)(0.073 \text{ N/m})}$$

$$= 5.5 \times 10^{-3} \text{ m} = 5.5 \text{ mm}.$$

We assume that the wire diameters are much smaller than their separation. This assumption is necessary because in deriving Eq. 34-6 we tacitly assumed that the magnetic field produced by one wire is uniform for all points within the second wire.

Is the equilibrium of the suspended wire stable or unstable against vertical displacements? This can be tested by displacing the wire vertically and examining how the forces on the wire change.

Suppose that the fine wire is suspended *below* the rigidly supported wire. How may it be made to "float"? Is the equilibrium against vertical displacements stable or unstable?

EXAMPLE 4

Two parallel wires a distance d apart carry equal currents i in opposite directions. Find the magnetic field for points between the wires and at a distance x from one wire.

Study of Fig. 34-11 shows that $\mathbf{B}_a$ due to the current i_a and $\mathbf{B}_b$ due to the current i_b point in the same direction at P. Each is given by Eq. 34-4 ($B = \mu_0 i/2\pi r$) so that

$$B = B_a + B_b = \frac{\mu_0 i}{2\pi}\left(\frac{1}{x} + \frac{1}{d-x}\right).$$

This relationship does not hold for points inside the wires because Eq. 34-4 is not valid there.

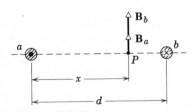

figure 34-11
Example 4

34-5
B FOR A SOLENOID

A *solenoid* is a long wire wound in a close-packed helix and carrying a current i. We assume that the helix is very long compared with its diameter. What is the nature of the field of **B** that is set up?

For points close to a single turn of the solenoid, the observer is not aware that the wire is bent in an arc. The wire behaves magnetically almost like a long straight wire, and the lines of **B** due to this single turn are almost concentric circles.

The solenoid field is the vector sum of the fields set up by all the

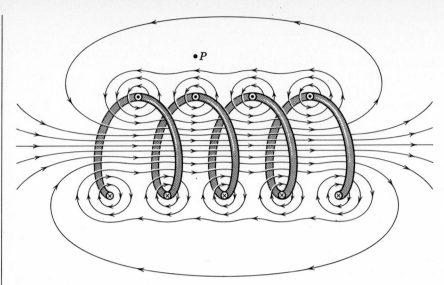

figure 34-12
A loosely wound solenoid.

turns that make up the solenoid. Figure 34-12, which shows a "solenoid" with widely spaced turns, suggests that the fields tend to cancel between the wires. It also suggests that, at points inside the solenoid and reasonably far from the wires, **B** is parallel to the solenoid axis. In the limiting case of adjacent square tightly packed wires, the solenoid becomes essentially a cylindrical current sheet and the requirements of symmetry then make the statement just given necessarily true. We assume that it is true in what follows.

For points such as P in Fig. 34-12 the field set up by the upper part of the solenoid turns (marked $\odot$) points to the left and tends to cancel the field set up by the lower part of the solenoid turns (marked $\otimes$), which points to the right. As the solenoid becomes more and more ideal, that is, as it approaches the configuration of an infinitely long cylindrical current sheet, the field **B** at outside points approaches zero. Taking the external field to be zero is not a bad assumption for a practical solenoid if its length is much greater than its diameter and if we consider only external points near the central region of the solenoid, that is, away from the ends. Figure 34-13 shows the lines of **B** for a real solenoid, which is far from ideal in that the length is not much greater than the diameter. Even here the spacing of the lines of **B** in the central plane shows that the external field is much weaker than the internal field.

Let us apply Ampère's law,

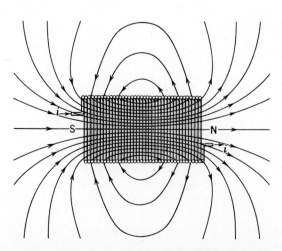

figure 34-13
A solenoid of finite length. The right end, from which lines of **B** emerge, behaves like the north pole of a compass needle. The left end behaves like the south pole.

figure 34-14
A section of an ideal solenoid, made of adjacent square turns, equivalent to an infinitely long cylindrical current sheet.

757   B FOR A SOLENOID   SEC. 34-5

$$\oint \mathbf{B} \cdot d\mathbf{l} = \mu_0 i,$$

to the rectangular path $abcd$ in the ideal solenoid of Fig. 34-14. We write the integral $\oint \mathbf{B} \cdot d\mathbf{l}$ as the sum of four integrals, one for each path segment:

$$\oint \mathbf{B} \cdot d\mathbf{l} = \int_a^b \mathbf{B} \cdot d\mathbf{l} + \int_b^c \mathbf{B} \cdot d\mathbf{l} + \int_c^d \mathbf{B} \cdot d\mathbf{l} + \int_d^a \mathbf{B} \cdot d\mathbf{l}.$$

The first integral on the right is Bh, where B is the magnitude of $\mathbf{B}$ inside the solenoid and h is the arbitrary length of the path from a to b. Note that path ab, though parallel to the solenoid axis, need not coincide with it.

The second and fourth integrals are zero because for every element of these paths $\mathbf{B}$ is at right angles to the path. This makes $\mathbf{B} \cdot d\mathbf{l}$ zero and thus the integrals are zero. The third integral, which includes the part of the rectangle that lies outside the solenoid, is zero because we have taken $\mathbf{B}$ as zero for all external points for an ideal solenoid.

Thus $\oint \mathbf{B} \cdot d\mathbf{l}$ for the entire rectangular path has the value Bh. The net current i that passes through the area bounded by the path of integration is not the same as the current i_0 in the solenoid because the path of integration encloses more than one turn. Let n be the number of turns per unit length; then

$$i = i_0(nh).$$

Ampère's law then becomes

$$Bh = \mu_0 i_0 nh$$

or
$$B = \mu_0 i_0 n. \qquad (34\text{-}7)$$

Although we derived Eq. 34-7 for an infinitely long ideal solenoid, it holds quite well for actual solenoids for internal points near the center of the solenoid. It shows that B does not depend on the diameter or the length of the solenoid and that B is constant over the solenoid cross section. A solenoid is a practical way to set up a known uniform magnetic field for experimentation, just as a parallel-plate capacitor is a practical way to set up a known uniform electric field.

The solenoid provides a good context in which to discuss Φ_B, the *flux* of the magnetic field $\mathbf{B}$. We discussed the flux Φ_E of the electric field $\mathbf{E}$ in Section 28-3, restricting ourselves largely, for reasons having to do with Gauss's law, to the flux for closed surfaces. We did, however, discuss the flux Φ_E for open surfaces; see Fig. 28-1.

We can, in a similar way, define the flux Φ_B for a magnetic field $\mathbf{B}$, for either a closed or an open surface. In either case it is given by

$$\Phi_B = \int \mathbf{B} \cdot d\mathbf{S},$$

in strict analogy to the discussion of Φ_E in Section 28-3. The SI unit of $\mathbf{B}$, as we have seen, is the *tesla* (1 tesla = 1 weber/meter^2) while that of Φ_B is simply the *weber* (abbr. Wb).

A solenoid is 1.0 m long and 3.0 cm in inner diameter. It has five layers of windings of 850 turns each and carries a current of 5.0 A.

(a) What is B at its center? From Eq. 34-7,

$$B = \mu_0 i_0 n = (4\pi \times 10^{-7} \text{ T·m/A})(5.0 \text{ A})(5 \times 850 \text{ turns/m})$$

$$= 2.7 \times 10^{-2} \text{ T} = 2.7 \times 10^{-2} \text{ Wb/m}^2.$$

We can use Eq. 34-7 even if the solenoid has more than one layer of windings because the diameter of the windings does not enter.

(b) What is the magnetic flux Φ_B for a cross section of the solenoid at its center? To the extent that $\mathbf{B}$ is constant, we can calculate the flux from

$$\Phi_B = \int \mathbf{B} \cdot d\mathbf{S} = BA,$$

where A is the effective cross-sectional area. Let us take A as the area of a circular disk whose diameter is the inner diameter of the windings (3.0 cm). The effective area can then be shown to be 7.1×10^{-4} m², and

$$\Phi_B = BA = (2.7 \times 10^{-2} \text{ Wb/m}^2)(7.1 \times 10^{-4} \text{ m}^2)$$

$$= 1.9 \times 10^{-5} \text{ Wb}.$$

EXAMPLE 5

A toroid. Figure 34-15 shows a toroid, which we may describe as a solenoid bent into the shape of a doughnut. Calculate $\mathbf{B}$ at interior points.

From symmetry the lines of $\mathbf{B}$ form concentric circles inside the toroid, as shown in the figure. Let us apply Ampère's law to the circular path of integration of radius r:

$$\oint \mathbf{B} \cdot d\mathbf{l} = \mu_0 i$$

or

$$(B)(2\pi r) = \mu_0 i_0 N,$$

where i_0 is the current in the toroid windings and N is the total number of turns. This gives

$$B = \frac{\mu_0}{2\pi} \frac{i_0 N}{r}.$$

In contrast to the solenoid, B is not constant over the cross section of a toroid. Show from Ampère's law that B equals zero for points outside an ideal toroid.

EXAMPLE 6

figure 34-15
Example 6. A toroid.

We can use Ampère's law to calculate magnetic fields only if the symmetry of the current distribution is high enough to permit the easy evaluation of the line integral $\oint \mathbf{B} \cdot d\mathbf{l}$. This requirement limits the usefulness of the law in practical problems. The law does not fail; it simply becomes difficult to apply in a useful way.

Similarly, in electrostatics, we use Gauss's law to calculate electric fields only if the symmetry of the charge distribution is high enough to permit the easy evaluation of the surface integral $\oint \mathbf{E} \cdot d\mathbf{S}$. We can, for example, use Gauss's law to find the electric field due to a long uni-

34-6
*THE BIOT-SAVART LAW**

---

* See "Ampère as a Contemporary Physicist" by R. A. Tricker, *Contemporary Physics*, August 1962; and "Electromagnetism as a Second Order Effect. III: The Biot-Savart Law" by W. G. V. Rosser, *Contemporary Physics*, October 1961, for more information about Ampère's law and the law of Biot and Savart.

formly charged rod but we cannot apply it usefully to an electric dipole, for the symmetry is not high enough in this case.

To compute **E** at a given point for an *arbitrary* charge distribution, we divided the distribution into *charge elements dq* and (see Section 27-4) we used Coulomb's law to calculate the field contribution $d\mathbf{E}$ due to each element at the point in question. We found the field **E** at that point by adding, that is, by integrating, the field contributions $d\mathbf{E}$ for the entire distribution.

We now describe a similar procedure for computing **B** at any point due to an arbitrary current distribution. We divide the current distribution into *current elements* and, using the law of Biot and Savart (which we describe below), we calculate the field contribution $d\mathbf{B}$ due to each current element at the point in question. We find the field **B** at that point by integrating the field contributions for the entire distribution.

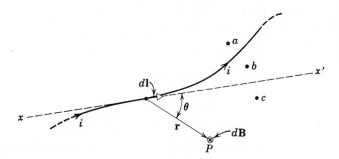

figure 34-16
The current element $d\mathbf{l}$ establishes a magnetic field contribution $d\mathbf{B}$ at point P.

Figure 34-16 shows an arbitrary current distribution consisting of a current i in a curved wire. The figure also shows a typical current element; it is a length $d\mathbf{l}$ of the conductor carrying a current i. Its direction is that of the tangent to the conductor (dashed line). A current element cannot exist as an isolated entity because a way must be provided to lead the current into the element at one end and out of it at the other. Nevertheless, we can think of an actual circuit as made up of a large number of current elements placed end to end.

Let P be the point at which we want to know the magnetic field $d\mathbf{B}$ associated with the current element. According to the Biot-Savart law, $d\mathbf{B}$ is given in magnitude by

$$dB = \frac{\mu_0 i}{4\pi} \frac{dl \, \sin \theta}{r^2}, \tag{34-8}$$

where **r** is a displacement vector from the element to P and θ is the angle between this vector and $d\mathbf{l}$. The direction of $d\mathbf{B}$ is that of the vector $d\mathbf{l} \times \mathbf{r}$. In Fig. 34-16, for example, $d\mathbf{B}$ at point P for the current element shown is directed into the page at right angles to the plane of the figure. Note that Eq. 34-8, being an inverse square law that describes the magnetic field due to a current element, may be viewed as the magnetic equivalent of Coulomb's law, which is an inverse square law that describes the electric field due to a charge element.

We may write the law of Biot and Savart in vector form as

$$d\mathbf{B} = \frac{\mu_0 i}{4\pi} \frac{d\mathbf{l} \times \mathbf{r}}{r^3}. \tag{34-9}$$

This formulation reduces to that of Eq. 34-8 when we express it in terms of magnitudes; it also gives complete information about the direction of $d\mathbf{B}$, namely, that it is the same as the direction of the vector $d\mathbf{l} \times \mathbf{r}$.

The resultant field at P is found by integrating Eq. 34-9, or

$$B = \int dB, \qquad (34\text{-}10)$$

where the integral is a vector integral.

A long straight wire. We illustrate the law of Biot and Savart by applying it to find $\mathbf{B}$ due to a current i in a long straight wire. We discussed this problem at length in connection with Ampère's law in Section 34-1.

Figure 34-17, a side view of the wire, shows a typical current element $d\mathbf{x}$. The magnitude of the contribution $d\mathbf{B}$ of this element to the magnetic field at P is found from Eq. 34-8, or

$$dB = \frac{\mu_0 i}{4\pi} \frac{dx \sin\theta}{r^2}.$$

The directions of the contributions $d\mathbf{B}$ at point P for all elements are the same, namely, into the plane of the figure at right angles to the page. Thus the vector integral of Eq. 34-10 reduces to a scalar integral, or

$$B = \int dB = \frac{\mu_0 i}{4\pi} \int_{x=-\infty}^{x=+\infty} \frac{\sin\theta\, dx}{r^2}.$$

Now, x, θ, and r are not independent, being related (see Fig. 34-17) by

$$r = \sqrt{x^2 + R^2}$$

and

$$\sin\theta\, [= \sin(\pi - \theta)] = \frac{R}{\sqrt{x^2 + R^2}},$$

so that the expression for B becomes

$$B = \frac{\mu_0 i}{4\pi} \int_{-\infty}^{+\infty} \frac{R\, dx}{(x^2 + R^2)^{3/2}}$$

$$= \frac{\mu_0 i}{4\pi R} \left| \frac{x}{(x^2 + R^2)^{1/2}} \right|_{x=-\infty}^{x=+\infty}$$

$$= \frac{\mu_0}{2\pi} \frac{i}{R}.$$

This is the result that we arrived at earlier for this problem (see Eq. 34-4). The law of Biot and Savart will always yield results that are consistent with Ampère's law and with experiment.

This problem reminds us of its electrostatic equivalent. We derived an expression for $\mathbf{E}$ due to a long charged rod, using Gauss's law (Section 28-8); we also solved this problem by integration methods, using Coulomb's law (Section 27-4).

EXAMPLE 7

figure 34-17
Example 7

A circular current loop. Figure 34-18 shows a circular loop of radius R carrying a current i. Calculate $\mathbf{B}$ for points on the axis.

The vector $d\mathbf{l}$ for a current element at the top of the loop points perpendicularly out of the page. The angle θ between $d\mathbf{l}$ and $\mathbf{r}$ is 90°, and the plane formed by $d\mathbf{l}$ and $\mathbf{r}$ is normal to the page. The vector $d\mathbf{B}$ for this element is at right angles to this plane and thus lies in the plane of the figure and at right angles to $\mathbf{r}$, as the figure shows.

Let us resolve $d\mathbf{B}$ into two components, one, $d\mathbf{B}_\parallel$, along the axis of the loop and another, $d\mathbf{B}_\perp$, at right angles to the axis. Only $d\mathbf{B}_\parallel$ contributes to the total magnetic field $\mathbf{B}$ at point P. This follows because the components $d\mathbf{B}_\parallel$ for all current elements lie on the axis and add directly; however, the components $d\mathbf{B}_\perp$ point in different directions perpendicular to the axis, and their result for the complete loop is zero, from symmetry. Thus

EXAMPLE 8

$$B = \int dB_\parallel,$$

where the integral is a simple scalar integration over the current elements.

For the current element shown in Fig. 34-18 we have, from the Biot-Savart law (Eq. 34-8),

$$dB = \frac{\mu_0 i}{4\pi} \frac{dl \sin 90°}{r^2}.$$

We also have

$$dB_\parallel = dB \cos \alpha.$$

Combining gives

$$dB_\parallel = \frac{\mu_0 i \cos \alpha \, dl}{4\pi r^2}.$$

Figure 34-18 shows that r and α are not independent of each other. Let us express each in terms of a new variable x, the distance from the center of the loop to the point P. The relationships are

$$r = \sqrt{R^2 + x^2}$$

and

$$\cos \alpha = \frac{R}{r} = \frac{R}{\sqrt{R^2 + x^2}}.$$

Substituting these values into the expression for $dB_\parallel$ gives

$$dB_\parallel = \frac{\mu_0 i R}{4\pi (R^2 + x^2)^{3/2}} \, dl.$$

Note that i, R, and x have the same values for all current elements. Integrating this equation, noting that $\int dl$ is simply the circumference of the loop ($= 2\pi R$), yields

$$B = \int dB_\parallel = \frac{\mu_0 i R}{4\pi (R^2 + x^2)^{3/2}} \int dl$$

$$= \frac{\mu_0 i R^2}{2(R^2 + x^2)^{3/2}}. \qquad (34\text{-}11)$$

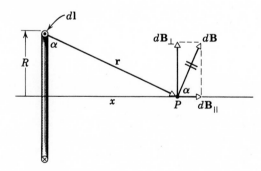

figure 34-18
Example 8. A ring of radius R carrying a current i.

If we put $x \gg R$ in Example 8 so that points close to the loop are not considered, Eq. 34-11 reduces to

$$B = \frac{\mu_0 i R^2}{2x^3}.$$

Recalling that πR^2 is the area A of the loop and considering loops with N turns, we can write this equation as

$$B = \frac{\mu_0}{2\pi} \frac{(NiA)}{x^3} = \frac{\mu_0}{2\pi} \frac{\mu}{x^3},$$

where μ is the *magnetic dipole moment* of the current loop. This reminds us of the result derived in Problem 25, Chapter 27 [$E = (1/2\pi\epsilon_0)(p/x^3)$], which is the formula for the *electric* field on the axis of an *electric* dipole.

Thus we have shown in two ways that we can regard a current loop as a magnetic dipole: It experiences a torque given by $\tau = \boldsymbol{\mu} \times \mathbf{B}$ when we place it in an *external* magnetic field (Eq. 33-11); it generates its own magnetic field given, for points on the axis, by the equation just developed.

Table 34-1 is a summary of the properties of electric and magnetic dipoles.

Table 34-1
Some dipole equations

Property	Dipole Type	Equation
Torque in an external field	electric	$\tau = \mathbf{p} \times \mathbf{E}$
	magnetic	$\tau = \boldsymbol{\mu} \times \mathbf{B}$
Energy in an external field	electric	$U = -\mathbf{p} \cdot \mathbf{E}$
	magnetic	$U = -\boldsymbol{\mu} \cdot \mathbf{B}$
Field at distant points along axis	electric	$E = \dfrac{1}{2\pi\epsilon_0}\dfrac{p}{x^3}$
	magnetic	$B = \dfrac{\mu_0}{2\pi}\dfrac{\mu}{x^3}$
Field at distant points along perpendicular bisector	electric	$E = \dfrac{1}{4\pi\epsilon_0}\dfrac{p}{x^3}$
	magnetic	$B = \dfrac{\mu_0}{4\pi}\dfrac{\mu}{x^3}$

EXAMPLE 9

In the Bohr model of the hydrogen atom the electron circulates around the nucleus in a path of radius 5.3×10^{-11} m at a frequency ν of 6.5×10^{15} Hz ($=$ rev/s). (*a*) What value of B is set up at the center of the orbit?

The current is the rate at which charge passes any point on the orbit and is given by

$$i = e\nu = (1.6 \times 10^{-19} \text{ C})(6.5 \times 10^{15} \text{ Hz}) = 1.0 \times 10^{-3} \text{ A}.$$

B at the center of the orbit is given by Eq. 34-11 with $x = 0$, or

$$B = \frac{\mu_0 i R^2}{2(R^2 + x^2)^{3/2}} = \frac{\mu_0 i}{2R}$$

$$= \frac{(4\pi \times 10^{-7} \text{ T} \cdot \text{m/A})(1.0 \times 10^{-3} \text{ A})}{(2)(5.3 \times 10^{-11} \text{ m})}$$

$$= 12 \text{ T}$$

(*b*) What is the equivalent magnetic dipole moment? From Eq. 33-10,

$$\mu = NiA = (1)(1.0 \times 10^{-3} \text{ A})(\pi)(5.3 \times 10^{-11} \text{ m})^2$$

$$= 8.8 \times 10^{-24} \text{ A} \cdot \text{m}^2.$$

questions

1. Can the path of integration around which we apply Ampère's law pass *through* a conductor?

2. Suppose we set up a path of integration around a cable that contains twelve wires with different currents (some in opposite directions) in each wire. How do we calculate i in Ampère's law in such a case?

3. Apply Ampère's law qualitatively to the three paths shown in Fig. 34-19.

figure 34-19
Question 3,
Problem 27

4. Discuss and compare Gauss's law and Ampère's law.

5. (a) Must Ampère's law (Eq. 34-1) be applied always around a *closed* loop? Can we, for example, apply it to a semicircular arc? (b) Can we apply Ampère's law to a closed *surface?*

6. Give details of three ways in which you can measure the magnetic field **B** at a point P, a perpendicular distance r from a long straight wire carrying a constant current i. Base them on: (a) firing a particle of charge q through point P with velocity **v**, parallel, say, to the wire; (b) measuring the force per unit length exerted on a second wire, parallel to the first wire and carrying a current i'; (c) measuring the torque exerted on a small magnetic dipole located a perpendicular distance r from the wire.

7. A current is set up in a long copper pipe. Is there a magnetic field (a) inside and (b) outside the pipe?

8. Equation 34-4 ($B = \mu_0 i/2\pi r$) suggests that a strong magnetic field is set up at points near a long wire carrying a current. Since there is a current i and a magnetic field **B**, why is there not a force on the wire in accord with the equation $\mathbf{F} = i\mathbf{l} \times \mathbf{B}$?

9. A beam of 20-MeV protons emerges from a cyclotron-like device. Is a magnetic field associated with these particles?

10. Does it necessarily follow from symmetry arguments alone that the lines of **B** around a long straight wire carrying a current i must be concentric circles?

11. A long straight wire of radius R carries a steady current i. In what sense, if any, does the magnetic field generated by this current depend on R?

12. A long straight wire carries a constant current i. Does Ampère's law (Eq. 34-1) hold for (a) a path of integration that encloses the wire but is not circular, (b) a path of integration that does not enclose the wire, and (c) a path that encloses the wire but does not all lie in one plane? Discuss.

13. Two long straight wires pass near one another at right angles. If the wires are free to move, describe what happens when currents are sent through them.

14. Is **B** constant in magnitude for points that lie on a given magnetic field line?

15. In electronics, wires that carry equal but opposite currents are often twisted together to reduce their magnetic effect at distant points. Why is this effective?

16. Two long parallel conductors carry equal currents i in the same direction. Sketch roughly the resultant lines of **B** due to the action of *both* currents. Does your figure suggest an attraction between the wires (in the same sense that Fig. 34-8 suggests an upward force on the wire in that figure)?

17. In Fig. 34-8 explain the relation between the figure and Eq. 33-6a ($\mathbf{F} = i\mathbf{l} \times \mathbf{B}$).

18. Test the "floating" wire of Example 3 for equilibrium under *horizontal* displacements. Consider that the wire floats above the rigidly supported wire and also below it. Summarize the equilibrium situation for both wire positions and for both vertical and horizontal displacements.

19. A current is sent through a vertical spring from whose lower end a weight is hanging; what will happen?

20. Comment on this statement: "The magnetic field **B** outside a long solenoid cannot be zero, if only for the reason that the helical nature of the windings produces a field for external points like that of a straight wire along the solenoid axis."

21. Does Eq. 34-7 $(B = \mu_0 in)$ hold for a solenoid of square cross section?

22. In your own words convince yourself that **B** = 0 for points outside an ideal solenoid, such as that of Fig. 34-14.

23. What is the direction of the magnetic fields at points a, b, and c in Fig. 34-16 set up by the particular current element shown?

24. In a circular loop of wire carrying a current i, is **B** uniform for all points within the loop?

25. Discuss analogies and differences between Coulomb's law and the Biot-Savart law.

26. Equation 34-9 gives the law of Biot and Savart in vector form. Write its electrostatic equivalent [that is, Eq. 27-6, or $dE = dq/(4\pi\epsilon_0 r^2)$] in vector form.

27. How might you measure the magnetic dipole moment of a compass needle?

28. What is the basis for saying that a current loop is a magnetic dipole?

29. A circular loop of wire lies on the floor of the room in which you are sitting and it carries a constant current i in a clockwise direction, as viewed from above. What is the direction of the magnetic dipole moment of this current loop?

30. As an exercise in vector representation, contrast and compare Fig. 18-12 which deals with fluid flow, with Fig. 34-7, which deals with the magnetic field. How strong an analogy can you make?

figure 34-20
Problem 1

problems

SECTION 34-1

1. Eight wires cut the page perpendicularly at the points shown in Fig. 34-20. A wire labeled with the integer k $(k = 1, 2, \ldots 8)$ bears the current ki_0. For those with odd k, the current flows up out of the page; for those with even k it flows down into the page. Evaluate $\oint \mathbf{B} \cdot \mathbf{dl}$ along the closed path shown in the direction indicated by the arrowhead.
   Answer: $-10\mu_0 i_0$ (Why the minus sign?).

2. Show that it is impossible for a uniform magnetic field **B** to drop abruptly to zero as one moves at right angles to it, as suggested by the horizontal arrow in Fig. 34-21 (see point a). In actual magnets fringing of the lines of **B** always occurs, which means that **B** approaches zero in a continuous and gradual way. (Hint: Apply Ampère's law to the rectangular path shown by the dashed lines.)

SECTION 34-2

3. A #10 bare copper wire (0.10 in. in diameter) can carry a current of 50 A without overheating. For this current, what is B at the surface of the wire?
   Answer: 7.9×10^{-3} T.

4. A surveyor is using a compass 20 ft below a power line in which there is a steady current of 100 A. Will this interfere seriously with the compass reading? The horizontal component of the earth's magnetic field at the site is 0.20 gauss.

5. If a point charge of magnitude $+q$ and speed v is a distance d from the axis of a long straight wire carrying a current i and is traveling perpendicular to

figure 34-21
Problem 2

the axis of the wire, what are the direction and magnitude of the force acting on it if the charge is moving (a) toward, or (b) away from the wire?

Answer: (a) $\dfrac{\mu_0}{2\pi}\dfrac{qvi}{r}$, antiparallel to i. (b) Same magnitude, parallel to i.

6. A long straight wire carries a current of 50 A. An electron, traveling at 1.0×10^7 m/s, is 5.0 cm from the wire. What force acts on the electron if the electron velocity is directed (a) toward the wire, (b) parallel to the wire, and (c) at right angles to the directions defined by (a) and (b)?

7. A long solid cylindrical copper wire of radius R carries a current i distributed uniformly over the cross section of the wire. Sketch roughly the magnitude of the magnetic field **B** as a function of the distance r from the axis of the wire for (a) $r < R$ and (b) $r > R$.

8. Four long copper wires are parallel to each other, their cross section forming a square 20 cm on edge. A 20-A current is set up in each wire in the direction shown in Fig. 34-22. What are the magnitude and direction of **B** at the center of the square?

9. A long coaxial cable consists of two concentric conductors with the dimensions shown in Fig. 34-23. There are equal and opposite currents i in the conductors. (a) Find the magnetic field B at r within the inner conductor $(r < a)$. (b) Find B between the two conductors $(a < r < b)$. (c) Find B within the outer conductor $(b < r < c)$. (d) Find B outside the cable $(r > c)$.

Answer: (a) $\dfrac{\mu_0 i r}{2\pi a^2}$. (b) $\dfrac{\mu_0 i}{2\pi r}$. (c) $\dfrac{\mu_0 i}{2\pi r}\left(\dfrac{c^2 - r^2}{c^2 - b^2}\right)$. (d) Zero.

10. Two long wires a distance d apart carry equal antiparallel currents i, as in Fig. 34-24. (a) Show that B at point P, which is equidistant from the wires, is given by

$$B = \frac{2\mu_0 i d}{\pi(4R^2 + d^2)}.$$

(b) In what direction does **B** point?

11. Figure 34-25 shows a hollow cylindrical conductor of radii a and b which carries a current i uniformly spread over its cross section. (a) Show that the magnetic field B for points inside the body of the conductor (that is, $a < r < b$) is given by

$$B = \frac{\mu_0 i}{2\pi(b^2 - a^2)} \cdot \frac{r^2 - a^2}{r}.$$

Check this formula for the limiting case of $a = 0$. (b) Make a rough plot of the general behavior of $B(r)$ from $r = 0$ to $r \to \infty$.

figure 34-22
Problem 8

figure 34-23
Problem 9

figure 34-24
Problem 10

figure 34-25
Problem 11

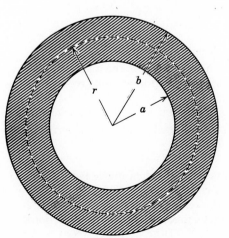

12. Two long straight wires a distance d (10 cm) apart each carry a current i (100 A). Figure 34-26 shows a cross section, with the wires running perpendicular to the page and point P lying on the perpendicular bisector of d. Find the magnitude and direction of the magnetic field at P when the current in the left-hand wire is out of the page and the current in the right-hand wire is (a) in the same direction and (b) in the opposite direction.

13. A conductor consists of an infinite number of adjacent wires, each infinitely long and carrying a current i. Show that the lines of **B** will be as represented in Fig. 34-27 and that B for all points in front of the infinite current sheet will be given by

$$B = \tfrac{1}{2}\mu_0 ni,$$

where n is the number of conductors per unit length. Derive both by direct application of Ampère's law and by considering the problem as a limiting case of Example 2.

14. A long straight conductor has a circular cross section of radius R and carries a current i. Inside the conductor there is a cylindrical hole of radius a whose axis is parallel to the conductor axis at a distance b from it. Use superposition ideas and obtain an expression for the magnetic field **B** inside the hole.

SECTION 34-3

15. A long wire carrying a current of 100 A is placed in a uniform external magnetic field of 50 gauss. The wire is at right angles to this external field. Locate the points at which the *resultant* magnetic field is zero.
Answer: $B = 0$ along a line parallel to the wire and 4.0 mm from it.

SECTION 34-4

16. In Problem 8 what is the force per unit length (N/m) acting on the lower left wire, in magnitude and direction?

17. Figure 34-28 shows a long wire carrying a current of 30 A. The rectangular loop carries a current of 20 A. Calculate the resultant force acting on the loop. Assume that $a = 1.0$ cm, $b = 8.0$ cm, and $l = 30$ cm.
Answer: 3.2×10^{-3} N, toward the wire.

18. Suppose, in Fig. 34-22, that the currents are all in the same direction. What is the force per unit length (N/m, magnitude and direction) on any one wire? In the analogous case of parallel motion of charged particles in a plasma, this is known as the pinch effect.

19. Two long parallel wires of negligible radius are a distance d apart. Let there be a current i in each wire (a) in the same direction and (b) in opposite directions. Letting r be the perpendicular distance from the center of one wire, find the magnitude B of the magnetic field in the region between the wires at points in the plane of the two wires.
Answer: (a) $\dfrac{\mu_0 i}{2\pi}\left[\dfrac{d-2r}{r(d-r)}\right]$. (b) $\dfrac{\mu_0 i}{2\pi}\left[\dfrac{d}{r(d-r)}\right]$.

SECTION 34-5

20. A 200-turn solenoid having a length of 25 cm and a diameter of 10 cm carries a current of 0.30 A. (a) What is the magnitude of the magnetic field **B** near the center of the solenoid? (b) What is the magnetic flux through an annular ring having an inside diameter of 2.0 cm and an outside diameter of 8.0 cm if the plane of the ring is perpendicular to the axis of the solenoid?

21. Express (a) the magnetic field **B** and (b) the magnetic flux Φ_B in terms of mass, length, time, and charge (M, L, T, Q).
Answer: (a) M/QT. (b) ML^2/QT.

22. A toroid having a 5.0 cm $\times$ 5.0 cm cross section and an inside radius of 15 cm has 500 turns of wire and carries a current of 0.80 A. (a) What is the magnitude of the magnetic field **B** in the center of the toroid (that is, at a radius of 17.5 cm)? (b) What is the magnetic flux through the cross section?

figure 34-26
Problem 12

figure 34-27
Problem 13

figure 34-28
Problem 17

23. A long straight wire of radius *a* carries a constant current *i* (*a*) Consider a concentric hypothetical circle of radius 2*a*, its plane being at right angles to the wire. What magnetic flux Φ_B passes through this circle? (*b*) If we were to double the current *i*, what do you imagine would happen to this flux?
Answer: (*a*) Zero. (*b*) Still zero; no lines of **B** pierce the circle in either case.

24. Derive the solenoid equation (Eq. 34-7) starting from the expression for the field on the axis of a circular loop (Example 8). (Hint: Subdivide the solenoid into a series of current loops of infinitesimal thickness and integrate.)

25. A long copper wire carries a current of 10 A. Calculate the magnetic flux per meter of wire through a plane surface *S* inside the wire, as in Fig. 34-29.
Answer: 1.0×10^{-6} Wb/m.

26. Two long, parallel #10 copper wires (diameter = 0.10 in.) carry currents of 10 A in opposite directions. (*a*) If their centers are 2.0 cm apart, calculate the flux per meter of wire length that exists in the space between the axes of the wires. (*b*) What fraction of the flux in (*a*) lies inside the wires? (*c*) Repeat the calculation of (*a*) for parallel currents.

SECTION 34-6

27. A long "hairpin" is formed by bending a piece of wire as shown in Fig. 34-19. If a 10-A current is set up, what are the direction and magnitude of **B** at point *a*? At point *b*? Take $R = 0.50$ cm.
Answer: (*a*) 1.0×10^{-3} T, out of figure. (*b*) 8.0×10^{-4} T, out of figure.

28. Use the Biot-Savart law to calculate the magnetic field **B** at *C*, the common center of the semicircular arcs *AD* and *HJ*, of radii R_2 and R_1 respectively, forming part of the circuit *ADJHA* carrying current *i*, as shown in Fig. 34-30.

29. Consider the circuit of Fig. 34-31. The curved segments are part of circles of radii *a* and *b*. The straight segments are along the radii. Find the magnetic field **B** at *P*, assuming a current *i* in the circuit.

Answer: $B = \dfrac{\mu_0 i \theta}{4\pi}\left(\dfrac{1}{b} - \dfrac{1}{a}\right)$, out of page.

30. The wire shown in Fig. 34-32 carries a current *i*. What is the magnetic field **B** at the center *C* of the semicircle arising from (*a*) each straight segment of length *l*, (*b*) the semicircular segment of radius *R*, and (*c*) the entire wire?

31. A straight conductor is split into identical semicircular turns as shown in Fig. 34-33. What is the magnetic field at the center *C* of the circular loop so formed?
Answer: Zero.

32. A circular loop of radius 10 cm carries a current of 15 A. At its center is placed a second loop of radius 1.0 cm, having 50 turns and a current of 1.0 A. (*a*) What magnetic field **B** does the large loop set up at its center? (*b*) What torque acts on the small loop? Assume that the planes of the two loops are at right angles and that **B** provided by the large loop is essentially uniform throughout the volume occupied by the small loop.

33. You are given a closed circuit with radii *a* and *b*, as shown in Fig. 34-34, carrying a current *i*. (*a*) What are the magnitude and direction of **B** at point *P*? (*b*) Find the dipole moment of the circuit.

Answer: (*a*) $\dfrac{\mu_0 i}{4}\left(\dfrac{1}{a} + \dfrac{1}{b}\right)$, into the page. (*b*) $\dfrac{i\pi}{2}(a^2 + b^2)$, into the page.

figure 34-29
Problem 25

figure 34-30
Problem 28

figure 34-31
Problem 29

figure 34-32
Problem 30

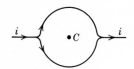

figure 34-33
Problem 31

figure 34-34
Problem 33

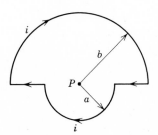

34. (a) A long wire is bent into the shape shown in Fig. 34-35, without cross-contact at P. The radius of the circular section is R. Determine the magnitude and direction of **B** at the center C of the circular portion when the current i flows as indicated. (b) The circular part of the wire is rotated without distortion about its (dashed) diameter perpendicular to the straight portion of the wire. The magnetic moment associated with the circular loop is now in the direction of the current in the straight part of the wire. Determine **B** at C in this case.

35. *Helmholtz coils.* Two 300-turn coils are arranged a distance apart equal to their radius, as in Fig. 34-36. For $R = 5.0$ cm and $i = 50$ A, plot B as a function of distance x along the common axis over the range $x = -5$ cm to $x = +5$ cm, taking $x = 0$ at point P. (Such coils provide an especially uniform field of B near point P.)

36. In Problem 35 let the separation of the coils be a variable z. Show that if $z = R$, then not only the first derivative (dB/dx) but also the second (d^2B/dx^2) of B is zero at point P. This accounts for the uniformity of B near point P for this particular coil separation.

37. A plastic disk of radius R has a charge q uniformly distributed over its surface. If the disk is rotated at an angular frequency ω about its axis, show that (a) the magnetic field at the center of the disk is

$$B = \frac{\mu_0 \omega q}{2\pi R}$$

and (b) the magnetic dipole moment of the disk is

$$\mu = \frac{\omega q R^2}{4}.$$

(Hint: The rotating disk is equivalent to an array of current loops; see Example 8.)

38. A messy loop of limp wire is placed on a table and anchored at points a and b as shown in Fig. 34-37. If a current i is now passed through the wire, will it try to form a circular loop or will it try to bunch up further?

39. A straight wire segment of length l carries a current i. (a) Show that the magnetic field **B** associated with this segment, at a distance R from the segment along a perpendicular bisector (see Fig. 34-38), is given in magnitude by

$$B = \frac{\mu_0 i}{2\pi R} \frac{l}{(l^2 + 4R^2)^{1/2}}.$$

(b) Does this expression reduce to an expected result as $l \to \infty$? *Answer:* (b) Yes.

40. A square loop of wire of edge a carries a current i. Show that the value of B at the center is given by

$$B = \frac{2\sqrt{2}\,\mu_0 i}{\pi a}.$$

41. A square loop of wire of edge a carries a current i. (a) Show that B for a point on the axis of the loop and a distance x from its center is given by

$$B = \frac{4\mu_0 i a^2}{\pi(4x^2 + a^2)(4x^2 + 2a^2)^{1/2}}.$$

(b) Does this reduce to the result of Problem 40 for $x = 0$? (c) Does the square loop behave like a dipole for points such that $x \gg a$? If so, what is its dipole moment? *Answer:* (b) Yes. (c) Yes; $\mu = ia^2$.

42. (a) Show that B at the center of a rectangle of length l and width d, carrying a current i, is given by

$$B = \frac{2\mu_0 i}{\pi} \frac{(l^2 + d^2)^{1/2}}{ld}.$$

figure 34-35
Problem 34

figure 34-36
Problems 35, 36

figure 34-37
Problem 38

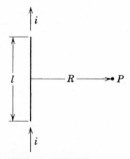

figure 34-38
Problem 39.

(b) What does B reduce to for $l \gg d$? Is this a result that you expect?

43. (a) A wire in the form of a regular polygon of n sides is just enclosed by a circle of radius a. If the current in this wire is i, show that the magnetic field **B** at the center of the circle is given in magnitude by

$$B = \frac{\mu_0 ni}{2\pi a} \tan{(\pi/n)}.$$

(b) Show that as $n \to \infty$ this result approaches that of a circular loop.

44. Calculate **B** approximately at point P in Fig. 34-39. Assume that $i = 10$ A and $a = 8.0$ cm.

45. You are given a length l of wire in which a current i may be established. The wire may be formed into a circle or a square. Which yields the larger value for B at the central point? See Problem 40.   *Answer:* The square.

46. (a) A current i flows in a straight wire of length L in the direction shown in Fig. 34-40a. Starting from the Biot-Savart law, determine the resulting magnetic field (**B**$_P$, **B**$_Q$, **B**$_R$, **B**$_S$, respectively-direction and magnitude in each case) at each of the four points P, Q, R, S (all coplanar with the wire). (b) Using the results of part (a), compute the magnetic field **B** (magnitude and direction) resulting at the point T from the current flowing as indicated in the six-sided rectilinear closed loop shown in Fig. 34-40b. (Everything drawn is meant to lie in the same plane and all angles are 90°.)

figure 34-39
Problem 44.

figure 34-40
Problem 46.

(a)

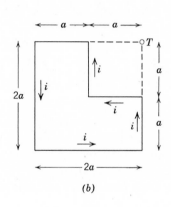

(b)

35

faraday's law of induction

For some physical laws it is hard to find experiments that lead in a direct and convincing way to the formulation of the law. Gauss's law, for example, emerged only slowly as the common factor with whose aid all electrostatic experiments could be interpreted and correlated. In Chapter 28 we found it best to state Gauss's law first and then to show that the underlying experiments were consistent with it.

Faraday's law of induction, which is one of the basic equations of electromagnetism (see Table 40-2), is different in that there are a number of simple experiments from which the law can be—and was—deduced directly. Such experiments were carried out by Michael Faraday in England in 1831 and by Joseph Henry in the United States at about the same time.

Faraday and Henry had several parallels in their lives. Both were apprentices at an early age. Faraday, at age 14, was apprenticed to a London bookbinder. He wrote, "There were plenty of books there and I read them." Henry, at age 13, was apprenticed to a watchmaker in Albany, New York.

In later years Faraday was appointed director of the Royal Institution in London, whose founding was due in large part to the American, Benjamin Thompson (Count Rumford). Henry, on the other hand, became secretary of the Smithsonian Institution in Washington, D.C., which was founded by an endowment from an Englishman, James Smithson.

Their greatest scientific overlap was that each discovered the law of electromagnetic induction, with which this chapter deals, independently and at about the same time. Even though Faraday published his results first, which gives him priority of discovery, the SI unit of inductance (see Chapter 36) is called the *henry* (abbr. H). On the other hand the SI unit of capacitance is, as we have seen, called the *farad* (abbr. F). In Section 38-1, in which we discuss oscillations in capacitative-inductive circuits, we shall see how appropriate it is to link the names of these two talented contemporaries in a single context.

Figure 35-1 shows the terminals of a coil connected to a galvanometer. Normally we would not expect this instrument to deflect because there seems to be no electromotive force in this circuit; but if we push a bar magnet toward the coil, with its north pole facing the coil, a remarkable thing happens. *While the magnet is moving,* the galvanometer deflects, showing that a current has been set up in the coil. If we hold the magnet stationary with respect to the coil, the galvanometer does not deflect. If we move the magnet *away* from the coil, the galvanometer again deflects, but in the opposite direction, which means that the current in the coil is in the opposite direction. If we use the south pole end of a magnet instead of the north pole end, the experiment works as described but the deflections are reversed. Further experimentation shows that *what matters is the relative motion of the magnet and the coil.* It makes no difference whether we move the magnet toward the coil or the coil toward the magnet.

The current that appears in this experiment is called an *induced current* and is said to be set up by an *induced electromotive force.* Note that there are no batteries anywhere in the circuit. Faraday was able to deduce from experiments like this the law that gives their magnitude and direction. Such emfs are very important in practice. The chances are good that the lights in the room in which you are reading this book are operated from an induced emf produced in a commercial electric generator.

In another experiment the apparatus of Fig. 35-2 is used. The coils are placed close together but at rest with respect to each other. When we close the switch S, thus setting up a steady current in the right-hand coil, the galvanometer deflects momentarily; when we open the switch, thus interrupting this current, the galvanometer again deflects momentarily, but in the opposite direction. No gross objects are moving in this experiment. In Faraday's words:

When the contact was made, there was a sudden and very slight effect at the galvanometer, and there was also a similar slight effect when the contact with the battery was broken. But whilst the voltaic current was continuing to pass through the one helix, no galvanometrical appearances nor any effect like induction upon the other helix could be perceived, although the active power of the battery was proved to be very great. . . .

Experiment shows that there will be an induced emf in the left coil of Fig. 35-2 whenever the current in the right coil is *changing.* It is the *rate at which the current is changing* and *not the size of the current* that is significant.

figure 35-1
Galvanometer G deflects while the magnet is moving with respect to the coil. Only their relative motion counts.

figure 35-2
Galvanometer G deflects momentarily when switch S is closed or opened. No motion is involved.

35-2
FARADAY'S LAW OF INDUCTION

Faraday had the insight to perceive that the change in the flux Φ_B for the left coil in the preceding experiments is the important common factor. This flux may be set up by a bar magnet or a current loop. Faraday's law of induction says that the induced emf ε in a circuit is equal (except for a minus sign) to the rate at which the flux through the circuit is changing. If the rate of change of flux is in webers/second, the emf ε will be in volts. In equation form

$$\varepsilon = -\frac{d\Phi_B}{dt}.$$
(35-1)

This is *Faraday's law of induction.* The minus sign is an indication of the *direction* of the induced emf, a matter we discuss in Section 35-3.*

If we apply Eq. 35-1 to a coil of N turns, an emf appears in every turn and these emfs are to be added. If the coil is so tightly wound that each turn can be said to occupy the same region of space, the flux through each turn will then be the same. The flux through each turn is also the same for (ideal) toroids and solenoids (see Section 34-5). The induced emf in all such devices is given by

$$\varepsilon = -N \frac{d\Phi_B}{dt} = -\frac{d(N\Phi_B)}{dt}, \qquad (35-2)$$

where $N\Phi_B$ measures the so-called *flux linkages* in the device.

Figures 35-1 and 35-2 suggest that there are at least two ways in which we can make the flux through a circuit change and thus induce an emf in that circuit. Speaking loosely, the coil that is connected to the galvanometer cannot tell in which of these experiments it is participating; it is aware only that the flux passing through its cross-sectional area is changing. The flux through a circuit can also be changed by changing its shape, that is, by squeezing or stretching it.

EXAMPLE 1

A long solenoid has 200 turns/cm and carries a current of 1.5 A; its diameter is 3.0 cm. At its center we place a 100-turn, close-packed coil of diameter 2.0 cm. This coil is arranged so that **B** at the center of the solenoid is parallel to its axis. The current in the solenoid is reduced to zero and then raised to 1.5 A in the other direction at a steady rate over a period of 0.050 s. What induced emf appears in the coil while the current is being changed?

The field B at the center of the solenoid is given by Eq. 34-7, or

$$B = \mu_0 ni = (4\pi \times 10^{-7}\,\text{T·m/A})(200 \times 10^2\,\text{turns/m})(1.5\,\text{A})$$

$$= 3.8 \times 10^{-2}\,\text{T}.$$

The area of the coil (not of the solenoid) is $3.1 \times 10^{-4}\,\text{m}^2$. The initial flux Φ_B through each turn of the coil is given by

$$\Phi_B = BA = (3.8 \times 10^{-2}\,\text{T})(3.1 \times 10^{-4}\,\text{m}^2) = 1.2 \times 10^{-5}\,\text{Wb}.$$

The flux goes from an initial value of 1.2×10^{-5} Wb to a final value of -1.2×10^{-5} Wb. The *change* in flux $\Delta\Phi_B$ for each turn of the coil during the 0.050-s period is thus twice the initial value. The induced emf is given by

$$\varepsilon = -\frac{N\Delta\Phi_B}{\Delta t} = -\frac{(100)(2 \times 1.2 \times 10^{-5}\,\text{Wb})}{0.050\,\text{s}} = -4.8 \times 10^{-2}\,\text{V} = -48\,\text{mV}.$$

The minus sign deals with the *direction* of the emf, as we explain below.

35-3
LENZ'S LAW

So far we have not specified the directions of the induced emfs. Although we can find these directions from a formal analysis of Faraday's law, we prefer to find them from the conservation-of-energy principle which, in this context, takes the form of Lenz's law, deduced by Heinrich Friedrich Lenz (1804–1865) in 1834: *The induced current will appear in such a direction that it opposes the change that produced it.* The minus sign in Faraday's law suggests this opposition. In mechanics

---

*Faraday, being untrained in mathematics, did not express his law of induction in equation form. In fact, in his three-volume "Experimental Researches in Electricity," a landmark work in the development of physics and chemistry, not a single equation appears!

the energy principle often allows us to draw conclusions about mechanical systems without analyzing them in detail. We use the same approach here.

Lenz's law refers to induced *currents*, which means that it applies only to closed conducting circuits. If the circuit is open, we can usually think in terms of what would happen if it *were* closed and in this way find the direction of the induced emf.

Consider the first of Faraday's experiments described in Section 35-1. Figure 35-3 shows the north pole of a magnet and a cross section of a nearby conducting loop. As we push the magnet toward the loop (or the loop toward the magnet) an induced current is set up in the loop. What is its direction?

figure 35-3
If we move the magnet toward the loop, the induced current points as shown, setting up a magnetic field that opposes the motion of the magnet.

A current loop sets up a magnetic field at distant points like that of a magnetic dipole, one face of the loop being a north pole, the opposite face being a south pole. The north pole, as for bar magnets, is that face *from* which the lines of **B** emerge. If, as Lenz's law predicts, the loop in Fig. 35-3 is to oppose the motion of the magnet toward it, the face of the loop toward the magnet must become a north pole. The two north poles — one of the current loop and one of the magnet — will repel each other. The right-hand rule shows that for the magnetic field set up by the loop to emerge from the right face of the loop, the induced current must be as shown. The current will be counterclockwise as we sight along the magnet toward the loop.

When we push the magnet toward the loop (or the loop toward the magnet), an induced current appears. In terms of Lenz's law this pushing is the "change" that produces the induced current, and, according to this law, the induced current will oppose the "push." If we pull the magnet away from the coil, the induced current will oppose the "pull" by creating a *south* pole on the right-hand face of the loop of Fig. 35-3. To make the right-hand face a south pole, the current must be opposite to that shown in Fig. 35-3. Whether we pull or push the magnet, its motion will always be automatically opposed.

The agent that causes the magnet to move, either toward the coil or away from it, will always experience a resisting force and will thus be required to do work. From the conservation-of-energy principle this work done on the system must be exactly equal to the thermal energy produced in the coil, since these are the only two energy transfers that occur in the system. If we move the magnet more rapidly, we will have to do work at a faster rate and the rate of production of thermal energy will increase correspondingly. If we cut the loop and then perform the experiment, there will be no induced current, no thermal energy, no force on the magnet, and no work required to move it. There will still be an emf in the loop, but, like a battery connected to an open circuit, it will not set up a current.

If the current in Fig. 35-3 were in the *opposite* direction to that shown, the face of the loop toward the magnet would be a south pole, which would *pull* the bar magnet toward the loop. We would only need to push the magnet slightly to start the process and then the action would be self-perpetuating. The magnet would accelerate toward the loop, increasing its kinetic energy all the time. At the same time thermal energy would appear in the loop at a rate that would increase with time. This would indeed be a something-for-nothing situation! Needless to say, it does not occur.

Let us apply Lenz's law to Fig. 35-3 in a different way. Figure 35-4 shows the lines of **B** for the bar magnet.* On this point of view the "change" is the increase in Φ_B through the loop caused by bringing the magnet nearer. The induced current opposes this change by setting up a field that tends to oppose the increase in flux caused by the moving magnet. Thus the field due to the induced current must point from left to right through the plane of the coil, in agreement with our earlier conclusion.

It is not significant here that the induced field opposes the *field* of the magnet but rather that it opposes the *change,* which in this case is the *increase* in Φ_B through the loop. If we withdraw the magnet, we reduce Φ_B through the loop. The induced field will now oppose this decrease in Φ_B (that is, the change) by *reenforcing* the magnet field. In each case the induced field opposes the change that gives rise to it.

figure 35-4
If we move the magnet toward the loop, we increase Φ_B through the loop.

figure 35-5
A rectangular loop is pulled out of a magnetic field with velocity **v**.

The example of Fig. 35-4, although easy to understand qualitatively, does not lend itself to quantitative calculations. Consider then Fig. 35-5, which shows a rectangular loop of wire of width *l,* one end of which is in a uniform field **B** pointing at right angles to the plane of the loop. This field **B** may be produced in the gap of a large electromagnet like that of Fig. 33-2. The dashed lines show the assumed limits of the magnetic field. The experiment consists in pulling the loop to the right at a constant speed *v.*

Note that the situation described by Fig. 35-5 does not differ in any

35-4
INDUCTION—A
QUANTITATIVE STUDY

---

* There are two fields of **B** in this problem—one connected with the current loop and one with the bar magnet. You must always be certain which one is meant.

essential particular from that of Fig. 35-4. In each case a conducting loop and a magnet are in relative motion; in each case the flux of the field of the magnet through the loop is being caused to change with time.

The flux Φ_B enclosed by the loop in Fig. 35-5 is

$$\Phi_B = Blx,$$

where lx is the area of that part of the loop in which B is not zero. We find the emf ε, from Faraday's law, or

$$\varepsilon = -\frac{d\Phi_B}{dt} = -\frac{d}{dt}(Blx) = -Bl\frac{dx}{dt} = Blv, \qquad (35\text{-}3)$$

where we have set $-dx/dt$ equal to the speed v at which the loop is pulled out of the magnetic field. Note that the only dimension of the loop that enters into Eq. 35-3 is the length l of the left end conductor. As we shall see later, the induced emf in Fig. 35-5 may be regarded as localized here. An induced emf such as this, produced by pulling a conductor through a magnetic field (or conversely), is sometimes called a *motional emf*.

The emf Blv sets up a current in the loop given by

$$i = \frac{\varepsilon}{R} = \frac{Blv}{R}, \qquad (35\text{-}4)$$

where R is the loop resistance. From Lenz's law, this current (and thus ε) must be clockwise in Fig. 35-5; it opposes the "change" (the decrease in Φ_B) by setting up a field that is parallel to the external field within the loop.

The current in the loop will cause forces $\mathbf{F}_1$, $\mathbf{F}_2$, and $\mathbf{F}_3$ to act on the three conductors, in accord with Eq. 33-6a, or

$$\mathbf{F} = i\mathbf{l} \times \mathbf{B}. \qquad (35\text{-}5)$$

Because $\mathbf{F}_2$ and $\mathbf{F}_3$ are equal and opposite, they cancel each other; $\mathbf{F}_1$, which is the force that opposes our effort to move the loop, is given in magnitude from Eqs. 35-4 and 35-5 as

$$F_1 = ilB\,\sin 90° = \frac{B^2l^2v}{R}.$$

The agent that pulls the loop must do work at the steady rate of

$$P = F_1v = \frac{B^2l^2v^2}{R}. \qquad (35\text{-}6)$$

From the principle of the conservation of energy, thermal energy must appear in the resistor at this same rate. We introduced the conservation-of-energy principle into our derivation when we wrote down the expression for the current (Eq. 35-4); recall that the relation $i = \varepsilon/R$ for single-loop circuits is a direct consequence of this principle. Thus we should be able to write down the expression for the rate of thermal energy production in the loop with the expectation that we will obtain a result identical with Eq. 35-6. Recalling Eq. 35-4, we put

$$P_J = i^2R = \left(\frac{Blv}{R}\right)^2 R = \frac{B^2l^2v^2}{R},$$

which is indeed the expected result. This example provides a quantitative illustration of the conversion of mechanical energy (the work done by an external agent) into electrical energy (associated with the induced emf) and finally into thermal energy.

Figure 35-6 shows a side view of the coil in the field. In Fig. 35-6a the coil is stationary; in Fig. 35-6b we are moving it to the right; in Fig. 35-6c we are moving it to the left. The lines of **B** in these figures represent the *resultant field* produced by the vector addition of the field **B**$_0$ due to the magnet and the field **B**$_i$ due to the induced current, if any, in the coil. These lines suggest convincingly that the agent moving the coil always experiences an opposing force.

figure 35-6
Side view of a rectangular loop in a magnetic field showing the loop (a) at rest, (b) being pulled out, and (c) being pushed in. The configuration of the lines suggests an opposing force in both (b) and (c).

Figure 35-7 shows a rectangular loop of resistance R, width l, and length a being pulled at constant speed v through a region of thickness d in which a uniform magnetic field **B** is set up by a magnet.

(a) Plot the flux Φ_B through the loop as a function of the coil position x. Assume that $l = 4$ cm, $a = 10$ cm, $d = 15$ cm, $R = 16$ Ω, $B = 2.0$ T, and $v = 1.0$ m/s.

The flux Φ_B is zero when the loop is not in the field; it is Bla when the loop is entirely in the field; it is Blx when the loop is entering the field and $Bl[a - (x - d)]$ when the loop is leaving the field. These conclusions, which you should verify, are shown graphically in Fig. 35-8a.

(b) Plot the induced emf ε.

The induced emf ε is given by $\varepsilon = -d\Phi_B/dt$, which we can write as

$$\varepsilon = -\frac{d\Phi_B}{dt} = -\frac{d\Phi_B}{dx}\frac{dx}{dt} = -\frac{d\Phi_B}{dx}v,$$

where $d\Phi_B/dx$ is the slope of the curve of Fig. 35-8a. $\varepsilon(x)$ is plotted in Fig. 35-8b. Lenz's law, from the same type of reasoning as that used for Fig. 35-5, shows that when the coil is entering the field, the emf ε acts counterclockwise as seen from above. Note that there is no emf when the coil is entirely in the magnetic field because the flux Φ_B through the coil is not changing with time, as Fig. 35-8a shows.

(c) Plot the rate P of thermal energy production in the loop.

EXAMPLE 2

figure 35-7
Example 2. A rectangular loop is caused to move with a velocity **v** through a magnetic field. The position of the loop is measured by x, the distance between the effective left edge of the field **B** and the right end of the loop.

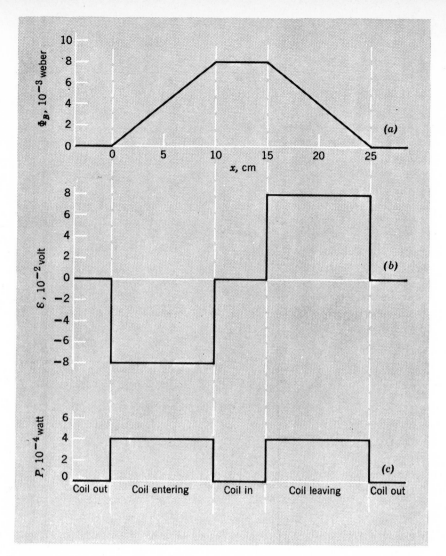

figure 35-8
Example 2. In practice the sharp corners would be rounded.

This is given by $P = \varepsilon^2/R$. It may be calculated by squaring the ordinate of the curve of Fig. 35-8b and dividing by R. The result is plotted in Fig. 35-8c.

If the fringing of the magnetic field, which cannot be avoided in practice (see Problem 34-2), is taken into account, the sharp bends and corners in Fig. 35-8 will be replaced by smooth curves. What changes would occur in the curves of Fig. 35-8 if the coil were open circuited?

EXAMPLE 3

A copper rod of length L rotates at angular frequency ω in a uniform magnetic field **B** as shown in Fig. 35-9. Find the emf ε developed between the two ends of the rod. (We might measure this emf if we place a conducting rail along the dashed circle in the figure and connect a voltmeter between the rail and point O.

If a wire of length $d\mathbf{l}$ is moved at velocity **v** at right angles to a field **B**, a motional emf $d\varepsilon$ will be developed (see Eq. 35-3) given by

$$d\varepsilon = Bv \, dl.$$

The rod of Fig. 35-9 may be divided into elements of length dl, the linear speed v of each element being ωl. Each element is perpendicular to **B** and is also moving in a direction at right angles to **B** so that, since the $d\varepsilon$'s are "in series",

$$\varepsilon = \int d\varepsilon = \int_0^L Bv \, dl = \int_0^L B(\omega l) \, dl = \tfrac{1}{2}B\omega L^2.$$

figure 35-9
Example 3

For a second approach, consider that at any instant the flux enclosed by the sector aOb in Fig. 35-9 is given by

$$\Phi_B = BA = B\left(\tfrac{1}{2}L^2\theta\right),$$

where $\tfrac{1}{2}L^2\theta$ can be shown to be the area of the sector. Differentiating gives

$$\frac{d\Phi_B}{dt} = \tfrac{1}{2}BL^2\frac{d\theta}{dt} = \tfrac{1}{2}B\omega L^2.$$

From Faraday's law, this is precisely the magnitude of ε, and agrees with the result just derived.

So far we have considered emfs induced by the relative motion of magnets and coils. In this section we assume that there is no physical motion of gross objects but that the magnetic field may vary with time. If we place a conducting loop in such a time-varying field, the flux through the loop will change and an induced emf will appear in the loop. This emf will set the charge carriers in motion, that is, it will induce a current.

From a microscopic point of view we can say, equally well, that the changing flux of **B** sets up an induced electric field **E** at various points around the loop. These induced electric fields are just as real as electric fields set up by static charges and will exert a force **F** on a test charge q_0 given by $\mathbf{F} = q_0\mathbf{E}$. Thus we can restate Faraday's law of induction in a loose but informative way as: *A changing magnetic field produces an electric field.*

To fix these ideas, consider Fig. 35-10, which shows a uniform magnetic field **B** at right angles to the plane of the page. We assume that **B** is increasing in magnitude at the same constant rate dB/dt at every point. This could be done by causing the current in the windings of the electromagnet that establishes the field to increase with time in the proper way.

The circle of arbitrary radius r shown in Fig. 35-10 encloses, at any instant, a flux Φ_B. Because this flux is changing with time, an induced emf given by $\varepsilon = -d\Phi_B/dt$ will appear around the loop. The electric fields **E** induced at various points of the loop must, from symmetry, be tangent to the loop. Thus the electric lines of force that are set up by the changing magnetic field are in this case concentric circles.

If we consider a test charge q_0 moving around the circle of Fig. 35-10, the work W done on it per revolution is, in terms of the definition of an emf, simply εq_0. From another point of view it is $(q_0 E)(2\pi r)$, where $q_0 E$ is the force that acts on the charge and $2\pi r$ is the distance over

35-5
TIME-VARYING MAGNETIC FIELDS

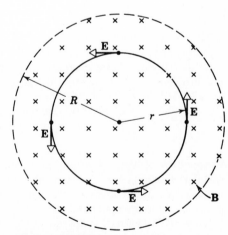

figure 35-10
The induced electric fields at four points produced by an increasing magnetic field. (The magnetic field cannot end abruptly at radius R but must approach zero gradually. This "fringing" does not change any of the arguments of this section.)

which the force acts. Setting the two expressions for W equal and canceling q_0 yields

$$\varepsilon = E2\pi r. \qquad (35\text{-}7)$$

In a more general case than that of Fig. 35-10 we must write

$$\varepsilon = \oint \mathbf{E} \cdot d\mathbf{l}. \qquad (35\text{-}8)$$

If this integral is evaluated for the conditions of Fig. 35-10, we obtain Eq. 35-7 at once. If Eq. 35-8 is combined with Eq. 35-1 ($\varepsilon = -d\Phi_B/dt$), we can write Faraday's law of induction as

$$\oint \mathbf{E} \cdot d\mathbf{l} = -\frac{d\Phi_B}{dt}, \qquad (35\text{-}9)$$

which is the form in which this law is expressed in Table 40-2.

EXAMPLE 4

Let B in Fig. 35-10 be increasing at the rate dB/dt. Let R be the radius of the cylindrical region in which the magnetic field is assumed to exist. What is the magnitude of the electric field $\mathbf{E}$ at any radius r? Assume that $dB/dt = 0.10$ T/s and $R = 10$ cm.

(a) For $r < R$, the flux Φ_B through the loop is

$$\Phi_B = B(\pi r^2).$$

Substituting into Faraday's law (Eq. 35-9),

$$\oint \mathbf{E} \cdot d\mathbf{l} = -\frac{d\Phi_B}{dt}$$

yields

$$(E)(2\pi r) = -\frac{d\Phi_B}{dt} = -(\pi r^2)\frac{dB}{dt}.$$

Solving for E yields

$$E = -\tfrac{1}{2}r\frac{dB}{dt}.$$

The minus sign is retained to suggest that the induced electric field $\mathbf{E}$ acts to *oppose* the change of the magnetic field. Note that $E(r)$ depends on dB/dt and not on B. Substituting numerical values, assuming $r = 5.0$ cm, yields, for the magnitude of E,

$$E = \tfrac{1}{2}r\frac{dB}{dt} = (\tfrac{1}{2})(5 \times 10^{-2} \text{ m})(0.10 \text{ T/s}) = 2.5 \times 10^{-3} \text{ V/m}.$$

(b) For $r > R$ the flux through the loop is

$$\Phi_B = \int \mathbf{B} \cdot d\mathbf{S} = B(\pi R^2).$$

This equation is true because $\mathbf{B} \cdot d\mathbf{S}$ is zero for those points of the loop that lie outside the effective boundary of the magnetic field.

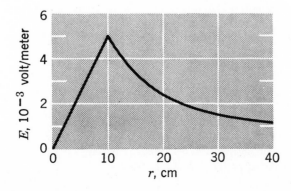

figure 35-11
Example 4. If the fringing of the field in Fig. 35-10 were to be taken into account, the result would be a rounding of the sharp cusp at $r = R$ ($= 10$ cm).

From Faraday's law (Eq. 35-9),

$$(E)(2\pi r) = -\frac{d\Phi_B}{dt} = -(\pi R^2)\frac{dB}{dt}.$$

Solving for E yields

$$E = -\frac{1}{2}\frac{R^2}{r}\frac{dB}{dt}.$$

These two expressions for $E(r)$ yield the same result, as they must, for $r = R$. Figure 35-11 is a plot of the magnitude of $\mathbf{E}(r)$ for the numerical values given.

In applying Lenz's law to Fig. 35-10, imagine that a circular conducting loop is placed concentrically in the field. Since Φ_B through this loop is increasing, the induced current in the loop will tend to oppose this "change" by setting up a magnetic field of its own that points up within the loop. Thus the induced current i must be counterclockwise, which means that the lines of the induced electric field **E**, which is responsible for the current, must also be counterclockwise. If the magnetic field in Fig. 35-10 were *decreasing* with time, the induced current and the lines of force of the induced electric field **E** would be clockwise, again opposing the *change* in Φ_B.

Figure 35-12 shows four of many possible loops to which Faraday's law may be applied. For loops 1 and 2, the induced emf ε is the same because these loops lie entirely within the changing magnetic field and, having the same area, thus have the same value of $d\Phi_B/dt$. Note that even though the emf ε $(= \oint \mathbf{E} \cdot d\mathbf{l})$ is the same for these two loops, the distribution of electric fields **E** around the perimeter of each loop, as indicated by the electric lines of force, is different. For loop 3 the emf is less because Φ_B and $d\Phi_B/dt$ for this loop are less, and for loop 4 the induced emf is zero.

The induced electric fields that are set up by the induction process are not associated with charges but with a changing magnetic flux. Although both kinds of electric fields exert forces on charges, there is a difference between them. The simplest manifestation of this difference is that lines of **E** associated with a changing magnetic flux can form closed loops (see Fig. 35-12); lines of **E** associated with charges cannot form closed loops but can always be drawn to start on a positive charge and end on a negative charge.

Equation 29-5, which defined the potential difference between two points a and b, is

$$V_b - V_a = \frac{W_{ab}}{q_0} = -\int_a^b \mathbf{E} \cdot d\mathbf{l}.$$

We have insisted that if potential is to have any useful meaning this integral (and W_{ab}) must have the same value for every path connecting a and b. This proved to be true for every case examined in earlier chapters.

An interesting special case comes up if a and b are the same point. The path connecting them is now a closed loop; V_a must be identical with V_b and this equation reduces to

$$\oint \mathbf{E} \cdot d\mathbf{l} = 0. \tag{35-10}$$

However, when changing magnetic flux is present, $\oint \mathbf{E} \cdot d\mathbf{l}$ is precisely *not* zero but is, according to Faraday's law (see Eq. 35-9), $-d\Phi_B/dt$. Electric fields associated with stationary charges are *conservative*, but those associated with changing magnetic fields are *nonconservative*; see Section 8-2. Electric potential, which can be defined only for a conservative force, *has no meaning for electric fields produced by induction.*

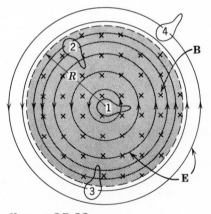

figure 35-12
Showing the circular lines of **E** from an increasing magnetic field. The four loops are imaginary paths around which an emf can be calculated.

The betatron is a device used to accelerate electrons to high speeds by allowing them to be acted upon by induced electric fields that are set up by a changing magnetic flux. It provides an excellent illustration of the "reality" of such induced fields and it is in this context that we discuss it. The energetic electrons can be used for fundamental research in physics or to produce penetrating X-rays which are useful in cancer therapy and in industry.

Figure 35-13 shows a 100-MeV betatron that was constructed at the General Electric Company. At this energy the electron speed is $0.999986c$, where c is the speed of light, so that relativistic mechanics must certainly be used in the analysis of its operation. Figure 35-14 shows a vertical cross section through the central part of the betatron to which the man in Fig. 35-13 is pointing.

The magnetic field in the betatron has several functions: (a) it guides the electrons in a circular path; (b) the changing magnetic field generates an electric field (Section 35-5) that accelerates the electrons in this path; (c) it keeps the radius of the orbit in which the electrons are moving essentially constant; (d) it introduces the electrons into the or-

figure 35-13
A 100-MeV betatron. M shows the magnet, C the magnetizing coils, and D the region in which the "doughnut" is located. (Courtesy General Electric Company.)

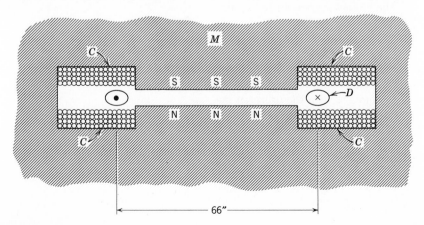

figure 35-14
Cross section of a betatron, showing magnet M, coils C, and "doughnut" D. Electrons emerge from the page at the left and enter it at the right. See Problem 33.

bit initially and removes them from the orbit after they have reached full energy; and finally (*e*) it provides a restoring force that resists any tendency for the electrons to leave their orbit, either vertically or radially. It is remarkable that it is possible to do all these things by proper shaping and control of the magnetic field.

The object marked *D* in Fig. 35-14 is an evacuated glass "doughnut" inside which the electrons travel. Their orbit is a circle at right angles to the plane of the figure. The electrons emerge from the plane at the left (·) and enter it at the right (×). In the General Electric machine the radius of the electron path is 33 in. The coils *C* and the 130-ton steel magnet shown in Fig. 35-13 provide the magnetic flux that passes through the plane of this orbit.

The current in coils *C* is made to alter periodically, 60 times/second to produce a changing flux through the orbit, shown in Fig. 35-15. Here Φ_B is taken as positive when **B** is pointing up, as in Fig. 35-14. If the electrons are to circulate in the direction shown, they must do so during the positive half-cycle, marked *ac* in Fig. 35-15. You should verify this (see Section 33-6). The electrons are accelerated by electric fields set up by the changing flux. The direction of these induced fields depends on the sign of $d\Phi_B/dt$ and must be chosen to accelerate, and not to decelerate, the electrons. Thus only half the positive half-cycle in Fig. 35-15 can be used for acceleration; it will prove to be *ab*.

The average value of $d\Phi_B/dt$ during the quarter-cycle *ab* is the slope of the dashed line, or

$$\frac{d\Phi_B}{dt} = \frac{1.8 \text{ Wb}}{4.2 \times 10^{-3} \text{ s}} = 430 \text{ V}.$$

From Faraday's law (Eq. 35-1), this is also the emf in volts. The electron will thus increase its energy by 430 eV every time it makes a trip around the orbit in the changing flux. If the electron gains only 430 eV of energy per revolution, it must make about 230,000 rev to gain its full 100 MeV. For an orbit radius of 33 in., this corresponds to a length of path of some 750 miles.

The betatron provides a good example of the fact that electric potential has no meaning for electric fields produced by induction. If a potential exists, it must be true that, as Eq. 35-10 shows, $\oint \mathbf{E} \cdot d\mathbf{l} = 0$ for any closed path. In the betatron, however, this integral, evaluated around the orbit, is precisely *not* zero but is, in our example, 430 volts. It must not be thought, of course, that the betatron violates the conservation-of-energy principle. The gain in kinetic energy of the circulating electron (430 eV/rev) must be supplied by an identifiable energy source. It comes, in fact, from the generator that energizes the magnet coils, thus providing the changing magnetic field. The energy is transmitted to the electron through the intermediary of this changing field.

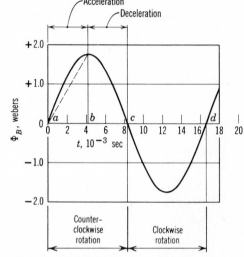

figure 35-15
The flux through the orbit of a betatron, during one cycle. Rotation of the electrons in the desired direction (counterclockwise as viewed from above in Fig. 35-14) is possible only during half-cycle *ac*.

EXAMPLE 5

In the betatron of Fig. 35-13, which is the "accelerating" quarter-cycle?

Let us assume that it is *ab* in Fig. 35-15, during which Φ_B through the orbit is *increasing*. If a conducting loop were placed to coincide with the orbit, an induced current would appear in the loop to oppose the tendency of Φ_B to increase. This means that a magnetic field would be set up that would oppose the field of the large magnet. Thus **E** would point outward at the right side of "doughnut" *D* in Fig. 35-14 and inward on the left side. The force (−*e***E**) acting on the electron is in the opposite direction to **E** because of the negative charge of the electron. Thus the tangential force acting on the electron is in the same direction as that at which it circulates in its orbit; this means that the speed of the electron will increase, as desired. You should go through this same analysis carefully, assuming (incorrectly, as it will turn out) that the accelerating half-cycle is *bc* in Fig. 35-15 rather than *ab*.

Faraday's law, in the form $\varepsilon = -d\Phi_B/dt$, gives correctly the induced emf no matter whether the change in Φ_B is produced by moving a coil, moving a magnet, changing the strength of a magnetic field, changing the shape of a conducting loop, or in other ways. However, observers who are in relative motion with respect to each other, even though they would all agree on the numerical value of the emf, would give different microscopic descriptions of the induction process. In electromagnetic systems, as well as in mechanical systems, it is important that the state of motion of the observer with respect to his environment be made perfectly clear.

Figure 35-16 shows a closed loop which is caused to move at velocity **v** with respect to a magnet that provides a uniform field **B** in the region shown. We consider first an observer, identified as *S*, who is *at rest with respect to the magnet* used to establish the field **B**; see Fig. 35-16*a*. The induced emf in this case is called a *motional emf* because the conducting loop is moving with respect to this observer.

Consider a positive charge carrier at the center of the left end of the conducting loop. To observer *S*, this charge, constrained to move to the right along with the loop, is a charge *q* moving with a velocity **v** in the magnetic field **B** and as such it experiences a sideways magnetic deflecting force given by Eq. 33-2 (**F** = *q***v** × **B**). This force causes the carriers to move upward along the conductor, so that they acquire a drift velocity $\mathbf{v}_d$ also, as shown in Fig. 35-16*a*.

The resultant equilibrium speed of the carrier is now **V**, which we find by adding **v** and $\mathbf{v}_d$ vectorially in Fig. 35-16*a*. The magnetic deflecting force $\mathbf{F}_m$ is, as always, at right angles to the resultant velocity **V** of the carrier and is given by

$$\mathbf{F}_m = q\mathbf{V} \times \mathbf{B}. \tag{35-13}$$

35-7
INDUCTION AND RELATIVE MOTION

figure 35-16

A closed conducting loop is in relative motion with respect to a magnet. (*a*) An observer *S*, fixed with respect to the magnet that produces the field **B**, sees the loop moving to the right. He also sees (see text) a *magnetic* force (Eq. 33-2) equal to $F_m \cos \theta$ acting upward on the positive charge carriers. (*b*) An observer *S'*, fixed with respect to the loop, sees the magnet moving toward the left. He also sees an *electric* force (see text) acting upward on the positive charge carriers. In both figures there is, of course, an average internal collision force (not shown; see Sections 31-4 and 31-5) that keeps the charge carriers from accelerating.

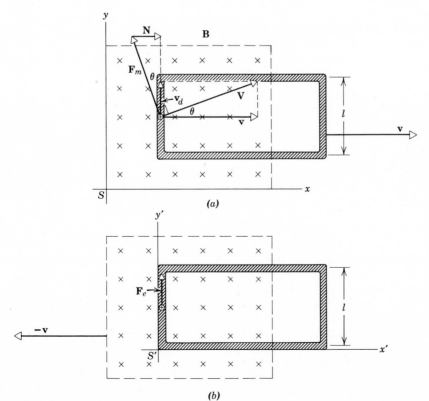

(*a*)

(*b*)

Now $\mathbf{F}_m$ acting alone would tend to push the carriers through the left wall of the conductor. Because this does not happen the conductor wall must exert a normal force $\mathbf{N}$ on the carriers (see Fig. 35-16a) of magnitude such that $\mathbf{v}_d$ lies parallel to the axis of the wire; in other words, $\mathbf{N}$ exactly cancels the horizontal component of $\mathbf{F}_m$, leaving only the component $\mathbf{F}_m \cos \theta$ that lies along the direction of the conductor. This component of force on the carrier is also canceled out, this time by the average force $\overline{\mathbf{F}}_i$ associated with the internal collisions that the carrier experiences as it drifts with (constant) speed v_d through the wire.

The kinetic energy of the charge carrier as it drifts through the wire remains constant. This is consistent with the fact that the resultant force acting on the charge carrier ($= \mathbf{F}_m + \overline{\mathbf{F}}_i + \mathbf{N}$) is zero. The work done by $\mathbf{F}_m$ is zero because magnetic forces, acting at right angles to the velocity of a moving charge, can do no work on that charge. Thus the (negative) work done on the carrier by the average internal collision force $\mathbf{F}_i$ must be exactly canceled by the (positive) work done on the carrier by the force $\mathbf{N}$. In the last analysis $\mathbf{N}$ is exerted by the agent who pulls the loop through the magnetic field, and the mechanical energy expended by this agent appears as thermal energy in the loop, as we have seen in Section 35-4.

Let us then calculate the work dW done on the carrier in time dt by the force $\mathbf{N}$; it is

$$dW = N(v\,dt) \tag{35-14}$$

in which $v\,dt$ is the distance that the loop (and the carrier) have moved to the right in Fig. 35-16a in time dt. We can write for N (see Eq. 35-13 and Fig. 35-16a).

$$N = F_m \sin \theta = (qVB)(v_d/V) = qBv_d. \tag{35-15}$$

Substituting Eq. 35-15 into Eq. 35-14 yields

$$\begin{aligned} dW &= (qBv_d)(v\,dt) \\ &= (qBv)(v_d\,dt) = qBv\,dl \end{aligned} \tag{35-16}$$

in which dl ($= v_d\,dt$) is the distance the carrier drifts along the conductor in time dt.

The work done on the carrier as it makes a complete circuit of the loop is found by integrating Eq. 35-16 around the loop and is

$$W = \oint dW = qBvl. \tag{35-17}$$

This follows because work contributions for the top and the bottom of the loops are opposite in sign and cancel and no work is done in those portions of the loop that lie outside the magnetic field.

An agent that does work on charge carriers, thus establishing a current in a closed conducting loop, can be viewed as an emf. We can write, making use of Eq. 35-17,

$$\varepsilon = \frac{W}{q} = \frac{qBvl}{q} = Blv, \tag{35-18}$$

which is, of course, the same result that we derived from Faraday's law of induction; see Eq. 35-3. Thus a motional emf is intimately connected with the sideways deflection of a charged particle moving through a magnetic field.

We now consider how the situation of Fig. 35-16 would appear to an observer S' who is *at rest with respect to the loop*. To this observer, the magnet is moving to the left in Fig. 35-16b with velocity $-\mathbf{v}$, and the charge q is at rest as far as its left-to-right motion is concerned. However S', like S, observes that the charge drifts clockwise around the loop and he measures the same emf ε that S measures. S' accounts for this, at the microscopic level, by postulating that an electric field $\mathbf{E}$ is induced in the loop by the action of the moving magnet. This induced field $\mathbf{E}$, which has the same origin as the induced fields that we discussed in Section 35-5, exerts a force on the charge carrier given by $q\mathbf{E}$.

The induced field $\mathbf{E}$, which exists in the end of the loop only, is associated with an emf ε and generates a current in the closed loop. Note that, in any closed loop in which there is a current, an internal electric field must exist at every point at which charges are moving. These electric fields, however, are set up by the emf, as in the case of a closed loop connected to a battery, and are *not* induced by the motion of the magnet. It is only this induced field **E** that we associate with the emf, through the relation (Eq. 35-8)

$$\varepsilon = \oint \mathbf{E} \cdot d\mathbf{l}$$

which reduced to

$$\varepsilon = El \qquad (35\text{-}19)$$

in this case. This is so because no *induced* electric field appears in the upper and lower bars (because of the nature of their motions), and none appears in the part of the loop outside the magnetic field.

The emfs given by Eqs. 35-18 and 35-19 must be identical because the relative motion of the loop and the magnet is identical in the two cases shown in Fig. 35-16. Equating these relations yields

$$El = Blv,$$

or

$$E = vB. \qquad (35\text{-}20a)$$

In Fig. 35-16*b* the vector **E** points upward along the axis of the left end of the conducting loop because this is the direction in which positive charges are observed to drift. The directions of **v** and **B** are clearly shown in this figure. We see, then, that Eq. 35-20*a* is consistent with the more general vector relation

$$\mathbf{E} = \mathbf{v} \times \mathbf{B}. \qquad (35\text{-}20b)$$

We have not proved Eq. 35-20*b* except for the special case of Fig. 35-16; nevertheless it proves to be true in general, that is, no matter what the angle between **v** and **B**.

We interpret Eq. 35-20*b* in the following way: Observer S fixed with respect to the magnet is aware only of a magnetic field. The force to him arises from the motion of the charges through **B**. Observer S' fixed on the charge carrier is aware of an electric field **E** also and attributes the force on the charge (at rest with respect to him initially) to the electric field. S says the force is of purely magnetic origin and S' says the force is of purely electric origin. From the point of view of S, the induced emf is given by $\oint (\mathbf{v} \times \mathbf{B}) \cdot d\mathbf{l}$. From the point of view of S', the same induced emf is given by $\oint \mathbf{E} \cdot d\mathbf{l}$, where **E** is the (induced) electric vector that he observes at points along the circuit.

For a third observer S'' who judges that both the magnet and the loop are moving, the force tending to move charges around the loop is neither purely electric nor purely magnetic, but a bit of each. In summary, in the equation

$$\mathbf{F}/q = \mathbf{E} + \mathbf{v} \times \mathbf{B},$$

different observers form different assessments of **E, B,** and **v** but, when these are combined, all observers form the same assessment of $\mathbf{F}/q$ and all obtain the same value for the induced emf in the loop (this depends only on the relative motion). That is, the total force is the same for all observers, but each observer forms a different estimate of the separate electric and magnetic forces contributing to the same total force.

The essential point is that what seems like a magnetic field to one observer may seem like a mixture of an electric field and a magnetic field to a second observer in a different inertial reference frame. Both observers would agree, however, on the gross measurable result, in the case of Fig. 35-16, the current in the loop. We are forced to conclude that magnetic and electric fields are *not* independent of each other and have no separate unique existence; they depend on the inertial frame.

Einstein started to think about relative motion at age 16 and published his famous paper on the theory of special relativity in 1905, when he was a 26-year-old patent examiner in the Swiss Patent Office in Berne. He was *not* led to this theory by early considerations of the nature of space and time but precisely by the problems raised in this section.

This is not only made clear by the title of his paper (translated from the German), "On the Electromagnetic Forces Acting on Moving Bodies," but is strongly reenforced by the opening lines of his paper, which are:

It is known that Maxwell's electrodynamics—as usually understood at the present time—when applied to moving bodies, leads to asymmetries which do not appear to be inherent in the phenomena. Take, for example, the reciprocal electrodynamic action of a magnet and a conductor. The observable phenomenon here depends only on the relative motion of the conductor and the magnet, whereas the customary view draws a sharp distinction between the two cases in which either the one or the other of these bodies is in motion. For if the magnet is in motion and the conductor at rest, there arises in the neighbourhood of the magnet an electric field with a certain definite energy, producing a current at the places where parts of the conductor are situated. But if the magnet is stationary and the conductor in motion, no electric field arises in the neighbourhood of the magnet. In the conductor, however, we find an electromotive force, to which in itself there is no corresponding energy, but which gives rise—assuming equality of relative motion in the two cases discussed—to electric currents of the same path and intensity as those produced by the electric forces in the former case.

If you want to pursue this matter further, please read Supplementary Topic V carefully and then refer to *Introduction to Special Relativity* (Chapter IV) by Robert Resnick, John Wiley & Sons (1968).

EXAMPLE 6

In Fig. 35-16 assume that $B = 2.0$ T, $l = 10$ cm, and $v = 1.0$ m/s. Calculate (a) the induced electric field observed by S' and (b) the emf induced in the loop.

(a) The electric field, which is apparent only to observer S', is associated with the moving magnet and is given in magnitude (see Eq. 35-20a) by

$$E = vB$$
$$= (1.0 \text{ m/s})(2.0 \text{ T})$$
$$= 2.0 \text{ V/m}.$$

(b) Observer S would calculate the induced (motional) emf from

$$\varepsilon = Blv$$
$$= (2.0 \text{ T})(1.0 \times 10^{-1} \text{m})(1.0 \text{ m/s})$$
$$= 0.20 \text{ V}.$$

Observer S' would not regard the emf as motional and would use the relationship

$$\varepsilon = El$$
$$= (2.0 \text{ V/m})(1.0 \times 10^{-1} \text{ m})$$
$$= 0.20 \text{ V}.$$

As must be the case, both observers agree as to the numerical value of the emf.

questions

1. In Figs. 35-1, 35-2, and 35-3, etc. we show, for simplicity, coils of only one turn. Explain the advantage of increasing the number of turns.
2. Are "induced" emfs and currents different in any way from emfs and currents provided by a battery connected to a conducting loop? Discuss.
3. Although we discussed these matters in earlier chapters, can you now explain more clearly in your own words the difference between a magnetic field **B** and the flux of a magnetic field Φ_B? Are they vectors or scalars? In what units may each be expressed? How are these units related? Are either or both (or neither) properties of a given point in space? Discuss fully.
4. A magnet is dropped from the ceiling along the axis of a copper loop lying flat on the floor. If the falling magnet is photographed with a time sequence

camera, what differences, if any, will be noted if (a) the loop is at room temperature and (b) the loop is packed in dry ice?

5. Two conducting loops face each other a distance d apart (Fig. 35-17). An observer sights along their common axis from left to right. If a clockwise current i is suddenly established in the larger loop, (a) what is the direction of the induced current in the smaller loop? (b) What is the direction of the force (if any) that acts on the smaller loop?

6. What is the direction of the induced emf in coil Y of Fig. 35-18 when (a) coil Y is moved toward coil X and (b) the current in coil X is decreased, without any change in the relative positions of the coils?

figure 35-17
Question 5

figure 35-18
Question 6

7. The north pole of a magnet is moved away from a metallic ring, as in Fig. 35-19. In the part of the ring farthest from the reader, which way does the current point?

8. *Eddy currents.* A sheet of copper is placed in a magnetic field as shown in Fig. 35-20. If we attempt to pull it out of the field or push it further in, an automatic resisting force appears. Explain its origin. (Hint: Currents, called eddy currents, are induced in the sheet in such a way as to oppose the motion.)

9. A current-carrying solenoid is moved toward a conducting loop as in Fig. 35-21. What is the direction of circulation of current in the loop as we sight toward it as shown?

10. If the resistance R in the left-hand circuit of Fig. 35-22 is increased, what is the direction of the induced current in the right-hand circuit?

figure 35-19
Question 7

figure 35-20
Question 8

Typical eddy current loop

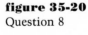

figure 35-21
Question 9

figure 35-22
Question 10

figure 35-23
Question 11

figure 35-24
Question 12

11. A loop, shown in Fig. 35-23, is removed from the magnet by pulling it vertically upward. (*a*) What is the direction of the induced current? (*b*) Is a force required to remove the loop? (*c*) Does the total amount of thermal energy produced in removing the loop depend on the time taken to remove it?

12. *Electromagnetic shielding.* Consider a conducting sheet lying in a plane perpendicular to a magnetic field **B**, as shown in Fig. 35-24. (*a*) If **B** suddenly changes, the full *change* in **B** is not immediately detected in region *P*. Explain. (*b*) If the resistivity of the sheet is zero, the *change* is not ever detected at *P*. Explain. (*c*) If **B** changes periodically at high frequency and the conductor is made of a material of low resistivity, the region near *P* is almost completely shielded from the *changes* in flux. Explain. (*d*) Is such a conductor useful as a shield from *static* magnetic fields? Explain.

13. *Magnetic damping.* A strip of copper is mounted as a pendulum about *O* in Fig. 35-25. It is free to swing through a magnetic field normal to the page. If the strip has slots cut in it as shown, it can swing freely through the field. If a strip without slots is substituted, the vibratory motion is strongly damped. Explain. (Hint: Use Lenz's law; consider the paths that the charge carriers in the strip must follow if they are to oppose the motion.)

14. What is the direction, if any, of the conventional current through resistor *R* in Fig. 35-26 (*a*) immediately after switch *S* is closed, (*b*) some time after switch *S* was closed, and (*c*) immediately after switch *S* is opened. (*d*) When switch *S* is held closed, which end of the long coil acts as a north pole? (*e*) How do the free charges in the coil containing *R* know about the flux within the long coil? What really gets them moving?

15. In Fig. 35-27 the movable wire is moved to the right, causing an induced current as shown. What is the direction of **B** in region *A*?

16. Account qualitatively for the configurations of the lines of **B** in Figs. 35-6 *b* and *c*.

17. How would Fig. 35-8 be changed if we took into account the necessary fringing of the magnetic field **B** in Fig. 35-7?

18. (*a*) In Fig. 35-10, need the circle of radius *r* be a conducting loop for **E** and ε to be present? (*b*) If the circle of radius *r* were not concentric (moved slightly to the left, say), would ε change? Would the configuration of **E** around the circle change? (*c*) For a concentric circle of radius *r*, with *r* > *R*, does an emf exist? Do electric fields exist?

19. A copper ring and a wooden ring of the same dimensions are placed so that there is the same changing magnetic flux through each. How do the induced electric fields in each ring compare?

20. In Fig. 35-12 how can the induced emfs around paths 1 and 2 be identical? The induced electric fields are much weaker near path 1 than near path 2, as the spacing of the lines of force shows. See also Fig. 35-11.

21. In a certain betatron the electrons rotate counterclockwise as seen from above. In what direction must the magnetic field point and how must it change with time while the electron is being accelerated?

22. Why can a betatron be used for acceleration only during one-quarter of a cycle?

figure 35-25
Question 13

figure 35-26
Question 14

figure 35-27
Question 15

23. To make the electrons in a betatron orbit spiral outward, would it be necessary to increase or to decrease the central flux? Assume that **B** at the orbit remains essentially unchanged.

24. A cyclotron (see Section 33-7) is a so-called *resonance device*. Does a betatron depend on resonance? Discuss.

25. In Fig. 35-16a we can see that a force $(F_m \cos \theta)$ acts on the charge carriers in the left branch of the loop. However, if there is to be a continuous current in the loop, and there is, a force of some sort must act on charge carriers in the other three branches of the loop to maintain the same drift speed v_d in these branches. What is its source? (Hint: Consider that the left branch of the loop was the only conducting element, the other three being nonconducting. Would not positive charge pile up at the top of the left half and negative charge at the bottom?)

26. Show that one volt = one weber/second.

problems

SECTION 35-2

1. A uniform magnetic field **B** is normal to the plane of a circular ring 10 cm in diameter made of #10 copper wire (diameter = 0.10 in.). At what rate must B change with time if an induced current of 10 A is to appear in the ring? *Answer:* 1.3 T/s.

2. You are given 50 cm of #18 copper wire (diameter = 0.040 in.). It is formed into a circular loop and placed at right angles to a uniform magnetic field that is increasing with time at the constant rate of 0.010 T/s. At what rate (in watts) is thermal energy generated in the loop?

3. A hundred turns of insulated copper wire are wrapped around an iron cylinder of cross-sectional area 0.001 m² and are connected to a resistor. The total resistance in the circuit is 10 Ω. If the longitudinal magnetic field in the iron changes from 1.0 T in one direction to 1.0 T in the opposite direction, how much charge flows through the circuit?   *Answer:* 2.0×10^{-2} C.

4. The current in the solenoid of Example 1 changes, not as in that Example, but according to $i = 3.0t + 1.0t^2$, where i is in amperes and t in seconds. (a) Plot quantitatively the induced emf in the coil from $t = 0$ to $t = 4.0$ s. (b) The resistance of the coil is 0.15 Ω. What is the instantaneous current in the coil at $t = 2.0$ s?

5. A small loop of area A is inside of, and has its axis in the same direction as, a long solenoid of n turns per unit length and current i. If $i = i_0 \sin \omega t$, find the emf ε in the loop.   *Answer:* $-\mu_0 nAi_0\omega \cos \omega t$.

6. *Alternating current generator.* A rectangular loop of N turns and of length a and width b is rotated at a frequency ν in a uniform magnetic field **B**, as in Fig. 35-28. (a) Show that an induced emf given by

$$\varepsilon = 2\pi\nu NbaB \sin 2\pi\nu t = \varepsilon_0 \sin 2\pi\nu t$$

appears in the loop. This is the principle of the commercial alternating-current generator. (b) Design a loop that will produce an emf with $\varepsilon_0 = 150$ V when rotated at 60 rev/s in a magnetic field of 0.50 T.

7. In Fig. 35-29 a closed single turn with a copper coil resistance of 5.0 Ω is placed *outside* a solenoid like that of Example 1. If the current in the solenoid is changed as in that example, (a) what current appears in the loop while the solenoid current is being changed? (b) How do the free charges

figure 35-28
Problem 6

in the loop "get the message" from the solenoid that they should start moving (to establish a current)? After all, the magnetic flux is entirely confined to the interior of the solenoid.
Answer: (a) 2.1×10^{-4} A.

8. Figure 35-30 shows a copper rod moving on conducting rails with velocity **v** parallel to a long straight wire carrying a current i. Calculate the induced emf ε in the rod, assuming that $v = 5.0$ m/s, $i = 100$ A, $a = 1.0$ cm, and $b = 20$ cm.

9. A circular loop of wire 10 cm in diameter is placed with its normal making an angle of 30° with the direction of a uniform 0.50-T magnetic field. The loop is "wobbled" so that its normal rotates about the field direction at the constant rate of 100 rev/min; the angle between the normal and the field direction (= 30°) remains unchanged during this process. What emf appears in the loop?   *Answer:* Zero.

10. A stiff wire bent into a semicircle of radius R is rotated with a frequency ν in a uniform magnetic field **B,** as shown in Fig. 35-31. What are the amplitude and frequency of the induced emf and of the induced current when the internal resistance of the meter M is R_M and the remainder of the circuit has negligible resistance?

11. Figure 35-32 shows a copper rod moving with velocity **v** parallel to a long straight wire carrying a current i. Calculate the induced emf in the rod, assuming that $v = 5.0$ m/s, $i = 100$ A, $a = 1.0$ cm, and $b = 20$ cm.
Answer: 3.0×10^{-4} V.

12. A uniform magnetic field **B** is changing in magnitude at a constant rate dB/dt. You are given a mass m of copper which is to be drawn into a wire of radius r and formed into a circular loop of radius R. Show that the induced current in the loop does not depend on the size of the wire or of the loop and, assuming **B** perpendicular to the loop, is given by

$$i = \frac{m}{4\pi\rho\delta} \frac{dB}{dt},$$

where ρ is the resistivity and δ the density of copper.

13. A circular loop of radius r (10 cm) is placed in a uniform magnetic field **B** (0.80 T) normal to the plane of the loop. The radius of the loop begins shrinking at a constant rate dr/dt (80 cm/s). (a) What is the emf ε induced in the loop? (b) At what constant rate would the area have to shrink to induce this same emf?   *Answer:* (a) 0.40 V. (b) 0.50 m²/s.

14. Prove that the electric field **E** in a charged parallel-plate capacitor cannot drop abruptly to zero as one moves at right angles to it, as suggested by the arrow in Fig. 35-33 (see point a). In actual capacitors fringing of the lines of force always occurs, which means that **E** approaches zero in a continuous and gradual way; see Problem 34-2. (Hint: Apply Faraday's law to the rectangular path shown by the dashed lines; see Eq. 35-8.

figure 35-29
Problem 7

figure 35-30
Problem 8

figure 35-31
Problem 10

figure 35-32
Problem 11

figure 35-33
Problem 14

15. An electromagnetic "eddy current" brake consists of a disk of conductivity σ and thickness t rotating about an axis through its center with a magnetic field **B** applied perpendicular to the plane of the disk over a small area a^2 (see Fig. 35-34). If the area a^2 is at a distance r from the axis, find an approximate expression for the torque tending to slow down the disk at the instant its angular velocity equals ω.   *Answer:* $\tau = B^2 a^2 r^2 \, \omega \, \sigma \, t$.

figure 35-34
Problem 15

SECTION 35-3

16. A small bar magnet is pulled rapidly through a conducting loop, along its axis. Sketch qualitatively (a) the induced current and (b) the rate of thermal energy production as a function of the position of the center of the magnet. Assume that the north pole of the magnet enters the loop first and that the magnet moves at constant speed. Plot the induced current as positive if it is clockwise as viewed along the path of the magnet.

17. A metal wire of mass m slides without friction on two rails spaced a distance d apart, as in Fig. 35-35. The track lies in a vertical uniform magnetic field **B**. (a) A *constant current* i flows from generator G along one rail, across the wire, and back down the other rail. Find the velocity (speed and direction) of the wire as a function of time, assuming it to be at rest at $t = 0$. (b) The generator is replaced by a battery with *constant emf* ε. The velocity of the wire now approaches a constant final value. What is this terminal speed? (c) What is the current in part (b) when the terminal speed has been reached?   *Answer:* (a) $Bidt/m$, away from G. (b) ε/Bd. (c) Zero.

figure 35-35
Problem 17

18. In Fig. 35-36 the magnetic flux through the loop perpendicular to the plane of the coil and directed into the paper is varying according to the relation

$$\Phi_B = 6t^2 + 7t + 1,$$

where Φ_B is in milliwebers (1 milliweber $= 10^{-3}$ weber) and t is in seconds. (a) What is the magnitude of the emf induced in the loop when $t = 2.0s$? (b) What is the direction of the current through R?

figure 35-36
Problem 18

SECTION 35-4

19. In the arrangement of Example 3 put B = 1.2 T and L = 5.0 cm. If $\varepsilon = 1.0$ V, what acceleration will a point at the end of the rotating rod experience? *Answer:* 2300 "g".

20. In Fig. 35-37 a conducting rod AB makes contact with the metal rails AD and BC which are 50 cm apart in a uniform magnetic field of 1.0 T perpendicular to the plane of the paper as shown. The total resistance of the circuit $ABCD$ is 0.4 Ω (assumed constant). (a) What is the magnitude and direction of the emf induced in the rod when it is moved to the left with a velocity of 8.0 m/s? (b) What force is required to keep the rod in motion? (c) Compare the rate at which mechanical work is done by the force **F** with the rate of development of thermal energy in the circuit.

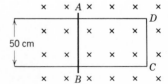

figure 35-37
Problem 20

21. In Fig. 35-38, $l = 2.0$ m and $v = 50$ cm/s. **B** is the earth's magnetic field, directed perpendicularly out of the page and having a magnitude 6.0×10^{-5} T at that place. The resistance of the circuit $ADCB$, assumed constant (explain how this may be achieved approximately), is $R = 1.2 \times 10^{-5}$ Ω. (a) What is the emf induced in the circuit? (b) What is the electric field in the wire AB? (c) What force does each electron in the wire experience due to the motion of the wire in the magnetic field? (d) What is the magnitude and direction of the current in the wire? (e) What force must an external agency exert in order to keep the wire moving with this constant velocity? (f) Compute the rate at which the external agency is doing work. (g) Compute the rate at which electrical energy is being converted into thermal energy.   *Answer:* (a) 6.0×10^{-5} V. (b) 3.0×10^{-5} V/m. (c) 4.8×10^{-24} N. (d) 5.0 A. (e) 6.0×10^{-4} N. (f) 3.0×10^{-4} W. (g) 3.0×10^{-4} W.

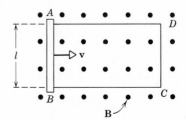

figure 35-38
Problem 21

22. A square wire of length l, mass m, and resistance R slides without friction down parallel conducting rails of negligible resistance, as in Fig. 35-39. The rails are connected to each other at the bottom by a resistanceless rail parallel to the wire, so that the wire and rails form a closed rectangular conduct-

ing loop. The plane of the rails makes an angle θ with the horizontal, and a uniform vertical magnetic field **B** exists throughout the region. (*a*) Show that the wire acquires a steady-state velocity of magnitude

$$v = \frac{mgR \sin \theta}{B^2 l^2 \cos^2 \theta}.$$

(*b*) Prove that this result is consistent with the conservation-of-energy principle. (*c*) What change, if any, would there be if **B** were directed down instead of up?

figure 35-39
Problem 22

SECTION 35-5

23. A long solenoid of radius r (2.5 cm) and turns per unit length n (100/cm) carries an initial current of i_0 (1.0 A). A single loop of wire of diameter D (10 cm) is placed around the solenoid, the axes coinciding. The current in the solenoid is reduced uniformly to i (0.50 A) over a period T (0.010 s). While the current is changing what is the induced emf ε in the surrounding loop? *Answer:* 1.2×10^{-3} V.

24. A circular coil of radius r (10 cm) is made of wire of resistance R (10 Ω). Perpendicular to the plane of the coil is a uniform magnetic field **B**. (*a*) At what constant rate must B increase so that there is a steady current i (0.010 A) in the circuit? (*b*) What power is dissipated in the resistor?

25. (*a*) For the arrangement of Fig. 34-28, what would be the current induced around the rectangular loop if the current in the wire decreased uniformly from 30 A to zero in 1.0 s? Assume no initial current in the loop and a resistance for the loop of 0.020 Ω. Take $a = 1.0$ cm, $b = 8.0$ cm, and $l = 30$ cm. (*b*) How much energy would be transferred to the loop in the 1.0-s interval? *Answer:* (*a*) 1.9×10^{-4} A. (*b*) 7.2×10^{-10} J.

26. The perpendicular field B through a one-turn circular loop of wire of negligible resistance changes with time as shown in Fig. 35-40. The loop is of radius r (10 cm) and is connected to a resistor R (10 Ω). (*a*) Plot the emf appearing across the resistor. (*b*) Plot the current i through the resistor R. (*c*) Plot the rate of thermal energy production in the resistor.

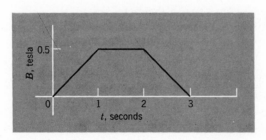

figure 35-40
Problem 26

27. A loop of wire of area A is connected to a resistance R. The loop is exposed to a time-varying **B** field (see Fig. 35-41). (*a*) Derive an expression for the net charge transferred through the resistor between $t = t_1$ and $t = t_2$. Show that your answer is proportional to the difference $\Phi_B(t_2) - \Phi_B(t_1)$, and is otherwise independent of the manner in which **B** is changing. (*b*) Suppose the change in flux, $\Phi_B(t_2) - \Phi_B(t_1)$ is zero. Does it then follow that no thermal energy production occurred during this time interval?

Answer: (*a*) $Q = -\dfrac{1}{R} [\Phi_B(t_2) - \Phi_B(t_1)]$. (*b*) No.

figure 35-41
Problem 27

28. Figure 35-42 shows two loops of wire having the same axis. The smaller loop is above the larger one, by a distance x which is large compared to the radius R of the larger loop. Hence, with current i flowing as indicated in the

larger loop, the consequent magnetic field is nearly constant throughout the plane area πr^2 bounded by the smaller loop. Suppose now that x is not constant but is changing at the constant rate $dx/dt = v$ (x increasing). (a) Determine the magnetic flux across the area bounded by the smaller loop as a function of x. (b) Compute the emf generated in the smaller loop at the instant when $x = NR$. (c) Determine the direction of the induced current flowing in the smaller loop if $v > 0$.

29. A wire is bent into three circular segments of radius r (10 cm) as shown in Fig. 35-43. Each segment is a quadrant of a circle, ab lying in the x-y plane, bc lying in the y-z plane, and ca lying in the z-x plane. (a) If a spatially uniform magnetic field **B** points in the x-direction, what is the magnitude of the emf ε developed in the wire when **B** increases at the rate of 3.0×10^{-3} T/s? (b) What is the direction of the current in the segment bc?
Answer: (a) 2.4×10^{-5} V. (b) From c to b.

30. Figure 35-44 shows a uniform magnetic field **B** confined to a cylindrical volume of radius R. **B** is decreasing in magnitude at a constant rate of 0.010 T/s. What is the instantaneous acceleration (direction and magnitude) experienced by an electron placed at a, at b, and at c? Assume $r = 5.0$ cm. (The necessary fringing of the field beyond R will not change your answer as long as there is axial symmetry about a perpendicular axis through b.)

31. A closed loop of wire consists of a pair of equal semicircles, radius 3.70 cm, lying in mutually perpendicular planes. The loop was formed by folding a circular loop along a diameter until the two halves became perpendicular. A uniform magnetic field **B** of magnitude 0.076 T is directed perpendicular to the fold diameter and makes equal angles (45°) with the planes of the semicircles as shown in Fig. 35-45a. (a) The magnetic field is reduced at a uniform rate to zero during a time interval 4.50×10^{-3} s. Determine the magnitude of the induced emf and the sense of the induced current in the loop during this interval. (b) How would the answers change if **B** is directed as shown in Fig. 35-45b, perpendicular to the direction first given for it but still perpendicular to the "fold-diameter?"
Answer: (a) 51×10^{-3} V. (b) $\varepsilon = 0$.

32. A uniform magnetic field **B** fills a cylindrical volume of radius R. A metal rod of length l is placed as shown in Fig. 35-46. If B is changing at the rate dB/dt, show that the emf that is produced by the changing magnetic field and that acts between the ends of the rod is given by

$$\varepsilon = \frac{dB}{dt} \frac{l}{2} \sqrt{R^2 - \left(\frac{l}{2}\right)^2} \, .$$

SECTION 35-6

33. Some measurements of the maximum magnetic field as a function of radius for the betatron described on page 781 are as follows:

r, cm	B, tesla	r, cm	B, tesla
0	0.400	81.2	0.409
10.2	0.950	83.7	0.400
68.2	0.950	88.9	0.381
73.2	0.528	91.4	0.372
75.2	0.451	93.5	0.360
77.3	0.428	95.5	0.340

figure 35-42
Problem 28

figure 35-43
Problem 29

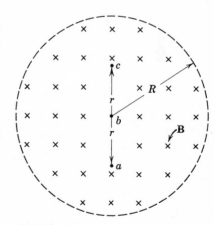

figure 35-44
Problem 30

figure 35-45
Problem 31

(a)   *(b)*

Show by graphical analysis that the relation $\overline{B} = 2B_R$ is satisfied at the orbit radius, $R = 84$ cm. (Hint: Note that $\overline{B} = \dfrac{1}{\pi R^2} \displaystyle\int_0^R B(r)(2\pi r)\,dr$ and evaluate the integral graphically.)

SECTION 35-7

34. (a) Estimate θ in Fig. 35-16a. Recall (see Section 31-1) that $v_d = 4 \times 10^{-2}$ cm/s in a typical case. Assume $v = 10$ cm/s. (b) It is clear that θ will be small. However, must we have $\theta \neq 0$ for the arguments presented in connection with this figure to be valid?

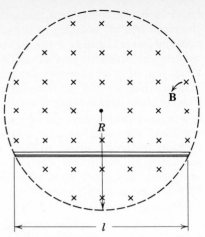

figure 35-46
Problem 32

36
inductance

If two coils are near each other, a current i in one coil will set up a flux Φ_B through the second coil. If this flux is changed by changing the current, an induced emf will appear in the second coil according to Faraday's law; see Fig. 35-2. However, two coils are not needed to show an inductive effect. An induced emf appears in a coil if the current *in that same coil* is changed. This is called *self-induction* and the electromotive force produced is called a *self-induced emf*. It obeys Faraday's law of induction just as other induced emfs do.

Consider first a "close-packed" coil, a toroid, or the central section of a long solenoid. In all three cases the flux Φ_B set up in each turn by a current i is essentially the same for every turn. Faraday's law for such coils (Eq. 35-2)

$$\varepsilon = -\frac{d(N\Phi_B)}{dt} \qquad (36\text{-}1)$$

shows that the number of *flux linkages* $N\Phi_B$ (N being the number of turns) is the important characteristic quantity for induction. For a given coil, provided no magnetic materials such as iron are nearby, this quantity is proportional to the current i, or

$$N\Phi_B = Li, \qquad (36\text{-}2)$$

in which L, the proportionality constant, is called the *inductance* of the device.

From Faraday's law (see Eq. 36-1) we can write the induced emf as

$$\varepsilon = -\frac{d(N\Phi_B)}{dt} = -L\frac{di}{dt}. \qquad (36\text{-}3a)$$

36-1
INDUCTANCE

Written in the form

$$L = -\frac{\mathcal{E}}{di/dt},$$ (36-3b)

induced

this relation may be taken as the defining equation for inductance for coils of all shapes and sizes, whether or not they are close-packed and whether or not iron or other magnetic material is nearby. It is analogous to the defining relation for capacitance, namely,

$$C = \frac{q}{V}.$$

If no iron or similar materials are nearby, L depends only on the geometry of the device. In an *inductor* (symbol 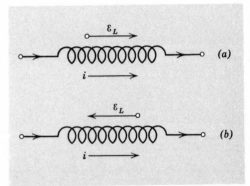) the presence of a *magnetic field* is the significant feature, corresponding to the presence of an *electric* field in a *capacitor*.

The SI unit of inductance, from Eq. 36-3b, is the volt·second/ampere. A special name, the *henry* (abbr. H), has been given to this combination of units, or

     1 henry = 1 volt·second/ampere.

As we saw in Section 35-1, the SI unit of inductance is named after Joseph Henry (1797–1878), an American physicist and a contemporary of Faraday.

We can find the direction of a self-induced emf from Lenz's law; see Section 35-3. Suppose that a steady current i, produced by a battery, exists in a coil. Let us suddenly *decrease* the (battery) emf in the circuit. The current i will start to *decrease* at once. This decrease in current, in the language of Lenz's law, is the "change" which the self-induction must oppose. To oppose the falling current, the induced emf must point in the *same* direction as the current, as shown in Fig. 36-1a.

However, if we suddenly *increase* the (battery) emf, the current i will start to *increase* at once. This increase in current is now the "change" which self-induction must oppose. To oppose the rising current the induced emf must point in the *opposite* direction to the current, as Fig. 36-1b shows. In each case the self-induced emf acts to oppose the *change* in the current. The minus sign in Eq. 36-3 shows that $\mathcal{E}$ and di/dt are opposite in sign, since L is always a positive quantity.

We were able to make a direct calculation of capacitance in terms of geometrical factors for a few special cases, such as the parallel-plate capacitor. In the same way, we can calculate the self-inductance L for a few special cases.

For a close-packed coil with no iron nearby, we have, from Eq. 36-2,

$$L = \frac{N\Phi_B}{i}.$$ (36-4)

Let us apply this equation to calculate L for a section of length l near the center of a long solenoid. The number of flux linkages in the length l of the solenoid is

$$N\Phi_B = (nl)(BA),$$

where n is the number of turns per unit length, B is the magnetic field inside the solenoid, and A is the cross-sectional area. From Eq. 34-7, B is given by

$$B = \mu_0 ni.$$

Combining these equations gives

figure 36-1
In (a) the current i is made to *decrease* and in (b) it is made to *increase*. The self-induced emf $\mathcal{E}_L$ opposes the *change* in each case.

36-2
CALCULATION OF INDUCTANCE

$$N\Phi_B = \mu_0 n^2 liA.$$

Finally, the inductance, from Eq. 36-4, is

$$L = \frac{N\Phi_B}{i} = \mu_0 n^2 lA. \qquad (36\text{-}5)$$

The inductance of a length l of a solenoid is proportional to its volume (lA) and to the square of the number of turns per unit length. Note that it depends on geometrical factors only. The proportionality to n^2 is expected. If we double the number of turns per unit length, not only is the *total* number of turns N doubled but the flux *through each turn* Φ_B is also doubled, an over-all factor of four for the flux linkages $N\Phi_B$, hence also a factor of four for the inductance (Eq. 36-4).

EXAMPLE 1

Derive an expression for the inductance of a toroid of rectangular cross section as shown in Fig. 36-2. Evaluate for $N = 10^3$, $a = 5.0$ cm, $b = 10$ cm, and $h = 1.0$ cm.

The lines of **B** for the toroid are concentric circles. Applying Ampère's law,

$$\oint \mathbf{B} \cdot d\mathbf{l} = \mu_0 i,$$

to a circular path of radius r yields

$$(B)(2\pi r) = \mu_0 i_0 N,$$

where N is the number of turns and i_0 is the current in the toroid windings; recall that i in Ampère's law is the *total* current that passes through the path of integration. Solving for B yields

$$B = \frac{\mu_0 i_0 N}{2\pi r}.$$

The flux Φ_B for the cross section of the toroid is

$$\Phi_B = \int \mathbf{B} \cdot d\mathbf{S} = \int_a^b (B)(h\ dr) = \int_a^b \frac{\mu_0 i_0 N}{2\pi r} h\ dr$$

$$= \frac{\mu_0 i_0 Nh}{2\pi} \int_a^b \frac{dr}{r} = \frac{\mu_0 i_0 Nh}{2\pi} \ln \frac{b}{a},$$

where $h\ dr$ is the area of the elementary strip shown (dashed lines) in the figure.

The inductance follows from Eq. 36-4, or

$$L = \frac{N\Phi_B}{i_0} = \frac{\mu_0 N^2 h}{2\pi} \ln \frac{b}{a}.$$

Substituting numerical values yields

$$L = \frac{(4\pi \times 10^{-7}\ \text{Wb/A} \cdot \text{m})(10^3)^2(1.0 \times 10^{-2}\ \text{m})}{2\pi} \ln \frac{10 \times 10^{-2}\ \text{m}}{5 \times 10^{-2}\ \text{m}}$$

$$= 1.4 \times 10^{-3}\ \text{Wb/A} = 1.4\ \text{mH}.$$

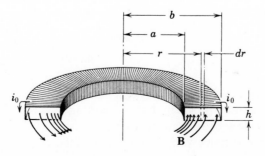

figure 36-2
Example 1. A cross section of a toroid, showing the current in the windings and the magnetic field.

In Section 32-8 we saw that if we suddenly introduce an emf ε, perhaps by using a battery, into a single loop circuit containing a resistor R and a capacitor C, the charge does not build up immediately to its final equilibrium value $(= C\varepsilon)$ but approaches it in an exponential fashion described by Eq. 32-15, or

$$q = C\varepsilon \, (1 - e^{-t/\tau_C}). \qquad (36\text{-}6)$$

The delay in the rise of the charge is described by the *capacitive time constant* τ_C, defined from

$$\tau_C = RC. \qquad (36\text{-}7)$$

If in this same circuit the battery emf ε is suddenly removed, the charge does not immediately fall to zero but approaches zero in an exponential fashion, described by Eq. 32-18b, or

$$q = C\varepsilon \, e^{-t/\tau_C}. \qquad (36\text{-}8)$$

The same time constant τ_C describes the fall of the charge as well as its rise.

An analogous delay in the rise (or fall) of the current occurs if we suddenly introduce an emf ε into (or remove it from) a single-loop circuit containing a resistor R and an inductor L. When the switch S in Fig. 36-3 is closed on a, for example, the current in the resistor starts to rise. If the inductor were not present, the current would rise rapidly to a steady value ε/R. Because of the inductor, however, a self-induced emf ε_L appears in the circuit and, from Lenz's law, this emf opposes the *rise* of current, which means that it opposes the battery emf ε in polarity. Thus the resistor responds to the difference between two emfs, a constant one ε due to the battery and a variable one ε_L $(= -L \, di/dt)$ due to self-induction. As long as this second emf is present, the current in the resistor will be less than ε/R.

As time goes on, the rate at which the current increases becomes less rapid and the self-induced emf ε_L, which is proportional to di/dt, becomes smaller. Thus a time delay is introduced, and the current in the circuit approaches the value ε/R asymptotically.

When the switch S in Fig. 36-3 is thrown to a, the circuit reduces to that of Fig. 36-4. Let us apply the loop theorem, starting at x in this figure and going clockwise around the loop. For the direction of current shown, x will be higher in potential than y, which means that we encounter a drop in potential of $-iR$ as we traverse the resistor. Point y is higher in potential than point z because, for an increasing current, the induced emf will oppose the *rise* of the current by pointing as shown. Thus as we traverse the inductor from y to z we encounter a drop in potential of $-L(di/dt)$. We encounter a rise in potential of $+\varepsilon$ in traversing the battery from z to x. The loop theorem thus gives

$$-iR - L\frac{di}{dt} + \varepsilon = 0$$

or

$$L\frac{di}{dt} + iR = \varepsilon. \qquad (36\text{-}9)$$

Equation 36-9 is a *differential equation* involving the variable i and its first derivative di/dt. We seek the function $i(t)$ such that when it and its first derivative are substituted in Eq. 36-9 the equation is satisfied.

Although there are formal rules for solving various classes of differential equations (and Eq. 36-9 can, in fact, be easily solved by direct

36-3
AN LR CIRCUIT

figure 36-3
An *LR* circuit.

figure 36-4
The circuit of Fig. 36-3 just after switch S is closed on a.

integration, after rearrangement) we often find it simpler to guess at the solution, guided by physical reasoning and by previous experience. We can test any proposed solution by substituting it in the differential equation and seeing whether this equation reduces to an identity.

The solution to Eq. 36-9 is, we assert,

$$i = \frac{\varepsilon}{R}\left(1 - e^{-Rt/L}\right). \tag{36-10}$$

To test this solution by substitution, we find the derivative di/dt, which is

$$\frac{di}{dt} = \frac{\varepsilon}{L}\,e^{-Rt/L}. \tag{36-11}$$

Substituting i and di/dt into Eq. 36-9 leads to an identity, as you can easily verify. Thus Eq. 36-10 is a solution of Eq. 36-9. Figure 36-5 shows how the potential difference V_R across the resistor $(= iR$; see Eq. 36-10) and V_L across the inductor $(= L\,di/dt$; see Eq. 36-11) vary with time for particular values of ε, L, and R. Compare this figure carefully with the corresponding figure for an RC circuit (Fig. 32-11).

We can rewrite Eq. 36-10 as

$$i = \frac{\varepsilon}{R}\left(1 - e^{-t/\tau_L}\right), \tag{36-12}$$

in which τ_L, the *inductive time constant*, is given by

$$\tau_L = L/R. \tag{36-13}$$

Note the correspondence between Eqs. 36-6 and 36-12.

To show that the quantity τ_L $(= L/R)$ has the dimensions of time, we put

$$\frac{1\ \text{henry}}{\text{ohm}} = \frac{1\ \text{henry}}{\text{ohm}}\left(\frac{1\ \text{volt} \cdot \text{second}}{1\ \text{henry} \cdot \text{ampere}}\right)\left(\frac{1\ \text{ohm} \cdot \text{ampere}}{1\ \text{volt}}\right) = 1\ \text{second}.$$

The first quantity in parentheses is a conversion factor based on the defining equation for inductance $[L = -\varepsilon/(di/dt)$; see Eq. 36-3$b$]. The second conversion factor is based on the relation $V = iR$.

The physical significance of the time constant follows from Eq. 36-12. If we put $t = \tau_L = L/R$ in this equation, it reduces to

$$i = \frac{\varepsilon}{R}\left(1 - e^{-1}\right) = (1 - 0.37)\,\frac{\varepsilon}{R} = 0.63\,\frac{\varepsilon}{R}.$$

Thus the time constant τ_L is that time at which the current in the circuit will reach a value within $1/e$ (about 37%) of its final equilibrium value (see Fig. 36-5).

t, milliseconds

(a)

t, milliseconds

(b)

figure 36-5
If in Fig. 36-3 we assume that $R = 2000\ \Omega$, $L = 4$ H, and $\varepsilon = 10$ V, then (a) shows the variation of V_R with t during the current buildup after switch S is closed on a, and (b) the variation of V_L with t. The time constant is $L/R = 2.0 \times 10^{-3}$ s. Compare this figure carefully with Fig. 32-11, the corresponding figure for an RC circuit.

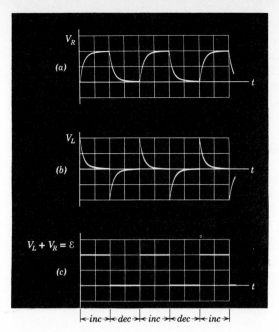

V_R	
(a)	
	t

V_L

(b)

t

$V_L + V_R = \varepsilon$

(c)

t

|← inc →|← dec →|← inc →|← dec →|← inc →|

figure 36-6
Oscilloscope photograph showing the variation with time of (a) the potential drop V_R across the resistor, (b) the potential drop V_L across the inductor, and (c) the applied emf ε. During the intervals marked *inc* the current is increasing; during those marked *dec* it is decreasing. Compare with Fig. 32-13. (Courtesy E. K. Hege, Rensselaer Polytechnic Institute.)

If the switch S in Fig. 36-3, having been left in position a long enough for the equilibrium current ε/R to be established, is thrown to $b*$, the effect is to remove the battery from the circuit. The differential equation that governs the subsequent decay of the current in the circuit can be found by putting $\varepsilon = 0$ in Eq. 36-9, or

$$L \frac{di}{dt} + iR = 0. \qquad (36\text{-}14)$$

You can show by the test of substitution that the solution of this differential equation is

$$i = \frac{\varepsilon}{R} e^{-t/\tau_L}. \qquad (36\text{-}15)$$

Just as for the RC circuit, the behavior of the circuit of Fig. 36-3 can be investigated experimentally, using a cathode-ray oscilloscope. If switch S in this figure is thrown periodically between a and b, the applied (battery) emf alternates between the values ε and zero. If the terminals of an oscilloscope are connected across b and c in Fig. 36-3, the oscilloscope will display the waveform of this applied emf on its screen, as in Fig. 36-6c.

If the terminals of the oscilloscope are connected across the resistor, the waveform displayed (Fig. 36-6a) will be that of the current in the circuit, since the potential drop across R, which determines the oscilloscope deflection, is given by $V_R = iR$. During the intervals marked *inc* in Fig. 36-6, the current is increasing and the waveform (see Eq. 36-12) is given by

$$V_R \ (= iR) = \varepsilon(1 - e^{-t/\tau_L}).$$

During the intervals marked *dec*, the current is decreasing and V_R (see Eq. 36-15) is given by

$$V_R \ (= iR) = \varepsilon e^{-t/\tau_L}.$$

Note that both the growth and the decay of the current are delayed.

If the oscilloscope terminals are connected across the inductor, the screen will show a plot of the potential difference across it as a function of time (Fig. 36-6b). While the current is increasing, the equation of the trace (see Eq. 36-11) should be

---

*The connection to b must be made before the connection to a is broken. A switch which does this is called a "make before break" switch.

$$V_L \left(= L\frac{di}{dt}\right) = \mathcal{E}e^{-t/\tau_L}.$$

When the current is decreasing, V_L is given in terms of the time derivative of Eq. 36-15 and is

$$V_L \left(= L\frac{di}{dt}\right) = -\mathcal{E}e^{-t/\tau_L}.$$

Note that V_L is opposite in sign when the current is increasing (di/dt positive) and when it is decreasing (di/dt negative), as is true also for the induced emf $\mathcal{E}_L[=-L(di/dt)=-V_L]$.

Examination of Fig. 36-6 shows that at any instant the sum of curves a and b always yields curve c. This is an expected consequence of the loop theorem, as Eq. 36-9 shows.

A solenoid has an inductance of 50 H and a resistance of 30 Ω. If it is connected to a 100-V battery, how long will it take for the current to reach one-half its final equilibrium value?

The equilibrium value of the current is reached as $t \to \infty$; from Eq. 36-12 it is $\mathcal{E}/R$. If the current has half this value at a particular time t_0, this equation becomes

$$\frac{1}{2}\frac{\mathcal{E}}{R} = \frac{\mathcal{E}}{R}(1 - e^{-t_0/\tau_L}).$$

Solving for t_0 yields

$$t_0 = \tau_L \ln 2 = 0.69 \frac{L}{R}.$$

Using the values given, this reduces to

$$t_0 = 0.69\tau_L = 0.69\left(\frac{50\text{ H}}{30\text{ Ω}}\right) = 1.2\text{ s}.$$

EXAMPLE 2

36-4
ENERGY AND THE MAGNETIC FIELD

When we lift a stone we do work, which we can get back again by lowering the stone. It is convenient to think of the work done to lift the stone being stored temporarily in the gravitational field between the earth and the lifted stone, and being withdrawn from this field when we lower the stone.

When we pull two unlike charges apart we like to say that the work we do is stored in the electric field between the charges. We can get it back from the field by letting the charges move closer together again.

In the same way we can store energy in a magnetic field. For example, two long, rigid, parallel wires carrying current in the same direction attract each other and we must do work to pull them apart. We can get this stored energy back at any time by letting the wires move back to their original positions.

To derive a quantitative expression for the storage of energy in the magnetic field, consider Fig. 36-4, which shows a source of emf $\mathcal{E}$ connected to a resistor R and an inductor L.

$$\mathcal{E} = iR + L\frac{di}{dt}, \tag{36-9}$$

is the differential equation that describes the growth of current in this circuit. We stress that this equation follows immediately from the loop theorem and that the loop theorem in turn is an expression of the principle of conservation of energy for single-loop circuits. If we multiply each side of Eq. 36-9 by i, we obtain

heat   *magnetic energy*

$$\dot{W} = \varepsilon i = i^2 R + Li\,\frac{di}{dt}, \qquad (36\text{-}16)$$

which has the following physical interpretation in terms of work and energy:

1. If a charge dq passes through the seat of emf ε in Fig. 36-4 in time dt, the seat does work on it in amount $\varepsilon\,dq$. The *rate* of doing work is $(\varepsilon\,dq)/dt$, or εi. Thus the left term in Eq. 36-16 is the *rate at which the seat of emf delivers energy to the circuit.*

2. The second term in Eq. 36-16 is the *rate at which energy appears as thermal (Joule) energy in the resistor.*

3. Energy that does not appear as thermal (Joule) energy must, by our hypothesis, be stored in the magnetic field. Since Eq. 36-16 represents a statement of the conservation of energy for LR circuits, the last term must represent the *rate dU_B/dt at which energy is stored in the magnetic field,* or

$$\frac{dU_B}{dt} = Li\,\frac{di}{dt}. \qquad (36\text{-}17)$$

We can write this as $\quad dU_B = Li\,di.$

Integrating yields

$$U_B = \int_0^{U_B} dU_B = \int_0^i Li\,di = \tfrac{1}{2}Li^2, \qquad (36\text{-}18)$$

which represents the total stored magnetic energy in an inductance L carrying a current i.

We can compare this relation with the expression for the energy associated with a capacitor C carrying a charge q, namely,

$$U_E = \frac{1}{2}\frac{q^2}{C}.$$

Here the energy is stored in an electric field. In each case the expression for the stored energy was derived by setting it equal to the work that must be done to set up the field.

EXAMPLE 3

A coil has an inductance of 5.0 H and a resistance of 20 Ω. If a 100-V emf is applied, what energy is stored in the magnetic field after the current has built up to its maximum value ε/R?

The maximum current is given by

$$i = \frac{\varepsilon}{R} = \frac{100\text{ V}}{20\ \Omega} = 5.0\text{ A}.$$

The stored energy is given by Eq. 36-18:

$$U_B = \tfrac{1}{2}Li^2 = \tfrac{1}{2}(5.0\text{ H})(5.0\text{ A})^2 = 63\text{ J}.$$

Note that the time constant for this coil $(= L/R)$ is 0.25 s. After how many time constants will *half* of this equilibrium energy be stored in the field?

EXAMPLE 4

A 3.0 H inductor is placed in series with a 10-Ω resistor, an emf of 3.0 V being suddenly applied to the combination. At 0.30 s (which is one inductive time constant) after the contact is made, (*a*) what is the rate at which energy is being delivered by the battery?

The current is given by Eq. 36-12, or

$$i = \frac{\varepsilon}{R}\left(1 - e^{-t/\tau_L}\right),$$

which at $t = 0.30$ s $(= \tau_L)$ has the value

$$i = \left(\frac{3.0 \text{ V}}{10 \text{ }\Omega}\right)(1 - e^{-1}) = 0.189 \text{ A}.$$

The rate P_ε at which energy is delivered by the battery is

$$P_\varepsilon = \varepsilon i$$

$$= (3.0 \text{ V})(0.189 \text{ A})$$

$$= 0.567 \text{ W}.$$

(b) At what rate does energy appear as thermal (Joule) energy in the resistor? This is given by

$$P_J = i^2 R$$

$$= (0.189 \text{ A})^2 (10 \text{ }\Omega)$$

$$= 0.357 \text{ W}.$$

(c) At what rate P_B is energy being stored in the magnetic field?

This is given by the last term in Eq. 36-16, which requires that we know di/dt. Differentiating Eq. 36-12 yields

$$\frac{di}{dt} = \left(\frac{\varepsilon}{R}\right)\left(\frac{R}{L}\right) e^{-t/\tau_L}$$

$$= \frac{\varepsilon}{L} e^{-t/\tau_L}.$$

At $t = \tau_L$ we have

$$\frac{di}{dt} = \left(\frac{3.0 \text{ V}}{3.0 \text{ H}}\right) e^{-1} = 0.37 \text{ A/s}.$$

From Eq. 36-17, the desired rate is

$$P_B = \frac{dU_B}{dt} = Li \frac{di}{dt}$$

$$= (3.0 \text{ H})(0.189 \text{ A})(0.37 \text{ A/s})$$

$$= 0.210 \text{ W}.$$

Note that as required by the principle of conservation of energy (see Eq. 36-16)

$$P_\varepsilon = P_J + P_B,$$

or

$$0.567 \text{ W} = 0.357 \text{ W} + 0.210 \text{ W}$$

$$= 0.567 \text{ W}.$$

We now derive an expression for the *density* of energy u_B in a magnetic field. Consider a length l near the center of a very long solenoid; Al is the volume associated with this length. The stored energy must lie entirely within this volume because the magnetic field outside such a solenoid is essentially zero. Moreover, the stored energy must be uniformly distributed throughout the volume of the solenoid because the magnetic field is uniform everywhere inside.

36-5
ENERGY DENSITY AND THE MAGNETIC FIELD

Thus, we can write $$u_B = \frac{U_B}{Al}$$

or, since $$U_B = \tfrac{1}{2}Li^2,$$

we have $$u_B = \frac{\tfrac{1}{2}Li^2}{Al}.$$

To express this in terms of the magnetic field, we can substitute for L in this equation, using the relation $L = \mu_0 n^2 lA$ (Eq. 36-5). Also we can solve Eq. 34-7 ($B = \mu_0 in$) for i and substitute in this equation. Doing so yields finally

$$u_B = \frac{1}{2}\frac{B^2}{\mu_0}. \qquad (36\text{-}19)$$

This equation gives the energy density stored at any point (in a vacuum or in a nonmagnetic substance) where the magnetic field is **B**. The equation is true for all magnetic field configurations, even though we derived it by considering a special case, the solenoid. Equation 36-19 is to be compared with Eq. 30-9,

$$u_E = \tfrac{1}{2}\epsilon_0 E^2, \qquad (36\text{-}20)$$

which gives the energy density (in a vacuum) at any point in an electric field. Note that both u_B and u_E are proportional to the square of the appropriate field quantity, B or E.

The solenoid plays a role with relationship to magnetic fields similar to the role the parallel-plate capacitor plays with respect to electric fields. In each case we have a simple device that can be used for setting up a uniform field throughout a well-defined region of space and for deducing, in a simple way, some properties of these fields.

A long *coaxial cable* (Fig. 36-7) consists of two concentric cylinders with radii a and b. Its central conductor carries a steady current i, the outer conductor providing the return path. (*a*) Calculate the energy stored in the magnetic field for a length l of such a cable.

In the space between the two conductors Ampère's law,

$$\oint \mathbf{B} \cdot d\mathbf{l} = \mu_0 i,$$

leads to $$(B)(2\pi r) = \mu_0 i$$

or $$B = \frac{\mu_0 i}{2\pi r}.$$

Ampère's law shows further that the magnetic field is zero for points outside the outer conductor (why?). Magnetic fields exist *inside* each of the conductors; although we can find their values from Ampère's law, we choose to ignore them, on the assumption that the cable dimensions are chosen so that most of the stored magnetic energy is in the space between the conductors.

The energy density for points between the conductors, from Eq. 36-19, is

$$u_B = \frac{1}{2\mu_0}B^2 = \frac{1}{2\mu_0}\left(\frac{\mu_0 i}{2\pi r}\right)^2 = \frac{\mu_0 i^2}{8\pi^2 r^2}.$$

Consider a volume element dV consisting of a cylindrical shell whose radii are r and $r + dr$ and whose length is l. The energy dU contained in it is

$$dU = u_B dV = \frac{\mu_0 i^2}{8\pi^2 r^2}(2\pi rl)(dr) = \frac{\mu_0 i^2 l}{4\pi}\frac{dr}{r}.$$

EXAMPLE 5

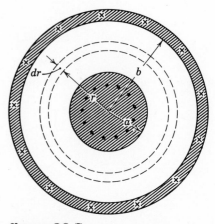

figure 36-7
Example 5. Cross section of a coaxial cable, showing steady equal but opposite currents in the central and outer conductors.

The total stored magnetic energy is found by integration, or

$$U = \int dU = \frac{\mu_0 i^2 l}{4\pi} \int_a^b \frac{dr}{r} = \frac{\mu_0 i^2 l}{4\pi} \ln \frac{b}{a},$$

which is the desired expression.

(b) What is the inductance of a length l of coaxial cable?

We can find the inductance L from Eq. 36-18 ($U = \frac{1}{2}Li^2$), which leads to

$$L = \frac{2U}{i^2} = \frac{\mu_0 l}{2\pi} \ln \frac{b}{a}.$$

You should also derive this expression directly from the definition of inductance, using the procedures of Example 1.

EXAMPLE 6

Compare the energy required to set up, in a cube 10 cm on edge (a) a uniform electric field of 1.0×10^5 V/m and (b) a uniform magnetic field of 1.0 T ($= 10^4$ gauss). Both these fields would be judged reasonably large but they are readily available in the laboratory.

(a) In the electric case we have, where V_0 is the volume of the cube,

$$U_E = u_E V_0 = \tfrac{1}{2} \epsilon_0 E^2 V_0$$

$$= (0.5)(8.9 \times 10^{-12} \text{ C}^2/\text{N} \cdot \text{m}^2)(10^5 \text{ V/m})^2 (0.1 \text{ m})^3$$

$$= 4.5 \times 10^{-5} \text{ J}.$$

(b) In the magnetic case, from Eq. 36-19, we have

$$U_B = u_B V_0 = \frac{B^2}{2\mu_0} V_o = \frac{(1.0 \text{ T})^2 (0.1 \text{ m})^3}{(2)(4\pi \times 10^{-7} \text{ T} \cdot \text{m/A})}$$

$$= 400 \text{ J}.$$

In terms of fields normally available in the laboratory, much larger amounts of energy can be stored in a magnetic field than in an electric one, the ratio being about 10^7 in this example. Conversely, much more energy is required to set up a magnetic field of reasonable laboratory magnitude than is required to set up an electric field of similarly reasonable magnitude.

36-6
MUTUAL INDUCTANCE

In Section 35-2 we saw that if two coils are close together, as in Fig. 35-2, a steady current i in one coil will set up a magnetic flux Φ linking the other coil. If we change i with time, an emf ε given by Faraday's law (Eq. 35-2) appears in the second coil; we called this process *induction*. We could better have called it *mutual induction*, to suggest the mutual interaction of the two coils and to distinguish it from *self-induction*, in which only one coil is involved as described in the preceding sections of this chapter.

Let us look a little more quantitatively at mutual induction. Figure 36-8 shows two circular close-packed coils near each other and sharing a common axis. There is a current i_1 in coil 1, set up by an external circuit not shown. This current produces a magnetic field suggested by the lines of $\mathbf{B}_1$ in the figure. Coil 2 stands alone as a closed circuit, with no external connections to a battery; a magnetic flux Φ_{21} passes through (links) it.

We define the *mutual inductance* M_{21} of coil 2 with respect to coil 1 as

$$M_{21} = \frac{N_2 \Phi_{21}}{i_1}. \qquad (36\text{-}21a)$$

Compare this with Eq. 36-4 $(L = N\Phi/i)$, the definition of (self-) inductance. We can recast Eq. 36-21a as

$$M_{21}i_1 = N_2\Phi_{21}. \qquad (36\text{-}21b)$$

If, by external means, we cause i_1 to vary with time, we have

$$M_{21}\frac{di_1}{dt} = N_2\frac{d\Phi_{21}}{dt}. \qquad (36\text{-}22)$$

The right side of this equation, from Faraday's law (Eq. 35-2), is, apart from a change in sign, just the emf ε_2 appearing in coil 2 due to the changing current in coil 1, or

$$\varepsilon_2 = -M_{21}\frac{di_1}{dt} \qquad (36\text{-}23a)$$

which you should compare with Eq. 36-3a $(\varepsilon = -L\,di/dt)$ for self-inductance.

Let us now interchange the roles of coils 1 and 2 in Fig. 36-8. That is, we set up a current i_2 in coil 2, by external circuitry not shown, and this produces a magnetic flux Φ_{12} that links coil 1 (from which the external circuitry has now been removed). If we change i_2 with time, we have, by the same argument given above,

$$\varepsilon_1 = -M_{12}\frac{di_2}{dt}; \qquad (32\text{-}23b)$$

compare Eq. 36-23a.

Thus we see that the emf in *either* coil is proportional to the rate of change of current in the *other* coil. The proportionality constants M_{21} and M_{12} seem to be different, but we assert, without proof, that they are in fact the same so that no subscripts are needed. This conclusion is in no way obvious. Thus we have

$$M_{21} = M_{12} = M \qquad (32\text{-}24)$$

and we can rewrite Eqs. 32-23 as

$$\varepsilon_2 = -M\,di_1/dt \quad \text{and} \quad \varepsilon_1 = -M\,di_2/dt. \qquad (32\text{-}25)$$

The induction is indeed *mutual.* The SI unit for M (compare Eqs. 36-3 and 36-23) is the henry.

The calculation of M, like that of L, depends on the geometry of the system. The simplest case is that in which *all* of the flux from one coil links the other coil. Example 7 shows such a situation.

figure 36-8
The arrangement used to define the mutual inductance M_{21} of coil 2 with respect to coil 1.

In the toroid of Example 1 (see Fig. 36-2) let us relabel the turns in the winding shown (coil 1) as N_1 and the current in this winding as i_1. Now let us superimpose a second winding (coil 2) over the first, having N_2 turns. The coils are electrically insulated from each other. In terms of the geometrical factors of Example 1, what is the mutual inductance M for the two windings?

From Eq. 36-21a we have

$$M = \frac{N_2\Phi_{21}}{i_1}$$

in which Φ_{21} is here identical with Φ, the total common flux in coils 1 and 2 due to i_1. From Example 1 we see that, with the noted changes in notation,

$$\Phi = \frac{\mu_0 i_1 N_1 h}{2\pi}\ln\frac{b}{a},$$

EXAMPLE 7

so that

$$M = \frac{\mu_0 N_1 N_2 h}{2\pi} \ln \frac{b}{a}.$$

Note that if $N_1 = N_2$, the inductances L_1 and L_2 of the two windings are virtually the same $(= L)$. A comparison with Example 1 shows that, in this case, $M = L$; see Problem 36 for a more general result.

questions

1. Under what conditions can we write Eq. 36-1 $[\varepsilon = d(N\Phi_B/dt)]$ as $\varepsilon = N(d\Phi_B/dt)$? Can you think of a physical situation in which changes in N alone with time would produce an induced emf?

2. In a case of mutual induction, as in Fig. 35-2, is self-induction also present? Discuss.

3. Is the inductance per unit length for a solenoid near its center (a) the same as, (b) less than, or (c) greater than the inductance per unit length near its ends?

4. Two solenoids, A and B, have the same diameter and length and contain only one layer of copper windings, with adjacent turns touching, insulation thickness being negligible. Solenoid A contains many turns of fine wire and solenoid B contains fewer turns of heavier wire. (a) Which solenoid has the larger inductance? (b) Which solenoid has the larger inductive time constant?

5. If the flux passing through each turn of a coil is the same, the inductance of the coil may be computed from $L = N\Phi_B/i$ (Eq. 36-4). How might one compute L for a coil for which this assumption is not valid?

6. Show that the dimensions of the two expressions for L, $N\Phi_B/i$ (Eq. 36-4) and $\varepsilon/(di/dt)$ (Eq. 36-3b), are the same.

7. You are given a length l of copper wire. How would you arrange it to obtain the maximum self-inductance?

8. You want to wind a coil so that it has resistance R but essentially no inductance. How would you do it?

9. Does the time required for the current in a particular LR circuit to build up to any given fraction of its equilibrium value depend on the value of the applied constant emf?

10. A steady current is set up in a coil with a very large inductive time constant. When the current is interrupted with a switch, a heavy arc tends to appear at the switch blades. Explain. (Note: Interrupting currents in highly inductive circuits can be dangerous.)

11. In an LR circuit like that of Fig. 36-4, can the self-induced emf ever be larger than the battery emf?

12. In an LR circuit like that of Fig. 36-4, is the current in the resistance *always* the same as the current in the inductance?

13. In the circuit of Fig. 36-3 the self-induced emf is a maximum at the instant the switch is closed on a. How can this be since there is no current in the inductance at this instant?

14. The switch in Fig. 36-3, having been closed on a for a "long" time, is thrown to b. What happens to the energy initially stored in the inductor?

15. A coil has a (measured) inductance L and a (measured) resistance R. Is its inductive time constant given by Eq. 36-13? Bear in mind that we derived that equation (see Fig. 36-3) for a situation in which the inductive and resistive elements were physically separated. Discuss.

16. In Section 36-3 we show that Eq. 36-10 is a solution of Eq. 36-9. Can you be sure that it is the *only* solution?

17. If a current in a source of emf is in the direction of the emf, the energy of the source decreases; if a current is in a direction opposite to the emf (as in

charging a battery), the energy of the source increases. Do these statements apply to the inductor in Fig. 36-1a and 36-1b?

18. Can you make an argument based on the manipulation of bar magnets to suggest that energy may be stored in a magnetic field?

19. Does Eq. 36-18 $(U = \frac{1}{2}Li^2)$ shed any light on the fact that (see Eq. 36-5) the inductance of a length l of a long solenoid is proportional to its volume?

20. Draw all the formal analogies that you can between a parallel-plate capacitor (for electric fields) and a long solenoid (for magnetic fields).

21. In a toroid is the energy density larger near the inner radius or near the outer radius?

22. Two coils are connected in series. Does their equivalent inductance depend on their geometrical relationship to each other?

23. You are given two similar flat circular coils of N turns each. For what geometry will their mutual inductance M be the greatest? For what geometry will it be the least? Assume that the coils are close together.

24. You are given two coils geometrically close together. Need they be electrically connected for them to exhibit mutual inductance? If they *are* electrically connected, can they still exhibit mutual inductance?

25. A flat circular coil is placed completely outside a long solenoid, near its center, the axes of the coil and the solenoid being parallel. Is there a mutual induction effect? Suppose the coil surrounds the solenoid? In each case justify your answer.

26. A circular coil of N turns surrounds a long solenoid. Is the mutual inductance greater when the coil is near the center of the solenoid or when it is near its end? Explain.

problems

SECTION 36-1

1. The inductance of a close-packed coil of 400 turns is 8.0 mH. What is the magnetic flux through the coil when the current is 5.0×10^{-3} A?
Answer: 1.0×10^{-7} Wb.

2. Each item (a) coulomb·ohm·meter/weber, (b) volt·second, (c) coulomb·ampere/farad, (d) kilogram·volt·meter^2/(henry·ampere)^2, (e) (henry/farad)$^{1/2}$ is equal to one of the items in the following list: meter, second, kilogram, dimensionless number, newton, joule, volt, ohm, watt, coulomb, ampere, weber, henry, farad. Give the equalities.

3. A 10-H inductor carries a steady current of 2.0 A. How can a 100-V self-induced emf be made to appear in the inductor?
Answer: Let the current change at 10 A/s.

SECTION 36-2

4. A solenoid is wound with a single layer of #10 copper wire (diameter, 0.10 in.). It is 4.0 cm in diameter and 2.0 m long. What is the inductance per unit length for the solenoid near its center? Assume that adjacent wires touch and that insulation thickness is negligible.

5. A long thin solenoid can be bent into a ring to form a toroid. Show that if the solenoid is long and thin enough, the equation for the inductance of a toroid (see Example 1) reduces to that for a solenoid of comparable length (Eq. 36-5).

6. *Inductors in Series* Two inductors L_1 and L_2 are connected in series and are separated by a large distance. (a) Show that the equivalent inductance L is $L_1 + L_2$. (b) Why must their separation be large?

7. Show that the self-inductance for a length l of a long wire associated with the flux *inside* the wire only is $\mu_0 l/8\pi$. Assume a uniform distribution of the current over the cross section of the wire.

8. *Inductors in Parallel*   Two inductors L_1 and L_2 are connected in parallel and separated by a large distance. (*a*) Show that the equivalent inductance L is given by

$$\frac{1}{L} = \frac{1}{L_1} + \frac{1}{L_2}.$$

(*b*) Why must their separation be large for this relationship to hold?

9. Two long parallel wires whose centers are a distance d apart carry equal currents in opposite directions. Show that, neglecting the flux within the wires themselves, the inductance of a length l of such a pair of wires is given by

$$L = \frac{\mu_0 l}{\pi} \ln \frac{d - a}{a},$$

where a is the wire radius. See Example 4, Chapter 34.

10. Calculate the self-inductance of two concentric hollow cylinders of radii a and b, and of length $l \gg a, b$. At one end the cylinders are connected by a flat conducting plate so that the current travels down in the inner cylinder and back in the outer. See Example 5 for hints.

figure 36-9
Problem 11

11. A very wide copper strip of width W is bent into a piece of slender tubing of radius R with two plane extensions, as shown in Fig. 36-9. A current i flows through the strip, distributed uniformly over its width. In this way a "one-turn solenoid" has been formed. (*a*) Find the magnitude of the magnetic field **B** in the tubular part (far away from the edges). (Hint: Assume that the field outside this one-turn solenoid is negligibly small.) (*b*) Find the inductance of this one-turn solenoid, neglecting the two plane extensions. *Answer:* (*a*) $\mu_0 i / W$. (*b*) $\pi \mu_0 R^2 / W$.

SECTION 36-3

12. The current in an LR circuit builds up to one-third of its steady-state value in 5.0 s. What is the inductive time constant?

13. How many "time constants" must we wait for the current in an LR circuit to build up to within 0.10 percent of its equilibrium value?   *Answer:* 6.9.

14. A 50-V potential difference is suddenly applied to a coil with $L = 50$ mH and $R = 180 \ \Omega$. At what rate is the current increasing after 0.001 s?

15. A wooden toroidal core with a square cross section has an inner radius of 10 cm and an outer radius of 12 cm. It is wound with one layer of #18 wire (diameter, 0.040 in.; "resistance", 160 ft/Ω). What are (*a*) the inductance and (*b*) the inductive time constant? Ignore the thickness of the insulation. *Answer:* (*a*) 2.8×10^{-4} H. (*b*) 2.7×10^{-4} s.

16. How long would it take for the potential difference across the resistor in an LR circuit ($L = 1.0$ H, $R = 1.0 \ \Omega$) to drop to 10% of its initial value?

17. A solenoid having an inductance of 6.0×10^{-6} H is connected in series to a 1.0×10^{3}-Ω resistor. (a) If a 10-V battery is switched across the pair, how long will it take for the current through the resistor to reach 80% of its final value? (b) What is the current through the resistor after one time constant? *Answer:* (a) 9.7×10^{-9} s. (b) 6.3×10^{-3} A.

18. The current in an LR circuit drops from 1.0 A at $t=0$ to 0.010 A one second later. If L is 10 H, find the resistance R in the circuit.

19. In the circuit shown in Fig. 36-10, $\varepsilon = 10$ V, $R_1 = 5.0$ Ω, $R_2 = 10$ Ω, and $L = 5.0$ H. For the two separate conditions (I) switch S just closed and (II) switch S closed for a very long time, calculate (a) the current i_1 through R_1, (b) the current i_2 through R_2, (c) the current i through the switch, (d) the potential difference across R_2, (e) the potential difference across L, and (f) di_2/dt.
Answer: I. (a) 2.0 A. (b) Zero. (c) 2.0 A. (d) Zero. (e) 10 V. (f) 2.0 A/s II. (a) 2.0 A. (b) 1.0 A. (c) 3.0 A. (d) 10 V. (e) Zero. (f) Zero.

20. In Fig. 36-11, $\varepsilon = 100$ V, $R_1 = 10$ Ω, $R_2 = 20$ Ω, $R_3 = 30$ Ω and $L = 2.0$ H. Find the values of i_1 and i_2 (a) immediately after switch S is closed; (b) a long time later; (c) immediately after switch S is opened again; (d) a long time later.

21. Show that the inductive time constant τ_L can also be defined as the time that would be required for the current in an LR circuit to reach its equilibrium value *if it continued to increase at its initial rate.*

22. The switch S in Fig. 36-3 is thrown from b to a. After one inductive time constant show that (a) the total energy transformed to thermal energy in the resistor is $0.168\varepsilon^2\tau_L/R$ and that (b) the energy stored in the magnetic field is $0.200\varepsilon^2\tau_L/R$. (c) Show that the equilibrium energy stored in the magnetic field is $0.500\varepsilon^2\tau_L/R$.

23. A coil with an inductance of 2.0 H and a resistance of 10 Ω is suddenly connected to a resistanceless battery with $\varepsilon = 100$ V. (a) What is the equilibrium current? (b) How much energy is stored in the magnetic field when this current exists in the coil? *Answer:* (a) 10 A (b) 100 J.

24. A coil with an inductance of 2.0 H and a resistance of 10 Ω is suddenly connected to a resistanceless battery with $\varepsilon = 100$ V. At 0.10 s after the connection is made, what are the rates at which (a) energy is being stored in the magnetic field, (b) thermal energy is appearing, and (c) energy is being delivered by the battery?

25. A given coil is connected in series with a 10,000-Ω resistor. When a 50-V battery is applied to the two, the current reaches a value of 2.0 mA after 5.0 ms. (a) Find the inductance of the coil. (b) What is the energy stored in the inductance at this same moment? *Answer:* (a) 98 H. (b) 2.0×10^{-4} J.

26. A long wire carries a current i uniformly distributed over a cross section of the wire. Show that the magnetic energy per unit length stored *within* the wire equals $\mu_0 i^2/16\pi$. Note that it does not depend on the wire diameter.

27. The coaxial cable of Example 5 has $a = 1.0$ mm, $b = 4.0$ mm, and $c = 5.0$ mm (c is the radius of the outer surface of the outer conductor). It carries a current of 10 A in the inner conductor and an equal but oppositely directed return current in the outer conductor. Calculate and compare the stored magnetic energy per meter of cable length (a) within the central conductor, (b) in the space between the conductors, and (c) within the outer conductor.
Answer: (a) 2.5×10^{-6} J/m. (b) 14×10^{-6} J/m. (c) 0.80×10^{-6} J/m.

28. Prove that when switch S in Fig. 36-3 is thrown from a to b all the energy stored in the inductor appears as thermal energy in the resistor.

SECTION 36-5
29. What is the energy density in the magnetic field near the center of the solenoid of Problem 23, Chapter 35? *Answer:* 63 J/m³.

30. A circular loop of wire 5.0 cm in radius carries a current of 100 A. What is the energy density at the center of the loop?

figure 36-10
Problem 19

figure 36-11
Problem 20

31. A length of #10 copper wire carries a current of 10 A. Calculate (a) the magnetic energy density and (b) the electric energy density at the surface of the wire. The wire diameter is 0.10 in. and its resistance per unit length is 1.0 ohm/1000 ft.

    Answer: (a) 0.99 J/m³. (b) 4.8×10^{-15} J/m³.

32. (a) What is the magnetic energy density of the earth's magnetic field of 5.0×10^{-5} T? (b) Assuming this to be relatively constant over distances small compared with the earth's radius and neglecting the variations near the magnetic poles, how much energy would be stored in a shell between the the earth's surface and 16 km above the surface?

33. (a) Find an expression for the energy density as a function of the radius for the toroid of Example 1. (b) Integrating the energy density over the volume of the toroid, find the total energy stored in the field of the toroid; assume $i = 0.50$ A. (c) Using Eq. 36-18 evaluate the energy stored in the toroid directly from the inductance and compare with (b).

    Answer: (a) $\dfrac{\mu_0 i^2 N^2}{8\pi^2 r^2}$. (b) 1.8×10^{-4} J. (c) 1.8×10^{-4} J.

34. What must be the magnitude of a uniform electric field if it is to have the same energy density as that possessed by a 0.50-tesla magnetic field?

35. What is the magnetic energy density at the center of a circulating electron in the hydrogen atom (see Example 9, Chapter 34)?

    Answer: 5.7×10^{7} J/m³.

SECTION 36-6
36. In Example 7 show that, if $N_1 \neq N_2$, the mutual inductance is given by

    $$M = \sqrt{L_1 L_2}.$$

    Do you suppose that this relation holds even if the situation is not like that of Example 7, that is, if it is not true that all the flux from one coil links all the turns of the other coil?

37. Two short cylindrical coils are connected in series; the coils are reasonably close together and share the same axis. (a) Show that the effective inductance of the combination is

    $$L = L_1 + L_2 \pm 2M.$$

    (b) What is the significance of the ± sign? Does it have anything to do with the relative sense (clockwise or counterclockwise) in which the coils are wound?

37
magnetic properties of matter

In electricity the *isolated charge q* is the simplest structure that can exist. If two such charges of opposite sign are placed near each other, they form an *electric dipole,* characterized by an electric dipole moment **p.** In magnetism isolated magnetic poles (usually called *magnetic monopoles*) which would correspond to isolated electric charges, apparently do not exist. The simplest magnetic structure is the *magnetic dipole,* characterized by a magnetic dipole moment **μ.** Table 34-1 summarizes some characteristics of electric and magnetic dipoles.

A current loop, a bar magnet, and a solenoid of finite length are examples of magnetic dipoles. We can identify their north poles (from which the lines of **B** *emerge*) by suspending them as compass needles and observing which end points north. We can find their magnetic dipole moments by placing the dipole in an external magnetic field **B,** measuring the torque τ that acts on it, and computing μ from Eq. 33-11, or

$$\tau = \mu \times \mathbf{B}. \qquad (37\text{-}1)$$

Alternatively, we can measure **B** due to the dipole at a point along its axis a (large) distance r from its center and compute μ from the expression in Table 34-1, or

$$B = \frac{\mu_0}{2\pi} \frac{\mu}{r^3}. \qquad (37\text{-}2)$$

Figure 37-1, which shows iron filings sprinkled on a sheet of paper under which there is a bar magnet, suggests that this dipole might be viewed as two "poles" separated by a distance d. However, all attempts to isolate these poles fail. If we break the magnet, as in Fig. 37-2, the fragments prove to be dipoles and not isolated poles. Where one north pole and one south pole existed there are now three of each. If we break up a magnet into the electrons, protons, and neutrons that make up its atoms, we will find that even these elementary particles are magnetic dipoles. Figure 37-3 contrasts the electric and the magnetic characters of the free electron.

All electrons have a characteristic *"spin" angular momentum* about a certain axis, which has the value of

$$L_s = 0.5272943 \times 10^{-34} \text{ joule·second.}$$

This is suggested by the vector $\mathbf{L}_s$ in Fig. 37-3b. Such a spinning charge can be viewed classically as being made up of infinitesimal current

figure 37-2
If we break a bar magnet, each fragment becomes a dipole. There will always be equal numbers of north and south poles, associated in pairs.

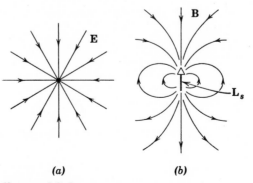

(a)   *(b)*

figure 37-3
(a) The lines of **E** and (b) the lines of **B** for an electron. The magnetic dipole moment of the electron, $\boldsymbol{\mu}_l$, is directed opposite to the spin angular momentum vector, $\mathbf{L}_s$.

loops. Each such loop is a tiny magnetic dipole, its moment being given by (see Eq. 33-10)

$$\mu = NiA, \qquad (37\text{-}3)$$

where i is the equivalent current in each infinitesimal loop and A is the loop area. The number of turns, N, is unity for each loop. The magnetic dipole moment of the spinning charge can be found by integrating over the moments of the infinitesimal current loops that make it up; see Problem 8.

Although this model of the spinning electron is too mechanistic and is not in accord with modern quantum physics, it remains true that the magnetic dipole moments of elementary particles are closely connected with their intrinsic angular momenta, or spins. Those particles and nuclei whose spin angular momentum is zero (the α-particle, the pion, the O^{16} nucleus, etc.) have no magnetic dipole moment. We must distinguish the "intrinsic" or "spin" magnetic moment of the electron from any additional magnetic moment it may have because of its *orbital* motion in an atom; see Example 2.

EXAMPLE 1

Devise a method for measuring μ for a bar magnet.

(a) Place the magnet in a uniform external magnetic field **B**, with μ making an angle θ with **B**. The magnitude of the torque acting on the magnet (see Eq. 37-1) is given by

$$\tau = \mu B \sin \theta.$$

Clearly we can find μ if we measure τ, B, and θ.

(b) A second technique is to suspend the magnet from its center of mass and to allow it to oscillate about its stable equilibrium position in the external field **B**. For *small* oscillations, $\sin \theta$ can be replaced by θ and the equation just given becomes

$$\tau = -(\mu B)\theta = -\kappa \theta,$$

where κ is a constant. The minus sign has been inserted to show that τ is a *restoring torque*. Since τ is proportional to θ, the condition for simple angular harmonic motion is met. The frequency ν is given by the reciprocal of Eq. 15-24, or

$$\nu = \frac{1}{2\pi} \sqrt{\frac{\kappa}{I}} = \frac{1}{2\pi} \sqrt{\frac{\mu B}{I}},$$

in which I is the rotational inertia. With this equation μ can be found from the measured quantities ν, B, and I.

EXAMPLE 2

An electron in an atom circulating in an assumed circular orbit of radius r behaves like a tiny current loop and has an *orbital magnetic dipole moment* usually represented by μ_l.*,† Derive a connection between μ_l and the *orbital angular momentum* L_l.

Newton's second law ($F = ma$) yields, if we substitute Coulomb's law for F,

$$\frac{1}{4\pi\epsilon_0} \frac{e^2}{r^2} = ma = \frac{mv^2}{r}$$

or

$$v = \sqrt{\frac{e^2}{4\pi\epsilon_0 mr}} \qquad (37\text{-}4)$$

---

* This must not be confused with the magnetic dipole moment μ_s of the electron spin, which is also present.
† Although this model is too mechanistic and not in the spirit of modern quantum physics, it is nevertheless instructive to examine it.

The angular velocity ω is given by

$$\omega = \frac{v}{r} = \sqrt{\frac{e^2}{4\pi\epsilon_0 mr^3}}.$$

The current for the orbit is the rate at which charge passes any given point, or

$$i = ev = e\left(\frac{\omega}{2\pi}\right) = \sqrt{\frac{e^4}{16\pi^3\epsilon_0 mr^3}}.$$

The orbital dipole moment μ_l is given from Eq. 37-3 if we put $N=1$ and $A=\pi r^2$, or

$$\mu_l = NiA = (1)\sqrt{\frac{e^4}{16\pi^3\epsilon_0 mr^3}}\,(\pi r^2) = \frac{e^2}{4}\sqrt{\frac{r}{\pi\epsilon_0 m}}. \tag{37-5}$$

The orbital angular momentum L_l is

$$L_l = (mv)r.$$

Combining with Eq. 37-4 leads to

$$L_l = \sqrt{\frac{e^2 mr}{4\pi\epsilon_0}}.$$

Finally, eliminating r between this equation and Eq. 37-5 yields

$$\mu_l = L_l\left(\frac{e}{2m}\right),$$

which shows that the orbital magnetic moment of an electron is proportional to its orbital angular momentum. Convince yourself that the vectors $\boldsymbol{\mu}_l$ and $\mathbf{L}_l$ point in opposite directions.

For $r = 5.3 \times 10^{-11}$ m, which corresponds to hydrogen in its normal state, we have, from Eq. 37-5,

$$\mu_l = \frac{e^2}{4}\sqrt{\frac{r}{\pi\epsilon_0 m}}$$

$$= \frac{(1.6 \times 10^{-19}\ \text{C})^2}{4}\sqrt{\frac{5.3 \times 10^{-11}\ \text{m}}{(\pi)(8.9 \times 10^{-12}\ \text{C}^2/\text{N}\cdot\text{m}^2)(9.1 \times 10^{-31}\ \text{kg})}}$$

$$= 9.2 \times 10^{-24}\ \text{A}\cdot\text{m}^2.$$

It is clear in Fig. 37-4 that, for the bar magnet and the solenoid, lines of $\mathbf{B}$ emerge from the top end and enter the bottom end. These are localized regions called the "north pole" (top end) and the "south pole" (bottom end). There are no sharply defined points so defined, as there are for the electrostatic dipole of Fig. 37-4c. Note that the lines of $\mathbf{B}$ and $\mathbf{E}$ in Fig. 37-4 are much alike (dipole field) at distances from the device that are much larger than its dimensions but that the behavior is much different for points very close to the device, including internal points.

The symmetry of nature has always been a guiding principle for physicists. For example, the existence of the (negative) electron suggested the existence of a positive electron, or positron, which was eventually discovered. In the same way the existence of a (positive) proton suggested that there might also be a negative proton and a large accelerator was built at the University of California (Berkeley) primarily to search for this particle; it was found. The discoveries of the positron and of the negative proton each were associated with Nobel prizes; see Appendix K.

With this motivation it is not surprising that physicists have long sought experimental evidence for the existence of magnetic monopoles.

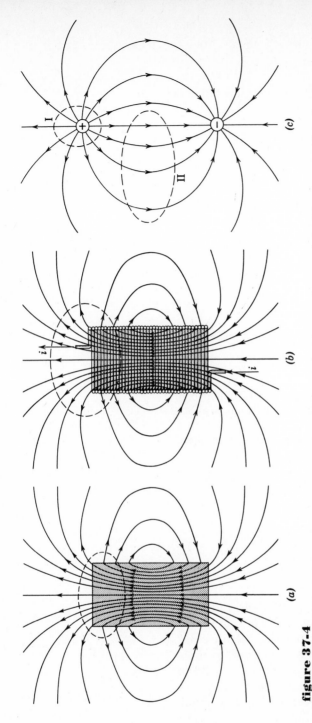

(a)   (b)   (c)

figure 37-4

Lines of **B** (a) for a bar magnet and (b) for a short solenoid. In each case the top of the figure is a north pole, the bottom being a south pole. (c) Lines of **E** for an electric dipole. At large enough distances all three fields vary like those for a dipole. The four dashed curves are intersections with the plane of the figure of closed Gaussian surfaces. Note that Φ_B equals zero for (a) or (b). Φ_E equals zero for surfaces like II in (c), which do not contain any charge, but Φ_E is not zero for surfaces like I. See Section 28-3.

Their absence, as we shall see in more detail in Chapter 40, represents a serious lack of symmetry between electricity and magnetism. Actually, magnetic monopoles were predicted to exist, on the basis of theory, by P. A. M. Dirac in 1931 and physicists have been searching for them constantly ever since.*

37-2
GAUSS'S LAW FOR MAGNETISM

Gauss's law for magnetism, which is one of the basic equations of electromagnetism (see Table 40-2), is a formal way of stating a conclusion that seems to be forced on us by the facts of magnetism, namely, that *isolated magnetic poles do not seem to exist*. This equation asserts that the flux Φ_B through any *closed* Gaussian surface must be zero, or

$$\Phi_B = \oint \mathbf{B} \cdot d\mathbf{S} = 0, \qquad (37\text{-}6)$$

where the integral is to be taken over the entire closed surface. We contrast this with Gauss's law for electricity, which is

$$\epsilon_0 \oint \mathbf{E} \cdot d\mathbf{S} = q. \qquad (37\text{-}7)$$

The fact that a zero appears at the right of Eq. 37-6, but not at the right of Eq. 37-7, means that in magnetism there seems to be no counterpart to the free charge q in electricity.

Figure 37-4a shows a Gaussian surface that encloses one end of a bar magnet. Note that the lines of **B** enter the surface inside the magnet and leave it outside the magnet. There is thus an inward (or negative) flux inside the magnet and an outward (or positive) flux outside it. The total flux for the whole surface is zero.

Figure 37-4b shows a similar surface for a solenoid of finite length which, like a bar magnet, is also a magnetic dipole. Here, too, Φ_B equals zero. Figures 37-4a and b show clearly that there are no "sources" of **B**; that is, there are no points from which lines of **B** emanate. Also, there are no "sinks" of **B**; that is, there are no points toward which **B** converges. In other words, *there are no free magnetic poles*.

Figure 37-4c shows a Gaussian surface (I) surrounding the positive end of an *electric* dipole. Here there *is* a net flux of the lines of **E**. There is a "source" of **E**; it is the charge q. If q is negative, we have a "sink" of **E** because the lines of **E** end on negative charges. For surfaces like surface II in Fig. 37-4c for which the charge inside is zero, the flux of **E** over the surface is also zero.

37-3
THE MAGNETISM OF THE EARTH

The magnet with which we are most familiar is that on which we live, the earth. The supposition that the earth is a large magnet, with magnetic poles and a magnetic equator, was first made by Sir William Gilbert (1544–1603), physician to Queen Elizabeth I. Gilbert made a small spherical *terrella* ("little earth") out of a naturally occurring magnetic loadstone (literally "leading stone" or compass) and traced its lines of magnetism. In those days of navigation and exploration there was a natural interest in the compass and in the earth's magnetism.

---

* See "Quest for the Magnetic Monopole" by Richard A. Carrigan, Jr., *The Physics Teacher*, October 1975, for more information.

Figure 37-5a is an idealized sketch of the lines of **B** associated with the earth's magnetic field, both at and above its surface. To a first approximation we can represent this field by imagining a strong bar magnet located at the center of the earth, as in Fig. 37-5b. Note that the earth's magnetic axis and its rotational axis RR do not coincide, being separated by about 15°.

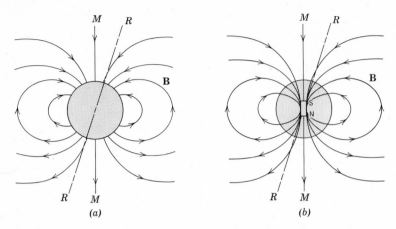

(a)   (b)

figure 37-5
(a) An idealized representation of the lines of **B** associated with the earth's magnetic field. MM is the earth's magnetic axis and RR is its rotational axis. (b) We can approximate the earth's external magnetic field by imagining that a srong bar magnet is located at the center of the earth.

The magnetic pole in the northern hemisphere is in arctic Canada. Note that it is a *south* magnetic pole because lines of **B** *converge* toward it, as in Fig. 37-4. There is a *north* magnetic pole in the southern hemisphere, in Antarctica; lines of **B** *emerge* from it.

There is, of course, no bar magnet buried at the center of the earth. The earth's magnetism must be related to the facts that the central core of the earth, whose radius is 55% of the earth's radius, is (a) liquid, (b) highly conducting, and (c) partakes of the earth's rotation. A dynamo effect, involving circulating currents in the earth's core whose mechanism is not yet fully understood, is almost certainly operating.*

Several other planets in our solar system, Mercury and Jupiter among them, also have magnetic fields. So do the sun and many other stars. There is also a magnetic field associated with our galaxy, that is, with the family of stars whose plane of symmetry is defined by the Milky Way. The galactic magnetic field is relatively weak (about 2 pT on the average) but its effects can be important because it extends over such large distances.

Two simple tools for exploring the earth's magnetic field are the compass needle and the dip needle, the latter being a gravitationally balanced magnetized needle whose axis of rotation is horizontal rather than vertical. In Tucson, Arizona, for example, the north pole of a compass needle (as of 1964) pointed about 13° east of geographic north. Such *declinations* must be known and taken into account when using the compass for navigation or direction finding. The horizontal component of the earth's magnetic field $\mathbf{B}_h$, to which the compass needle responds, is, at Tucson, 26 μT (=0.26 gauss).

Now let us orient the (horizontal) rotational axis of a dip needle at Tucson until it is at right angles to the horizontal component of **B**. Can you convince yourself that the needle will now point in the direction of **B**? It does, and we find that, at this site and at this date, the north end of the needle points downward toward the earth, the needle making an angle φ_i of about 59° (the *inclination*) with a horizontal plane containing

* See "The Earth's Magnetism" by S. K. Runcorn, *Scientific American*, September 1955.

Color Plate 1

At the right we see an aurora viewed from Sky Lab 3, from somewhere over the Indian Ocean. The bright band extending to the left is the normal "airglow", a phenomenon not visible from earth to the naked eye. It represents the emission of radiation from atoms and molecules in the upper atmosphere that absorb energy from sunlight during the daytime. The full moon illuminates a largely clouded earth. A piece of Sky Lab barely shows at the upper left. (This photograph was taken by astronaut Owen K. Garriott by a hand-held camera in Sky Lab. We are indebted to him and to the National Aeronautics and Space Administration for permission to publish the photograph.)

its rotational axis. This shows that, as we expect from Fig. 37-5a, lines of **B** are *entering* the earth's surface at this point.

The earth's magnetic field is neither as regular nor as static as the idealized field in Fig. 37-5 suggests. Also, there are observable phenomena, going far beyond deflections of the compass needle, that would not occur if the earth had no magnetic field. Consider the following:

1. *Local variations.* The earth's magnetic field has important local variations, caused by differences in the magnetic properties of the rocks that make up the earth's crust and by the presence of concentrated magnetic ore bodies.

2. *Changes with time.* The mean magnetic declination and inclination vary measurably from year to year at any given location. For example, between the years 1600 and 1800 the measured magnetic declination at London varied in a continuous way from 11° east to 24° west. The north magnetic pole (as of 1948) was measured to be moving north west at about 8 km/y.

This wandering of the earth's magnetic axis, and thus the variation with time of local declinations and inclinations, has resulted in a new archeological specialty, *archeomagnetism*, by means of which dates may be assigned to ancient kilns, ovens, and hearths. The principle is based on the fact that most clays, from which such structures were made, contain small amounts of magnetic materials, whose orientation is frozen into position by heating during normal use. By comparing the present direction of the earth's magnetic field with that of the "frozen in" direction of the magnetization, an approximate archeological date may be established.

On a longer (geological) time scale there is evidence that the earth's magnetic axis has actually completely reversed in direction as many as nine times during the last 4×10^6 years.* The evidence rests on measurements of the (weak) magnetism frozen into rocks of known geological age at the time of their formation.

3. *Interactions with the solar wind.* The sun emits a steady stream of ionized hydrogen atoms (protons) and electrons, that sweeps through the solar system at supersonic speeds. This ever present "solar wind" interacts strongly in several ways with the earth's magnetic field.† (a) Occasional sharp increases in the intensity of the solar wind produce terrestrial *magnetic storms*, which seriously interfere with long-distance radio communication. (b) The protons and electrons of the solar wind, acted upon by forces given by Eq. 33-2 ($\mathbf{F} = q_0\mathbf{v} \times \mathbf{B}$), spiral along the lines of **B**, moving back and forth between the magnetic north and south polar regions. These "trapped" electrons and protons form the so called *Van Allen radiation belts.* They were discovered by James A. Van Allen, of the State University of Iowa, in early satellite experiments.‡ (c) The trapped solar wind particles, as they interact with the earth's atmosphere, produce the dazzling spectacle of the *aurora*, which is most prominent at about ±75° geomagnetic latitude; see Plate I, opposite.**

* "Reversals of the Earth's Magnetic Field" by Allan Cox, G. Brent Dalrymple, and Richard R. Doell, *Scientific American*, February 1967.
† "The Solar Wind" by E. N. Parker, *Scientific American*, April 1964.
‡ See "Radiation Belts" by Brian J. O'Brien, *Scientific American*, May 1963. See also "The Magnetosphere" by J. A. Ratcliffe, *Contemporary Physics*, vol. 18, 1977.
** "Aurora and Airglow" by C. T. Elvey and Franklin E. Roach, *Scientific American*, September 1955.

From data given earlier in this section find (*a*) the vertical component $\mathbf{B}_v$ of the earth's magnetic field and (*b*) the resultant magnetic field $\mathbf{B}$ at Tucson.

Figure 37-6 shows the situation. We have

(*a*)

$$B_v = B_h \tan \varphi_i$$

$$= (26 \ \mu\text{T})(\tan 59°)$$

$$= 44 \ \mu\text{T} \ (= 0.44 \ \text{gauss}),$$

and

(*b*)

$$B = \sqrt{B_h{}^2 + B_v{}^2}$$

$$= \sqrt{(26 \ \mu\text{T})^2 + (44 \ \mu\text{T})^2}$$

$$= 51 \ \mu\text{T} \ (= 0.51 \ \text{gauss}).$$

Note that the magnetic declination at Tucson plays no role in this problem.

EXAMPLE 3

figure 37-6
Example 3.

37-4
PARAMAGNETISM

Magnetism as we know it in our daily experience is an important but special branch of the subject called *ferromagnetism*; we discuss this in Section 37-6. Here we discuss a weaker form of magnetism called *paramagnetism*.

For most atoms and ions, the magnetic effects of the electrons, including both their spins and orbital motions, exactly cancel so that the atom or ion is not magnetic. This is true for the rare gases such as neon and for ions such as Cu^+, which makes up ordinary copper.* For other atoms or ions the magnetic effects of the electrons do *not* cancel, so that the atom as a whole has a magnetic dipole moment $\boldsymbol{\mu}$. Examples are found among the transition elements, such as Mn^{++}; the rare earths, such as Gd^{+++}; and the actinide elements, such as U^{++++}.

If we place a sample of N atoms, each of which has a magnetic dipole moment $\boldsymbol{\mu}$, in a magnetic field, the elementary atomic dipoles tend to line up with the field. This tendency to align is called *paramagnetism*. For perfect alignment, the sample as a whole would have a magnetic dipole moment of $N\boldsymbol{\mu}$. However, the aligning process is seriously interfered with by thermal agitation effects. The importance of thermal agitation may be measured by comparing two energies: one $(= \frac{3}{2}kT)$ is the mean kinetic energy of translation of a gas atom at temperature T; the other $(= 2\mu B)$ is the difference in energy between an atom lined up with the magnetic field and one pointing in the opposite direction. As Example 4 shows, the effect of the collisions at ordinary temperatures and fields is very great. The sample acquires a magnetic moment when placed in an external magnetic field, but this moment is usually very much smaller than the maximum possible moment $N\mu$.

A paramagnetic gas, whose atoms (see Example 2) have a magnetic dipole moment of about $10^{-23} \ \text{A} \cdot \text{m}^2$, is placed in an external magnetic field of magnitude 1.0 T $(= 10^4 \ \text{gauss})$. At room temperature $(T = 300 \ \text{K})$ calculate and compare U_T, the mean kinetic energy of translation $(= \frac{3}{2}kT)$, and U_B, the magnetic energy $(= 2\mu B)$:

EXAMPLE 4

*Cu^+ indicates a copper atom from which one electron has been removed; Al^{+++} indicates an aluminum atom from which three electrons have been removed, etc.

$$U_T = \tfrac{3}{2}kT = (\tfrac{3}{2})(1.38 \times 10^{-23} \text{ J/K})(300 \text{ K}) = 6 \times 10^{-21} \text{ J},$$

$$U_B = 2\mu B = (2)(10^{-23} \text{ A} \cdot \text{m}^2)(1.0 \text{ T}) = 2 \times 10^{-23} \text{ J}.$$

Because U_T equals 300 U_B, we see that energy exchanges in collisions can interfere seriously with the alignment of the dipoles with the external field.

If we place a specimen of a paramagnetic substance in a nonuniform magnetic field, such as that near the pole of a strong magnet, it will be attracted toward the region of higher field, that is, toward the pole. We can understand this by drawing an analogy with the corresponding electric case of Fig. 37-7, which shows a dielectric specimen (a sphere) in a nonuniform electric field. The net electric force points to the right in the figure and is

$$F_e = q(E_0 + \Delta E) - q(E_0 - \Delta E) = q(2\Delta E),$$

which we can write as

$$F_e = \frac{(q\,\Delta x)}{\Delta x}2\Delta E = p\left(\frac{2\Delta E}{\Delta x}\right) \cong p\left(\frac{dE}{dx}\right)_{\text{max}}$$

Here p (= $q\,\Delta x$) is the induced electric dipole moment of the sphere. In the differential limit of a very small sphere $(2\Delta E/\Delta x)$ approaches $(dE/dx)_{\text{max}}$, the gradient of the electric field at the center of the sphere.

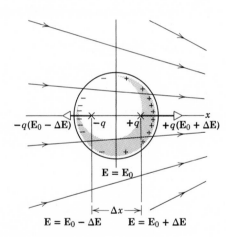

figure 37-7
A dielectric sphere in a nonuniform electric field. The effective induced charges are represented by the point charges $+q$ and $-q$

In the corresponding magnetic case we have, by analogy,

$$F_m = \mu\left(\frac{dB}{dx}\right)_{\text{max}} \tag{37-8}$$

Thus, by measuring the magnetic force F_m that acts on a small paramagnetic specimen when we place it in a nonuniform magnetic field whose field gradient $(dB/dx)_{\text{max}}$ is known, we can learn its magnetic dipole moment μ. The *magnetization* **M** of the specimen is defined as the magnetic moment per unit volume, or

$$\mathbf{M} = \frac{\boldsymbol{\mu}}{V},$$

where V is the volume of the specimen. It is a vector because $\boldsymbol{\mu}$, the dipole moment of the specimen, is a vector.

In 1895 Pierre Curie (1859–1906) discovered experimentally that the magnetization M of a paramagnetic specimen is directly proportional to

B, the effective value of magnetic field in which the specimen is placed, and inversely proportional to the temperature, or

$$M = C \frac{B}{T},$$ (37-9)

in which C is a constant. This equation is known as *Curie's law*. The law is physically reasonable in that increasing B tends to align the elementary dipoles in the specimen, that is, to increase M, whereas increasing T tends to interfere with this alignment, that is, to decrease M. Curie's law is well verified experimentally, provided that the ratio B/T does not become too large.

M cannot increase without limit, as Curie's law implies, but must approach a value M_{max} ($= \mu N/V$) corresponding to the complete alignment of the N dipoles contained in the volume V of the specimen. Figure 37-8 shows this saturation effect for a sample of $CrK(SO_4)_2 \cdot 12H_2O$. The chromium ions are responsible for all the paramagnetism of this salt, all the other elements being paramagnetically inert. To achieve 99.5% saturation, it is necessary to use applied magnetic fields as high as 50,000 gauss ($= 5.0$ T) and temperatures as low as 1.3 K. Note that for more readily achievable conditions, such as $B = 10,000$ gauss ($= 1.0$ T) and $T = 10$ K, the abscissa in Fig. 37-8 is only 1.0 so that Curie's law would appear to be well obeyed for this and for all lower values of B/T.

figure 37-8
The ratio M/M_{max} for a paramagnetic salt (chromium potassium alum) in various magnetic fields and at various temperatures. The curve through the experimental points is a theoretical curve calculated from modern quantum physics. (From measurements by W. E. Henry.)

The curve that passes through the experimental points in this figure is calculated from a theory based on modern quantum physics; it is in excellent agreement with experiment.

37-5
DIAMAGNETISM

In 1846 Michael Faraday discovered that a specimen of bismuth brought near to the pole of a strong magnet is *repelled*. He called such substances *diamagnetic* (in contrast with paramagnetic specimens, which are *attracted*). Diamagnetism, present in all substances, is such a feeble effect that its presence is masked in substances made of atoms that have a net magnetic dipole moment, that is, in paramagnetic or ferromagnetic substances.

Figures 37-9a and b show an electron circulating in a diamagnetic atom at angular frequency ω_0 in an assumed circular orbit of radius r.

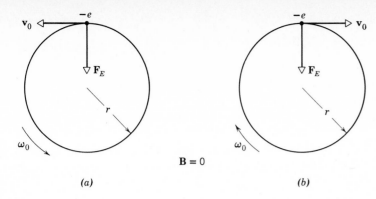

(a)

(b)

$\mathbf{B} = 0$

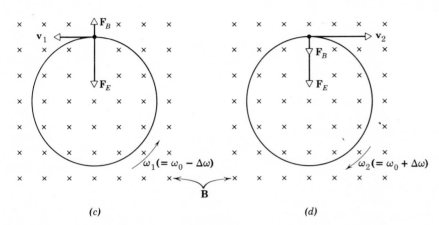

(c)

(d)

figure 37-9
(a) An electron circulating in an atom. (b) An electron circulating in the opposite direction. (c) A magnetic field is introduced, *decreasing* the linear speed of the electron in (a), that is, $v_1 < v_0$. (d) The magnetic field *increases* the linear speed of the electron in (b), that is, $v_2 > v_0$. See problem 15 for a different way of looking at diamagnetism. Our treatment of diamagnetism, though mechanistic and classical, nevertheless yields results in reasonable agreement with experiment.

Each electron is moving under the action of a centripetal force $\mathbf{F}_E$ of electrostatic origin where, from Newton's second law,

$$F_E = ma = m\omega_0^2 r. \qquad (37\text{-}10)$$

Each revolving electron has an orbital magnetic moment, but for the atom as a whole the orbits are randomly oriented so that there is no *net* magnetic effect. In Fig. 37-9a, for example, the magnetic dipole moment $\boldsymbol{\mu}_l$ points into the page; in Fig. 37-9b it points out and the suggested net effect for the two orbits shown is cancellation. This cancellation is also suggested at the left in Fig. 37-10.

If we apply an external field $\mathbf{B}$ as in Figs. 37-9c and d, an *additional* force, given by $-e(\mathbf{v} \times \mathbf{B})$, acts on the electron. This magnetic force acts always at right angles to the direction of motion; its magnitude is

$$F_B = evB = e(\omega r)B. \qquad (37\text{-}11)$$

Show that in Fig. 37-9c $\mathbf{F}_B$ and $\mathbf{F}_E$ point in opposite directions and that in Fig. 37-9d they point in the same direction. Note that since the centripetal force changes when we turn on the magnetic field (the radius can be shown to remain constant), the angular velocity must also change; thus ω in Eq. 37-11 differs from ω_0 in Eq. 37-10.

Applying Newton's second law to Figs. 37-9c and d, and allowing for both directions of circulation, yields for the *resultant* forces on the electrons

$$F_E \pm F_B = ma = m\omega^2 r.$$

Substituting Eqs. 37-10 and 37-11 into this equation yields

(a)   (b)

figure 37-10
The magnetic moments of the two oppositely circulating electrons in an atom cancel when there is no external magnetic field, as in (a), but do *not* cancel when a field is applied, as in (b). Note that the resultant moment in (b) points in the *opposite* direction to $\mathbf{B}$. Compare carefully with Fig. 37-9.

$$m\omega_0{}^2 r \pm e\omega r B = m\omega^2 r$$

or

$$\omega^2 \mp \left(\frac{eB}{m}\right)\omega - \omega_0{}^2 = 0. \tag{37-12}$$

We could solve this quadratic equation for ω, the new angular velocity. Rather than doing this, we take advantage of the fact (presented without proof; see Problem 14) that ω differs only slightly from ω_0, even in the strongest external magnetic fields. Thus

$$\omega = \omega_0 + \Delta\omega, \tag{37-13}$$

where $\Delta\omega \ll \omega_0$. Substituting this equation into Eq. 37-12 yields

$$[\omega_0{}^2 + 2\omega_0\,\Delta\omega + (\Delta\omega)^2] \mp [\beta\omega_0 + \beta\Delta\omega] - \omega_0{}^2 = 0,$$

where β is a convenient abbreviation for eB/m. The two terms $\omega_0{}^2$ cancel each other; the terms $(\Delta\omega)^2$ and $\beta\Delta\omega$ are small compared to the remaining terms and may be set equal to zero with only small error. This leads, as an excellent approximation, to

$$\Delta\omega \cong \mp \tfrac{1}{2}\beta = \pm \frac{eB}{2m} \tag{37-14}$$

or, from Eq. 37-13, $\qquad \omega = \omega_0 \pm \dfrac{eB}{2m}.$

Thus the effect of applying a magnetic field is to increase or decrease (depending on the direction of circulation) the angular velocity. This, in turn, increases or decreases the orbital magnetic moment of the circulating electron (see Example 2).

In Fig. 37-9c the angular velocity is reduced (because the centripetal force is reduced) so that the magnitude of the magnetic moment is reduced. In Fig. 37-9d, however, the angular velocity is increased so that the magnitude of $\boldsymbol{\mu}_l$ is increased. These effects are shown on the right in Fig. 37-10, where it will be noted that the two magnetic moments *no longer cancel*.

We see that if we apply a magnetic field **B** to a diamagnetic substance, a magnetic moment will be *induced* whose direction (out of the plane of Fig. 37-9) is *opposite* to **B**; see also Fig. 37-10. This is precisely the reverse of paramagnetism, in which the (*permanent*) magnetic dipoles tend to point in the *same* direction as the applied field.

We can now understand why a diamagnetic specimen is repelled when brought near to the pole of a strong magnet. If the pole is a north pole, there exists a nonuniform magnetic field **B** pointing away from the pole. If a sphere made of a diamagnetic material (bismuth, say) is brought near to this pole, the magnetization **M** that is induced in it points toward the pole, that is, *opposite to* **B.** Thus the side of the sphere closest to the magnet behaves like a north pole and is *repelled* by the nearby north pole of the magnet. For a paramagnetic sphere, the vector **M** points *along the direction of* **B** and the side of the sphere nearest to the magnet is a south pole, which is *attracted* to the north pole of the magnet.

EXAMPLE 5

Calculate the *change* in magnetic moment for a circulating electron, as described in Example 2, if a magnetic field **B** of 2.0 T (= 20,000 gauss) acts at right angles to the plane of the orbit.

We obtain μ from Eq. 37-3, or

$$\mu = NiA = (1)(ev)(\pi r^2) = (1)\left(\frac{e\omega}{2\pi}\right)(\pi r^2) = \tfrac{1}{2}er^2\omega.$$

The *change* in μ is
$$\Delta\mu = \tfrac{1}{2}er^2\,\Delta\omega$$

or, from Eq. 37-14,
$$\Delta\mu = \pm\,\tfrac{1}{2}er^2\left(\frac{eB}{2m}\right) = \pm\frac{e^2Br^2}{4m}.$$

Substituting numbers yields
$$\Delta\mu = \pm\frac{(1.6\times10^{-19}\text{ C})^2(2.0\text{ T})(5.3\times10^{-11}\text{ m})^2}{(4)(9.1\times10^{-31}\text{ kg})}$$

$$= \pm4.0\times10^{-29}\text{ A}\cdot\text{m}^2.$$

In Example 2 the moment μ_l was 9.2×10^{-24} A·m², so that the change induced by even a strong external magnetic field is rather small, the ratio $\Delta\mu/\mu_l$ being about 4×10^{-6}.

37-6
FERROMAGNETISM

For three elements (Fe, Co, and Ni) and for a variety of alloys of these and other elements, a special effect occurs which permits a specimen to achieve a high degree of magnetic alignment in spite of the randomizing tendency of the thermal motions of the atoms. In such materials, described as *ferromagnetic*, a special form of interaction called *exchange coupling* occurs between adjacent atoms, coupling their magnetic moments together in rigid parallelism.* If the temperature is raised above a certain critical value, called the *Curie temperature*, the exchange coupling suddenly disappears and the materials become simply paramagnetic. For iron the Curie temperature is 1043 K. Ferromagnetism is evidently a property not only of the individual atom or ion but also of the interaction of each atom or ion with its neighbors in the crystal lattice (see Fig. 21-5) of the solid.

Figure 37-11 shows a *magnetization curve* for a specimen of iron. To obtain such a curve, we form the specimen, assumed initially unmagnetized, into a ring and wind a toroidal coil around it, as in Fig. 37-12, to form a so-called *Rowland ring*.† When a current i is set up in the coil, *if the iron core is not present*, a magnetic field is set up within the toroid given by (see Eq. 34-7)

$$B_0 = \mu_0 ni, \qquad (37\text{-}15)$$

where n is the number of turns per unit length for the toroid. Although this formula was derived for a long solenoid, we can apply it to a toroid

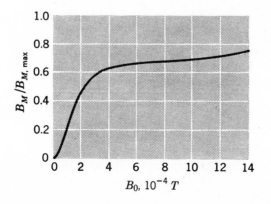

figure 37-11
A magnetization curve for iron.

*Exchange coupling, a purely quantum effect, cannot be "explained" in terms of classical physics.
† See Section 33-1 for further information about H. A. Rowland.

figure 37-12
A Rowland ring, showing a
secondary coil S.

if $d \ll r$ in Fig. 37-12. Because of the iron core, the actual value of **B** in the toroidal space will exceed **B**$_0$, by a large factor in many cases, since the elementary atomic dipoles in the core line up with the applied field **B**$_0$, thereby setting up their own magnetic field. Thus we can write

$$B = B_0 + B_M \qquad (37\text{-}16)$$

where B_M is the magnetic field contribution due to the core; it is proportional to the magnetization M of the core. Often $B_M \gg B_0$.

The field B_0 is proportional to the current in the toroid and we can calculate it readily, using Eq. 37-15; B can be measured in a way that we describe below. An experimental value for B_M can be derived from Eq. 37-16. It has a maximum value $B_{M.\max}$ corresponding to complete alignment of the atomic dipoles in the iron. Thus we can plot, as in Fig. 37-11, the fractional degree of alignment $(= B_M/B_{M.\max})$ as a function of B_0. For this specimen a value of 96.5% saturation is achieved at $B_0 = 0.13$ T $(= 1300$ gauss; this point is about 16 ft to the right of the origin in the figure); increasing B_0 to 1.0 T $(= 10,000$ gauss; about 120 ft to the right in Fig. 37-11) increases the fractional saturation only to 97.7%.

The use of iron in transformers, electromagnets, etc., greatly increases the strength of the magnetic field that can be generated by a given current in a given set of windings. That is, very often, $B_M \gg B_0$ in Eq. 37-16. However, in the design of electromagnets generating really intense magnetic fields (3 - 25 T, say) it proves poor practise to incorporate iron in the limited space available in the magnet core. The controlling design factor proves to be the rate at which cooling water can be pumped through this core to carry away the large amount of Joule heat generated in the energizing coils. A particular (iron-free) 25-T magnet at MIT, though only 36 in. in diameter, dissipates 16 MW and requires that cooling water course through its core at the rate of 2000 gal/min.

To measure B in the system of Fig. 37-12, let the current in the toroid windings be increased from zero to i. The flux through the secondary coil S will change by BA, where A is the area of the toroid. While the flux is changing, an induced emf will appear in coil S, according to Faraday's law. For simplicity, we assume that the current in the toroid is so adjusted that B increases linearly with time for an interval Δt, as shown in Fig. 37-13a. The emf in coil S during this interval, from Faraday's law,† will then be

† We ignore the minus sign because we are concerned only with the magnitude of ε.

$$\varepsilon = N\frac{\Delta\Phi_B}{\Delta t} = \frac{NBA}{\Delta t},$$

where N is the number of turns in coil S. This emf will set up a current i_s in coil S given by

$$i_s = \frac{\varepsilon}{R} = \frac{NBA}{R\,\Delta t}$$

or

$$B = \frac{(i_s\,\Delta t)R}{NA} = \frac{qR}{NA},$$

in which R is the resistance of coil S and $i_s\,\Delta t$ is the charge q that passes through this coil during time Δt. If a so-called *ballistic galvanometer* is connected to S, its deflection will be a measure of the charge q. Thus it is possible to find B for any value of the current i in the toroid windings. A more detailed analysis shows that it is not necessary that the curve $B(t)$ in Fig. 37-13a be linear during the interval Δt.

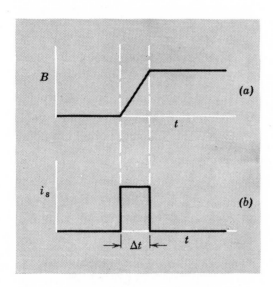

figure 37-13
(a) B for a Rowland ring as the current in the windings is increased from zero during an interval Δt. (b) The corresponding induced current in the secondary coil. Both curves are idealized; in practice, the sharp corners would be rounded off.

Magnetization curves for ferromagnetic materials *do not retrace themselves* as we increase and then decrease the toroid current. Figure 37-14 shows the following operations with a Rowland ring: (1) starting with the iron unmagnetized (point a), increase the toroid current until B_0 ($= \mu_0 ni$) has the value corresponding to point b; (2) reduce the current in the toroid winding back to zero (point c); (3) reverse the toroid current and increase it in magnitude until point d is reached; (4) reduce the current to zero again (point e); (5) reverse the current once more until point b is reached again. The lack of retraceability shown in Fig. 37-14 is called *hysteresis*. Note that at points c and e the iron core is magnetized, even though there is no current in the toroid windings; this is the familiar phenomenon of *permanent magnetism*.

We explained the magnetization curve for paramagnetism (see Fig. 37-8) in terms of the mutually opposing tendencies of alignment with the external field and of randomization because of thermal agitation. In ferromagnetism, however, we have assumed that adjacent atomic dipoles are locked in rigid parallelism. Why, then, does the magnetic moment of the specimen not reach its saturation value for very low—even zero—values of B_0? The current interpretation is to assume the existence within the specimen of *domains*, that is, of local regions

figure 37-14
A magnetization curve (ab) for a specimen of iron and an associated hysteresis loop ($ebcde$).

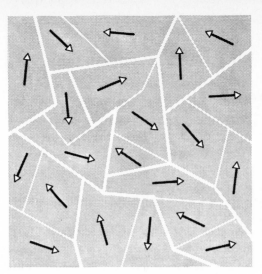

figure 37-15
The separate magnetic domains in an unmagnetized polycrystalline ferromagnetic sample are oriented to produce little external effect. Each domain, however, is made up of completely aligned atomic dipoles, as suggested by the arrows. The white boundaries define the crystals that make up the solid; the light boundaries define the domains within the crystals.

within which there is essentially perfect alignment. The domains themselves, however, as Fig. 37-15 suggests, are not parallel at moderately low values of $\mathbf{B}_0$.

Figure 37-16 shows some domain photographs, taken by sprinkling a colloidal suspension of finely powdered iron oxide on a properly etched single crystal of iron. The domain boundaries, which are thin regions in which the alignment of the elementary dipoles changes from a certain direction in one domain to an entirely different direction in the other, are the sites of intense but highly localized and nonuniform magnetic fields. The suspended colloidal particles are attracted to these regions. Although the atomic dipoles in the individual domains are completely aligned, the specimen as a whole may have a very small resultant magnetic moment. This is the state of affairs in an unmagnetized iron nail.

figure 37-16
Domain patterns for a single crystal of iron containing 3.8% silicon. The white lines show the boundaries between the domains. These boundaries are regular rather than irregular, as in Fig. 37-15, because the specimen is a single crystal. In Fig. 37-15, the specimen is not a single crystal but is made up of many crystallites or grains. (Courtesy H. J. Williams, Bell Telephone Laboratories.)

As we magnetize a piece of iron by placing it in an external magnetic field, two effects take place. One is a growth in size of the domains that are favorably oriented at the expense of those that are not, as in Fig. 37-17. Second, the direction of orientation of the dipoles within a domain may swing around as a unit, becoming closer to the field direction. Hysteresis comes about because the domain boundaries do not move completely back to their original positions when the external field $\mathbf{B}_0$ is removed.

Two other types of magnetism, closely related to ferromagnetism, are *antiferromagnetism* and *ferrimagnetism* (note spelling). In antiferro-

0.01 mm

(a)   (b)   (c)

figure 37-17
(a) A boundary between two domains, with the magnetization in each domain shown by the white arrows. (b) If an external magnetic field pointing from left to right is imposed on the specimen, the upper domain will grow at the expense of the lower. The domain boundary will move down as the elementary dipoles reverse themselves. (c) The process continues. The boundary has moved across a region in which there is a crystal imperfection. (Courtesy H. J. Williams, Bell Telephone Laboratories.)

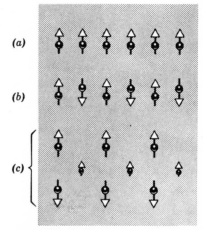

(a)

(b)

(c)

figure 37-18
Showing how elementary magnetic dipoles are oriented by the interatomic exchange coupling in (a) ferromagnetism, (b) antiferromagnetism, and (c) ferrimagnetism. The actual arrangements are, of course, three-dimensional.

37-7
NUCLEAR MAGNETISM

magnetic substances, of which MnO_2 is an example, the exchange coupling to which we referred on page 825 serves to lock adjacent ions into rigid *antiparallelism* (see Fig. 37-18). Such materials exhibit very little gross external magnetism. However, if they are heated above a certain temperature, called the *Néel temperature,* the exchange coupling ceases to act and the material becomes paramagnetic. In ferrimagnetic substances, of which iron ferrite is an example, two different kinds of magnetic ions are present. In iron ferrite the two ions are Fe^{++} and Fe^{+++}. The exchange coupling locks the ions into a pattern like that of Fig. 37-18c, in which the external effects are intermediate between ferromagnetism and antiferromagnetism. Here, too, the exchange coupling disappears if the material is heated above a certain characteristic temperature.

Many nuclei have magnetic dipoles, and the possibility arises that a specimen of matter may exhibit gross external magnetic effects associated with its nuclei. However, nuclear magnetic moments are several orders of magnitude smaller than those associated with the electronic motions in an atom or ion. The magnetic moment of an electron associated with its spin, for example, exceeds that of the proton (the nucleus of hydrogen) by a factor of 660.

Gross external effects for nuclear magnetism are smaller than the corresponding (ionic) paramagnetic effects by the *square* of ratios of this order of magnitude, because (a), *all else being equal*, the external magnetism is reduced by such a ratio, but (b) the very fact that the magnetic dipole moment of the nucleus is smaller means that (see Example 4) the thermal vibrations are proportionally (to a good approximation) more effective in reducing the degree of alignment of the elementary dipoles in an external magnetic field; thus all else is *not* equal and the ratio enters twice.

Techniques such as the Rowland ring (see Fig. 37-12) are far too insensitive to detect nuclear magnetism. We describe here a *nuclear resonance technique* by means of which nuclear magnetism can readily reveal itself. This method is also vastly useful for studying paramagnetism, ferromagnetism, antiferromagnetism, and ferrimagnetism, in all of which cases the magnetic effects are associated not with the nuclei but with the atomic electrons. The nuclear-resonance technique was developed in 1946 by E. M. Purcell and his co-workers at Harvard. Simultaneously and independently, F. Bloch and his co-workers at Stanford discovered a very similar method. For these achievements the two physicists shared a Nobel prize.

We focus our attention on the problem of measuring the magnitude μ of the magnetic moment of the proton. In principle, this can be done by placing a specimen containing protons in an external magnetic field **B** and by measuring the energy $(= 2\mu B)$ required to turn the protons end for end. A rigorously correct description of the procedures cannot be given without using quantum physics. The description given, although based entirely on classical physics, nevertheless leads to the correct conclusions.

(a)

(b)

figure 37-19
(a) A spinning proton precessing in an external magnetic field and (b) a spinning top precessing in an external gravitational field. $\mathbf{L}_p$ and $\mathbf{L}_t$ are the two angular momentum vectors.

Figure 37-19a shows a spinning proton with its axis making an angle θ with a uniform external magnetic field **B.** Figure 37-19b shows a spinning top with its axis making an angle θ with a uniform external gravitational field **g.** In each case there is a torque that tends to align the axis of the spinning object with the field. For the proton (see Eq. 33-11) it is given by

$$\tau_p = \mu B \sin\theta. \qquad (37\text{-}17a)$$

For the top it is given by $\quad \tau_t = mgr \sin\theta, \qquad (37\text{-}17b)$

where r locates the center of mass of the top and m is its mass.

In Section 13-2, we saw that the spinning top precesses about a vertical axis with an angular frequency given by

$$\omega_t = \frac{mgr}{L_t}, \qquad (37\text{-}18a)$$

in which L_t is the spin angular momentum of the top.

The proton, which has a quantized spin angular momentum L_p, will also precess about the direction of the (magnetic) field because of the action of the (magnetic) torque. You should derive the expression for the frequency of precession, being guided by the derivation of Section 13-2, but using the magnetic torque (see Eq. 37-17a) instead of the gravitational torque (see Eq. 37-17b). The relation is

$$\omega_p = \frac{\mu B}{L_p}. \qquad (37\text{-}18b)$$

EXAMPLE 6

831 NUCLEAR MAGNETISM SEC. 37-7

What is the precession frequency of a proton in a magnetic field of 0.50 T (= 5000 gauss)?

The quantities μ and L_p in Eq. 37-18b are 1.4×10^{-26} A·m² and 0.53×10^{-34} J·s. This equation then yields

$$\nu_p = \frac{\omega_p}{2\pi} = \frac{\mu B}{2\pi L_p} = \frac{(1.4 \times 10^{-26} \text{ A·m}^2)(0.50 \text{ T})}{(2\pi)(0.53 \times 10^{-34} \text{ J·s})} = 2.1 \times 10^7 \text{ Hz}.$$

This frequency (=21 MHz) is in the radio-frequency range.

It is possible to change the energy of any system in periodic motion if we allow an external influence to act on it at the same frequency as that of its motion. This is the familiar *resonance* condition. As an "external influence" for the precessing proton, we use a small alternating magnetic field $\mathbf{B}_{osc}$ arranged to be at right angles to the steady field $\mathbf{B}$. This oscillating field combines vectorially with the steady field so that the *resultant* field rocks back and forth between the limits shown by the dashed lines in Fig. 37-20. Typical values for B and for the amplitude of $\mathbf{B}_{osc}$ are 5000 gauss and one gauss, respectively, so that the rocking angle α in the figure is quite small. If the angular frequency ω_0 of the oscillating field is chosen equal to the angular precession frequency ω_p of the proton, it turns out that the precessing proton can absorb energy. An increase in energy means an increase in θ in Fig. 37-19a.

The resonance condition

$$\omega_0 = \frac{\mu B}{L_p} \tag{37-19}$$

can be used to measure μ. We place the spinning proton in a known field $\mathbf{B}$, apply a "perturbing field" at right angles to it, and vary the angular frequency ω_0 of this perturbing field until resonance occurs. It is possible to tell when Eq. 37-19 is satisfied because, at resonance, many spinning protons will tend to turn end for end in the field, absorbing energy which can be detected by appropriate electronic techniques.

figure 37-20
In the nuclear magnetic resonance method a small oscillating magnetic field $\mathbf{B}_{osc}$ is placed at right angles to a steady field $\mathbf{B}$.

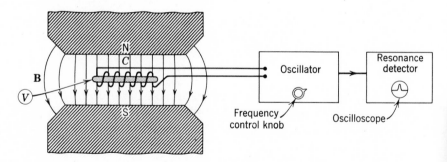

figure 37-21
An arrangement to observe nuclear resonance. The oscillating field is horizontal within the coil.

Figure 37-21 is a schematic diagram of an experimental arrangement. The protons, present as hydrogen nuclei in a small vial V of water, are immersed in a strong steady magnetic field caused by the electromagnet whose pole faces N and S are shown. A rapidly alternating current in the small coil C provides the (horizontal) weak, perturbing magnetic field $\mathbf{B}_{osc}$. This current is provided by a radio-frequency oscillator whose angular frequency ω_0 can be varied; an electronic "resonance detector," also connected to the oscillator, serves to indicate when energy is being drained from the oscillator and used to "flip the protons." In principle, the oscillator angular frequency ω_0 is varied until the resonance detector shows that Eq. 37-19 is satisfied (see Fig. 37-22). The magnetic moment μ can then be determined by measuring B and ω_0. Surprisingly enough, magnetic moments can be measured in this and similar ways to a much greater accuracy than we can measure μ for a bar magnet. For the proton we have

$$\mu_p = 1.410617 \times 10^{-26} \text{ A·m}^2.$$

figure 37-22
An oscilloscope photograph of a proton resonance peak showing energy absorbed from the oscillator (vertical axis) versus oscillator frequency (horizontal axis). (From Bloembergen et al.

In Chapter 30 we saw that if a dielectric is placed in an electric field, polarization charges will appear on its surface. These surface charges, which find their origin in the elementary electric dipoles (permanent or induced) that make up the dielectric, set up a field of their own that modifies the original field. For the simple case discussed in Chapter 30 — a dielectric slab in a parallel-plate capacitor — this complication can readily be handled in terms of the electric field vector **E** and some knowledge of the electric properties of the slab material, such as the dielectric constant. For more complex problems we asserted that it was useful to introduce two other (subsidiary) electric vectors, the *electric polarization* **P** and the *electric displacement* **D.** Table 30-2 shows some of the characteristics of these three vectors.

In magnetism we find a similar situation. If magnetic materials are placed in an external magnetic field, the elementary magnetic dipoles (permanent or induced) will act to set up a field of their own that will modify the original field. For the simple case discussed in this chapter — a Rowland ring with a ferromagnetic core — this complication can readily be handled in terms of the magnetic field vector **B** and some knowledge of the magnetic properties of the ring material, such as that provided by Fig. 37·11. For more complex problems we find it useful to introduce two other (subsidiary) magnetic vectors, the *magnetization* **M** and the *magnetic field strength* **H.** We do so largely so that the student who takes a second course in electromagnetism will have some familiarity with them.

Consider a Rowland ring carrying a current i_0 in its windings and designed so that its core, assumed to be iron, can be removed. The magnetic field **B,** measured by the methods of Section 37-6, will be much greater when the core is in place than when it is not, assuming that the current in the windings remains unchanged.

Physically we can understand the large value of B in the iron core in terms of the alignment of the elementary dipoles in the iron. A hypothetical slice out of the iron core, as in Fig. 37-23b, has a magnetic moment $d\mu$ equal to the vector sum of all of the elementary dipoles contained in it. We define our first subsidiary vector, the *magnetization* **M,** as the magnetic moment per unit volume of the core material. For the slice of Fig. 37-23b we have

$$d\mu = \mathbf{M}(A\ dl),$$

where $(A\ dl)$ is the volume of the slice, A being the cross section of the core.

37-8
THREE MAGNETIC VECTORS

figure 37-23
(a) A Rowland ring with an iron core. (b) A slice of the core, showing its magnetic moment $d\mu$ caused by the alignment of the elementary magnetic dipoles in the iron.

(a)   *(b)*

When we discussed Ampère's law in Chapter 34, we assumed that no magnetic materials were present. If we apply this law, namely,

$$\oint \mathbf{B} \cdot d\mathbf{l} = \mu_0 i, \tag{37-20}$$

to the circular path of integration shown in Fig. 37-23a, we obtain

$$(B)(2\pi r_0) = \mu_0(N_0 i_0), \tag{37-21}$$

has when the core is in place if we increase the current in the windings by an amount $i_{M,0}$. The magnetization of the iron core is thus *equivalent in its effect on* **B** to such a hypothetical current increase. We choose to modify Ampère's law by arbitrarily inserting a *magnetizing current* term i_M on the right, obtaining

$$\oint \mathbf{B} \cdot d\mathbf{l} = \mu_0(i + i_M). \tag{37-22}$$

If we give i_M a suitable value when the iron core is in place, it is clear that Ampère's law, in this new form, can remain valid. It remains to relate this (largely hypothetical) magnetizing current to something more physical, the magnetization **M**.*

Applying Eq. 37-22 to the iron ring of Fig. 37-23a yields

$$(B)(2 \pi r_0) = \mu_0(N_0 i_0 + N_0 i_{M,0}). \tag{37-23}$$

We can relate $i_{M,0}$ to the magnetization **M** if we recall (Eq. 33-10) that the magnetic moment of a magnetic dipole in the form of a current loop is given by

$$\mu = NiA,$$

where N is the number of turns in the loop, i is the loop current, and A is the loop area. Let us use this equation to find what increase $i_{M,0}$ in current in the windings around the slice of Fig. 37-23b would produce a magnetic moment equivalent to that actually produced by the alignment of elementary dipoles in the slice. We have

$$M(A \; dl) = \left(N_0 \frac{dl}{2\pi r_0}\right)(i_{M,0})(A),$$

the quantity in the first parentheses on the right being the number of turns associated with the slice of thickness dl. This reduces to

$$N_0 i_{M,0} = M(2\pi r_0). \tag{37-24}$$

Substituting this into Eq. 37-23 yields

$$(B)(2\pi r_0) = \mu_0(N_0 i_0) + \mu_0(M)(2\pi r_0). \tag{37-25}$$

We now choose to generalize from the special case of the Rowland ring by writing Eq. 37-25 as

$$\oint \mathbf{B} \cdot d\mathbf{l} = \mu_0 i + \mu_0 \oint \mathbf{M} \cdot d\mathbf{l}$$

or

$$\oint \left(\frac{\mathbf{B} - \mu_0 \mathbf{M}}{\mu_0}\right) \cdot d\mathbf{l} = i.$$

The quantity $(\mathbf{B} - \mu_0 \mathbf{M})/\mu_0$ occurs so often in magnetic situations that we give it a special name, the *magnetic field strength* **H**, or

$$\mathbf{H} = \frac{\mathbf{B} - \mu_0 \mathbf{M}}{\mu_0}$$

which we write as

$$\mathbf{B} = \mu_0 \mathbf{H} + \mu_0 \mathbf{M}. \tag{37-26}$$

* It is possible to give reality to the magnetizing current by viewing it as a real current that flows around the magnet at its surface, being the resultant macroscopic effect of all the microscopic current loops that constitute the atomic electron orbits. This *Amperian current* viewpoint, however, does not take the magnetization due to electron spin readily into account. Since we do not attempt to measure magnetizing currents experimentally, other than through their (postulated) magnetic effects, we prefer to view the magnetizing current as a convenient formalism.

Ampère's law can now be written in the simple form

$$\oint \mathbf{H} \cdot d\mathbf{l} = i, \tag{37-27}$$

which holds in the presence of magnetic materials and in which i is the *true current only*, that is, it does not include the magnetizing current. This reminds us that the electric displacement vector $\mathbf{D}$ permitted us to write Gauss's law for the case in which dielectric materials are present, in a form involving *free charges only*, that is, not polarization charges; see Table 30-2.

We state without proof (see Problems 23 and 24) that at a boundary between two media (1) the component of $\mathbf{H}$ tangential to the surface has the same value on each side of the surface* and (2) the component of $\mathbf{B}$ perpendicular to the surface has the same value on each side of the surface. These *boundary conditions* are of great value in solving complex problems.

To find H in our Rowland ring, let us apply Ampère's law in the generalized form of Eq. 37-27. We have

$$(H)(2\pi r_0) = N_0 i_0,$$

where i_0 is the (true) current in the windings. This gives

$$H = \left(\frac{N_0}{2\pi r_0}\right) i_0 = n i_0, \tag{37-28}$$

in which n is the number of turns per unit length. Since we have not introduced any information describing the core into Eq. 37-27, the value of H computed from Eq. 37-28 is independent of the core material.

B can be measured experimentally by the method of Section 37-6 and M can then be calculated from Eq. 37-26. Note in passing (see Eq. 37-15) that the abscissa B_0 in Fig. 37-11 is proportional to $H(= \mu_0 H)$, the ordinate being proportional to B. Curves such as this and that of Fig. 37-14 are called *B-H curves*.

Let us assume that we have made measurements of $\mathbf{H}$, $\mathbf{B}$, and $\mathbf{M}$ for a wide variety of magnetic materials, using either the technique described or an equivalent one. For *paramagnetic* and *diamagnetic* materials we would find, as an experimental result, that $\mathbf{B}$ is directly proportional to $\mathbf{H}$, or

$$\mathbf{B} = \kappa_m \mu_0 \mathbf{H}, \tag{37-29}$$

in which κ_m, the *permeability* of the magnetic medium, is a constant for a given temperature and density of the material. Eliminating $\mathbf{B}$ between Eqs. 37-29 and 37-26 allows us to write

$$\mathbf{M} = (\kappa_m - 1)\mathbf{H}, \tag{37-30}$$

which is another expression of the linear or proportional character of paramagnetic and diamagnetic materials.

For a vacuum, in which there are no magnetic dipoles present to be aligned, the magnetization $\mathbf{M}$ must be zero. Putting $\mathbf{M} = 0$ in Eq. 37-26 leads to

$$\mathbf{B} = \mu_0 \mathbf{H} \qquad \text{(in vacuum).} \tag{37-31}$$

Comparison with Eq. 37-29 shows that a vacuum must be described by $\kappa_m = 1$. Equation 37-30 verifies that the magnetization vanishes if we

---

* Assuming that there are no true currents at the surface, as there are in the Rowland ring in Fig. 37-23a, for example.

Table 37-1
Three magnetic vectors

Name	Symbol	Associated with	Boundary Condition
Magnetic induction*	**B**	All currents	Normal component continuous
Magnetic field strength	**H**	True currents only	Tangential component continuous†
Magnetization (magnetic dipole moment per unit volume)	**M**	Magnetization currents only	Vanishes in a vacuum

Defining equations for **B**		$\mathbf{F} = q\mathbf{v} \times \mathbf{B}$	Eq. 33-2
	or	$= i\mathbf{l} \times \mathbf{B}$	Eq. 33-6a
General relation among the three vectors		$\mathbf{B} = \mu_0\mathbf{H} + \mu_0\mathbf{M}$	Eq. 37-26
Ampère's law when magnetic materials are present		$\oint \mathbf{H} \cdot d\mathbf{l} = i$ (i = true current only)	Eq. 37-27
Empirical relations for certain magnetic materials**		$\mathbf{B} = \kappa_m \mu_0 \mathbf{H}$	Eq. 37-29
		$\mathbf{M} = (\kappa_m - 1)\mathbf{H}$	Eq. 37-30

*We have usually called **B** simply "the magnetic field." Here we call it by its alternative name, "magnetic induction," to avoid confusion with "magnetic field strength," the name for **H**.
** For paramagnetic and diamagnetic materials only, if κ_m is to be independent of **H**.
† Assuming no true currents exist at the boundary.

put κ_m equal to unity. For paramagnetic materials κ_m is slightly greater than unity. For diamagnetic materials it is slightly less than unity; Eq. 37-30 shows that this requires **M** and **H** to be oppositely directed, a fact implicit in Section 37-5.

In ferromagnetic materials the relationship between **B** and **H** is far from linear, as Figs. 37-11 and 37-14 show. Experimentally, κ_m proves to be a function not only of the value of H but also, because of hysteresis, of the magnetic and thermal history of the specimen.*

An interesting special case of ferromagnetism is the permanent magnet, for which **H**, **M**, and **B** all have nonvanishing values inside the magnet even though there is no true current. Figure 37-24 shows typical lines of **B** and **H** associated with such a magnet. The lines of **B** may be drawn as continuous loops, the boundary condition (2), mentioned above, being satisfied where the lines enter and leave the magnet.

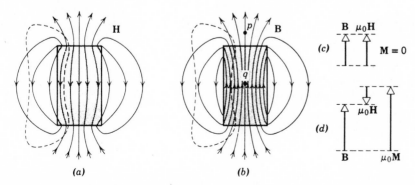

(a)   (b)

figure 37-24

(a) The lines of **H** and (b) the lines of **B** for a permanent magnet. Note that the lines of **H** change direction at the boundary. The closed dashed curves are paths of integration around which Ampère's law may be applied. The relation $\mathbf{B} = \mu_0\mathbf{H} + \mu_0\mathbf{M}$ is shown to be satisfied for (c) a particular outside point p and (d) a particular inside point q.

*In the dielectric case there are waxy materials, called *ferroelectrics*, for which the relationship between **D** and **E** is nonlinear, which exhibit hysteresis, and from which quasi-permanent electric dipoles (*electrets*) can be constructed. However, most commonly useful dielectric materials are linear, whereas most commonly useful magnetic materials are nonlinear.

Equation 37-22 shows that the vector **B** is associated with the *total* current, both true and magnetizing. In Fig. 37-24b $\oint \mathbf{B} \cdot d\mathbf{l}$ around any loop such as that shown by the dashed curve is not zero and must be associated with a hypothetical magnetizing current i_M, imagined to circulate around the magnet at its surface; actual or true currents (i) do not exist in this problem. Figure 37-24a shows that **H** reverses direction at the boundary. Since **H** (see Eq. 37-27) is associated with true currents only, we must have $\oint \mathbf{H} \cdot d\mathbf{l} = 0$ around any loop such as that shown by the dashed lines. The reversal of **H** at the boundary makes this possible. Note that **M** and **H** point in opposite directions within the magnet. Table 37-1 summarizes the properties of the three vectors **B**, **H**, and **M**.

EXAMPLE 7

In the Rowland ring the (true) current i_0 in the windings is 2.0 A and the number of turns per unit length (n) in the toroid is 10 turns/cm. B, measured by the technique of Section 37-6, is 1.0 T. Calculate (a) H, (b) M, and (c) the magnetizing current $i_{M,0}$ both when the core is in place and when it is removed. (d) For these particular operating conditions, what is κ_m?

(a) H is independent of the core material and may be found from Eq. 37-28:

$$H = ni$$

$$= (10 \text{ turns/cm})(2.0 \text{ A})$$

$$= 2.0 \times 10^3 \text{ A/m}.$$

(b) M is zero when the core is removed. With the core in place, we may solve Eq. 37-26 for **M**, obtaining for the magnitude of **M**,

$$M = \frac{B - \mu_0 H}{\mu_0}$$

$$= \frac{(1.0 \text{ T}) - (4\pi \times 10^{-7} \text{ T} \cdot \text{m/A})(2.0 \times 10^3 \text{ A/m})}{(4\pi \times 10^{-7} \text{ T} \cdot \text{m/A})}$$

$$= 7.9 \times 10^5 \text{ A/m}.$$

(c) The effective magnetizing current follows from Eq. 37-24:

$$i_{M,0} = M \left(\frac{2\pi r_0}{N_0}\right) = \frac{M}{n}$$

$$= \frac{7.9 \times 10^5 \text{ A/m}}{1.0 \times 10^3 \text{ turns/m}}$$

$$= 790 \text{ A}.$$

An *additional* current of this amount in the windings would produce the same value of B in the absence of a core as that obtained by alignment of the elementary dipoles, with the core in place.

(d) The permeability can be found from Eq. 37-29, or

$$\kappa_m = \frac{B}{\mu_0 H}$$

$$= \frac{1.0 \text{ T}}{(4\pi \times 10^{-7} \text{ T} \cdot \text{m/A})(2.0 \times 10^3 \text{ A/m})}$$

$$= 397.$$

We emphasize that this value of κ_m holds only for the special conditions of this experiment.

questions

1. Two iron bars are identical in appearance. One is a magnet and one is not. How can you tell them apart? You are not permitted to suspend either bar as a compass needle or to use any other apparatus.

2. Two iron bars always attract, no matter the combination in which their ends are brought near each other. Can you conclude that one of the bars must be unmagnetized?

3. The neutron, which has no charge, has a magnetic dipole moment. Is this possible on the basis of classical electromagnetism, or does this evidence alone indicate that classical electromagnetism has broken down?

4. Do all permanent magnets have to have identifiable north and south poles? Consider geometries other than the straight bar magnet.

5. If magnetic monopoles are shown to exist, is it conceivable that there might be a whole family of them, with different masses, magnetic pole strengths, intrinsic angular momenta, etc.? Compare Appendix F.

6. A certain short iron rod is found, by test, to have a north pole at each end. You sprinkle iron filings over the rod. Where (in the simplest case) will they cling? Make a rough sketch of what the lines of B must look like, both inside and outside the rod. See "A Three-Pole Bar Magnet?" by Jerry D. Wilson, *The Physics Teacher*, September 1976.

7. Consider these two situations. (*a*) A (hypothetical) magnetic monopole is pulled through a single-turn conducting loop along its axis, at a constant speed. (*b*) A short bar magnet (a magnetic dipole) is similarly pulled. Compare qualitatively the net amounts of charge transferred through any cross section of the loop during these two processes.

8. Cosmic rays are charged particles that strike our atmosphere from some external source. We find that more low-energy cosmic rays reach the earth near the north and south magnetic poles than at the (magnetic) equator. Why is this so?

9. How might the magnetic dipole moment of the earth be measured?

10. Give three reasons for believing that the flux Φ_B of the earth's magnetic field is greater through the boundaries of Alaska than through those of Texas.

11. You are a manufacturer of compasses. (*a*) Describe ways in which you might magnetize the needle. (*b*) The end of the needle that points north is usually painted a characteristic color. Without suspending the needle in the earth's field, how might you find out which end of the needle to paint? (*c*) Is the painted end a north or a south magnetic pole?

12. Can you think of a mechanism by which a magnetic storm, that is, a violent perturbation of the earth's magnetic field, can interfere with radio communication?

13. Convince yourself, in terms of the relation $\mathbf{F} = q_0\mathbf{v} \times \mathbf{B}$ (Eq. 33-2) that electrons and protons of the solar wind that are trapped in the earth's Van Allen radiation belts will indeed spiral around the lines of $\mathbf{B}$ and be reflected backward near the earth's north and south magnetic poles (the magnetic mirror effect). In what way do the motions of the trapped electrons and protons differ?

14. Aurorae are most frequently observed, not at the north and south magnetic poles, but at magnetic latitudes about 23° away from these poles (passing through Hudson Bay, for example, in the northern geomagnetic hemisphere). Can you think of any reason, however qualitative, why the auroral activity should not be strongest at the poles themselves?

15. Is the magnetization at saturation for a paramagnetic substance very much different from that for a saturated ferromagnetic substance of about the same size?

16. The magnetization induced in a given diamagnetic sphere by a given external magnetic field does not vary with temperature, in sharp contrast to the situation in paramagnetism. Is this understandable in terms of the description that we have given of the origin of diamagnetism?

17. Explain why a magnet attracts an unmagnetized iron object such as a nail.

18. Does any net force or torque act on (a) an unmagnetized iron bar or (b) a permanent bar magnet when placed in a uniform magnetic field?

19. A nail is placed at rest on a smooth table top near a strong magnet. It is released and attracted to the magnet. What is the source of the kinetic energy it has just before it strikes the magnet?

20. Compare the magnetization curves for a paramagnetic substance (see Fig. 37-8) and for a ferromagnetic substance (see Fig. 37-11). What would a similar curve for a diamagnetic substance look like? Do you think that it would show saturation effects in strong applied fields (say 10 T)?

21. Why do iron filings line up with a magnetic field, as in Fig. 37-1? After all, they are not intrinsically magnetized.

22. Distinguish between the precession frequency and the cyclotron frequency of a proton in a magnetic field.

23. In our discussion of nuclear magnetism we said that energy absorption occurs because the dipoles are turned end for end. However, a given dipole might initially be lined up either with the field or against it. In the first case there would be an *absorption* of energy, but in the second case there would be a *release* of energy, each amount being $2\mu B$. Why do we observe a *net* absorption? These two events would seem to cancel.

24. Discuss similarities and differences in Tables 30-2 and 37-1.

25. In what sense do a parallel plate capacitor filled with a dielectric and a Rowland ring (see Fig. 37-12) with an iron core show formal similarities as far as E and B (and their related vectors) are concerned? Discuss in terms of Tables 30-2 and 37-1.

26. A Rowland ring (see Fig. 37-12) carries a constant current. If we cut a small slot in the iron core, leaving an air gap, what changes occur in **B**, **H**, and **M** within the core?

27. You are given a bar magnet, with its north pole pointing up and its south pole pointing down. What are the directions of **B**, **H**, and **M** for points (a) inside the magnet near its center, (b) outside and just above the magnet, and (c) outside and just below the magnet?

problems

SECTION 37-1

1. The dipole moment of a small, single-turn, current loop is 2.0×10^{-4} A·m². What is the magnetic field (due to the dipole) on the axis of the dipole 8.0 cm away from the loop?   *Answer:* 7.8×10^{-8} T, pointing along the axis.

2. A simple bar magnet is suspended by a string, as shown in Fig. 37-25. If a uniform magnetic field **B** directed parallel to the ceiling is then established, sketch the resulting orientation of string and magnet.

3. Calculate (a) the electric field **E** and (b) the magnetic field **B** at a point 0.10 nm away from a proton, measured along its axis of spin. The magnetic moment of the proton is 1.4×10^{-26} A·m².
Answer: (a) 1.4×10^{11} V/m. (b) 2.8×10^{-3} T.

4. Show that, classically, a spinning positive charge will have a spin magnetic moment that points in the same direction as its spin angular momentum.

5. An electron has a spin angular momentum L of 0.53×10^{-34} J·s and a magnetic moment μ of 9.3×10^{-24} A·m². Compare μ/L and e/m for the electron. *Answer:* They are equal.

6. A total charge q is distributed uniformly on a dielectric ring of radius r. If the ring is rotated about an axis perpendicular to its plane and through its center at an angular speed ω, find the magnitude and direction of its resulting magnetic moment.

7. Show, by sketching the magnetic field of a magnetic dipole, that (a) if the dipole moments of two nearby dipoles are parallel, they will not tend to remain that way and (b) if they are anti-parallel, they *will* tend to remain that way.

figure 37-25
Problem 2

In each case consider the torques on the second dipole in the field of the first.

8. Assume that the electron is a small sphere of radius R, its charge and mass being spread uniformly throughout its volume. Such an electron has a "spin" angular momentum L of 0.53×10^{-34} J·s and a magnetic moment μ of 9.3×10^{-24} A·m². Show that $e/m = 2\mu/L$. Is this prediction in agreement with experiment? (Hint: The spherical electron must be divided into infinitesimal current loops and an expression for the magnetic moment found by integration. This model of the electron is too mechanistic to be in the spirit of quantum physics.)

SECTION 37-3

9. From the data given in the text find (a) the average vertical component of the earth's magnetic field at Tucson in 1964 and (b) the average magnitude of **B**. *Answer:* (a) 44μT (=0.44 gauss). (b) 51μT (=0.51 gauss).

10. The earth has a magnetic dipole moment of 8.0×10^{22} A·m². (a) What current would have to be set up in a single turn of wire going around the earth at its magnetic equator if we wished to set up such a dipole? (b) Could such an arrangement be used to cancel out the earth's magnetism at points in space well above the earth's surface? (c) On the earth's surface?

SECTION 37-4

11. (a) What is the magnetic moment due to the orbital motion of an electron in an atom when the orbital angular momentum is one quantum unit ($= h/2\pi = 1.05 \times 10^{-34}$ J · s). (b) The intrinsic spin magnetic moment of an electron is 0.928×10^{-23} A·m². What is the difference in the magnetic potential energy U between the states in which the magnetic moment is aligned with and aligned in the opposite direction to an external magnetic field of 1.2 T? (c) What absolute temperature would be required so that the energy difference in (b) would equal the mean thermal energy $kT/2$? *Answer:* (a) 9.2×10^{-24} A · m². (b) 2.2×10^{-23} J. (c) 3.2 K.

12. At what temperature will the average thermal energy of a paramagnetic gas be equal to the magnetic energy in a field of 0.50 T (=5000 gauss) if the dipole moments of the atoms are about 10^{-23} A·m²?

SECTION 37-5

13. An electron travels in a circular orbit about a fixed positive point charge in the presence of a uniform magnetic field **B** directed normal to the plane of its motion. The electric force has precisely N times the magnitude of the magnetic force on the electron. (a) Determine the two possible angular speeds of the electron's motion. (b) Evaluate these speeds numerically if $B = 0.427$ T (=4.27×10^3 gauss) and $N = 100$.

   Answer: (a) $(N \pm 1)\dfrac{eB}{m}$. (b) 7.43×10^{12} rad/s; 7.57×10^{12} rad/s.

14. Prove that $\Delta\omega \ll \omega_0$ in Eq. 37-13.

15. Can you give an explanation of diamagnetism based on Faraday's law of induction? In Figs. 37-9a and b, for example, what inductive effects can be expected as the magnetic field is built up from zero to the final value **B**?

SECTION 37-6

16. The dipole moment associated with an atom of iron in an iron bar is 1.8×10^{-23} A·m². Assume that all the atoms in the bar, which is 5.0 cm long and has a cross-sectional area of 1.0 cm², have their dipole moments aligned. (a) What is the dipole moment of the bar? (b) What torque must be exerted to hold this magnet at right angles to an external field of 1.5 T (=15,000 gauss)? The density of iron is 7.9 g/cm³.

17. A Rowland ring is formed of ferromagnetic material. It is circular in cross section, with an inner radius of 5.0 cm and an outer radius of 6.0 cm and is wound with 400 turns of wire. (a) What current must be set up in the windings to attain $B_0 = 2.0 \times 10^{-4}$ T in Fig. 37-11? (b) A secondary coil wound around the toroid has 50 turns and has a resistance of 8.0 Ω. If, for this value

of B_0, we have $B_M = 800B_0$, how much charge moves through the secondary coil when the current in the toroid windings is turned on?

Answer: (a) 0.14 A. (b) 7.9×10^{-5} C.

18. *Dipole-dipole interaction.* The exchange coupling mentioned in Section 37-6 as being responsible for ferromagnetism is *not* the mutual magnetic interaction energy between two elementary magnetic dipoles. To show this (a) compute B a distance a (=10 nm) away from a dipole of moment μ (= 1.8×10^{-23} A·m²); (b) compute the energy (= $2\mu B$) required to turn a second similar dipole end for end in this field. What do you conclude about the strength of this dipole-dipole interaction? Compare with the results of Example 4. (Note: For the same distance, the field in the median plane of a dipole is only half as large as on the axis; see Eq. 37-2.)

SECTION 37-7

19. Assume that the hydrogen nuclei (protons) in 1.0 g of water could all be aligned. What magnetic field B would be produced 5.0 cm from the sample, along its alignment axis? 	*Answer:* 7.5×10^{-6} T.

20. It is possible to measure e/m for the electron by measuring (a) the cyclotron frequency ν_c of electrons in a given magnetic field and (b) the precession frequency ν_p of protons in the same field. Show that the relation is

$$\frac{e}{m} = \frac{\nu_c}{\nu_p} \frac{\mu_s}{L_s}.$$

SECTION 37-8

21. The magnetic energy density can be shown to be given in its most general form as

$$\mu_B = \tfrac{1}{2}\mathbf{B} \cdot \mathbf{H}.$$

Does this reduce to a familiar result for a vacuum? 	*Answer:* Yes.

22. An iron magnet containing iron of relative permeability 5000 has a flux path 1.0 m long in the iron and an air gap 0.01 m long each with cross-sectional areas of 0.02 m². What current is necessary in a 500 turn coil wrapped around the iron to give a flux density in the air gap of 1.8 T?

23. *Boundary condition for* **H.** Prove that at the boundary between two media the tangential component of **H** has the same value on each side of the surface, assuming that there is no current at the surface. (Hint: Construct a closed rectangular loop, the two opposite longer sides being parallel to the surface, with one side in each medium. Use Ampère's law in the form that applies when magnetic materials are present.)

24. *Boundary condition for* **B.** Prove that at the boundary between two media the normal component of **B** has the same value on each side of the surface. (Hint: Construct a closed Gaussian surface shaped like a flat pillbox with one face in each medium and apply Gauss's law for magnetism.)

38
electromagnetic oscillations

The LC system in Fig. 38-1 (assumed resistanceless*) resembles the mass-spring system of Fig. 8-4 (assumed frictionless*) in that, among other things, each system has a characteristic frequency of oscillation. The analogy actually goes far beyond this, as we will see quantitatively in Section 38-3. Let us first, however, treat LC oscillations from a physical but semiquantitative point of view.

Assume that initially the capacitor C in Fig. 38-1a carries a charge q_m and the current i in the inductor is zero. At this instant the energy stored in the capacitor is given by Eq. 30-7, or

$$U_E = \frac{1}{2} \frac{q_m^2}{C}. \qquad (38\text{-}1)$$

The energy stored in the inductor, given by

$$U_B = \tfrac{1}{2} L i^2, \qquad (38\text{-}2)$$

is zero because the current is zero. The capacitor now starts to discharge through the inductor, positive charge carriers moving counterclockwise, as shown in Fig. 38-1b. This means that a current i, given by dq/dt and pointing down in the inductor, is established.

As q decreases, the energy stored in the electric field in the capacitor also decreases. This energy is transferred to the magnetic field that appears around the inductor because of the current i that is building up

---

* A common term to describe the closely related situations of Figs. 8-4 and 38-1 is "undamped." If friction is present (as in Fig. 15-19) or resistance (as in Fig. 38-3), we describe the situation as "damped."

there. Thus the electric field decreases, the magnetic field builds up, and energy is transferred from the former to the latter.

At a time corresponding to Fig. 38-1c, all the charge on the capacitor will have disappeared. The electric field in the capacitor will be zero, the energy stored there having been transferred entirely to the magnetic field of the inductor. According to Eq. 38-2, there must then be a current—and indeed one of maximum value—in the inductor. Note that even though q equals zero, the current (which is dq/dt) is *not* zero at this time.

figure 38-1
Showing eight stages in a cycle of oscillation of a resistanceless LC circuit. The bar graphs below each figure show the stored magnetic and electric energy. The vertical arrows on the inductor axis show the current. Compare this figure in detail with Fig. 8-4, to which it exactly corresponds.

The large current in the inductor in Fig. 38-1c continues to transport positive charge from the top plate of the capacitor to the bottom plate, as shown in Fig. 38-1d; energy now flows from the inductor back to the capacitor as the electric field builds up again. Eventually, the energy will have been transferred completely back to the capacitor, as in Fig. 38-1e. The situation of Fig. 38-1e is like the initial situation, except that the capacitor is charged oppositely.

The capacitor will start to discharge again, the current now being clockwise, as in Fig. 38-1f. Reasoning as before, we see that the circuit eventually returns to its initial situation and that the process continues at a definite frequency ν (measured, say, in hertz, or cycles/second) to which corresponds a definite *angular* frequency ω ($= 2\pi\nu$ and measured, say, in radians/second). Once started, such LC oscillations (in the ideal case described, in which the circuit contains no resistance) continue indefinitely, energy being shuttled back and forth between the electric field in the capacitor and the magnetic field in the inductor. Any configuration in Fig. 38-1 can be set up as an initial condition. The oscillations will then continue from that point, proceeding clockwise around the figure. Compare these oscillations carefully with those of the mass-spring system described in Fig. 8-4.

To measure the charge q as a function of time, we can measure the variable potential difference $V_C(t)$ that exists across capacitor C. The relation

$$V_C = \left(\frac{1}{C}\right) q$$

shows that V_C is proportional to q. To measure the current we can insert a small resistance R in the circuit and measure the potential difference across it. This is proportional to i through the relation

$$V_R = (R)i.$$

We assume here that R is so small that its effect on the behavior of the circuit is negligible. Both q and i, or more correctly V_C and V_R, which are proportional to them, can be displayed on a cathode-ray oscilloscope. This instrument can plot automatically on its screen graphs proportional to $q(t)$ and $i(t)$, as in Fig. 38-2.

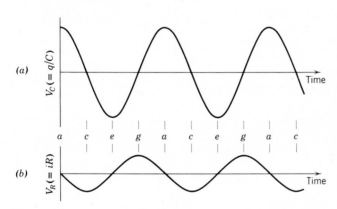

figure 38-2
A drawing of an oscilloscope screen showing potential differences proportional to (a) the charge and (b) the current, in the circuit of Fig. 38-1, as a function of time. The letters indicate corresponding phases of oscillation in that figure. Note that because $i = dq/dt$ the lower curve is proportional to the derivative of the upper. Can you verify this?

EXAMPLE 1

A 1.0-μF capacitor is charged to 50 V. The charging battery is then disconnected and a 10-mH coil is connected across the capacitor, so that LC oscillations occur. What is the maximum current in the coil? Assume that the circuit contains no resistance.

The maximum stored energy in the capacitor must equal the maximum stored energy in the inductor, from the conservation-of-energy principle. This leads, from Eqs. 38-1 and 38-2, to

$$\frac{1}{2}\frac{q_m^2}{C} = \tfrac{1}{2}Li_m^2,$$

where i_m is the *maximum* current and q_m is the *maximum* charge. Note that the maximum current and the maximum charge do not occur at the same time but one-fourth of a cycle apart; see Figs. 38-1 and 38-2. Solving for i_m and substituting CV_0 for q_m gives

$$i_m = V_0 \sqrt{\frac{C}{L}} = (50 \text{ V}) \sqrt{\frac{1.0 \times 10^{-6} \text{ F}}{10 \times 10^{-3} \text{ H}}} = 0.50 \text{ A}.$$

In an actual *LC* circuit the oscillations will not continue indefinitely because there is always some resistance present that will drain energy from the electric and magnetic fields and dissipate it as thermal (Joule) energy; the circuit will warm up. The oscillations, once started, will die away, as in Fig. 38-3. Compare this figure with Fig. 15-19, which shows the decay of the mechanical oscillations of a mass-spring system caused by frictional damping.

It is possible to have sustained electromagnetic oscillations if arrangements are made to supply, automatically and periodically (once a cycle, say), enough energy from an outside source to compensate for that dissipated as thermal energy.* We are reminded of a clock escapement, which is a device for feeding energy from a spring or a falling weight into an oscillating pendulum, thus compensating for frictional losses that would otherwise cause the oscillations to die away. *LC* oscillators, whose frequency may be varied between certain limits, are commercially available as packaged units over a wide range of frequencies, extending from low audio-frequencies (lower than 10 Hz) to microwave frequencies (higher than 10 GHz).

figure 38-3
A photograph of an oscilloscope trace showing how the oscillations in an *RCL* circuit die away because energy is dissipated as thermal energy in the resistor. The figure is a plot of the potential difference across the resistor as a function of time.

38-2
ANALOGY TO SIMPLE HARMONIC MOTION

Figure 8-4 shows that in a mass-spring system performing simple harmonic motion, as in an oscillating *LC* circuit, two kinds of energy occur. One is potential energy of the compressed or extended spring; the other is kinetic energy of the moving mass. These are given by the familiar formulas in the first column of Table 38-1. The table suggests that a

Table 38-1
Energy in oscillating systems

Mechanical (Fig. 8-4)		Electromagnetic (Fig. 38-1)	
spring	$U_P = \frac{1}{2}kx^2$	capacitor	$U_E = \frac{1}{2}\frac{q^2}{C}$
mass	$U_K = \frac{1}{2}mv^2$	inductor	$U_B = \frac{1}{2}Li^2$
	$v = dx/dt$		$i = dq/dt$

capacitor is in some formal way like a spring and an inductor is like a mass and that certain electromagnetic quantities "correspond" to certain mechanical ones, namely,

q corresponds to x,
i corresponds to v,
C corresponds to $1/k$,
L corresponds to m.

Comparison of Fig. 38-1, which shows the oscillations of a resistance-less *LC* circuit, with Fig. 8-4, which shows the oscillations in a frictionless mass-spring system, indicates how close the correspondence is. Note how v and i correspond in the two figures; also x and q. Note, too, how in each case the energy alternates between two forms, magnetic

* From another point of view it is possible to supply an electronically generated "negative resistance" to the circuit, large enough in magnitude to just cancel the actual resistance. See "Negative Resistor to Provide Self-Oscillation in RLC Circuits" by Edwin A. S. Lewis, *American Journal of Physics,* December 1976.

and electric for the LC system, and kinetic and potential for the mass-spring system.

In Section 15-3 we saw that the natural angular frequency of oscillation of a (frictionless) mass-spring system is

$$\omega = 2\pi\nu = \sqrt{\frac{k}{m}}.$$

The method of correspondences suggests that to find the natural frequency for a (resistanceless) LC circuit, k should be replaced by $1/C$ and m by L, obtaining

$$\omega = 2\pi\nu = \sqrt{\frac{1}{LC}}. \tag{38-3}$$

This formula is indeed correct, as we show in the next section.

38-3 ELECTROMAGNETIC OSCILLATIONS— QUANTITATIVE

We now derive an expression for the frequency of oscillation of a (resistanceless) LC circuit, our derivation being based on the conservation-of-energy principle. The total energy U present at any instant in an oscillating LC circuit is given by

$$U = U_B + U_E = \tfrac{1}{2}Li^2 + \frac{1}{2}\frac{q^2}{C},$$

which expresses the fact that at any arbitrary time the energy is stored partly in the magnetic field in the inductor and partly in the electric field in the capacitor. If we assume the circuit resistance to be zero, there is no energy transfer to thermal energy and U remains constant with time, even though i and q vary. In more formal language, dU/dt must be zero. This leads to

$$\frac{dU}{dt} = \frac{d}{dt}\left(\tfrac{1}{2}Li^2 + \frac{1}{2}\frac{q^2}{C}\right) = Li\frac{di}{dt} + \frac{q}{C}\frac{dq}{dt} = 0. \tag{38-4}$$

Now, q and i are not independent variables, being related by

$$i = \frac{dq}{dt}.$$

Differentiating yields

$$\frac{di}{dt} = \frac{d^2q}{dt^2}.$$

Substituting these two expressions into Eq. 38-4 leads to

$$L\frac{d^2q}{dt^2} + \frac{1}{C}q = 0. \tag{38-5}$$

This is the differential equation that describes the oscillations of a (resistanceless) LC circuit. To solve it, note that Eq. 38-5 is mathematically of *exactly* the same form as Eq. 15-6,

$$m\frac{d^2x}{dt^2} + kx = 0, \tag{15-6}$$

which is the differential equation for the mass-spring oscillations. Fundamentally, it is by comparing these two equations that the correspondences on p. 844 arise.

The solution of Eq. 15-6 proved to be

$$x = A \cos{(\omega t + \phi)}, \tag{15-8}$$

where $A\ (= x_m)$ is the amplitude of the motion and ϕ is an arbitrary *phase angle.* Since q corresponds to x, we can write the solution of Eq. 38-5 as

$$q = q_m \cos (\omega t + \phi), \tag{38-6}$$

where ω is the still unknown angular frequency of the electromagnetic oscillations.

We can test whether Eq. 38-6 is indeed a solution of Eq. 38-5 by substituting it and its second derivative in that equation. To find the second derivative, we write

$$\frac{dq}{dt} = i = -\omega q_m \sin (\omega t + \phi) \tag{38-7a}$$

and

$$\frac{d^2q}{dt^2} = -\omega^2 q_m \cos (\omega t + \phi). \tag{38-7b}$$

Substituting q and d^2q/dt^2 into Eq. 38-5 yields

$$-L\omega^2 q_m \cos (\omega t + \phi) + \frac{1}{C} q_m \cos (\omega t + \phi) = 0.$$

Canceling $q_m \cos (\omega t + \phi)$ and rearranging leads to

$$\omega = \sqrt{\frac{1}{LC}}.$$

Thus, if ω is given the constant value $1/\sqrt{LC}$, Eq. 38-6 is indeed a solution of Eq. 38-5. This expression for ω agrees with Eq. 38-3, which we arrived at by the method of correspondences.

The phase angle ϕ in Eq. 38-6 is determined by the conditions that prevail at $t = 0$. If the initial condition is as represented by Fig. 38-1a, then we put $\phi = 0$ in order that Eq. 38-6 may predict $q = q_m$ at $t = 0$. What initial condition in Fig. 38-1 is implied if we select $\phi = 90°$?

EXAMPLE 2

(a) In an oscillating LC circuit, what value of charge, expressed in terms of the maximum charge, is present on the capacitor when the energy is shared equally between the electric and the magnetic field? (b) How much time is required for this condition to arise, assuming the capacitor to be fully charged initially? Assume that $L = 10$ mH and $C = 1.0$ μF.

(a) The stored energy and the *maximum* stored energy in the capacitor are, respectively,

$$U_E = \frac{q^2}{2C} \quad \text{and} \quad U_{E,m} = \frac{q_m^2}{2C}.$$

Substituting $U_E = \frac{1}{2}U_{E,m}$ yields

$$\frac{q^2}{2C} = \frac{1}{2}\frac{q_m^2}{2C} \quad \text{or} \quad q = \frac{1}{\sqrt{2}} q_m.$$

(b) To find the time, we write, assuming $\phi = 0$ in Eq. 38-6,

$$q = q_m \cos \omega t = \frac{1}{\sqrt{2}} q_m,$$

which leads to

$$\omega t = \cos^{-1} \frac{1}{\sqrt{2}} = \frac{\pi}{4} \quad \text{or} \quad t = \frac{\pi}{4\omega}.$$

The angular frequency ω is found from Eq. 38-3, or

$$\omega = \sqrt{\frac{1}{LC}} = \sqrt{\frac{1}{(10 \times 10^{-3} \text{ H})(1.0 \times 10^{-6} \text{ F})}} = 1.0 \times 10^4 \text{ rad/s}.$$

The time t is then

$$t = \frac{\pi}{4\omega} = \frac{\pi}{(4)(1.0 \times 10^4 \text{ rad/s})} = 7.9 \times 10^{-5} \text{ s} \ (=79 \ \mu\text{s}).$$

What is the frequency ν in Hz?

The stored electric energy in the LC circuit, using Eq. 38-6, is

$$U_E = \frac{1}{2} \frac{q^2}{C} = \frac{q_m^2}{2C} \cos^2 (\omega t + \phi), \qquad (38\text{-}8)$$

and the magnetic energy, using Eq. 38-7a, is

$$U_B = \tfrac{1}{2} L i^2 = \tfrac{1}{2} L \omega^2 q_m^2 \sin^2 (\omega t + \phi).$$

Substituting the expression for ω (see Eq. 38-3) into this last equation yields

$$U_B = \frac{q_m^2}{2C} \sin^2 (\omega t + \phi). \qquad (38\text{-}9)$$

Figure 38-4 shows plots of $U_E(t)$ and $U_B(t)$ for the case of $\phi = 0$. Note that (a) the maximum values of U_E and U_B are the same ($= q_m^2/2C$); (b) at any instant the sum of U_E and U_B is a constant ($= q_m^2/2C$); and (c) when U_E has its maximum value, U_B is zero and conversely. This analysis supports the qualitative analysis of Section 38-1. Compare this discussion with that given in Section 15-4 for the energy transfers in a mass-spring system.

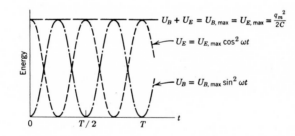

$U_B + U_E = U_{B,\text{max}} = U_{E,\text{max}} = \frac{q_m^2}{2C}$

$U_E = U_{E,\text{max}} \cos^2 \omega t$

$U_B = U_{B,\text{max}} \sin^2 \omega t$

figure 38-4
The stored magnetic ($-\cdot-\cdot-$) and electric ($-\,-\,-\,-$) energies in the circuit of Fig. 38-1. Note that their sum is a constant. $T \ (=1/\nu)$ is the period of the oscillation.

EXAMPLE 3

The RCL circuit. (a) Derive an expression for the quantity $q(t)$ for a single-loop circuit containing a resistance R as well as an inductance L and a capacitance C. (b) After how long a time will the charge oscillations decay to half-amplitude if $L = 10$ mH, $C = 1.0 \ \mu\text{F}$, and $R = 0.10 \ \Omega$?

(a) If U is the total stored field energy, we have, as before,

$$U = \tfrac{1}{2} L i^2 + \frac{1}{2} \frac{q^2}{C}.$$

U is no longer constant but rather

$$\frac{dU}{dt} = -i^2 R,$$

the minus sign signifying that the stored energy U *decreases* with time, being converted to thermal (Joule) energy at the rate $i^2 R$. Combining these two equations leads to

$$L i \frac{di}{dt} + \frac{q}{C} \frac{dq}{dt} = -i^2 R.$$

Substituting dq/dt for i and d^2q/dt^2 for di/dt leads finally to

$$L\frac{d^2q}{dt^2} + R\frac{dq}{dt} + \frac{1}{C}q = 0,$$

which is the differential equation that describes the damped LC oscillations. If we put $R = 0$, this equation reduces, as expected, to Eq. 38-5.

Compare this differential equation for damped LC oscillations with Eq. 15-37, or

$$m\frac{d^2x}{dt^2} + b\frac{dx}{dt} + kx = 0, \qquad (15\text{-}37)$$

which describes damped mass-spring oscillations. Once again the equations are mathematically identical, the resistance R corresponding to the mechanical damping constant b and q corresponding to x.

The solution of the RCL circuit follows at once, by correspondence, from the solution of Eq. 15-37. It is (see Eqs. 15-38 and 15-39) for R reasonably small, and for an initial condition in which the capacitor has a maximum charge (that is, $\phi = 0$),

$$q = q_m e^{-Rt/2L}\cos\omega't, \qquad (38\text{-}10)$$

where

$$\omega' = \sqrt{\frac{1}{LC} - \left(\frac{R}{2L}\right)^2}. \qquad (38\text{-}11)$$

Note that Eq. 38-10, which can be described as a cosine function with an exponentially decreasing amplitude, is the equation of the decay curve of Fig. 38-3. Note, too (see Eq. 38-11), that the presence of resistance reduces the oscillation frequency. These two equations reduce to familiar results as $R \to 0$.

(b) The oscillation amplitude will have decreased to half the initial amplitude when the amplitude factor $e^{-Rt/2L}$ in Eq. 38-10 has the value one-half, or

$$\tfrac{1}{2} = e^{-Rt/2L}.$$

At what time t will this occur? What is ω'?

We have

$$t = \frac{2L}{R}\ln 2 = \frac{(2)(10 \times 10^{-3}\text{ H})(0.69)}{0.10\ \Omega} = 0.14\text{ s.}$$

The angular frequency, from Eq. 38-11, is

$$\omega' = \sqrt{\frac{1}{(10 \times 10^{-3}\text{ H})(1.0 \times 10^{-6}\text{ F})} - \left(\frac{0.10\ \Omega}{2 \times 10 \times 10^{-3}\text{ H}}\right)^2}$$

$$= \sqrt{10^8\ (\text{rad/s})^2 - 25\ (\text{rad/s})^2} = 1.0 \times 10^4\text{ rad/s.}$$

Note that the second term is quite small, so that in this case, as in many practical cases, the resistance has a negligible effect on the frequency. Can you show that 0.14 s, the time at which the oscillations decrease to half-amplitude, corresponds to about 220 cycles of oscillation? The damping is much less severe than that shown in Fig. 38-3.

38-4 LUMPED AND DISTRIBUTED ELEMENTS

In this section we put electromagnetic oscillations aside for the moment and return to mechanical oscillating systems, so that we can develop the concepts of lumped and distributed elements. In Section 38-5 we will return to electromagnetic systems and make a similar comparison.

In the oscillating mass-spring system in Fig. 8-4 the two kinds of energy involved appear in quite separate parts of the system, the potential energy being associated with the spring and the kinetic energy with the moving mass. A closed organ pipe (see Fig. 20-6b) is a mechanical oscillating system in which the two kinds of energy are not separated in space. Kinetic energy, associated with moving elements of air, and potential energy, associated with compressions and rarifications of elements of air, are both present throughout the volume of the pipe. We say that such an oscillating system has *distributed* rather than *lumped* (as in the mass-spring system) elements.

We see at once one difference between lumped and distributed mechanical oscillating systems. A lumped system has a single frequency of oscillation, given for the mass-spring system by Eq. 15-11; see Section 15-5 for other examples of lumped oscillating mechanical systems. On the other hand, distributed systems, such as the organ pipes in Fig. 20-6 and the vibrating string of Fig. 20-4, have a number of discrete oscillation frequencies (harmonics), the values for these two cases being given in Section 20-5.

Now let us examine in more detail, from the point of view of distributed elements, the behavior of a closed organ pipe, which we will idealize as an *acoustic resonant cavity*.

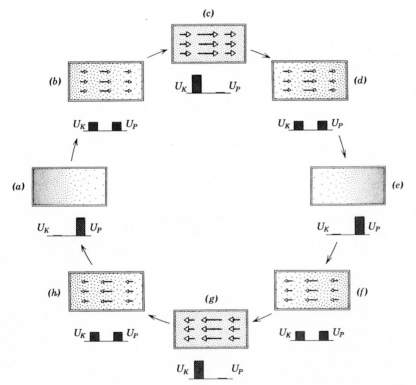

figure 38-5
Showing eight stages in a cycle of oscillation of a cylindrical acoustic resonant cavity, suggestive of a closed organ pipe. The bar graphs below each figure show the kinetic and potential energy. The arrows represent the directed velocities of small volume elements of the gas.

Figure 38-5, a series of "snapshots" taken one-eighth of a cycle apart, shows the pressure and velocity variations in the fundamental mode of a particular acoustic resonant cavity. There is a pressure node at the center and a pressure antinode at each end. There is a velocity* node at each end and a velocity antinode at the center. When the pressure variation is the greatest, the velocity is zero (Figs. 38-5a and e). When the pressure is uniform, the velocities have their maximum values (Figs. 38-5c, and g).

The energy in the acoustic resonator alternates between kinetic

---

* The velocity of interest here is the directed velocity $\mathbf{v}_{\text{gas}}$ of small volume elements of the gas which are, however, large enough to contain a great number of molecules. The thermal velocities of the molecules have no directional preference and are ignored.

energy associated with the moving gas and potential energy associated with the compression and rarefaction of the gas. In Figs. 38-5c and g the energy is all kinetic and in Figs. 38-5a and e it is all potential. In intermediate phases it is part kinetic and part potential.

The kinetic energy of a small mass Δm of the gas, which is moving parallel to the cylinder axis with a speed v_{gas}, is $\frac{1}{2}\Delta m v_{gas}^2$. The *kinetic energy density*, that is, the kinetic energy per unit volume, is

$$u_K = \frac{1}{2}\frac{\Delta m}{\Delta V} v_{gas}^2 = \frac{1}{2}\rho_0 v_{gas}^2,$$

in which ΔV is the volume of the gas element and ρ_0 is the mean density of the gas.

The potential energy per unit volume in the gas, that is, the *potential energy density*, associated with the compressions and rarefactions of the gas is given (we state) by

$$u_P = \frac{1}{2} B \left(\frac{\Delta\rho}{\rho_0}\right)^2.$$

Here B is the bulk modulus* of elasticity of the gas and $\Delta\rho/\rho_0$, which is positive for a compression and negative for a rarefaction, is the fractional change in gas density at any point.

The angular frequency of oscillation for the cavity of Fig. 38-5, in the fundamental (or lowest frequency) mode, shown in that figure, is found from

$$\omega_1 = 2\pi\nu = \frac{2\pi v}{\lambda} = \frac{\pi v}{l},$$

where v is the speed of sound in the gas and l is the length of the cavity. From Eq. 20-1 we may write v as $\sqrt{B/\rho_0}$. Note that in the above we have put $\lambda = 2l$, corresponding to the fundamental mode. What are the angular frequencies ω_2, ω_3, etc., of the higher frequency modes?

38-5
ELECTROMAGNETIC RESONANT CAVITY

Consider now a second closed cylinder, of radius a and length l, whose walls are made of copper or some other good conductor. A system of oscillating electric and magnetic fields can be set up in such a cavity. Such an *electromagnetic resonant cavity* is a distributed electromagnetic oscillator, in contrast to an LC circuit, which is a lumped system. As for the acoustic resonator, many modes of oscillation with discrete frequencies are possible; we describe only the fundamental mode, which has the lowest frequency. The cavity oscillations can be set up by suitably connecting the cavity, through a small hole in its side wall, to a source of electromagnetic radiation such as a magnetron. If the cavity dimensions are of the order of a few centimeters, the resonant frequencies will be of the order of 10^{10} Hz. This is far higher than the acoustic frequencies in cavities of the same size, reflecting the fact that the speed of electromagnetic waves in free space $(= 3 \times 10^8$ m/s) is much greater than the speed of sound in air (about 350 m/s).

Figure 38-6, which is a series of "snapshots" taken one-eighth of a cycle apart, shows, by the horizontal lines, how the electric field **E** varies with time in the cavity. The electric lines of force originate on positive charges at one end of the cylinder and terminate on negative charges at the other end. As **E** changes with time, eventually reversing itself, currents flow from one end of the cylinder to the other on the inner cylinder wall. At any point in the cavity, energy is stored in the electric field in an amount per unit volume given by Eq. 30-9, or

$$u_E = \frac{1}{2}\epsilon_0 E^2. \tag{38-12}$$

* See p. 435.

figure 38-6
Showing eight stages in a cycle of oscillation of a cylindrical electromagnetic resonant cavity. The bar graphs below each figure show the stored electric and magnetic energy. The dots and crosses represent circular lines of **B**; the horizontal lines represent **E**. Compare with Fig. 38-5; both figures are examples of distributed elements.

Figure 38-6 also shows, by the dots and crosses, how the magnetic field **B** varies with time. The magnetic lines form circles about the cylinder axis. Note that the magnetic field has a maximum value when the electric field is zero, and conversely. At any point in the cavity energy is stored in the magnetic field in an amount per unit volume given by Eq. 36-19, or

$$u_B = \frac{1}{2\mu_0} B^2. \qquad (38\text{-}13)$$

Thus, as in the LC circuit, energy is shuttled back and forth between the electric and the magnetic fields. However, these fields no longer occupy completely separate regions of space. In Section 40-5 we clarify just *why* the electric and magnetic fields in Fig. 38-6 interact in the way that we have described.

We state without proof that the angular frequency of oscillation for the electromagnetic cavity in Fig. 38-6 is, in the fundamental mode shown in that figure,

$$\omega_1 = \frac{2.41c}{a},$$

in which a is the cavity radius and c is the speed of electromagnetic radiation in free space. We will see in Section 41-8 that c may be written in terms of electromagnetic quantities as $1/\sqrt{\mu_0\epsilon_0}$. As the field patterns in Fig. 38-6 suggest, the resonant frequency of oscillation of the cavity, for the mode of oscillation shown, depends only on the radius of the cavity and not on its length.

Table 38-2 summarizes some characteristics of the four oscillating systems that we have discussed so far. For lumped systems it gives expressions for the two kinds of energy involved and for the (single) oscillation frequency. For the distributed systems it gives expressions for the two kinds of energy density involved and for the oscillation frequency in the fundamental mode. You should study carefully all the correspondences, similarities, and differences that occur in the table.

Table 38-2
Four oscillating systems

	Mechanical Systems (frictionless)	Electromagnetic Systems (resistanceless)
Lumped systems	Mass + spring $U_K = \frac{1}{2}mv^2$ $U_P = \frac{1}{2}kx^2$ $\omega = \sqrt{\dfrac{k}{m}}$ (Fig. 8-4)	LC circuit $U_B = \frac{1}{2}Li^2$ $U_E = \frac{1}{2}(1/C)q^2$ $\omega = \sqrt{\dfrac{1}{LC}}$ (Fig. 38-1)
Distributed systems	Acoustic cavity $u_K = \frac{1}{2}\rho_0 v_{gas}^2$ $u_P = \frac{1}{2}B(\Delta\rho/\rho_0)^2$ $\omega_1 = \dfrac{\pi v}{l}; \quad v = \sqrt{\dfrac{B}{\rho_0}}$ (Fig. 38-5)	Electromagnetic cavity $u_B = \frac{1}{2}(1/\mu_0)B^2$ $u_E = \frac{1}{2}\epsilon_0 E^2$ $\omega_1 = \dfrac{2.41c}{a}; \quad c = \sqrt{\dfrac{1}{\epsilon_0\mu_0}}$ (Fig. 38-6)

EXAMPLE 4

In the cavity of Fig. 38-6, what is the relationship between the "average" value of E throughout the cavity, measured at the instant corresponding to Fig. 38-6a, to the "average" value of B, measured at the instant corresponding to Fig. 38-6c?

At the first instant the energy is all electric and at the second it is all magnetic. The total energy U, found by integrating the energy density over the volume of the cavity, must be the same at these two instants, or

$$U = \int u_{E,m} \, dV = \int u_{B,m} \, dV,$$

where dV is a volume element in the cavity and $u_{E,m}$ and $u_{B,m}$ are the *maximum* values of u_E and of u_B at the site of this volume element; these maximum values occur one-fourth of a cycle apart, as Fig. 38-6 shows. Substituting Eqs. 38-12 and 38-13 leads to

$$\int \frac{\epsilon_0 E_m^2}{2} \, dV = \int \frac{B_m^2}{2\mu_0} \, dV$$

or

$$\mu_0\epsilon_0 \int E_m^2 \, dV = \int B_m^2 \, dV.$$

The quantity $\int E_m^2 \, dV$ can be written as $\overline{E_m^2}V$, where V is the cavity volume and $\overline{E_m^2}$ is the average value of E_m^2 throughout the cavity. Treating B_m in the same way leads to

$$\mu_0\epsilon_0 \overline{E_m^2} = \overline{B_m^2}$$

or, taking square roots, $\quad \sqrt{\overline{B_m^2}} = \sqrt{\mu_0\epsilon_0} \sqrt{\overline{E_m^2}}.$

We can represent $\sqrt{\overline{B_m^2}}$ by B_{rms}, a so-called "root-mean-square" value of B_m. In computing B_{rms}, note that the averaging is done throughout the volume of the cavity, at the instant corresponding to Fig. 38-6c. It is not a time average for a particular point in the cavity. Doing the same for E yields

$$E_{rms}/B_{rms} = \frac{1}{\sqrt{\mu_0\epsilon_0}}$$

$$= [(4\pi \times 10^{-7} \text{ T} \cdot \text{m/A})(8.9 \times 10^{-12} \text{ C}^2/\text{N} \cdot \text{m}^2)]^{-1/2}$$

$$= 3.00 \times 10^8 \text{ m/s}.$$

If this reminds you of the speed of light, it is no coincidence because that is exactly what it is; we will explore this in detail in Section 41-8.

If E_{rms} above equals 1.0×10^4 V/m, a reasonable value, then

$$B_{rms} = (1.0 \times 10^4 \text{ V/m})/(3.00 \times 10^8 \text{ m/s})$$

$$= 3.3 \times 10^{-5} \text{ T} = 0.33 \text{ gauss.}$$

What is the total stored energy in the cavity under these conditions, assuming the cavity to be 10 cm long and 3.0 cm in diameter?

Recall that in Example 6, Chapter 36, we showed the energy density for a magnetic field of "ordinary" laboratory magnitude (say, 1 T) to be enormously greater than that for an electric field of "ordinary" magnitude (say, 10^5 V/m). This fact is consistent with the present example.

questions

1. Why doesn't the *LC* circuit of Fig. 38-1 simply stop oscillating when the capacitor has been completely discharged?

2. How might you start an *LC* circuit into oscillation with its initial condition being represented by Fig. 38-1c? Devise a switching scheme to bring this about.

3. In an oscillating *LC* circuit, assumed resistanceless, what determines (a) the frequency and (b) the amplitude of the oscillations?

4. In connection with Figs. 38-1c and g, explain how there can be a current in the inductor even though there is no charge on the capacitor.

5. In Fig. 38-1 is it possible to have (a) an *LC* circuit without resistance, (b) an inductor without inherent capacitance, or (c) a capacitor without inherent inductance? Discuss the practical validity of the *LC* circuit of Fig. 38-1, in which each of the above possibilities is ignored. See "Self-resonant Effects in Coils and Capacitors: an Experiment" by Samuel Derman, *The Physics Teacher*, September 1976.

6. In Fig. 38-1, what changes are called for if the oscillations are to proceed *counterclockwise* around the figure?

7. In Fig. 38-1, what phase angles ϕ in Eq. 38-6 correspond to the eight circuit situations shown?

8. All practical *LC* circuits have to contain *some* resistance and thus are *RCL* circuits. However, one can buy a packaged audio oscillator in which the output maintains a constant amplitude indefinitely and does not decay, as in Fig. 38-3. How can this happen? (*Hint:* Consider the analogy to the pendulum clock, in which falling weights are involved.)

9. Can you see any physical reason for assuming that *R* is "small" in Eqs. 38-10 and 38-11? (*Hint:* Consider what might happen if the damping *R* were so large that Eq. 38-10 could not even go through one cycle of oscillation before *q* was reduced to essentially zero. Could this happen? If so, what do you imagine Fig. 38-3 would look like?)

10. Tabulate as many mechanical or electric systems as you can think of that possess a natural frequency, along with the formula for that frequency if given in the text.

11. Discuss the periodic flow of energy, if any, from point to point in an acoustic resonant cavity.

12. Can a given circuit element (a capacitor, say) behave like a "lumped" element under some circumstances and like a "distributed" element under others?

13. List as many (a) lumped and (b) distributed mechanical oscillating systems as you can.

14. Are oscillating systems (mechanical, say) *either* lumped *or* distributed? That is, is there no middle ground? (a) Consider a lumped system such as an

idealized mass-spring arrangement. How might you change it physically to make it more distributed? (b) Consider a distributed system such as a vibrating string. How might you change it physically to make it more lumped?

15. A coil has a measured inductance L. In a practical case it also has a capacitance C, adjacent windings behaving as "plates." It can be made to oscillate at a certain frequency without attaching it to an external capacitance. Is this a case of distributed elements? Do you suppose that it can oscillate at more than one frequency? Discuss.

16. A violin string is an oscillating mechanical system with distributed elements. Give some qualitative details as to why this is so. For example, where are the kinetic and potential energies to be found? (Compare Figs. 19-16 and 38-5.)

17. An air-filled acoustic resonant cavity and an electromagnetic resonant cavity of the same size have resonant frequencies that are in the ratio of 10^6 or so. Which has the higher frequency and why?

18. What constructional difficulties would you encounter if you tried to build an LC circuit of the type shown in Fig. 38-1 to oscillate (a) at 0.01 Hz, or (b) at 10^{10} Hz?

19. Electromagnetic cavities are often silver-plated on the inside. Why?

problems

SECTION 38-1

1. Find the capacitance of an LC circuit if the maximum charge on the capacitor is 1.0 μC and the total energy is 1.4×10^{-4} J.   *Answer:* 3.6×10^{-9} F.

2. A 1.5-mH inductor in an LC circuit stores a maximum energy of 1.0×10^{-5} J. What is the peak current?

3. In an oscillating LC circuit $L = 1.0$ mH, $C = 4.0$ μF and the maximum charge on C is 3.0 μC. Find the maximum current.   *Answer:* 47 mA.

4. An oscillating LC circuit consisting of a 1.0 nF ($= 1.0$ nanofarad $= 1.0 \times 10^{-9}$ F) capacitor and a 3.0-mH coil carries a peak voltage of 3.0 V. (a) What is the maximum charge on the capacitor? (b) What is the peak current through the circuit? (c) What is the maximum energy stored in the magnetic field of the coil?

5. In an oscillating LC circuit, (a) in terms of the maximum charge on the capacitor, what value of charge is present when the energy is shared equally between the electric and the magnetic fields? (b) What fraction of a period must elapse following the time the capacitor is fully charged for this condition to arise?   *Answer:* (a) $q = q_m/\sqrt{2}$. (b) $t = T/8$.

6. At some instant in an oscillating LC circuit, three-fourths of the total energy is stored in the magnetic field of the inductor. (a) In terms of the maximum charge on the capacitor, what is the charge on the capacitor at this instant? (b) In terms of the maximum current in the inductor, what is the current in the inductor at this instant?

SECTION 38-2

7. Given a 1.0-mH inductor, how would you make it oscillate at 1.0 MHz ($= 1.0 \times 10^6$ Hz)?
   Answer: Connect a 25-pF capacitor across it and use it as the resonant element in an oscillator.

8. You are given a 10-mH inductor and two capacitors, of 5.0- and 2.0-μF capacitance. (a) Can you find four resonant frequencies that can be obtained by connecting these elements in various ways? (b) Are there more than four such frequencies?

9. An LC circuit has an inductance $L = 3.0$ mH and a capacitance $C = 10$ μF. (a) Calculate the angular frequency ω of oscillation. (b) Find the period T of the oscillation. (c) At time $t = 0$ the capacitor is charged to 200 μC, and the current is zero. Sketch roughly the charge on the capacitor as a function of time.   *Answer:* (a) 5.8×10^3 rad/s. (b) 1.1×10^{-3} s.

10. How long will it take for an uncharged 4.0-pF capacitor in an LC circuit to charge if its final voltage is 1.0 mV and the maximum current is 50 mA?

11. An inductor is connected across a capacitor whose capacitance can be varied by turning a knob. We wish to make the frequency of the LC oscillations vary linearly with the angle of rotation of the knob, going from 2×10^5 Hz to 4×10^5 Hz as the knob turns through 180°. If $L = 1.0$ mH, plot C as a function of angle for the 180° rotation.
    Answer: 0°, 45°, 90°, 135°, and 180° correspond respectively to 6.4, 4.1, 2.8, 2.1, and 1.6×10^{-10} F.

12. A variable capacitor with a range from 10 to 365 pF is used with a coil to tune the input to a radio. (a) What ratio of maximum to minimum frequencies may be tuned with such a capacitor? (b) If this capacitor is to tune from 0.54 to 1.60 MHz, the ratio computed in (a) is too large. By adding a capacitor in parallel to the variable capacitor this range may be adjusted. How large should this capacitor be and what inductance should be chosen in order to tune the desired range of frequencies?

13. An oscillating LC circuit is designed to operate at a peak current i (30 mA). The inductance L (0.042 H) is fixed and the frequency is varied by changing C. (a) If the capacitor has a maximum peak voltage V_m (50 V), can the circuit safely operate at a frequency ν of 1.0 MHz? (b) What is the maximum safe operating frequency? (c) What is the minimum capacitance?
    Answer: (a) 7900 V; no. (b) 6300 Hz. (c) 1.5×10^{-8} F.

14. A 10-kg mass oscillates on a spring that, when extended 2.0 cm from equilibrium, has a restoring force of 5.0 N. (a) Find the capacitance of the analogous LC system with $L = 1.0 \times 10^{-3}$ H. (b) Would it be a simple matter to construct the analogous circuit?

15. Initially the 900-μF capacitor is charged to 100 V, and the 100-μF capacitor is uncharged in Fig. 38-7. (a) Describe in detail how one may charge the 100-μF capacitor to 300 V using S_1 and S_2 appropriately. (b) Describe in detail the mass + spring mechanical analogy of this problem.
    Answer: Let T_2 be the period of the inductor and 900 μF capacitor and T_1 the period of inductor and 100 μF capacitor. Then (a) close S_2, wait $T_2/4$; quickly close S_1, then open S_2; wait $T_1/4$ and then open S_1.

figure 38-7
Problem 15

SECTION 38-3

16. A circuit has $L = 10$ mH and $C = 1.0$ μF. How much resistance must be inserted in the circuit to reduce the (undamped) resonant frequency by 0.01%?

17. In an oscillating LC circuit $L = 3.0$ mH and $C = 2.7$ μF. At $t = 0$ the charge $q = 0$ and the current $i = 2.0$ A. (a) What is the maximum charge that will appear on the capacitor? (b) In terms of the period T of oscillation, how much time after $t = 0$ will elapse until the energy stored in the capacitor will be increasing at its greatest rate? (c) What is this greatest rate at which the energy stored in the capacitor increases?
    Answer: (a) 1.8×10^{-4} C. (b) $T/8$. (c) 67 W.

18. In a damped LC circuit, find the time required for the maximum energy present in the capacitor during one oscillation to fall to half the maximum energy present during the first oscillation. Assume $q = q_m$ at $t = 0$ (i.e., use Eq. 38-10).

19. Derive the differential equation for an LC circuit (Eq. 38-5) using the loop theorem.

20. Show that, for low damping, the current in a damped LC circuit is given approximately by

$$i = -q_m \omega' e^{-Rt/2L} \sin (\omega' t + \phi),$$

in which

$$\phi = \tan^{-1} \frac{R}{2L\omega'}.$$

Start from Eq. 38-10.

21. Suppose that in an oscillating RCL circuit the amplitude of the charge oscillations drops to one-half its initial value after n cycles. Show that the fractional reduction in the frequency of resonance, caused by the presence of the resistor, is given to a close approximation by

$$\frac{\omega - \omega'}{\omega} = \frac{0.0061}{n^2},$$

which is independent of L, C, or R. Apply to the decay curve of Fig. 38-3.

22. "Q" *for a circuit.* In the damped LC circuit of Example 3 show that the fraction of the energy lost per cycle of oscillation, $\Delta U/U$, is given to a close approximation by $2\pi R/\omega L$. The quantity $\omega L/R$ is often called the "Q" of the circuit (for "quality"). A "high-Q" circuit has low resistance and a low fractional energy loss per cycle $(= 2\pi/Q)$.

SECTION 38-5

23. What would be the dimensions of a cylindrical electromagnetic resonant cavity (like that described in the text) operating, in the fundamental mode, at 60 Hz, the frequency of household alternating current?
Answer: Radius $= 1.9 \times 10^3$ km, independent of its length.

24. Sketch diagrams like those shown in Figure 38-6 showing a cycle of oscillation of a cylindrical electromagnetic resonant cavity operating, not in the fundamental mode, but in the first overtone.

39
alternating currents

Thus far we have discussed the response, that is, the generated currents, when emfs that vary with time in different ways are applied to circuits containing elements of resistance R, capacitance C, and inductance L, in various combinations.

In Chapter 32 we discussed the (steady) currents generated when we apply (steady) emfs to purely resistive networks. In Section 32-8 we discussed the response of a single-loop RC circuit when a "square wave" emf is applied, as in Fig. 32-13. In Section 36-3 we described the behavior of a single-loop LR network when a similar square wave is applied, as in Fig. 36-6. Finally, in Chapter 38, we analyzed the responses of single-loop LC and RCL circuits in which *no* emf is applied other than the transient emf needed to start the "free oscillations"; see Figs. 38-1, 38-2*b*, and 38-3.

This chapter deals with *alternating currents* set up in single-loop RCL circuits, such as that of Fig. 39-1, by an emf that varies with time as

$$\mathcal{E} = \mathcal{E}_m \sin \omega t, \qquad (39\text{-}1)$$

$\omega(= 2\pi\nu$, where ν is measured in hertz) being a fixed angular frequency. An emf of this type might be established, for example, by an alternating current generator in a commercial power plant, where usually $\nu = 60$ Hz. (See Problem 6, Chapter 35.) The symbol for a source of alternating emf such as that described by Eq. 39-1 is ———⟨∿⟩——— Such a device is called an alternating current generator or an *ac generator*.*

Alternating currents are important for two reasons: (1) On the practical side, modern technology and indeed the life-style in technologically

---

* The term 'alternating current' is used loosely here: strictly this is an alternating voltage.

advanced nations, would be very different if emfs such as those given by Eq. 39-1, and the alternating currents to which they correspond, were not available. Power distribution systems, radio, television, satellite communication systems, computor systems, and so on, would be either much less effective or impossible without alternating emfs and the alternating currents generated by them. (2) On the theoretical side, if we know the response of *any RCL* circuit (no matter how many elements or loops are involved) to the emf of Eq. 39-1, we can find the response, that is, the currents generated, to *any* arbitrary emf, no matter how complicated its waveform. We rely here on the facts that we can write any complex waveform as the sum of separate sine (and cosine) terms in a Fourier series and that we can apply the principle of superposition. See Section 19-4 and Fig. 19-5 for an analogy.

A central aim of this chapter is to find the alternating current i in the circuit of Fig. 39-1 in terms of ε_m, ω, R, C, and L. Note that, for the conditions assumed in Fig. 39-1,* the current in all parts of the loop is the same (as for single-loop direct current circuits in Chapter 32) and we may safely assume that i is given by

$$i = i_m \sin(\omega t - \phi) \tag{39-2}$$

in which ω is the angular frequency of the applied alternating emf of Eq. 39-1.

In Eq. 39-2 i_m is the current amplitude and ϕ is the phase angle between the alternating current of Eq. 39-2 and the alternating emf of Eq. 39-1. It is our job to express i_m and ϕ in Eq. 39-2 in terms of ε_m, ω, R, C, and L.†

To do so, let us break down the problem suggested by Fig. 39-1 into three separate problems, in which we consider R, C, and L separately and in turn.‡ We start with R.

1. A resistive circuit. Figure 39-2a shows a circuit containing a resistive element only, acted on by the alternating emf of Eq. 39-1. From the loop theorem and from the definition of resistance we can write

$$V_R = \varepsilon_m \sin \omega t \quad \text{(loop theorem)} \tag{39-3a}$$
$$V_R = i_R R \quad \text{(definition of } R\text{)} \tag{39-3b}$$

or

$$i_R = \left(\frac{\varepsilon_m}{R}\right) \sin \omega t. \tag{39-3c}$$

Comparison of Eqs. 39-3a and 39-3c shows that the time-varying quantities V_R and i_R are *in phase*, that is, they reach their maximum values at the same time. As expected from Eq. 39-2 they also have the same angular frequency ω. We show both of these things in Fig. 39-2b, a plot of Eqs. 39-3a and 39-3c.

39-2
RCL ELEMENTS, CONSIDERED SEPARATELY

$(\varepsilon = \varepsilon_m \sin \omega t)$

figure 39-1
A single-loop *RCL* circuit contains an ac generator. V_R, V_C, and V_L are the time-varying potential differences across the resistor, the capacitor, and the inductor, respectively.

---

* The condition is, as Fig. 39-1 implies, that we can localize R, C, and L in separate, physically identifiable parts of the circuit. We discuss the basis of this assumption (that of *lumped impedances*) fully in Chapter 40. The lower the frequency the more justifiable this assumption becomes.

† As we shall see (Eq. 39-14), ϕ actually does not depend on ε_m.

‡ We could apply the loop theorem (see Section 32-2) to the circuit of Fig. 39-1 and solve the differential equation that results but this would involve us in mathematical difficulties beyond the scope of this book. We adopt an indirect but equally valid and more physically informative approach.

(ε = ε_m sin ωt)

(a)

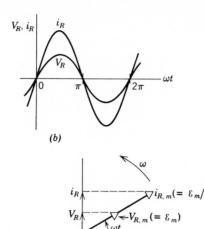

(b)

(c)

figure 39-2

(a) A single-loop resistive circuit containing an ac generator. (b) The current and the potential difference across the resistor are in phase. In Eq. 39-2, $\phi = 0$. (c) A phasor diagram shows the same thing. The arrows on the vertical axis are instantaneous values.

(ε = ε_m sin ωt)

(a)

(b)

(c)

Figure 39-2c shows another way of looking at the situation. It is sometimes called a *phasor diagram*, in which the phasors, represented by the open arrows, rotate counterclockwise with an angular frequency ω about the origin. The phasors have these properties: (a) The length of the phasor is proportional to the *maximum* value of the alternating quantity involved, thus ε_m for V_R (Eq. 39-3a) and (ε_m/R) for i_R (Eq. 39-3c), and (b) the projection of the phasors on the vertical axis gives the *instantaneous* values of the alternating currents involved. Thus the arrows on the vertical axis represent the time-varying quantities V_R and i_R, as in Eqs. 39-3a and 3c, respectively. That V_R and i_R are in phase follows from the fact that their phasors lie along the same line in Fig. 39-2c.

Follow the rotation of the phasors in Fig. 39-2c and convince yourself that this phasor diagram completely and correctly describes Eqs. 39-3a and 3c.

2. *A capacitive circuit.* Figure 39-3a shows a circuit containing a capacitive element only, acted on by the alternating emf of Eq. 39-1. From the loop theorem and from the definition of capacitance we can write

$$V_C = \varepsilon_m \sin \omega t \text{ (loop theorem)} \qquad (39\text{-}4a)$$

and

$$V_C = q/C \text{ (definition of } C\text{).} \qquad (39\text{-}4b)$$

From these relations we have

$$q = \varepsilon_m C \sin \omega t$$

or

$$i_C \left(= \frac{dq}{dt}\right) = \omega C \varepsilon_m \cos \omega t. \qquad (39\text{-}4c)$$

Comparison of Eqs. 39-4a and 39-4c shows that the time-varying

figure 39-3

(a) A single-loop capacitive circuit containing an ac generator. (b) The potential difference across the capacitor *lags* the current by a quarter-cycle. In Eq. 39-2, $\phi = -90°$. (c) A phasor diagram shows the same thing. The arrows on the vertical axis are instantaneous values.

quantities V_C and i_C are one-quarter cycle out of phase. We show this in Fig. 39-3b, a plot of Eqs. 39-4a and 39-4c. We see that V_C lags i_C, that is, as time goes on, V_C reaches its maximum *after* i_C does, by one-quarter cycle.

We show this with equal clarity in the phasor diagram of Fig. 39-3c. As the phasors rotate in the assumed counterclockwise direction it is clear that the $V_{C,m}$ phasor *lags* behind the $i_{C,m}$ phasor by one-quarter cycle.

The phase angle ϕ between V_C and i_C in Fig. 39-3 is –90°. To show this put $\phi = -90°$ in Eq. 39-2 and expand. We obtain

$$i = i_m \cos \omega t,$$

in agreement with Eq. 39-4c.

For reasons of symmetry of notation we choose to rewrite Eq. 39-4c as

$$i_C = \left(\frac{\varepsilon_m}{X_C}\right) \cos \omega t \qquad (39\text{-}5a)$$

in which we must have

$$X_C = \frac{1}{\omega C}. \qquad (39\text{-}5b)$$

We call X_C the *capacitive reactance*. Comparing Eqs. 39-3c and 39-5a we see that the current amplitudes are (ε_m/R) and (ε_m/X_C), respectively. We see from this that the unit of X_C must be the ohm. Can you prove this explicitly, starting from Eq. 39-5b?

Also, let us recognize that ε_m in Eq. 39-4a is the maximum value of V_C $(= V_{C,m})$. Thus, from Eq. 39-5a we can write

$$V_{C,m} = i_{C,m} X_C. \qquad (39\text{-}6a)$$

This suggests that when *any* alternating current, of amplitude i_m and angular frequency ω, exists in a capacitor the maximum potential difference across the capacitor (*regardless of the complexity of the circuit in which the capacitor is involved*) is given by

$$V_{C,m} = i_m X_C. \qquad (39\text{-}6b)$$

EXAMPLE 1

In Fig. 39-3a, let $C = 150$ μF, $\nu = 60$ Hz, and $\varepsilon_m = 300$ V. Find (a) $V_{C,m}$, (b) X_C, and (c) $i_{C,m}$.

(a) From Eq. 39-4a,

$$V_{C,m} = 300 \text{ V}.$$

(b) From Eq. 39-5b,.

$$X_C = \frac{1}{\omega C} = \frac{1}{2\pi\nu C}$$
$$= \frac{1}{(2\pi)(60 \text{ Hz})(150 \times 10^{-6} \text{ F})}$$
$$= 18 \ \Omega.$$

(c) From Eq. 39-6a,

$$i_{C,m} = \frac{V_{C,m}}{X_C}$$
$$= \frac{300 \text{ V}}{18 \ \Omega} = 17 \text{ A}.$$

How would $i_{C,m}$ change if you doubled the frequency? Does this seem intuitively reasonable?

3. *An inductive circuit.* Figure 39-4a shows a circuit containing an inductive element only, acted on by the alternating emf of Eq. 39-1. From the loop theorem and the definition of inductance we can write

$$V_L = \mathcal{E}_m \sin \omega t \quad \text{(loop theorem)} \tag{39-7a}$$

and

$$V_L = L(di/dt) \quad \text{(definition of } L\text{)}. \tag{39-7b}$$

From these relations we see that

$$di = (\mathcal{E}_m/L) \sin \omega t \, dt$$

or

$$i_L = \int di = -(\mathcal{E}_m/\omega L) \cos \omega t. \tag{39-7c}$$

Comparison of Eqs. 39-7a and 7c shows that the time-varying quantities V_L and i_L are a quarter-cycle out of phase. We show this in Fig. 39-4b, a plot of Eqs. 39-7a and 39-7c. We see that V_L leads i_L, that is, as time goes on V_L reaches its maximum *before* i_L does, by a quarter-cycle.

We show this with equal clarity in the phasor diagram of Fig. 39-4c. As the phasors rotate in the counterclockwise direction it is clear that the $V_{L,m}$ phasor precedes (that is, *leads*) the $i_{L,m}$ phasor by a quarter-cycle.

The phase angle ϕ between V_L and i_L in Fig. 39-4 is $+90°$. To show this put $\phi = +90°$ in Eq. 39-2 and expand. We obtain

$$i = -i_m \cos \omega t,$$

in agreement with Eq. 39-7c.

Again, for reasons of compactness of notation we choose to rewrite Eq. 39-7c as

$$i_L = -(\mathcal{E}_m/X_L) \cos \omega t \tag{39-8a}$$

in which we must have

$$X_L = \omega L. \tag{39-8b}$$

We call X_L the *inductive reactance.* As for the capacitive reactance (Eq. 39-5) we see that the SI unit for X_L must also be the ohm. Can you prove this explicitly, starting from Eq. 39-8b?

Also, let us recognize that $\mathcal{E}_m$ in Eq. 39-7a is the maximum value of $V_L (= V_{L,m})$. Thus from Eq. 39-7a we can write

$$V_{L,m} = i_{L,m} X_L. \tag{39-9a}$$

This suggests that when *any* alternating current of amplitude i_m and angular frequency ω exists in an inductor, the maximum potential difference across the inductor (*regardless of the complexity of the circuit in which the inductor is involved*) is given by

$$V_{L,m} = i_m X_L. \tag{39-9b}$$

In Fig. 39-4 let $L = 60$ mH, $\nu = 60$ Hz, and $\mathcal{E}_m = 300$ V. Find (a) $V_{L,m}$, (b) X_L and (c) $i_{L,m}$.

EXAMPLE 2

(a) From Eq. 39-7a,

$$V_{L,m} = \mathcal{E}_m = 300 \text{ V}.$$

(b) From Eq. 39-8b,

$$X_L = \omega L = 2\pi\nu L$$
$$= (2\pi)(60 \text{ Hz})(60 \times 10^{-3} \text{ H})$$
$$= 23 \text{ } \Omega.$$

($\mathcal{E} = \mathcal{E}_m \sin \omega t$)

(a)

(b)

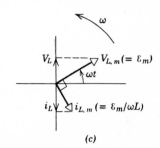

(c)

figure 39-4
(a) A single-loop inductive circuit containing an ac generator. (b) The potential difference across the inductor *leads* the current by a quarter-cycle. In Eq. 39-2, $\phi = +90°$. (c) A phasor diagram shows the same thing. The arrows on the vertical axis are instantaneous values.

(c) From Eq. 39-9a,

$$i_{L,m} = \frac{V_{L,m}}{X_L}$$

$$= \frac{300 \text{ V}}{23 \text{ }\Omega} = 13 \text{ A}.$$

How would $i_{L,m}$ change if you doubled the frequency? Does your answer seem intuitively reasonable?

Having finished our analysis of separate single-loop R, C, and L circuits, we now return to Fig. 39-1, in which all three elements are present. The emf is given by Eq. 39-1

$$\varepsilon = \varepsilon_m \sin \omega t \qquad (39\text{-}1)$$

and the (single) current in the circuit has the form shown in Eq. 39-2

$$i = i_m \sin (\omega t - \phi) \qquad (39\text{-}2)$$

in which we have yet to determine i_m and ϕ.

We start by applying the loop theorem to the circuit of Fig. 39-1, obtaining

$$\varepsilon = V_R + V_C + V_L \qquad (39\text{-}10)$$

These are all sinusoidally time-varying quantities, their maximum values being, in order, ε_m (see Eq. 39-1), $V_{R,m}$ ($= i_m R$), $V_{C,m}$ ($= i_m X_C$; see Eq. 39-6b), and $V_{L,m}$ ($= i_m X_L$; see Eq. 39-9b).

Although Eq. 39-10 is true at any instant of time we cannot easily use it to find i_m and ϕ in Eq. 39-2 because of the phase differences that exist among the separate terms. We therefore turn to the phasor diagram of Fig. 39-5a, which shows the maximum values of i, V_R, V_C, and V_L. Check Section 39-2 to see that the phase differences are correct, that is, V_R is *in phase* with the current (Fig. 39-2), V_C *lags* the current by a quarter-cycle (Fig. 39-3), and V_L *leads* the current by a quarter-cycle (Fig. 39-4).

As in Section 39-2, the projections of the phasors on the vertical axis give the *instantaneous* values of the quantities involved. Thus the *algebraic* sum of V_R, V_C, and V_L on the vertical axis equals ε, as Eq. 39-10 requires.

On the other hand, we assert that the *vector* sum of the phasor amplitudes $V_{R,m}$, $V_{C,m}$, and $V_{L,m}$ yields a phasor whose amplitude is the ε_m of Eq. 39-1. The projection of ε_m on the vertical axis would, of course, be the time-varying ε of Eq. 39-1, that is, it would be $V_R + V_C + V_L$ as Eq. 39-10 asserts. Note that, in vector operations, the (algebraic) sum of the projections of any number of vectors on a given straight line is equal to the projection on that line of the (vector) sum of those vectors.

We can find ε_m from Fig. 39-5b, in which we have formed the phasor $V_{L,m} - V_{C,m}$. This is at right angles to $V_{R,m}$ and we have

$$\varepsilon_m = \sqrt{V_{R,m}^2 + (V_{L,m} - V_{C,m})^2}$$

$$= \sqrt{(i_m R)^2 + (i_m X_L - i_m X_C)^2}$$

$$= i_m \sqrt{R^2 + (X_L - X_C)^2} \qquad (39\text{-}11)$$

We call the quantity multiplying i_m the *impedance Z* of the circuit of Fig. 39-1. Thus we can write

39-3
THE SINGLE-LOOP RCL CIRCUIT

(a)

(b)

figure 39-5

(a) A phasor diagram corresponding to the circuit of Fig. 39-1. Note the phasor amplitudes and the instantaneous values shown on the vertical axis. (b) The same diagram, showing the relationship between the emf ξ and the current i of Eqs. 39-1 and 39-2, respectively. The phase angle ϕ of Eq. 39-2 is clearly shown.

$$i_m = \frac{\varepsilon_m}{Z}, \qquad (39\text{-}12)$$

which reminds us of the relation $i = \varepsilon/R$ (Eq. 32-2) for single-loop resistive networks with steady emfs. The SI unit of impedance is evidently the ohm.

We can write out Eq. 39-12 in full detail (see Eqs. 39-11, 39-5b, and 39-8b) as

$$i_m = \frac{\varepsilon_m}{\sqrt{R^2 + (\omega L - 1/\omega C)^2}}. \qquad (39\text{-}13)$$

Thus we have solved the first problem posed in the second paragraph of Section 39-2; we have expressed i_m in terms of ε_m, ω, R, C, and L. It remains to find the phase angle ϕ between i and ε; compare Eqs. 39-1 and 39-2.

We show the angle ϕ in Fig. 39-5b and we can write

$$\tan \phi = \frac{V_{L,m} - V_{C,m}}{V_{R,m}}$$

$$= \frac{i_m(X_L - X_C)}{i_m R}$$

$$= \frac{X_L - X_C}{R}. \qquad (39\text{-}14)$$

Thus we have solved the second problem posed in the second paragraph of Section 39-2; we have expressed ϕ in terms of ω, R, C, and L. As we said earlier, ϕ does not depend on ε_m. Increasing ε_m increases i_m (see Eq. 39-13) but it does not change ϕ; the *scale* of the operation is changed but not its *nature*.

We drew Fig. 39-5b arbitrarily with $X_L > X_C$; that is, we assumed the circuit of Fig. 39-1 to be more inductive than capacitive. For this assumption ε_m *leads* i_m (although not by so much as a quarter-cycle as it did in the purely inductive circuit of Fig. 39-4). The phase angle ϕ in Eq. 39-14 (and thus in Eq. 39-2) is positive but less than $+90°$.

If, on the other hand, we had $X_C > X_L$, the circuit would be more capacitive than inductive and ε_m would *lag* i_m (although not by as much as a quarter-cycle, as it did in the purely capacitive circuit of Fig. 39-3). Consistent with this change from leading to lagging, the angle ϕ in Eq. 39-14 (and thus in Eq. 39-2) would automatically become negative.

EXAMPLE 3

In Fig. 39-1 let $R = 4.0\ \Omega$, $C = 150\ \mu$F, $L = 60$ mH, $\nu = 60$ Hz, and $\varepsilon_m = 300$ V. Find (a) X_C, (b) X_L, (c) Z, (d) i_m, and (e) ϕ.

(a) $X_C = 18\ \Omega$ as in Example 1. Note that X_C depends only on C and ω (Eq. 39-5b) and *not* on the nature of the circuit in which C is an element.

(b) $X_L = 23\ \Omega$, as in Example 2. Note that X_L depends only on L and ω (Eq. 39-8b) and *not* on the nature of the circuit in which L is an element.

(c) From Eq. 39-11 we have

$$Z = \sqrt{R^2 + (X_L - X_C)^2}$$
$$= \sqrt{(4.0\ \Omega)^2 + (23\ \Omega - 18\ \Omega)^2}$$
$$= 6.4\ \Omega.$$

Note that this circuit is more inductive than capacitive, that is, $X_L > X_C$, as in Fig. 39-5.

(d) From Eq. 39-12 we have

$$i_m = \frac{\varepsilon_m}{Z} = \frac{300\ V}{6.4\ \Omega} = 47\ A.$$

(e) From Eq. 39-14 we have

$$\tan\phi = \frac{X_L - X_C}{R}$$
$$= \frac{23\ \Omega - 18\ \Omega}{4.0\ \Omega} = 1.25$$

or $\phi = +51°$. Because $X_L > X_C$, ϕ is positive and ε_m *leads* i_m, as Fig. 39-5b suggests, but, as we expect, by less than +90°.

Power is an important practical matter. If you turn on the lights in a room, you are dealing directly with the subject matter of this section. We start by realizing that power dissipation in *RCL* circuits (Fig. 39-1 is an example on which we will focus), occurs *only* in the resistive element R; there is no mechanism for dissipating power in purely capacitive or purely inductive elements.

Let us therefore consider the purely resistive single-loop circuit of Fig. 39-2a, from two points of view. First let us replace the alternating emf by a *steady* emf of magnitude ε_m. The (steady) power dissipated in R (see Eq. 31-17) is given by

$$P = \frac{\varepsilon_m^2}{R}\ \text{(steady emf)}. \qquad (39\text{-}15)$$

If the alternating emf of Eq. 39-1 applies, we have, however

$$P(t) = \frac{(\varepsilon_m \sin \omega t)^2}{R}\ \text{(alternating emf)}. \qquad (39\text{-}16)$$

Our real concern is not so much for $P(t)$, which is sometimes zero, but for the average power $\overline{P(t)}\ (= P_{av})$, or

$$P_{av} = \frac{\varepsilon_m^2}{R}\ \overline{(\sin \omega t)^2}. \qquad (39\text{-}17)$$

Figure 39-6 shows (a) a plot of $\sin \omega t$ and (b) a plot of $(\sin \omega t)^2$. Note that $\overline{\sin \omega t} = 0$ (the positive parts just cancel the negative parts). However, $\overline{(\sin \omega t)^2} = \frac{1}{2}$ (the parts above the line "$\frac{1}{2}$" just cancel those below that line; there are no negative values. Thus,

$$P_{av} = \frac{1}{2}\frac{\varepsilon_m^2}{R} = \left(\frac{\varepsilon_m}{\sqrt{2}}\right)^2\frac{1}{R}. \qquad (39\text{-}18)$$

We choose to call $\varepsilon_m/\sqrt{2}$ the *root-mean-square* (abbr. rms) value of $\varepsilon\ (= \varepsilon_{rms})$. The notation is justified. First we *squared* ε_m, then we averaged it (or took the *mean* of it) over an integral number of cycles (the averaging factor, in this case of sinusoidal functions, being $\frac{1}{2}$), and

39-4
POWER IN ALTERNATING CURRENT CIRCUITS

(a)

(b)

figure 39-6
(a) A plot of $\sin \omega t$, showing that its average value over an integral number of cycles is zero. The parts of the curve above the horizontal axis just cancel those below. (b) A plot of $(\sin \omega t)^2$, showing that its average value over an integral number of cycles is $\frac{1}{2}$. The parts of the curve above the horizontal line "$\frac{1}{2}$" just cancel those below it; the curve has no negative values.

finally we took the *square root.** We can thus write Eq. 39-18 as

$$P_{\text{av}} = \frac{\mathcal{E}_{\text{rms}}^2}{R}. \qquad (39\text{-}19)$$

Equation 39-19 looks much like Eq. 31-17 and the message is that if we use rms quantities for $\mathcal{E}$, for V, and for i, the *average* power dissipation, which is all that usually matters, will be the same for alternating current circuits as for direct current circuits with a constant emf. Alternating current instruments, such as ammeters and voltmeters, are almost always too sluggish mechanically to follow the actual cycles of oscillation. They are deliberately calibrated to read $\mathcal{E}_{\text{rms}}$, V_{rms}, and i_{rms}. Thus, if you plug an alternating current voltmeter into a household electric outlet and it reads "120 V," the maximum value of the potential difference at the outlet is $\sqrt{2}$ (120 V) or 170 V. The sole reason for using rms values in alternating current circuits is to let us use the familiar direct current power relationships of Section 31-5. Thus, we summarize:

$$\mathcal{E}_{\text{rms}} = \mathcal{E}_m/\sqrt{2}, \quad V_{\text{rms}} = V_m/\sqrt{2}, \quad \text{and} \quad i_{\text{rms}} = i_m/\sqrt{2}. \qquad (39\text{-}20)$$

Because the proportionality factor $(= 1/\sqrt{2})$ is the same in each case, we can write the important Eq. 39-13 as

$$i_{\text{rms}} = \frac{\mathcal{E}_{\text{rms}}}{\sqrt{R^2 + (\omega L - 1/\omega C)^2}} \qquad (39\text{-}21)$$

and, indeed, this is the form that we almost always use.

EXAMPLE 4

(a) A steady emf $(\mathcal{E}_0 = 120 \text{ V})$ is applied to a single-loop resistive circuit with $R = 150 \ \Omega$. What is the power dissipation? (b) The steady emf $\mathcal{E}_0$ is replaced by an alternating emf $(\mathcal{E} = \mathcal{E}_m \sin \omega t)$. If the average power is to remain unchanged, what must $\mathcal{E}_m$ be?

(a)
$$P = \frac{\mathcal{E}_0^2}{R} = \frac{(120 \text{ V})^2}{150 \ \Omega} = 96 \text{ W}.$$

(b) From Eq. 39-18 and the conditions of the problem,

$$P_{\text{av}} (= 96 \text{ W}) = \left(\frac{\mathcal{E}_m}{\sqrt{2}}\right)^2 \frac{1}{R}$$

or

$$\mathcal{E}_m = \sqrt{2 P_{\text{av}} R}$$
$$= \sqrt{(2)(96 \text{ W})(150 \ \Omega)}$$
$$= 170 \text{ V}.$$

From Eq. 39-20 we have

$$\mathcal{E}_{\text{rms}} = \mathcal{E}_m/\sqrt{2} = (170 \text{ V})/\sqrt{2}$$
$$= 120 \text{ V},$$

just as we expect.

Now we turn from power considerations for the resistive circuit of Fig. 39-2 to those for the more general circuit of Fig. 39-1, in which all three elements, R, C, and L, appear. For the instantaneous power we can write (compare Eq. 31-15)

---

*We have encountered rms quantities before, namely, in the rms speeds of molecules (see Section 23-4; also, see p. 852).

$$P(t) = \varepsilon(t)\,i\,(t)$$

$$= [\varepsilon_m \sin \omega t]\,[i_m \sin (\omega t - \phi)]. \qquad (39\text{-}22)$$

If we expand the factor $\sin(\omega t - \phi)$ by a trignometric identity that is evident by comparison with Eq. 39-23, we obtain

$$P(t) = (\varepsilon_m\,i_m)(\sin \omega t)(\sin \omega t \cos \phi - \cos \omega t \sin \phi)$$

$$= \varepsilon_m\,i_m \sin^2 \omega t \cos \phi - \varepsilon_m\,i_m \sin \omega t \cos \omega t \sin \phi. \qquad (39\text{-}23)$$

If we now find $\overline{P(t)}$ $(= P_{av})$, which is our principal practical concern, we have

$$P_{av} = \tfrac{1}{2}\varepsilon_m\,i_m \cos \phi + 0. \qquad (39\text{-}24)$$

This comes about because, as we have seen, $\overline{\sin^2 \omega t} = \tfrac{1}{2}$ and we assert (see Problem 9) that $\overline{\sin \omega t \cos \omega t} = 0$. Because $\varepsilon_{rms} = \varepsilon_m/\sqrt{2}$ and $i_{rms} = i_m/\sqrt{2}$ (Eq. 39-20) we can write Eq. 39-24 as

$$P_{av} = \varepsilon_{rms}\,i_{rms} \cos \phi \qquad (39\text{-}25)$$

in which we call $\cos \phi$ the *power factor*. Figure 39-2, which shows a purely resistive load, is characterized by $\phi = 0$ (and thus $\cos \phi = 1$)so that Eq. 39-25 becomes

$$P_{av} = \varepsilon_{rms}\,i_{rms} \text{ (resistive load)}. \qquad (39\text{-}26)$$

Subscripts aside, this is just the relation that we would write for direct current circuits (compare Eq. 31-15).

EXAMPLE 5

Let us use the same parameters for Fig. 39-1 that we did in Example 3, namely, $R = 4.0\,\Omega$, $C = 150\mu F$, $L = 60\,mH$, $\nu = 60\,Hz$, and $\varepsilon_m = 300\,V$. Find (a) ε_{rms}, (b) i_{rms}, (c) ϕ, (d) $\cos\phi$, and (e) P_{av}.

(a) $\varepsilon_{rms} = \varepsilon_m/\sqrt{2} = (300\,V)/\sqrt{2} = 210\,V$.

(b) $i_{rms} = i_m/\sqrt{2} = (47\,A)/\sqrt{2} = 33\,A$; see Example 3, part (d).

(c) $\phi = 51°$, as in Example 3, part (e).

(d) The power factor, $\cos \phi$, is $\cos 51°$ or 0.63.

(e) $P_{av} = \varepsilon_{rms}\,i_{rms} \cos \phi$

$= (210\,V)(33\,A)(0.63)$

$= 4.4\,kW.$

39-5
RESONANCE IN ALTERNATING CURRENT CIRCUITS

In this section we return to the circuit of Fig. 39-1 and consider the effect, as far as the current i_{rms} is concerned, of varying the angular frequency ω of the applied emf of Eq. 39-1, assuming that ε_m, R, C, and L remain fixed.

We have stressed the analogy between mechanical systems, such as mass-spring arrangements (Fig. 8-4) and electromagnetic systems (Fig. 38-1). With this in mind we are entitled to look upon the *RCL* circuit of Fig. 39-1 as possessing a "natural" frequency of oscillation, and to view the circuit as acted upon by an external influence, in this case the applied alternating emf, given by $\varepsilon = \varepsilon_m \sin \omega t$ (Eq. 39-1), in which ω is the angular frequency of the "driving force." We expect a maximum "response," defined here by the i_{rms} of Eq. 39-21, when the angular frequency ω of the driving force just equals the natural frequency of oscillation ω_0 for the free oscillations of the circuit. Let us see whether these analogies and expectations turn out to be true.

From Eq. 39-21 we see that the maximum value of i_{rms} occurs when $X_L = X_C$ and has the value

$$i_{rms,max} = \frac{\varepsilon_{rms}}{R}, \qquad (39\text{-}27)$$

that is, i_{rms} is limited only by the circuit resistance. If $R \to 0$, $i_{rms,max} \to \infty$. Putting $X_L = X_C$ yields

$$\omega L = \frac{1}{\omega C}$$

or

$$\omega = \frac{1}{\sqrt{LC}}. \qquad (39\text{-}28)$$

The quantity on the right (see Eq. 38-3) is just the natural angular frequency ω_0 of the circuit of Fig. 39-1, with ε and R removed. That is:

the maximum value of i_{rms} occurs when the frequency ω of the driving force [the $\varepsilon(t)$ of Eq. 39-1], is exactly equal to the natural frequency ω_0 of the *undamped* $(R = 0)$ circuit of Fig. 39-1.

This condition

$$\omega = \omega_0 \qquad (39\text{-}29)$$

is called *resonance.* We met it earlier in connection with the cyclotron (see Section 33-7) and in many equivalent mechanical situations.

Figure 39-7 is a plot of i_{rms} versus ω (see Eq. 39-21), for fixed values of ε_m, C, and L, but for three different values of R. Note how rapidly the sharpness of the resonance peak broadens as R increases.

Figure 39-7 suggests to us the common experience of tuning a radio set. What we do here when we turn the tuning knob is to adjust the natural frequency ω_0 of an internal circuit to the frequency ω of the signal transmitted by the station antenna, until Eq. 39-29 is satisfied. In a metropolitan area, where there are many incoming signals whose fre-

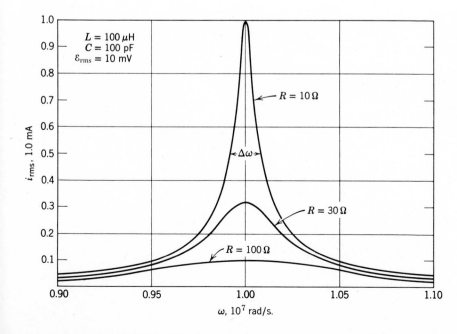

figure 39-7
Showing resonance in the *RCL* circuit of Fig. 39-1 for three different values of *R*. Along the horizontal axis we plot the (variable) angular frequency ω of the impressed emf. The arrow marked "$\Delta\omega$" on the curve for $R = 10\ \Omega$ is called the "full width of the curve at half maximum" or, more concisely, the *half width.*

quencies are not too far apart, the sharpness (or "quality"; see Problem 17) of tuning becomes very important.

Now let us consider the phenomenon of resonance from the point of view of the phasor diagram of Fig. 39-5b. If $X_L = X_C$, as resonance requires, the phasors $V_{L,m}$ $(=i_m X_L)$ and $V_{C,m}$ $(=i_m X_C)$ just cancel each other and the phase angle $\phi = 0$. The fact that $\phi = 0$ if $X_L = X_C$ is explicitly shown to be true by Eq. 39-14. If $\phi > 0$, as in Fig. 39-5b, we have $X_L > X_C$ and the circuit is predominantly inductive. On the other hand, if $\phi < 0$, we have $X_C > X_L$ and the circuit is predominantly capacitive.

EXAMPLE 6

In Fig. 39-7 (a) show that the angular frequency of the impressed emf at which resonance occurs is indeed 1.0×10^7 rad/s; (b) show also that the maximum value of i_{rms} at resonance, for $R = 10\ \Omega$, is 1.0 mA.

(a) At resonance $X_L = X_C$ or

$$\omega L = \frac{1}{\omega C}$$

which leads to

$$\omega = \frac{1}{\sqrt{LC}} = \frac{1}{\sqrt{(100\ \mu\mathrm{H})(100\ \mathrm{pF})}}$$

$$= 1.0 \times 10^7 \text{ rad/s.}$$

(b) At resonance Eq. 39-27 applies, or

$$i_{\mathrm{rms.max}} = \frac{\mathcal{E}_{\mathrm{rms}}}{R}$$

$$= \frac{10\ \mathrm{mV}}{10\ \Omega} = 1.0\ \mathrm{mA.}$$

39-6
ALTERNATING CURRENT RECTIFIERS AND FILTERS

We often have available a source of alternating emf (given, say, by Eq. 39-1, $\mathcal{E} = \mathcal{E}_m \sin\omega t$) and we want to derive from it, by electronic means, a constant potential difference. For example, in television sets, sound reproducing systems, etc., the available electric input is usually an alternating emf, often described by 120 V $(= \mathcal{E}_{\mathrm{rms}})$ and 60 Hz $(= \omega/2\pi)$. From this we need to derive one or more constant potential differences (50 V, 300 V, 1500 V, etc.) to operate the electronic circuitry of the device. This process is called *rectification* (literally, "making straight") and the devices that make it possible are called *rectifiers.*

Rectifiers are nonohmic devices that have the property (see Section 31-3) that their resistance depends on the polarity of the applied potential difference. In this section we assume an ideal rectifier, which has $R = 0$ for a given polarity and $R \to \infty$ when that polarity is reversed.

Physically, rectifiers may be semiconducting solid-state devices or vacuum tube diodes. The symbol for a rectifier is ———▷|———, left to right being the direction of "easy conduction."

Figure 39-8 suggests how rectifiers work. To establish a baseline consider Fig. 39-8a, which is identical to Fig. 39-2a. A source of alternating emf $\mathcal{E}$ $(= \mathcal{E}_m \sin \omega t)$ is connected with a resistive load R with no rectifiers in the circuit. The arrows indicate the current directions for the polarity shown. When the polarity of $\mathcal{E}$ reverses, the current arrows also reverse.

If we connect a cathode ray oscilloscope between points b and c, it

(a)

(b)

(c)

figure 39-8
An ac generator $\mathcal{E}$ $(= \mathcal{E}_m \sin \omega t)$ acts on the three circuits shown. The potential difference V_{bc} is displayed on the right. Note that $\overline{V_{bc}}$ increases from zero to $\mathcal{E}_m/\pi$ to $2\mathcal{E}_m/\pi$.

displays the waveform shown on the right. Note that $\overline{V_{bc}} = 0$ in this case, the positive halves of the sine wave just cancelling the negative halves. No rectification occurs, which is not surprising because there are no rectifiers in the circuit.

If we connect a dc voltmeter* between points b and c in Fig. 39-8a, it would read zero because the coil assembly is too sluggish to follow the waveform shown in Fig. 39-8a. For a typical dc voltmeter, at 60 Hz, say, the pointer will oscillate rapidly with small amplitude about its zero position. Why?

In Fig. 39-8b we insert a rectifier in the path ab, the direction of "easy conduction" being from a to b. When the polarity reverses, the circuit is virtually open because the (ideal) rectifier has infinite resistance.

If we connect a cathode ray oscilloscope between points b and c, it displays the waveform shown on the right. It is clear that $\overline{V_{bc}} > 0$. In fact,

$$\overline{V_{bc}} = \frac{\mathcal{E}_m}{2\pi} \left\{ \int_0^\pi \sin \omega t \, d(\omega t) + \int_\pi^{2\pi} 0 \, d(\omega t) \right\} = \frac{\mathcal{E}_m}{\pi}. \qquad (39\text{-}30)$$

The arrangement of Fig. 39-8b is called a *half-wave rectifier*. It does not yet produce a truly constant potential difference but it is a step in

---

* We can construct such a voltmeter by putting a large resistor R_0 in series with the coil of the sensitive galvanometer shown in Fig. 33-9. The large series resistor R_0 ($\gg R$ in Fig. 39-8a) is needed to insure that touching the voltmeter terminals to points b and c in Fig. 39-8a will not appreciably change the current distribution in that circuit.

that direction. If we connect a dc voltmeter between points b and c in Fig. 39-8b, it will give a definite reading. As for Fig. 39-8a there will be rapid small-amplitude oscillations of the pointer about its equilibrium position.

In Fig. 39-8c we introduce four rectifiers. For the polarity shown the current follows the path indicated by the arrows. When the polarity of the emf reverses, the current becomes counterclockwise and its path through the rectifier network is *dbca. Note that the direction of the current through R remains unchanged;* it is *always* from b to c, regardless of the polarity of the source of emf.

If we connect a cathode ray oscilloscope between points b and c, it displays the waveform shown on the right. Convince yourself that $\overline{V_{bc}}$ is just twice the value given for the half-wave rectifier of Fig. 39-8b, that is,

$$\overline{V_{bc}} = 2\,\mathcal{E}_m/\pi. \tag{39-31}$$

The arrangement of Fig. 39-8c is called a *full-wave rectifier.* If we connect a dc voltmeter between points b and c, it will deflect twice as far as it did in connection with Fig. 39-8b. Oscillations of the pointer will still occur but they will have smaller amplitude. Why?

EXAMPLE 7

In Fig. 39-8 the emf is $\mathcal{E} = \mathcal{E}_m \sin \omega t$, in which $\mathcal{E}_m = 300$ V and $R = 15\ \Omega$. What average power P_{av} is dissipated in the resistor for the three cases shown?

(*a*) For Fig. 39-8a we have, from Eqs. 39-18 and 39-19,

$$P_{av} = \frac{\mathcal{E}_m{}^2}{2R} \left(= \frac{\mathcal{E}_{rms}{}^2}{R} \right)$$

$$= \frac{(300\ \text{V})^2}{(2)(15\ \Omega)} = 3.0\ \text{kW}.$$

(*b*) Because the circuit of Fig. 39-8b is nonconducting (open circuit) for half the time, P_{av} is just half the value shown in (a) above, or 1.5 kW.

(*c*) In Fig. 39-8c P_{av} has the same value (= 3.0 kW) as in (a) above. Note that P_{av} does not depend on the *direction* of i or V; both quantities occur as squares.

The waveform of Fig. 39-8c is *still* not a nontime-varying potential difference. We can, however, resolve it into a steady component and a series of sinusoidally alternating potential differences with various angular frequencies ω, amplitudes V, and relative phases ϕ. We often call these ac components *the ripple.* Qualitatively we can see that this division of the waveform of Fig. 39-8c is reasonable, just by inspection. Quantitatively we could (but won't) derive it by Fourier analysis of the waveform.

What we need to do now is to connect the potential difference between points b and c of the circuit of Fig. 39-8 c, that is, the waveform of that figure, through a *filter circuit,* which has these properties: (1) it passes the dc component from input to output with negligible reduction in value, and (2) it greatly decreases the amplitude of the time-varying components, that is, of the ripple.

Figure 39-9 shows a simple filter circuit. It contains an ideal inductor L (that is, it has no resistive or capacitive properties) and an ideal capacitor C (that is, it has no resistive or inductive properties). The input V_{in} to the filter may be steady or sinusoidally oscillating. To explore the behavior of the filter we will look at these two cases separately.

For $V_{in} = $ a constant we see that,

$$V_{out} = V_{in} = \text{the same constant.}$$

Neither L nor C has any effect. In fact, L might be replaced by a straight wire (shorted out) and C removed from (clipped out of) the circuit, with no noticeable effect on V_{out}.

For an ac input, however, the situation is quite different. We assume from the beginning that both L and C are "large" so that X_L ($=\omega L$) $\gg X_C$ ($=1/\omega C$). If ω and C are large enough, $X_C \to 0$ and the capacitor acts as a virtual short circuit for ac components, even though it has no effect on the dc component.

Assume

$$V_{in} = V_{in,m} \sin \omega t. \qquad (39\text{-}32)$$

For the current we can put

$$i = i_m \sin (\omega t + \phi). \qquad (39\text{-}2)$$

From Eq. 39-13, with $R = 0$ and ε_m replaced $V_{in,m}$, we have

$$i = \frac{V_{in,m}}{X_L - X_C} \sin (\omega t + \phi). \qquad (39\text{-}33)$$

Because we have already assumed that $X_L \gg X_C$ we can write this as

$$i \cong \frac{V_{in,m}}{X_L} \sin (\omega t + \phi). \qquad (39\text{-}34)$$

To find the phase angle ϕ we turn to Eq. 39-14, or

$$\tan \phi = \frac{X_L - X_C}{R}. \qquad (39\text{-}14)$$

With $X_L \gg X_C$ and $R \to 0$ we have $\tan \phi \to +\infty$ or $\phi = +90°$. Thus Eq. 39-33 becomes (see Eq. 39-14)

$$i = \frac{V_{in,m}}{X_L} \cos \omega t. \qquad (39\text{-}35)$$

Note that a cosine function is just 90° out of phase with a sine function.

The output V_{out} is given by q/C so that we must find q for the capacitor. From Eq. 39-35 we have

$$q = \int_0^t i \,(dt) = \frac{1}{\omega} \int_0^{\omega t} i \, d(\omega t)$$

$$= \left(\frac{1}{\omega}\right)\left(\frac{V_{in,m}}{X_L}\right) \int_0^{\omega t} \cos \omega t \, d(\omega t)$$

$$= \left(\frac{1}{\omega}\right)\left(\frac{V_{in,m}}{X_L}\right) \Big| \sin \omega t \Big|_0^{\omega t}$$

$$= \left(\frac{1}{\omega}\right)\left(\frac{V_{in,m}}{X_L}\right) \sin \omega t. \qquad (39\text{-}36)$$

Because $X_C = 1/\omega C$ we have

$$V_{out} = \frac{q}{C} = (X_C)\left(\frac{V_{in,m}}{X_L}\right) \sin \omega t. \qquad (39\text{-}37)$$

Comparison with Eq. 39-32 gives for the ac attenuation

$$V_{out,m}/V_{in,m} = X_C/X_L. \qquad (39\text{-}38)$$

Because we have chosen $X_L \gg X_C$ we see that the attenuation of the "ripple components" is large. Much more effective filters than those of Fig. 39-9 are possible, as amateur builders of sound reproducing equipment are well aware (see Problem 23).

figure 39-9
A simple filter circuit designed to pass dc potential differences from input to output without appreciable attenuation but to attenuate time-varying components greatly. Convince yourself that if L, C, and ω all have "large" values, the filter element will be more effective in reducing the amplitudes of ac input components.

In Fig. 39-9 let $V_{in} = V_{in,m} \sin \omega t$, in which $V_{in,m} = 300$ V, and $\omega/2\pi = 60$ Hz. Let $L = 10$ H and $C = 300$ μF. Find $V_{out,m}$.

EXAMPLE 8

From Eq. 39-37, we have

$$V_{out,m} = (V_{in,m})(X_C)\left(\frac{1}{X_L}\right)$$

$$= (V_{in,m})\left(\frac{1}{\omega C}\right)\left(\frac{1}{\omega L}\right)$$

$$= \frac{V_{in,m}}{\omega^2 LC}$$

$$= \frac{(300 \text{ V})}{(2\pi \times 60 \text{ Hz})^2 (10 \text{ H})(300 \times 10^{-6} \text{ F})}$$

$$= 0.70 \text{ V}.$$

In this case the attenuation factor is

$$(V_{out,m}/V_{in,m}) = (0.70 \text{ V})/(300 \text{ V})$$

or 2.3×10^{-3}. Convince yourself that this is the ratio X_C/X_L, in which we have assumed $X_L \gg X_C$.

In dc circuits the power dissipation in a resistive load is given by Eq. 31-15 ($P = iV$). This means that, for a given power requirement, we have our choice of a relatively large current i and a relatively small potential difference V or just the reverse, provided that their product remains constant. In the same way, for ac circuits the average power dissipation is given by Eq. 39-25 ($P_{av} = i_{rms} V_{rms}$) and we have the same choice as to the relative values of i_{rms} and V_{rms}.*

In electric power distribution systems it is clear that at both the generating end (the electric power plant) and the receiving end (the home or factory) it is desirable, both for reasons of safety and the efficient design of equipment, to deal with relatively low voltages. For example, no one wants an electric toaster or a child's electric train to operate at, say, 10 kV.

On the other hand, in the transmission of electric energy from the generating plant to the consumer, we want the *lowest* possible current (and thus the *largest* possible potential difference) so as to minimize the i^2R ohmic losses in the transmission line. $V_{rms} = 350$ kV is not uncommon. Thus there is a fundamental "mismatch" between the requirements for efficient transmission on the one hand and efficient and safe generation and consumption on the other. We need a device that can, as design considerations require, raise (or lower) the potential difference in a circuit, keeping the product iV essentially constant. The alternating current *transformer* of Fig. 39-10 is such a device. It has no direct current counterpart of equivalent simplicity, which is why dc distribution systems, strongly advocated by Edison, have now been essentially totally replaced by ac systems, strongly advocated by Tesla and others.

Within the last few decades the use of dc transmission over very long distances or for special situations such as underwater or underground transmission cables has undergone a revival. In the Soviet Union, for example, there is a 300-mile line from Volgograd to the Donbass operating at a dc potential difference of 800 kV. The power is generated as ac, stepped up to the transmission voltage, converted to dc, reconverted

39-7
THE TRANSFORMER

---

* We assume a resistive load so that $\cos \phi = 1$ in Eq. 39-25.

to ac at the receiving point, and stepped down to lower voltages for transmission. The advantage of high-voltage dc transmission under these circumstances is that, for dc, the capacitive and inductive properties of the transmission line can be ignored. Note that transformers, the subject of this section, are still very much required.*

In Fig. 39-10 two coils are shown wound around a soft iron core. The *primary* winding, of N_1 turns, is connected with an alternating current generator whose emf $\varepsilon_1(t)$ is given by $\varepsilon_1 = \varepsilon_m \sin \omega t$. The *secondary* winding, of N_2 turns, is on open circuit as long as switch S is open, which we assume for the present. Thus there is no secondary current. We assume further that the resistances of the primary and secondary windings and also the magnetic "losses" in the iron core are negligible. Actually, well-designed, high-capacity transformers can have energy losses as low as one percent so that our assumption of an ideal transformer is not unreasonable.

figure 39-10
An ideal transformer, showing two coils wound on the same soft iron core. An ac generator is connected with the primary winding. The secondary winding may be on open circuit (if switch S is open) or connected with a resistive load R (if switch S is closed).

For the above conditions the primary winding is a pure inductance; compare Fig. 39-4. Thus the (very small) primary current, called the *magnetizing current* $i_{mag}(t)$, lags the primary potential difference $V_1(t)$ by 90°; the power factor ($=\cos \phi$ in Eq. 39-25) is zero, and thus no power is delivered from the generator to the transformer.

However, the small alternating primary current $i_{mag}(t)$ induces an alternating magnetic flux $\Phi_B(t)$ in the iron core and we assume that all this flux links the turns of the secondary windings. From Faraday's law of induction (see Eq. 35-2) the emf *per turn* ε_T is the same for both the primary and secondary windings. Thus, on an rms basis, we can write

$$(\varepsilon_T)_{rms} = \left(-\frac{d\Phi_B}{dt}\right)_{rms} = \frac{V_{1,rms}}{N_1} = \frac{V_{2,rms}}{N_2}$$

or

$$V_{2,rms} = V_{1,rms}(N_2/N_1). \qquad (39-39)$$

If $N_2 > N_1$, we speak of a *step-up transformer*; if $N_2 < N_1$, we speak of a *step-down transformer*.

In all of the above we have assumed an open circuit secondary so that no power is transmitted through the transformer. If we now close switch S in Fig. 39-10, however, we have a more practical situation in which the secondary winding is connected with a resistive load R. In the general case the load would also contain inductive and capacitive elements but we confine ourselves to this special case.

Several things happen when we close switch S. (1) A current $i_2(t)$ appears in the secondary circuit, with a corresponding power dissipation $i_{2,rms}^2 R$ ($=V_{2,rms}^2/R$) in the resistive load. (2) This current induces its

* See L. O. Bartold and H. G. Pfeiffer, "High-Voltage Power Transmission," *Scientific American*, May 1964.

own alternating magnetic flux in the iron core and this flux induces (from Faraday's law and Lenz's law) an *opposing* emf in the primary windings. The two windings appear now as a fully coupled mutual inductance; see Section 36-6. (3) $V_1(t)$, however, cannot change in response to this opposing emf because it must *always* equal $\varepsilon(t)$ as provided by the generator; closing switch S cannot change this fact. (4) For this reason a *new* resultant current $i_1(t)$ must appear in the primary windings, its magnitude and phase angle being just that needed to cancel the opposing emf generated in the primary windings by $i_2(t)$. In particular the phase angle ϕ between $i_1(t)$ and $\varepsilon(t)$ for an ideal transformer must approach 0°, so that the power factor in Eq. 39-25, $\cos\phi$, must approach unity.

All of the above is consistent with energy conservation. When we close switch S, power is dissipated in the resistive load. This requires that the generator provide an equal power to the (ideal) transformer, or, assuming $\phi = 0°$ ($\cos\phi = 1$),

$$\varepsilon_{rms}i_{1,rms} = V^2_{2,rms}/R. \qquad (39\text{-}40)$$

This relation expresses the fact that, for an ideal transformer with a resistive load, the power provided by the generator on the primary side just equals that dissipated in the resistive load on the secondary side.

EXAMPLE 9

A transformer on a utility pole operates at $V_{1,rms} = 8.0$ kV (see Fig. 39-10) on the primary side and supplies electric energy to a number of nearby houses at $V_{2,rms} = 120$ V.

(*a*) What is the turns ratio N_1/N_2? (*b*) If the average power consumption in the houses for a given time interval is 70 kW, what are the rms currents in the primary and secondary windings of the transformer? Assume an ideal transformer, a resistive load, and a power factor of unity. (*c*) What is the equivalent resistive load R in the secondary circuit?

(*a*) From Eq. 39-39,

$$N_1/N_2 = V_{1,rms}/V_{2,rms}$$

$$= 8.0 \text{ kV}/120 \text{ V}$$

$$= 67.$$

(*b*) From Eq. 39-25 we have, with $\cos\phi = 1$

$$i_{2,rms} = P_{av}/(V_{2,rms})(\cos\phi)$$

$$= (70 \text{ kW})/(120 \text{ V})(1)$$

$$= 580 \text{ A},$$

and also

$$i_{1,rms} = P_{av}/(V_{1,rms})(\cos\phi)$$

$$= (70 \text{ kW})/(8.0 \text{ kV})(1)$$

$$= 8.8 \text{ A}.$$

Note that, as required for an ideal transformer,

$$i_{1,rms}V_{1,rms} = i_{2,rms}V_{2,rms} = 70 \text{ kW}.$$

(*c*) Here we have

$$R = (V_{2,rms})^2/P_{av}$$

$$= (120 \text{ V})^2/70 \text{ kW}$$

$$= 0.21 \ \Omega.$$

Show that the same result follows by solving Eq. 39-40 for R.

questions

1. In the relation $\omega = 2\pi\nu$ we measure ω in radians/second and ν in hertz or cycles/second. The radian is a measure of angle. What connection do angles have with alternating currents?

2. Problem 35-6 suggests how an alternating emf such as that described by Eq. 39-1 can be generated. If the output of such an ac generator is connected to an *RCL* circuit such as that of Fig. 39-1, what is the ultimate source of the power dissipated in the circuit? In what part of the circuit does the power dissipation occur?

3. In the circuit of Fig. 39-1, why is it safe to assume that (a) the alternating current of Eq.39-2 has the same angular frequency ω as the alternating emf of Eq. 39-1, and (b) that the phase angle ϕ in Eq. 39-2 does not vary with time? What would happen if either of these (true) statements were false?

4. Is there an analogy, however loose, between the facts that (a) the phase angle ϕ in Eq. 39-2 does not depend on the value of ω in Eq. 39-1, and (b) the oscillation frequency of a mass-spring system does not depend on the amplitude of the oscillation? Distinguish between the *nature* of a physical event and its *scale.*

5. How does a phasor differ from a vector? We know, for example, that emfs, potential differences, and currents are *not* vectors. How then can we justify constructions such as Fig. 39-5?

6. In the purely resistive circuit of Fig. 39-2 how does the maximum value of the alternating current i_m vary with the angular frequency of the applied emf?

7. Would any of the discussion of Section 39-2 be invalid if the phasor diagrams were to rotate in the clockwise direction, rather than the counterclockwise direction which we assumed?

8. Does it seem intuitively reasonable that the capacitive reactance $(= 1/\omega C)$ should vary inversely with the angular frequency, whereas the inductive reactance $(=\omega L)$ varies directly with this quantity?

9. During World War II, at a large research laboratory in this country, an alternating current generator was located a mile or so from the laboratory building which it served. A technician increased the speed of the generator to compensate for what he called "the loss of frequency along the transmission line" connecting the generator with the laboratory building. Comment.

10. Discuss in your own words what it means to say that a potential difference "leads" or "lags" an alternating current.

11. If, as we stated in Section 39-3, a given circuit is "more inductive than capacitive," that is, that $X_L > X_C$, does this mean, for a fixed angular frequency, that (a) L is relatively "large" and C is relatively "small," or (b) L and C are both relatively "large"? (c) For fixed values of L and C does this mean that ω is relatively "large" or relatively "small"?

12. Assume that in Fig. 39-1 we let $\omega \rightarrow 0$. Does Eq. 39-13 approach an expected value? Discuss.

13. How is it that in Fig. 39-2 the current phasor points in the direction of the voltage phasor but in Figs. 39-3, 39-4, and 39-5 the current phasors point in three different directions? The applied emf (Eq. 39-1) is the same in all four cases.

14. Consider this statement: "If $X_L > X_C$, then, regardless of the frequency, we must have $L > 1/C$." What is wrong with this statement?

15. Do Kirchoff's loop theorem (see Section 32-2) and Kirchoff's junction theorem (see Section 32-5) apply to multiloop ac circuits as well as to multiloop dc circuits?

16. In Example 5 what would be the effect on P_{av} if you increased (a) R, (b) C, and (c) L? How would ϕ in Eq. 39-25 change in these three cases?

17. Do commercial power station engineers like to have a low power factor (see Eq. 39-25) or a high one, or does it make any difference to them? Between what values can the power factor fluctuate? What determines the power factor; is it characteristic of the generator, of the transmission line,

of the circuit to which the transmission line is connected, or of some combination of these?

18. If you know the power factor (=cos ϕ in Eq. 39-25) for a given RCL circuit, can you tell whether or not the applied alternating emf is leading or lagging the current? If so, how? If not, why not?

19. If $R = 0$ in the circuit of Fig. 39-1, there can be no Joule internal energy dissipation in the circuit. However, an alternating emf and an alternating current are still present. Discuss the energy flow in the circuit under these conditions.

20. If you want to reduce your electric bill, do you hope for a small or a large power factor (=cos ϕ in Eq. 39-25) or does it make any difference? If it does, is there anything that you can do about it? Discuss.

21. If $R = 0$ in Fig. 39-1, how would the instantaneous current vary with time? What value would the phase angle ϕ have? What would be the average power dissipation?

22. In Eq. 39-25 is ϕ the phase angle between $\mathcal{E}(t)$ and $i(t)$ or between $\mathcal{E}_{rms}$ and i_{rms}? Discuss.

23. Resonance in RCL circuits, as judged by Eq. 39-29 and Fig. 39-7, occurs when the angular frequency of the alternating emf (the driving force) is exactly equal to the natural angular frequency of the (undamped) LC circuit. In Section 15-10, however, we saw that resonance for damped mass-spring systems, as judged by Eq. 15-41 and Fig. 15-20, occurs when the angular frequency of the driving force is close to, *but not exactly equal to*, the natural angular frequency of the (undamped) mass-spring system. Is there a failure of the principle of correspondence here? (*Hint:* Do the quantities displayed on the vertical axes of Figs. 15-20 and 39-7 truly "correspond"? See Problem 15.)

24. Convince yourself that filter circuits, such as those shown in Fig. 39-9 become, for fixed values of L and C, more effective as the angular frequency ω of V_{in} increases. What does "more effective" mean in this connection?

25. In Fig. 39-8c why does the current approaching junction c not divide equally between paths cd and ca? Both are potentially conducting paths.

26. Sketch roughly the waveforms of Fig. 39-8b,c if the rectifiers are not "ideal," that is, if the resistance in the forward direction, though small, is not actually zero, and if the resistance in the backward direction, though large, is not infinitely large.

27. Sketch an $i - V$ curve (see Figs. 31-4, 31-5, and 31-6) for the idealized rectifier assumed in Section 39-6. Include both polarities of the applied potential difference V (and of the corresponding current i) in your plot.

28. A doorbell transformer is designed for a primary rms input of 120 V and a secondary rms output of 6.0 V. What would happen if the primary and secondary connections were accidentally interchanged during installation? Would you have to wait for someone to push the doorbell to find out? Discuss.

29. You are given a transformer enclosed in a wooden box, its primary and secondary terminals being available at two opposite faces of the box. How could you find its turns ratio without opening the box?

problems

SECTION 39-1

1. A commercial alternating current generator is characterized by $\nu = 60$ Hz. What is the angular frequency ω and in what units is it expressed?
Answer: $\omega = 2\pi\nu = 380$ rad/s.

SECTION 39-2

2. An 0.50-μF capacitor is connected, as shown in Fig. 39-3a, to an ac generator with $\mathcal{E}_m = 300$ V. What is the amplitude i_m of the resulting alternating current if the angular frequency ω is (a) 100 rad/s, and (b) 1000 rad/s?

3. A 45-mH inductor has a reactance X_L of 1300Ω. What must be (a) the applied angular frequency ω and (b) the applied frequency ν for this to be true? (c) If, as in Fig. 39-4a, an alternating emf with $\mathcal{E}_m = 300$ V is applied, what is the amplitude i_m of the alternating current that results? *Answer:* (a) 2.9×10^4 rad/s. (b) 4.6 kHz. (c) 0.23A.

4. A 1.5-μF capacitor has a reactance X_C of 12 Ω. What must be (a) the applied frequency ν and (b) the applied angular frequency ω for this to be true? (c) If, as in Fig. 39-3a, an alternating emf with $\mathcal{E}_m = 300$ V is applied, what is the amplitude i_m of the alternating current that results?

5. (a) At what frequency ν would a 6.0-mH inductor and a 10-μF capacitor have the same reactance? (b) What would this reactance be? (c) How would this frequency compare with the natural resonant frequency of free oscillations if the components were connected as a (resistanceless) *LC* oscillator, as in Fig. 38-1? *Answer:* (a)650 Hz. (b) 24 Ω. (c) They are equal.

SECTION 39-3
6. Recalculate Example 3 for $C = 20$ μF, all other given quantities remaining the same. Will $\mathcal{E}_m$ lead or lag i_m in this new situation? Draw a rough diagram corresponding to Fig. 39-5.

7. Redraw (roughly) Figs. 39-5a and 5b for the cases of $X_C > X_L$ and $X_C = X_L$.

SECTION 39-4
8. We have seen that $\overline{\sin^2 \omega t} = \frac{1}{2}$ (average value). Find the average value of $\sin^2(\omega t + \phi)$ where ϕ is a (constant) phase angle?

9. Show that (see Eq. 39-23) $\overline{\sin \omega t \cos \omega t} = 0$ by using the trignometric identity $2 \sin \omega t \cos \omega t = \sin 2\omega t$. Also, plot $\sin \omega t \cos \omega t$ roughly and show graphically that its average value is zero.

10. The average value of $\mathcal{E}_m \sin \omega t$ over many cycles (see Fig. 39-6a) is zero. We have also seen that $\mathcal{E}_{\text{rms}} = \mathcal{E}_m/\sqrt{2}$. How does the average value of $\mathcal{E}_m \sin \omega t$, averaged over one-half of a cycle only, compare with $\mathcal{E}_m$ and with $\mathcal{E}_{\text{rms}}$?

11. In an *RCL* circuit such as that of Fig. 39-1, assume $R = 5.0$ Ω, $L = 60$ mH, $\nu = 60$ Hz, and $\mathcal{E}_m = 300$ V. For what values of C would P_{av} for the circuit be (a) a maximum and (b) a minimum? (c) What are these maximum and minimum values and what are the corresponding phase angles and power factors? (d) If $\cos \phi$ in Eq. 39-25 is negative, it would seem that P_{av} in Eq. 39-25 would be negative. What can this mean? Can you find a value of C above that will give a negative value for $\cos \phi$? Discuss.
Answer: (a) 117μF. *(b)* Infinity. *(c)* 9000W, 420W; 0°, 78°; 1.0, 0.22.

SECTION 39-5
12. An *RCL* circuit, such as that of Fig. 39-1, has $R = 20$ Ω, $C = 20$ μF, and $L = 1.0$ H. (a) At what angular frequency ω of the ac generator will the circuit resonate with maximum response? (b) At what angular frequencies will the response be one-half the maximum value? We define "response" to be measured by the rms current in the circuit, as in Fig. 39-7.

13. A resistor-capacitor-inductor circuit R_1, C_1, L_1, connected as in Fig. 39-1, exhibits resonance at the same frequency as a second, separate, combination R_2, C_2, L_2. If the two combinations are connected in series in a single circuit, at what frequency would the combined circuit resonate?

Answer: $\omega = \sqrt{\dfrac{1}{L_1 C_1}} = \sqrt{\dfrac{1}{L_2 C_2}}$, independent of R_1 and R_2.

14. In Fig. 39-7 show that for frequencies higher than the resonant frequency the circuit is predominantly inductive and for frequencies lower than the resonant frequency it is predominantly capacitive. What does this statement mean? How do you interpret it in terms of Fig. 39-5b?

15. Show that the amplitude of the *charge* (not current) oscillations in an *RCL* circuit such as that of Fig. 39-1 is

$$q_m = \frac{\mathcal{E}_m}{\sqrt{(\omega^2 L - 1/C)^2 + (\omega R)^2}}$$

(*a*) For what value of ω will q_m be a maximum? (*b*) Does this result shed any light on the comparison of Figs. 15-20 and 39-7, suggested in Question 23?

Answer: (*a*) $\omega = \sqrt{\dfrac{1}{LC} - \dfrac{R^2}{2L^2}}$. (Assume $R^2 < 2L/C$).

16. Figure 39-11 shows an ac generator connected, through terminals a and b, to a "black box" containing an *RCL* circuit whose elements and arrangements we do not know. An alternating current, given by $i = i_m \sin(\omega t + \phi)$ appears in the lead-in wires. (*a*) What is the power factor? (*b*) Is the circuit in the box capacitive or inductive in nature? (*c*) Does the emf lag the current or lead it? (*d*) What average power P_{av} is delivered to the box by the source of emf if $\varepsilon_m = 750$ V and $i_m = 12$ A? (*e*) Why don't you need to know the angular frequency ω to answer the preceding question? (*f*) If you wanted the circuit to resonate, in the sense of Fig. 39-7, what is the nature of the circuit element that you would connect between terminals a and b? (*g*) At resonance, what values would ϕ and P_{av} have?

$\varepsilon = \varepsilon_m \sin \omega t$

figure 39-11
Problem 16

17. Show that the fractional half-width of the resonance curves of Fig. 39-7 is given, to a close approximation, by

$$\frac{\Delta\omega}{\omega} = \frac{\sqrt{3}R}{\omega L},$$

in which ω is the resonant frequency and $\Delta\omega$ is the width of the resonance peak at $i = \frac{1}{2}i_m$. Note (see Problem 38-22) that this expression may be written as $\sqrt{3}/Q$ which shows clearly that a "high-Q" circuit has a sharp resonance peak, that is, a small $\Delta\omega/\omega$.

SECTION 39-6

18. In Fig. 39-8 what are (*a*) $V_{bc,rms}$ and (*b*) P_{av} for the resistor R in each of the cases shown?

19. In Fig. 39-8, shade the areas, in all three cases, that cancel to form $\overline{V_{bc}}$.

20. For the full-wave rectified waveform of Fig. 39-8*c*, what is the fundamental (lowest) frequency ν_0 of the "ac ripple"? Assume that, for the ac generator, $\nu = \omega/2\pi = 60$ Hz.

21. In Fig. 39-12 in which $R_1 \gg R_2$ and both V_{in} and V_{out} are constant potential differences, show that the attenuation factor $V_{out}/V_{in} \cong R_2/R_1$. Compare with Fig. 39-9 and Eq. 39-38, in which the (ac) attenuation factor is (approximately) X_C/X_L. Discuss analogies and differences.

figure 39-12
Problem 21

22. (*a*) Show that the attenuation factor $V_{out,m}/V_{in,m}$ (see Eq. 39-38 and Fig. 39-9) can be written, for $\omega \gg \omega_0$, as $(\omega_0/\omega)^2$. Here ω is the input angular frequency and ω_0 is the resonant angular frequency $(= \sqrt{1/LC})$ of the *LC* filter combination. (*b*) Show that $\omega \gg \omega_0$ corresponds to $X_L \gg X_C$. Is this reasonable?

23. Show by qualitative argument that the three-stage filter of Fig. 39-13 is more effective than the one-stage filter of Fig. 39-9. Sketch its dc equivalent, that is, replace L by R_1 and C by R_2, with $R_1 \gg R_2$. Derive the dc attenuation factor V_{out}/V_{in}, both of these quantities being dc potentials. *Answer:* $(R_2/R_1)^3$.

Input

figure 39-13
Problem 23

24. Show that, in low current, high-voltage applications (such as power supplies for television picture tubes) we can replace the inductor L of Fig. 39-9 by a "large" resistor R and still achieve substantial reduction of the ac component of V_{in} without *too* much reduction of the dc component.

SECTION 39-7

25. An ac transmission line transfers energy at the rate $P_{av} = 5.0$ MW from a generating plant to a factory. (*a*) What current i_{rms} is present in the line if the transmission voltage V_{rms} is 120 V? (*b*) If $V_{rms} = 80$ kV? (*c*) What is the ratio of the thermal (Joule) energy losses in the line for these two cases? Assume that the power factor $\cos \phi = 1$. *Answer:* (*a*) 42 kA, (*b*) 63A, (*c*) 4.5×10^5.

26. A transformer has 500 primary turns and 10 secondary turns. (*a*) If $V_{1,rms}$ for the primary is 120 V, what is $V_{2,rms}$ for the secondary, assumed on open

circuit? (b) If the secondary now has a resistive load R of 15Ω, what are $i_{1,\text{rms}}$ and $i_{2,\text{rms}}$? Assume an ideal transformer with $\phi = 0$.

27. The output of a full-wave rectifier (see Fig. 39-8c) is fed into an (ideal) transformer with a step-up ratio of 2:1. Sketch roughly the waveform appearing at the secondary, assumed to be on open circuit.

28. In Fig. 39-10 compare these quantities for (a) switch S open, and (b) switch S closed: $\Phi_B(t)$, $\mathcal{E}(t)$, $V_1(t)$, $V_2(t)$, $V_{1,\text{rms}}$, $V_{2,\text{rms}}$, $i_1(t)$, $i_2(t)$, $i_{1,\text{rms}}$, and $i_{2,\text{rms}}$. Assume an ideal transformer, with $\phi = 0$.

29. In Fig. 39-10 show that $i_1(t)$ in the primary circuit remains unchanged if a resistance $R'[=R(N_1/N_2)^2]$ is connected directly across the generator, the transformer and the secondary circuit being removed. That is,

$$i_1(t) = \frac{\mathcal{E}(t)}{R'}.$$

In this sense we see that a transformer not only "transforms" potential differences and currents but also resistances. In the more general case, in which the secondary load in Fig. 39-10 contains capacitive and inductive elements as well as resistive, we say that a transformer transforms *impedances*. See Problem 31 for an example.

30. In Problem 7 of Chapter 32 we asserted (see Fig. 32-3a) that, for dc circuits, the power dissipated in an external resistance R is a maximum when $R = r$, where r is the internal resistance of the (dc) source of emf. In Fig. 39-14 show that, in the same way, the average power dissipation P_{av} in R is a maximum when $R = r$, in which r is the internal resistance of the ac generator. In the text we have tacitly assumed that $r = 0$; here we assume that $r \neq 0$.

figure 39-14
Problem 30

31. *Impedance Matching.* We have seen in Problem 29 that a transformer can serve as a resistance (generally, impedance) transforming device. We saw also, in Problem 30 that (see Fig. 39-14) the transfer of power from an ac generator (internal resistance r) to a resistive load R is a maximum when $R = r$. Suppose that, in Fig. 39-14, $r = 1.0$ kΩ, $R = 10$ Ω, $\omega/2\pi = 60$ Hz, and $\mathcal{E}_{\text{rms}} = 120$ V. Design a transformer, to be interposed between the ac generator and the load, that will insure maximum power transfer to R. Assume an ideal transformer with $\phi = 0$. Such a technique is used, for example, when it is necessary to transfer power efficiently from a (high-impedance) audio amplifier to a (low-impedance) loudspeaker.
Answer: $N_1/N_2 = 10$.

40

maxwell's equations

In classical mechanics and thermodynamics we sought to identify the smallest, most compact set of equations or laws that would define the subject as completely as possible. In mechanics we found this in Newton's three laws of motion (Sections 5-2, 5-4, and 5-5) and in the associated force laws, such as Newton's law of gravitation (Section 16-2). In thermodynamics we found it in the three laws described in Sections 21-2, 22-7, and 25-4.

In this chapter we seek to do the same thing for electromagnetism, proceeding in several steps. First we display, in Table 40-1, a *tentative* set of such equations. After studying this table, we will conclude from arguments of symmetry that these equations are not yet complete and that there may be (and indeed *is*) a missing term in one of them.

40-1
THE BASIC EQUATIONS OF ELECTROMAGNETISM

Table 40-1
Tentative* basic equations of electromagnetism

Symbol	Name	Equation	Text Reference
I	Gauss's law for electricity	$\epsilon_0 \oint \mathbf{E} \cdot d\mathbf{S} = q$	Eq. 28-6
II	Gauss's law for magnetism	$\oint \mathbf{B} \cdot d\mathbf{S} = 0$	Eq. 37-6
III	Faraday's law of induction	$\oint \mathbf{E} \cdot d\mathbf{l} = -\dfrac{d\Phi_B}{dt}$	Eq. 35-9
IV	Ampère's law	$\oint \mathbf{B} \cdot d\mathbf{l} = \mu_0 i$	Eq. 34-1

* "Tentative" suggests, as we shall see below, that Eq. IV is not yet complete and requires an additional term; see Table 40-2.

The missing term will prove to be no trifling correction but will round out the complete description of electromagnetism and, beyond this, will establish optics as an integral part of electromagnetism. In particular it will allow us to prove that the speed of light c in free space is related to purely electric and magnetic quantities by

$$c = \frac{1}{\sqrt{\epsilon_0 \mu_0}}. \qquad (40\text{-}1)$$

It will also lead us to the concept of the electromagnetic spectrum, which lies behind the experimental discovery of radio waves.

We have seen how the principle of symmetry permeates physics and how it has often led to new insights or discoveries. For example, (a) if body A attracts body B with a force **F**, then body B attracts body A with a force $-\mathbf{F}$ (it does), and (b) if there is a negative electron, there may well be a positive electron (there is), etc.

Let us examine Table 40-1 from this point of view. First we say that when we are dealing with symmetry considerations alone (that is, not making quantitative calculations) we can ignore ϵ_0 and μ_0. These constants result from our choice of unit systems and play no role in arguments of symmetry. There are in fact unit systems in which $\epsilon_0 = \mu_0 = 1$.

With this in mind we see that the left sides of the equations in Table 40-1 are completely symmetrical, in pairs. Equations I and II are surface integrals of **E** and **B,** respectively, over closed surfaces. Equations III and IV are line integrals of **E** and **B,** respectively, around closed loops.

On the right side of these equations, things are *not* symmetrical and, in fact, there are two kinds of asymmetries, which we shall discuss separately.

1. The first asymmetry, which is not really the concern of this chapter, deals with the apparent fact that although there are isolated centers of charge (electrons and protons, say) there seem not to be isolated centers of magnetism (magnetic monopoles; see Section 37-1). Thus we account for the "q" on the right of Eq. I and for the "0" on the right of Eq. II. In the same way the term $\mu_0 i \,(= \mu_0 \, dq/dt)$ appears on the right of Eq. IV but no similar term (a current of magnetic monopoles) appears on the right of Eq. III. The resolution of *this* asymmetry depends on the not as yet discovered magnetic monopole. Considerations of symmetry have motivated physicists to search for the magnetic monopole in great earnest and in many ways. It is as though nature were hinting and guiding physicists in their explorations.

2. The second asymmetry, with which this chapter deals, sticks out like a sore thumb. On the right side of Eq. III (Faraday's law of induction; see Eq. 35-9) we find the term $-d\Phi_B/dt$ and we interpret this law loosely by saying:

If you change a magnetic field ($d\Phi_B/dt$), you produce an electric field ($\oint \mathbf{E} \cdot d\mathbf{l}$).

We learned this in Section 35-1 where we showed that if you shove a bar magnet through a closed conducting loop, you do indeed induce an electric field, and thus a current, in that loop.

From the principle of symmetry we are entitled to suspect that the symmetrical relation holds, that is:

If you change an electric field ($d\Phi_E/dt$), you produce a magnetic field ($\oint \mathbf{B} \cdot d\mathbf{l}$).

This symmetry principle indeed meets the test of experiment and we discuss it fully in the next section. This supposition supplies us with the important "missing" term in Eq. IV in Table 40-1.

40-2
INDUCED MAGNETIC FIELDS

Here we discuss in detail the evidence for the supposition of the previous section, namely:

"A changing electric field induces a magnetic field."

Although we will be guided by considerations of symmetry alone, we will also point to direct experimental verification.

Figure 40-1a shows a uniform electric field **E** filling a cylindrical region of space. It might be produced by a circular parallel-plate capacitor, as suggested in Fig. 40-1b. We assume that E is increasing at a steady rate dE/dt, which means that charge must be supplied to the capacitor plates at a steady rate; to supply this charge requires a steady current i into the positive plate and an equal steady current i out of the negative plate.

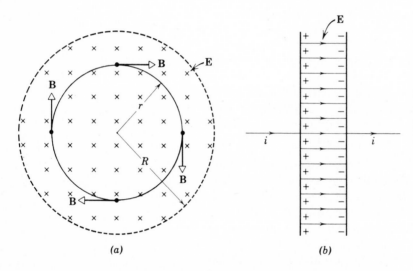

(a)   (b)

figure 40-1
(a) Showing the induced magnetic fields **B** at four points, produced by a changing electric field **E**. The electric field is increasing in magnitude. Compare Fig. 35-10.
(b) Such a changing electric field may be produced by charging a parallel-plate capacitor as shown.

Experiment shows (see Problem 4) that *a magnetic field is set up by this changing electric field.* Figure 40-1a shows **B** for four arbitrary points. Figure 40-1 suggests a beautiful example of the symmetry of nature. A changing *magnetic* field induces an *electric* field (Faraday's law); now we see that a changing *electric* field induces a *magnetic* field.

To describe this new effect quantitatively, we are guided by analogy with Faraday's law of induction,

$$\oint \mathbf{E} \cdot d\mathbf{l} = -\frac{d\Phi_B}{dt},$$ (40-2)

which asserts that an electric field (left term) is produced by a changing magnetic field (right term). For the symmetrical counterpart we might write*

$$\oint \mathbf{B} \cdot d\mathbf{l} = \mu_0 \epsilon_0 \frac{d\Phi_E}{dt}.$$ (40-3)

Equation 40-3 asserts that a magnetic field (left term) can be produced

---

* Our system of units requires that we insert the constants ϵ_0 and μ_0 in Eq. 40-3. In some unit systems they would not appear.

by a changing electric field (right term). Compare carefully Fig. 35-10, which illustrates the production of an electric field by a changing magnetic field, with Fig. 40-1a. In each case the appropriate flux Φ_B or Φ_E is *increasing*. However, experiment shows that the lines of **E** in Fig. 35-10 are *counterclockwise*, whereas those of **B** in Fig. 40-1a are *clockwise*. This difference requires that the minus sign of Eq. 40-2 be omitted from Eq. 40-3.

In Section 34-1 we saw that a magnetic field can also be set up by a current in a wire. We described this quantitatively by Ampère's law:

$$\oint \mathbf{B} \cdot d\mathbf{l} = \mu_0 i,$$

in which i is the conduction current passing through the loop around which the line integral is taken. Thus there are at least two ways of setting up a magnetic field: (a) by a changing electric field and (b) by a current. In general, both possibilities must be allowed for, or*

$$\oint \mathbf{B} \cdot d\mathbf{l} = \mu_0 \epsilon_0 \frac{d\Phi_E}{dt} + \mu_0 i. \tag{40-4}$$

Maxwell is responsible for this important generalization of Ampère's law. It is a central and vital contribution, as we have pointed out earlier.

In Chapter 34 we assumed that no changing electric fields were present so that the term $d\Phi_E/dt$ in Eq. 40-4 was zero. In the discussion just given we assumed that there were no conduction currents in the space containing the electric field. Thus the term i in Eq. 40-4 is zero. We see now that each of these situations is a special case.

A parallel-plate capacitor with circular plates is being charged as in Fig. 40-1. (a) Derive an expression for the induced magnetic field at various radii r. Consider both $r < R$ and $r > R$.

From Eq. 40-3,

$$\oint \mathbf{B} \cdot d\mathbf{l} = \mu_0 \epsilon_0 \frac{d\Phi_E}{dt},$$

we can write, for $r < R$,

$$(B)(2\pi r) = \mu_0 \epsilon_0 \frac{d}{dt} [(E)(\pi r^2)] = \mu_0 \epsilon_0 \pi r^2 \frac{dE}{dt}.$$

Solving for B yields $B = \frac{1}{2} \mu_0 \epsilon_0 r \frac{dE}{dt}$ $(r < R)$.

For $r > R$, Eq. 40-3 yields

$$(B)(2\pi r) = \mu_0 \epsilon_0 \frac{d}{dt} [(E)(\pi R^2)] = \mu_0 \epsilon_0 \pi R^2 \left(\frac{dE}{dt}\right),$$

or $B = \frac{\mu_0 \epsilon_0 R^2}{2r} \frac{dE}{dt}$ $(r > R)$.

EXAMPLE 1

---

* Actually, there is a third way of setting up a magnetic field, by the use of magnetized bodies. In Section 37-8 we saw that this could be accounted for by inserting a *magnetizing current* term i_M on the right side of Ampère's law. This law would then read, in its full generality,

$$\oint \mathbf{B} \cdot d\mathbf{l} = \mu_0 \epsilon_0 \frac{d\Phi_E}{dt} + \mu_0 i + \mu_0 i_M.$$

In all that follows we assume that no magnetic materials are present so that $i_M = 0$.

(b) Find B at $r = R$ for $dE/dt = 10^{12}$ V/m · s and for $R = 5.0$ cm.
At $r = R$ the two equations for B reduce to the same expression, or

$$B = \tfrac{1}{2}\mu_0\epsilon_0 R \frac{dE}{dt}$$

$$= (\tfrac{1}{2})(4\pi \times 10^{-7}\ \text{T} \cdot \text{m/A})(8.9 \times 10^{-12}\ \text{C}^2/\text{N} \cdot \text{m}^2)(5.0 \times 10^{-2}\ \text{m})(10^{12}\ \text{V/m} \cdot \text{s})$$

$$= 2.8 \times 10^{-7}\ \text{T} = 0.0028\ \text{gauss}.$$

This shows that the induced magnetic fields in this example are so small that they can scarcely be measured with simple apparatus, in sharp contrast to induced *electric* fields (Faraday's law), which can be demonstrated easily. This experimental difference is in part due to the fact that induced emfs can easily be multiplied by using a coil of many turns. No technique of comparable simplicity exists for magnetic fields. In experiments involving oscillations at very high frequencies dE/dt above can be very large, resulting in significantly larger values of the induced magnetic field.

40-3 DISPLACEMENT CURRENT

Equation 40-4 shows that the term $\epsilon_0\ d\Phi_E/dt$ has the dimensions of a current. Even though no motion of charge is involved, there are advantages in giving this term the name *displacement* current.* Thus we can say that a magnetic field can be set up either by a conduction current i or by a displacement current i_d $(=\epsilon_0 d\Phi_E/dt)$, and we can rewrite Eq. 40-4 as†

$$\oint \mathbf{B} \cdot d\mathbf{l} = \mu_0(i_d + i). \tag{40-5}$$

The concept of displacement current permits us to retain the notion that *current is continuous,* a principle established for steady conduction currents in Section 31-1. In Fig. 40-1b, for example, a current i enters the positive plate and leaves the negative plate. The *conduction* current is *not* continuous across the capacitor gap because no charge is transported across this gap. However, the displacement current i_d in the gap will prove to be exactly i, thus retaining the concept of the continuity of current.

To calculate the displacement current, recall (see Eq. 30-2) that E in the gap is given by

$$E = \frac{q}{\epsilon_0 A}.$$

Differentiating gives $\quad \dfrac{dE}{dt} = \dfrac{1}{\epsilon_0 A}\dfrac{dq}{dt} = \dfrac{1}{\epsilon_0 A}\,i.$

The displacement current i_d is by definition

$$i_d = \epsilon_0 \frac{d\Phi_E}{dt} = \epsilon_0 \frac{d(EA)}{dt} = \epsilon_0 A \frac{dE}{dt}.$$

Eliminating dE/dt between these two equations leads to

---

* The word "displacement" was introduced for historical reasons that need not concern us here.
† We may write this more generally, taking the presence of magnetic materials into account, as

$$\oint \mathbf{B} \cdot d\mathbf{l} = \mu_0(i_d + i + i_M).$$

See the footnote on p. 883.

$$i_d = (\epsilon_0 A)\left(\frac{1}{\epsilon_0 A}\, i\right) = i,$$

which shows that the displacement current in the gap is identical with the conduction current in the lead wires.

EXAMPLE 2

What is the displacement current for the situation of Example 1? From the definition of displacement current,

$$i_d = \epsilon_0 \frac{d\Phi_E}{dt} = \epsilon_0 \frac{d}{dt}\,[(E)(\pi R^2)] = \epsilon_0 \pi R^2 \frac{dE}{dt}$$

$$= (8.9 \times 10^{-12}\ \mathrm{C^2/N \cdot m^2})(\pi)(5.0 \times 10^{-2}\ \mathrm{m})^2(10^{12}\ \mathrm{V/m \cdot s})$$

$$= 0.070\ \mathrm{A}.$$

Even though this displacement current is reasonably large, it produces only a small magnetic field (see Example 1) because it is spread out over a large area. In contrast, the *conduction* current i in the lead wires (also = 0.070 A, because $i = i_d$) can produce magnetic effects close to the (thin) wires that are easily detectable by a compass needle. The discussion of Section 34-2 shows that the magnetic effect is greatest at the surface of the wire. The capacitor of Fig. 40-1 behaves like a "conductor" of radius 5.0 cm, carrying a (displacement) current of 0.070 A. Its largest magnetic effect is at the capacitor edge, that is, at $r = 5.0$ cm, and is given by Eq. 34-4 ($B = \mu_0 i/2\pi r$).

40-4
MAXWELL'S EQUATIONS

Equation 40-4 completes our presentation of the basic equations of electromagnetism, called *Maxwell's equations.** They are summarized in Table 40-2, which rounds out the "tentative" Table 40-1 by supplying the "missing" term in Eq. IV of that table. All equations of physics that serve, as these do, to correlate experiments in a vast area and to predict new results have a certain beauty about them and can be appreciated, by those who understand them, on an aesthetic level. This is true for Newton's laws of motion, for the laws of thermodynamics, for the theory of relativity, and for the theories of quantum physics. As for Maxwell's equations, the physicist Ludwig Boltzmann (quoting a line from Goethe) wrote, "Was it a god who wrote these lines. . . ." In more recent times J. R. Pierce, in a book chapter entitled 'Maxwell's Wonderful Equations' says, "To anyone who is motivated by anything beyond the most narrowly practical, it is worth while to understand Maxwell's equations simply for the good of his soul." The scope of these equations is remarkable, including as it does the fundamental operating principles of all large-scale electromagnetic devices such as motors, synchrotrons, television, and microwave radar.†

We suggested in Section 40-1 that Maxwell's equations (as they appear in

* In the preceding sections we have leaned heavily on symmetry arguments in developing Maxwell's equations as displayed in Table 40-2. Actually Maxwell did not base his work on such arguments. The British physicist Oliver Heaviside (1850–1925) seems to have been the first to point out the symmetry between **E** and **B** that these equations display. See "Maxwell, Displacement Current, and Symmetry" by Alfred M. Bork, *American Journal of Physics*, November, 1963, for an interesting historical account.
† For a recent spectacular example of an application of Maxwell's equations, see "Electromagnetic Flight" by Henry H. Kolm and Richard D. Thornton, *Scientific American*, October 1973. The article states: "The future of high-speed ground transportation may well lie not with wheeled trains but with vehicles that 'fly' a foot or so above a guideway, lifted and propelled by electromagnetic forces."

Table 40-2
The basic equations of electromagnetism (Maxwell's equations) *

Number	Name	Equation	Describes	Crucial Experiment	General Text Reference
I	Gauss's law for electricity	$\epsilon_0 \oint \mathbf{E} \cdot d\mathbf{S} = q$	Charge and the electric field	1. Like charges repel and unlike charges attract, as the inverse square of their separation.	Chapter 28
				1'. A charge on an insulated conductor moves to its outer surface.	
II	Gauss's law for magnetism	$\oint \mathbf{B} \cdot d\mathbf{S} = 0$	The magnetic field	2. It has thus far not been possible to verify the existence of a magnetic monopole.	Section 37-2
III	Faraday's law of induction	$\oint \mathbf{E} \cdot d\mathbf{l} = -\dfrac{d\Phi_B}{dt}$	The electrical effect of a changing magnetic field	4. A bar magnet, thrust through a closed loop of wire, will set up a current in the loop.	Chapter 35
IV	Ampère's law (as extended by Maxwell)	$\oint \mathbf{B} \cdot d\mathbf{l}$ $= \mu_0 \left(\epsilon_0 \dfrac{d\Phi_E}{dt} + i \right)$	The magnetic effect of a changing electric field or of a current	3. The speed of light can be calculated from purely electromagnetic measurements.	Section 41-8
				3'. A current in a wire sets up a magnetic field near the wire.	Chapter 34

* Written on the assumption that no dielectric or magnetic material is present.

Table 40-2) bear the same relationship to electromagnetism that Newton's laws of motion do to mechanics. There is, however, an important difference. Einstein presented his special theory of relativity (see Supplementary Topic V) in 1905, roughly 200 years after Newton's laws appeared and about 40 years after Maxwell's equations. As it turns out, Newton's laws had to be drastically modified in cases in which the relative speeds approached that of light. However, *no changes whatever were required in Maxwell's equations.* In the language of Supplementary Topic V we say that "Maxwell's equations are invariant under a Lorentz transformation but Newton's laws of motion are not."

Now that we have identified Maxwell's equations as the basic equations of electromagnetism, we want to test them in several situations. In this section we address ourselves to the electromagnetic cavity oscillations of Section 38-5 (see Fig. 38-6). In Chapter 41 we will show that Maxwell's equations predict the existence of electromagnetic waves, that visible light is such a wave, and that the speed c of electromagnetic radiation (light) is indeed given by Eq. 40-1.

A completely formal treatment of the cavity oscillations of Fig. 38-6, which is beyond our scope here, would start from Maxwell's equations and would end with mathematical expressions for the variation of **B** and **E** with time and with position in the cavity for all modes of oscillation of the cavity. We confine ourselves to the fundamental mode only, shown in Fig. 38-6, for which we *postulated* the variations of **B** and **E** given in that figure; we will show that these postulated fields are completely consistent with Maxwell's equations.

Figure 40-2 presents two views of the cavity of Fig. 38-6d, in which both electric and magnetic fields are present. Study of Fig. 38-6 reveals that **B** is *decreasing* in magnitude and **E** is *increasing* at this phase of the cycle of oscillation. Let us apply Faraday's law,

$$\oint \mathbf{E} \cdot d\mathbf{l} = -\frac{d\Phi_B}{dt},$$

40-5
MAXWELL'S EQUATIONS AND CAVITY OSCILLATIONS

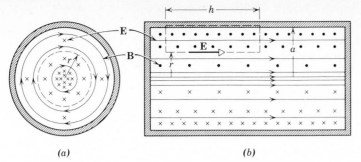

(a)   (b)

figure 40-2
Two cross sections of an electromagnetic resonant cavity at a phase of oscillation corresponding to Fig. 38-6d. (a) The dashed circle is a path suitable for applying Ampère's law. (b) The dashed rectangle is a path suitable for applying Faraday's law.

to the rectangle of dimensions h and $a - r$. There is a definite flux Φ_B through the rectangular path in question, and this flux is decreasing with time because **B** is decreasing. The line integral above is

$$\oint \mathbf{E} \cdot d\mathbf{l} = hE(r),$$

in which $E(r)$ is the value of E at a radius r from the axis of the cavity.

Note that **E** equals zero for the upper leg of the integration path (which lies in the cavity wall) and that **E** and $d\mathbf{l}$ are at right angles on the two side legs. Combining these equations yields

$$E(r) = -\frac{1}{h}\frac{d\Phi_B}{dt}. \tag{40-6}$$

Equation 40-6 shows that $E(r)$ depends on the rate at which Φ_B through the path shown is changing with time and that it has its maximum value when $d\Phi_B/dt$ is a maximum. This occurs when **B** is zero, that is, when **B** is changing its direction; recall that a sine or cosine is changing most rapidly, that is, it has the steepest tangent, at the instant it crosses the axis between positive and negative values. Thus the electric field pattern in the cavity will have its *maximum* value when the magnetic field is *zero* everywhere, consistent with Figs. 38-6a and 38-6e and with the concept of the interchange of energy between electric and magnetic fields. You can show, by applying Lenz's law, that the electric field in Fig. 40-2b indeed points to the *right*, as shown, if the magnetic field is *decreasing.*

Figure 40-2a shows an end view of the cavity; the electric lines of force are entering the page at right angles to the page and the magnetic lines form clockwise circles. Let us apply Ampère's law in the form

$$\oint \mathbf{B} \cdot d\mathbf{l} = \mu_0\epsilon_0 \frac{d\Phi_E}{dt} + \mu_0 i, \tag{40-4}$$

to the circular path of radius r shown in the figure. No charge is transported through the ring so that the conduction current i in Eq. 40-4 is zero. The line integral on the left is $(B)(2\pi r)$ so that the equation reduces to

$$B(r) = \frac{\mu_0\epsilon_0}{2\pi r}\frac{d\Phi_E}{dt}. \tag{40-7}$$

Equation 40-7 shows that the magnetic field $B(r)$ is proportional to the rate at which the electric flux Φ_E through the ring is changing with time. The field $B(r)$ has its maximum value when $d\Phi_E/dt$ is at its maximum; this occurs when $\mathbf{E} = 0$, that is, when **E** is reversing its direction. Thus we see that **B** has its *maximum* value when **E** is *zero* for all points in the cavity. This is consistent with Figs. 38-6c and 38-6g and with the concept of the interchange of energy between electric and magnetic forms. A comparison with Fig. 40-1a, which like Fig. 40-2a, corresponds to an increasing electric field, shows that the lines of **B** are indeed clockwise, as viewed along the direction of the electric field.

Comparison of Eqs. 40-6 and 40-7 suggests the complete interdependence of **B** and **E** in the cavity. As the magnetic field changes with time, it induces the electric field in a way described by Faraday's law. The electric field, which also changes with time, induces the magnetic field in a way described by Maxwell's extension of Ampère's law. The oscillations, once established, sustain each other and would continue indefinitely were it not for losses due to thermal energy in the cavity walls or leakage of energy from openings that might be present in the walls. In Chapter 41 we show that a similar interplay of **B** and **E** occurs not only in standing electromagnetic waves in cavities but also in traveling electromagnetic waves, such as radio waves or visible light.

EXAMPLE 3

In Fig. 40-2 analyze the currents (both conduction and displacement) that occur in the cavity (both in its conducting walls and within its volume). Show the relationship between these currents and the electric and magnetic fields and also show that, considering both conduction and displacement currents together, it is reasonable to conclude that current is continuous around closed loops.

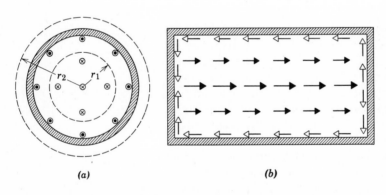

(a)   (b)

figure 40-3
The cavity of Fig. 40-2, showing (a) the conduction current coming up the walls and displacement current going down the cavity volume and (b) the displacement current (black arrowheads) in the volume of the cavity and the conduction currents (white arrowheads) at the walls. The arrows in each case represent current densities. Note the continuity of current (conduction + displacement), that is, it is possible to form closed current loops.

Figure 40-3 shows two views of the cavity, at an instant corresponding to that of Fig. 40-2. For simplicity, we do not show the **E** and **B** fields; the arrows represent currents. Because E is increasing in Figs. 40-2 and 40-3, the positive charge on the left end cap must be increasing. Thus there must be conduction currents in the walls pointing from right to left in Fig. 40-3b. These currents are also shown by the dots (representing the tips of arrows) near the cavity walls in Fig. 40-3a.

Bearing in mind that $\epsilon_0 \, d\Phi_E/dt$ is a displacement current, we can write Eq. 40-7 as

$$B(r) = \frac{\mu_0}{2\pi r}\left(\epsilon_0 \frac{d\Phi_E}{dt}\right) = \frac{\mu_0}{2\pi r} \, i_d.$$

This equation stresses that **B** in the cavity is associated with a displacement current; compare Eq. 34-4. Applying the right-hand rule in Fig. 40-2a shows that the displacement current i_d must be directed into the plane of the figure if it is to be associated with the clockwise lines of **B** that are present.

The displacement current is represented in Fig. 40-3b by the arrows that point to the right and in Fig. 40-3a by the crosses that represent arrows entering the page. Study of Fig. 40-3 shows that the current is continuous, being directed up the walls as a conduction current and then back down through the volume of the cavity as a displacement current. If we apply Ampère's law as extended by Maxwell,

$$\oint \mathbf{B} \cdot d\mathbf{l} = \mu_0(i_d + i), \qquad (40\text{-}8)$$

to the circular path of radius r_1 in Fig. 40-3a, we see that **B** at that path is due entirely to the displacement current, the conduction current *i* *within the path* being zero.

For the path of radius r_2, the *net* current enclosed is zero because the conduction current in the walls is exactly equal and opposite to the displacement current in the cavity volume. Since *i* equals i_d in magnitude, but is oppositely directed, it follows from Eq. 40-8 that *B* must be zero for all points outside the cavity, in agreement with observation.

questions

1. In your own words explain why Eq. III of Table 40-1 can be interpreted by saying, "A changing magnetic field can generate an electric field."

2. If (as is true) there are unit systems in which ϵ_0 and μ_0 do not appear, how can Eq. 40-1 be true?

3. If a uniform flux Φ_E through a plane circular ring decreases with time, is the induced magnetic field (as viewed along the direction of **E**) clockwise or counterclockwise? Does it make any difference (a) whether the electric field is uniform or (b) whether the plane of the ring is perpendicular to the electric field?

4. Compare Tables 40-1 and 40-2. Is it enough to rely on the principle of symmetry alone or do we really need experimental verification for the "missing" term in Eq. IV?

5. Why is it so easy to show that "a changing magnetic field produces an electric field" but so hard to show in a simple way that "a changing electric field produces a magnetic field"?

6. In Fig. 40-1a consider a circle with $r > R$. How can a magnetic field be induced around this circle, as Example 1 shows? After all, there is no electric field at the location of this circle and $dE/dt = 0$ here.

7. In Fig. 40-1a, **E** is into the figure and is increasing in magnitude. Find the direction of **B** if (a) **E** is into the figure and decreasing, (b) **E** is out of the figure and increasing, (c) **E** is out of the figure and decreasing, and (d) **E** remains constant.

8. In Fig. 38-1c a displacement current is needed to maintain continuity of current in the capacitor. How can one exist, considering that there is no charge on the capacitor?

9. At what parts of the cycle will (a) the conduction current and (b) the displacement current in the cavity of Fig. 38-6 be zero?

10. In Figs. 40-1a and 40-1b what is the direction of the displacement current i_d? In this same figure, can you find a rule relating the directions of **B** and **E**?

11. What advantages are there in calling the term $\epsilon_0 \, d\Phi_E/dt$ in Eq. IV, Table 40-2 a displacement *current?*

12. Why are the magnetic effects of conduction currents in wires so easy to detect but the magnetic effects of displacement currents in capacitors so hard to detect?

13. Discuss the time variation during one complete cycle of the charges that appear at various points on the inner walls of the oscillating electromagnetic cavity of Fig. 38-6.

14. Would you expect that the arrangement of the magnetic and electric fields in Fig. 40-2 is the only possible arrangement? If there are other arrangements, would you expect them to have higher or lower frequencies than those shown in Fig. 40-2?

15. In connection with Fig. 40-3, in what sense can the end caps be considered as capacitor plates? In what sense can the cylindrical walls be considered as an inductor? (*Note:* Figure 40-3 is clearly a case of distributed elements but there must be a smooth transition between distributed and lumped elements.)

16. (a) In Fig. 40-2a is it possible to apply Faraday's law usefully to the dashed circle? (b) Is it possible to apply Ampère's law usefully to the dashed rectangle of Fig. 40-2b? Discuss.

problems

SECTION 40-2

1. (a) Convince yourself that, for the conditions of Fig. 40-1, the induced magnetic field is greatest at the edge of the capacitor. (b) Make a rough, non-numerical, plot of $B(r)$ from $r = 0$ to $r \gg R$.

2. You are given a 1.0-μF parallel-plate capacitor. How would you establish an (instantaneous) displacement current of 1.0 A in the space between its plates?

3. Prove that the displacement current in a parallel-plate capacitor can be written as

$$i_d = C \frac{dV}{dt}.$$

4. In 1929 M. R. Van Cauwenerghe succeeded in measuring directly, for the first time, the displacement current i_d between the plates of a parallel-plate capacitor to which an alternating potential difference was applied, as suggested by Fig. 40-1. He used circular plates whose effective radius was 40 cm and whose capacitance was 1.0×10^{-10} F. The applied potential difference had a maximum value V_m of 174 kV at a frequency of 50 Hz. (a) What maximum displacement current is present between the plates? (See Problem 3.) (b) Why is the applied potential difference chosen to be as high as it is? [The delicacy of these measurements is such that they were only performed in a direct manner more than 60 years after Maxwell enunciated the concept of displacement current! The reference is *Journal de Physique*, No. 8, p. 303, (1929)].

SECTION 40-3

5. Figure 40-4 shows the plates P_1 and P_2 of a circular parallel-plate capacitor of radius R. They are connected as shown to long straight wires in which a constant conduction current i exists. A_1, A_2, and A_3 are hypothetical circles of radius r, two of them outside the capacitor and one between the plates.

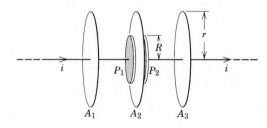

figure 40-4
Problem 5

Write expressions for the magnetic field B at the circumference of each of these circles.

Answer: $\frac{\mu_0 i}{2\pi r}$, in all cases.

6. In Example 1 show that the *displacement current density* j_d is given, for $r < R$, by

$$j_d = \epsilon_0 \frac{dE}{dt}.$$

7. A parallel-plate capacitor has square plates 1.00 m on a side as in Fig. 40-5. There is a charging current of 2.00 A flowing into (and out of) the capacitor. (a) What is the displacement current through the region between the plates?

Edge view   Top view

figure 40-5
Problem 7

(b) What is dE/dt in this region? (c) What is the displacement current through the square dashed path between the plates? (d) What is $\int \mathbf{B} \cdot d\mathbf{l}$ around this square dashed path?

Answer: (a) 2.00 A. (b) 2.3×10^{11} V/m·s. (c) 0.50 A. (d) 6.3×10^{-7} T·m.

8. The capacitor in Fig. 40-6 consisting of two circular plates with area $A = 0.10$ m² is connected to a source of potential $\varepsilon = \varepsilon_m \sin \omega t$, where $\varepsilon_m = 200$ V and $\omega = 100$ rad/s. The maximum value of the displacement current is $i_d = 8.9 \times 10^{-6}$ A. Neglect fringing of the electric field at the edges of the plates. (a) What is the maximum value of the current i? (b) What is the maximum value of $d\Phi_E/dt$, where Φ_E is the electric flux through the region between the plates? (c) What is the separation d between the plates? (d) Find the maximum value of the magnitude of $\mathbf{B}$ between the plates at a distance $R = 0.10$ m from the center.

figure 40-6
Problem 8

9. In Example 1 how does the displacement current through a hypothetical concentric circular loop of radius r vary with r? Consider both (a) $r < R$ and (b) $r > R$. Answer: (a) $\pi \epsilon_0 r^2 (dE/dt)$. (b) $\pi \epsilon_0 R^2 (dE/dt)$.

10. A parallel-plate capacitor with circular plates 20.0 cm in diameter is being charged as in Fig. 40-1a. The displacement current density throughout the region is uniform, into the paper in the diagram, and has a value of 20.0 A/m². (a) Calculate the magnetic field B at a distance $R = 5.0$ cm from the axis of symmetry of the region. (b) Calculate dE/dt in this region.

11. Identify the Maxwell equation that is equivalent to or includes: (a) Electric lines of force end only on electric charges. (b) The displacement current. (c) Under static conditions, there cannot be any charge inside a conductor. (d) A changing electric field must be accompanied by a magnetic field. (e) The net magnetic flux through a closed surface is always zero. (f) A changing magnetic field must be accompanied by an electric field. (g) Magnetic flux lines have no ends. (h) The net electric flux through a closed surface is proportional to the total charge inside. (i) An electric charge is always accompanied by an electric field. (j) There are no magnetic monopoles. (k) An electric current is always accompanied by a magnetic field. (l) Coulomb's law. (m) The electrostatic field is conservative.

Answer: See Table 40-2. (a) I. (b) IV. (c) I. (d) IV. (e) II. (f) III. (g) II. (h) I. (i) I. (j) II. (k) IV. (l) I. (m) III.

12. Collect and tabulate expressions for the following four quantities, considering both $r < R$ and $r > R$. Copy down the derivations side by side and study them as interesting applications of Maxwell's equations to problems having cylindrical symmetry. (a) $B(r)$ for a current i in a long wire of radius R (see Section 34-2). (b) $E(r)$ for a long uniform cylinder of charge of radius R (see Section 28-8; also Problem 27, Chapter 28). (c) $B(r)$ for a parallel-plate capacitor, with circular plates of radius R, in which E is changing at a constant rate (see Section 40-2). (d) $E(r)$ for a cylindrical region of radius R in which a uniform magnetic field B is changing at a constant rate (see Section 35-5).

13. A long cylindrical conducting rod with radius a is centered on the x-axis as shown in Fig. 40-7. A narrow saw cut is made in the rod at $x = b$. A conduction current i, increasing with time and given by $i = \alpha t$, flows toward the right in the rod; α is a (positive) proportionality constant. At $t = 0$ there is no charge on the cut faces near $x = b$. (a) Find the magnitude of the charge on these faces, as a function of time. (b) Use Eq. I in Table 40-2 to find E in the gap as a function of time. (c) Sketch the lines of $\mathbf{B}$ for $r < a$, where r is the distance from the x-axis. (d) Use Eq. IV in Table 40-2 to find $B(r)$ in the gap for $r < a$. (e) Compare the above answer with $B(r)$ in the rod for $r < a$.

Answer: (a) $\frac{1}{2}\alpha t^2$. (b) $\alpha t^2/2\pi\epsilon_0 a^2$. (c) As you sight along the axis of the rod from left to right in Fig. 40-7 the lines of $\mathbf{B}$ form clockwise circles, both in the rod and in the gap. (d) $\mu_0 \alpha t r/2\pi a^2$. (e) Same as (d).

figure 40-7
Problem 13

14. Using the definitions of flux Φ, volume charge density ρ, and current density $\mathbf{j}$, write the four Maxwell equations of Table 40-2 in such a manner that all the fluxes, currents, and charges appear as volume or surface integrals.

SECTION 40-4

15. Assume that the existence of magnetic monopoles is firmly established by experiment. (a) How would you modify the equations of Table 40-2? Let q_m be the expression for the strength of the presumed magnetic monopole, analogous to the basic electric charge e. (b) What SI units would q_m have?

    Answer: (a) $\oint \mathbf{B} \cdot d\mathbf{S} = q_m$, $\oint \mathbf{E} \cdot d\mathbf{l} = -\dfrac{d\Phi_B}{dt} - \dfrac{dqm}{dt}$; other equations are unchanged. (b) $V \cdot s$.

16. *A self-consistency property of two of the Maxwell equations* (numbers III and IV in Table 40-2). Two adjacent closed paths *abcda* and *efcbe* share the common edge *bc* as shown in Fig. 40-8. (a) We may apply $\oint \mathbf{E} \cdot d\mathbf{l} = -d\Phi_B/dt$ to each of these closed paths separately. Show that from this alone, $\oint \mathbf{E} \cdot d\mathbf{l} = -d\Phi_B/dt$ is *automatically* satisfied for the composite closed path *abefcda*. (b) Repeat using $\dfrac{1}{\mu_0} \oint \mathbf{B} \cdot d\mathbf{l} = i + \epsilon_0 \dfrac{d\Phi_E}{dt}$.

figure 40-8
Problem 16

17. *A self-consistency property of two of the Maxwell equations* (numbers I and II in Table 40-2). Two adjacent parallelepipeds share a common face as shown in Fig. 40-9. (a) We may apply $\epsilon_0 \oint \mathbf{E} \cdot d\mathbf{S} = q$ to each of the two closed surfaces separately. Show that, from this alone, it follows that $\epsilon_0 \oint \mathbf{E} \cdot d\mathbf{S} = q$ is *automatically* satisfied for the composite closed surface. (b) Repeat using $\oint \mathbf{B} \cdot d\mathbf{S} = 0$.

figure 40-9
Problem 17

SECTION 40-5

18. A cylindrical electromagnetic cavity 5.0 cm in diameter and 7.0 cm long is oscillating in the mode shown in Fig. 38-6. (a) Assume that, for points on the axis of the cavity, $E_m = 10^4$ V/m. For such axial points what is the maximum rate $(dE/dt)_m$ at which E changes? (b) Assume that the average value of $(dE/dt)_m$, for all points over a cross section of a cavity, is about one-half the value found above for axial points. On this assumption, what is the maximum value of B at the cylindrical surface of the cavity?

19. in microscopic terms the principle of continuity of current may be expressed as

$$\oint (\mathbf{j} + \mathbf{j}_d) \cdot d\mathbf{S} = 0,$$

in which $\mathbf{j}$ is the conduction current density and $\mathbf{j}_d$ is the displacement current density. The integral is to be taken over any closed surface; the equation essentially says that whatever current flows into the enclosed volume must also flow out. (a) Apply this equation to the surface shown by the dashed lines in Fig. 40-10 shortly after switch S is closed. (b) Apply it to various surfaces that may be drawn in the cavity of Fig. 40-3, including some that cut the cavity walls.

figure 40-10
Problem 19

41
electromagnetic
waves

In Chapter 40 we stated without proof that Maxwell's equations predict the existence of electromagnetic waves whose speed in a vacuum is given by Eq. 40-1, or*

$$c = \frac{1}{\sqrt{\mu_0 \epsilon_0}}.$$ (40-1)

We have called c the "speed of light" but, because neither λ nor ν (recall that $c = \lambda \nu$) appears in Eq. 40-1, c is the speed of electromagnetic waves in general, no matter what their wavelengths (λ) or their frequencies (ν).

We can classify waves as standing waves or as traveling waves. In Section 40-5 (see Figs. 40-2 and 40-3) we studied *standing* electromagnetic waves in an electromagnetic resonant cavity. In this chapter we study *traveling* electromagnetic waves, either confined within the conducting walls of a transmission line (Sections 41-4, 5, 6) or in free space (Sections 41-3, 7, 8).

Figure 41-1 suggests the range of the electromagnetic spectrum as we now know it.† From Maxwell's equations we conclude that all these waves have the same nature and speed and that they differ only in frequency, and thus in wavelength. The names attached to the various regions of the spectrum are identified with the experimental techniques

41-1
INTRODUCTION

41-2
THE ELECTROMAGNETIC SPECTRUM

---

* See Section 41-8 and Supplementary Topic VI for a proof of this relationship.
† *Spectrum* is a Latin word that means "ghost" or "apparition." It was first used in this way by Isaac Newton in 1671 to describe the wavering, rainbowlike image formed on the wall of a darkened room when he held a prism in the path of a beam of sunlight entering through a small hole in his window shade.

figure 41-1
The electromagnetic spectrum. Note that both scales are logarithmic. The terms μm and nm are *micrometer* and *nanometer,* respectively; both are accented on the first syllable.

for producing and detecting the waves in question. For the AM and the FM-TV bands the frequency ranges are also a matter of legal definition and are sharply defined.

There are no gaps in the spectrum. For example, we can produce electromagnetic waves with $\nu \cong 3 \times 10^{11}$ Hz, either by microwave techniques (microwave oscillators) or by infrared techniques (heated sources). Also, there are no realized upper or lower limits to the frequency or wavelength scales. As an example at one end of the spectrum, electromagnetic waves with $\nu \cong 10^{-2}$ Hz, which corresponds to a period of about 100 s and to a wavelength of about 5000 earth radii, have been detected at the earth's surface.*

It is hard to realize the extent to which we are bathed in electromagnetic waves. The sun is our predominant source, in the sense that its radiations define the environment to which we as a species have adapted.

Let us also consider earth-originated sources of electromagnetic radiation. We are criss-crossed by radio and TV signals. Microwaves from radar systems, from telephone relay systems, and so forth, may reach us. There are electromagnetic waves from light bulbs, from heated engine blocks in automobiles, from X-ray machines, from fireflies, from lightning flashes, from γ-radiation from the radioactive materials in the earth, and so on.

Many electromagnetic waves reach us from extraterrestrial sources and in fact essentially all that we know about the universe comes to us in this way. Electromagnetic waves from outside the earth in the visual range have, of course, been observed since the dawn of the human race. Three technological developments have greatly extended our horizons as far as the study of electromagnetic waves from space is concerned. They are:

The Telescope (see Section 44-6)

This instrument was first used for astronomical observations by Galileo in 1610. With it he (*a*) discovered the mountains and craters on the moon, which had previously been assumed to be a perfectly spherical body; (*b*) discovered that the Milky Way was composed of multitudes of individual stars; (*c*) discovered the four innermost satellites of Jupiter, revealing a model of the Copernican solar system; (*d*) observed the phases of the planet Venus, an important support for the Copernican theory of the solar system; (*e*) shed some light on the distinctive character of Saturn (his telescope could not resolve the rings of Saturn but at least he claimed it to differ from other planets) and (*f*) made a step forward in understanding sunspots. It would be hard to claim, for a single person and in so short a span of years, such major discoveries with astro-

---

* See "The Longest Electromagnetic Waves" by James R. Heirtzler, *Scientific American,* March 1962.

nomical telescopes. Ground-based optical astronomy reaches its present potential with the construction of the 6.0m (= 236 in.) reflecting telescope in the Soviet Union.*

In October, 1957 the USSR satellite Sputnik I (mass = 83 kg) orbited the earth. From this beginning has developed the major space effort with which we are all familiar. The UHURU satellite, launched by the United States in 1970, initiated the study of X-ray emmission from extraterrestrial objects.† Since that time many other satellites have studied electromagnetic waves from space in this and other regions of the electromagnetic spectrum.

Orbiting Satellites and Other Spacecraft

A Bell Laboratories engineer, Karl G. Jansky,‡ was investigating electromagnetic disturbances to transoceanic telephone traffic when he realized, in 1931, that a source of extraterrestrial signals at radio frequencies was present. This was the beginning of the science of radioastronomy,

The Development of Radioastronomy

figure 41-2
One of eight similar components of the 5-km radio telescope array at the Mullard Radio Astronomy Observatory near Cambridge, England. At a wavelength of 6 cm this array has a resolving power of 2 seconds of arc, comparable with that of large optical telescopes. See "Radio Astronomy and Cosmology" by John B. Irwin, *Sky and Telescope,* December 1976.

* See "The Soviet 6-meter Altazimuth Reflector" by Bazart K. Ioannisiani, *Sky and Telescope,* November 1977.
† See "Some Recent Advances in X-Ray Astronomy" by Alan P. Lightman, *Sky and Telescope,* October 1976. Satellites made it possible to study electromagnetic radiations from space which otherwise would be absorbed by the earth's atmosphere.
‡ The unit of energy flux in radioastronomy is called the *jansky* in his honor.

based on high performance radiotelescopes such as the 1000-ft diameter parabolic reflector at Arecibo, Puerto Rico. The radiotelescope array at the Mullard Radio Astronomy Observatory, near Cambridge, England, is another example. This 5-km array, which became operational in 1972, employs four fixed and four movable parabolic reflectors of the type shown in Fig. 41-2.

41-3
ELECTROMAGNETIC WAVES FROM SPACE

In this section we consider just four of the important discoveries about the nature of our universe made by studying electromagnetic waves from space. We restrict ourselves to the radio frequency region of the spectrum.

1. Electromagnetic radiation with $\lambda = 21.1$ cm ($\nu = 1420$ MHz) is emitted by neutral hydrogen atoms* that populate the spaces between the stars that comprise our galaxy.† The density of hydrogen atoms in interstellar space is only about 1 atom/cm³ and the mean time spent by a hydrogen atom capable of radiating before it does so is about 10^7 y. Nevertheless our galaxy is so huge that this electromagnetic radiation is readily detectable.

For hydrogen atoms at rest with respect to the terrestrial detector of their radiations the precise wavelength is 21.1061 cm ($\nu = 1420.406$ MHz). Our galaxy, however, is not a rigid structure, its outer regions move with respect to the inner core. Just as for sound, we can measure the radial velocities of light-emitting sources by the Doppler effect; see Sections 20-7 and 42-5. By observing these frequency shifts we can learn a lot about the structure and the internal motions of our galaxy. In recent years electromagnetic radiation from molecules in the interstellar regions of our galaxy has been detected. We may list formaldehyde (HCHO), ammonia (NH_3), and carbon monoxide (CO).

2. Quasi-stellar radio sources (*quasars*) were discovered in 1962. These are optical objects associated normally with large radioemissions. What makes them so interesting is that they have very large "red shifts" in their spectra which, with our present understanding of astrophysics, would lead us to believe that they are at enormous distances from us, well outside our own galaxy. On the other hand, if they are really so far away, they must have enormous energy outputs for which we cannot account. Their nature is under active study today.

3. The primeval fireball radiation.‡ Most astrophysicists today believe that the universe originated about 7×10^9 years ago in the explosion of a very highly concentrated matter-radiation complex. This is the so-called "Big Bang" theory of the origin of the universe. The debris from this explosion expanded outward uniformly in all directions (hence the expanding universe concept) and the matter gradually condensed to form galaxies, stars, and planets as we know them today.

In 1965 the Princeton University physicist Robert Dicke recognized that, if the remnants of the electromagnetic radiation from this primeval fireball, expected to be detectable by radiotelescopes, could be found, it would provide impressive evidence in support of the Big-Bang theory.

* See "Radio Emission from Interstellar Neutral Hydrogen," by R. D. Davies, *Contemporary Physics*, August 1961.
† Our galaxy is a disc-shaped cluster of stars about 10^5 light years in diameter. The Milky Way defines the plane of the disc. Our sun is located about halfway from the galactic center to the galactic edge. There are many other galaxies in our universe; every new high-performance telescope reveals more.
‡ See "The Primeval Fireball" by P. J. E. Peebles and D. T. Wilkinson, *Scientific American*, June 1976.

Unknown to Dicke two Bell Laboratory scientists, A. Penzias and R. Wilson, had actually discovered this electromagnetic radiation from space without realizing its cosmological significance. They made their discovery during engineering studies of communication satellite problems. Within a few years the reality of the primeval fireball radiation was well established.

This episode reminds us that Jansky, also of the Bell Laboratories, founded the science of radio astronomy in 1931. In each case a strictly engineering endeavor led to unexpected insights into our understanding of the universe.*

4. In 1968 a group of British radio astronomers at Cambridge University, headed by Antony Hewish, discovered the first of a series of objects that emit strong short bursts of electromagnetic radiation separated by the astronomically incredibly short time period of about one second. These objects were called *pulsars.*†

It is now generally agreed that pulsars are ordinary stars in which, during the process of shrinking under their own attractive gravitational forces, the electrons of the star combine with the protons, leaving a compact ball of neutrons. Typical *neutron stars* (see Problems 4-23 and 16-5) have these properties: (a) radii of about 10 km (the solar radius is about 7×10^5 km), (b) densities of about 10^{16} kg/m³ (the mean solar density is about 10^3 kg/m³), and (c) periods of rotation of about 10^{-2} s (the solar rotation period is about 10^6 s). The observed pulses seem to arise from a radio-emitting spot on the surface of the neutron star. The emitted electromagnetic rotation sweeps by a terrestrial observer as the star rotates; the analogy to the revolving beam from a lighthouse beacon is clear.

41-4 TRANSMISSION LINES

In this section we turn our attention from the passive observation of electromagnetic waves that reach us to the deliberate generation of such waves and their transmission from point to point by a *transmission line.* Figure 41-3 shows one type of line, a *coaxial cable,* its input end being connected to a switch S. For the time being we assume that the cable is infinitely long and that the cable elements have zero resistance. (The Atlantic cable, first laid successfully in 1866, is a coaxial cable, although it is far from resistanceless.)

When switch S is closed on b, the central and the outer conductors are at the same potential. If we then throw the switch to a, a potential difference V suddenly appears between these elements. This potential

figure 41-3
An electromagnetic signal can be sent along the transmission line (coaxial cable) by throwing switch S from b to a.

* See "The Roots of Solid-State Research at Bell Labs" by Lillian Hartmann Hoddeson, *Physics Today,* March 1977, for a fascinating historical account of the interaction of science and technology in a particular industrial laboratory. In 1937 and 1956 Bell Laboratory scientists were awarded Nobel Prizes.
† See "Pulsars" by A. Hewish, *Scientific American,* October 1968, and "The Nature of Pulsars" by J. P. Ostriker, *Scientific American,* January 1971. Hewish received a shared Nobel Prize in 1974 for this discovery.

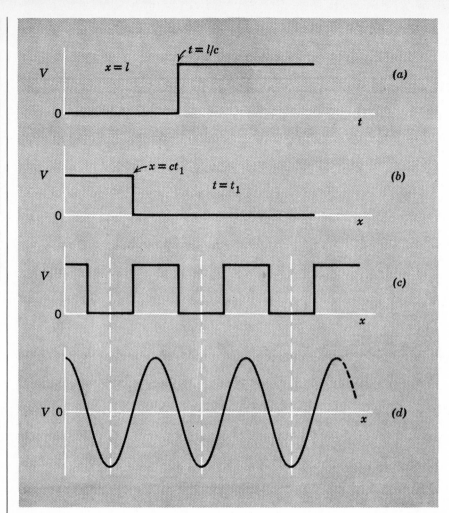

figure 41-4
(*a*) The variation with time of the
potential difference between the
conductors of a coaxial cable at a
distance *l* from the input end. (*b*) An
instantaneous "snapshot" of the
pulse in the cable at a certain time t_1.
(*c*) The waveform if switch *S* in Fig.
41-3 is periodically thrown between
a and *b*. (*d*) The waveform if we
replace switch *S* by an electromagnetic
oscillator with a sinusoidal output.

difference does not appear instantaneously all along the line but is propagated with a finite speed *c* that will turn out to be exactly that of light, assuming a resistanceless line. Figure 41-4*a* shows that the potential difference between the conductors at a distance *l* along the line suddenly rises, at a time given by $t = l/c$, from zero to a value determined by the battery emf. We can also consider the variation of *V* with position *x* along the line at a given time t_1 after closing the switch. Figure 41-4*b* shows such an instantaneous "snapshot." It, too, suggests a traveling "wavefront" moving along the line at speed *c*. At $t = t_1$ the signal has not yet reached points where $x > ct_1$.

If we throw switch *S* periodically from *b* to *a* and back again, a wave disturbance like that of Fig. 41-4*c* is propagated. This suggests that if we replace the battery and switch arrangement by an electromagnetic oscillator with a sinusoidal output of frequency *ν* a wave like that of Fig. 41-4*d* will be propagated.

A traveling wave in a resistanceless transmission line will exhibit a wavelength λ given by

$$\lambda = \frac{c}{\nu}.$$

If the oscillator frequency is 60 Hz, the common commercial power frequency, the wavelength is 5×10^6 m, which is about 3000 miles. At this low frequency traveling waves are not apparent in any line of normal length. By the time the polarity of the oscillator has changed appreciably, the energy fed into the line at the oscillator end has been delivered to the load.

Frequencies in the radio or the microwave range are much higher and the wavelengths correspondingly smaller. Commercial television frequencies as established by the Federal Communications Commission range from 54 to 980 MHz. In terms of wavelength this is a range of 5.6 to 0.31 m. At these wavelengths the patterns of potential difference in the transmission lines used to send television signals across the country can be described aptly as traveling waves. Microwaves, used in radar systems and for communication purposes, have even smaller wavelengths, in the range of about 20 cm to about 0.5 mm.

These considerations suggest another way of viewing the difference between lumped and distributed circuit elements. A system is "distributed" if the wavelength is about the same size as, or less than, the dimensions of the system. If the wavelength is much larger than the dimensions of the system, we are dealing with lumped components. A transmission line 50 m long would be a lumped system for electromagnetic radiation at 60 Hz ($\lambda = 5 \times 10^6$ m) but a distributed system at 100 MHz ($\lambda = 3$ m). In a lumped system the circuit analysis is normally carried out in terms of lumped system parameters such as L, C, and R; in a distributed system the analysis is often carried out in terms of the fields that are set up and the charges and currents that are related to them.

EXAMPLE 1

A potential difference given by

$$V_0 = V_m \sin \omega t$$

is applied between the terminals of a long resistanceless transmission line; the frequency $\nu(=\omega/2\pi)$ is 3×10^9 Hz ($=3$ GHz; see Table 1-2). Write an equation for $V(t)$ at a point P which is 1.5 wavelengths down the line from the oscillator.

The general equation for a wave traveling in the x direction (see Eq. 19-10a) is

$$V = V_m \sin (\omega t - kx),$$

where $k(=2\pi/\lambda)$ is the wave number. At $x = 0$ this gives correctly the time variation of the input terminal potential difference. At $x = 1.5\lambda$ we have

$$V_P = V_m \sin \left[\omega t - \left(\frac{2\pi}{\lambda}\right)(1.5\lambda)\right] = V_m \sin (\omega t - 3\pi)$$

$$= -V_m \sin \omega t.$$

Thus V_P is always equal in magnitude to V_0 but is opposite in sign. What is the wavelength in this example?

41-5
COAXIAL CABLE— FIELDS AND CURRENTS

Figures 41-5a and 41-5b are "snapshots" of the electric and magnetic field configurations in a coaxial cable. The electric field is radial and the magnetic field forms concentric lines about the central conductor. The entire pattern moves along the line, assumed resistanceless, at speed c.

The field patterns in this figure obey the *boundary condition* required for a line that is assumed to be resistanceless, namely, that **E** for all points on either conducting surface has no tangential component (see Example 7, Chapter 28). We can find the field patterns mathematically from Maxwell's equations by imposing this requirement. The configuration

figure 41-5
(a) The electric and magnetic fields in a coaxial cable, showing a wave traveling to the right at speed c. (b) A cross-sectional view at a plane through xx in (a); the wave is emerging from the page. (c) Conduction currents in the walls (open arrows) and displacement currents in the space between the conductors (filled arrows) associated with the wave in (a); the arrows in each case represent current density vectors.

shown is the simplest of many different wave patterns that can travel along the line. The coaxial cable, unlike the electromagnetic cavity of Fig. 38-6, is not a resonant device. The angular frequency ω of waves that travel along it can be varied continuously, as is the case for all traveling waves, such as transverse waves in a long stretched cord.

Figure 41-5c shows the currents in the cable at the instant corresponding to Figs. 41-5a and 41-5b. The arrows parallel to the cable axis represent conduction currents in the central and the outer conductors. The vertical arrows with filled heads represent displacement currents that exist in the space between the conductors. Note that the conduction current and the displacement current arrows form closed loops, preserving the concept of the continuity of current.

Verify that the displacement current represented in Fig. 41-5c is consistent with the pattern of **B** and **E** shown in Fig. 41-5a.

Consider a small surface element $\Delta \mathbf{S}$ shown edge-on in Fig. 41-5; it is shown as viewed from above in Fig. 41-6. This hypothetical element is stationary with respect to the cable while the field configuration moves through it at speed c. Figure 41-6a shows the electric lines of force in and near this element. It is clear from symmetry that, at the instant shown, the net flux Φ_E through this area is zero. However, even though Φ_E is zero in magnitude, it is, at this instant, *changing at its most rapid rate*, since **E** at the element ΔS is at the very moment of reversing its direction as the wave moves through. Thus the displacement current, which is given by

$$i_d = \epsilon_0 \frac{d\Phi_E}{dt},$$

also has its maximum value.

Figure 41-6b shows the magnetic field in the vicinity of the element of area. Let us apply the generalized form of Ampère's law,

$$\oint \mathbf{B} \cdot d\mathbf{l} = \mu_0(i + i_d),$$

to the element. The conduction current i is zero since no charge is transported

EXAMPLE 2

(a)   (b)

figure 41-6
The area element S in Fig. 41-5 enlarged and viewed from above, showing the adjacent (a) electric and (b) magnetic fields.

through ΔS. The displacement current i_d is not zero, having, in fact, its maximum value. Thus, since the right side of this equation $(=\mu_0 i_d)$ does not vanish, the left side must not vanish. Study of Fig. 41-6b shows that $\oint \mathbf{B} \cdot d\mathbf{l}$ around the boundary of this square has, indeed, a nonzero value. Thus the field and displacement current configurations of Fig. 41-5a–c are consistent. We have not discussed the direction of i_d; that is, does it point into the plane of Fig. 41-6, as Fig. 41-5c asserts, or out of it? We leave this as a question. You may be guided by considering the direction of the displacement current in Figs. 40-1a and 40-1b.

EXAMPLE 3

Show that the conduction currents in Fig. 41-5c are appropriately related to the magnetic field pattern.

Figure 41-7 shows a cross section of the cable at a plane through xx in Fig. 41-5a. Let us apply Ampère's law,

$$\oint \mathbf{B} \cdot d\mathbf{l} = \mu_0 (i + i_d),$$

to the ring of radius r. The displacement current through this ring is zero, this current being at right angles to the central conductor. The conduction current i in the central conductor, shown by the $\times$'s in the figure, does pass through this ring so that the equation becomes

$$(B)(2\pi r) = \mu_0 i,$$

or

$$B = \frac{\mu_0 i}{2\pi r}.$$

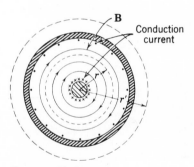

figure 41-7
Example 3. The coaxial cable of Fig. 41-5b, showing the conduction currents in the central and outer conductors. The wave is emerging from the page.

Note that $\mathbf{B}$ is related to the current in the central conductor by the usual right-hand rule and that the expression for B is that found earlier (Eq. 34-4) for a long straight wire carrying a steady current.

If we apply Ampère's law to the large ring of radius r', we must put the *net* conduction current equal to zero because the current in the outer conductor, shown by the dots, is equal and opposite to that in the inner conductor. This means that $\mathbf{B}$ must be zero for points outside the cable, in agreement with experiment.

It is interesting to compare the electromagnetic oscillations in a typical *traveling* wave, such as that of Fig. 41-5, with those in a cavity resonator, such as that of Fig. 38-6. The latter oscillations are an electromagnetic *standing* wave. In a traveling wave $\mathbf{E}$ and $\mathbf{B}$ are in phase, which means that at a given position along the transmission line they reach their maxima at the same time. However, Fig. 38-6 shows that at a given position in a cavity resonator $\mathbf{E}$ and $\mathbf{B}$ reach their maxima one-fourth of a cycle apart; they are 90° out of phase.

A complete analogy exists in mechanical systems. In the acoustic resonator of Fig. 38-5 the time variations of pressure and velocity for the standing acoustic wave are also 90° out of phase, in exact correspondence to the electromagnetic cavity oscillations of Fig. 38-6. The acoustic analogy to a transmission line (Fig. 41-8) would be an infinitely long gas-filled tube, one end being connected to an acoustic oscillator such as a loud-speaker. The entire configuration of Fig. 41-8 moves to the right with speed v. The pressure variations, suggested by the dots, and the instantaneous velocities, suggested by the arrows, are *in phase*, just as are **E** and **B** in the coaxial cable of Fig. 41-5.

Loudspeaker

figure 41-8
An acoustic transmission line, showing a sound wave traveling to the right. The small arrows with filled heads show the directed drift velocities for small volume elements of the gas. Compare Fig. 41-5.

It is possible to send electromagnetic waves through a hollow metal pipe that has no central conductor. We assume that the inner walls of such a pipe, or *waveguide* as it is called, are resistanceless and that the cross section is rectangular.

Figure 41-9 shows a typical electric and magnetic field pattern. We imagine that a microwave oscillator is connected to the left end and sends electromagnetic energy down the guide. Figure 41-9(*i*) shows a side view of the guide and Fig. 41-9(*ii*), a top view; Fig. 41-9(*iii*) shows the cross section. As for the coaxial cable, the field patterns are such that **E** has no tangential component for any point on the inner surface of the guide. The fields **E** and **B** are in phase, again like the coaxial cable.

41-6
WAVEGUIDE

figure 41-9
A waveguide, showing (*i*) a side view of the lines of **E,** (*ii*) a top view of the lines of **B,** and (*iii*) a cross-sectional view of the lines of **E.** In (*iii*) the wave is emerging from the page. For simplicity the lines of **B** are not shown in (*i*) and (*iii*), nor are the lines of **E** shown in (*ii*).

As for all traveling waves, the angular frequency ω of electromagnetic waves traveling down a guide can be varied continuously. In a waveguide of given dimensions, however, there exists, for every mode of transmission, that is, for every pattern of **E** and **B,** a so-called *cutoff frequency* ω_0. A given guide will not transmit waves in a given mode if their frequency is below the cutoff value for that mode in that guide. The field patterns of Fig. 41-9 show the *dominant mode* for a rectangular guide; this is the mode with the lowest cutoff frequency. Given the frequency ω of electromagnetic waves to be transmitted, it is common practice to select a guide whose dimensions are such that ω is larger than the cutoff frequency ω_0 for the dominant mode but smaller than the cutoff frequencies of all other modes. Under these conditions the dominant mode of propagation is the only one possible.

In a (resistanceless) coaxial cable the wave patterns travel at speed c. In the acoustic transmission line of Fig. 41-8 (assumed "resistanceless") the waves also travel at a speed v, which is the same as the propagation speed in an infinite medium. In a waveguide, however, the speed is *not c*. In waveguides we must distinguish between (*a*) the *phase speed* v_{ph}, which is the speed at which the wave patterns of Fig. 41-9 travel, and (*b*) the *group speed* v_{gr}, which is the speed at which electromagnetic energy or information-carrying "signals" travel along the guide. These speeds, which are identical for electromagnetic waves in a coaxial cable and for acoustic waves in a tube, are different for waves in a waveguide.

The phase speed is not directly measurable. The wave pattern is a repetitive structure, and there is no way to distinguish one wave maximum from another. We can observe the waves entering one end of the guide and leaving at the other, but there is no way to identify a particular wave maximum so that we can time its passage down the guide. We can put a "signal" on the wave by increasing the power level of the oscillator for a short time. We can time this power pulse as it passes through the guide, but there is no guarantee that it travels at the same speed as the wave pattern and, indeed, it does not. The speed of such signals or markers is the speed at which *energy* is propagated, that is, the group speed.

From Maxwell's equations we can show that the phase speed and the group speed for the mode of Fig. 41-9 are

$$v_{ph} = \frac{c}{\sqrt{1 - \left(\frac{\lambda}{2a}\right)^2}} \qquad (41\text{-}1)$$

and

$$v_{gr} = c\sqrt{1 - \left(\frac{\lambda}{2a}\right)^2}, \qquad (41\text{-}2)$$

in which a is the width of the guide and λ the free-space wavelength. Note that as $a \to \infty$, which corresponds to free-space conditions, $v_{ph} = v_{gr} = c$.

The phase speed v_{ph} is *greater* than the velocity of light, the group speed v_{gr} being correspondingly less. In relativity theory we learn that no speed at which signals or energy travel can be faster than that of light. However, signals or energy *cannot* be transmitted down a guide at the phase speed v_{ph}; they travel with speed v_{gr} which is *always* less than c so there is no conflict with the theory of relativity.

The wavelength λ in Eqs. 41-1 and 41-2 is the wavelength that would be measured for the oscillations in free space, that is,

$$\lambda = \frac{c}{\nu}, \qquad (41\text{-}3)$$

where c is the speed in free space and ν is the frequency. For waves of a given frequency, the wavelength exhibited in a guide (λ_g) must differ from the free-space wavelength λ because the speed v_{ph} has changed. The so-called *guide wavelength* λ_g is given by

$$\lambda_g = \frac{v_{ph}}{\nu} = \frac{v_{ph}}{c/\lambda} = \lambda \frac{v_{ph}}{c}.$$

From Eq. 41-1 this yields

$$\lambda_g = \frac{\lambda}{\sqrt{1 - \left(\frac{\lambda}{2a}\right)^2}}. \qquad (41\text{-}4)$$

Thus the guide wavelength, which is the wavelength exhibited by the field patterns in Fig. 41-9, is larger than the free-space wavelength.

EXAMPLE 4

What must be the width a of a rectangular guide such that the energy of electromagnetic radiation whose free-space wavelength is 3.0 cm travels down the guide (a) at 95% of the speed of light? (b) At 50% of the speed of light?

From Eq. 41-2 we have

$$v_{gr} = 0.95c = c\sqrt{1 - \left(\frac{\lambda}{2a}\right)^2}.$$

Solving for a yields $a = 4.8$ cm; repeating for $v_{gr} = 0.50c$ yields $a = 1.7$ cm.

This illustrates the cutoff phenomenon described above. If $\lambda = 2a$, then $v_{gr} = 0$ and energy cannot travel down the guide. For the radiation considered in this example $\lambda = 3.0$ cm, so that the guide must have a width a of *at least* $\frac{1}{2} \times 3.0$ cm $= 1.5$ cm if it is to transmit this wave. The guide whose width we calculated in (a) above can transmit radiations whose free-space wavelength is 2×4.8 cm $= 9.6$ cm *or less*.

The acoustic transmission line of Fig. 41-8 cannot be infinitely long. Its far end may be sealed by a solid cap or left open, or it may have a flange, a horn, or some similar device mounted on it. If the far end is not sealed, energy will escape into the medium beyond. We call this *acoustic radiation*. In general, some energy will also be reflected back down the transmission line. If acoustic radiation is desirable, the designer's task is to fashion a termination (that is, an "acoustic antenna") for the transmission line such that the smallest possible fraction of the incident energy will be reflected back down the line. Such a termination might take the form of a flared horn. Acoustic radiation, of course, requires a medium such as air in order to be propagated.

An electromagnetic transmission line such as a coaxial cable or a waveguide can also be terminated in many ways, and energy can escape from the end of the line into the space beyond. In contrast to sound waves, a physical medium is not required. Thus electromagnetic energy can be radiated from the end of the transmission line, to form a traveling electromagnetic wave in free space.

Figure 41-10 shows an effective termination for a coaxial cable; it consists of two wires arranged as shown and is called an *electric dipole antenna*. The potential difference between the two conductors alternates sinusoidally as the wave reaches them, the effect being that of an electric dipole whose dipole moment **p** varies with time, both in magnitude and direction.

41·7
RADIATION

figure 41-10
An electric dipole antenna on the end of a coaxial cable.

figure 41-11
Showing the fields of **E** and **B** radiated away from an oscillating electric dipole. We show only the fields at distances from the dipole that are large compared with its dimensions. The electromagnetic wave sweeping through distant point P is a plane wave, moving in the x-direction.

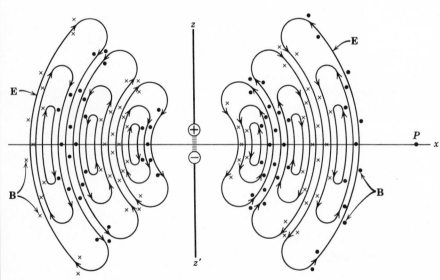

Figure 41-11 shows an electromagnetic wave generated by such an oscillating electric dipole. The figure is a section through a figure of revolution about the dipole axis zz' and the wave moves out in any direction from the dipole with speed c. The fields shown are those that exist at distances from the dipole which are large compared with the dipole dimensions. We call this the *radiation field* and it is the focus of our concern here. The fields close to the dipole (the *near field*) are more complex but they have no interest for us here because they die out rapidly and we want to study only the fields at large distances.

Figure 41-12 shows eight cyclical "snapshots" of the fields of **E** and **B** sweeping past an observer at point P in Fig. 41-11. We assume that P is so far distant from the oscillating dipole that the wave fronts passing

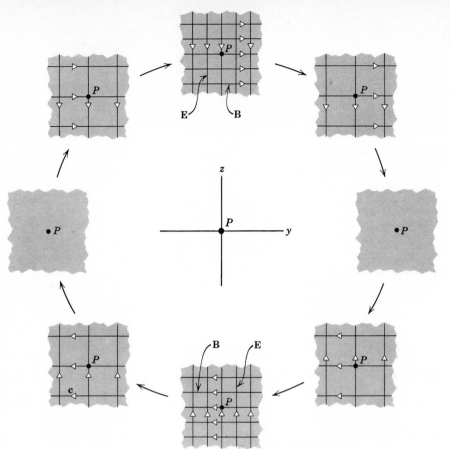

figure 41-12
Showing eight cyclical "snapshots"
of the plane electromagnetic wave
radiated from the oscillating dipole
of Fig. 41-11 through point P. The
direction of the wave (the x-direction)
is out of the plane of the figure.

through and near P describe a plane wave; see Section 19-2. The speed c
of the wave in free space is given by $c = \nu\lambda$, which we can write as

$$c = \frac{\omega}{k},\qquad(41\text{-}5)$$

where ω, the angular frequency, and k, the wave number, are related to
the frequency ν and the wavelength λ by

$$\omega = 2\pi\nu \quad \text{and} \quad k = 2\pi/\lambda.$$

41-8
TRAVELING WAVES AND MAXWELL'S EQUATIONS

In earlier sections we have postulated the existence of certain magnetic
and electric field distributions, in resonant cavities, coaxial cables, and
waveguides, and we have shown that these postulated distributions are
consistent with Maxwell's equations, as are the distributions of conduction and displacement currents associated with the fields. If you
pursue your studies of electromagnetism, you will learn how to derive
mathematical expressions for $\mathbf{E}$ and $\mathbf{B}$ by subjecting Maxwell's equations
to the boundary conditions appropriate to the problem at hand. In this
section we continue our program by showing that the postulated pat-
terns of $\mathbf{E}$ and $\mathbf{B}$ for a traveling electromagnetic wave are completely
consistent with Maxwell's equations. In doing so, we will be able to
show that the speed of such waves in free space is that of visible light
and thus that visible light is itself an electromagnetic wave.

If the observer at P in Fig. 41-11 is, as we have assumed, at a considerable distance from the source, the *wavefronts* described by the
electric and magnetic fields that reach him (see Fig. 41-12) will be

figure 41-13
A plane electromagnetic wave traveling to the right at speed c. Lines of **B** are parallel to the y axis; those of **E** are parallel to the z axis. The shaded rectangles on the right refer to Fig. 41-14.

planes and the wave that moves past him will be a *plane wave*. Figure 41-13 shows a three-dimensional "snapshot" of a plane wave traveling in the x direction. The lines of **E** are parallel to the z axis and those of **B** are parallel to the y axis. The values of **B** and **E** for this wave depend only on x and t (not on y or z). We postulate that they are given in magnitude by

$$B = B_m \sin(kx - \omega t) \qquad (41\text{-}6)$$

and

$$E = E_m \sin(kx - \omega t). \qquad (41\text{-}7)$$

Figure 41-14 shows two sections through the three-dimensional diagram of Fig. 41-13. In Fig. 41-14a the plane of the page is the xz plane and in Fig. 41-14b it is the xy plane. Note that, as for the traveling waves in a coaxial cable (Fig. 41-5a) and in a waveguide (Fig. 41-9), **E** and **B** are in phase, that is, at any point through which the wave is moving they reach their maximum values at the same time.

The shaded rectangle of dimensions dx and h in Fig. 41-14a is fixed in space. As the wave passes over it, the magnetic flux Φ_B through the rectangle will change, which will give rise to induced electric fields around

(a)

(b)

figure 41-14
The wave of Fig. 41-13 viewed (a) in the xz plane and (b) in the xy plane.

the rectangle, according to Faraday's law of induction. These induced electric fields are, in fact, simply the electric component of the traveling wave.

Let us apply Lenz's law to this induction process. The flux Φ_B for the shaded rectangle of Fig. 41-14a is *decreasing* with time because the wave is moving through the rectangle to the right and a region of weaker magnetic fields is moving into the rectangle. The induced field will act to oppose this change, which means that if we imagine that the boundary of the rectangle is a conducting loop, a *counterclockwise* induced current would appear in it. This current would produce a field of **B** that, within the rectangle, would point out of the page, thus opposing the decrease in Φ_B. There is, of course, no conducting loop, but the net induced electric field **E** does indeed act counterclockwise around the rectangle because $E + dE$, the magnitude of **E** at the right edge of the rectangle, is greater than E, the magnitude of **E** at the left edge. Thus the electric field configuration is entirely consistent with the concept that it is induced by the changing magnetic field.

For a more detailed analysis let us apply Faraday's law of induction, or

$$\oint \mathbf{E} \cdot d\mathbf{l} = -\frac{d\Phi_B}{dt}, \tag{41-8}$$

going counterclockwise around the shaded rectangle of Fig. 41-14a. There is no contribution to the integral from the top or bottom of the rectangle because **E** and $d\mathbf{l}$ are at right angles here. The integral then becomes

$$\oint \mathbf{E} \cdot d\mathbf{l} = [(E + dE)(h)] - [(E)(h)] = dE\, h.$$

The flux Φ_B for the rectangle is

$$\Phi_B = (B)(dx\, h),$$

where B is the magnitude of **B** at the rectangular strip and $dx\, h$ is the area of the strip. Differentiating gives

$$\frac{d\Phi_B}{dt} = h\, dx\, \frac{dB}{dt}.$$

From Eq. 41-8 we then have

$$dE\, h = -h\, dx\, \frac{dB}{dt},$$

or

$$\frac{dE}{dx} = -\frac{dB}{dt}. \tag{41-9}$$

Actually, both B and E are functions of x and t; see Eqs. 41-6 and 41-7. In evaluating dE/dx, we assume that t is constant because Fig. 41-14a is an "instantaneous snapshot." Also, in evaluating dB/dt we assume that x is constant since what is required is the time rate of change of B at a particular place, the strip in Fig. 41-14a. The derivatives under these circumstances are called *partial derivatives*, and a somewhat different notation is used for them; see Example 2, p. 414 and p. 636. In this notation Eq. 41-9 becomes

$$\frac{\partial E}{\partial x} = -\frac{\partial B}{\partial t}. \tag{41-10}$$

The minus sign in this equation is appropriate and necessary, for, although E is increasing with x at the site of the shaded rectangle in Fig. 41-14a, B is decreasing with t. Since $E(x,t)$ and $B(x,t)$ are known (see

Eqs. 41-6 and 41-7), Eq. 41-10 reduces to

$$kE_m \cos{(kx - \omega t)} = \omega B_m \cos{(kx - \omega t)},$$

or (see Eq. 41-5) $$\frac{\omega}{k} = \frac{E_m}{B_m} = c. \tag{41-11a}$$

Thus the speed of the wave c is the ratio of the amplitudes of the electric and the magnetic components of the wave. From Eqs. 41-6 and 41-7 we see that the ratio of amplitudes is the same as the ratio of instantaneous values, or

$$E = cB. \tag{41-11b}$$

This important result will be useful in later sections.

We now turn our attention to Fig. 41-14b, in which the flux Φ_E for the shaded rectangle is decreasing with time as the wave moves through it. According to Equation IV, Table 40-2 (with $i = 0$, because there are no conduction currents in a traveling electromagnetic wave),

$$\oint \mathbf{B} \cdot d\mathbf{l} = \mu_0 \epsilon_0 \frac{d\Phi_E}{dt}, \tag{41-12}$$

this changing flux will induce a magnetic field at points around the periphery of the rectangle. This induced magnetic field is simply the magnetic component of the electromagnetic wave. Thus, as in the cavity resonator of Section 38-5, the electric and the magnetic components of the wave are intimately connected with each other, each depending on the time rate of change of the other.

Comparison of the shaded rectangles in Fig. 41-14 shows that for each the appropriate flux, Φ_B or Φ_E, is *decreasing* with time. However, if we proceed counterclockwise around the upper and lower shaded rectangles, we see that $\oint \mathbf{E} \cdot d\mathbf{l}$ is *positive*, whereas $\oint \mathbf{B} \cdot d\mathbf{l}$ is *negative*. This is as it should be. If you compare Figs. 35-10 and 40-1a, you will be reminded that although the fluxes Φ_B and Φ_E in those figures are changing with time in the same way (both are increasing) the lines of the induced fields, $\mathbf{E}$ and $\mathbf{B}$, respectively, circulate in opposite directions.

The integral in Eq. 41-12, evaluated by proceeding counterclockwise around the shaded rectangle of Fig. 41-14b, is

$$\oint \mathbf{B} \cdot d\mathbf{l} = [-(B + dB)(h)] + [(B)(h)] = -h \, dB,$$

where B is the magnitude of $\mathbf{B}$ at the left edge of the strip and $B + dB$ is its magnitude at the right edge.

The flux Φ_E through the rectangle of Fig. 41-14b is

$$\Phi_E = (E)(h \, dx).$$

Differentiating gives $$\frac{d\Phi_E}{dt} = h \, dx \, \frac{dE}{dt}.$$

Thus we can write Eq. 41-12 as

$$-h \, dB = \mu_0 \epsilon_0 \left(h \, dx \, \frac{dE}{dt} \right)$$

or, substituting partial derivatives,

$$-\frac{\partial B}{\partial x} = \mu_0 \epsilon_0 \frac{\partial E}{\partial t}. \tag{41-13}$$

Again, the minus sign in this equation is appropriate and necessary, for, although B is increasing with x at the site of the shaded rectangle in Fig. 41-14b, E is decreasing with t.

Combining this equation with Eqs. 41-6 and 41-7 yields

$$-kB_m \cos{(kx - \omega t)} = -\mu_0 \epsilon_0 \omega E_m \cos{(kx - \omega t)},$$

or (see Eq. 41-5)
$$\frac{E_m}{B_m} = \frac{k}{\mu_0 \epsilon_0 \omega} = \frac{1}{\mu_0 \epsilon_0 c}. \tag{41-14}$$

Eliminating E_m/B_m between Eqs. 41-11a and 41-14 yields

$$c = \frac{1}{\sqrt{\mu_0 \epsilon_0}}. \tag{41-15}$$

Substituting numerical values yields

$$c = \frac{1}{\sqrt{(4\pi \times 10^{-7}\ \text{T·m/A})(8.9 \times 10^{-12}\ \text{C}^2/\text{N·m}^2)}}$$
$$= 3.0 \times 10^8\ \text{m/s}, \tag{41-16}$$

which is the speed of light in free space! This emergence of the speed of light from purely electromagnetic considerations is the crowning achievement of Maxwell's electromagnetic theory. Maxwell made this prediction before radio waves were known and before it was realized that light was electromagnetic in nature. His prediction led to the concept of the electromagnetic spectrum and to the discovery of radio waves by Heinrich Hertz in 1890. It made it possible to discuss optics as a branch of electromagnetism and to derive its fundamental laws from Maxwell's equations.

A conclusion as basic as Eq. 41-15 must be subject to rigorous experimental testing and indeed this has been done. Of the three quantities in this equation, one, μ_0, has an exact assigned value of $4\pi \times 10^{-7}$ T·m/A. The other two, c and ϵ_0, are measurable quantities. Our confidence in Maxwell's equations, bolstered by many successful predictions and agreements with experiment, is such that we now use the measured speed of light (2.99792458×10^8 m/s; see Appendix B) to determine ϵ_0, by way of Eq. 41-15. This gives a much better value than we could measure by the methods of Section 30-2.

Curiously enough Maxwell himself did not view the propagation of electromagnetic waves and electromagnetic phenomena in general, in anything like the terms suggested by, say, Fig. 41-13. Like all physicists of his day he believed firmly that space was permeated by a subtle substance called the *ether* and that electromagnetic phenomena could be accounted for in terms of rotating vortices in this aether.

It is a tribute to Maxwell's genius that, with such mechanical models in his mind, he was able to deduce the laws of electromagnetism that bear his name. These laws, as we have pointed out, not only required no change when Einstein's special theory of relativity came on the scene almost half-century later but, indeed, were strongly confirmed by that theory. Today we no longer believe in the aether concept; see Supplementary Topic V, "Special Relativity—A Summary of Conclusions."

41-9 THE POYNTING VECTOR

One of the important characteristics of an electromagnetic wave is that it can transport energy from point to point. As we show below, we can describe the rate of energy flow per unit area in a plane electromagnetic wave by a vector **S**, called the *Poynting vector* after John Henry Poynting (1852–1914), who first pointed out its properties. We define **S** from

$$S = \frac{1}{\mu_0} E \times B. \qquad (41\text{-}17)$$

In SI units **S** is expressed in watts/meter²; the direction of **S** gives the direction in which the energy moves. The vectors **E** and **B** refer to their instantaneous values at the point in question. If we apply Eq. 41-17 to the traveling plane electromagnetic wave of Fig. 41-13, it is clear that **E** × **B**, hence **S**, point in the direction of propagation. Note, too, that **S** points parallel to the axis for all points in the coaxial cable of Fig. 41-5.

We get meaningful results if we extend the Poynting vector concept to other electromagnetic situations involving either traveling or standing electromagnetic waves, as we will see in Examples 5 and 6. If we extend it to circuit situations involving steady or almost steady currents and lumped circuit elements, we are led to some interesting conclusions, which we explore in Problems 28, 31, and 32.

Analyze energy flow in the cavity of Fig. 38-6, using the Poynting vector.

Study of Fig. 41-15 shows that when the energy is all electric (Figs. 41-15a and 41-15e) it is concentrated along the axis, because this is the region in which **E** has its maximum value. When the energy is all magnetic (Figs. 41-15c and 41-15g), it is concentrated near the walls. Thus the energy surges back and forth periodically between the central region and the region near the walls. The figure

EXAMPLE 5

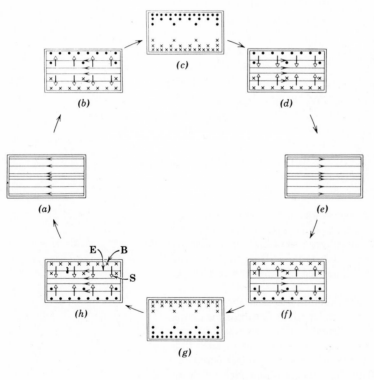

figure 41-15
Example 5. Energy surges back and forth periodically between the central region of the cavity and the region near the walls, as indicated by the Poynting vector **S** (vertical arrows).

figure 41-16
A plane wave is traveling to the right at speed c; compare Fig. 41-14. The dashed rectangle in this figure represents a three-dimensional box, of area A and thickness dx, that extends at right angles to the plane of the figure.

shows, by the open arrows, the direction of **S** at various points in the cavity and at various times in the cycle. Note that **S** equals zero for Figs. 41-15*a, c, e,* and *g,* which is appropriate because at these instants of time the field configurations are momentarily stationary and energy is not flowing. A pendulum bob at the end of its swing forms a mechanical analogy. Verify from Eq. 41-17 that these arrows point in the correct directions.

We can use Figure 41-16 to derive the Poynting relation for the special case of a traveling plane electromagnetic wave. It shows a cross section of a traveling plane wave, along with a thin "box" of thickness dx and area A. The box, a mathematical construction, is fixed with respect to the axes while the wave moves through it. At any instant the energy stored in the box, from Eqs. 30-9 and 36-19, is

$$dU = dU_E + dU_B = (u_E + u_B)(A \; dx)$$

$$= \left(\tfrac{1}{2}\epsilon_0 E^2 + \frac{1}{2\mu_0} B^2\right)A \; dx, \qquad (41\text{-}18)$$

where $A \; dx$ is the volume of the box and E and B are the instantaneous values of the field vectors in the box.

Using Eq. 41-11*b* $(E = cB)$ to eliminate *one* of the E's in the first term in Eq. 41-18 and *one* of the B's in the second term leads to

$$dU = \left[\tfrac{1}{2}\epsilon_0 E(cB) + \frac{1}{2\mu_0} B \left(\frac{E}{c}\right)\right] A \; dx$$

$$= \frac{(\mu_0\epsilon_0 c^2 + 1)(EBA \; dx)}{2\mu_0 c}.$$

From Eq. 41-15, however, $\mu_0\epsilon_0 c^2 = 1$, so that

$$dU = \frac{EBA \; dx}{\mu_0 c}.$$

This energy dU will pass through the right face of the box in a time dt equal to dx/c. Thus the energy per unit area per unit time, which is S, is given by

$$S = \frac{dU}{dt \; A} = \frac{EBA \; dx}{\mu_0 c \; (dx/c) \; A} = \frac{1}{\mu_0} EB.$$

This is exactly the prediction of the more general relation Eq. 41-17 for a traveling plane wave.

This relation refers to values of S, E, and B at any instant of time. We are usually more interested in the *average* value of S, taken over one or more cycles of the wave. An observer making intensity measurements on a wave moving past him would measure this average value $\overline{S}$. We can easily show (see Example 6) that $\overline{S}$ is related to the *maximum* values of E and B by

$$\overline{S} = \frac{1}{2\mu_0} E_m B_m.$$

EXAMPLE 6

An observer is at a distance r from a point light source whose power output is P_0. Calculate the magnitudes of the electric and the magnetic fields. Assume that the source is monochromatic, that it radiates uniformly in all directions, and that at distant points it behaves like the traveling plane wave of Fig. 41-13.

The power that passes through a sphere of radius r is $(\overline{S})(4\pi r^2)$, where $\overline{S}$ is the *average* value of the Poynting vector at the surface of the sphere. This power must equal P_0, or

$$P_0 = \overline{S} 4\pi r^2.$$

From the definition of **S** (Eq. 41-17), we have

$$\overline{S} = \overline{\left(\frac{1}{\mu_0} EB\right)}.$$

Using the relation $E = cB$ (Eq. 41-11b) to eliminate B leads to

$$\overline{S} = \frac{1}{\mu_0 c} \overline{E^2}.$$

The average value of E^2 over one cycle is $\frac{1}{2}E_m^2$, since E varies sinusoidally (see Eq. 41-7). This leads to

$$P_0 = \left(\frac{E_m^2}{2\mu_0 c}\right)(4\pi r^2),$$

or

$$E_m = \frac{1}{r}\sqrt{\frac{P_0 \mu_0 c}{2\pi}}.$$

For $P_0 = 10^3$ W and $r = 1.0$ m this yields

$$E_m = \frac{1}{(1.0\text{ m})}\sqrt{\frac{(10^3\text{ W})(4\pi \times 10^{-7}\text{ Wb/A·m})(3 \times 10^8\text{ m/s})}{2\pi}}$$

$$= 240\text{ V/m}.$$

The relationship $E_m = cB_m$ (Eq. 41-11a) leads to

$$B_m = \frac{E_m}{c} = \frac{240\text{ V/m}}{3 \times 10^8\text{ m/s}} = 8 \times 10^{-7}\text{ T}.$$

Note that E_m is appreciable as judged by ordinary laboratory standards but that B_m (= 0.008 gauss) is quite small.

---

questions

1. If you are asked what fraction of the electromagnetic spectrum lies in the visible range, how would you respond?

2. Project Seafarer was an ambitious program to construct an enormous antenna, buried underground on a site about 4000 square miles in area. Its purpose was to transmit signals to submarines while they were deeply submerged. If the effective wavelength was, say, about 10^4 earth radii, what would be the frequency and the period of the radiations emitted? Ordinarily electromagnetic radiations do not penetrate very far into conductors such as sea water. Can you think of any reason why such ELF (extremely low frequency) radiations should penetrate more effectively? Think of the limiting case of zero frequency. (Why not transmit signals at zero frequency?)

3. How would you characterize electromagnetic radiation that has a frequency of 10 kHz? 10^{20} Hz? or a wavelength of 500 nm? 10 km? 0.50 nm?

4. Electromagnetic waves reach us from the farthest depths of the universe. Do they tell us what the universe is like now, what it was like at some time in the distant past, or something in between? Discuss.

5. *The Dark Night Sky Paradox:** Perhaps the simplest astronomical observation that you can make is this: When the sun sets, the sky becomes dark. This is true and seems obvious but an argument can be made that it should not be so. Consider:

   "Assuming an infinite universe, uniformly populated by stars more or less like our sun, we can say that a straight line projected from the observer in any direction will eventually hit a star. The *distances R* of most of these stars will be very great indeed so that the stars illuminate him only weakly, the illumination varying as $1/R^2$. On the other hand, the *number* of distant stars located within a spherical shell whose radii are R and $R + \Delta R$ increases as R^2 (assuming that ΔR is constant). Can you prove this last statement? These two effects seem to cancel precisely. Thus the night sky should be virtually infinitely bright, the observer being illuminated by an infinity of suns."

   Can you see any flaw in this argument? Think of the finite speed of light, the large scale of the universe, the expanding universe concept, the finite

* Usually called *Olbers' paradox.*

lifetime of stars, and so on. See "The Dark Night Sky Paradox" by E. R. Harrison, *American Journal of Physics,* February 1977, for an excellent historical review and a lucid explanation.

6. When ordinary stars condense to form neutron stars (pulsars), why does the angular rotation rate become so large?

7. In the coaxial cable of Fig. 41-3, what are the directions of the conduction current (*a*) in the central conductor and (*b*) in the outer conductor, shortly after the switch is thrown to position *a*? Consider points that have been reached by the wavefront of Fig. 49-4*b* and those that have not.

8. What is the relation between the wavelength in the cable of Fig. 41-5 and that in free space for a coaxial cable?

9. What is the direction of the displacement current in Fig. 41-6? Give an argument to support your answer.

10. Compare a coaxial cable and a waveguide, used as a transmission line. Point out both similarities and differences.

11. Can traveling waves with a continuous range of wavelengths be sent down (*a*) a coaxial cable and (*b*) a waveguide? Can standing waves with a continuous range of wavelengths be set up in a resonant cavity? Develop mechanical or acoustical analogies to support your answers.

12. If a certain wavelength is larger than the cutoff wavelength for a guide in its dominant mode, can energy be sent down it in any other mode?

13. Explain why the term $\epsilon_0 \, d\Phi_E/dt$ is needed in Ampère's equation to understand the propagation of electromagnetic waves.

14. In the equation $c = 1/\sqrt{\mu_0 \epsilon_0}$ (Eq. 40-1), how can c always have the same value if μ_0 is *arbitrarily* assigned and ϵ_0 is measured?

15. Is it conceivable that electromagnetic theory might some day be able to predict the value of c (3×10^8 m/s), not in terms of μ_0 and ϵ_0, but directly and numerically without recourse to any measurements?

16. Figure 41-17 shows a magnetic dipole activated by an oscillating LC circuit. Discuss the nature of the traveling wave at a distant point P.

figure 41-17
Question 16

17. In a coaxial cable is the energy transported in the conductors, through the agency of the currents, or in the space between them, through the agency of the fields?

SECTION 41-2

1. In Fig. 41-1, showing the electromagnetic spectrum, the frequency ν and the wavelength λ are related by $c = \lambda\nu$. If the frequency intervals in the figure are evenly spaced, show that the wavelength intervals are also necessarily equally spaced. Bear in mind that both scales are logarithmic.

problems

SECTION 41-3

2. The earth's mean radius is 6.4×10^6 m and the mean earth-sun distance is 1.5×10^8 km. What fraction of the electromagnetic radiation emitted by the

sun is intercepted by the disc of the earth? Assume an inverse square law for the decrease of intensity, measured perhaps in watts/meter², with distance.

3. The intensity of direct solar radiation that was unabsorbed by the atmosphere on a lazy summer day is 100 W/m². How close would you have to stand to a 1.0-kW electric heater to feel the same intensity? Assume that the heater is 100% efficient and radiates equally in all directions.
Answer: 0.89 m.

4. Our closest stellar neighbor, α-Centauri, is 4.3 light years away. It has been suggested that TV programs from our planet (The Tonight Show, for example) have reached this star and may have been viewed by the hypothetical inhabitants of a hypothetical planet orbiting this star. The moon is 3.8×10^5 km from earth and perhaps its hypothetical inhabitants (who always hide when a space ship approaches) also watch these programs. From the inverse square law, what would be the ratio of the intensities of such signals, measured perhaps in W/m², for the moon and for α-Centauri?

SECTION 41-5

5. A coaxial cable is made of a center wire of radius a surrounded by a thin metal tube of radius b. The space between the conductors is evacuated. (a) Find the capacitance per unit length of this coaxial cable (*Hint:* Imagine equal but opposite charges to be on the wire and the tube). (b) Find the inductance per unit length of this coaxial cable (*Hint:* Imagine a current i flowing down the center wire and back along the tube.)
Answer: (a) $2\pi\epsilon_0/\ln b/a$. (b) $(\mu_0/2\pi) \ln b/a$.

6. Using Gauss's law, sketch the instantaneous charges that appear on the conductors of the coaxial cable of Fig. 41-5 and show that this pattern of charges is appropriately related to the conduction currents shown in Fig. 41-5c.

7. A resonant cavity is constructed by closing each end of the coaxial cable of Fig. 41-5 with a metal cap. The cavity contains three half-waves. Describe the patterns of **E** and **B** that occur, assuming the same mode of oscillation as that shown in Fig. 41-5. (Hint: Remember that **E** can have no tangential component at a conducting surface and that **B** and **E** must be 90° out of phase.)

SECTION 41-6

8. (a) For a rectangular guide of width 3.0 cm, what must the free-space wavelength of radiation be if it is to require 1.0 μs (= 10^{-6} s) for energy to traverse a 100-m length of guide? (b) What is the phase speed under these circumstances?

9. For a rectangular guide of width 3.0 cm, plot the phase speed, the group speed, and the guide wavelength as a function of the free-space wavelength. Assume the dominant mode.

10. Under what conditions will the guide wavelength in the guide of Fig. 41-9 be double the free-space wavelength?

11. (a) What guide wavelength does 10-cm radiation (free-space wavelength) exhibit in a rectangular guide whose width is 6.0 cm? Assume the dominant mode. (b) What is the cutoff wavelength for this guide?
Answer: (a) 18 cm. (b) 12 cm.

SECTION 41-7

12. (a) At a distance of 80 miles from a radio transmitter, how much later would you observe a wave emitted from the antenna? (b) If the radiation were a radio wave emitted from the sun? (c) If it were emitted from a stellar radio source 380 light years distant?

SECTION 41-8

13. An electromagnetic wave is traveling in the negative y-direction. At a particular position and time, the electric field is along the positive z-axis and has a magnitude of 100 V/m. What are the direction and magnitude of the magnetic field?   *Answer:* 3.3×10^{-7} T, in the negative x-direction.

14. The magnetic field equations for an electromagnetic wave in free space are $B_x = B \sin(ky + \omega t)$, $B_y = B_z = 0$. (a) What is the direction of propagation? (b) Write the electric field equations.

15. How does the displacement current density vary with space and time in a traveling plane electromagnetic wave?

16. (a) Starting with Eqs. 41-10 and 41-13, show that $E(x, t)$ satisfies the "wave equation"

$$\frac{\partial^2 E}{\partial x^2} = \mu_0 \epsilon_0 \frac{\partial^2 E}{\partial t^2}.$$

(b) What "wave equation" does $B(x, t)$ satisfy?

17. Show that (a) through (d) below satisfy Eqs. 41-10 and 41-13. In each of these, A is a constant. (a) $E = Ac(x - ct)$, $B = A(x - ct)$. (b) $E = Ac(x + ct)^{15}$, $B = -A(x + ct)^{15}$. (c) $E = Ace^{(x-ct)}$, $B = Ae^{(x-ct)}$. (d) $E = Ac \ln(x + ct)$, $B = -A \ln(x + ct)$. (e) Generalize these examples to show that $E = Acf(x - ct)$, $B = Af(x - ct)$ is a solution where f is *any* function of $(x - ct)$. What is the corresponding situation for functions of $(x + ct)$?

18. Show that the directions of the fields **E** and **B** in Fig. 41-11 are consistent with the direction of propagation of the radiation.

19. Show that the directions of the fields E and B in the various "patches" in Fig. 41-12 are consistent with the direction of propagation of the radiation.

SECTION 41-9

20. Sunlight strikes the earth, outside its atmosphere, with an intensity of 2.0 cal/cm²-min. Calculate E_m and B_m for sunlight, assuming it to be a wave like that of Fig. 41-12.

21. A #10 copper wire (diameter, 0.10 in.; resistance per 1000 ft, 1.00 Ω) carries a current of 25 A. Calculate (a) **E**, (b) **B**, and (c) **S** for a point on the surface of the wire.
Answer: (a) **E** = 8.2×10^{-2} V/m, parallel to the wire. (b) **B** = 3.9×10^{-3} T, tangent to the wire and perpendicular to its axis. (c) **S** = 260 W/m², radially inward.

22. Prove that for any point in an electromagnetic wave such as that of Fig. 41-13 the density of energy stored in the electric field equals that stored in the magnetic field.

23. A plane radio wave has $E_m \cong 10^{-4}$ V/m. Calculate (a) B_m and (b) the intensity of the wave, as measured by $\overline{S}$.
Answer: (a) 3.3×10^{-13} T. (b) 1.3×10^{-11} W/m².

24. If a coaxial cable has resistance, energy must flow from the fields into the conducting surfaces to provide Joule heating. How must the electric lines of force of Fig. 41-5a be modified in this case? (*Hint:* The Poynting vector near the surface must have a component pointing toward the surface.)

25. Analyze the flow of energy in the waveguide of Fig. 41-9, using the Poynting vector.

26. A coaxial cable (inner radius a, outer radius b) is used as a transmission line between a battery ε and a resistor R, as shown in Fig. 41-18. (a) Calculate **E, B** for $a < r < b$. (b) Calculate the Poynting vector **S** for $a < r < b$. (c) By suitably integrating the Poynting vector, show that the total power flowing across the annular cross section $a < r < b$ is ε^2/R. Is this reasonable?

figure 41-18
Problem 26

(d) Show that the direction of **S** is always from the battery to the resistor, no matter which way the battery is connected.

27. An airplane flying at a distance of 10 km from a radio transmitter receives a signal of intensity 10 μW/m². Calculate (a) the (rms) electric field at the airplane due to this signal; (b) the (rms) magnetic field at the airplane; (c) the total power radiated by the transmitter, assuming the transmitter to radiate isotropically and the earth to be a perfect absorber. *Answer:* (a) 6.1×10^{-2} V/m. (b) 2.1×10^{-10} T. (c) 13 kW.

28. A cube of edge a has its edges parallel to the x-, y-, and z-axes of a rectangular coordinate system. A uniform electric field **E** is parallel to the y-axis and a uniform magnetic field **B** is parallel to the x-axis. Calculate (a) the rate at which, according to the Poynting vector point of view, energy may be said to pass through each face of the cube and (b) the net rate at which the energy stored in the cube may be said to change.

29. Consider the possibility of standing waves:

$$E = E_m (\sin \omega t)(\sin kx)$$
$$B = B_m (\cos \omega t)(\cos kx).$$

(a) Show that these satisfy Eqs. 41-10 and 41-13 if E_m is suitably related to B_m and ω suitably related to k. What are these relationships? (b) Find the (instantaneous) Poynting vector. (c) Show that the time average power flow across any area is zero. (d) Describe the flow of energy in this problem. *Answer:* (a) Eq. 41-11a must hold for standing waves as well as for traveling waves. (b) $S = (E_m^2/4\mu_0 c) \sin 2\omega t \sin 2kx$.

30. Figure 41-19 shows a long resistanceless transmission line, delivering power from a battery to a resistive load. A steady current i exists as shown. (a) Sketch qualitatively the electric and magnetic fields around the line, and (b) show that, according to the Poynting vector point of view, energy travels from the battery to the resistor through the space around the line and not through the line itself. (*Hint:* Each conductor in the line is an equipotential surface, since the line has been assumed to have no resistance.)

31. Figure 41-20 shows a cylindrical resistor of length l, radius a, and resistivity ρ, carrying a current i. (a) Show that the Poynting vector **S** at the surface of the resistor is everywhere directed normal to the surface, as shown. (b) Show that the rate P at which energy flows into the resistor through its cylindrical surface, calculated by integrating the Poynting vector over this surface, is equal to the rate at which thermal energy is produced; that is,

$$\int \mathbf{S} \cdot d\mathbf{A} = i^2 R,$$

where $d\mathbf{A}$ is an element of area of the cylindrical surface. This shows that, according to the Poynting vector point of view, the energy that appears in a

figure 41-19
Problem 30

figure 41-20
Problem 31

resistor as thermal energy does not enter it through the connecting wires but through the space around the wires and the resistor. (*Hint:* **E** is parallel to the axis of the cylinder, in the direction of the current; **B** forms concentric circles around the cylinder, in a direction given by the right-hand rule.)

32. Figure 41-21 shows a parallel-plate capacitor being charged. (*a*) Show that the Poynting vector **S** points everywhere radially into the cylindrical volume. (*b*) Show that the rate P at which energy flows into this volume, calculated by integrating the Poynting vector over the cylindrical boundary of this volume, is equal to the rate at which the stored electrostatic energy increases; that is, that

$$\int \mathbf{S} \cdot d\mathbf{A} = Ad \, \frac{d}{dt} \left(\tfrac{1}{2} \epsilon_0 E^2 \right),$$

where Ad is the volume of the capacitor and $\tfrac{1}{2}\epsilon_0 E^2$ is the energy density for all points within that volume. This analysis shows that, according to the Poynting vector point of view, the energy stored in a capacitor does not enter it through the wires but through the space around the wires and the plates. (*Hint:* To find **S** we must first find **B**, which is the magnetic field set up by the displacement current during the charging process; see Fig. 40-1. Ignore fringing of the lines of **E**.)

33. A long hollow nonconducting cylinder (radius R, length l) carries a uniform charge per unit area of σ on its surface. An externally applied torque causes the cylinder to rotate at constant acceleration $\omega(t) = \alpha t$ about the cylinder axis. (*a*) Find **B** within the cylinder (treat it as a solenoid). (*b*) Find **E** at the inner surface of the cylinder. (*c*) Find **S** at the inner surface of the cylinder. (*d*) Show that the flux of **S** entering the interior volume of the cylinder is equal to $\dfrac{d}{dt}\left(\dfrac{\pi R^2 l}{2\mu_0} B^2 \right).$

Answer: (*a*) $B = \sigma R \omega \mu_0.$ (*b*) $E = \tfrac{1}{2}\mu_0 \sigma R^2 \alpha.$ (*c*) $S = \tfrac{1}{2}\mu_0 \sigma^2 R^3 \alpha^2 t.$

figure 41-21
Problem 32

42
the nature and propagation of light

This chapter connects what is true about the entire electromagnetic spectrum (see Fig. 41-1) to visible light, which is, of course, a part of that spectrum. In chapters that immediately follow we will focus on visible light. This is important to do because almost all of us, having adapted by genetic selection to the sun, have two electromagnetic receptors that function in this range and, indeed, define the range. These, of course, are our eyes. We must always remember that whatever we say about the visible part of the spectrum holds, at its foundations, for all other parts of the spectrum as well. The differences are largely in the nature of the means of production and detection in the various ranges.

Here we define "light" as radiation that can affect the eye. Figure 42-1, which shows the relative eye sensitivity of an assumed *standard*

figure 42-1
The relative eye sensitivity of an assumed *standard observer* at different wavelengths for a specified level of illumination.

918

observer to radiations of various wavelengths, shows that the center of the visible region is about 5.55×10^{-7} m or 555 nm. Light of this wavelength produces the sensation of yellow-green.*

In optics we often use the micrometer (abbr. μm) the nanometer (abbr. nm), and the Ångstrom (abbr. Å) as units of wavelength. They are defined from

$$1 \text{ micrometer} = 10^{-6} \text{ meter}$$

$$1 \text{ nanometer} = 10^{-9} \text{ meter}$$

$$1 \text{ Ångstrom} = 10^{-10} \text{ meter.}$$

Thus we can say that the center of the visible region is 0.555 μm, 555 nm, or 5550 Å.

The limits of the visible spectrum are not well defined because the eye sensitivity curve approaches the axis asymptotically at both long and short wavelengths. If we take the limits, arbitrarily, as the wavelengths at which the eye sensitivity has dropped to 1% of its maximum value, these limits are about 430 and 690 nm, less than a factor of two in wavelength. The eye can detect radiation beyond these limits if it is intense enough. In many experiments in physics one can use photographic plates or light-sensitive electronic detectors in place of the human eye.

42-2
ENERGY AND MOMENTUM

Energy is carried by electromagnetic waves from the sun to the earth or from an open fire to a hand placed nearby. We described the transport of energy by such a wave in free space in Section 41-9 by the Poynting vector **S**, or

$$\mathbf{S} = \frac{1}{\mu_0} \mathbf{E} \times \mathbf{B}, \qquad (42\text{-}1)$$

where **E** and **B** are the instantaneous values of the electric and magnetic field vectors.

Less familiar is the fact that electromagnetic waves may also transport linear momentum. In other words, it is possible to exert a pressure (a *radiation pressure*†) on an object by shining a light on it. Such forces must be small in relation to forces of our daily experience because we do not ordinarily notice them. We do not, after all, fall over backward when we raise a window shade in a dark room and let sunlight shine on us. Radiation pressure effects are, however, important in the life cycles of stars because of the incredibly high temperatures (2×10^7 K for our sun) that we associate with stellar interiors.

The first measurement of radiation pressure was made in 1901–1903 by Nichols and Hull at Dartmouth College and by Lebedev in Russia, about thirty years after the existence of such effects had been predicted theoretically by Maxwell.

* See "The Retinex Theory of Color Vision" by Edwin H. Land, *Scientific American*, December 1977, and "Color and Perception: the Work of Edwin Land in the Light of Current Concepts" by M. H. Wilson and R. W. Brocklebank, *Contemporary Physics*, December 1961, for a fascinating discussion of the problems of perception and the distinction between color as a characteristic of light and color as a perceived property of objects.
† See "Radiation Pressure" by G. E. Henry, *Scientific American*, June 1957; see also "The Pressure of Laser Light" by Arthur Ashkin, *Scientific American*, February 1972.

Let a parallel beam of light fall on an object for a time t, the incident light being *entirely absorbed* by the object. If energy U is absorbed during this time, the momentum p delivered to the object is given, according to Maxwell's prediction, by

$$p = \frac{U}{c} \qquad \text{(total absorption)}, \qquad (42\text{-}2a)$$

where c is the speed of light. The direction of **p** is the direction of the incident beam. If the light energy U is *entirely reflected*, the momentum delivered will be twice that given above, or

$$p = \frac{2U}{c} \qquad \text{(total reflection)}. \qquad (42\text{-}2b)$$

In the same way, twice as much momentum is delivered to an object when a perfectly elastic tennis ball is bounced from it as when it is struck by a perfectly inelastic ball (a lump of putty, say) of the same mass and speed. If the light energy U is partly reflected and partly absorbed, the delivered momentum will lie between U/c and $2U/c$.

EXAMPLE 1

A parallel beam of light with an energy flux S of 10 W/cm² falls for 1 hr on a perfectly reflecting plane mirror of 1.0-cm² area. (a) What momentum is delivered to the mirror in this time and (b) what force acts on the mirror?

(a) The energy that is reflected from the mirror is

$$U = (10 \text{ W/cm}^2)(1.0 \text{ cm}^2)(3600 \text{ s}) = 3.6 \times 10^4 \text{ J}.$$

The momentum delivered after 1 hour's illumination is

$$p = \frac{2U}{c} = \frac{(2)(3.6 \times 10^4 \text{ J})}{3 \times 10^8 \text{ m/s}} = 2.4 \times 10^{-4} \text{ kg m/s}.$$

(b) From Newton's second law, the average force on the mirror is equal to the average rate at which momentum is delivered to the mirror, or

$$F = \frac{p}{t} = \frac{2.4 \times 10^{-4} \text{ kg m/s}}{3600 \text{ s}} = 6.7 \times 10^{-8} \text{ N}.$$

This is a small force.

Nichols and Hull, in 1903, measured radiation pressures and verified Eq. 42-2, using a torsion balance technique. They allowed light to fall on mirror M in Fig. 42-2; the radiation pressure caused the balance arm to turn through a measured angle θ, twisting the torsion fiber F. Assuming a suitable calibration for their torsion fiber, the experimenters could arrive at a numerical value for this pressure. Nichols and Hull measured the intensity of their light beam by allowing it to fall on a blackened metal disk of known absorptivity and by measuring the temperature rise of this disk. In a particular run these experimenters measured a radiation pressure of 7.01×10^{-6} N/m²; for their light beam, the value predicted, using Eq. 42-2, was 7.05×10^{-6} N/m², in excellent agreement. Assuming a mirror area of 1 cm², this represents a force on the mirror of only 7×10^{-10} N, about 100 times smaller than the force calculated in Example 1.

The success of the experiment of Nichols and Hull was the result in large part of the care they took to eliminate spurious deflecting effects caused by changes in the speed distribution of the molecules in the gas

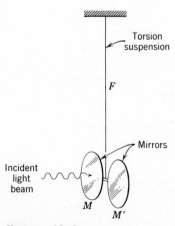

figure 42-2
Suggesting the experiment of Nichols and Hull, used to measure radiation pressure.

surrounding the mirror. These changes were brought about by the small rise in the temperature of the mirror as it absorbed light energy from the incident beam. This "radiometer effect" is responsible for the spinning action of the familiar toy radiometers when placed in a beam of sunlight. In a perfect vacuum such effects would not occur, but in the best vacuums available in 1903 radiometer effects were present and had to be taken specifically into account in the design of the experiment.

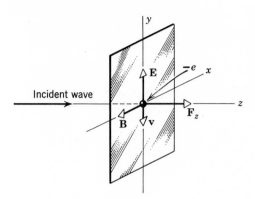

figure 42-3
An incident plane light wave falls on an electron in a thin resistive sheet. Instantaneous values of **E**, **B**, the electron velocity **v**, and the radiation force **F**$_z$ are shown.

To demonstrate the transport of momentum from Maxwell's equations in a particular case, let a plane electromagnetic wave traveling in the z-direction fall on a large thin sheet of a material of high resistivity as in Fig. 42-3. A small part of the incident energy will be absorbed within the sheet, but most of it will be transmitted if the sheet is thin enough.*

The incident wave vectors **E** and **B** vary with time at the sheet as

$$\mathbf{E} = \mathbf{E}_m \sin \omega t \qquad (42\text{-}3)$$

and

$$\mathbf{B} = \mathbf{B}_m \sin \omega t \qquad (42\text{-}4)$$

where **E** is parallel to the $\pm y$ axis and **B** is parallel to the $\pm x$ axis.

In Section 31-4 we saw that the effect of a (constant) electric force ($= -eE$) on a conduction electron in a metal was to make it move with a (constant) drift speed v_d. The electron behaves as if it is immersed in a viscous fluid, the electric force acting on it being counterbalanced by a "viscous" force, which may be taken as proportional to the electron speed. Thus for a constant field E, after equilibrium is established,

$$eE = bv_d, \qquad (42\text{-}5)$$

where b is a resistive damping coefficient. The electron equilibrium speed, dropping the subscript d, is thus

$$v = \frac{eE}{b}. \qquad (42\text{-}6)$$

If the applied electric field varies with time and if the variation is slow enough, the electron speed can continually readjust itself to the changing value of E so that its speed continues to be given essentially by its equilibrium value (Eq. 42-6) at all times. These readjustments are more rapidly made the more viscous the medium, just as a stone falling in air reaches a constant equilibrium rate of descent only relatively slowly but one falling in a viscous oil does so quite rapidly. We assume that the sheet in Fig. 42-3 is so viscous, that is, that its resistivity is so high, that Eq. 42-6 remains true even for the rapid oscillations of E in the incident light beam.

As the electron vibrates parallel to the y-axis, it experiences a *second* force

---

* Some of the incident energy will also be reflected, but the reflected wave is of such low intensity that we can ignore it in the derivation that follows.

due to the *magnetic* component of the wave. This force $\mathbf{F}_z$ $(= -e\mathbf{v} \times \mathbf{B})$ points in the z-direction, being at right angles to the plane formed by $\mathbf{v}$ and $\mathbf{B}$, that is, the xy-plane. The instantaneous magnitude of $\mathbf{F}_z$ is given by

$$F_z = evB = \frac{e^2EB}{b}.\qquad(42\text{-}7)$$

$\mathbf{F}_z$ always points in the positive z-direction because $\mathbf{v}$ and $\mathbf{B}$ reverse their directions simultaneously; this force is, in fact, the mechanism by which the radiation pressure acts on the sheet of Fig. 42-3.

From Newton's second law, F_z is the rate dp_e/dt at which the incident wave delivers momentum to each electron in the sheet, or

$$\frac{dp_e}{dt} = \frac{e^2EB}{b}.\qquad(42\text{-}8)$$

Momentum is delivered at this rate to every electron in the sheet and thus to the sheet itself. It remains to relate the momentum transfer to the sheet to the absorption of energy within the sheet.

The electric field component of the incident wave does work on each oscillating electron at an instantaneous rate (see Eq. 42-6) given by

$$\frac{dU_e}{dt} = F_Ev = (eE)\left(\frac{eE}{b}\right) = \frac{e^2E^2}{b}.$$

Note that the magnetic force $\mathbf{F}_z$, always being at right angles to the velocity $\mathbf{v}$, does no work on the oscillating electron. Equation 41-11b shows that for a plane wave in free space B and E are related by

$$E = Bc.$$

Substituting above for *one* of the E's leads to

$$\frac{dU_e}{dt} = \frac{e^2EBc}{b}.\qquad(42\text{-}9)$$

This equation represents the rate, per electron, at which energy is absorbed from the incident wave.

Comparing Eqs. 42-8 and 42-9 shows that

$$\frac{dp_e}{dt} = \frac{1}{c}\frac{dU_e}{dt}.$$

Integrating yields
$$\int_0^t \frac{dp_e}{dt}\, dt = \frac{1}{c}\int_0^t \frac{dU_e}{dt}\, dt,$$

or
$$p_e = \frac{U_e}{c},\qquad(42\text{-}10)$$

where p_e is the momentum delivered to a single electron in any given time t and U_e is the energy absorbed by that electron in the same time interval. Multiplying each side by the number of free electrons in the sheet leads to Eq. 42-2a.

Although we derived this relation (Eq. 42-10) for a particular kind of absorber, no characteristics of the absorber—for example, the resistive damping coefficient b—remain in the final expression. This is as it should be because Eq. 42-10 is a general property of radiation absorbed by *any* material.

42-3 THE SPEED OF LIGHT*

Light travels so fast that there is nothing in our daily experience to suggest that its speed is not infinite. It calls for considerable insight even to ask, "How fast does light travel?" Galileo asked himself this question and actually tried to answer it experimentally. His book, *Two New Sciences*, published in 1638, is written in the form of a conversation

* See "The Speed of Light" by J. H. Rush, *Scientific American*, August 1955.

among three persons called Salviati, Sagredo, and Simplicio. Here is part of what they say about the speed of light.

Simplicio: Everyday experience shows that the propagation of light is instantaneous; for when we see a piece of artillery fired, at a great distance, the flash reaches our eyes without lapse of time; but the sound reaches the ear only after a noticeable interval.

Sagredo: Well, Simplicio, the only thing I am able to infer from this familiar bit of experience is that sound, in reaching our ear, travels more slowly than light; it does not inform me whether the coming of the light is instantaneous or whether, although extremely rapid, it still occupies time. . . .

Salviati, who speaks with Galileo's voice, then describes a possible method (actually carried out) for measuring the speed of light. He and an assistant stand facing each other some distance apart, at night. Each carries a lantern which can be covered or uncovered at will. Galileo started the experiment by uncovering his lantern. When the light reached the assistant he uncovered his own lantern, whose light was then seen by Galileo. Galileo tried to measure the time between the instant at which he uncovered his own lantern and the instant at which the light from his assistant's lantern reached him. For a one-mile separation we now know that the round trip travel time would be only 11 μs. This is much less than human reaction times, so the method fails.

To measure a large velocity directly, we must either measure a small time interval or use a long base line. This situation suggests that astronomy, which deals with great distances, might be able to provide an experimental value for the speed of light; this proved to be true. Although it would be desirable to time the light from the sun as it travels to the earth, there is no way of knowing when the light that reaches us at any instant left the sun; we must use subtler astronomical methods.

Note, however, that microwave pulses are quite regularly reflected from the moon; this gives a 7.68×10^8-m base line (there and back) for timing purposes. The speed of light (and of microwaves) is so well known now from other experiments that we use these measurements to measure the lunar distance accurately. Microwave signals have also been reflected from Venus.

In 1675 Ole Roemer, a Danish astronomer working in Paris, made some observations of the moons of Jupiter (see Problem 20) from which a speed of light of 2×10^8 m/s may be deduced. About fifty years later James Bradley, an English astronomer, made some astronomical observations of an entirely different kind from which a value of 3.0×10^8 m/s may be deduced.

In 1849 Hippolyte Louis Fizeau (1819–1896), a French physicist, first measured the speed of light by a nonastronomical method, obtaining a value of 3.13×10^8 m/s. Figure 42-4 shows Fizeau's apparatus. Let us first ignore the toothed wheel. Light from source S is made to converge

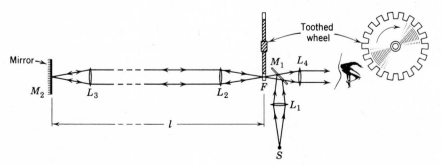

figure 42-4
Fizeau's apparatus for measuring the speed of light.

by lens L_1, is reflected from mirror M_1, and forms in space at F an image of the source. Mirror M_1 is a so-called "half-silvered mirror"; its reflecting coating is so thin that only half the light that falls on it is reflected, the other half being transmitted.

Light from the image at F enters lens L_2 and emerges as a parallel beam; after passing through lens L_3 it is reflected back along its original direction by mirror M_2. In Fizeau's experiment the distance l between M_2 and F was 8630 meters or 5.36 miles. When the light strikes mirror M_1 again, some will be transmitted, entering the eye of the observer through lens L_4.

The observer will see an image of the source formed by light that has traveled a distance $2l$ between the wheel and mirror M_2 and back again. To time the light beam a marker of some sort must be put on it. This is done by "chopping" it with a rapidly rotating toothed wheel. Suppose that during the round-trip travel time of $2l/c$ the wheel has turned just enough so that, when the light from a given "burst" returns to the wheel, point F is covered by a tooth. The light will hit the face of the tooth that is toward M_2 and will not reach the observer's eye.

If the speed of the wheel is exactly right, the observer will not see any of the bursts because each will be screened by a tooth. The observer measures c by increasing the angular speed ω of the wheel from zero until the image of source S disappears. Let θ be the angular distance from the center of a gap to the center of a tooth. The time needed for the wheel to rotate a distance θ is the round-trip travel time $2l/c$. In equation form,

$$\frac{\theta}{\omega} = \frac{2l}{c} \quad \text{or} \quad c = \frac{2\omega l}{\theta}. \quad (42\text{-}11)$$

This "chopped beam" technique, suitably modified, is used today to measure the speeds of neutrons and other particles.

EXAMPLE 2

The wheel used by Fizeau had 720 teeth. What is the smallest angular speed at which the image of the source will vanish?

The angle θ is 1/1440 rev; solving Eq. 42-11 for ω gives

$$\omega = \frac{c\theta}{2l} = \frac{(3.00 \times 10^8 \text{ m/s})(1/1440 \text{ rev})}{(2)(8630 \text{ m})} = 12.1 \text{ rev/s}.$$

The French physicist Foucault (1819–1868) greatly improved Fizeau's method by substituting a rotating mirror for the toothed wheel. The American physicist Albert A. Michelson (1852–1931) conducted an extensive series of measurements of c, extending over a fifty-year period, using this technique.

We must view the speed of light within the larger framework of the speed of electromagnetic radiation in general. It is a significant experimental confirmation of Maxwell's theory of electromagnetism that the speed in free space of waves in all parts of the electromagnetic spectrum has the same value c. Table 42-1 shows some selected measurements that have been made of the speed of electromagnetic radiation since Galileo's day. It stands as a monument to man's persistence and ingenuity. Note in the last column how the uncertainty in the measurement has improved through the years. Note also the international character of the effort and the variety of methods.

Table 42-1
The speed of electromagnetic radiation in free space (some selected measurements)*

Date	Experimenter	Country	Method	Speed (km/s)	Uncertainty (km/s)
1600(?)	Galileo	Italy	Lanterns and shutters	"If not instantaneous, it is extraordinarily rapid"	
1675	Roemer	France	Moons of Jupiter	200,000	
1729	Bradley	England	Aberration of starlight	304,000	
1849	Fizeau	France	Toothed wheel	313,300	
1862	Foucault	France	Rotating mirror	298,000	500
1876	Cornu	France	Toothed wheel	299,990	200
1880	Michelson	U.S.A.	Rotating mirror	299,910	50
1883	Newcomb	England	Rotating mirror	299,860	30
1906	Rosa and Dorsey	U.S.A.	Electromagnetic theory	299,781	10
1923	Mercier	France	Standing waves on wires	299,782	15
1926	Michelson	U.S.A.	Rotating mirror	299,796	4
1928	Karolus and Mittelstaedt	Germany	Kerr cell	299,778	10
1932	Michelson, Pease, and Pearson	U.S.A.	Rotating mirror	299,774	11
1941	Anderson	U.S.A.	Kerr cell	299,776	14
1950	Bergstrand	Sweden	Geodimeter	299,792.7	0.25
1950	Essen	England	Microwave cavity	299,792.5	3
1950	Bol and Hansen	U.S.A.	Microwave cavity	299,789.3	0.4
1951	Aslakson	U.S.A.	Shoran radar	299,794.2	1.9
1952	Rank, Ruth, and Ven der Sluis	U.S.A.	Molecular spectra	299,776	7
1952	Froome	England	Microwave interferometer	299,792.6	0.7
1954	Florman	U.S.A.	Microwave interferometer	299,795.1	1.9
1957	Bergstrand	Sweden	Geodimeter	299,792.85	0.16
1958	Froome	England	Microwave interferometer	299,792.50	0.10
1965	Kolibayev	U.S.S.R.	Geodimeter	299,792.6	0.06
1967	Grosse	West Germany	Geodimeter	299,792.5	0.05
1973	Evenson, Wells, Peterson, Danielson, and Day	U.S.A.	Laser techniques	299,792.4574	0.0012

* See "Some Recent Determinations of the Velocity of Light, III" by Joseph F. Mulligan, *American Journal of Physics,* October 1976. Here the 1974 measurements of Blayney et al. at the National Physical Laboratory in England are described. They are in very close agreement with the results of Evenson et al. (1973).

The task of arriving at a single "best" value for any physical quantity, such as c, from many independent measurements is usually difficult because it involves a careful evaluation of each measurement and a complex averaging process, which takes into account other physical quantities with which the quantity in question may be associated. In the case of c, however, the matter is straightforward. The latest (1973) compilers† of the "best" values of the physical constants state: "All past measurements of c have been rendered obsolete by Evenson et al.'s recent measurement . . ."; this measurement is the last entry in Table 42-1.

These workers, at the National Bureau of Standards in Boulder, Colorado, measured the frequency ν of a certain radiation emitted by a helium-neon laser by comparing it directly with the oscillation frequency of the cesium clock, which is used to define the second (see

† See footnote to Appendix B.

Appendix A). Then, using accurate measurements of the wavelength of this radiation made by several groups of workers, they computed c from the relation $c = \lambda\nu$ and deduced the value displayed in Appendix B, namely,

$$c = (299{,}792.4574 \pm 0.0012) \text{ km/s.}$$

The largest source of uncertainty in this measurement is that of the definition of the meter (needed to find the wavelength) in terms of radiations from the krypton-86 atom (see Appendix A).

It is now clear that the best measurements of c are *not* made by timing the passage of light over a measured distance, as by Fizeau in 1849 and by Michelson et al. in 1932 (see Table 42-1). They are made by measuring the frequency ν and the wavelength λ of the light and computing c from $c = \lambda\nu$. This holds true for either traveling waves, as in the measurements of Evenson and his collaborators, described above, or for standing waves, an example of which we now describe. These standing wave experiments will also remind us that c in free space has the same value throughout the entire electromagnetic spectrum and is not confined to visible light.

We describe here the "microwave cavity method" used by Essen in England and by Bol and Hansen in the U.S.A. It employs standing electromagnetic waves confined to a cavity rather than traveling waves in free space.

It is possible to convert a section of waveguide such as that of Fig. 41-9 into a resonant cavity by closing it with two metal caps; see Fig. 42-5. The pattern of oscillations in the cavity is closely related to that in the guide and exhibits the same "guide wavelength" λ_g. The guide wavelength is related to the cavity length l by

$$\lambda_g = \frac{2l}{n} \qquad n = 1,\, 2,\, 3,\, \cdots, \qquad (42\text{-}12)$$

which is the same relationship used for acoustic waves in closed pipes; n (=3 for Fig. 42-5) gives the number of half-waves contained in the cavity.

The procedure is to measure λ_g for such a cavity, which has been tuned to resonance, and then, using Eq. 41-4,

$$\lambda_g = \frac{\lambda}{\sqrt{1 - \left(\frac{\lambda}{2a}\right)^2}}, \qquad (42\text{-}13)$$

calculate the free-space wavelength λ. From the measured resonant frequency, the speed c can be found from $c = \lambda\nu$.

figure 42-5
A resonant cavity constructed from a section of waveguide; compare Fig. 41-9. For simplicity the lines of **E** are not shown in (a) and those of **B** are not shown in (b).

EXAMPLE 3

Essen of the National Physical Laboratory in England made a resonant cavity measurement of the speed of electromagnetic waves in 1950; see Table 42-1. His cavity was made of a circular waveguide rather than a rectangular one; it can be shown that, for the oscillation pattern used by him, the geometrical factor $2a$ in Eq. 42-13 must be replaced by $1.64062R$, where R is the guide radius. The cavity radius was 3.25876 cm; the cavity length was 15.64574 cm and it proved to resonate at 9.498300×10^9 Hz. At resonance it was determined that there were eight half-waves in the cavity. What value of c results?

From Eq. 42-12, computing only an approximate result,

$$\lambda_g = \frac{2l}{n} = \frac{(2)(15.6 \text{ cm})}{8} = 3.90 \text{ cm.}$$

Substituting into Eq. 42-13, suitably modified for a circular waveguide, yields

$$3.90 \text{ cm} = \frac{\lambda}{\sqrt{1 - \left(\dfrac{\lambda}{(1.64)(3.26 \text{ cm})}\right)^2}}.$$

Solving this equation for λ yields $\lambda = 3.15$ cm. Finally, we have

$$c = \lambda\nu = (3.15 \text{ cm})(9.50 \times 10^9 \text{ Hz}) = 2.99 \times 10^8 \text{ m/s}.$$

For practical reasons Essen analyzed his data in a more roundabout way than that given. His final result, based on many measurements under different conditions and carried to much greater accuracy than that illustrated in the above example, was 299,792.5 km/s with an uncertainty of 3.0 km/s.

When we say that the speed of sound in dry air at 0°C is 331.7 m/s, we imply a reference frame fixed with respect to the air mass. When we say that the speed of light in free space is 2.99792458×10^8 m/s, what reference frame is implied? It cannot be the medium through which the light wave travels because, in contrast to sound, no medium is required.

The concept of a wave requiring no medium was abhorrent to the physicists of the nineteenth century, influenced as they then were by a false analogy between light waves and sound waves or other purely mechanical disturbances. These physicists postulated the existence of an *ether,* which was a tenuous substance that filled all space and served as a medium of transmission for light. The ether was required to have a vanishingly small density to account for the fact that it could not be observed by any known means in an evacuated space.

The ether concept, although it proved useful for many years, did not survive the test of experiment. In particular, careful attempts to measure the speed of the earth through the ether always gave the result of zero.† Physicists were not willing to believe that the earth was permanently at rest in the ether and that all other bodies in the universe were in motion through it. Other hypotheses about the nature of the propagation of light also proved unsatisfactory for one reason or another.

Einstein in 1905 resolved the difficulty of understanding the propagation of light by making a bold postulate: If a number of observers are moving (at uniform velocity) with respect to each other and to a source of light and if each observer measures the speed of the light emerging from the source, *they will all obtain the same value.* This is the fundamental assumption of Einstein's special theory of relativity. It does away with the need for an ether by asserting that the speed of light is the same in *all* reference frames; none is singled out as fundamental. The theory of relativity, derived from this postulate, has been subject to many experimental tests, from which agreement with the predictions of theory has always emerged. These agreements, extending over half a century, lend strong support to Einstein's basic postulate about light propagation.

Figure 42-6 focuses specifically on the fundamental problem of light propagation. A source of light, at rest in reference frame S', emits a

42·4
MOVING SOURCES AND OBSERVERS*

figure 42-6
Observers S and S', who are in relative motion, each observe a light pulse P. The pulse is emitted from a source, not shown, that is at rest in the S' frame of reference.

---

* For further information about the material of this and the next section see Supplementary Topic V.
† See Section 45-7, which describes the crucial experiment of Michelson and Morley.

light pulse P whose speed v' is measured by an observer at rest in this same frame. From the point of view of an observer in reference frame S, frame S' and its associated observer are moving in the positive x-direction at speed u. Question: What speed v would observer S measure for the light pulse P? Einstein's hypothesis asserts that *each* observer would measure the same speed c, or that

$$v = v' = c.$$

This hypothesis contradicts the classical law of addition of velocities (see Section 4-6), which asserts that

$$v = v' + u. \tag{42-14}$$

This law, which is so familiar that it seems (incorrectly) to be intuitively true, is in fact based on observations of gross moving objects in the world about us. Even the fastest of these—an earth satellite, say—is moving at a speed that is quite small compared to that of light. The body of experimental evidence that underlies Eq. 42-14 thus represents a severely restricted area of experience, namely, experiences in which $v' \ll c$ and $u \ll c$. If we assume that Eq. 42-14 holds for all particles regardless of speed, we are making a gross extrapolation. Einstein's theory of relativity predicts that this extrapolation is indeed not valid and that Eq. 42-14 is a limiting case of a more general relationship that holds for light pulses and for material particles, whatever their speed, or

$$v = \frac{v' + u}{1 + v'u/c^2}. \tag{42-15}$$

Equation 42-15 is quite indistinguishable from Eq. 42-14 at low speeds, that is, when $v' \ll c$ and $u \ll c$; see Example 4.

If we apply Eq. 42-15 to the case in which the moving object is a light pulse, and if we put $v' = c$, we obtain

$$v = \frac{c + u}{1 + cu/c^2} = c.$$

This is consistent, as it must be, with the fundamental assumption on which the derivation of Eq. 42-15 is based; it shows that *both* observers measure the same speed c for light. Equation 42-14 predicts (incorrectly) that the speed measured by S will be $c + u$. Figure 42-7 shows that the (correct) Eq. 42-15 and the (approximate) Eq. 42-14 cannot be distin-

figure 42-7
The speed of a particle P, as seen by observer S in Fig. 42-6, for the special case of $v' = u$. All speeds are expressed as a ratio to c, the speed of light. The vertical arrow corresponds to about 0.25×10^8 m/s. ($= 5 \times 10^7$ mi/h.)

guished from each other at speeds that are small compared with the speed of light.

EXAMPLE 4

Suppose that $v' = u = 1.0 \times 10^5$ m/s $(=0.0003c = 230{,}000$ mi/h). What error is made in using Eq. 42-14 rather than Eq. 42-15 to calculate v?

Equation 42-14 gives

$$v = v' + u = 1.0 \times 10^5 \text{ m/s} + 1.0 \times 10^5 \text{ m/s} = 2.0 \times 10^5 \text{ m/s}.$$

Equation 42-15 gives

$$v = \frac{v' + u}{1 + v'u/c^2}$$

$$= \frac{1.0 \times 10^5 \text{ m/s} + 1.0 \times 10^5 \text{ m/s}}{1 + \left(\dfrac{1.0 \times 10^5 \text{ m/s}}{3.0 \times 10^8 \text{ m/s}}\right)^2}$$

$$= \frac{2.0 \times 10^5 \text{ m/s}}{1.00000011}.$$

Even at 230,000 mi/hr the error in Eq. 42-14 is immeasurably small.

EXAMPLE 5

Two electrons are ejected in opposite directions from radioactive atoms in a sample of radioactive material. Let each electron have a speed, as measured by a laboratory observer, of $0.6c$ (this corresponds to a kinetic energy of 130 keV). What is the speed of one electron as seen from the other?

Equation 42-14 gives

$$v = v' + u = 0.6c + 0.6c = 1.2c.$$

Equation 42-15 gives

$$v = \frac{v' + u}{1 + v'u/c^2} = \frac{0.6c + 0.6c}{1 + (0.6c)^2/c^2} = 0.88c.$$

This example shows that for speeds that are comparable to c, Eqs. 42-14 and 42-15 yield rather different results. A wealth of indirect experimental evidence points to the latter result as being correct.

42-5
DOPPLER EFFECT

We have seen that the same speed is measured for light no matter what the relative speeds of the light source and the observer are. The measured frequency and wavelength will change, but always in such a way that their product, which is the velocity of light, remains constant. Such frequency shifts are called *Doppler shifts,* after Johann Doppler (1803–1853), who first predicted them.

In Section 20-7 we showed that if a source of *sound* is moving away from an observer at a speed u, the frequency heard by the observer (see Eq. 20-10, which we have rearranged and in which we have substituted u for v_s) is

$$\nu' = \nu \frac{1}{1 + u/v}. \qquad \begin{cases} 1. \text{ sound wave} \\ 2. \text{ observer fixed in medium} \\ 3. \text{ source receding from observer} \end{cases} \qquad (42\text{-}16)$$

In this equation ν is the frequency heard when the source is at rest and v is the speed of sound.

If the source is at rest in the transmitting medium but the observer is moving away from the source at speed u, the observed frequency (see Eq. 20-9b, in which u has been substituted for v_0) is

$$\nu' = \nu\left(1 - \frac{u}{v}\right) \cdot \quad \begin{cases} \text{1. sound wave} \\ \text{2. source fixed in medium} \\ \text{3. observer receding from source} \end{cases} \qquad (42\text{-}17)$$

Even if the relative separation speeds u of the source and the observer are the same, the frequencies predicted by Eqs. 42-16 and 42-17 are different. This is not surprising, because a sound source moving through a medium in which the observer is at rest is physically different from an observer moving through that medium with the source at rest, as comparison of Figs. 20-10 and 20-11 shows.

We might be tempted to apply Eqs. 42-16 and 42-17 to light, substituting c, the speed of light, for v, the speed of sound. For light, as contrasted with sound, however, it has proved impossible to identify a medium of transmission relative to which the source and the observer are moving. This means that "source receding from observer" and "observer receding from source" are physically identical situations and must exhibit *exactly the same* Doppler frequency. As applied to light, either Eq. 42-16 or Eq. 42-17 or both must be incorrect. The Doppler frequency predicted by the theory of relativity is, in fact,

$$\nu' = \nu\frac{1 - u/c}{\sqrt{1 - (u/c)^2}} \cdot \quad \begin{cases} \text{1. light wave} \\ \text{2. source and observer} \\ \quad \text{separating} \end{cases} \qquad (42\text{-}18)$$

In all three of the foregoing equations we obtain the appropriate relations for the source and the observer *approaching* each other if we replace u by $-u$.

Equations 42-16, 42-17, and 42-18 are not so different as they seem if the ratio u/c is small enough. This was made clear in Example 3, Chapter 20, for the first two of these equations. Let us expand Eqs. 42-16, 42-17, and 42-18 by the binomial theorem, as in the example referred to. The equations then become, substituting c for v,

$$\nu' = \nu\left[1 - \frac{u}{c} + \left(\frac{u}{c}\right)^2 + \cdots\right], \qquad (42\text{-}16a)$$

$$\nu' = \nu\left(1 - \frac{u}{c}\right), \qquad (42\text{-}17a)$$

and

$$\nu' = \nu\left[1 - \frac{u}{c} + \frac{1}{2}\left(\frac{u}{c}\right)^2 + \cdots\right] \cdot \qquad (42\text{-}18a)$$

The ratio u/c for all available monochromatic light sources, even those of atomic dimensions, is small. This means that successive terms in these equations become small rapidly and, depending on the accuracy required, only a limited number of terms need be retained.

Under nearly all circumstances the differences among these three equations are not important. Nevertheless, it is of extreme interest to carry out at least one experiment precisely enough to serve as a test of Eq. 42-18a and thus, in part, of the theory of relativity.

H. E. Ives and G. R. Stilwell carried out such a precision experiment in 1938. They sent a beam of hydrogen atoms, generated in a gas discharge, down a tube at speed u, as in Fig. 42-8a. They could observe light emitted by these atoms in a direction opposite to **u** (atom 1, for example) using a mirror, and also in a direction parallel to **u** (atom 2, for example). With a precision spectrograph, they could photograph a particular characteristic spectrum line in this light, obtain-

figure 42-8
The Ives-Stilwell experiment.

ing, on a frequency scale, the lines marked ν_1' and ν_2' in Fig. 42-8b. It is also possible to photograph, on the same photographic plate, a line corresponding to light emitted from *resting* atoms; such a line appears as ν in Fig. 42-9b. A fundamental measured quantity in this experiment is $\Delta\nu/\nu$, defined from

$$\frac{\Delta\nu}{\nu} = \frac{\Delta\nu_2 - \Delta\nu_1}{\nu},$$
(42-19)

(see Fig. 42-8b). It measures the extent to which the frequency of the light from resting atoms fails to lie halfway between the frequencies ν_1' and ν_2'. Table 42-2 shows that the measured results agree with the formula predicted by the theory of relativity (Eq. 42-18a) and not with the classical formula borrowed from the theory of sound propagation in a material medium (Eq. 42-16a).

Table 42-2
The Ives-Stilwell experiment*

$\frac{\Delta\nu}{\nu}$, 10^{-5}	Speed of moving atoms ($=u$), 10^6 m/s			
	0.865	1.01	1.15	1.33
Theoretical value according to classical theory (Eq. 42-16a)	1.67	2.26	2.90	3.94
Theoretical value according to the theory of relativity (Eq. 42-18a)	0.835	1.13	1.45	1.97
Experimental value	0.762	1.1	1.42	1.9

* See Eq. 42-19; the table shows only part of the data taken by Ives and Stilwell.

Ives and Stilwell did not present their experimental results as evidence for the support of Einstein's theory of relativity but rather gave them an alternative theoretical explanation. Modern observers, looking not only at their excellent experiment but at the whole range of experimental evidence, now give the Ives-Stilwell experiment the interpretation we have described for it above.

The Doppler effect for light finds many applications in astronomy, where it is used to determine the speeds at which luminous heavenly bodies are moving toward us or receding from us. Such Doppler shifts measure only the radial or line-of-sight components of the relative velocity. All galaxies† for which such measurements have been made

† See "The Red-Shift" by Allen R. Sandage, *Scientific American*, September 1956.

(a)   (b)

figure 42-9

(a) The central spot is a nebula in the constellation Corona Borealis; it is 130,000,000 light years distant. (b) The central streak shows the distribution in wavelength of the light emitted from this nebula. The two vertical dark bands show the presence of calcium. The horizontal arrow shows that these calcium lines occur at longer wavelengths than those for terrestrial light sources containing calcium, the length of the arrow representing the wavelength shift. Measurement of this shift shows that the galaxy is receding from us at 13,400 mi/s. The lines above and below the central streak represent light from a terrestrial source, used to establish a wavelength scale. (Courtesy Mount Wilson and Mount Palomar Observatories.)

(Fig. 42-9) appear to be receding from us, the recession velocity being greater for the more distant galaxies; these observations are the basis of the expanding-universe concept.

EXAMPLE 6

Certain characteristic wavelengths in the light from a galaxy in the constellation Virgo are observed to be increased in wavelength, as compared with terrestrial sources, by about 0.4%. What is the radial speed of this galaxy with respect to the earth? Is it approaching or receding?

If λ is the wavelength for a terrestrial source, then

$$\lambda' = 1.004\lambda.$$

Since we must have $\lambda'\nu' = \lambda\nu = c$, we can write this as

$$\nu' = 0.996\nu.$$

This frequency shift is so small that, in calculating the source velocity, it makes no practical difference whether we use Eq. 42-16, 42-17, or 42-18. Using Eq. 42-17 we obtain

$$\nu' = 0.996\nu = \nu\left(1 - \frac{u}{c}\right).$$

Solving yields $u/c = 0.004$, or $u = (0.004)(3 \times 10^8 \text{ m/s}) = 1.2 \times 10^6$ m/s or 2.7×10^6 mi/h. The galaxy is *receding*; had u turned out to be negative, the galaxy would have been moving toward us.

---

questions

1. How might an eye-sensitivity curve like that of Fig. 42-1 be measured?
2. Why are danger signals in red, when the eye is most sensitive to yellow-green?
3. Comment on this definition of the limits of the spectrum of visible light given by a physiologist: "The limits of the visible spectrum occur when the eye is no better adapted than any other organ of the body to serve as a detector."
4. The human body can "detect" electromagnetic radiations in parts of the electromagnetic spectrum apart from the visual range. Give examples for the infrared, ultraviolet, X-ray, and gamma ray regions. Some of the "detection processes" are not very pleasant.

5. In connection with Fig. 42-1 (a) do you think it possible that the wavelength of maximum sensitivity could vary if the intensity of the light is changed? (b) What might the curve of Fig. 42-1 look like for a group of color blind people who could not, for example, distinguish red from green? (See "The Science of Color" by S. J. Edwards, *Physics Education*, June 1975.)

6. Suppose that human eyes were insensitive to visible light but were very sensitive to infrared light. What environmental changes would be needed if you were to (a) walk down a long corridor and (b) drive a car? Would the phenomenon of color exist? How would traffic lights have to be modified?

7. How can an object absorb light energy without absorbing momentum?

8. Name two historic experiments, in addition to the radiation pressure measurements of Nichols and Hull, in which a torsion balance was used. Both are described in this book, one in Part 1 and one in Part 2.

9. A searchlight sends out a parallel beam of light. Does the searchlight experience any force associated with the emission of light?

10. In Section 42-2 we stated that the force on the mirror in the radiation pressure experiment of Nichols and Hull (see Fig. 42-2) was about 7×10^{-10} N. Identify an object whose weight at the earth's surface is about this magnitude (Table 1-4 might help).

11. Discuss this (1972) statement by Lewis M. Branscomb, Director of the National Bureau of Standards. "Scientists have looked forward to the possibility of using one gauge—one 'yardstick so to speak'—not only for the three dimensions of space but for the fourth dimension of time as well. To interchange clocks and rules scientists must know the speed with which light travels. . . . With this demonstration that both the space (wavelength) and time (frequency) dimensions of a single light source can be measured with prodigious accuracy, this goal is now within our grasp." What does he mean? (See *Science News*, February 1972.)

12. As you recline in a beach chair in the sun why are you so very conscious of the thermal energy that is delivered to you, but totally unresponsive to the linear momentum delivered from the same source?

13. Some advocate that the speed of light, now known to such high accuracy, be proclaimed as a defined constant rather than a measured one. What are the implications of this in terms of the present definitions of the meter and the second (Appendix A)?

14. When a parallel beam of light falls on an object the momentum transfers are given by Eqs. 42-2. Do these equations still hold if the light source is moving rapidly toward or away from the object at, perhaps, a speed of 0.1 c?

15. How could Galileo test experimentally that reaction times were an overwhelming source of error in his attempt to measure the speed of light, described on p. 923?

16. It has been suggested that the velocity of light may change slightly in value as time goes on. Can you find any evidence for this in Table 42-1?

17. Can you think of any "every day" observation (that is, without experimental apparatus) to show that the speed of light is not infinite? Think of lightning flashes, possible discrepancies between the predicted time of sunrise and the observed time, radio communications between earth and astronauts in orbiting space ships, and so on? Discuss.

18. A friend asserts that Einstein's postulate (that the speed of light is not affected by the uniform motion of the source or the observer) must be discarded because it violates "common sense." How would you answer him?

19. Why is the rotating mirror method for measuring the speed of light better than the toothed wheel method of Fizeau? (See Fig. 42-4.)

20. In a vacuum, does the speed of light depend on (a) the wavelength, (b) the frequency, (c) the intensity, (d) the speed of the source, or (e) the speed of the observer?

21. Can a galaxy be so distant that its recession speed equals c? If so, how can we see the galaxy? That is, will its light ever reach us?

22. How do the Doppler effects for light and for sound differ? In what ways are they the same?

23. Gamma rays are electromagnetic radiation emitted from radioactive nuclei. In free space, do they travel with the same speed as visible light? Does their speed depend on the speed of the nucleus that emits them?

problems

SECTION 42-1

1. (a) At what wavelengths does the eye sensitivity have half its maximum value? (b) What are the frequency and the period of the light for which the eye is most sensitive? See Fig. 42-1.
Answer: (a) 510 and 610 nm. (b) 5.5×10^{14} Hz and 1.8×10^{-15} s.

SECTION 42-2

2. Radiation from the sun striking the earth has an intensity of 1400 W/m². Assuming that the earth behaves like a flat disk at right angles to the sun's rays and that all the incident energy is absorbed, calculate the force on the earth due to radiation pressure. Compare it with the force due to the sun's gravitational attraction.

3. What is the radiation pressure 1.0 m away from a small 500-W light bulb? Assume that the surface on which the pressure is exerted faces the bulb and is perfectly absorbing and that the bulb radiates uniformly in all directions.
Answer: 1.3×10^{-7} Pa.

4. Prove, for a plane wave at normal incidence on a plane surface, that the radiation pressure on the surface is equal to the energy density in the beam outside the surface. This relation holds no matter what fraction of the incident energy is reflected.

5. Prove, for a stream of bullets striking a plane surface at right angles, that the "pressure" is *twice* the (kinetic) energy density in the stream above the surface; assume that the bullets are completely "absorbed" by the surface. Contrast this with the behavior of light. (See preceding problem.)

6. Show that for complete absorption of a parallel beam of light, the radiation pressure on the absorbing object is given by $p = S/c$, where S is the magnitude of the Poynting vector and c is the speed of light in free space.

7. It has been proposed that a spaceship might be propelled in the solar system by radiation pressure, using a large sail made of aluminum foil. How large must the sail be if the radiation force is to be equal in magnitude to the sun's gravitational attraction? Assume that the mass of the *ship + sail* is 1500 kg, that the sail is perfectly reflecting, and that the sail is oriented at right angles to the sun's rays. The sun's mass is 1.97×10^{30} kg. (Incidentally, NASA had plans to build such a "space clipper" to intercept Halley's Comet on its next swing into the inner solar system in 1986. The basic design was a sail made of polymer film coated with an aluminum film a few atoms thick. The sail area would be about 100 acres. The space clipper ship would have been well instrumented but unmanned.)
Answer: 9.2×10^5 m² (3200 ft on edge).

8. A particle in the solar system is under the combined influence of the sun's gravitational attraction and the radiation force due to the sun's rays. Assume that the particle is a sphere of density 1.0 g/cm³ and that all of the incident light is absorbed. (a) Show that all particles with radius less than some critical radius, R_0, will be blown out of the solar system. (b) Calculate R_0. (c) Does R_0 depend on the distance from the earth to the sun? (See the appendices for the necessary constants.)

9. Show that $c\,\epsilon_0 \mathbf{E} \times \mathbf{B}$ has the dimensions of $\dfrac{\text{momentum}}{\text{area-time}}$, whereas $\dfrac{1}{\mu_0}\mathbf{E} \times \mathbf{B}$ has

the dimensions of $\dfrac{\text{energy}}{\text{area-time}}$. (The vector $c\,\epsilon_0\mathbf{E}\times\mathbf{B}$ may be used for computing momentum flow in the same manner that $\mathbf{S}=\dfrac{1}{\mu_0}\mathbf{E}\times\mathbf{B}$ is used to compute energy flow.)

10. A small spaceship whose mass, with occupant, is 1500 kg is drifting in outer space, where no gravitational field exists. If it shines a searchlight, which radiates 10^4 W, into space, what speed would the ship attain in one day because of the reaction force associated with the momentum carried away by the light beam?

11. A plane electromagnetic wave propagating through a vacuum is described by the following electric and magnetic fields: $E_x = E_0 \sin{(kz - \omega t)}$, $B_y = \dfrac{E_0}{c}\sin{(kz - \omega t)}$, $E_y = E_z = B_x = B_z = 0$. (a) What is the direction of propagation of the wave? (b) Find the average power per unit area. (c) What is the rate at which momentum is delivered to a perfectly absorbing surface of area A that is normal to the direction of propagation?
Answer: (a) +z-direction. (b) $E_0^2/2\mu_0 c$. (c) $\epsilon_0 E_0^2 A/2$.

12. A plane electromagnetic wave, with wavelength 3.0 m, travels in free space in the +x-direction with its electric vector **E**, of amplitude 300 V/m, directed along the y-axis. (a) What is the frequency ν of the wave? (b) What is the amplitude of the **B** field associated with the wave? (c) If $E = E_m \sin{(kx - \omega t)}$, what are the values of k and ω for this wave? (d) What is the time-averaged rate of energy flow per unit area associated with this wave? (e) If the wave fell upon a perfectly absorbing sheet of area A, what momentum would be delivered to the sheet per second and what is the radiation pressure exerted on the sheet?

SECTION 42-3

13. On page 926 the uncertainty in the measurement of the speed of light c is given as ±0.0012 km/s. In Appendix B it is given as ±0.004 parts per million. Show that these uncertainties are consistent.

14. For the value of the speed of light quoted on p. 926 (and in Appendix B) what is the uncertainty of the measurement, in feet/second? How many feet does light travel in one second?

15. Suppose that light is timed over a one-mile base line and its speed is measured to the accuracy quoted on p. 926. How large an error in the length of the base line could be tolerated, assuming other sources of error to be negligible?
Answer: 3×10^{-4} in.

16. *Bradley's method for determining the speed of light.* Consider a star located on a line through the sun, drawn perpendicular to the plane of the earth's orbit about the sun. The star's distance is much greater than the radius of the earth's orbit. Show that, due to the finite speed of light, a telescope through which the star is seen must be tilted at an angle of 20.5″ to the perpendicular, in the direction the earth is moving. This phenomenon, called *aberration, is* noticeable and was first explained by James Bradley in 1729. (Hint: draw an analogy to a person running through rain while holding an umbrella.)

17. Suppose that we were able to establish radio communication with the hypothetical inhabitants of a hypothetical planet orbiting our nearest star, α-Centauri, which is 4.2 light years from us. How long would it take to receive a reply to a message? Repeat for the Great Nebula in Andromeda, one of our closest extragalactic neighbors but 2×10^6 light years distant. What do these considerations lead you to conclude about the nature of our possible communication with extragalactic peoples?
Answer: 8.4 y; 4×10^6 y.

18. In Table 42-1, which of the four determinations of c reported by Michelson and his collaborators best agrees with the currently accepted value of c as displayed in Appendix B? Take the uncertainties of measurement into ac-

count. You might make a graph showing, on a suitable scale, the three values and their uncertainties displayed as error bars.

19. The uncertainty of the distance to the moon, as measured by the reflection of radar waves from it, is about 0.8 km. Assuming that this uncertainty is associated only with the measurement of the elapsed time, what uncertainty in this time is implied?　　　　　　　　　　*Answer:* 5.3 μs.

20. Roemer's method for measuring the speed of light consisted in observing the apparent times of revolution of one of the moons of Jupiter. The true period of revolution is 42.5 hr. (*a*) Taking into account the finite speed of light, how would you expect the apparent time of revolution to alter as the earth moves in its orbit from point *x* to point *y* in Fig. 42-10? (*b*) What observations would be needed to compute the speed of light? Neglect the motion of Jupiter in its orbit. Figure 42-10 is not drawn to scale.

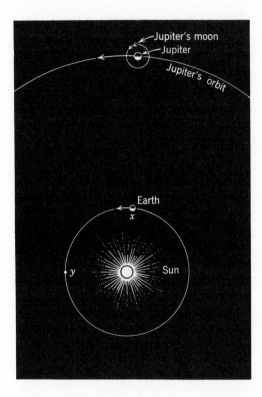

figure 42-10
Problem 20

SECTION 42-4
21. Assume that $v' = u$ in Eqs. 42-14 and 42-15. At what speed u would these expressions differ by 1.0%?　　　　　　　　　　*Answer:* 0.1 *c*.

SECTION 42-5
22. Show that, for the 21.1-cm line so much used by radioastronomers, a Doppler frequency shift in kHz can be converted to a radial velocity in km/s by multiplying by 0.211.

23. A rocketship is receding from the earth at a speed of 0.2*c*. A light in the rocketship appears blue to passengers on the ship. What color would it appear to be to an observer on the earth? (See Fig. 42-1.)
Answer: Yellow-orange.

24. The difference in wavelength between an incident microwave beam and one reflected from an approaching or receding car is used to determine automobile speeds on the highway. (*a*) Show that if *v* is the speed of the car and ν the frequency of the incident beam, the change of frequency is approximately $2v\nu/c$, where *c* is the speed of the electromagnetic radiation. (*b*) For microwaves of frequency 2450 MHz, what is the change of frequency per mi/hr of speed?

25. The period of rotation of the sun at its equator is 24.7 days; its radius is 7.0×10^8 m. What Doppler wavelength shifts are expected for characteristic wavelengths in the vicinity of 550 nm emitted from the edge of the sun's disk?   *Answer:* 3.8×10^{-3} nm.

26. An earth satellite, transmitting on a frequency of 40 MHz (exactly), passes directly over a radio receiving station at an altitude of 250 miles and at a speed of 18,000 mi/hr. Plot the change in frequency, attributable to the Doppler effect, as a function of time, counting $t = 0$ as the instant the satellite is over the station. (*Hint:* The speed u in the Doppler formula is not the actual velocity of the satellite but its component in the direction of the station. Use the nonrelativistic formula (Eq. 42-16a) and neglect the curvature of the earth and of the satellite orbit.)

27. The "red shift" of radiation from a distant nebula consists of the light (H$_\gamma$), known to have a wavelength of 4340 Å when observed in the laboratory, appearing to have a wavelength of 6562 Å. (a) What is the speed of the nebula in the line of sight relative to the earth? (b) Is it approaching or receding?   *Answer:* (a) 1.2×10^8 m/s. (b) Receding.

28. Show that, for slow speeds, the Doppler shift can be written in the approximate form

$$\frac{\Delta\lambda}{\lambda} = \frac{u}{c},$$

    where $\Delta\lambda$ is the change in wavelength.

29. In the experiment of Ives and Stilwell the speed u of the hydrogen atoms in a particular run was 8.61×10^5 m/s. Calculate $\Delta\nu/\nu$, on the assumptions that (a) Eq. 42-18a is correct and (b) that Eq. 42-16a is correct; compare your results with those given in Table 42-2 for this speed. Retain the first three terms only in Eqs. 42-18a and 42-16a.
    Answer: (a) $\Delta\nu/\nu = 0.825 \times 10^{-5}$. (b) $\Delta\nu/\nu = 1.65 \times 10^{-5}$.

43
reflection and refraction-plane waves and plane surfaces

So far we have considered the electromagnetic spectrum, including visible light, only in unimpeded free space. Here we examine it as reflected from plane surfaces such as glass or water and especially as it behaves when it passes through such transparent materials. As we shall see in the next chapter, without the consideration of these properties, such devices as cameras, telescopes, spectacles, microscopes, etc., could not be explained. As always, what we say in this chapter for visible light has its counterpart in other regions of the electromagnetic spectrum.

In Fig. 43-1a a light beam falling on a water surface is both reflected from the surface and bent (that is, *refracted*) as it enters the water.* We represent the incident beam in Fig. 43-1b by a single line, the *incident ray*, parallel to the direction of propagation. We assume the incident beam in Fig. 43-1b to be a *plane wave*, the wavefronts being normal to the incident ray. The reflected and refracted beams are also represented

43-1
REFLECTION AND REFRACTION

figure 43-1
(a) A photograph showing reflection and refraction at an air-water interface. (b) A representation using rays.

*A common and spectacular example of reflection and refraction is the rainbow. See "The Theory of the Rainbow" by H. Moysés Nussenzveig, *Scientific American*, April 1977.

by rays. The angles of *incidence* (θ_1), of *reflection* (θ_1'), and of *refraction* (θ_2) are measured between the normal to the surface and the appropriate ray, as shown in the figure.

By experiment, we find these laws governing reflection and refraction:

1. The reflected and the refracted rays lie in the plane formed by the incident ray and the normal to the surface at the point of incidence, that is, the plane of Fig. 43-1b.

2. For reflection:

$$\theta_1' = \theta_1. \tag{43-1}$$

3. For refraction:

$$\frac{\sin \theta_1}{\sin \theta_2} = n_{21}, \tag{43-2}$$

where n_{21} is a constant called the *index of refraction* of medium 2 with respect to medium 1. Table 43-1 shows the indices of refraction for some common substances with respect to a vacuum for a wavelength (sodium light) of 589 nm (= 5890 Å).

The index of refraction of one medium with respect to another generally varies with wavelength, as Fig. 43-2 shows. Because of this fact

figure 43-2
The index of refraction of fused quartz with respect to a vacuum.

Table 43-1
Some indices of refraction*
[**for** λ = 589 nm (= 5890 Å)]

Medium	Index of Refraction
Water	1.33
Ethyl alcohol	1.36
Carbon bisulfide	1.63
Air (1 atm and 20°C)	1.0003
Methylene iodide	1.74
Fused quartz	1.46
Glass, crown	1.52
Glass, dense flint	1.66
Sodium chloride	1.53

* Measured with respect to a vacuum. The index with respect to air (except, of course, the index of air itself) will be negligibly different in most cases.

figure 43-3
Sunlight from a hole F in a screen is refracted by prism ABC, forming a spectrum on screen *MN*; from Newton's *Opticks* (1704).

refraction, unlike reflection, can be used to analyze a beam of light into its component wavelengths. Figure 43-3, taken from Newton's *Opticks,* shows how Newton, using glass prism *ABC*, formed a spectrum of sunlight entering his window through a small hole at *F.*

The law of reflection was known to Euclid. That of refraction was discovered experimentally by Willebrod Snell (1591–1626) and deduced from the early corpuscular theory of light by René Descartes (1596–1650). The law of refraction is known as Snell's law or (in France) as Descartes' law.

We can derive the laws of reflection and refraction from Maxwell's equations, which means that these laws should hold for all regions of the electromagnetic spectrum. Figure 43-4*a* shows an experimental setup for investigating the reflection of microwaves from a large metal sheet. Figure 43-4*b* shows the reading of the detector as a function of the angular position of the mirror. The existence of a reflected beam at the proper angle confirms the law of reflection for microwaves. There is ample experimental evidence that Eqs. 43-1 and 43-2 correctly describe the behavior of reflected and refracted beams in all parts of the electromagnetic spectrum.

It is common knowledge that a polished steel surface will form a well-defined reflected beam if an incident beam falls on it, but a sheet of paper will reflect

figure 43-4
(*a*) An apparatus to test the law of reflection for microwaves. (*b*) A reflected beam (for $\lambda \cong 10$ cm) appears for the expected orientation of the mirror.

light more or less in all directions (diffuse reflection). It is largely by diffuse reflection that we see nonluminous objects around us. The difference between diffuse and specular (that is, mirrorlike) reflection is a matter of surface roughness; a reflected beam will be formed only if the average depth of the surface irregularities of the reflector is substantially less than the wavelength of the incident light. This criterion of surface roughness has different implications in different regions of the electromagnetic spectrum. The bottom of a cast-iron skillet, for example, is a good reflector for microwaves of wavelength 0.5 cm but is not a good reflector for visible light (that is, one cannot shave or put on makeup by it).

A second requirement for the existence of a reflected beam is that the transverse dimensions of the reflector must be substantially larger than the wavelength of the incident beam. If a beam of visible light falls on a polished metal disk the size of a dime, a reflected beam will be formed. However, if the same disk is placed in a beam of short radio waves with, say, $\lambda = 1.0$ m, radiation will be scattered in all directions from it, and no well-defined unidirectional beam will appear. We investigate this phenomenon of *diffraction* in Chapter 46. The requirements that surfaces be "smooth" and "large" also apply to the formation of refracted beams. If these two requirements are not met the description of reflection and refraction in terms of rays, whose behavior is governed by Eqs. 43-1 and 43-2, is not valid.

EXAMPLE 1

Figure 43-5 shows an incident ray i striking a plane mirror MM' at angle of incidence θ. Trace this ray.

The reflected ray makes an angle θ with the normal at b and falls as an incident ray on mirror $M'M''$. Its angle of incidence θ' on this mirror is $\pi/2 - \theta$. A second reflected ray r' makes an angle θ' with the normal erected at b'. Rays i and r' are antiparallel for any value of θ. To see this, note that

$$\phi = \pi - 2\theta' = \pi - 2\left(\frac{\pi}{2} - \theta\right) = 2\theta.$$

Two lines are parallel if their opposite interior angles for an intersecting line (ϕ and 2θ) are equal.

Repeat the problem if the angle between the mirrors is 120° rather than 90°.

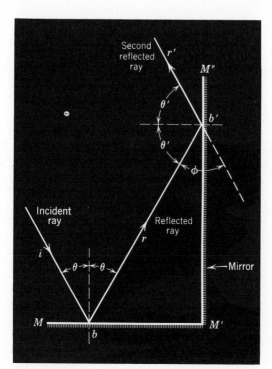

figure 43-5
Example 1.

EXAMPLE 2

An incident ray in air falls on the plane surface of a block of quartz and makes an angle of 30° with the normal. This beam contains two wavelengths, 400 and 500 nm. The indices of refraction for quartz with respect to air (n_{qa}) at these wavelengths are 1.4702 and 1.4624, respectively. What is the angle between the two refracted beams?

From Eq. 43-2 we have, for the 400-nm beam,

$$\sin \theta_1 = n_{qa} \sin \theta_2,$$

or

$$\sin 30° = (1.4702) \sin \theta_2,$$

which leads to

$$\theta_2 = 19.88°.$$

For the 500-nm beam we have

$$\sin 30° = (1.4624) \sin \theta_2',$$

or

$$\theta_2' = 19.99°.$$

The angle $\Delta\theta$ between the beams is 0.11°, the shorter wavelength component being bent through the larger angle, that is, having the smaller angle of refraction.

EXAMPLE 3

An incident ray falls on one face of a glass prism in air as in Fig. 43-6. The angle θ is so chosen that the emerging ray also makes an angle θ with the normal to the other face. Derive an expression for the index of refraction of the prism material with respect to air.

Note that $\angle abc = \alpha$, the two angles having their sides mutually perpendicular. Therefore,

$$\alpha = \tfrac{1}{2}\phi \text{ in which } \phi \text{ is the prism angle.} \tag{43-3}$$

The deviation angle ψ is the sum of the two opposite interior angles in triangle aed, or

$$\psi = 2(\theta - \alpha).$$

Substituting $\tfrac{1}{2}\phi$ for α and solving for θ yields

$$\theta = \tfrac{1}{2}(\psi + \phi). \tag{43-4}$$

At point a, θ is the angle of incidence and α the angle of refraction. The law of refraction (see Eq. 43-2) is

$$\sin \theta = n_{ga} \sin \alpha,$$

in which n_{ga} is the index of refraction of the glass with respect to air.

From Eqs. 43-3 and 43-4 this yields

$$\sin \frac{\psi + \phi}{2} = n_{ga} \sin \frac{\phi}{2}$$

figure 43-6
Example 3.

$$or \qquad n_{ga} = \frac{\sin \frac{1}{2}(\psi + \phi)}{\sin (\phi/2)}, \qquad (43\text{-}5)$$

which is the desired relation. This equation holds only for θ so chosen that the light ray passes symmetrically through the prism. For this condition the deviation angle ψ is a minimum; if θ is either increased or decreased, a larger deviation will be produced. ψ is called the *angle of minimum deviation*.

A theory of light would not be accepted if it were not able to predict the well-established laws of reflection and refraction. We could derive these laws from Maxwell's equations, but mathematical complexity prevents us from doing so here. Fortunately, we can derive these and several other laws of optics on the basis of a simpler but less comprehensive theory of light, put forward by the Dutch physicist Christian Huygens in 1678. This theory simply assumes that light is a wave rather than, say, a stream of particles. It says nothing about the nature of the wave and, in particular—since Maxwell's theory of electromagnetism appeared only after the lapse of a century—gives no hint of the electromagnetic character of light. Huygens did not know whether light was a transverse wave or a longitudinal one; he did not know the wavelengths of visible light; he had little knowledge of the speed of light. Nevertheless, his theory was a useful guide to experiment for many years and remains useful today for pedagogic and certain other practical purposes. We must not expect it to yield the same wealth of detailed information that Maxwell's more complete electromagnetic theory does.

Huygens' theory is based on a geometrical construction, called *Huygens' principle*, that allows us to tell where a given wavefront will be at any time in the future if we know its present position; it is: *All points on a wavefront can be considered as point sources for the production of spherical secondary wavelets. After a time t the new position of the wavefront will be the surface of tangency to these secondary wavelets.*

We illustrate this by a trivial example: Given a wavefront (ab in Fig. 43-7) in a plane wave in free space, where will the wavefront be a

43-2
HUYGENS' PRINCIPLE

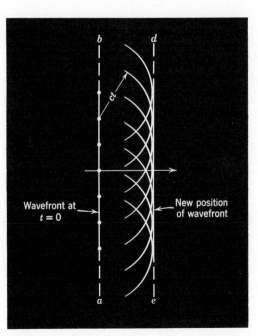

figure 43-7
The propagation of a plane wave in free space is described by the Huygens construction. Note that the ray (horizontal arrow) representing the wave is perpendicular to the wavefronts.

time t later? Following Huygens' principle, we let several points on this plane (see dots) serve as centers for secondary spherical wavelets. In a time t the radius of these spherical waves is ct, where c is the speed of light in free space. We represent the plane of tangency to these spheres at time t by de. As we expect, it is parallel to plane ab and a perpendicular distance ct from it. Thus plane wavefronts are propagated as planes and with speed c. Note that the Huygens method involves a three-dimensional construction and that Fig. 43-7 is the intersection of this construction with the plane of the page.

We might expect that, contrary to observation, a wave should be radiated backward as well as forward from the dots in Fig. 43-7. This result is avoided by assuming that the intensity of the spherical wavelets is not uniform in all directions but varies continuously from a maximum in the forward direction to a minimum of zero in the back direction. This is suggested by the shading of the circular arcs in Fig. 43-7. Huygens' method can be applied quantitatively to *all* wave phenomena; see Problem 14. The method was put on a firm mathematical footing by Augustin Fresnel (1788–1827).

43-3 HUYGENS' PRINCIPLE AND THE LAW OF REFLECTION

Figure 43-8a shows three wavefronts in a plane wave falling on a plane mirror MM'. For convenience they are chosen to be one wavelength apart. Note that θ_1, the angle between the wavefronts and the mirror, is the same as the angle between the incident ray and the normal to the mirror. In other words, θ_1 is the *angle of incidence.* The three wavefronts are related to each other by the Huygens construction, as in Fig. 43-7.

In Fig. 43-8b a Huygens wavelet centered on point a will expand to include point l after a time λ/c. Light from point p in this same wavefront cannot move beyond the mirror but must expand upward as a spherical Huygens wavelet. Setting a compass to radius λ and swinging an arc about p provides a semicircle to which the reflected wavefront must be tangent. Since point l must lie on the new wavefront, this tangent must pass through l. Note that the angle θ_1' between the wavefront and the mirror is the same as the angle between the reflected ray and the normal to the mirror. In other words, θ_1' is the *angle of reflection.*

Consider right triangles alp and $a'lp$. They have side lp in common and side al $(=\lambda)$ is equal to side $a'p$. The two right triangles are thus congruent and we may conclude that

$$\theta_1 = \theta_1',$$

as required by the law of reflection. If you recall that the Huygens construction is three-dimensional and that the arcs shown represent segments of spherical surfaces, you will be able to convince yourself that the reflected ray lies in the plane formed by the incident ray and the normal to the mirror, that is, the plane of Fig. 43-8. This is also a requirement of the law of reflection; see p. 939.

43-4 HUYGENS' PRINCIPLE AND THE LAW OF REFRACTION

Figure 43-9 shows four stages in the refraction of three successive wavefronts in a plane wave falling on an interface between air (medium 1) and glass (medium 2). For convenience, we assume that the incident wavefronts are separated by λ_1, the wavelength as measured in medium 1. Let the speed of light in air be v_1 and that in glass be v_2. We assume that

$$v_2 < v_1. \qquad (43\text{-}6)$$

figure 43-8
The reflection of a plane wave from a plane mirror as described by the Huygens construction.

figure 43-9
The refraction of a plane wave at a plane interface as described by Huygens construction; we do not show the reflected wave, for simplicity. Note the change in wavelength on refraction.

This assumption about the speeds is vital to the derivation that follows. It was not possible to test it experimentally because of technical difficulties until 1850, when the assumption was shown by Foucault to be correct.

The wavefronts in Fig. 43-9a are related to each other by the Huygens construction of Fig. 43-7. As in Fig. 43-8, θ_1 is the angle of incidence. In Fig. 43-9b consider the time $(=\lambda_1/v_1)$ during which a Huygens wavelet from point e moves to include point c. Light from point h, traveling through glass at a reduced speed (recall the assumption of Eq. 43-6) will move a shorter distance

$$\lambda_2 = \lambda_1 \frac{v_2}{v_1} \qquad (43\text{-}7)$$

during this time. The refracted wavefront must be tangent to an arc of this radius centered on h. Since c lies on the new wavefront, the tangent must pass through this point, as shown. Note that θ_2, the angle between the refracted wavefront and the air-glass interface, is the same as the angle between the refracted ray and the normal to this interface. In other words, θ_2 is the *angle of refraction*. Note, too, that the wavelength in glass (λ_2) is less than the wavelength in air (λ_1).

For the right triangles hce and hce' we may write

$$\sin \theta_1 = \frac{\lambda_1}{hc} \qquad \text{(for } hce\text{)}$$

and

$$\sin \theta_2 = \frac{\lambda_2}{hc} \qquad \text{(for } hce'\text{)}.$$

Dividing and using Eq. 43-7 yields

$$\frac{\sin \theta_1}{\sin \theta_2} = \frac{\lambda_1}{\lambda_2} = \frac{v_1}{v_2} = \text{a constant.} \qquad (43\text{-}8)$$

The law of refraction, as stated in Eq. 43-2, is

$$\frac{\sin \theta_1}{\sin \theta_2} = n_{21}, \qquad (43\text{-}2)$$

so that n_{21} is now revealed as the ratio of the speeds of light in the two media, or

$$n_{21} = \frac{v_1}{v_2}. \qquad (43\text{-}9)$$

We may rewrite Eq. 43-8 as

$$\left(\frac{c}{v_1}\right) \sin \theta_1 = \left(\frac{c}{v_2}\right) \sin \theta_2, \qquad (43\text{-}10)$$

in which c is the speed of light in free space. The quantities (c/v_1) and (c/v_2) (see Eq. 43-9) are the indices of refraction of medium 1 and of medium 2, respectively, with respect to a vacuum. Introducing the symbols n_1 and n_2 for these quantities allows us to write the law of refraction as

$$n_1 \sin \theta_1 = n_2 \sin \theta_2. \qquad (43\text{-}11)$$

If we assume that the medium above the glass in Fig. 43-9 is a vacuum rather than air, the speed v_1 becomes c and the wavelength, called λ_1 in Fig. 43-9, assumes a value λ that is characteristic of the wave in free space. Equation 43-7 may thus be written

$$\lambda_2 = \lambda \frac{v_2}{c} = \frac{\lambda}{n_2} \qquad (43\text{-}12a)$$

or

$$\lambda_n = \frac{\lambda}{n}. \qquad (43\text{-}12b)$$

This shows specifically that the wavelength of light in a material medium is less than the wavelength of the same wave in a vacuum. Figure 43-9 shows clearly the difference in wavelength in the two media.

The application of Huygens' principle to refraction requires that if a light ray is bent toward the normal in passing from air to an optically dense medium then the speed of light in that optically dense medium (glass, say) must be *less* than that in air; see Eq. 43-6. This requirement holds for all wave theories of light. For the early particle theory of light put forward by Newton, refraction can be explained only if the speed of light in the medium in which light is bent toward the normal (the optically dense medium) is *greater* than that in air. The dense medium was thought to exert attractive forces on the light "corpuscles" as they neared the surface, speeding them up and changing their direction to cause them to make a smaller angle with the normal. Figure 43-10 shows a figure from a 1637 work of René Descartes, in which he makes an analogy between the refraction of light and the motion of a tennis ball on entering a medium in which it moves more slowly.

An experimental comparison of the speed of light in water and in air is decisive, therefore, between the wave and corpuscular theories of light. Such a measurement was first carried out by Foucault in 1850; he showed conclusively

figure 43-10
According to the early (incorrect) corpuscular theory of light, ABI is the path of a ray on entering a medium in which its speed is less. The (correct) wave theory predicts that for the ray shown the speed in the lower medium (below BE) must be greater; from Descartes' La Dioptrique (1637).

that light travels more slowly in water than in air, thus ruling out the corpuscular theory of Newton.

Let light rays in an optically dense medium (glass, say) fall on a surface on the other side of which is a less optically dense medium (air, say); see Fig. 43-11. As the angle of incidence θ is increased, a situation is reached (see ray e) at which the refracted ray points along the surface, the angle of refraction being 90°. For angles of incidence larger than this *critical angle* θ_c no refracted ray exists, giving rise to a phenomenon called *total internal reflection.*

We find the critical angle by putting $\theta_2 = 90°$ in the law of refraction (see Eq. 43-11):

$$n_1 \sin \theta_c = n_2 \sin 90°,$$

or

$$\sin \theta_c = \frac{n_2}{n_1}. \qquad (43\text{-}13)$$

For glass and air $\sin \theta_c = (1.00/1.50) = 0.667$, which yields $\theta_c = 41.8°$. Total internal reflection does not occur when light originates in the medium of lower index of refraction.

figure 43-11
Showing the total internal reflection of light from a source S; the *critical angle* is θ_e.

Light can be "piped" from one point to another with little loss by allowing it to enter one end of a rod of transparent plastic. The light will undergo total internal reflection at the boundary of the rod and will follow its contour, emerging at its far end. Images may be transferred from one location to another, using a bundle of fine glass fibers, each fiber transmitting a small fraction of the image.*

Fiber optics techniques make possible many useful optical devices for transmitting and transforming luminous images. Figure 43-12 shows a short fiber bundle constructed so that the fibers taper in diameter along its length. The wide end is shown placed over the letter *S* in the printed word "OPTICS." We see, with the aid of a mirror placed above the bundle, that a letter *S*, reduced in size, has been transmitted by total internal reflection through the bundle to its narrow end.

* See "Fiber Optics" by N. S. Kapany, *Scientific American,* November 1960.

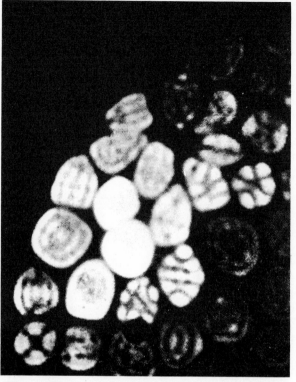

figure 43-12
A bundle of tapered fibers (below) is placed over the letter S. Above, with the aid of a mirror, we see that the image, reduced in size, is transmitted to the top of the bundle by total internal reflection in the individual fibers. (Courtesy Dr. N. S. Kapany, Optics Technology, Inc.)

figure 43-13
A photomicrograph of light emerging from the end of a bundle of fibers. The fiber diameters approach the wavelength of light so that each fiber acts like an "optical waveguide." We have here convincing visual evidence that light is an electromagnetic wave. (Courtesy of Dr. N. S. Kapany, Optics Technology, Inc.)

Figure 43-13 is an enlarged view of a cross section of such a fiber bundle in which the diameters of the individual fibers are made so small that they are of the order of magnitude of the wavelength of light. This condition violates the spirit of our assumption of p. 941, namely, that the transverse dimensions of reflecting and refracting surfaces would be large compared to the wavelength of light. Consequently, a description of the reflection and refraction of light in terms of rays, as in Figs. 43-1b and 43-11, is not possible. Figure 43-13 is readily interpreted, however, on the basis of the electromagnetic wave theory of light and provides convincing pictorial supporting evidence for that theory. The fibers behave like waveguides (see Section 41-6), and the patterns of darkness and light represent the distribution of the **E** and **B** vectors for various modes of oscillation of the electromagnetic waves traveling down the "guides."

As of 1977 several experimental, field condition, tests of telephone communication* by optical fiber are underway. In a test in Chicago 24 hairlike fibers formed, with their protective sheath, a cable 0.50 in. in diameter and 1.5 miles long. The incoming voice signal is transformed into coded light pulses which are sent along the fiber at 4.47 pulses/second; the pulses are reconstituted at the far end to form a replica of the input signal. The design capability of the cable is 8064 simultaneous two-way conversations. The chemical purity requirements of the fiber materials and the design requirements of the electrical ⇌ optical input and output devices at the ends of the cable push solid-state and materials research technologies close to their present-day limits.

* See "Communication by Optical Fiber" by J. S. Cook, *Scientific American*, November 1973; "Telephoning by Light, I: The Breakthroughs" by John H. Douglas, *Science News*, July 1975; "Light-Wave Communications" by W. S. Boyle, *Scientific American*, August 1977.

EXAMPLE 4

Figure 43-14a shows a triangular prism of glass, a ray incident normal to one face being totally reflected. If θ_1 is 45°, what can you conclude about the index of refraction n of the glass?

The angle θ_1 must be equal to or greater than the critical angle θ_c where θ_c is given by Eq. 43-13:

$$\sin \theta_c = \frac{n_2}{n_1} = \frac{1}{n},$$

in which, for all practical purposes, the index of refraction of air $(=n_2)$ is set equal to unity. Suppose that the index of refraction of the glass is such that total internal reflection just occurs, that is, that $\theta_c = 45°$. This would mean

$$n = \frac{1}{\sin 45°} = 1.41.$$

Thus the index of refraction of the glass must be *equal to or larger than* 1.41. If it were less, total internal reflection would not occur.

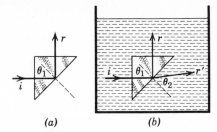

figure 43-14
Examples 4, 5

EXAMPLE 5

What happens if the prism in Example 4 (assume that $n = 1.50$) is immersed in water $(n = 1.33)$? See Fig. 43-14b.

The new critical angle, given by Eq. 43-13, is

$$\sin \theta_c = \frac{n_2}{n_1} = \frac{1.33}{1.50} = 0.887,$$

which corresponds to $\theta_c = 62.5°$. The actual angle of incidence $(=45°)$ is less than this so that we do *not* have total internal reflection.

There is a reflected ray, with an angle of reflection of 45°, as Fig. 43-14b shows. There is also a refracted ray, with an angle of refraction given by

$$n_1 \sin \theta_1 = n_2 \sin \theta_2$$

$$(1.50)(\sin 45°) = (1.33) \sin \theta_2,$$

which yields $\theta_2 = 52.9°$. Show that as $n_2 \to n_1$, $\theta_c \to 90°$.

Maxwell's equations permit us to calculate how the incident energy is divided between the reflected and the refracted beams. Figure 43-15 shows the theoretical prediction for (a) a light beam in air falling on a glass-air interface and (b) a light beam in glass falling on such an interface. Figure 43-15a shows that for angles of incidence up to about 50°, less than 10% of the light energy is reflected. At grazing incidence, however (that is, angles of incidence near 90°), the surface becomes an excellent reflector. We are all familiar with the high reflecting power of a wet road for light from automobile headlights that strikes near grazing incidence.

Figure 43-15b shows clearly that at a certain critical angle (41.8° in this case; see Eq. 43-13) *all* the light is reflected. For angles of incidence appreciably below this value, about 4% of the energy is reflected.

43-6
FERMAT'S PRINCIPLE

In 1650 Pierre Fermat discovered a remarkable principle which we can express in these terms: *A light ray traveling from one point to another will follow a path such that, compared with nearby paths, the time required is either a minimum or a maximum or will remain unchanged (that is, it will be stationary).*

We can readily derive the laws of reflection and refraction from this

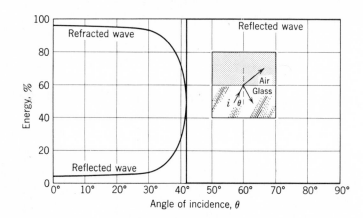

figure 43-15
(a) The percent of the energy reflected and refracted when an incident wave in air falls on glass (n = 1.50). The same for the incident wave in glass, showing total internal reflection.

principle. Figure 43-16 shows two fixed points A and B and a ray APB connecting them.* The total length l of this ray is

$$l = \sqrt{a^2 + x^2} + \sqrt{b^2 + (d - x)^2},$$

where x locates the point P at which the ray touches the mirror.

According to Fermat's principle, P will have a position such that the time of travel of the light must be a minimum (or a maximum or must remain unchanged). Expressed in another way, the total length l of the ray must be a minimum (or a maximum or must remain unchanged). In either case, the methods of the calculus require that dl/dx be zero. Taking this derivative yields

$$\frac{dl}{dx} = (\tfrac{1}{2})(a^2 + x^2)^{-1/2}(2x) + \tfrac{1}{2}[b^2 + (d - x)^2]^{-1/2}(2)(d - x)(-1) = 0,$$

which we can rewrite as

$$\frac{x}{\sqrt{a^2 + x^2}} = \frac{d - x}{\sqrt{b^2 + (d - x)^2}}.$$

Comparison with Fig. 43-16 shows that we can rewrite this as

$$\sin \theta_1 = \sin \theta_1',$$

or

$$\theta_1 = \theta_1',$$

which is the law of reflection.

To prove the law of refraction from Fermat's principle, consider Fig.

* We assume that ray APB lies in the plane of the figure; see Problem 26.

figure 43-16
A ray from A passes through B after reflection at P.

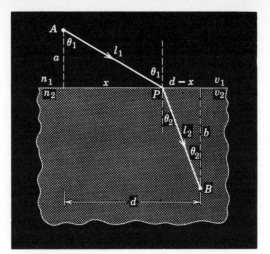

figure 43-17
A ray from A passes through B after refraction at P.

43-17, which shows two points *A* and *B* in two different media and a ray *APB* connecting them. The time *t* is given by

$$t = \frac{l_1}{v_1} + \frac{l_2}{v_2}.$$

Using the relation $n = c/v$ we can write this as

$$t = \frac{n_1 l_1 + n_2 l_2}{c} = \frac{l}{c}.$$

The quantity $l (= n_1 l_1 + n_2 l_2)$ is called the *optical path length* of the ray. Equation 43-12b $(\lambda_n = \lambda/n)$ shows that the optical path length is equal to the length that this same number of waves would have if the medium were a vacuum. We must not confuse the optical path length with the geometrical path length, which is $l_1 + l_2$.

Fermat's principle requires that *l* be a minimum (or a maximum or must remain unchanged) which, in turn, requires that *x* be so chosen that $dl/dx = 0$. The optical path length is

$$l = n_1 l_1 + n_2 l_2 = n_1 \sqrt{a^2 + x^2} + n_2 \sqrt{b^2 + (d - x)^2}.$$

Differentiating yields

$$\frac{dl}{dx} = n_1(\tfrac{1}{2})(a^2 + x^2)^{-1/2}(2x) + n_2(\tfrac{1}{2})[b^2 + (d - x)^2]^{-1/2}(2)(d - x)(-1) = 0,$$

which we can recast as

$$n_1 \frac{x}{\sqrt{a^2 + x^2}} = n_2 \frac{d - x}{\sqrt{b^2 + (d - x)^2}}.$$

Comparison with Fig. 43-17 shows that this, in turn, we can write as

$$n_1 \sin \theta_1 = n_2 \sin \theta_2,$$

which is the law of refraction.

In each of the examples of this section the time required, or, what is equivalent, the optical path length, proves to be a *minimum*.

questions

1. Describe what your immediate environment would be like if all objects were totally absorbing. Sitting in a chair in a room, could you see anything? If a person entered the room could you see her?

2. Would you expect sound waves to obey the laws of reflection and of refraction obeyed by light waves? Discuss the propagation of spherical and of cylindrical waves, using Huygens' principle. Does Huygens' principle apply to sound waves in air? If Huygens' principle predicts the laws of reflection and refraction, why is it necessary or desirable to view light as an electromagnetic wave, with all its attendant complexity?

3. A street light, viewed by reflection across a body of water in which there are ripples, appears very elongated. Explain.

4. The light beam in Fig. 43-1a is broadened on entering the water. Explain.

5. By what percent does the speed of blue light in fused quartz differ from that of red light?

6. Can (a) reflection phenomena or (b) refraction phenomena be used to determine the wavelength of light?

7. How can one determine the indices of refraction of the media in Table 43-1 relative to water, given the data in that table?

8. You are given a cube of glass. How can you find the speed of light (from a sodium light source) in this cube?

9. Describe and explain what a fish sees as he looks in various directions above his "horizon."

10. How did Foucault's measurement of the speed of light in water decide between the wave and the particle theories of light?

11. Why does a diamond "sparkle" more than a glass imitation cut to the same shape?

12. Is it plausible that the wavelength of light should change in passing from air into glass but that its frequency should not? Explain.

13. Light has (a) a wavelength, (b) a frequency, and (c) a speed. Which, if any, of these quantities remains unchanged when light passes from a vacuum into a slab of glass?

14. In reflection and refraction why do the reflected and refracted rays lie in the plane defined by the incident ray and the normal to the surface? Can you think of any exceptions?

15. What causes mirages? Does it have anything to do with the fact that the index of refraction of air is not constant but varies with its density? See "Mirages" by Alistair B. Fraser and William B. Mach, *Scientific American*, January, 1976.

16. Design a periscope, taking advantage of total internal reflection. What are the advantages compared with silvered mirrors?

17. What characteristics must a material have in order to serve as an efficient "light pipe"?

18. A certain toothbrush has a red plastic handle into which rows of nylon bristles are set. The tops of the bristles (but not their sides) appear red. Explain.

19. Discuss the formation of rainbows. See "The Theory of the Rainbow" by H. Moysés Nussenzveig, *Scientific American*, April 1977.

20. Why are optical fibers potentially more effective carriers of information than, say, microwaves or cables? Think of the frequencies involved.

21. What does "optical path length" mean? Can the optical path length ever be less than the geometrical path length? Ever greater?

22. A solution of copper sulphate appears blue when we view it through transmitted light. Does this mean that a copper sulphate solution absorbs blue light selectively? Discuss.

23. If you are an English literature major and interested in James Joyce, what would the letters *alp* and *hce* in Figures 43-8 and 43-9 mean to you?

problems

1. The wavelength of yellow sodium light in air is 5890 Å. (*a*) What is its frequency? (*b*) What is its wavelength in glass whose index of refraction is 1.52? (*c*) From the results of (*a*) and (*b*) find its speed in this glass. *Answer:* (*a*) 5.1×10^{14} Hz. (*b*) 3880 Å. (*c*) 1.98×10^8 m/s.

2. Prove that if a mirror is rotated through an angle α, the reflected beam is rotated through an angle 2α. Is this result reasonable for $\alpha = 45°$?

3. The astronomer Claudius Ptolemy, who flourished in Alexandria, Egypt during the second century A.D., gave the following measured values for the angle on incidence θ_1 and the angle of refraction θ_2 for a light beam passing from air to water:

θ_1	θ_2	θ_1	θ_2
10°	7°45′	50°	35°0′
20°	15°30′	60°	40°30′
30°	22°30′	70°	45°30′
40°	29°0′	80°	50°0′

   Are these data consistent with Snell's law? If so, what index of refraction results? These data are interesting as the oldest recorded physical measurements. *Answer:* Yes; $n = 1.34$ vs. today's value of 1.33.

4. What is the speed in fused quartz of light of wavelength 550 nm (= 5500 Å)? (See Fig. 43-2.)

5. The speed of yellow sodium light in a certain liquid is measured to be 1.92×10^8 m/s. What is the index of refraction of this liquid, with respect to air, for sodium light? *Answer:* 1.56.

6. Suppose that the speed of light in air has been measured with an uncertainty of, say, 1 km/s. In calculating the speed in vacuum, suppose that it is not certain whether n for air is 1.00029 or 1.00030. (*a*) How much extra uncertainty is introduced into the calculated value for c? (*b*) Estimate how accurately n should be known for this purpose.

7. A bottom-weighted vertical pole extends 2.0 m above the bottom of a swimming pool and 0.5 m above the water. Sunlight is incident at 45°. What is the length of the shadow of the pole on the bottom of the pool? *Answer:* 1.4 m.

8. A 60° prism is made of fused quartz. A ray of light falls on one face, making an angle of 45° with the normal. Trace the ray through the prism graphically with some care, showing the paths traversed by rays representing (*a*) blue light, (*b*) yellow-green light, and (*c*) red light. (See Figs. 43-2 and 43-6.)

9. Prove that a ray of light incident on the surface of a sheet of plate glass of thickness t emerges from the opposite face parallel to its initial direction but displaced sideways, as in Fig. 43-18. Show that, for small angles of incidence θ, this displacement is given by

$$x = t\theta \frac{n-1}{n}$$

   where n is the index of refraction and θ is measured in radians.

10. Assume that the index of refraction of the earth's atmosphere varies, with altitude only, from the value one at the "edge" of the atmosphere to some larger value at the surface of the earth. (*a*) Neglecting the earth's curvature, show that the apparent angle of a star from the zenith direction is independent of how the refractive index of the atmosphere varies with altitude and depends only on the value of n at the earth's surface. (*Hint:* Compare a uniform atmosphere with one consisting of layers of increasing refractive index.) (*b*) How does the earth's curvature affect the analysis?

figure 43-18
Problem 9

11. Two perpendicular mirrors form the sides of a vessel filled with water, as shown in Fig. 43-19. A light ray is incident from above, normal to the water surface. (a) Show that the emerging ray is parallel to the incident ray. Assume that there are two reflections at the mirror surfaces. (b) Repeat the analysis for the case of oblique incidence, the ray lying in the plane of the figure. (c) Using three mirrors, state and prove the three-dimensional analog to this problem.

figure 43-19
Problem 11

12. Show that for a thin prism (ϕ small) and light not far from normal incidence (θ_1 small) the deviation angle is independent of the angle of incidence and is equal to $(n - 1)\ \phi$ (see Fig. 43-6).

13. In Fig. 43-6 show by graphical ray tracing, using a protractor, that if θ for the incident ray is *either* increased *or* decreased, the deviation angle ψ is increased. The symmetrical situation shown in this figure is called the *position of minimum deviation*.

SECTION 43-2

14. One end of a stick is dragged through water at a speed v which is greater than the speed u of water waves. Applying Huygens' construction to the water waves, show that a conical wavefront is set up and that its half-angle α is given by

$$\sin \alpha = u/v.$$

This is familiar as the bow wave of a ship or the shock wave caused by an object moving through air with a speed exceeding that of sound, as in Fig. 20-12.

15. When an electron moves through a medium at a speed exceeding the speed of light in that medium, it radiates electromagnetic energy (the Cerenkov effect, see Section 20-7). What minimum speed must an electron have in a liquid of refractive index 1.54 in order to radiate?
Answer: 1.9×10^8 m/s.

SECTION 43-5

16. A ray of light is incident normally on the face ab of a glass prism ($n = 1.52$), as shown in Fig. 43-20. (a) Assuming that the prism is immersed in air, find the largest value for the angle ϕ so that the ray is totally reflected at face ac. (b) Find ϕ if the prism is immersed in water.

figure 43-20
Problem 16

17. A light ray falls on a square glass slab as in Fig. 43-21. What must the index of refraction of the glass be if total internal reflection occurs at the vertical face?　　　　　　　　　　　　　　　　　　　　　　*Answer:* $n > 1.22$.

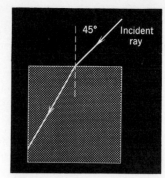

18. A given monochromatic light ray, initially in air, strikes the 90° prism at P (see Fig. 43-22) and is refracted there and at Q to such an extent that it just grazes the right-hand prism surface after it emerges into air at Q. (a) Determine the index of refraction, relative to air, of the prism for this wavelength in terms of the angle of incidence θ_1 which gives rise to this situation. (b) Give a numerical upper bound for the index of refraction of the prism. (c) Show, by a ray diagram, what happens if the angle of incidence at P is slightly greater than θ_1, is slightly less than θ_1.

figure 43-21
Problem 17

19. A glass prism with an apex angle of 60° has $n = 1.60$. (a) What is the smallest angle of incidence for which a ray can enter one face of the prism and emerge from the other? (b) What angle of incidence would be required for the ray to pass through the prism symmetrically, as in Fig. 43-6?
Answer: (a) 36°. (b) 53°.

20. A point source is 80 cm below the surface of a body of water. Find the diameter of the largest circle at the surface through which light can emerge from the water.

figure 43-22
Problem 18

21. A drop of liquid may be placed on a semicircular slab of glass as in Fig. 43-23. (a) Show how to determine the index of refraction of the liquid by observing total internal reflection. The index of refraction of the glass is unknown and must also be determined. Is the range of indices of refraction that can be measured in this way restricted in any sense? (b) In reality, how practical is this method?

figure 43-23
Problem 21

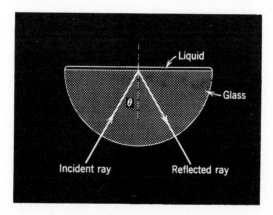

22. A point source of light is placed a distance h below the surface of a large deep lake. (a) Show that the fraction f of the light energy that escapes directly from the water surface is independent of h and is given by

$$f = \tfrac{1}{2}(1 - \sqrt{1 - 1/n^2})$$

where n is the index of refraction of water. (*Note:* Absorption within the water and reflection at the surface—except where it is total—have been neglected.) (b) Evaluate this fraction for $n = 1.33$.

23. Figure 43-24 shows a *constant-deviation prism*. Although made of one piece of glass, it is equivalent to two 30° 60° 90° prisms and one 45° 45° 90° prism. White light is incident in the direction i. θ_1 is changed by rotating the prism so that, in turn, light of any desired wavelength may be made to follow the path shown, emerging at r. Show that, if $\sin \theta_1 = \tfrac{1}{2}n$, then $\theta_2 = \theta_1$ and beams i and r are at right angles.

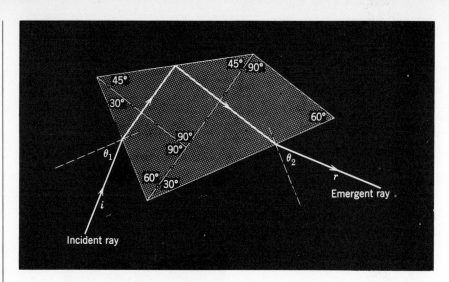

figure 43-24
Problem 23

24. A plane wave of white light traveling in fused quartz strikes a plane surface of the quartz, making an angle of incidence θ. Is it possible for the internally reflected beam to appear (a) bluish or (b) reddish? Roughly what value of θ must be used? (*Hint:* White light will appear bluish if wavelengths corresponding to red are removed from the spectrum.)

25. A glass cube has a small spot at its center. (a) What parts of the cube face must be covered to prevent the spot from being seen, no matter what the direction of viewing? (b) What fraction of the cube surface must be so covered? Assume a cube edge of 1.0 cm and an index of refraction of 1.50. (Neglect the subsequent behavior of an internally reflected ray.)
    Answer: (a) Cover the center of each face with an opaque disk of radius 0.45 cm. (b) About 0.63.

SECTION 43-6

26. Using Fermat's principle, prove that the reflected ray, the incident ray, and the normal lie in one plane.

27. Prove that the optical path lengths for reflection and refraction in Figs. 43-16 and 43-17 are minimal when compared with other nearby paths connecting the same two points.

28. Light of free space wavelength 600 nm (= 6000 Å) travels 1.6×10^{-4} cm in a medium of index of refraction 1.5. Find (a) the optical path length, (b) the wavelength in the medium, and (c) the phase difference after moving that distance, with respect to light traveling the same distance in free space.

44

reflection and refraction-spherical waves and spherical surfaces

In Chapter 43 we described the reflection and refraction of plane waves at plane surfaces. In this chapter we consider the more general case of spherical waves falling on spherical reflecting and refracting surfaces. All of the results of Chapter 43 will emerge as special cases of the results of this chapter, because we can view a plane as a spherical surface with an infinite radius of curvature.

Both in Chapter 43 and in this chapter we make extensive use of *rays*. Although a ray is a convenient construction, it proves impossible to isolate one physically. Figure 44-1a shows schematically a plane wave of wavelength λ falling on a slit of width $a = 5\lambda$. We find that the light flares out into the geometrical shadow of the slit, a phenomenon called *diffraction*, which we will study in Chapter 46. Figures 44-1b ($a = 3\lambda$) and 44-1c ($a = \lambda$) show that diffraction becomes more pronounced as $a/\lambda \rightarrow 0$ and that attempts to isolate a single ray from the incident plane wave are futile.

Figure 44-2 shows water waves in a shallow ripple tank, produced by tapping the water surface periodically with the edge of a stick. We see that the plane wave so generated also flares out by diffraction when it encounters a gap in a barrier placed across it. Diffraction is characteristic of waves of all types. We can hear around corners, for example, because of the diffraction of sound waves.

The diffraction of waves at a slit (or at an obstacle such as a wire) is expected from Huygens' principle. Consider the portion of the wavefront that arrives at the position of the slit in Fig. 44-1. We can view every point on it as the site of an expanding spherical Huygens' wavelet. The "bending" of light into the region of the geometrical shadow is as-

44-1
GEOMETRICAL OPTICS AND WAVE OPTICS

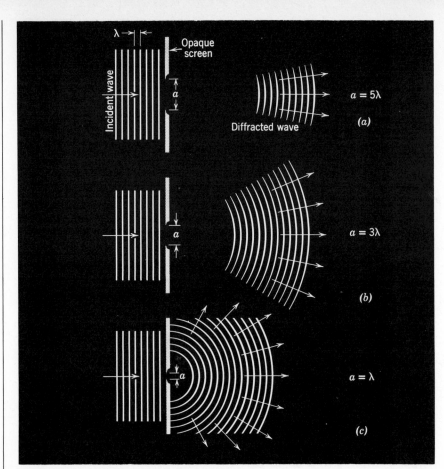

figure 44-1
An attempt to isolate a ray by reducing the slit width a fails because of diffraction, which becomes more pronounced as a/λ approaches zero. Certain details of the diffracted waves are omitted for simplicity (we will discuss them fully in Chapter 46; see also Fig. 44-3). The main point is clear, however: for a fixed wavelength λ the diffracted wave broadens as we reduce the slit width a.

sociated with the blocking off of Huygens' wavelets from those parts of the incident wavefront that lie behind the slit edges.

Figure 44-3 was made by allowing parallel light to fall on a slit placed 50 cm in front of a photographic plate. In Fig. 44-3a the slit width was about 6 μm. The central band of light is much wider than this, showing that light has "flared out" into the geometric shadow of the slit. In addition, many secondary maxima, omitted from Fig. 44-1 for simplicity, appear. Figure 44-3b shows what happens when the slit width is reduced by a factor of two. The central maximum becomes wider, in agreement with Fig. 44-1. Figure 44-3c shows the effect of reducing the slit width by an additional factor of 7, to 0.4 μm. The central maximum is now much wider and the secondary maxima, whose intensities relative to the central maximum have been deliberately overemphasized by long exposure, are very evident.

Diffraction can be ignored if the ratio a/λ is large enough, a being a measure of the smallest sideways dimension of the slit or obstacle. If $a \gg \lambda$, light appears to travel in straight lines which we can represent by rays that obey the laws of reflection (Eq. 43-1) and refraction (Eq. 43-2). In Chapter 43 this condition, called *geometrical optics*, prevailed, the lateral dimensions of all mirrors, prisms, etc., being much greater than the wavelength. We also assume in this chapter that the conditions for geometrical optics are *also* satisfied.

If the requirement for geometrical optics is not met, we cannot describe the behavior of light by rays but must take its wave nature specifically into account. This subject is called *wave optics*; it includes geometrical optics as an important limiting case in much the same

figure 44-2
Diffraction of water waves at a slit in a ripple tank. Note that the slit width is about the same size as the wavelength. Compare with Fig. 44-1c. (Courtesy of Educational Services Incorporated.)

figure 44-3
(a) The intensity of light diffracted from a slit of width $a \cong 6$ μm and falling on a screen 50 cm beyond. (b) The slit width is reduced by a factor of two. (c) The slit width is further reduced by an additional factor of seven. Note that secondary maxima, made prominent in this case by deliberate overexposure, appear on either side of the central maximum. These secondary maxima have been omitted from Fig. 44-1 for simplicity.

sense that we can view an ideal gas as a limiting case of a real gas; both abstractions are exceedingly useful. We will treat wave optics in later chapters.

Figure 44-4 shows a point source of light O, the *object*, placed a distance o in front of a plane mirror. The light falls on the mirror as a spherical wave represented in the figure by rays emanating from O.* At the point at which each ray strikes the mirror we construct a reflected ray. If we

44-2
SPHERICAL WAVES— PLANE MIRROR

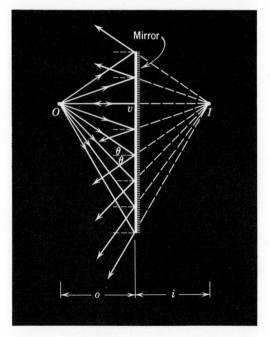

figure 44-4
A point object O forms a virtual image I in a plane mirror. The rays *appear* to emanate from I, but actually light does not pass through this point.

---

* In our discussion of reflection from mirrors in Chapter 43 (see Fig. 43-8) we assumed an incident *plane* wave; the incident rays are parallel to each other in that case. Here we have a *point* source and the rays striking the mirror are *diverging* from that point source.

extend the reflected rays backward, they intersect at a point I which is the same distance behind the mirror that the object O is in front of it; I is called the *image* of O.

Images may be *real* or *virtual*. In a real image light actually passes through the image point; in a virtual image the light *behaves* as though it diverges from the image point, although, in fact, it does not pass through this point; see Fig. 44-4. Images in plane mirrors are always virtual. We know from daily experience how "real" such a virtual image appears to be and how definite is its location in the space behind the mirror, even though this space may, in fact, be occupied by a brick wall.

figure 44-5
Two rays from Fig. 44-4; ray Oa makes an arbitrary angle θ with the normal.

Figure 44-5 shows two rays from Fig. 44-4. One strikes the mirror at v, along a perpendicular line. The other strikes it at an arbitrary point a, making an angle of incidence θ with the normal at that point. Elementary geometry shows that the angles aOv and aIv are also equal to θ. Thus the right triangles $aOva$ and $aIva$ are congruent and

$$o = -i, \tag{44-1}$$

in which we arbitrarily introduce the minus sign to show that I and O are on opposite sides of the mirror. Equation 44-1 does not involve θ, which means that *all* rays striking the mirror pass through I when extended backward, as we have seen above. Beyond assuming that the mirror is truly plane and that the conditions for geometrical optics hold, we have made no approximations in deriving Eq. 44-1. A point object produces a point image in a plane mirror, with $o = -i$, no matter how large the angle θ in Fig. 44-5.

Because of the finite diameter of the pupil of the eye, only rays that lie fairly close together can enter the eye after reflection at a mirror. For the eye position shown in Fig. 44-6 only a small patch of the mirror near point a is effective in forming the image; the rest of the mirror may be covered up or removed. If we move our eye to another location, a different patch of the mirror will be effective; the location of the virtual image I will remain unchanged, however, as long as the object remains fixed.

If the object is an extended source such as the head of a person, a virtual image is also formed. From Eq. 44-1, every point of the source has an image point that lies an equal distance directly behind the plane of

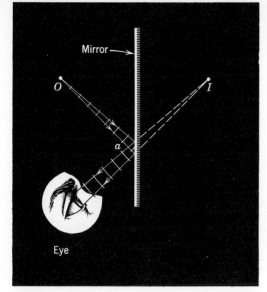

figure 44-6
A "pencil" of rays from O enters the eye after reflection at the plane mirror. Only a small portion of the mirror near a is effective. The small arcs represent portions of spherical wavefronts.

figure 44-7
A plane mirror reverses left and right. The object O is a left hand; the image I is a right hand. Try it in a mirror.

the mirror. Thus the image reproduces the object point by point. Most of us prove this every day by looking into a mirror.

Images in plane mirrors differ from objects in that left and right are interchanged. The image of a printed page is different from the page itself. Similarly, if a top is made to spin clockwise, the image, viewed in a vertical mirror, will seem to spin counterclockwise. Figure 44-7 shows an image of a left hand, constructed by using point-by-point application of Eq. 44-1; the image has the symmetry of a right hand.*

EXAMPLE 1

How tall must a vertical mirror be if a person 6 ft high is to be able to see her entire length? Assume that her eyes are 4 in. below the top of her head.

Figure 44-8 shows the paths followed by light rays leaving the top of the

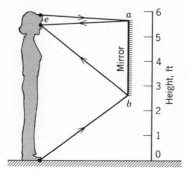

figure 44-8
Example 1. A person can view her full-length image in a mirror that is only half her height.

---

* See "The Overthrow of Parity" by Philip Morrison, *Scientific American*, April 1957, for a discussion of the distinction in nature between right and left.

woman's head and the tips of her toes. These rays, chosen so that they will enter the eye e after reflection, strike the vertical mirror at points a and b, respectively. The mirror need occupy only the region between these two points. Calculation shows that b is 2 ft, 10 in. and a is 5 ft, 10 in. above the floor. The length of the mirror is thus 3.0 ft, or half the height of the person. Note that this height is independent of the distance between the person and the mirror. Mirrors that extend below point b only show reflections of the floor between the person and the mirror.

Two plane mirrors are placed at right angles, and a point object O is located on the perpendicular bisector, as shown in Fig. 44-9a. Locate the images. Images I_1 and I_2 are formed in mirrors ab and cd, respectively. There is also a third image; it may be considered to be the image of I_1 in mirror cd or the image of I_2 in mirror ab. The three images and the object O lie on a circle whose center is on the line of intersection of the mirrors and whose plane is at right angles to that line.

EXAMPLE 2

figure 44-9
Example 2. (a) Real object O has three virtual images. (b) A typical bundle of rays used to view I_3. The light originates from O.

In viewing I_3 the light entering the observer's eye is reflected *twice* after leaving the source. Figure 44-9b shows a typical bundle of rays. In viewing I_1 or I_2, the light is reflected only once, as in Fig. 44-6.

In Fig. 44-10 a spherical light wave from a point object O falls on a concave spherical mirror whose radius of curvature is r.* A line through O and the center of curvature C makes a convenient reference axis.

A ray from O that makes an arbitrary angle α with this axis intersects the axis at I after reflection from the mirror at a. A ray that leaves O along the axis will be reflected back along itself at v and will also pass through I. Thus, for these two rays at least, I is the image of O; it is a *real* image because light actually passes through I. Let us find the location of I.

figure 44-10
Two rays from O converge after reflection in a spherical concave mirror, forming a real image at I.

A useful theorem is that the exterior angle of a triangle is equal to the sum of the two opposite interior angles. Applying this to triangles $OaCO$ and $OaIO$ in Fig. 44-10 yields

$$\beta = \alpha + \theta$$

and

$$\gamma = \alpha + 2\theta.$$

Eliminating θ between these equations leads to

$$\alpha + \gamma = 2\beta. \tag{44-2}$$

In radian measure we can write angles α, β, and γ as

$$\alpha \cong \frac{av}{vO} = \frac{av}{o}$$

$$\beta = \frac{av}{vC} = \frac{av}{r} \tag{44-3}$$

$$\gamma \cong \frac{av}{vI} = \frac{av}{i}.$$

Note that only the equation for β is exact, for the reason that the center of curvature of arc av is at C and not at O or I. However, the equations for α and for γ are approximately correct if these angles are sufficiently small. *In all that follows we assume that the rays diverging from the object make only a small angle α with the axis of the mirror.* We call such rays, which lie close to the mirror axis, *paraxial rays.* We did not find it necessary to make such an assumption for plane mirrors. Sub-

---

* A spherical shell, viewed from inside, is everywhere *concave*; viewed from outside it is everywhere *convex*. In this chapter we will always judge concave and convex from the point of view of an observer sighting along the direction of the incident light.

stituting these equations into Eq. 44-2 and canceling av yields

$$\frac{1}{o} + \frac{1}{i} = \frac{2}{r}, \tag{44-4}$$

in which o is the *object distance* and i is the *image distance.* Both these distances are measured from the *vertex* of the mirror, which is the point v at which the axis intercepts the mirror.

Significantly, Eq. 44-4 does not contain α (or β, γ, or θ), so that it holds for all rays that strike the mirror provided that they are sufficiently paraxial. In an actual case the rays can be made as paraxial as one likes by putting a circular diaphragm in front of the mirror, centered about the vertex v; this will impose a certain maximum value of α.

As α in Fig. 44-10 is permitted to become larger, it will become less true that a point object will form a point image; the image will become extended and fuzzy. No sharp criterion for deciding whether a given ray is paraxial can be laid down. If the maximum permitted value of α is reduced, the rays will become more paraxial and the image will become sharper. Unfortunately, the image will also become fainter because less total light energy will be reflected from the mirror. A compromise must often be made between image brightness and image quality.

As for plane mirrors, the image (real or virtual) in a spherical mirror can be seen only if the eye is located so that light rays from the object can enter it after reflection. In Fig. 44-11 a bundle of light rays is shown entering the eye in position x; only the small patch of mirror near a is effective for this eye position. If the observer moves his eye to position y, the image will vanish for him because the mirror does not exist near point a'.

Although Eq. 44-4 was derived for the special case in which the object is located beyond the center of curvature, it is generally true, no matter where the object is located. It is also true for convex mirrors, as in Fig. 44-12.

In applying Eq. 44-4, we must be careful to follow a consistent convention of signs for o, i, and r. As the basis for the sign conventions to be used in this book, we start from this statement:

> In Fig. 44-10, in which light *diverges* from a *real* object, falls on a *concave* mirror, and *converges* after reflection to form a *real* image, the quantities o, i, and r in Eq. 44-4 and f in Eq. 44-5 are given positive numerical values.

We used Fig. 44-10 to derive Eq. 44-4 and you should associate them in your mind as an aid in getting the signs correct.

Let us fix our minds on the side of the mirror from which incident light comes. Because mirrors are opaque, the light, after reflection, must remain on this side, and if an image is formed here it will be a *real* image. Therefore, we call the side of the mirror from which the light comes the *R-side* (for *real image*). We call the back of the mirror the *V-side* (for *virtual image*), because images formed on this side of the mirror must be virtual, no light being actually present on this side.

In the indented statement above we associated real images with positive image distances. This suggests our first sign convention:

1. The image distance i is positive if the image (real) lies on the R-side of the mirror, as in Fig. 44-10; i is negative if the image (virtual) lies on the V-side, as in Fig. 44-12.

If the mirror in Fig. 44-10, which is concave as viewed from the direction of the incident light, is made convex, the rays will *diverge* after re-

figure 44-11
The eye must be properly located to
see (real) image *I*. An observer at *x*
can see the image. One at *y* cannot.

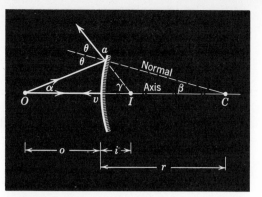

figure 44-12
Two rays from *O* diverge after
reflection in a spherical convex
mirror, forming a virtual image at *I*,
the point from which they appear to
originate. Compare Fig. 44-10.

flection and will form a *virtual* image, as Fig. 44-12 shows. Thus the
indented statement above suggests our second sign convention:

2. The radius of curvature *r* is positive if the center of curvature of the
mirror lies on the R-side, as in Fig. 44-10; *r* is negative if the center of
curvature lies on the V-side, as in Fig. 44-12.

You should not commit these sign conventions to memory but should
deduce them in each case from the basic statement on p. 964, using
Figs. 44-10, 12 as mnemonic aids.

(a)

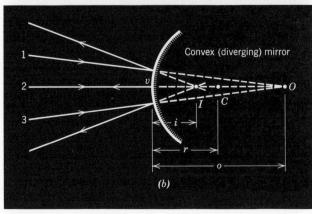

(b)

figure 44-13
(*a*) *Converging* rays (see 1, 2, and 3)
fall on a *concave* mirror. A *virtual*
object *O* produces a *real* image *I*;
note that no light passes through *O*
but it does pass through *I*. Also
note that *o* is negative but *i, r* (and
thus *f*), are positive. (*b*) Again
converging rays (see 1, 2, and 3) fall
on a *convex* mirror. A *virtual* object
O produces a *virtual* image *I*. Note
again that no light passes through
O or *I*. Here *o, i, r* (and thus *f*), are
all negative. Compare these figures
with Fig. 44-10, in which a *real*
object (*diverging* rays) produces a
real image on reflection from a
concave mirror. Here *o, i, r* (and
thus *f*), are all positive. See also
Fig. 44-12, in which a *real* object
(*diverging* rays) produces a *virtual*
image in a *convex* mirror. Here *o* is
positive but *i, r,* (and thus *f*), are
negative. These four figures show all
possibilities for arrangements of
virtual/real objects and virtual/real
images in mirrors. Compare them
carefully. Figure 44-23 shows a
similar comparison for thin lenses.

For all cases that we have considered so far, including plane mirrors, we have assumed that the incident light striking the mirror was *diverging* from a point source (the *object*) when it struck the mirror. In such cases, as we have seen, we have taken the object distance *o* in Eq. 44-4 as *positive*.

It is possible, by arrangements of mirrors and/or lenses, to make *converging* light fall on the mirror in question. In such cases we declare the *object* to be *virtual* and we assign a *negative* sign to the object distance *o* in Eq. 44-4. Figure 44-13 is an example. No matter what the source of light, *O*, which lies on the V-side, is a virtual object and *o*, the object distance, is taken as negative.

figure 44-14

(*a*) The focal point for a concave spherical mirror, showing both the rays and the wavefronts. *F* and *C* lie on the R-side, the focal point is real, and the focal length *f* of the mirror is positive (as is *r*). (*b*) The same, except that the incident light makes an angle *α* with the mirror axis; the rays are focused at a point in the *focal plane F′ F″*. (*c*) Same as (*a*) except that the mirror is convex; *F* and *C* lie on the V-side of the mirror. The focal point is virtual and the focal length *f* is negative (as is *r*).

EXAMPLE 3

967 SPHERICAL WAVES–SPHERICAL MIRROR SEC. 44-3

A convex mirror has a radius of curvature of 20 cm. If a point source is placed 14 cm away from the mirror, as in Fig. 44-12, where is the image?

A rough graphical construction, applying the law of reflection at a in the figure, shows that the image will be on the V-side of the mirror and thus will be virtual. We may verify this quantitatively and analytically from Eq. 44-4, noting that r is negative here because the center of curvature of the mirror is on its V-side. We have

$$\frac{1}{o} + \frac{1}{i} = \frac{2}{r}$$

or

$$\frac{1}{+14 \text{ cm}} + \frac{1}{i} = \frac{2}{-20 \text{ cm}},$$

which yields $i = -5.8$ cm, in agreement with the graphical prediction. The negative sign for i reminds us that the image is on the V-side of the mirror and thus is virtual.

When *parallel* light falls on a mirror (Fig. 44-14), we call the image point (real or virtual) the *focal point F* of the mirror. The focal length f is the distance between F and the vertex. If we put $o \to \infty$ in Eq. 44-4, thus insuring parallel incident light, we have

$$i = \tfrac{1}{2}r = f.$$

Equation 44-4 can then be rewritten

$$\frac{1}{o} + \frac{1}{i} = \frac{1}{f}, \qquad (44\text{-}5)$$

where f, like r, is taken as positive for mirrors whose centers of curvature are on the R-side (that is, for *concave*, or *converging* mirrors; see Fig. 44-14a) and negative for those whose centers of curvature are on the V-side (that is, for *convex*, or *diverging* mirrors; see Fig. 44-13c). Figure 44-14b shows an incident plane wave that makes a small angle α with the mirror axis. The rays are focused at a point in the *focal plane* of the mirror. This is a plane at right angles to the mirror axis at the focal point.

We now consider objects that are not points. Figure 44-15 shows a candle in front of (a) a concave mirror and (b) a convex mirror. We choose

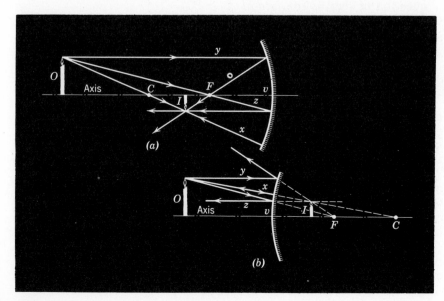

figure 44-15
The image of an extended object in (a) a concave mirror and (b) a convex mirror is located graphically. Any two of the three special rays shown are sufficient.

to draw the mirror axis through the foot of the candle and, of course, through the center of curvature. We can find the image of any off-axis point, such as the tip of the candle, graphically, using the following facts:

1. A ray that strikes the mirror after passing (either directly or upon being extended) through the center of curvature C returns along itself (ray x in Fig. 44-15). Such rays strike the mirror at right angles.

2. A ray that strikes the mirror parallel to its axis passes (or will pass when extended) through the focal point (ray y).

3. A ray that strikes the mirror after passing (either directly or upon being extended) through the focal point emerges parallel to the axis (ray z).

figure 44-16
A particular ray for the arrangement of Fig. 44-15, used to show that the *lateral magnification m* is given by $-i/o$.

Figure 44-16 shows a ray (dve) that originates on the tip of the object candle of Fig. 44-15a, is reflected from the mirror at point v, and passes through the tip of the image candle. The law of reflection demands that this ray make equal angles θ with the mirror axis as shown. For the two similar right triangles in the figure we can write

$$\frac{ce}{bd} = \frac{vc}{vb}.$$

The quantity on the left (apart from a question of sign) is the *lateral magnification m* of the mirror. Since we want to represent an *inverted* image by a *negative* magnification, we arbitrarily define m for this case as $-(ce/bd)$. Since $vc = i$ and $vb = o$, we have at once

$$m = -\frac{i}{o}. \tag{44-6}$$

This equation gives the magnification for spherical and plane mirrors under all circumstances. For a plane mirror, $o = -i$ and the predicted magnification is $+1$ which, in agreement with experience, indicates an *erect* image the same size as the object.

Images in spherical mirrors suffer from several "defects" that arise because the assumption of paraxial rays is never completely justified. In general, a point source will not produce a point image; see Problem 14. Apart from this, distortion arises because the magnification varies somewhat with distance from the mirror axis, Eq. 44-6 being strictly correct only for paraxial rays. Superimposed on these defects are *diffraction effects* which come about because the basic assumption of geometrical optics, that light travels in straight lines, must always be considered an approximation.

Figure 44-17 shows a point source O near a convex spherical *refracting* surface of radius of curvature r. The surface separates two media whose indices of refraction differ, that of the medium in which the incident light falls on the surface being n_1 and that on the other side of the surface being n_2.

44-4
SPHERICAL REFRACTING SURFACE

figure 44-17
Two rays from O converge after refraction at a spherical surface, forming a real image at I. Here o (diverging rays), i, and r are all positive; see Example 4.

From O we draw a line through the center of curvature C of the refracting surface, thus establishing a convenient axis which intercepts the surface at vertex v. From O we draw a ray that makes a small but arbitrary angle α with the axis and strikes the refracting surface at a, being refracted according to

$$n_1 \sin \theta_1 = n_2 \sin \theta_2.$$

The refracted ray intersects the axis at I. A ray from O that travels along the axis will not be bent on entering the surface and will also pass through I. Thus, for these two rays at least, I is the image of O.

As in the derivation of the mirror equation, we use the theorem that the exterior angle of a triangle is equal to the sum of the two opposite interior angles. Applying this to triangles $COaC$ and $ICaI$ yields

$$\theta_1 = \alpha + \beta \qquad (44\text{-}7)$$

and
$$\beta = \theta_2 + \gamma. \qquad (44\text{-}8)$$

As α is made small, angles β, γ, θ_1, and θ_2 in Fig. 44-17 also become small. We assume that α, hence all these angles, *are arbitrarily small*. We made this same paraxial ray assumption for spherical mirrors. Replacing the sines of the angles by the angles themselves — since the angles are required to be small — permits us to write the law of refraction as

$$n_1 \theta_1 \cong n_2 \theta_2. \qquad (44\text{-}9)$$

Combining Eqs. 44-8 and 44-9 leads to

$$\beta = \frac{n_1}{n_2} \theta_1 + \gamma.$$

Eliminating θ_1 between this equation and Eq. 44-7 leads, after rearrangement, to

$$n_1 \alpha + n_2 \gamma = (n_2 - n_1)\beta. \qquad (44\text{-}10)$$

In radian measure the angles α, β, and γ in Fig. 44-17 are

$$\alpha \cong \frac{av}{o}$$

$$\beta = \frac{av}{r} \qquad\qquad (44\text{-}11)$$

$$\gamma \cong \frac{av}{i}.$$

Only the second of these equations is exact. The other two are approximate because I and O are *not* the centers of circles of which av is an arc. However, for paraxial rays (α small enough) the inaccuracies in Eq. 44-11 can be made as small as desired.

Substituting Eqs. 44-11 into Eq. 44-10 leads readily to

$$\frac{n_1}{o} + \frac{n_2}{i} = \frac{n_2 - n_1}{r}. \qquad\qquad (44\text{-}12)$$

This equation holds whenever light is refracted from point objects at spherical surfaces, assuming only that the rays are paraxial. As with the mirror formula, care must be taken to use Eq. 44-12 with consistent signs for o, i, and r. Once again we establish our sign conventions by physical reasoning from a particular case, that of Fig. 44-17:

> In Fig. 44-17, in which light *diverges* from a *real* object, falls on a *convex* refracting surface, and *converges* after refraction to form a *real* image, the quantities o, i, and r in Eq. 44-12 have positive numerical values.

We used Fig. 44-17 to derive Eq. 44-12, and you should associate it in your mind as an aid in getting the signs correct. This basic statement is quite similar to that which we made for mirrors on p. 964.

We fix our attention on the side of the refracting surface from which the incident light falls on the surface. In contrast to mirrors, the light energy *passes through* a refracting surface to the other side, and if a real image is formed it must appear on the far side, which we call the R-side. The side from which the incident light comes is called the V-side because virtual images must appear here. Figure 44-18 suggests this important distinction between reflection and refraction.

In the indented statement above we associated real images with positive-image distances. Thus we are led to the sign convention:

1. The image distance i is positive if the image (real) is on the R-side of the refracting surface, as in Fig. 44-17; i is negative if the image (virtual) lies on the V-side, as in Fig. 44-19.

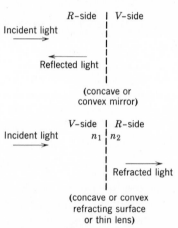

figure 44-18
Real images are formed on the same side as the incident light for mirrors but on the opposite side for refracting surfaces and thin lenses. This is so because the incident light is reflected back by mirrors but is transmitted through by refracting surfaces. Note that o is positive if rays *diverge* from a *real* object O.

figure 44-19
Two rays from O diverge after refraction at a spherical surface, forming a virtual image at I. Here o is positive (the rays from O are diverging) but i and r are negative.

The refracting surface in Fig. 44-17 is convex. If it is made concave (still assuming that $n_2 > n_1$), the rays will *diverge* after refraction and form a *virtual* image, as Fig. 44-19 shows. Thus we are led to our second sign convention:

2. The radius of curvature r is positive if the center of curvature of the refracting surface lies on the R-side, as in Fig. 44-17; r is negative if the center of curvature is on the V-side, as in Fig. 44-19.

The sign conventions for refracting surfaces are the same as for mirrors (p. 964), the fundamental difference between the two situations being absorbed in the definitions of R-side and V-side in Fig. 44-18. This difference is easily remembered on physical grounds.

EXAMPLE 4

Locate the image for the geometry shown in Fig. 44-17, assuming the radius of curvature to be 10 cm, n_2 to be 2.0, and n_1 to be 1.0. Let the object be 20 cm to the left of v.

From Eq. 44-12,

$$\frac{n_1}{o} + \frac{n_2}{i} = \frac{n_2 - n_1}{r},$$

we have

$$\frac{1.0}{+20 \text{ cm}} + \frac{2.0}{i} = \frac{2.0 - 1.0}{+10 \text{ cm}}.$$

Note that r is positive because the center of curvature of the surface lies on the R-side. This relation yields $i = +40$ cm in agreement with the graphical construction. The light energy actually passes through I so that the image is real, as indicated by the positive sign for i.

---

figure 44-20
Two rays from O appear to originate from I (virtual image) after refraction at a spherical surface.

EXAMPLE 5

An object is immersed in a medium with $n_1 = 2.0$, being 15 cm from the spherical surface whose radius of curvature is -10 cm, as in Fig. 44-20, r is negative because C lies on the V-side. Locate the image.

Figure 44-20 shows a ray traced through the surface by applying the law of refraction at point a. A second ray from O along the axis emerges undeflected at v. We find the image I by extending these two rays backward; it is virtual.

From Eq. 44-12,

$$\frac{n_1}{o} + \frac{n_2}{i} = \frac{n_2 - n_1}{r},$$

we have

$$\frac{2.0}{+15 \text{ cm}} + \frac{1.0}{i} = \frac{1.0 - 2.0}{-10 \text{ cm}},$$

which yields $i = -30$ cm, in agreement with Fig. 44-20 and with the sign conventions. Note that n_1 always refers to the medium on the side of the surface from which the light comes.

What is the relationship between i and o if the refracting surface is plane?

A plane surface has an infinite radius of curvature. Putting $r \to \infty$ in Eq. 44-12 leads to

$$i = -o\,\frac{n_2}{n_1}.$$

Figure 44-21 illustrates the situation graphically (a) for an object in air as seen from below water and (b) for an object in water with air above. This shows that a diver, looking upward at, say, an overhanging tree branch, will think it higher than it is by the factor 1.33/1.00. Similarly, an observer in air will think that objects on the bottom of a water tank are closer to the surface than they actually are, in the ratio 1.00/1.33. These considerations, being based on Eq. 44-12, hold

EXAMPLE 6

figure 44-21
Refraction at a plane surface at near-normal incidence, showing a pencil of rays and the corresponding wavefronts entering the pupil. (a) Source in air and (b) source in water.

only for paraxial rays, which means that the incident rays can make only a small angle with the normal; this angle has been exaggerated in the figure for clarity. Note again that n_1 is always identified with the medium that lies on the side of the surface containing the incident light.

In most refraction situations there is more than one refracting surface. This is true even for a spectacle lens, the light passing from air into glass and then from glass into air. In microscopes, telescopes, cameras,* etc., there are often many more than two surfaces.

Figure 44-22a shows a thick glass "lens" of length l whose surfaces are ground to radii r' and r''. A point object O' is placed near the left surface as shown. A ray leaving O' along the axis is not deflected on entering or leaving the lens because it falls on each surface along a normal.

A second ray leaving O', at an arbitrary angle α with the axis, strikes the surface at point a', is refracted, and strikes the second surface at point a''. The ray is again refracted and crosses the axis at I'', which, being the intersection of two rays from O', is the image of point O', formed after refraction at two surfaces.

44-5
THIN LENSES

* For insight into the difficult task of designing real high-performance lenses see "The Photographic Lens" by William H. Price, *Scientific American,* August 1976.

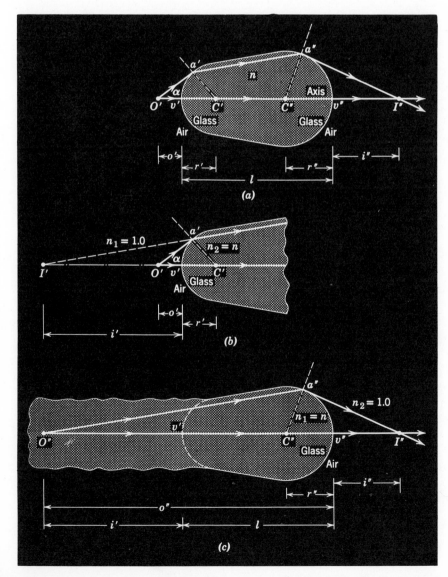

figure 44-22
(a) Two rays from O' intersect at I'' (real image) after refraction at two spherical surfaces. (b) The first surface and (c) the second surface shown separately. The quantities α and n have been exaggerated for clarity.

Figure 44-22b shows the first surface, which forms a virtual image of O' at I'. To locate I', we use Eq. 44-12,

$$\frac{n_1}{o} + \frac{n_2}{i} = \frac{n_2 - n_1}{r}.$$

Putting $n_1 = 1.0$ and $n_2 = n$ and bearing in mind that the image distance is negative (that is, $i = -i'$ in Fig. 44-22b), we obtain

$$\frac{1}{o'} - \frac{n}{i'} = \frac{n-1}{r'}. \tag{44-13}$$

In this equation i' will be a positive number because we have arbitrarily introduced the minus sign appropriate to a virtual image.

Figure 44-22c shows the second surface. Unless an observer at point a'' were aware of the existence of the first surface, he would think that the light striking that point originated at point I' in Fig. 44-22b and that the region to the left of the surface was filled with glass. Thus the (virtual) image I' formed by the first surface serves as a real object O'' for the second surface. The distance of this object from the second surface is

$$o'' = i' + l. \tag{44-14}$$

In applying Eq. 44-12 to the second surface, we insert $n_1 = n$ and $n_2 = 1.0$ because the object behaves as if it were imbedded in glass. If we use Eq. 44-14, Eq. 44-12 becomes

$$\frac{n}{i' + l} + \frac{1}{i''} = \frac{1-n}{r''}. \tag{44-15}$$

Let us now assume that the thickness l of the "lens" in Fig. 44-22 is so small that we can neglect it in comparison with other linear quantities in this figure (such as o', i', o'', i'', r', and r''). In all that follows we make this *thin-lens approximation*. Putting $l = 0$ in Eq. 44-15 leads to

$$\frac{n}{i'} + \frac{1}{i''} = -\frac{n-1}{r''}. \tag{44-16}$$

Adding Eqs. 44-13 and 44-16 leads to

$$\frac{1}{o'} + \frac{1}{i''} = (n-1)\left(\frac{1}{r'} - \frac{1}{r''}\right).$$

Finally, calling the original object distance simply o and the final image distance simply i leads to

$$\frac{1}{o} + \frac{1}{i} = (n-1)\left(\frac{1}{r'} - \frac{1}{r''}\right). \tag{44-17}$$

This equation holds only for paraxial rays and only if the lens is so thin that it essentially makes no difference from which surface of the lens the quantities o and i are measured. In Eq. 44-17 r' refers to the first surface struck by the light as it traverses the lens and r'' to the second surface.

The sign conventions for Eq. 44-17 are the same as those for mirrors and for single refracting surfaces. Because the lens is assumed to be thin, we refer to the R-side and the V-side of the lens itself (see Fig. 44-18) rather than those of its separate surfaces. The sign conventions are suggested by Fig. 44-23.

1. The image distance i is positive if the image (real) lies on the R-side

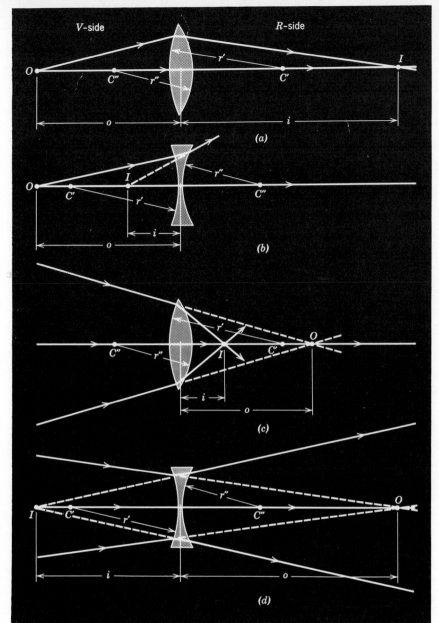

(a)

(b)

(c)

(d)

figure 44-23

(a) A real image and a real object. Both o and i are positive. (b) A virtual image and a real object. Note that o is positive but i is negative. (c) A real image and a virtual object. Note that o is negative but i is positive. (d) A virtual image and a virtual object. Both o and i are negative. Note that the sign of o is determined by whether the incident rays are diverging (first two cases) or converging (second two cases) when they strike the thin lens. For diverging incident rays the object is real; for converging rays it is virtual. The sign of i is determined by whether I is on the R-side of the lens (in which case I is real and i is positive, as in the first and third figures) or by whether I is on the V-side of the lens (in which case I is virtual and i is negative, as in the second and fourth figures).

of the lens, as in Figs. 44-23a and 44-23c. It is negative if the image (virtual) lies on the V-side of the lens, as in Figs. 44-23b and 44-23d.

2. The object distance o is positive if *diverging* rays fall on the lens, as in Figs. 44-23a and 44-23b; the object is real in such cases. The object distance o is negative if *converging* rays fall on the lens, as in Figs. 44-23c and 44-23d; the object is virtual in such cases.

3. The radii of curvature r' and r" refer, respectively, to the first and to the second surfaces to be struck by the incident light. Each of these quantities is positive if the corresponding centers of curvature C' and C" lie on the R-side of the lens; they are negative otherwise.

Figure 44-24a and 44-24c shows parallel light from a distant object falling on a thin lens. The image location is called the *second focal point*

(a)

(b)

(c)

figure 44-24

(a) Parallel light passes through the second focal point F_2 of a converging lens. (b) The incident light makes an angle α with the lens axis, the rays being focused in the focal plane $F'F''$. (c) Parallel light, passing through a diverging lens, seems to originate at the second focal point F_2. C' and C'' are centers of curvature for the lens surfaces; F_1 is the first focal point.

F_2 of the lens. The distance from F_2 to the lens is called the *focal length f*. The *first focal point* for a thin lens (F_1 in figure) is the object position for which the image is at infinity. For thin lenses the first and second focal points are on opposite sides of the lens and are equidistant from it.

We can find the focal length from Eq. 44-17 by inserting $o \rightarrow \infty$ and $i = f$. This yields

$$\frac{1}{f} = (n-1)\left(\frac{1}{r'} - \frac{1}{r''}\right). \qquad (44\text{-}18)$$

This relation is called the *lens maker's equation* because it allows us to compute the focal length of a lens in terms of the radii of curvature and the index of refraction of the material. Combining Eqs. 44-17 and 44-18 allows us to write the thin-lens equation as

$$\frac{1}{o} + \frac{1}{i} = \frac{1}{f}. \qquad (44\text{-}19)$$

Figure 44-24b shows parallel incident rays that make a small angle α with the lens axis; they are brought to a focus in the *focal plane $F'F''$*, as shown. This is a plane normal to the lens axis at the focal point.

In Fig. 44-24a we note that all rays in the figure contain the same number of wavelengths; in other words, they have the same *optical path*

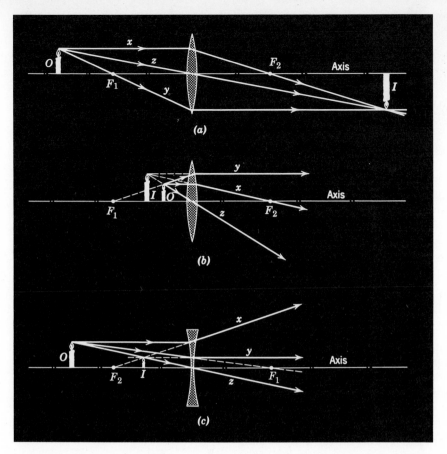

figure 44-25
Showing the graphical location of images for three thin lenses. We treat here real objects only, that is, diverging rays falling on the lens.

lengths; see Section 43-6. The optical path lengths are the same because the wavefronts are surfaces over which the wave disturbance has the same constant value and because all the rays shown pass through the same number of wavefronts.

The lenses of Fig. 44-24 have radii of curvature of magnitude 40 cm and are made of glass with $n = 1.65$. Compute their focal lengths.

Since C' lies on the R-side of the lens in Fig. 44-24a, r' is positive $(= +40 \text{ cm})$. Since C'' lies on the V-side, r'' is negative $(= -40 \text{ cm})$. Substituting in Eq. 44-18 yields

$$\frac{1}{f} = (n-1)\left(\frac{1}{r'} - \frac{1}{r''}\right) = (1.65 - 1)\left(\frac{1}{+40 \text{ cm}} - \frac{1}{-40 \text{ cm}}\right),$$

or
$$f = +31 \text{ cm.}$$

A positive focal length indicates that in agreement with Fig. 44-24a the focal point F_2 is on the R-side of the lens and parallel incident light converges after refraction to form a real image.

In Fig. 44-24c C' lies on the V-side of the lens so that r' is negative $(= -40 \text{ cm})$. Since r'' is positive $(= +40 \text{ cm})$, Eq. 44-18 yields

$$f = -31 \text{ cm.}$$

A negative focal length indicates that in agreement with Fig. 44-24c the focal point F_2 is on the V-side of the lens and incident light diverges after refraction to form a virtual image.

EXAMPLE 7

figure 44-26
Two rays for the situation of Fig. 44-25a.

We can locate the image of an extended object such as a candle (Fig. 44-25) graphically by using the following three facts:

1. A ray parallel to the axis and falling on the lens passes, either directly or when extended, through the second focal point F_2 (ray x in Fig. 44-25).

2. A ray falling on a lens after passing, either directly or when extended, through the first focal point F_1 will emerge from the lens parallel to the axis (ray y).

3. A ray falling on the lens at its center will pass through undeflected. There is no deflection because the lens, near its center, behaves like a thin piece of glass with parallel sides. The direction of the light rays is not changed and the sideways displacement can be neglected because the lens thickness has been assumed to be negligible (ray z).

Figure 44-26, which represents part of Fig. 44-25a, shows a ray passing from the tip of the object through the center of curvature to the tip of the image. For the similar triangles abc and dec we may write

$$\frac{de}{ab} = \frac{dc}{ac}.$$

The right side of this equation is i/o and the left side is $-m$, where m is the *lateral magnification*. The minus sign is required because we wish m to be negative for an inverted image. This yields

$$m = -\frac{i}{o}, \qquad (44\text{-}20)$$

which holds for all types of thin lenses and for all object distances.

EXAMPLE 8

A converging thin lens has a focal length of +24 cm. An object is placed 9.0 cm from the lens as in Fig. 44-25b. Describe the image.

From Eq. 44-19,

$$\frac{1}{o} + \frac{1}{i} = \frac{1}{f},$$

we have

$$\frac{1}{+9.0 \text{ cm}} + \frac{1}{i} = \frac{1}{+24 \text{ cm}},$$

which yields $i = -14.4$ cm, in agreement with the figure. The minus sign means that the image is on the V-side of the lens and is thus virtual.

The lateral magnification is given by

$$m = -\frac{i}{o} = -\frac{-14.4 \text{ cm}}{+9.0 \text{ cm}} = +1.6,$$

again in agreement with the figure. The plus signifies an erect image.

Images formed by lenses suffer from defects similar to those discussed for mirrors on p. 968. There are effects connected with the failure of a point object to form a point image, with the variation of magnification with distance from the lens axis, and with diffraction. For lenses, but not for mirrors, there are also *chromatic aberrations* associated with the fact that the refracting properties of the lens vary with wavelength because the index of refraction of the lens material does. If a point object on the lens axis emits white light, the image, neglecting other lens defects, will be a series of colored points spread out along the axis. We have all seen the colored images produced by inexpensive lenses. A great deal of ingenious optical engineering goes into the design of lenses (more commonly lens systems) in which the various lens defects are minimized. The lens surfaces are normally not spherical nor are the lenses "thin."

Note that we can write the mirror formula (Eq. 44-5) and the (identical) thin lens formula (Eq. 44-19) in this form:

$$\frac{1}{o/|f|} + \frac{1}{i/|f|} = \pm 1 \tag{44-21}$$

in which $|f|$, the absolute value of the focal length f, is always positive. On the right side of the equation we choose $+1$ for a converging lens or a concave mirror and -1 for a diverging lens or a convex mirror. See Problem 41.

Figure 44-27 is a graphical representation of Eq. 44-21, for both mirrors and lenses, Fig. 44-27a being for concave mirrors and converging lenses and Fig. 44-27b for convex mirrors and diverging lenses.

(a)

(b)

figure 44-27

(a) A representation of $i/|f|$ and $o/|f|$ for concave mirrors and converging lenses. Note that (lower left quadrant) a virtual object cannot produce a virtual image. The numbers near the x's are the magnifications (see Eq. 44-20). A positive sign means an erect image, a negative sign an inverted image. Compare previous figures.

(b) A representation of $i/|f|$ and $o/|f|$ for convex mirrors and diverging lenses. Note that (upper right quadrant) a real object cannot produce a real image. Can you verify this? Once again, the numbers near the x's are the magnifications (Eq. 44-20). See "Image Formation in Lenses & Mirrors, a Complete Representation" by Albert A. Bartlett, *The Physics Teacher,* May 1976.

The human eye is a remarkably effective organ* but its range can be extended in many ways by a host of optical instruments such as spectacles, simple magnifiers, motion picture projectors, cameras (including TV cameras), microscopes, telescopes, etc. In many cases these devices extend the scope of our vision beyond the visible range; satellite-borne infrared cameras and X-ray microscopes are examples.

In almost all cases of modern sophisticated optical instruments the mirror and thin lens formulas (Eqs. 44-5 and 44-19) hold only as an approximation. The rays may not be paraxial, as anyone who has used a camera knows; in astronomical telescopes, however, the rays are indeed paraxial. In typical laboratory microscopes the lens cannot be considered "thin" in the sense in which that term was defined in Section 44-5. In most optical instruments lenses are compound, that is, they are made of several components, cemented together, the interfaces rarely being exactly spherical. This is done to improve image quality and brightness and to relax greatly the dependence on paraxial rays.

In what follows we describe three optical instruments, assuming, for simplicity of illustration, that the thin lens formula applies.

a. The Simple Magnifier. The normal human eye can focus a sharp image of an object on the retina if the object O is located anywhere from infinity, the stars, say, to a certain point called the near point P_n, which we take to be about 25 cm from the eye. If you move the object within the near point the perceived retinal image becomes fuzzy. The location of the near point normally varies with age. We have all heard stories about people who claim not to need glasses but who read their newspapers at arm's length; their near points are receding! Find your own near point by moving this page closer to your eyes, considered separately, until you reach a position at which the image begins to become indistinct.

Figure 44-28a shows an object O placed at the near point P_n. The size

44-6
OPTICAL INSTRUMENTS

figure 44-28
(a) An object O of height h is placed at the near point P_n of the human eye. If moved any closer, it would fail to form a clear image on the retina. (b) A converging lens (that is, a simple magnifier) is placed close to the eye and the object O is moved from P_n to F_1. Not drawn to scale.

* Insect's eyes are even more versatile than human eyes. (Have you ever tried to catch a fly?) See "The Compound Eye of Insects" by G. Adrian Horridge, *Scientific American*, July 1977.

of the perceived image on the retina is measured by the angle θ. In Fig. 44-28b we insert a converging lens of focal length f just in front of the eye and move the object O to the first focal point F_1 of the lens. The eye now perceives an image at infinity, the angle of the image rays being θ', where $\theta' > \theta$. The *angular magnification* m_θ, not to be confused with the *lateral magnification* m given by Eq. 44-20, can be found from

$$m_\theta = \theta'/\theta$$

where $\qquad \theta \cong h/25 \text{ cm} \quad \text{and} \quad \theta' \cong h/f,$

or

$$m_\theta \cong 25 \text{ cm}/f. \qquad (44\text{-}22)$$

Note that, as expected, if $f = 25$ cm, then $m_\theta = 1$, that is, $\theta' = \theta$. Lens aberrations limit the angular magnifications for a simple converging lens to a few orders of magnitude. This is enough, however, for stamp collectors and for actors portraying Sherlock Holmes. More sophisticated magnifier designs have appreciably greater angular magnifications.

b. A Compound Microscope. Figure 44-29 shows a thin lens version of a compound microscope, used for viewing small objects that are very close to the instrument. The object O, of height h, is placed just outside the first focal point F_1 of the objective lens, whose focal length is f_{ob}. A real, inverted image I of height h' is formed by the objective, the lateral magnification being given by Eq. 44-20, or

$$m = \frac{h'}{h} = -\frac{s \tan \theta}{f_{ob} \tan \theta} = -\frac{s}{f_{ob}}. \qquad (44\text{-}23)$$

As usual, the minus sign indicates an inverted image.

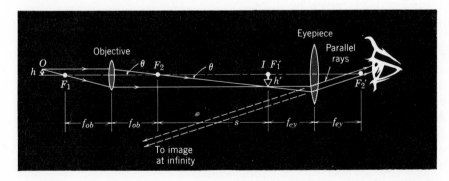

figure 44-29
A simplified version of a compound microscope, using "thin" lenses. Some dimensions have been distorted for clarity. Two rays only are shown.

The distance s (sometimes called the tube length) is so chosen that the image I falls on the first focal point F_1' of the eyepiece, which then acts as a simple magnifier as in subsection a above. Parallel rays enter the eye and a final image I' forms at infinity. The final magnification M is given by the product of the linear magnification m for the objective lens, given by Eq. 44-23, and the angular magnification m_θ for the eyepiece, given by Eq. 44-22, or

$$M = m \times m_\theta = -\left(\frac{s}{f_{ob}}\right)\left(\frac{25 \text{ cm}}{f_{ey}}\right). \qquad (44\text{-}24)$$

c. An Astronomical Telescope. Like microscopes, telescopes come in a large variety of forms. The form we describe here is the simple re-

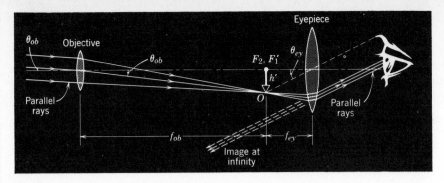

figure 44-30

A simplified version of an astronomical telescope, using "thin" lenses. Some dimensions have been distorted for clarity. Only three rays are shown.

fracting telescope* that consists of an objective lens and an eyepiece, both represented in Fig. 44-30 by thin lenses, although in practice, as for microscopes, they will each be compound lens systems.

At first glance it may seem that the lens arrangements for telescopes (Fig. 44-30) and for microscopes (Fig. 44-29) are similar. However, telescopes are designed to view large objects, such as galaxies, stars, and planets, at large distances, whereas microscopes are designed for just the opposite purpose. Note also that in Fig. 44-30 the second focal point of the objective F_2 coincides with the first focal point of the eyepiece F_1', but in Fig. 44-29 these points are separated by a distance s (sometimes called the *tube length*).

In Fig. 44-30 parallel rays from a distant object strike the objective lens, making an angle θ_{ob} with the telescope axis and forming a real, inverted image at the common focal point F_2, F_1'. This image acts as an object for the eyepiece and a (still inverted) virtual image is formed at infinity. The rays defining the image make an angle θ_{ey} with the telescope axis.

The angular magnification m_θ of the telescope is given by θ_{ey}/θ_{ob}. For paraxial rays we can write $\theta_{ob} = h'/f_{ob}$ and $\theta_{ey} = -h'/f_{ey}$ or

$$m_\theta = -\frac{f_{ob}}{f_{ey}}. \tag{44-25}$$

Magnification is only one of the design factors of an astronomical telescope and is indeed easily achieved (How?). There is also *light gathering power* which determines how bright the image is. This is important when viewing faint objects such as distant galaxies and is accomplished by making the objective lens diameter as large as possible. There is *field of view*. An instrument designed for galactic observation (narrow field of view) must be quite different from one designed for the observation of meteors (wide field of view). Also, there are lens and mirror aberrations including *spherical aberration* (that is, lenses and mirrors with truly spherical surfaces do not form sharp images) and *chromatic aberration* (that is, for simple lenses the focal length varies with wavelength so that fuzzy images are formed, displaying unnatural colors). There is also *resolving power* which describes the ability of any optical instrument to distinguish between two objects (stars, say) whose

---

* For an example of the diversity of astronomical telescopes, see "The MMT Observatory on Mount Hopkins" by Nathaniel P. Carleton and Thomas E. Hoffman, *Sky and Telescope*, July 1976. "MMT" stands for Multiple Mirror Telescope.

* See also *Physics Today*, p. 18, April 1977 for a preview of a 2.4-meter (aperture) Space Telescope, planned to orbit the earth in 1983. The device is 46 ft long and will weigh 17,000 lb. It will be launched into orbit by a space shuttle, which will return every two or three years to make repairs and change or add instruments.

angular separation is small. We will discuss this fully in Section 46-5. This by no means exhausts the design parameters of astronomical telescopes. We could also make a similar listing for compound microscopes and, indeed, for any high performance optical instrument.

EXAMPLE 9

Figure 44-31 shows a simple astronomical telescope like that of Fig. 44-30 with the exceptions that (a) the incident parallel rays are parallel to the telescope axis, and (b) the incident rays fill the objective lens, whose diameter is d_{ob}, as, they normally would. Find the diameter d of the *emergent pencil*, as it is called, which contains all the information that is available for entering the eye.

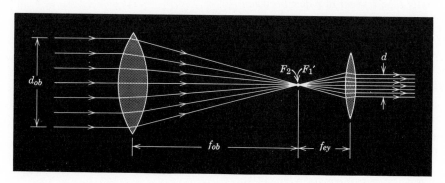

figure 44-31
The formula for d (Eq. 44-26) holds even if the incident rays are not parallel to the telescope axis, as in Fig. 44-30. The rays must be close to the axis, however, or most of the quantitative considerations of this section do not hold true.

In Fig. 44-31 we have, from similar triangles,

$$\frac{d_{ob}/2}{f_{ob}} = \frac{d/2}{f_{ey}},$$

or (see Eq. 44-25)

$$d = d_{ob}\left(\frac{f_{ey}}{f_{ob}}\right) = -\frac{d_{ob}}{m_\theta}. \qquad (44\text{-}26)$$

Note that m_θ is inherently negative so that d is positive, as it must be.

We must compare d with d_p, the diameter of the pupil of the eye. The pupil diameter is not constant but varies with illumination, between the limits of about 2 mm (sunlight) to 9 mm (darkness). The ideal condition is $d = d_p$. For example, if $d > d_p$ some light does not enter the eye and is wasted.

Actually these considerations are modified in practice. For most major optical instruments we use photographic recording. Also, telescopes (and cameras) in orbiting satellites and space probes often send data to earth electronically, where it is processed by computor techniques into photographic form. The emergent pencil concept remains important however, not only for visual observation but for many lens systems design considerations.

questions

1. Can you think of a simple test or observation to prove that the law of reflection is the same for all wavelengths, under conditions in which geometrical optics prevails?

2. We all know that when we look into a mirror right and left are reversed. Our right hand will seem to be a left hand; if we part our hair on the left it will seem to be parted on the right, etc. Can you think of a system of mirrors that would let us see ourselves as others see us? If so draw it and prove your point by sketching some typical rays.

3. We have seen that a single reflection in a plane mirror reverses right and left. When we drive down a highway, for example, the letters on the highway signs are reversed as seen through the rear view mirror. And yet, as

seen through this same mirror, you still seem to be driving down the right lane. Why does the mirror reverse the signs and not the lanes? Or does it? Discuss.

4. A young women peers closely at her face in a high-quality plane mirror. She sees an image of her face in perfect focus. (Ignore the right-left reversal). Why the "perfect focus"? After all her nose is closer to the mirror than her earlobes are.

5. Note (Fig. 44-8) that any portion of the mirror that extends below b or above a is totally wasted if you wish to see yourself at full length in a mirror. If you have a "full length" mirror available convince yourself that this is true by taping a sheet of newspaper across the mirror at points a and b, leaving only the region between a and b exposed. Then, by stepping backward and forward, convince yourself that this theorem is true no matter what your distance from the mirror. Finally tape a sheet of paper *between* a and b and note what you see in the mirror.

6. If a mirror reverses right and left, why doesn't it reverse up and down?

7. Devise a system of plane mirrors that will let you see the back of your head. Trace the rays to prove your point.

8. If converging rays fall on a plane mirror, is the image virtual?

9. In many city buses a convex mirror is suspended over the door, in full view of the driver. Why not a plane or a concave mirror?

10. Dentists and dental hygienists use a small mirror with a long handle attached to examine your teeth. Is the mirror concave, convex, or plane, and why?

11. What approximations were made in deriving the mirror equation (Eq. 44-4):

$$\frac{1}{o} + \frac{1}{i} = \frac{2}{r}?$$

12. Under what conditions will a spherical mirror, which may be concave or convex, form (a) a real image, (b) an inverted image, and (c) an image smaller than the object?

13. Is it possible to see a perfect reflector? That is, if a perfectly reflecting concave spherical mirror is placed in a dark room and illuminated by, say, a luminous point object placed on its optical axis, could you see the surface of the mirror? Discuss.

14. Can a virtual image be projected onto a screen? Photographed? If you put a piece of paper at the site of a virtual object (assuming a high-intensity light beam) will it ignite after sufficient exposure? Consider these two questions for real images and real objects and note the differences, if any. Discuss.

15. The human eye has often been likened to a camera and, indeed, this comparison is valid. There is a difference however; the eye has no shutter (the eyelids do not serve this purpose). If you leave the shutter of a camera open and scan it across the horizon, the developed image on the film will be a blur. On the other hand, if you scan the horizon with your eye you will see every object distinctly. Have you any ideas to explain the difference between these two situations? See "Visual Motion Perception" by Gunnar Johansson, *Scientific American*, June 1975.

16. In connection with Question 15 consider this further difference between the eye and the camera. If you photograph, say, an elm tree, the image on the film can easily be overexposed beyond the point of recognition. On the other hand, if you simply look at an elm tree your perceived image remains constant for hours on end. Why this difference?

17. For the human eye we assert that the main focusing device is the cornea, that is, the curved outer surface of the eye and that the "lens" of the eye serves to make minor focussing adjustments. True or false? What function does the iris have? Can you think of advantages that the camera has over the human eye? There are some.

18. Does the apparent depth of an object below water depend on the angle of view of the observer in air? Explain and illustrate with ray diagrams.

19. An unsymmetrical thin lens forms an image of a point object on its axis. Is the image location changed if the lens is reversed?

20. Why has a lens two focal points and a mirror only one?

21. Under what conditions will a thin lens, which may be converging or diverging, form (a) a real image, (b) an inverted image, and (c) an image smaller than the object?

22. A skin diver wants to use an air-filled plastic bag as a converging lens for underwater use. Sketch a suitable cross section for the bag.

23. What approximations were made in deriving the thin lens equation (Eq. 44-19):

$$\frac{1}{o} + \frac{1}{i} = \frac{1}{f}?$$

24. Under what conditions will a thin lens have a lateral magnification (a) of -1 and (b) of $+1$?

25. How does the focal length of a glass lens for blue light compare with that for red light, assuming the lens is (a) diverging and (b) converging?

26. Does the focal length of a lens depend on the medium in which the lens is immersed? Is it possible for a given lens to act as a converging lens in one medium and a diverging lens in another medium?

27. Are the following statements true for a glass lens in air? (a) A lens that is thicker at the center than at the edges is a converging lens for parallel light. (b) A lens that is thicker at the edges than at the center is a diverging lens for parallel light. Explain and illustrate, using wavefronts.

28. Under what conditions would the lateral magnification ($m = -i/o$) for lenses and mirrors become infinite? Is there any practical significance to such a condition?

29. Light rays are reversible. Discuss the situation in terms of objects and images if all rays in Figs. 44-10, 44-14, 44-17, 44-19, 44-23, and 44-25 are reversed in direction.

30. In connection with Fig. 44-24a, we pointed out that all rays originating on the same wavefront in the incident wave have the same optical path length to the image point. Discuss this in connection with Fermat's principle (Section 43-6).

31. Is the focal length of a spherical mirror affected by the medium in which it is immersed? . . . of a thin lens . . . ? What's the difference?

32. What is the significance of the origin of coordinates in Figs. 44-27a and 44-27b?

33. In Fig. 44-23 what are the signs of o, i, r', and r'' in the four cases shown?

34. In Fig. 44-27a how do you interpret $o/|f| = +1$ and $i/|f| = +1$? Draw a ray diagram to illustrate these two situations, for thin converging lenses. In Fig. 44-27b answer the same question for $o/|f| = -1$ and $i/|f| = -1$, for thin diverging lenses.

35. Why is the magnification of a simple magnifier (see the derivation leading to Eq. 44-22) defined in terms of angles rather than image/object size?

36. Ordinary spectacles do not magnify but a simple magnifier does. What then, is the function of spectacles?

37. The "f-number" of a camera lens (see Problem 47) is its focal length divided by its aperture, that is, its effective diameter. Why is this useful to know in photography? How can the f-number of the lens by changed? How is exposure time related to f-number?

38. Does it matter whether (a) an astronomical telescope, (b) a compound microscope, (c) a simple magnifier, (d) a camera, including a TV camera, or (e) a projector, including a slide projector and a motion picture projector produce erect or inverted images? What about real or virtual images? Discuss each case.

39. Why does chromatic aberration occur in simple lenses but not in mirrors?

40. The unaided human eye produces a real but *inverted* image on the retina. (*a*) Why then don't we perceive objects such as people and trees as upside down? (*b*) We don't, of course, but suppose that we wore special glasses so that we did. If you then turned this book upside down, could you read this question with the same facility that you do now? Discuss.

41. In the movies, when the director wishes to portray a scene as viewed through a pair of binoculars, a mask with a horizontal figure-eight opening usually appears on the screen. What is wrong with this?

42. Why are all recent large astronomical telescopes of the reflecting rather than the refracting variety? Think of mechanical mounting problems for lenses and mirrors, the difficulties of shaping (that is, "figuring") the various optical surfaces involved, problems with small flaws in the optical glass blanks used to make lenses and mirrors, etc.

problems

SECTION 44-2

1. A small object is 10 cm in front of a plane mirror. If you stand behind the object, 30 cm from the mirror, and look at its image, for what distance must you focus your eyes?   *Answer:* 40 cm.

2. Suppose you wished to photograph an object seen in a plane mirror. If the object is 5.0 m to your right and 1.0 m closer to the plane of the mirror than you, for what distance must you focus the lens of your camera?

3. A point object is 10 cm away from a mirror while the eye of an observer (pupil diameter 5.0 mm) is 20 cm away. Assuming both the eye and the point to be on the same line perpendicular to the surface, find the area of the mirror used in observing the reflection of the point.   *Answer:* 2.2 mm².

4. Two plane mirrors make an angle of 90° with each other. What is the largest number of images of an object placed between them that can be seen by a properly placed eye? The object need *not* lie on the mirror bisector.

5. Solve Example 2 if the angle between the mirrors is (*a*) 45°, (*b*) 60°, (*c*) 120°, the object always being placed on the bisector of the mirrors.
   Answer: (*a*) 7, (*b*) 5, (*c*) 2.

6. How many images of himself can an observer see in a room whose ceiling and two adjacent walls are mirrors? Explain.

7. A small object *O* is placed one-third of the way between two parallel plane mirrors as in Fig. 44-32. Trace appropriate bundles of rays for viewing the four images that lie closest to the object.

8. In Fig. 44-6 you rotate the mirror 30° counterclockwise, leaving the point object *O* in place. Is the (virtual) image point displaced? If so, where is it? Can the eye still see the image without being moved? Sketch a figure showing the new situation, with *O* and the eye remaining unchanged in position.

9. You are peering into a peephole into an illuminated box. A plane mirror is on the interior wall facing you. A square wire framework is rotating slowly in a counterclockwise direction (as seen from above) about a vertical axis.

figure 44-32
Problem 7

The distance from the peephole to the mirror is 15 cm. (a) How far from the mirror should the vertical axis of rotation of the square be located so that its image in the mirror would appear at half size? (b) Compare the directions of rotation of the object and image. (c) Describe in your own words what you would see, assuming that you have not been told that the box contains a mirror.

10. Extend Fig. 44-9 to three dimensions by adding a mirror perpendicular to the common axis of the two mirrors shown. This forms a *corner reflector*, much used in optical, microwave, and other applications. It has the property that an incident ray is returned, after three reflections, along the same direction. Can you prove this?

SECTION 44-3

11. For clarity, the rays in figures like Fig. 44-10 are not drawn paraxial enough for Eq. 44-4 to hold with great accuracy. With a ruler, measure r and o in this figure and calculate, from Eq. 44-4, the predicted value of i. Compare this with the measured value of i.

12. Fill in this table, each column of which refers to a spherical mirror and a real object. Check your results by graphical analysis. Distances are in centimeters; if a number has no plus or minus sign in front of it, find the correct sign.

	a	b	c	d	e	f	g	h
Type	Concave						Convex	
f	20		+20			20		
r				−40			40	
i				−10			4	
o	+10	+10	+30	+60				+24
m		+1		−0.5		+0.10		0.50
Real image?		no						
Erect image?								no

Answer: For alternate vertical columns: (a) +, +40, −20, +2, no, yes. (c) Concave, +40, +60, −2, yes, no. (e) Convex, −20, +20, +0.5, no, yes. (g) −20, −, −, +5, no, yes.

13. A short linear object of length l lies on the axis of a spherical mirror, a distance o from the mirror. (a) Show that its image will have a length l' where

$$l' = l\left(\frac{f}{o-f}\right)^2.$$

(b) Show that the *longitudinal magnification* $m'(=l'/l)$ is equal to m^2 where m is the lateral magnification discussed in Section 44-3. (c) Is there any condition such that, neglecting all mirror defects, the image of a small cube would also be a cube? *Answer:* (c) Yes; object at center of curvature.

14. Redraw Fig. 44-33 on a large sheet of paper and trace carefully the reflected rays, using the law of reflection. Is a point focus formed? Discuss.

15. A thin flat plate of partially reflecting glass is a distance b from a convex mirror. A point source of light S is placed a distance a in front of the plate (see Fig. 44-34) so that its image in the partially reflecting plate coincides with its image in the mirror. If $b = 7.5$ cm and the focal length of the mirror is $f = -30$ cm, find a and draw the ray diagram. *Answer:* 23 cm.

figure 44-33
Problem 14

figure 44-34
Problem 15

16. Verify that Eq. 44-4 is consistent with the situations of Figs. 44-5, 10, 12 (see Example 3), 13, 14, and 15. In some cases only a qualitative answer is possible.

17. Modify Fig. 44-12 in such a way (possibly by adding a second mirror to the left of the one shown) that the object for Fig. 44-12 is virtual. Trace the rays.

SECTION 44-4

18. A penny lies on the bottom of a swimming pool 10 ft. deep. What is its apparent depth as viewed from above the water? The index of refraction of water is 1.33.

19. Fill in the following table, each column of which refers to a spherical surface separating two media with different indices of refraction. Distances are measured in centimeters. The object is real in all cases.

	a	b	c	d	e	f	g	h
n_1	1.0	1.0	1.0	1.0	1.5	1.5	1.5	1.5
n_2	1.5	1.5	1.5		1.0	1.0	1.0	
o	+10	+10		+20	+10		+70	+100
i		−13	+600	−20	−6	−7.5		+600
r	+30		+30	−20		−30	+30	−30
Real image?								

Draw a figure for each situation and construct the appropriate rays graphically. Assume a point object.
Answer: For alternate vertical columns: (a) −18, no. (c) +71, yes. (e) +30, no. (g) −26, no.

20. A layer of water ($n = 1.33$) 2.0 cm thick floats on carbon tetrachloride ($n = 1.46$) 4.0 cm thick. How far below the water surface, viewed at normal incidence, does the bottom of the tank seem to be?

21. As an example of the importance of the paraxial ray assumption, consider this problem. You place a coin at the bottom of a swimming pool filled with water ($n = 1.33$) to a depth of 8.0 ft. What is the apparent depth of the coin below the surface when viewed (a) at near normal incidence (that is, by paraxial rays) and (b) by rays that leave the coin making an angle of 30° with the normal (that is, definitely not paraxial rays)? What do you conclude? *Answer:* (a) 6.0 ft. (b) 3.9 ft.

22. Define and locate the first and second focal points (see p. 975) for a single spherical refracting surface such as that of Fig. 44-17.

23. A parallel incident beam falls on a solid glass sphere at normal incidence. Locate the image in terms of the index of refraction n and the sphere radius r.

Answer: Assuming the light incident from the left, $i = \dfrac{2-n}{2(n-1)}r$, to the right

of the right edge of the sphere if n < 2, as it is for glass.

SECTION 44-5

24. A double-convex lens is to be made of glass with an index of refraction of 1.50. One surface is to have twice the radius of curvature of the other and the focal length is to be 6.0 cm. What are the radii?

25. A lens is made of glass having an index of refraction of 1.5. One side of the lens is flat and the other convex with a radius of curvature of 20 cm. (a) Find the focal length of the lens. (b) If an object is placed 40 cm to the left of the

figure 44-35
Problem 26

lens, where will the image be located? (c) Would either of these answers change if we rotated the lens through 360°?

Answer: (a) +40 cm. (b) Image to the right at +∞. (c) No.

26. Using the lensmaker's equation (Eq. 44-18), decide which of the thin lenses in Fig. 44-35 are converging and which diverging for parallel incident light.

27. An object is placed at a center of curvature of a double-concave lens, both of whose radii of curvature have the same magnitude. (a) What are the signs of the two radii of curvature? (b) Find the location of the image in terms of the radius of curvature r and the index of refraction, n, of the glass. (c) Describe the nature of the image. (d) Verify your result with a ray diagram.

Answer: (a) r' is negative and r'' is positive; see Fig. 44-23b and 44-23c.

(b) $i = -\dfrac{r}{2n-1}$. (c) Virtual and erect.

28. Show that the focal length f' for a thin lens whose index of refraction is n and which is immersed in a medium, water, say, whose index of refraction is n' is given by

$$\frac{1}{f'} = \frac{n-n'}{n'}\left(\frac{1}{r'} - \frac{1}{r''}\right).$$

29. Reproduce Fig. 44-27a from first principles, that is, from Eq. 44-19. How do you know (a) that the lens is diverging or converging? (b) That the image is real or virtual? (c) That the object is real or virtual? (d) That the lateral magnification is > 1 or < 1?

30. Fill in this table, each column of which refers to a thin lens, to the extent possible. Check your results by graphical analysis. Distances are in centimeters; if a number (except in row n) has no plus sign or minus sign in front of it, find the correct sign.

	a	b	c	d	e	f	g	h	i
Type	converging								
f	10	+10	10	10					
r'					+30	−30	−30		
r''					−30	+30	−60		
i									
o	+20	+5	+5	+5	+10	+10	+10	+10	+10
n					1.5	1.5	1.5		
m			>1	<1				0.5	0.5
Real image?									yes
Erect image?								yes	

Draw a figure for each situation and construct the appropriate rays graphically. The object is real in all cases.

Answer: Alternate vertical columns (an X means that the quantity cannot be found from the data given): (*a*) +, X, X, +20, X, −1, yes, no. (*c*) Converging, +, X, X, −10, X, no, yes. (*e*) Converging, +30, −15, +1.5, no, yes. (*g*) Diverging, −120, −9.2, +0.92, no, yes. (*i*) Converging, +3.3, X, X, +5, X, −, no.

31. A luminous object and a screen are a fixed distance *D* apart. (*a*) Show that a converging lens of focal length *f* will form a real image on the screen for two positions that are separated by

$$d = \sqrt{D(D - 4f)}.$$

(*b*) Show that the ratio of the two image sizes for these two positions is

$$\left(\frac{D - d}{D + d}\right)^2.$$

32. A converging lens with a focal length of +20 cm is located 10 cm to the left of a diverging lens having a focal length of −15 cm. If a real object is located 40 cm to the left of the first lens, locate and describe completely the image formed.

33. Two thin lenses of focal length f_1 and f_2 are in contact. Show that they are equivalent to a single thin lens with a focal length given by

$$f = \frac{f_1 f_2}{f_1 + f_2}.$$

34. Show that the distance between a real object and its real image formed by a thin converging lens is always greater than or equal to four times the focal length of the lens.

35. The formula

$$\frac{1}{o} + \frac{1}{i} = \frac{1}{f}$$

is called the *Gaussian* form of the thin lens formula. Another form of this formula, the *Newtonian* form, is obtained by considering the distance *x* from the object to the first focal point and the distance *x'* from the second focal point to the image. Show that

$$xx' = f^2.$$

36. An erect object is placed a distance in front of a converging lens equal to twice the focal length f_1 of the lens. On the other side of the lens is a converging mirror of focal length f_2 separated from the lens by a distance $2(f_1 + f_2)$. (*a*) Find the location, nature and relative size of the final image. (*b*) Draw the appropriate ray diagram. See Fig. 44-36.

figure 44-36
Problem 36

37. Two thin lenses, one having $f = +12.0$ cm and the other having $f = −10.0$ cm, are separated by 7.0 cm. A small object is placed 43.5 cm from the center of

the lens system on the principal axis first on one side and next on the other side. Find the location of the final image in each case.

Answer: For object nearer to the converging lens the image is 8.5 cm to the opposite side of the center of the lens system. For an object nearer the diverging lens, the image is 63.5 cm to the opposite side of the center of the lens system.

38. (a) Show that a thin converging lens of focal length f followed by a thin diverging lens of focal length −f will bring parallel light to a focus beyond the second lens provided that the separation of the lenses L satisfies $0 < L < f$. (b) Does this property change if the lenses are interchanged? (c) What happens when $L = 0$?

39. An object is placed 1.0 m in front of a converging lens, of focal length 0.5 m, which is 2.0 m in front of a plane mirror. (a) Where is the final image, measured from the lens, that would be seen by an eye looking toward the mirror through the lens? (b) Is the final image real or virtual? (c) Is the final image erect or inverted? (d) What is the lateral magnification?

Answer: (a) 0.60 m on the side of the lens away from the mirror. (b) Real. (c) Erect. (d) +0.20.

40. An object is 20 cm to the left of a lens with a focal length of +10 cm. A second lens of focal length +12.5 cm is 30 cm to the right of the first lens. (a) Using the image formed by the first lens as the object for the second, find the location and relative size of the final image. (b) Verify your conclusions by drawing the lens system to scale and constructing a ray diagram. (c) Describe the final image.

41. Show that Eq. 44-21 is correct.

42. An object is placed 1.0 m in front of a converging lens, of focal length 0.50 m, which is 2.0 m in front of a plane mirror. (a) Where is the final image, measured from the lens, that would be seen by an eye looking toward the mirror through the lens? (b) Is the final image real or virtual? (c) Is the final image erect or inverted? (d) What is the lateral magnification?

SECTION 44-6

43. In connection with Fig. 44-28b, (a) show that if the object O is moved from the first focal point F_1 toward the eye, the image moves in from infinity and the angle θ' (and thus the angular magnification m_θ) is increased. (b) If you continue this process, at what image location will m_θ have its maximum useable value? (c) Show that the maximum useable value of m_θ is $1 + (25\ \text{cm})/f$. (d) Show that in this situation the angular magnification is equal to the linear magnification.

44. A microscope of the type shown in Fig. 44-29 has a focal length for the objective lens of 4.0 cm and for the eyepiece lens of 8.0 cm. The distance between the lenses is 25 cm. (a) What is the distance s in Fig. 44-29? (b) To reproduce the conditions of Fig. 44-29 how far beyond F_1 in that figure should the object be placed? (c) What is the lateral magnification m of the objective? (d) What is the angular magnification m_θ of the eyepiece? (e) What is the overall magnification M of the microscope?

45. *The eye—the basic optical instrument:* Figure 44-37a suggests a normal human eye. Parallel rays, entering a relaxed eye gazing at infinity, produce a real, inverted image on the retina. The eye thus acts as a converging lens. Most of the refraction occurs at the outer surface of the eye, the *cornea*. Assume a focal length f for the eye of 2.50 cm.

In Fig. 44-37b the object is moved in to a distance o (=40.0 cm) from the

(a)

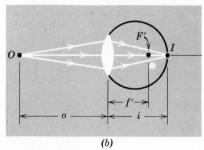

(b)

figure 44-37
Problem 45

eye. To form an image on the retina the effective focal length of the eye must be reduced to f'. This is done by the action of the ciliary muscles that change the shape of the lens and thus the effective focal length of the eye.

(a) Find f' from the above data. (b) Would the effective radii of curvature of the lens become larger or smaller in the transition from a to b in Fig. 44-37? (In the figure the structure of the eye is only roughly suggested and Fig. 44-37b is not to scale.)   *Answer:* (a) 2.35 cm. (b) Smaller.

(a)

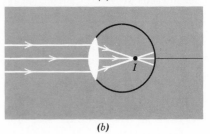

(b)

figure 44-38
Problem 46

46. In an eye that is *farsighted* the eye focuses parallel rays so that the image would form behind the retina, as in Fig. 44-38a. In an eye that is *nearsighted* the image is formed in front of the retina, as in Fig. 44-38b. (a) How would you design a corrective lens for each eye defect? Make a ray diagram for each case. (b) If you need spectacles only for reading, are you nearsighted or farsighted? (c) What is the function of bifocal spectacles, in which the upper parts and lower parts have different focal lengths? (d) Some musicians in symphony orchestras (and others as well) wear *trifocals*. What occupational optical problem is involved here?

47. *The camera:* Figure 44-39 shows an idealized camera focused on an object at infinity. A real, inverted image I is formed on the film, the image distance i being equal to the (fixed) focal length f (=5.0 cm, say) of the lens system. In Fig. 44-39b the object O is closer to the camera, the object distance o being, say, 100 cm. To focus an image I on the film, we must extend the lens away from the camera (why?). (a) Find i' in Fig. 44-39b. (b) By how much must the lens be moved? Note that the camera differs from the eye (see Problem 45) in this respect. In the camera, f remains constant and the image distance i must be adjusted by moving the lens. For the eye the image distance i remains constant and the focal length f is adjusted by distorting the lens. Compare Figs. 44-37 and 44-39 carefully.
Answer: (a) 5.3 cm. (b) 3.0 mm.

figure 44-39
Problem 47

48. *The reflecting telescope:* Isaac Newton, having convinced himself (erroneously as it turned out) that chromatic aberration was an inherent property of refracting telescopes, invented the reflecting telescope, shown schematically in Fig. 44-40. He presented his second model of this telescope, which has a magnifying power of 38, to the Royal Society, which still has it.

In Fig. 44-40 incident light falls, closely parallel to the telescope axis, on the objective mirror M. After reflection from small mirror M' (the figure is not to scale), the rays form a real, inverted image in the focal plane through F. This image is then viewed through an eyepiece.

(a) Show that the angular magnification m_θ is also given by Eq. 44-25, or

$$m_\theta = -f_{ob}/f_{ey}$$

where f_{ob} is the focal length of the objective mirror and f_{ey} that of the eyepiece. (b) The 200-in. mirror of the reflecting telescope at Mt. Palomar in California has a focal length of 16.8 m. Estimate the size of the image formed

figure 44-40
Problem 48

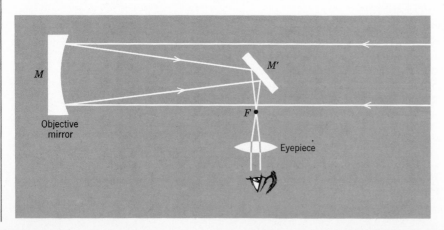

in the focal plane of this mirror when the object is a meter stick 2.0 km away. Assume parallel incident rays. (c) The mirror of a reflecting astronomical telescope has an effective radius of curvature ("effective" because such mirrors are ground to a parabolic rather than a spherical shape, to eliminate spherical aberration defects) of 10 m. To give an angular magnification of 200, what must be the focal length of the eyepiece?

45
interference

In Section 19-7 we saw that if two mechanical waves of the same frequency travel in approximately the same direction and have a phase difference that remains constant with time, they may combine so that their energy is not distributed uniformly in space but is a maximum at certain points and a minimum (possibly even zero) at others. The demonstration of such *interference* effects for light by Thomas Young in 1801 first established the wave theory of light on a firm experimental basis. Young was able to deduce the wavelength of light from his experiments, the first measurement of this important quantity.

Young allowed sunlight to fall on a pinhole S_0 punched in a screen A in Fig. 45-1. The emerging light spreads out by diffraction (see Section 44-1) and falls on pinholes S_1 and S_2 punched into screen B. Again diffraction occurs and two overlapping spherical waves expand into the space to the right of screen B.

The condition for geometric optics, namely, that $a \gg \lambda$ where a is the diameter of the pinholes, is definitely *not* met in this experiment. The pinholes do not cast geometrical shadows but act as sources of expanding Huygens' wavelets. We are dealing here (and in the three succeeding chapters) with *wave optics* rather than with geometrical optics.

Figure 45-2, taken from an 1803 paper of Young, shows the region between screens B and C. The blackening represents the minima of the wave disturbance; the white space between represents the maxima. If you hold the page with your eye close to the left edge and look at a grazing angle along the figure, you will see that along lines marked by x's there is cancellation of the wave; between them there is reinforcement. If we place a screen anywhere across the superimposed waves, we expect to find alternate bright and dark spots on it. Figure 45-3 shows a

figure 45-1
Showing how Thomas Young produced an interference pattern by allowing diffracted waves from pinholes S_1 and S_2 to overlap on screen C.

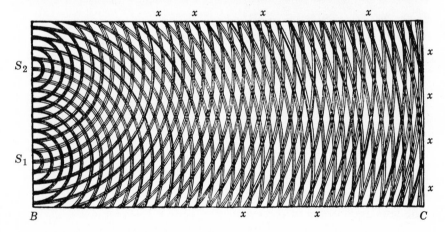

figure 45-2
Thomas Young's original drawing showing interference effects in overlapping waves. Place the eye near the left edge and sight at a grazing angle along the figure. (From Thomas Young, *Phil. Transactions*, 1803.) See also *Great Experiments in Physics*, p. 93, Morris H. Shamos, ed., Holt and Company, N.Y., 1959, for a readable annotated account of this experiment.

figure 45-3
Interference fringes for monochromatic light, made with an arrangement like that of Fig. 45-1, using long narrow slits rather than pinholes.

photograph of such *interference fringes*; in keeping with modern technique, long narrow slits rather than pinholes were used in preparing this figure.

Interference is not limited to light waves but is a characteristic of all wave phenomena. Figure 45-4, for example, shows the interference pattern of water waves in a shallow *ripple tank*. The waves are generated by two vibrators that tap the water surface in synchronism, producing two expanding spherical waves.

Let us analyze Young's experiment quantitatively, assuming that the incident light consists of a single wavelength only. In Fig. 45-5 P is an arbitrary point on the screen, a distance r_1 and r_2 from the narrow slits S_1 and S_2, respectively. Let us draw a line from S_2 to b in such a way that the lines PS_2 and Pb are equal. If d, the slit spacing, is much smaller than the distance D between the two screens (the ratio d/D in the figure has been exaggerated for clarity), S_2b is then almost perpendicular to both r_1 and r_2. This means that angle S_1S_2b is almost equal to angle PaO, both

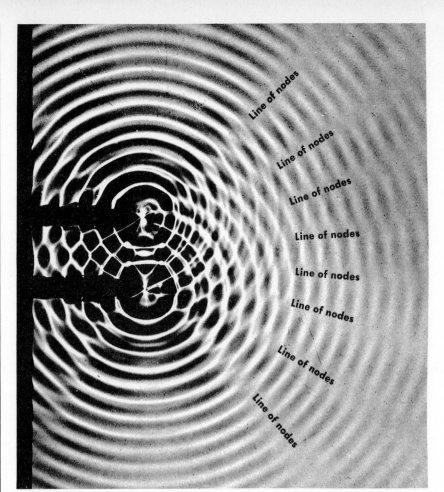

figure 45-4
The interference of water waves in a ripple tank. There is destructive interference along the lines marked "Line of nodes" and constructive interference between these lines. (Courtesy Physical Science Study Committee.)

angles being marked θ in the figure. This is equivalent to saying that the lines r_1 and r_2 may be taken as parallel.

We often put a lens in front of the two slits, as in Fig. 45-6, the screen C being in the focal plane of the lens. Under these conditions light focused at P must have struck the lens parallel to the line Px, drawn from P through the center of the (thin) lens. Under these conditions rays r_1 and r_2 are strictly parallel even though the requirement $D \gg d$ is not met. The lens L may in practice be the lens and cornea of the eye, screen C being the retina.

The two rays arriving at P in Figs. 45-5 or 45-6 from S_1 and S_2 are in phase at the source slits, both being derived from the same wavefront in the incident plane wave. Because the rays have different optical path lengths, they arrive at P with a phase difference. The number of wavelengths contained in $S_1 b$, which is the path difference, determines the nature of the interference at P.

To have a *maximum* at P, $S_1 b$ ($= d \sin \theta$) must contain an integral number of wavelengths, or

$$S_1 b = m\lambda \qquad m = 0, 1, 2, \ldots,$$

which we can write as

$$d \sin \theta = m\lambda \qquad m = 0, 1, 2, \ldots \text{ (maxima)}. \qquad (45\text{-}1)$$

Note that each maximum above O in Figs. 45-5 and 45-6 has a symmetrically located maximum below O. There is a central maximum described by $m = 0$.

figure 45-5
Rays from S_1 and S_2 combine at P.
The wave fronts of light falling on
screen B have been taken as parallel.
Actually, $D \gg d$, the figure being
distorted for clarity. We represent
the midpoint of the slits by a.

figure 45-6
A lens is normally used to produce
interference fringes; compare with
Fig. 45-5. The figure is again distorted
for clarity in that $f \gg d$ in practice.

For a *minimum* at P, $S_1 b$ $(=d \sin \theta)$ must contain a half-integral
number of wavelengths, or

$$d \sin \theta = (m + \tfrac{1}{2})\lambda \qquad m = 0, 1, 2, \ldots \text{(minima)}. \qquad (45\text{-}2)$$

If a lens is used as in Fig. 45-6, it may seem that a phase difference
should develop between the rays beyond the plane represented by $S_2 b$,
the path lengths between this plane and P being clearly different. In
Section 44-5, however, we saw that for such parallel rays focused by a
lens the *optical path lengths* are identical. Two rays with the same optical path lengths contain the same number of wavelengths, so that no
phase difference will result because of the light passing through the lens.

EXAMPLE 1

The double-slit arrangement in Fig. 45-5 is illuminated with light from a mercury vapor lamp so filtered that only the strong green line ($\lambda = 546$ nm or 5460 Å)
is effective. The slits are 0.10 mm apart, and the screen on which the interference pattern appears is 20 cm away. What is the angular position of the first
minimum? Of the tenth maximum?

At the first minimum we put $m = 0$ in Eq. 45-2, or

$$\sin \theta = \frac{(m + \tfrac{1}{2})\lambda}{d} = \frac{(\tfrac{1}{2})(546 \times 10^{-9} \text{ m})}{0.10 \times 10^{-3} \text{ m}} = 0.0027.$$

This value for $\sin \theta$ is so small that we can take it to be the value of θ, expressed
in radians; expressed in degrees it is 0.16°.

At the tenth maximum (not counting the central maximum) we must put
$m = 10$ in Eq. 45-1. Doing so and calculating as before leads to an angular position of 3.8°. For these conditions we see that the angular spread of the first
dozen or so fringes is small.

EXAMPLE 2

In Example 1 what is the linear distance on screen C between adjacent maxima?
If θ is small enough, we can use the approximation

$$\sin \theta \cong \tan \theta \cong \theta.$$

From Fig. 45-5 we see that

$$\tan \theta = \frac{y}{D}.$$

Substituting this into Eq. 45-1 for $\sin \theta$ leads to

$$y = m \frac{\lambda D}{d} \qquad m = 0, 1, 2, \ldots \text{(maxima)}.$$

The positions of any two adjacent maxima are given by

$$y_m = m \frac{\lambda D}{d}$$

and

$$y_{m+1} = (m+1) \frac{\lambda D}{d}.$$

We find their separation Δy by subtracting

$$\Delta y = y_{m+1} - y_m = \frac{\lambda D}{d}$$

$$= \frac{(546 \times 10^{-9} \text{ m})(20 \times 10^{-2} \text{ m})}{0.10 \times 10^{-3} \text{ m}} = 1.09 \text{ mm}.$$

As long as θ in Figs. 45-5 and 45-6 is small, the separation of the interference fringes is independent of m; that is, the fringes are evenly spaced. Note that if the incident light contains more than one wavelength the separate interference patterns, which will have different fringe spacings, will be superimposed.

We can use Eq. 45-1 to measure the wavelength of light; to quote Thomas Young:

From a comparison of various experiments, it appears that the breadth of the undulations [that is, the wavelength] constituting the extreme red light must be supposed to be, in air, about one 36 thousandth of an inch, and those of the extreme violet about one 60 thousandth; the mean of the whole spectrum, with respect to the intensity of light, being one 45 thousandth.

Young's value for the average effective wavelength present in sunlight (1/45,000 in.) can be written as 570 nm, which agrees rather well with the wavelength at which the eye sensitivity is a maximum, 555 nm (see Fig. 42-1). It must not be supposed that Young's work was received without criticism. One of his contemporaries, a firm believer in the corpuscular theory of light, wrote in part:

We wish to raise our feeble voice against innovations that can have no other effect than to check the progress of science, and renew all those wild phantoms of the imagination which Bacon and Newton put to flight from her temple. This paper contains nothing that deserves the name of either experiment or discovery.

Needless to say, posterity has decided in favor of Young.

45-2
COHERENCE

Analysis of the derivation of Eqs. 45-1 and 45-2 shows that a fundamental requirement for the existence of well-defined interference fringes on screen C in Figs. 45-5, 6 is that the light waves that travel from S_1 and S_2 to any point P on this screen must have a sharply defined phase difference ϕ that remains constant with time. If this condition is satisfied, a stable, well-defined fringe pattern will appear. At certain points P, ϕ

will be given, independent of time, by $n\pi$ where $n = 1, 3, 5, \ldots$ so that the resultant intensity will be strictly zero and will remain so throughout the time of observation. At other points ϕ will be given by $n\pi$ where $n = 0, 2, 4 \ldots$ and the resultant intensity will be a maximum. Under these conditions the two beams emerging from slits S_1 and S_2 are said to be completely *coherent*.

Let the source in Fig. 45-5 be removed and let slits S_1 and S_2 be replaced by two completely independent light sources, such as two fine incandescent wires placed side by side in a glass envelope. No interference fringes will appear on screen C but only a relatively uniform illumination. We can interpret this if we make the reasonable assumption that for completely independent light sources the phase difference between the two beams arriving at P will vary with time in a random way. At a certain instant conditions may be right for cancellation and a short time later (perhaps 10^{-8} s) they may be right for re-enforcement. This same random phase behavior holds for all points on screen C with the result that this screen is uniformly illuminated. The intensity at any point is equal to the sum of the intensities that each source S_1 and S_2 produces separately at that point. Under these conditions the two beams emerging from S_1 and S_2 are said to be completely *incoherent*.

Note that for completely *coherent* light beams one (1) combines the amplitudes vectorially, taking the (constant) phase difference properly into account, and then (2) squares this resultant amplitude to obtain a quantity proportional to the resultant intensity. For completely *incoherent* light beams, on the other hand, one (1) squares the individual amplitudes to obtain quantities proportional to the individual intensities, and then (2) adds the individual intensities to obtain the resultant intensity. This procedure is in agreement with the experimental fact that for completely independent light sources the resultant intensity at every point is always greater than the intensity produced at that point by either light source acting alone.

It remains to investigate under what experimental conditions coherent or incoherent beams may be produced and to give an explanation for coherence in terms of the mode of production of the radiation. Consider first a parallel beam of microwave radiation emerging from an antenna connected by a coaxial cable to an oscillator based on an electromagnetic resonant cavity. The cavity oscillations (see Section 38-5) are completely periodic with time and produce, at the antenna, a completely periodic variation of **E** and **B** with time. The radiated wave at large enough distances from the antenna is well represented by Fig. 41-13. Note that (1) the wave has essentially infinite extent in time, including both future times ($t > 0$, say) and past times ($t < 0$; see Fig. 45-7a. At any point, as the wave passes by, the wave disturbance (i.e., **E** or **B**) varies with time in a perfectly periodic way. (2) The wavefronts at points

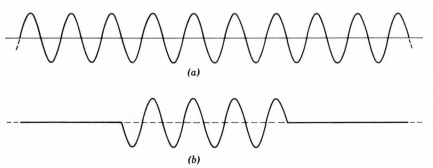

(a)

(b)

figure 45-7
(a) A section of an infinite wave and (b) a wavetrain.

far removed from the antenna are parallel planes of essentially infinite extent at right angles to the propagation direction. At any instant of time the wave disturbance varies with distance along the propagation direction in a perfectly periodic way.

Two beams generated from a single traveling wave like that of Fig. 41-13 will be completely coherent. One way to generate two such beams is to put an opaque screen containing two slits in the path of the beam. The waves emerging from the slits will always have a constant phase difference at any point in the region in which they overlap and interference fringes will be produced. Coherent radio beams can also be readily established, as can coherent elastic waves in solids, liquids and gases. The two prongs of the vibrating tapper in Fig. 45-4, for example, generate two coherent waves in the water of the ripple tank.

The technique of producing two beams from a single beam (and thus from a single source) tests specifically whether the wavefronts in the single parallel beam are truly planes, that is, whether all points in a plane at right angles to the direction of propagation have the same phase at any given instant. By dividing the beam in another way it is possible to test whether the beam is truly periodic over a large number of cycles of oscillation. This can be done, as we show in detail in Section 45-7, by inserting in the beam at 45° to it a thin sheet of material possessing the property that two beams are produced, one (which will be at right angles to the incident beam) by reflection and a second (which will be in the direction of the incident beam) by transmission. In the visible region such a sheet, called a *half-silvered mirror,* may be formed from a glass plate by depositing on it an appropriately thin film of silver. By appropriate use of mirrors (see Section 45-7) these two sub-beams can be recombined into a single beam traveling in a chosen direction. If the beams travel different distances before they are recombined, we are comparing in the combined beam a sample of the original beam with another sample an arbitrarily large number of cycles away. If the original beam is truly periodic in space and time, the two sub-beams will be completely coherent and interference fringes will be produced when they are recombined.

If we turn from microwave sources to common sources of visible light, such as incandescent wires or an electric discharge passing through a gas, we become aware of a fundamental difference. In both of these sources the fundamental light emission processes occur in individual atoms and these atoms do not act together in a cooperative (that is, *coherent*) way. The act of light emission by a single atom takes, in a typical case, about 10^{-8} s and the emitted light is properly described as a *wavetrain* (Fig. 45-7b) rather than as a wave (Fig. 45-7a). For emission times such as these the wavetrains are a few meters long.

Interference effects from ordinary light sources may be produced by putting a very narrow slit (S_0 in Fig. 45-1) directly in front of the source. This insures that the wavetrains that strike slits S_1 and S_2 in screen B in this figure originate from the same small region of the source. The diffracted beams emerging from S_1 and S_2 thus represent the same population of wavetrains and are coherent with respect to each other. If the phase of the light emitted from S_0 changes, this change is transmitted simultaneously to S_1 and S_2. Thus, at any point on screen C, a constant phase difference is maintained between the beams from these two slits and a stationary interference pattern occurs.

If the width of slit S_0 in Fig. 45-1 is gradually increased, it will be observed experimentally that the maxima of the interference fringes become reduced in intensity and that the intensity in the fringe minima is no longer strictly zero. In other words, the fringes become less distinct. If S_0 is opened extremely wide, the lowering of the maximum intensity and the raising of the minimum intensity will be such that the fringes disappear, leaving only a uniform illumination. Under these conditions we say that the beams from S_1 and S_2 pass continuously from a condition of complete coherence to one of complete incoherence. When not at either of these two limits, the beams are said to be partially coherent.

Partial coherence can also be demonstrated in two beams that are produced, as described on p. 1000, by inserting a "half-silvered mirror" in a beam at an angle of 45° to the direction of propagation. The two beams so produced, by reflection and transmission, traverse paths of different lengths before they are recombined. If the path difference is small compared with the average length of a wavetrain, the interference fringes will be sharply defined and will go essentially to zero at their minima. If the path difference is deliberately made longer, the fringes will become less distinct, and finally, when the path difference is larger than the average length of a wavetrain, the fringes will disappear altogether. Thus it is possible once again to progress smoothly in an experimental arrangement from complete coherence, through partial coherence, to complete incoherence.

The lack of coherence of the light from ordinary sources such as glowing wires is due to the fact that the emitting atoms do not act cooperatively (that is, *coherently*). Since 1960 it has proved possible to construct sources of visible light in which the atoms *do* act cooperatively and in which the emitted light is highly coherent. Such devices are called *lasers**; their light output is extremely monochromatic, intense, and highly collimated. The coherence of the emitted light can be demonstrated by placing a screen with two holes in it in the beam emerging from the laser. An interference pattern results, as Fig. 45-8 shows. These methods permit, for the first time, a degree of control of visible light approaching that possible for radio and for microwaves. The practical applications of lasers, which include the amplification of weak light signals, the use of light beams as highly efficient carriers of in-

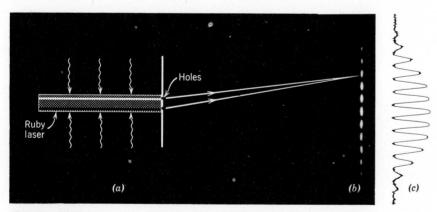

figure 45-8
A screen with two small holes is placed against the end of a laser. (a) The light passing through the holes forms an interference pattern on a strip of photographic film placed at (b). The fact that the pattern is formed shows that the light emitted from the laser is coherent across the beam cross section. At (b) is shown the image as formed on a photographic strip exposed in this manner. At (c) is an intensity plot of the film made by measuring the degree to which the film has been exposed. (Courtesy of D. F. Nelson and R. J. Collins, Bell Telephone Laboratories.)

* Laser, is a coined word, meaning "*l*ight *a*mplification through *s*timulated *e*mission of *r*adiation." For an early report, see "Optical Masers" by Arthur L. Schawlow, *Scientific American,* June 1961.

formation from point to point,* and the production of high temperatures by intense local heating, are being actively exploited.†

Let us assume that the electric field components of the two waves in Fig. 45-5 vary with time at point P as

$$E_1 = E_0 \sin \omega t \qquad (45\text{-}3)$$

and

$$E_2 = E_0 \sin (\omega t + \phi) \qquad (45\text{-}4)$$

where $\omega \, (=2\pi\nu)$ is the angular frequency of the waves and ϕ is the phase difference between them. Note that ϕ depends on the location of point P, which, in turn, for a fixed geometrical arrangement, is described by the angle θ in Figs. 45-5 and 45-6. We assume that the slits are so narrow that the diffracted light from each slit illuminates the central portion of the screen uniformly. This means that near the center of the screen E_0 is independent of the position of P, that is, of the value of θ.

The resultant wave disturbance‡ at P is found from

$$E = E_1 + E_2 \qquad (45\text{-}5)$$

and (see Eq. 19-16) proves to be

$$E = E_\theta \sin (\omega t + \beta), \qquad (45\text{-}6a)$$

where

$$\beta = \tfrac{1}{2}\phi \qquad (45\text{-}6b)$$

and

$$E_\theta = 2E_0 \cos \beta = E_m \cos \beta. \qquad (45\text{-}6c)$$

E_m, the maximum possible amplitude for E_θ, is equal to twice the amplitude of the combining waves $(=2E_0)$, corresponding to complete reinforcement. You should verify Eq. 45-6 carefully. The amplitude E_θ of the resultant wave disturbance, which determines the intensity of the interference fringes, will turn out to depend strongly on the value of θ, that is, on the location of point P in Figs. 45-5 and 45-6.

In Section 19-6 we showed that the intensity of a wave I, measured perhaps in watts/meter², is proportional to the square of its amplitude. For the resultant wave then, ignoring the proportionality constant,

$$I_\theta \propto E_\theta{}^2. \qquad (45\text{-}7)$$

This relationship seems reasonable if we recall (Eq. 30-9) that the energy density in an electric field is proportional to the *square* of the electric field strength. This is true for rapidly varying electric fields, such as those in a light wave, as well as for static fields.

The ratio of the intensities of two light waves is the ratio of the squares of the amplitudes of their electric fields. If I_θ is the intensity of the resultant wave at P, and I_0 is the intensity that a single wave acting alone would produce, then

$$\frac{I_\theta}{I_0} = \left(\frac{E_\theta}{E_0}\right)^2. \qquad (45\text{-}8)$$

---

* See "Light-Wave Communications" by W. S. Boyle, *Scientific American*, August 1977.
† For an example of one of the many applications of lasers, see "Processing Materials with Lasers" by Edward N Breinan, Bernard H. Kear, and Conrad M. Banas, *Physics Today*, November 1976.
‡ The electric field **E** in the light wave rather than the magnetic field **B** is normally identified with the "wave disturbance" because the effects of **B** on the human eye and on various light detectors are exceedingly small. Note too that, although Eq. 45-5 should be a vector equation, in most cases of interest the **E** vectors in the two interfering waves are closely parallel so that an algebraic equation suffices.

Combining with Eq. 45-6c leads to

$$I_\theta = 4I_0 \cos^2 \beta = I_m \cos^2 \beta. \qquad (45\text{-}9)$$

Note that the intensity of the resultant wave at any point P varies from zero [for a point at which $\phi \, (=2\beta) = \pi$, say] to I_m, which is four times the intensity I_0 of each individual wave [for a point at which $\phi \, (=2\beta) = 0$, say]. Let us compute I_θ as a function of the angle θ in Figs. 45-5 or 45-6.

The phase difference ϕ in Eq. 45-4 is associated with the path difference $S_1 b$ in Fig. 45-5 or 45-6. If $S_1 b$ is $\frac{1}{2}\lambda$, ϕ will be π; if $S_1 b$ is λ, ϕ will be 2π, etc. This suggests that

$$\frac{\text{phase difference}}{2\pi} = \frac{\text{path difference}}{\lambda},$$

$$\phi = \frac{2\pi}{\lambda} \, (d \, \sin \theta),$$

or, finally, from Eq. 45-6b,

$$\beta = \tfrac{1}{2}\phi = \frac{\pi d}{\lambda} \sin \theta. \qquad (45\text{-}10)$$

We can substitute this expression for β into Eq. 45-9 for I_θ, yielding the latter quantity as a function of θ. For convenience we collect here the expressions for the amplitude and the intensity in double-slit interference.

[Eq. 45-6c]	$E_\theta = E_m \cos \beta$	interference	(45-11a)
[Eq. 45-9]	$I_\theta = I_m \cos^2 \beta$	from narrow	(45-11b)
		slits (that is,	
[Eq. 45-10]	$\beta \, (=\tfrac{1}{2}\phi) = \dfrac{\pi d}{\lambda} \sin \theta$	$a \ll \lambda$)	(45-11c)

To find the positions of the intensity maxima, we put

$$\beta = m\pi \qquad m = 0, 1, 2, \ldots$$

in Eq. 45-11b. From Eq. 45-11c this reduces to

$$d \sin \theta = m\lambda \qquad m = 0, 1, 2, \ldots \text{ (maxima)},$$

which is the equation derived in Section 45-1 (Eq. 45-1). To find the intensity minima we write

$$\frac{\pi d \, \sin \theta}{\lambda} = (m + \tfrac{1}{2})\pi \qquad m = 0, 1, 2, \ldots \text{ (minima)},$$

which reduces to the previously derived Eq. 45-2.

Figure 45-9 shows the intensity pattern for double-slit interference.

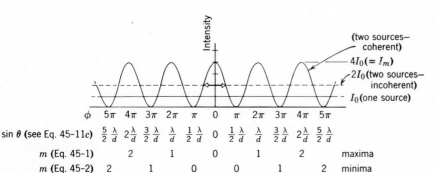

figure 45-9
The intensity pattern for double-slit interference. The heavy arrow in the central peak represents the half-width of the peak. This figure is constructed on the assumption that the two interfering waves each illuminate the central portion of the screen uniformly, that is, I_0 is independent of position as shown.

The horizontal solid line is I_0; this describes the (uniform) intensity pattern on the screen if one of the slits is covered up. If the two sources were incoherent, the intensity would be uniform over the screen and would be $2I_0$; see the horizontal dashed line in Fig. 45-9. For coherent sources we expect the energy to be merely redistributed over the screen, because energy is neither created nor destroyed by the interference process. Thus the *average* intensity in the interference pattern should be $2I_0$, as for incoherent sources. This follows at once if, in Eq. 45-11b, we substitute one-half for the cosine-squared term and if we recall that $I_m = 4I_0$. We have seen several times that the average value of the square of a sine or a cosine term over one or more half-cycles is one-half.

In Section 45-3 we combined two time-varying wave disturbances, namely,

$$E_1 = E_0 \sin \omega t \qquad (45\text{-}3)$$

and

$$E_2 = E_0 \sin (\omega t + \phi), \qquad (45\text{-}4)$$

which have the same angular frequency ω and amplitude E_0 but which have a phase difference ϕ between them. In this case the result (Eqs. 45-11a and 45-11c) is easily obtained algebraically.

In later chapters we will want to add larger numbers of wave disturbances, often an infinite number, with infinitesimal individual amplitudes. Since analytic methods become more difficult in such cases we describe a graphical method, illustrating it by rederiving Eq. 45-11a.

A sinusoidal wave disturbance such as that represented by Eq. 45-3 can be represented graphically, using a rotating vector. In Fig. 45-10a a vector of magnitude E_0 is allowed to rotate about the origin in a counterclockwise direction with an angular frequency ω. Following electrical engineering practice we call such a rotating vector a *phasor*; see Section 39-2. The alternating wave disturbance E_1 (Eq. 45-3) is represented by the projection of this phasor on the vertical axis.

A second wave disturbance E_2, which has the same amplitude E_0 but a phase difference ϕ with respect to E_1,

$$E_2 = E_0 \sin (\omega t + \phi), \qquad (45\text{-}4)$$

can be represented graphically (Fig. 45-10b) as the projection on the vertical axis of a second phasor of magnitude E_0 which makes an angle ϕ with the first phasor. As this figure shows, the sum E of E_1 and E_2 is the sum of the projections of the two phasors on the vertical axis. This is revealed more clearly if we redraw the phasors, as in Fig. 45-10c, placing the foot of one arrow at the head of the other, maintaining the proper phase difference, and letting the whole assembly rotate counterclockwise about the origin.

In Fig. 45-10c E can also be regarded as the projection on the vertical axis of a phasor of length E_θ, which is the vector sum of the two phasors of magnitude E_0. Note that the (algebraic) sum of the projections of the two phasors is equal to the projection of the (vector) sum of the two phasors.

In most problems in optics we are concerned only with the *amplitude* E_θ of the resultant wave disturbance and not with its time variation. This is because the eye and other common measuring instruments respond to the resultant intensity of the light (that is, to the square of the amplitude) and cannot detect the rapid time variations that characterize visible light. For sodium light ($\lambda = 589$ nm $= 5890$ Å), for example, the

45-4
ADDING WAVE DISTURBANCES

(a)

(b)

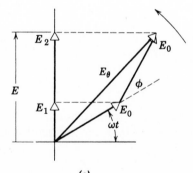

(c)

figure 45-10

(a) A time-varying wave disturbance E_1 is represented by a rotating vector or *phasor.* (b) Two wave disturbances E_1 and E_2, with a phase difference ϕ between them, are so represented. These two phasors can represent the two wave disturbances in the double-slit problem; see Eqs. 45-3 and 45-4. (c) Another way of drawing (b).

$\nu(=\omega/2\pi)$ is 5.1×10^{14} Hz. Often, then, we need not consider the rotation of the phasors but can confine our attention to finding the magnitude of the resultant phasor.

Figure 45-11a shows the phasors for double-slit interference at time $t = 0$; compare Fig. 45-10c. We see that

$$E_\theta = 2E_0 \cos \beta = E_m \cos \beta,$$

in which, from the theorem that the exterior angle of a triangle (ϕ) is equal to the sum of its two opposite interior angles $(\beta + \beta)$,

$$\beta = \tfrac{1}{2}\phi.$$

This is exactly the result arrived at earlier algebraically; compare Eqs. 45-11a and 45-11c.

In a more general case we might want to find the resultant of a number (>2) of sinusoidally varying wave disturbances. The general procedure is the following:

1. Construct a series of phasors representing the functions to be added. Draw them end to end, maintaining the proper phase relationships between adjacent phasors.

2. Construct the vector sum of this array. Its length gives the amplitude of the resultant. The angle between it and the first phasor is the phase of the resultant with respect to this first phasor. The projection of this phasor on the vertical axis gives the time variation of the resultant wave disturbance.

Find graphically the resultant $E(t)$ of the following wave disturbances:

$$E_1 = 10 \sin \omega t$$

$$E_2 = 10 \sin (\omega t + 15°)$$

$$E_3 = 10 \sin (\omega t + 30°)$$

$$E_4 = 10 \sin (\omega t + 45°).$$

Figure 45-12 in which E_0 equals 10, shows the assembly of four phasors that

(a)

(b)

figure 45-11
(a) A construction to find the amplitude E_θ of two wave disturbances of amplitude E_0 and phase difference ϕ. (b) The maximum possible amplitude for these wave disturbances occurs for $\phi = 0$ and has the value $E_m = 2E_0$.

EXAMPLE 3

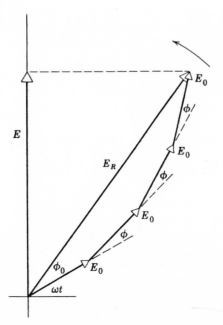

figure 45-12
Example 3. Four wave disturbances are added graphically, using the method of phasors.

represents these functions. Their vector sum, by graphical measurement, has an amplitude E_R of 38 and a phase ϕ_0 with respect to E_1 of 23°. In other words,

$$E(t) = E_1 + E_2 + E_3 + E_4 = 38 \sin(\omega t + 23°).$$

Check this result by trigonometric calculation.

The colors of soap bubbles, oil slicks, and other thin films are the result of interference. Figure 45-13 shows interference effects in a thin vertical film of soapy water illuminated by monochromatic light.

Figure 45-14 shows a film of uniform thickness d and index of refraction n, the eye being focused on spot a. The film is illuminated by a broad source of monochromatic light S. There exists on this source a point P such that two rays, identified by the single and double arrows, respectively, can leave P and enter the eye as shown, after passing through point a. These two rays follow different paths in going from P to the eye, one being reflected from the upper surface of the film, the other from the lower surface. Whether point a appears bright or dark depends on the nature of the interference between the two waves that diverge from a. These waves are coherent because they both originate from the same point P on the light source.

If the eye looks at another part of the film, say a', the light that enters the eye must originate from a different point P' of the extended source, as suggested by the dashed lines in Fig. 45-14.

For near-normal incidence ($\theta \cong 0$ in Fig. 45-14) the geometrical path difference for the two rays from P will be close to $2d$. We might expect the resultant wave reflected from the film near a to be an interference maximum if the distance $2d$ is an integral number of wavelengths. This statement must be modified for two reasons.

First, the wavelength must refer to the wavelength of the light in the film λ_n and not to its wavelength in air λ; that is, we are concerned with optical path lengths rather than geometrical path lengths. The wavelengths λ and λ_n (see Eq. 43-12b) are related by

$$\lambda_n = \lambda/n. \qquad (45\text{-}12)$$

To bring out the second point, let us assume that the film is so thin that $2d$ is very much less than a wavelength. The phase difference between the two waves would be close to zero on our assumption, and we would expect such a film to appear bright on reflection. However, it appears dark. This is clear from Fig. 45-13, in which the action of gravity produces a wedge-shaped film, extremely thin at its top edge. As drainage continues, the dark area increases in size. To explain this and many similar phenomena, we assume that one or the other of the two rays of Fig. 45-14 suffers an abrupt phase change of π ($=180°$) associated either with reflection at the air-film interface or transmission through it. As it turns out, the ray reflected from the upper surface suffers this phase change. The other ray is not changed abruptly in phase, either on transmission through the upper surface or on reflection at the lower surface.

In Section 19-9 we discussed phase changes on reflection for transverse waves in strings. To extend these ideas, consider the composite string of Fig. 45-15, which consists of two parts with different masses per unit length, stretched to a given tension. If a pulse moves to the right in Fig. 45-15a, approaching the junction, there will be a reflected and a transmitted pulse, the reflected pulse being *in phase* with the incident pulse. In Fig. 45-15b the situation is reversed, the incident pulse now

45-5
INTERFERENCE FROM THIN FILMS

figure 45-13
A soapy water film on a wire loop, viewed by reflected light. The black segment at the top is not a tear. It arises because the film, by drainage, is so thin here that there is destructive interference between the light reflected from its front surface and that reflected from its back surface. We shall see that these two waves differ in phase by 180°.

figure 45-14
Interference by reflection from a thin film, assuming an extended source S.

1007   INTERFERENCE FROM THIN FILMS   SEC. 45-5

being in the lighter string. In this case the reflected pulse will differ in phase from the incident pulse by π (=180°). In each case the transmitted pulse will be in phase with the incident pulse.

Figure 45-15*a* suggests a light wave in glass, say, approaching a surface beyond which there is a less optically dense medium (one of lower index of refraction) such as air. Figure 45-15*b* suggests a light wave in air approaching glass. To sum up the optical situation, when reflection occurs from an interface beyond which the medium has a *lower* index of refraction, the reflected wave undergoes *no phase change;* when the medium beyond the interface has a *higher* index, there is a phase change of π.* The transmitted wave does not experience a change of phase in either case.

We are now able to take into account both factors that determine the nature of the interference, namely, differences in optical path length and phase changes on reflection. For the two rays of Fig. 45-14 to combine to give a *maximum* intensity, assuming normal incidence, we must have

$$2d = (m + \tfrac{1}{2})\lambda_n \qquad m = 0, 1, 2, \ldots.$$

The term $\tfrac{1}{2}\lambda_n$ is introduced because of the phase change on reflection, a phase change of 180° being equivalent to half a wavelength. Substituting λ/n for λ_n yields finally

$$2dn = (m + \tfrac{1}{2})\lambda \qquad m = 0, 1, 2, \ldots \text{ (maxima).} \qquad (45\text{-}13)$$

The condition for a *minimum* intensity is

$$2dn = m\lambda \qquad m = 0, 1, 2, \ldots \text{ (minima).} \qquad (45\text{-}14)$$

* These statements, which can be proved rigorously from Maxwell's equations (see also Section 45-6), must be modified for light falling on a less dense medium at an angle such that total internal reflection occurs. They must also be modified for reflection from metallic surfaces.

figure 45-15
Phase changes on reflection at a junction between stretched composite string. (*a*) Incident pulse in heavy string and (*b*) incident pulse in light string.

These equations hold when the index of refraction of the film is either greater or less than the indices of the media on *each* side of the film. Only in these cases will there be a relative phase change of 180° for reflections at the two surfaces. A water film in air and an air film in the space between two glass plates provide examples of cases to which Eqs. 45-13 and 45-14 apply. Example 5 provides a case in which they do not apply.

If the film thickness is not uniform, as in Fig. 45-13, where the film is wedge-shaped, constructive interference will occur in certain parts of the film and destructive interference will occur in others. Lines of maximum and of minimum intensity will appear—these are the interference fringes. They are called *fringes of constant thickness*, each fringe being the locus of points for which the film thickness d is a constant. If the film is illuminated with white light rather than monochromatic light, the light reflected from various parts of the film will be modified by the various constructive or destructive interferences that occur. This accounts for the brilliant colors of soap bubbles and oil slicks.

Only if the film is "thin," which implies that d is no more than a few wavelengths of light, will fringes of the type described, that is, fringes that appear localized on the film and associated with a variable film thickness, be possible. For very thick films (say $d \cong 1$ cm), the path difference between the two rays of Fig. 45-14 will be many wavelengths and the phase difference at a given point on the film will change rapidly as we move even a small distance away from a. For "thin" films, however, the phase difference at a also holds for reasonably nearby points; there is a characteristic "patch brightness" for any point on the film, as Fig. 45-13 shows. Interference fringes can be produced for thick films; they are not localized on the film but are at infinity. See Section 45-7.

EXAMPLE 4

A water film ($n = 1.33$) in air is 320 nm thick. If it is illuminated with white light at normal incidence, what color will it appear to be in reflected light?

By solving Eq. 45-13 for λ,

$$\lambda = \frac{2dn}{m + \frac{1}{2}} = \frac{(2)(320 \text{ nm})(1.33)}{m + \frac{1}{2}} = \frac{850 \text{ nm}}{m + \frac{1}{2}} \quad \text{(maxima)}.$$

From Eq. 45-14 the minima are given by

$$\lambda = \frac{850 \text{ nm}}{m} \quad \text{(minima)}.$$

Maxima and minima occur for the following wavelengths:

m	0 (max)	1 (min)	1 (max)	2 (min)	2 (max)
λ, nm	1700	850	570	425	340

Only the maximum corresponding to $m = 1$ lies in the visible region (see Fig. 42-1); light of this wavelength appears yellow-green. If white light is used to illuminate the film, the yellow-green component will be enhanced when viewed by reflection.

EXAMPLE 5

*Nonreflecting glass.** Lenses are often coated with thin films of transparent substances such as MgF_2 ($n = 1.38$) in order to reduce the reflection from the glass surface, using interference. How thick a coating is needed to produce a minimum reflection at the center of the visible spectrum (550 nm)?

---

* See "Optical Interference Coatings" by Philip Baumeister and Gerald Pineus, *Scientific American*, December 1970.

We assume that the light strikes the lens at near-normal incidence (θ is exaggerated for clarity in Fig. 45-16), and we seek destructive interference between rays r and r_1. Equation 45-14 does not apply because in this case a phase change of 180° is associated with *each* ray, for at *both* the upper and lower surfaces of the MgF$_2$ film the reflection is from a medium of greater index of refraction.

There is no net change in phase produced by the two reflections, which means that the optical path difference for destructive interference is $(m + \frac{1}{2})\lambda$ (compare Eq. 45-13), leading to

$$2dn = (m + \tfrac{1}{2})\lambda \qquad m = 0, 1, 2, \ldots \text{ (minima)}.$$

Solving for d and putting $m = 0$ yields

$$d = \frac{(m + \frac{1}{2})\lambda}{2n} = \frac{\lambda}{4n} = \frac{550 \text{ nm}}{(4)(1.38)} = 100 \text{ nm}.$$

figure 45-16
Example 5. Unwanted reflections from glass can be reduced by coating the glass with a thin transparent film.

figure 45-17
Example 6. Apparatus for observing Newton's rings.

EXAMPLE 6

Newton's rings. Figure 45-17 shows a lens of radius of curvature R resting on an accurately plane glass plate and illuminated from above by light of wavelength λ. Figure 45-18 shows that circular interference fringes (Newton's rings) appear, associated with the variable thickness air film between the lens and the plate. Find the radii of the circular interference maxima.

Here it is the ray from the *bottom* of the (air) film rather than from the top that undergoes a phase change of 180°, for it is the one reflected from a medium of higher refractive index. The condition for a maximum remains unchanged (Eq. 45-13), however, and is

$$2d = (m + \tfrac{1}{2})\lambda \qquad m = 0, 1, 2, \ldots, \qquad (45\text{-}15)$$

the index of refraction of the air film being assumed to be unity. From Fig. 45-17

figure 45-18
Example 6. Newton's rings. (Courtesy of Bausch and Lomb Optical Co.)

we can write

$$d = R - \sqrt{R^2 - r^2} = R - R\left[1 - \left(\frac{r}{R}\right)^2\right]^{1/2}$$

If $r/R \ll 1$, we can expand the square bracket by the binomial theorem, keeping only two terms, or

$$d = R - R\left[1 - \frac{1}{2}\left(\frac{r}{R}\right)^2 + \cdots\right] \cong \frac{r^2}{2R}.$$

Combining with Eq. 45-15 yields

$$r = \sqrt{(m + \tfrac{1}{2})\lambda R} \qquad m = 0, 1, 2, \ldots,$$

which gives the radii of the bright rings. If white light is used, each spectrum component will produce its own set of circular fringes, the sets all overlapping.

45-6
OPTICAL REVERSIBILITY AND PHASE CHANGES ON REFLECTION

G. G. Stokes (1819–1903) used the principle of optical reversibility to investigate the reflection of light at an interface between two media. The principle states that if there is no absorption of light, a light ray that is reflected or refracted will retrace its original path if its direction is reversed. This reminds us that any mechanical system can run backward as well as forward, provided there is no absorption of energy because of friction, etc.

Figure 45-19a shows a wave of amplitude E reflected and refracted at a surface separating media 1 and 2, where $n_2 > n_1$. The amplitude of the reflected wave is $r_{12}E$, in which r_{12} is an *amplitude reflection coefficient*. The amplitude of the refracted wave is $t_{12}E$, where t_{12} is an *amplitude transmission coefficient*.

We consider only the possibility of phase changes of 0 or 180°. If $r_{12} = +0.5$, for example, we have a reduction in amplitude on reflection by one-half and no change in phase. For $r_{12} = -0.5$ we have a phase change of 180° because

$$E \sin (\omega t + 180°) = -E \sin \omega t.$$

Figure 45-19b suggests that if we reverse these two rays, they should combine to produce the original ray reversed in direction. Ray $r_{12}E$, identified by the single arrows in the figure, is reflected and refracted, producing the rays of amplitudes $r_{12}^2 E$ and $r_{12}t_{12}E$. Ray $t_{12}E$, identified by the triple arrows, is also reflected and refracted, producing the rays of amplitudes $t_{12}t_{21}E$ and $t_{12}r_{21}E$ as shown. Note that r_{12} describes a ray in medium 1 reflected from medium 2, and r_{21} describes a ray in medium 2 reflected from medium 1. Similarly, t_{12} describes a ray that passes from medium 1 to medium 2; t_{21} describes a ray that passes from medium 2 to medium 1.

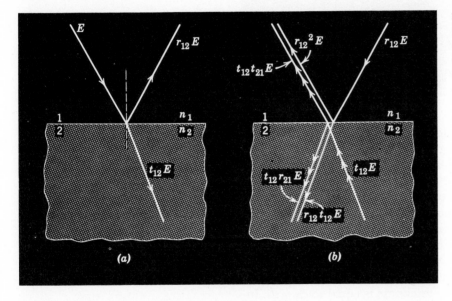

(a) (b)

figure 45-19
(a) A ray is reflected and refracted at an air-glass interface. (b) The optically reversed situation; the two rays in the lower left must cancel.

The two rays in the upper left of Fig. 45-19b must be equivalent to the incident ray of Fig. 45-19a, reversed; the two rays in the lower left of Fig. 45-19b must cancel. This second requirement leads to

$$r_{12}t_{12}E + t_{12}r_{21}E = 0,$$

or

$$r_{12} = -r_{21}.$$

This result tells us that if we compare a wave reflected from medium 1 with one reflected from medium 2, they behave differently in that one or the other undergoes a phase change of 180°. We must rely on experiment or analysis by Maxwell's equations to show that, as we pointed out earlier, the ray reflected from the more optically dense medium is the one that experiences the phase change of 180°.

45-7
MICHELSON'S INTERFEROMETER*

An *interferometer* is a device that can be used to measure lengths or changes in length with great accuracy by means of interference fringes. We describe the form originally built by A. A. Michelson (1852–1931) in 1881.

Consider light that leaves point P on extended source S (Fig. 45-20) and falls on half-silvered mirror M. This mirror has a silver coating just

* See "Michelson: America's first Nobel Prize Winner in Science" by R. S. Shankland, *The Physics Teacher*, January 1977. See also "Michelson and his Interferometer" by R. S. Shankland, *Physics Today*, April 1974.

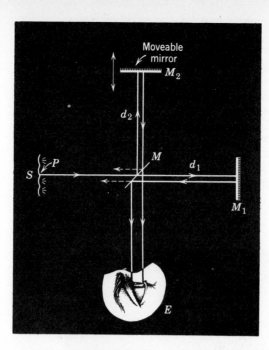

figure 45-20
Michelson's interferometer, showing
the path of a particular ray originating
at point P of an extended source S.

thick enough to transmit half the incident light and to reflect half; in
the figure we have assumed that this mirror, for convenience, possesses
negligible thickness. At M the light divides into two waves. One proceeds by transmission toward mirror M_1; the other proceeds by reflection toward M_2. The waves are reflected at each of these mirrors and are
sent back along their directions of incidence, each wave eventually entering the eye E. Because the waves are coherent, being derived from the
same point on the source, they will interfere.

If the mirrors M_1 and M_2 are exactly perpendicular to each other, the
effect is that of light from an extended source S falling on a uniformly
thick slab of air, between glass, whose thickness is equal to $d_2 - d_1$. Interference fringes appear, caused by small changes in the angle of incidence of the light from different points on the extended source as it
strikes the equivalent air film. For *thick* films a path difference of one
wavelength can be brought about by a very small change in the angle
of incidence.

If M_2 is moved backward or forward, the effect is to change the thickness of the equivalent air film. Suppose that the center of the (circular)
fringe pattern appears bright and that M_2 is moved just enough to cause
the first bright circular fringe to move to the center of the pattern. The
path of the light beam striking M_2 has been changed by one wavelength.
This means (because the light passes twice through the equivalent air
film) that the mirror must have moved one-half a wavelength.

The interferometer is used to measure changes in length by counting
the number of interference fringes that pass the field of view as mirror
M_2 is moved. Length measurements made in this way can be accurate
if large numbers of fringes are counted.

Michelson measured the length of the standard meter, kept in Paris,
in terms of the wavelength of certain monochromatic red light emitted
from a light source containing cadmium. He showed that the standard
meter was equivalent to 1,553,163.5 wavelengths of the red cadmium
light. For this he received the Nobel prize in 1907.

Physicists have long speculated on the advantages of discarding the
standard meter bar as the basic standard of length and of *defining* the

meter in terms of the wavelength of some carefully chosen monochromatic radiation. This would make the primary length standard readily available in laboratories all over the world. It would improve the accuracy of length measurements, since one would no longer need to compare an unknown object with a standard object (the meter bar), using interferometer techniques, but could measure the unknown object *directly* and in an absolute sense, using these techniques. There is the additional advantage that if the standard meter bar were destroyed, it could never be replaced, whereas light sources and interferometers will (presumably) always be available.

In 1961 such an atomic standard of length was adopted by international agreement. Quoting from an article* describing the event:

The wavelength of the orange-red light of krypton-86 has replaced the platinum iridium bar as the world standard of length. Formerly the wavelength of this light was defined as a function of the length of the meter bar. Now the meter is defined as a multiple (1,650,763.73) of the wavelength of the light.

The light from krypton-86 was used in preference to that from other atomic sources because it produces sharper interference fringes in the interferometer over the long optical paths sometimes used in length measurement.

45-8
MICHELSON'S INTERFEROMETER AND LIGHT PROPAGATION†

In Section 42-4 we presented Einstein's hypothesis, now well verified, that in free space light is propagated with the same speed c no matter what the relative velocity of the source and the observer may be. We pointed out that this hypothesis contradicted the views of nineteenth-century physicists regarding wave propagation. It was difficult for these physicists, trained as they were in the classical physics of the time, to believe that a wave could be propagated without a medium. If such a medium could be established, the speed c of light would naturally be construed as the speed *with respect to that medium,* just as the speed of sound always refers to a medium such as air.

Although no medium for light propagation was obvious, the physicists postulated one, called the *ether*‡ and hypothesized that its properties were such that it was undetectable by ordinary means such as weighing.

In 1881 (24 years before Einstein's hypothesis) A. A. Michelson set himself the task of forcing the ether, assuming that it existed, to submit to direct physical verification. In particular, Michelson, later joined by E. W. Morley, tried to measure the speed u with which the earth moves through the ether. Michelson's interferometer was their instrument of choice for this now-famous Michelson-Morley experiment.

The earth together with the interferometer moving with velocity **u** through the ether is equivalent to the interferometer at rest with the ether streaming through it with velocity −**u,** as shown in Fig. 45-21. Consider a wave moving along the path MM_1M and one moving along MM_2M. The first corresponds classically to a man rowing a boat a distance d downstream and the same distance upstream; the second corresponds to rowing a boat a distance d across a stream and back.

* *Scientific American,* p. 75, December 1960.
† See Supplementary Topic V.
‡ More fully, the *luminiferous* (or light-carrying) *ether.*

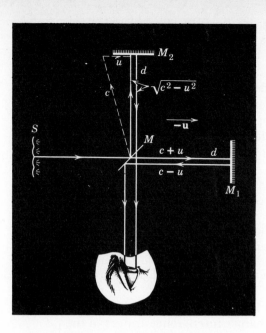

figure 45-21
The "ether" is streaming with velocity −**u** through Michelson's interferometer. The wave speeds shown are on the basis of the (incorrect) ether hypothesis.

On the ether hypothesis the speed of light on the path MM_1 is $c + u$; on the return path M_1M it is $c - u$. The time required for the complete trip is

$$t_1 = \frac{d}{c+u} + \frac{d}{c-u} = d\,\frac{2c}{c^2-u^2} = \frac{2d}{c}\,\frac{1}{1-(u/c)^2}.$$

The speed of light, *on the ether hypothesis*, for path MM_2 is $\sqrt{c^2 - u^2}$, as Fig. 45-21 suggests. This same speed holds for the return path M_2M, so that the time required for this complete path is

$$t_2 = \frac{2d}{\sqrt{c^2-u^2}} = \frac{2d}{c}\,\frac{1}{\sqrt{1-(u/c)^2}}.$$

The difference of time for the two paths is

$$\Delta t = t_1 - t_2$$
$$= \frac{2d}{c}\left\{\left[1 - \left(\frac{u}{c}\right)^2\right]^{-1} - \left[1 - \left(\frac{u}{c}\right)^2\right]^{-1/2}\right\}$$

Assuming $u/c \ll 1$, we can expand the quantities in the square brackets by using the binomial theorem, retaining only the first two terms. This leads to

$$\Delta t = \frac{2d}{c}\left\{\left[1 + \left(\frac{u}{c}\right)^2 + \cdots\right] - \left[1 + \frac{1}{2}\left(\frac{u}{c}\right)^2 + \cdots\right]\right\}$$

$$= \frac{2d}{c}\left\{\frac{1}{2}\left(\frac{u}{c}\right)^2\right\} = \frac{du^2}{c^3}. \qquad (45\text{-}16)$$

Now let the entire interferometer be rotated through 90°. This will interchange the roles of the two light paths, MM_1M now being the "cross-stream" path and MM_2M the "down- and up-stream" path. The time difference between the two waves entering the eye is also reversed; this changes the phase difference between the combining waves and alters the positions of the interference maxima. *The experiment consists of looking for a shift of the interference fringes as the apparatus is rotated.*

The *change* in time difference is $2\Delta t$, which corresponds to a fringe shift of $2\Delta t/T$ where $T(=\lambda/c)$ is the period of vibration of the light. The expected maximum shift in the number of fringes on a 90° rotation (see Eq. 45-16) is

$$\Delta N = \frac{2\Delta t}{T} = \frac{2\Delta t c}{\lambda} = \frac{2d}{\lambda}\left(\frac{u}{c}\right)^2 \qquad (45\text{-}17)$$

In the Michelson-Morley interferometer let $d = 11$ m (obtained by multiple reflection in the interferometer) and $\lambda = 5.9 \times 10^{-7}$ m. If u is assumed to be roughly the orbital speed of the earth, then $u/c \cong 10^{-4}$. The expected maximum fringe shift when the interferometer is rotated through 90° is then

$$\Delta N = \frac{2d}{\lambda}\left(\frac{u}{c}\right)^2 = \frac{(2)(11\text{ m})}{5.9 \times 10^{-7}\text{ m}}(10^{-4})^2 = 0.4.$$

Even though a shift of only about 0.4 of a fringe was expected, Michelson and Morley were confident that they could observe a shift of 0.01 fringe. *They found from their experiment, however, that there was no observable fringe shift!*

The analogy between a light wave in the supposed ether and a boat moving in water, which seemed so evident in 1881, is simply incorrect. The derivation based on this analogy is incorrect for light waves. When the analysis is carried through on Einstein's hypothesis, the observed negative result is clearly predicted, the speed of light being c for all paths. The motion of the earth around the sun and the rotation of the interferometer have, in Einstein's view, no effect whatever on the speed of the light waves in the interferometer.

It should be made clear that although Einstein's hypothesis is completely consistent with the negative result of the Michelson-Morley experiment, this experiment standing alone cannot serve as a proof for Einstein's hypothesis. Einstein said that no number of experiments, however large, could prove him right but that a single experiment could prove him wrong. Our present-day belief in Einstein's hypothesis rests on consistent agreement in a large number of experiments designed to test it. The "single experiment" that might prove Einstein wrong has never been found.

questions

1. Is Young's experiment an interference experiment or a diffraction experiment, or both?

2. Do interference effects occur for sound waves? Recall that sound is a longitudinal wave and that light is a transverse wave.

3. In Young's double-slit interference experiment, using a monochromatic laboratory light source, why is screen A in Fig. 45-1 necessary? What would happen if one gradually enlarged the hole in this screen?

4. What changes occur in the pattern of interference fringes if the apparatus of Fig. 45-5 is placed under water?

5. If interference between light waves of different frequencies is possible, one should observe light beats, just as one obtains sound beats from two sources of sound with slightly different frequencies. Discuss how one might experimentally look for this possibility.

6. Why are parallel slits preferable to the pinholes that Young used in demonstrating interference?

7. Is coherence important in reflection and refraction?

8. If your source of light is a laser beam, you do not need (see Fig. 45-8) the equivalent of screen A in Fig. 45-1. Why?

9. Defend this statement: Fig. 45-7a is a sine (or cosine) wave but Fig. 45-7b is not. Indeed you cannot assign a unique frequency to the curve of Fig. 45-7b. Why not? (Hint: think of Fourier analysis.)

10. Most of us are familiar with rotating or oscillating radar antennas which produce rotating or oscillating beams of microwave radiation. It is also possible to produce an oscillating beam of microwave radiation *without* any mechanical motion of the transmitting antenna. This is done by periodically changing the phase of the radiation as it emerges from various sections of the (long) transmitting antenna. Convince yourself that, by constructive interference from various parts of the fixed antenna, an oscillating microwave beam could indeed be so produced. A very large radar installation of this type has been installed by the United States in the Aleutian Islands to monitor USSR missile experiments.

11. Describe the pattern of light intensity on screen C in Fig. 45-5 if one slit is covered with a red filter and the other with a blue filter, the incident light being white.

12. Define carefully, and distinguish between, the angles θ and ϕ that appear in Eq. 45-10.

13. If one slit in Fig. 45-5 is covered, what change occurs in the intensity of light in the center of the screen?

14. In Young's double-slit experiment suppose that screen A in Fig. 45-1 contained *two* very narrow parallel slits instead of one. (a) Show that if the spacing between these slits is properly chosen, the interference fringes can be made to disappear. (b) Under these conditions, would you call the beams emerging from slits S_1 and S_2 in screen B coherent? They do not produce interference fringes. (c) Discuss what would happen to the interference fringes in the case of a single slit in screen A if the slit width were gradually increased.

15. What are the requirements for a maximum intensity when viewing a thin film by *transmitted* light?

16. In a Newton's rings experiment, is the central spot, as seen by reflection, dark or light? Explain.

17. Why must the film of Fig. 45-14 be "thin" for us to see an interference pattern of the type described?

18. Why do coated lenses (see Example 5) look purple by reflected light?

19. A person wets his eyeglasses to clean them. As the water evaporates he notices that for a short time the glasses become markedly more nonreflecting. Explain.

20. A lens is coated to reduce reflection, as in Example 5. What happens to the energy that had previously been reflected? Is it absorbed by the coating?

21. Very small changes in the angle of incidence do not change the interference conditions much for "thin" films but they do change them for "thick" films. Why?

22. In connection with the phase change on reflection at an interface between two transparent media, do you think that phase shifts other than 0 or π are possible? Do you think that phase shifts can be calculated rigorously from Maxwell's equations?

23. Consider the following objects that produce colors when exposed to sunlight: (1) soap bubbles, (2) rose petals, (3) the inner surface of an oyster shell (irridescence), (4) thin oil slicks, (5) non-reflecting coatings on camera lenses, and (6) peacock tail feathers. All but one of these are purely interference phenomena, no pigments being involved. Which one is it? Discuss.

24. Why does a film (soap bubble, oil slick, etc.) have to be "thin" to display interference effects? Or does it? How thin is "thin"? Discuss.

25. An *optical flat* is a slab of glass that has been ground flat to within a small fraction of a wavelength. How may it be used to test the flatness of a second slab of glass?

26. The directional characteristics of a certain radar antenna as a receiver of radiation are known. What can be said about its directional characteristics as a transmitter?

27. A person in a dark room, looking through a small window, can see a second person standing outside in bright sunlight. The second person cannot see the first person. Is this a failure of the principle of optical reversibility? Assume no absorption of light.

28. Why is it necessary to rotate the interferometer in the Michelson-Morley experiment?

29. How is the negative result of the Michelson-Morley experiment interpreted according to Einstein's theory of relativity?

30. An automobile directs its headlights onto the side of a barn. Why are interference fringes not produced?

31. In principle, could you construct an acoustical version of Michelson's interferometer and use it to measure wind velocities? Discuss some of its design difficulties. Are there simpler ways to measure wind velocities?

problems

SECTION 45-1

1. Design a double-slit arrangement that will produce interference fringes 1° apart on a distant screen. Assume sodium light ($\lambda = 589$ nm $= 5890$ Å). *Answer:* Slit separation must be 0.034 mm.

2. In a double-slit arrangement the distance between slits is 5.0 mm and the slits are 1.0 m from the screen. Two interference patterns can be seen on the screen, one due to light of 480 nm and the other 600 nm. What is the separation on the screen between the third-order interference fringes of the two different patterns?

3. Sodium light ($\lambda = 589$ nm) falls on a double slit of separation $d = 2.0$ mm. D in Fig. 45-5 is 4 cm. What percent error is made in locating the tenth bright fringe if it is *not* assumed that $D \gg d$? *Answer:* 0.03%

4. In a double-slit arrangement the slits are separated by a distance equal to 100 times the wavelength of the light passing through the slits. (a) What is the angular separation between the first and second maxima? (b) What is the linear distance between the first and second maxima if the screen is at a distance of 50 cm from the slits?

5. A double-slit arrangement produces interference fringes for sodium light ($\lambda = 589$ nm) that are 0.20° apart. For what wavelength would the angular separation be 10 percent greater? *Answer:* 650 nm

6. Sodium light ($\lambda = 589$ nm) falls on a double slit of separation $d = 0.20$ mm. A thin lens ($f = +1.0$ m) is placed near the slit as in Fig. 45-6. What is the linear fringe separation on a screen placed in the focal plane of the lens?

7. In the front of a lecture hall, a coherent beam of monochromatic light from a helium-neon laser ($\lambda = 640$ nm $= 6400$ Å) illuminates a double slit. From there it travels a distance d (20 m) to a mirror at the back of the hall, and returns the same distance to a screen. (a) In order that the distance between interference maxima be s (10 cm) what should be the distance between the two slits? (b) State briefly what you will see if the lecturer slips a thin sheet of cellophane over one of the slits. The optical path length through the cellophane is 2.5 wavelengths longer than the equivalent air path length. *Answer:* (a) 0.26 mm. (b) In place of the central maximum you get a minimum.

8. In Young's interference experiment in a large ripple tank (see Fig. 45-4) the coherent vibrating sources are placed 12.0 cm apart. The distance between

maxima 2.0 m away is 18.0 cm. If the speed of ripples is 25.0 cm/s, find the frequency of the vibrators.

9. A thin flake of mica ($n = 1.58$) is used to cover one slit of a double-slit arrangement. The central point on the screen is occupied by what used to be the seventh bright fringe. If $\lambda = 550$ nm, what is the thickness of the mica?

Answer: 6.6×10^{-3} mm

10. One slit of a double-slit arrangement is covered by a thin glass plate of refractive index 1.4, and the other by a thin glass plate of refractive index 1.7. The point on the screen where the central maximum fell before the glass plates were inserted is now occupied by what had been the fifth bright fringe before. Assume $\lambda = 480$ nm and that the plates have the same thickness t and find the value of t.

11. A double-slit arrangement produces interference fringes for sodium light ($\lambda = 589$ nm) that are 0.20° apart. What is the angular fringe separation if the entire arrangement is immersed in water?

Answer: 0.15°.

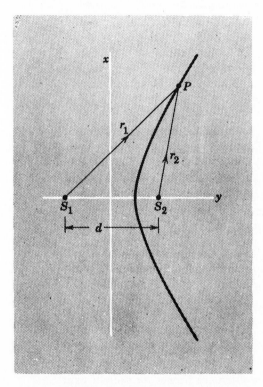

figure 45-22
Problem 12

12. Two point sources in Fig. 45-22 emit coherent waves. Show that curves, such as that given, over which the phase difference for rays r_1 and r_2 is a constant, are hyperbolas. (*Hint:* A constant phase difference implies a constant difference in length between r_1 and r_2.)

13. In Fig 45-23, the source emits monochromatic light of wavelength λ. S is a narrow slit in an otherwise opaque screen I. A plane mirror, whose surface includes the axis of the lens shown, is located a distance h below S. Screen II is at the focal plane of the lens. (a) Find the condition for maximum and minimum brightness of fringes on screen II in terms of the usual angle θ, the wavelength λ, and the distance h. (b) Do fringes appear only in region A (above the axis of the lens), only in region B (below the axis of the lens), or in both regions A and B? Explain. (*Hint:* Consider the image of S formed by the mirror.)

Answer: (a) 2 h sin θ = mλ (minimum); 2 h sin θ = (m + $\frac{1}{2}$)λ (maximum).
(b) Only in region A.

figure 45-23
Problem 13

14. Design, semi-quantitatively, a Young-type double slit interference experiment (see Fig. 45-1) for sound waves in air. Assume a frequency of 500 Hz and a speed of 330 m/s. Discuss some of the design parameters such as the nature of the source, the size of screen B, the width, height, and separation of the two slits, the "fringe" separation, the nature of the fringe detector, etc.

SECTION 45-3

15. Light of wavelength 600 nm is incident normally on a system of two parallel narrow slits separated by 0.60 mm. Sketch the intensity in the pattern observed on a distant screen as a function of angle θ, as in Fig. 45-5, for the range of values $0 \leqslant \theta \leqslant 0.0040$ radians.

16. Show that the half-width $\Delta\theta$ of the double-slit interference fringes (see arrow in Fig. 45-9) is given by

$$\Delta\theta = \frac{\lambda}{2d}$$

if θ is small enough so that $\sin\theta \cong \theta$.

17. One of the slits of a double-slit system is wider than the other, so that the amplitude of the light reaching the central part of the screen from one slit, acting alone, is twice that from the other slit, acting alone. Derive an expression for I_θ in terms of θ, corresponding to Eqs. 45-11b and 45-11c.

Answer: $I_\theta = \frac{1}{9} I_m \left[1 + 8 \cos^2\left(\frac{\pi d \sin\theta}{\lambda}\right) \right]$

18. S_1 and S_2 in Fig. 45-24 are effective point sources of radiation, excited by the same oscillator. They are coherent and in phase with each other. Placed 4.0 m apart, they emit equal amounts of power in the form of 1.0-m wavelength electromagnetic waves. (a) Find the positions of the first (that is, the nearest), the second, and the third maxima of the received signal, as the detector is moved out along Ox. (b) Is the intensity at the nearest minimum equal to zero? Justify your answer.

figure 45-24
Problem 18

SECTION 45-4

19. Find the sum of the following quantities (a) by the vector method and (b) analytically:

$$y_1 = 10 \sin \omega t$$

$$y_2 = 8 \sin (\omega t + 30°).$$

Answer: $y = 17 \sin (\omega t + 13°)$.

20. Add the following quantities graphically, using the vector method:

$$y_1 = 10 \sin \omega t$$

$$y_2 = 15 \sin (\omega t + 30°)$$

$$y_3 = 5 \sin (\omega t - 45°).$$

21. Consider the problem of determining the sum

$$A_1 \sin (\omega t + \phi_1) + A_2 \sin (\omega t + \phi_2) + \cdots + A_n \sin (\omega t + \phi_n)$$

from the phasor diagram.
(a) Show that the sum may always be written in the form

$$B \sin \omega t + C \cos \omega t.$$

(b) Show that $B^2 + C^2 \leq (A_1 + A_2 + \cdots + A_n)^2$.
(c) When does the equality sign in (b) hold?
Answer: (c) When all phase angles ϕ_i are equal.

22. We wish to coat a flat slab of glass ($n = 1.50$) with a transparent material ($n = 1.25$) so that light of wavelength 600 nm (in vacuum) incident normally is not reflected. How can this be done?

23. A thin film 4.0×10^{-5} cm thick is illuminated by white light normal to its surface. Its index of refraction is 1.5. What wavelengths within the visible spectrum will be intensified in the reflected beam?
Answer: 480 nm (blue).

24. White light reflected at perpendicular incidence from a soap bubble has, in the visible spectrum, a single interference maximum (at $\lambda = 600$ nm) and a single minimum at the violet end of the spectrum. If $n = 1.33$ for the film, calculate its thickness.

25. A plane wave of monochromatic light falls normally on a uniformly thin film of oil which covers a glass plate. The wavelength of the source can be varied continuously. Complete destructive interference of the reflected light is observed for wavelengths of 500 and 700 nm and for no other wavelengths in between. If the index of refraction of the oil is 1.30 and that of the glass is 1.50, find the thickness of the oil film.
Answer: 670 nm.

26. A plane monochromatic light wave in air falls at normal incidence on a thin film of oil which covers a glass plate. The wavelength of the source may be varied continuously. Complete destructive interference in the reflected beam is observed for wavelengths of 500 and 700 nm and for no other wavelength in between. The index of refraction of glass is 1.50. Show that the index of refraction of the oil must be less than 1.50.

27. A thin film of acetone (refractive index 1.25) is floated on a thick glass plate (refractive index 1.50). Plane light waves of variable wavelengths are incident normal to the film. When one views the reflected wave it is noted that complete destructive interference occurs at 600 nm and constructive interference at 700 nm. Calculate the thickness of the acetone film.
Answer: 840 nm.

28. White light reflected at perpendicular incidence from a soap film has, in the visible spectrum, an interference maximum at 600 nm (= 6000 Å) and a minimum at 450 nm (= 4500 Å), with no minimum in between. If $n = 1.33$ for the film, what is the film thickness, assumed uniform?

29. In Example 5 assume that there is zero reflection for light of wavelength 550 nm at normal incidence. Calculate the factor by which the reflection is diminished by the coating at 450 and at 650 nm.
Answer: The intensity is diminished by 88% at 450 nm and by 94% at 650 nm.

30. A tanker has dumped a large quantity of kerosene ($n = 1.2$) into the Gulf of Mexico, creating a large slick on top of the water ($n = 1.3$). (a) If you are looking straight down from an airplane onto a region of the slick where its thickness is 460 nm, for which wavelength(s) of visible light is the reflection the greatest? (b) If you are scuba-diving directly under this same region of the slick, for which wavelengths of visible light is the transmitted intensity the strongest?

31. Light of wavelength 630 nm is incident normally on a thin wedge-shaped film of index of refraction 1.5. There are ten bright and nine dark fringes over the length of film. By how much does the film thickness change over this length?　　　*Answer:* 1.9×10^{-6} m (= 1.9 μm).

32. A broad source of light ($\lambda = 680$ nm) illuminates normally two glass plates 12 cm long that touch at one end and are separated by a wire 0.048 mm in diameter at the other (Fig. 45-25). How many bright fringes appear over the 12-cm distance?

figure 45-25
Problem 32

33. An oil drop $(n = 1.20)$ floats on a water $(n = 1.33)$ surface and is observed from above by reflected light (see Fig. 45-26). (a) Will the outer (thinnest) regions of the drop correspond to a bright or a dark region? (b) Approximately how thick is the oil film where one observes the third blue region from the outside of the drop? (c) Why do the colors gradually disappear as the oil thickness becomes larger?   *Answer:* (a) Bright. (b) 594 nm.

figure 45-26
Problem 33

34. The diameter of the tenth bright ring in a Newton's rings apparatus changes from 1.40 to 1.27 cm as a liquid is introduced between the lens and the plate. Find the index of refraction of the liquid.

35. In a Newton's rings experiment the radius of curvature R of the lens is 5.0 m and its diameter is 2.0 cm. (a) How many rings are produced? (b) How many rings would be seen if the arrangement were immersed in water $(n = 1.33)$? Assume that $\lambda = 589$ nm.   *Answer:* (a) 34. (b) 46.

36. In the Newton's rings experiment, use the result of Example 6 to show that the difference in radius between adjacent rings is

$$\Delta r = r_{m+1} - r_m \cong \tfrac{1}{2} \sqrt{\frac{\lambda R}{m}}.$$

Assume $m \gg 1$ and use the binomial theorem. Is this result qualitatively consistent with Fig. 45-18?

37. In the Newton's rings experiment use the result of Example 6 and Problem 36 to show that the *area* between adjacent rings is given, for $m \gg 1$, by

$$A_m = \pi \lambda R \ (= \text{a constant}).$$

Is this result qualitatively consistent with Fig. 45-18?

SECTION 45-7

38. If mirror M_2 in Michelson's interferometer is moved through 0.233 mm, 792 fringes are counted. What is the wavelength of the light?

39. A thin film with $n = 1.40$ for light of wavelength 589 nm $(= 5890$ Å) is placed in one arm of a Michelson interferometer. If a shift of 7.0 fringes occurs, what is the film thickness?   *Answer:* 5200 nm.

40. A Michelson interferometer is used with a sodium discharge tube as a light source. The yellow sodium light consists of two wavelengths, 589.0 (=5890 Å) and 589.6 nm (=5896 Å). It is observed that the interference pattern disappears and reappears periodically as one moves mirror M_2 in Fig. 45-20. (a) Explain this effect. (b) Calculate the change in path difference between two successive reappearances of the interference pattern.

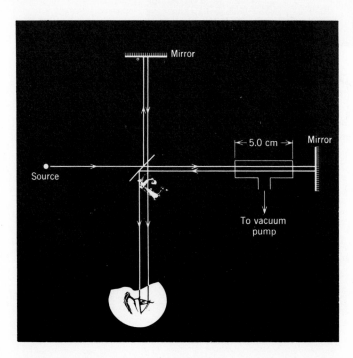

figure 45-27
Problem 41

41. An air-tight chamber 5.0 cm long with glass windows is placed in one arm of a Michelson interferometer as indicated in Fig. 45-27. Light of $\lambda = 5000$ Å is used. The air is slowly evacuated from the chamber using a vacuum pump. While the air is being removed, 60 fringes are observed to pass through the view. From these data, find the index of refraction of air at atmospheric pressure.   *Answer:* 1.0003

42. Write an expression for the intensity observed in Michelson's interferometer (Fig. 45-20) as a function of the position of the moveable mirror. Measure the position of the mirror from the point at which $d_1 = d_2$.

46
diffraction

Diffraction, which is illustrated in Fig. 44-3, is the bending of light around an obstacle such as the edge of a slit. We can see the diffraction of light by looking through a crack between two fingers at a distant light source such as a tubular neon sign or by looking at a street light through a cloth umbrella. Usually diffraction effects are small and we must look for them carefully. Also, most sources of light have an extended area so that a diffraction pattern produced by one point of the source will overlap that produced by another. Finally, common sources of light are not monochromatic. The patterns for the various wavelengths overlap and again the effect is less apparent.

Diffraction was discovered by Francesco Grimaldi (1618–1663), and the phenomenon was known both to Huygens and to Newton. Newton did not see in it any justification for a wave theory for light. Huygens, although he believed in a wave theory, did not believe in diffraction! He imagined his secondary wavelets to be effective only at the point of tangency to their common envelope, thus denying the possibility of diffraction (see Section 43-3). In his words:

And thus we see the reasons why light . . . proceeds only in straight lines in such a way that it does not illuminate any object except when the path from the source to the object is open along such a line.

Jean Augustin Fresnel (1788–1827) correctly applied Huygens' principle to explain diffraction. In these early days the light waves were believed to be mechanical waves in an all-pervading ether. We have seen (Section 41-8) how Maxwell showed that light waves were not mechanical in nature but electromagnetic. Einstein rounded out our

46-1
INTRODUCTION

modern view of light waves by eliminating the need to postulate an ether (see Section 42-4).

Figure 46-1 shows a general diffraction situation. Surface A is a wavefront that falls on B, which is an opaque screen containing an aperture of arbitrary shape; C is a diffusing screen that receives the light that passes through this aperture. We can calculate this pattern of light intensity on C by subdividing the wavefront into elementary areas dS, each of which becomes a source of an expanding Huygens' wavelet. The light intensity at an arbitrary point P is found by superimposing the wave disturbances (that is, the **E** vectors) caused by the wavelets reaching P from all these elementary radiators.

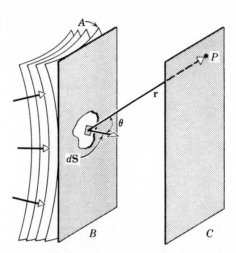

figure 46-1
Light is diffracted at the aperture in screen B and illuminates screen C. The intensity at P is found by dividing the wavefront at B into elementary radiators dS and combining their effects at P.

The wave disturbances reaching P differ in amplitude and in phase because (a) the elementary radiators are at varying distances from P, (b) the light leaves the radiators at various angles to the normal to the wavefront (see Section 43-2), and (c) some radiators are blocked by screen B; others are not. Diffraction calculations — simple in principle — may become difficult in practice. The calculation must be repeated for every point on screen C at which we wish to know the light intensity. We followed exactly this program in calculating the double-slit intensity pattern in Section 45-3. The calculation there was simple because we assumed only two elementary radiators, the two narrow slits.

Figure 46-2a shows the general case of *Fresnel diffraction*, in which the light source and/or the screen on which the diffraction pattern is displayed are a finite distance from the diffracting aperture; the wavefronts that fall on the diffracting aperture in this case and that leave it to illuminate any point P of the diffusing screen are not planes; the corresponding rays are not parallel.

A simplification results if source S and screen C are moved to a large distance from the diffraction aperture, as in Fig. 46-2b. This limiting case is called *Fraunhofer diffraction*. The wavefronts arriving at the diffracting aperture from the distant source S are planes, and the rays associated with these wavefronts are parallel to each other. Fraunhofer conditions can be established in the laboratory by using two converging lenses, as in Fig. 46-2c. The first of these converts the diverging wave from the source into a plane wave. The second lens causes plane waves leaving the diffracting aperture to converge to point P. All rays that

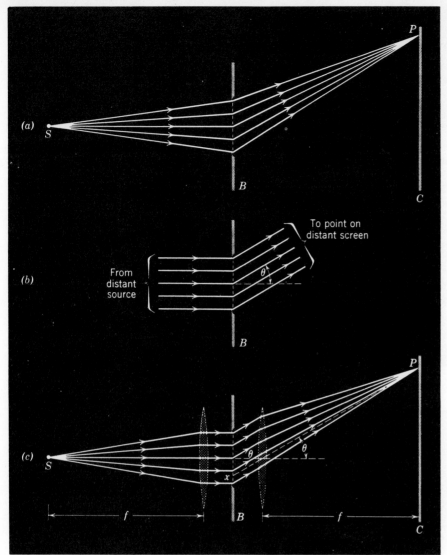

figure 46-2
(*a*) Fresnel diffraction. (*b*) Source *S* and screen *C* are moved to a large distance, resulting in Fraunhofer diffraction. (*c*) Fraunhofer diffraction conditions produced by lenses, leaving source *S* and screen *C* in their original positions.

illuminate *P* will leave the diffracting aperture parallel to the dashed line *Px* drawn from *P* through the center of this second (thin) lens. We assumed Fraunhofer conditions for Young's double-slit experiment in Section 45-1 (see Fig. 45-5).

Although Fraunhofer diffraction is a limiting case of the more general Fresnel diffraction, it is an important limiting case and is easier to handle mathematically. This book deals only with Fraunhofer diffraction.

46-2
SINGLE SLIT

Figure 46-3 shows a plane wave falling at normal incidence on a long narrow slit of width *a*. Let us focus our attention on the central point P_0 of screen *C*. The parallel rays extending from the slit to P_0 all have the same optical (though not geometrical) path lengths, as we saw in Section 44-5. Since they are in phase at the plane of the slit, they will still be in phase at P_0, and the central point of the diffraction pattern that appears on screen *C* has a maximum intensity.

We now consider another point on the screen. Light rays which reach

figure 46-3
Conditions at the central maximum of the diffraction pattern. The slit extends a distance above and below the figure, this distance being much greater than the slit width a.

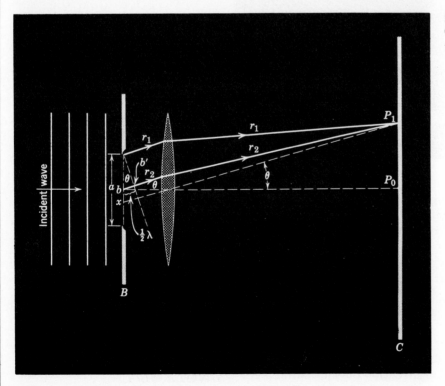

figure 46-4
Conditions at the first minimum of the diffraction pattern.

P_1 in Fig. 46-4 leave the slit at an angle θ as shown. (Note that the ray represented by the dashed line xP_1 is drawn to pass through the center of the lens and is thus undeflected; this ray determines θ.) Ray r_1 originates at the top of the slit and ray r_2 at its center. If θ is chosen so that the distance bb' in the figure is one-half a wavelength, r_1 and r_2 will be out of phase and will produce no effect at P_1.* In fact, every ray from the upper half of the slit will be canceled by a ray from the lower half, originating at a point $a/2$ below the first ray. The point P_1, the first minimum of the diffraction pattern, will have zero intensity (compare Fig. 44-3).

The condition shown in Fig. 46-4 is

$$\frac{a}{2} \sin \theta = \frac{\lambda}{2},$$

or
$$a \sin \theta = \lambda. \tag{46-1}$$

As we stated earlier (see Fig. 44-1), the central maximum becomes wider as the slit is made narrower. If the slit width is as small as one wavelength ($a = \lambda$), the first minimum occurs at $\theta = 90°$ ($\sin \theta = 1$ in Eq. 46-1), which implies that the central maximum fills the entire forward hemisphere. We assumed a condition approaching this in our discussion of Young's double-slit interference experiment in Section 45-1.

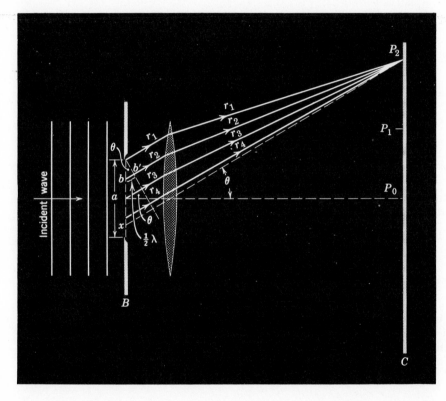

figure 46-5
Conditions at the second minimum of the diffraction pattern.

In Fig. 46-5 the slit is divided into four equal zones, with a ray leaving the top of each zone. Let θ be chosen so that the distance bb' is one-half a wavelength. Rays r_1 and r_2 will then cancel at P_2. Rays r_3 and r_4 will also be half a wavelength out of phase and will also cancel. Consider four other rays, emerging from the slit a given distance below the four rays above. The two rays below r_1 and r_2 will cancel uniquely, as will the two rays below r_3 and r_4. We can proceed across the entire slit and conclude again that no light reaches P_2; we have located a second point of zero intensity.

---

* Whatever phase relation exists between r_1 and r_2 at the plane represented by the sloping dashed line in Fig. 46-4 that passes through b' also exists at P_1, not being affected by the lens (see Section 44-5).

The condition described (see Fig. 46-5) requires that

$$\frac{a}{4} \sin \theta = \frac{\lambda}{2},$$

or

$$a \sin \theta = 2\lambda.$$

By extension, the general formula for the minima in the diffraction pattern on screen C is

$$a \sin \theta = m\lambda \qquad m = 1, 2, 3, \ldots \text{ (minima)}. \qquad (46\text{-}2)$$

There is a maximum approximately halfway between each adjacent pair of minima. You should consider the simplification that has resulted in this analysis by examining Fraunhofer (Fig. 46-2c) rather than Fresnel conditions (Fig. 46-2a).

EXAMPLE 1

A slit of width a is illuminated by white light. For what value of a will the first minimum for red light ($\lambda = 650$ nm $= 6500$ Å) fall at $\theta = 30°$?

At the first minimum we put $m = 1$ in Eq. 46-2. Doing so and solving for a yields

$$a = \frac{m\lambda}{\sin \theta} = \frac{(1)(650 \text{ nm})}{\sin 30°} = 1300 \text{ nm}.$$

Note that the slit width must be twice the wavelength in this case.

EXAMPLE 2

In Example 1 what is the wavelength λ' of the light whose first diffraction maximum (not counting the central maximum) falls at $\theta = 30°$, thus coinciding with the first minimum for red light?

This maximum is about halfway between the first and second minima. We can find it without too much error by putting $m = 1.5$ in Eq. 46-2, or

$$a \sin \theta \cong 1.5\lambda'.$$

From Example 1, however, $\quad a \sin \theta = \lambda.$

Dividing gives $\quad \lambda' = \dfrac{\lambda}{1.5} = \dfrac{650 \text{ nm}}{1.5} = 430 \text{ nm}.$

Light of this color is violet. The second maximum for light of wavelength 430 nm will *always* coincide with the first minimum for light of wavelength 650 nm, no matter what the slit width. If the slit is relatively narrow, the angle θ at which this overlap occurs will be relatively large.

46-3
SINGLE SLIT—QUALITATIVE

In Section 46-2 we located the positions of the maxima and minima for a single-slit diffraction pattern. We now wish to find an expression for the intensity of the *entire* diffraction pattern as a function of the diffraction angle θ. We do so qualitatively in this section and quantitatively in the next section.

Figure 46-6 shows a slit of width a divided into N parallel strips of width Δx. Each strip acts as a radiator of Huygens' wavelets and produces a characteristic wave disturbance at point P, whose position on the screen, for a particular arrangement of apparatus, can be described by the angle θ.

If the strips are narrow enough—which we assume—all points on a given strip have essentially the same optical path length to P, and therefore all the light from the strip will have the same phase when it ar-

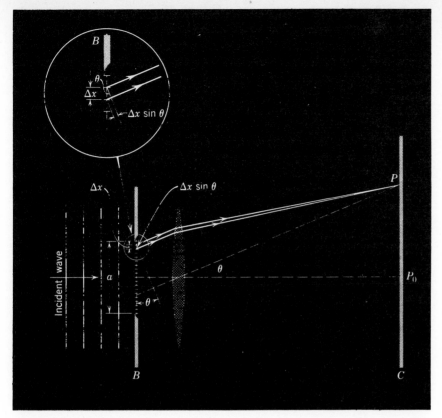

figure 46-6
We divide a slit of width a into N
strips of width Δx. The insert shows
conditions at the second strip more
clearly. In the differential limit the
slit is divided into an infinite
number of strips (that is, $N \to \infty$)
of differential width dx. For clarity
in this and the following figure, we
take $N = 18$.

rives at P. We may take the amplitudes ΔE_0 of the wave disturbances at
P from the various strips may be taken as equal if θ in Fig. 46-6 is not
too large.

We limit our considerations to points that lie in, or very close to, the plane
of Fig. 46-6. It can be shown that this procedure is valid for a slit whose length is
much greater than its width a. We made this same assumption tacitly both
earlier in this chapter and in Chapter 45; see Figs. 45-5 and 46-3, for example.

The wave disturbances from adjacent strips have a constant phase
difference $\Delta\phi$ between them at P given by

$$\frac{\text{phase difference}}{2\pi} = \frac{\text{path difference}}{\lambda},$$

or
$$\Delta\phi = \left(\frac{2\pi}{\lambda}\right)(\Delta x \sin\theta), \qquad (46\text{-}3)$$

where $\Delta x \sin\theta$ is, as the figure insert shows, the path difference for rays
originating at the top edges of adjacent strips. Thus, at P, N vectors with
the same amplitude ΔE_0, the same frequency, and the same phase difference $\Delta\phi$ between adjacent members combine to produce a resultant
disturbance. We ask, for various values of $\Delta\phi$ [that is, for various points
P on the screen, corresponding to various values of θ (see Eq. 46-3)], what
is the amplitude E_θ of the resultant wave disturbance? We find the
answer by representing the individual wave disturbances ΔE_0 by phasors
and calculating the resultant phasor amplitude, as described in Section
45-4.

At the center of the diffraction pattern θ equals zero, and the phase
shift between adjacent strips (see Eq. 46-3) is also zero. As Fig. 46-7a
shows, the phasor arrows in this case are laid end to end and the ampli-

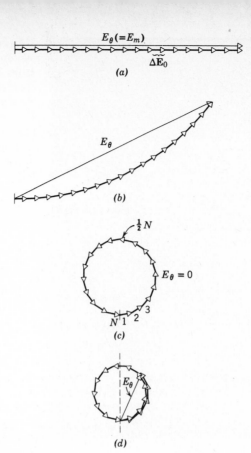

figure 46-7
Conditions at (a) the central maximum, (b) a direction slightly removed from the central maximum, (c) the first minimum, and (d) the first maximum beyond the central maximum for single-slit diffraction. This figure corresponds to $N = 18$ in Fig. 46-6.

tude of the resultant has its maximum value E_m. This corresponds to the center of the central maximum.

As we move to a value of θ other than zero, $\Delta\phi$ assumes a definite nonzero value (again see Eq. 46-3), and the array of arrows is now as shown in Fig. 46-7b. The resultant amplitude E_θ is less than before. Note that the length of the "arc" of small arrows is the same for both figures and indeed for all figures of this series. As θ increases further, we reach a situation (Fig. 46-7c) in which the chain of arrows curls around through 360°, the tip of the last arrow touching the foot of the first arrow. This corresponds to $E_\theta = 0$, that is, to the first minimum. For this condition the ray from the top of the slit (1 in Fig. 46-7c) is 180° out of phase with the ray from the center of the slit ($\frac{1}{2}N$ in Fig. 46-7c). These phase relations are consistent with Fig. 46-4, which also represents the first minimum.

As θ increases further, the phase shift continues to increase, and the chain of arrows coils around through an angular distance greater than 360°, as in Fig. 46-7d, which corresponds to the first maximum beyond the central maximum. This maximum is much smaller than the central maximum. In making this comparison, recall that the arrows marked E_θ in Fig. 46-7 correspond to the *amplitudes* of the wave disturbance and not to the *intensity*. The amplitudes must be squared to obtain the corresponding relative intensities (see Eq. 45-7).

The "arc" of small arrows in Fig. 46-8 shows the phasors representing, in amplitude and phase, the wave disturbances that reach an arbitrary point P on the screen of Fig. 46-6, corresponding to a particular angle θ. The resultant amplitude at P is E_θ. If we divide the slit of Fig. 46-6 into

46-4
*SINGLE SLIT—
QUANTITATIVE*

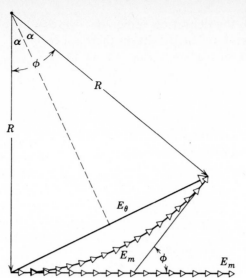

infinitesimal strips of width dx, the arc of arrows in Fig. 46-8 approaches the arc of a circle, its radius R being indicated in that figure. The length of the arc is E_m, the amplitude at the center of the diffraction pattern, for at the center of the pattern the wave disturbances are all in phase and this "arc" becomes a straight line as in Fig. 46-7*a*.

The angle ϕ in the lower part of Fig. 46-8 is revealed as the difference in phase between the infinitesimal vectors at the left and right ends of the arc E_m. This means that ϕ is the phase difference between rays from the top and the bottom of the slit of Fig. 46-6. From geometry we see that ϕ is also the angle between the two radii marked R in Fig. 46-8. From this figure we can write

$$E_\theta = 2R \sin \frac{\phi}{2}.$$

In radian measure ϕ, from the figure, is

$$\phi = \frac{E_m}{R}.$$

Combining yields

$$E_\theta = \frac{E_m}{\phi/2} \sin \frac{\phi}{2},$$

or

$$E_\theta = E_m \frac{\sin \alpha}{\alpha}, \qquad (46\text{-}4)$$

in which

$$\alpha = \frac{\phi}{2}. \qquad (46\text{-}5)$$

From Fig. 46-6, recalling that ϕ is the phase difference between rays from the top and the bottom of the slit and that the path difference for these rays is $a \sin \theta$, we have

$$\frac{\text{phase difference}}{2\pi} = \frac{\text{path difference}}{\lambda},$$

or

$$\phi = \left(\frac{2\pi}{\lambda}\right)(a \sin \theta).$$

Combining with Eq. 46-5 yields

$$\alpha = \frac{\phi}{2} = \frac{\pi a}{\lambda} \sin \theta. \qquad (46\text{-}6)$$

Equation 46-4, taken together with the definition of Eq. 46-6, gives the amplitude of the wave disturbance for a single-slit diffraction pattern at any angle θ. The intensity I_θ for the pattern is proportional to the square of the amplitude, or

$$I_\theta = I_m \left(\frac{\sin \alpha}{\alpha}\right)^2. \qquad (46\text{-}7)$$

For convenience we display together, and renumber, the formulas for the amplitude and the intensity in single-slit diffraction.

[Eq. 46-4] $\qquad\qquad E_\theta = E_m \dfrac{\sin \alpha}{\alpha}$ $\qquad\qquad\qquad\qquad (46\text{-}8a)$

[Eq. 46-7] $\qquad\qquad I_\theta = I_m \left(\dfrac{\sin \alpha}{\alpha}\right)^2$ single-slit diffraction $\qquad (46\text{-}8b)$

[Eq. 46-6] $\qquad \alpha \left(= \tfrac{1}{2}\phi\right) = \dfrac{\pi a}{\lambda} \sin \theta$ $\qquad\qquad\qquad (46\text{-}8c)$

Figure 46-9 shows plots of I_θ for several values of the ratio a/λ. Note that the pattern becomes narrower as we increase a/λ; compare this figure with Figs. 44-1 and 44-3.

Minima occur in Eq. 46-8b when

$$\alpha = m\pi \qquad m = 1, 2, 3, \ldots. \qquad (46\text{-}9)$$

Combining with Eq. 46-8c leads to

$$a \sin \theta = m\lambda \qquad m = 1, 2, 3, \ldots \text{ (minima)},$$

which is the result derived in the preceding section (Eq. 46-2). In that section, however, we derived *only* this result, obtaining no quantitative information about the intensity of the diffraction pattern at places in which it was not zero. Here (Eqs. 46-8) we have complete intensity information.

EXAMPLE 3

Intensities of the secondary diffraction maxima. Calculate, approximately, the relative intensities of the secondary maxima in the single-slit Fraunhofer diffraction pattern.

The secondary maxima lie approximately halfway between the minima and are found (compare Eq. 46-9) from

$$\alpha \cong \left(m + \tfrac{1}{2}\right)\pi \qquad m = 1, 2, 3, \ldots.$$

Substituting into Eq. 46-8b yields

$$I_\theta = I_m \left[\frac{\sin \left(m + \tfrac{1}{2}\right)\pi}{\left(m + \tfrac{1}{2}\right)\pi}\right]^2,$$

which reduces to $\qquad\qquad \dfrac{I_\theta}{I_m} = \dfrac{1}{\left(m + \tfrac{1}{2}\right)^2 \pi^2}.$

This yields, for $m = 1, 2, 3, \ldots$, $I_\theta/I_m = 0.045, 0.016, 0.0083$, etc. The successive maxima decrease rapidly in intensity.

EXAMPLE 4

Width of the central diffraction maximum. Derive the *half-width* $\Delta\theta$ of the central maximum in a single-slit Fraunhofer diffraction (see Fig. 46-9b). The half-width is the angle between the two points in the pattern where the intensity is one-half that at the center of the pattern.

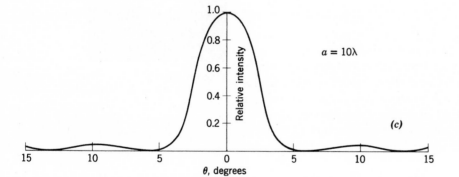

figure 46-9
The relative intensity in single-slit diffraction for three values of the ratio a/λ. The arrow in (b) shows the half-width $\Delta\theta$ of the central maximum.

Point x in Fig. 46-9b is so chosen that $I_\theta = \frac{1}{2}I_m$, or, from Eq. 46-8b,

$$\frac{1}{2} = \left(\frac{\sin \alpha_x}{\alpha_x}\right)^2.$$

This equation cannot be solved analytically for α_x. We can solve it graphically, as accurately as we wish, by plotting the quantity $(\sin \alpha_x/\alpha_x)^2$ as ordinate versus α_x as abscissa and noting the value of α_x at which the curve intersects the line "one-half" on the ordinate scale (see Problem 11). However, if only an approximate answer is desired, it is often quicker to use trial-and-error methods.

We know that α equals π at the first minimum; we guess that α_x is perhaps $\pi/2$ (= 90° = 1.57 rad). Trying this in Eq. 46-8b yields

$$\frac{I_\theta}{I_m} = \left[\frac{\sin (\pi/2)}{\pi/2}\right]^2 = 0.4.$$

This intensity ratio is *less* than 0.5, so that α_x must be *less* than 90°. After a few more trials we find easily enough that

$$\alpha_x = 1.40 \text{ rad} = 80°$$

does yield a ratio close to the correct value of 0.5.

We now use Eq. 46-8c to find the corresponding angle θ:

$$\alpha_x = \frac{\pi a}{\lambda} \sin \theta_x = 1.40,$$

or, noting that $a/\lambda = 5$ for Fig. 46-9b,

$$\sin \theta_x = \frac{1.40\lambda}{\pi a} = \frac{1.40}{5\pi} = 0.0892.$$

The half-width $\Delta\theta$ of the central maximum (see Fig. 46-9b) is given by

$$\Delta\theta = 2\theta_x = 2 \sin^{-1} 0.0892 = 2 \times 5.1° = 10.2°,$$

which is in agreement with the figure.

46-5
DIFFRACTION AT A CIRCULAR APERTURE

Diffraction will occur when a wavefront is partially blocked off by an opaque object such as a metal disk or an opaque screen containing an aperture. Here we consider diffraction at a circular aperture of diameter d, the aperture constituting the boundary of a circular converging lens.

Our previous treatment of lenses was based on geometrical optics, diffraction being specifically assumed not to occur. A rigorous analysis would be based from the beginning on wave optics, since geometrical optics is always an approximation, although often a very good one. Diffraction phenomena would emerge in a natural way from such a wave-optical analysis.

Figure 46-10 shows the image of a distant point source of light (a star) formed on a photographic film placed in the focal plane of the converging lens of a telescope.* It is not a point, as the (approximate) geometrical optics treatment suggests, but a circular disk surrounded by several progressively fainter secondary rings. Comparison with Fig. 44-3c leaves little doubt that we are dealing with a diffraction phenomenon in which, however, the aperture is a circle rather than a long narrow slit. The ratio d/λ, where d is the diameter of the lens (or of a circular aperture placed in front of the lens), determines the scale of the diffraction pattern, just as the ratio a/λ does for a slit.

Analysis shows that the first minimum for the diffraction pattern of a circular aperture of diameter d, assuming Fraunhofer conditions, is given by

$$\sin \theta = 1.22 \frac{\lambda}{d}. \qquad (46\text{-}10)$$

This is to be compared with Eq. 46-1, or

$$\sin \theta = \frac{\lambda}{a},$$

which locates the first minimum for a long narrow slit of width a. The factor 1.22 emerges from the mathematical analysis when we integrate over the elementary radiators into which the circular aperture may be divided.

In actual lenses the image of a distant point object will be somewhat larger than that shown in Fig. 46-10 and may not have radial symmetry. This is caused by the various lens "defects" mentioned in Section 44-5. However, even if all of these defects could be eliminated by suitable shaping of the lens surfaces or by intro-

figure 46-10
The image of a star formed by a converging lens is a diffraction pattern. Note the central maximum, sometimes called the Airy disk (after Sir George Airy, who first solved the problem of diffraction at a circular aperture in 1835), and the circular secondary maximum. Other secondary maxima occur at larger radii but are too faint to be seen.

---

*Diffraction phenomena also occur in microscopes and other optical instruments. They are, in fact, characteristic of the entire electromagnetic spectrum.

ducing correcting lenses, the diffraction pattern of Fig. 46-10 would remain. It is an inherent property of the lens aperture and of the wavelength of light used.

The fact that lens images are diffraction patterns is important when we wish to distinguish two distant point objects whose angular separation is small. Figure 46-11 shows the visual appearances and the corresponding intensity patterns for two distant point objects (stars, say) with small angular separations and approximately equal central intensities. In (a) the objects are not resolved; that is, they cannot be distinguished from a single point object. In (b) they are barely resolved and in (c) they are fully resolved.*

In Fig. 46-11b the angular separation of the two point sources is such that the maximum of the diffraction pattern of one source falls on the first minimum of the diffraction pattern of the other. This is called *Rayleigh's criterion*. This criterion, though useful, is arbitrary; other criteria for deciding when two objects are resolved are sometimes used. From Eq. 46-10, two objects that are barely resolvable by Rayleigh's criterion must have an angular separation θ_R of

$$\theta_R = \sin^{-1}\frac{1.22\lambda}{d}.$$

Since the angles involved are rather small, we can replace $\sin\theta_R$ by θ_R, or

$$\theta_R = 1.22\frac{\lambda}{d}, \tag{46-11}$$

in which θ_R is expressed in radians. If the angular separation θ between the objects is greater than θ_R, we can resolve the two objects; if it is less, we cannot. The angle θ_R is the smallest angular separation for which resolution is possible, using Rayleigh's criterion.

EXAMPLE 5

A converging lens 3.0 cm in diameter has a focal length f of 20 cm. (a) What angular separation must two distant point objects have to satisfy Rayleigh's criterion? Assume that $\lambda = 550$ nm.

From Eq. 46-11,

$$\theta_R = 1.22\frac{\lambda}{d} = \frac{(1.22)(5.5 \times 10^{-7}\text{ m})}{3.0 \times 10^{-2}\text{ m}} = 2.2 \times 10^{-5}\text{ rad}.$$

(b) How far apart are the centers of the diffraction patterns in the focal plane of the lens? The linear separation is

$$x = f\theta = (20\text{ cm})(2.2 \times 10^{-5}\text{ rad}) = 4400\text{ nm}.$$

This is 8.0 wavelengths of the light employed.

When we wish to use a lens to resolve objects of small angular separation, it is desirable to make the central disk of the diffraction pattern as small as possible. This can be done (see Eq. 46-11) by increasing the lens diameter or by using a shorter wavelength. One reason for constructing large telescopes is to produce *sharper* images so that we can examine celestial objects in finer detail. The images are also *brighter*, not only because the energy is concentrated into a smaller diffraction

---

*Diffraction effects set the so-called "bottom line" on resolution. However, other effects combine to hamper resolution. For atmospheric effects on the observation of stars see "Toward the Diffraction Limit" by Dietrick E. Thomsen, *Science News*, August 1975.

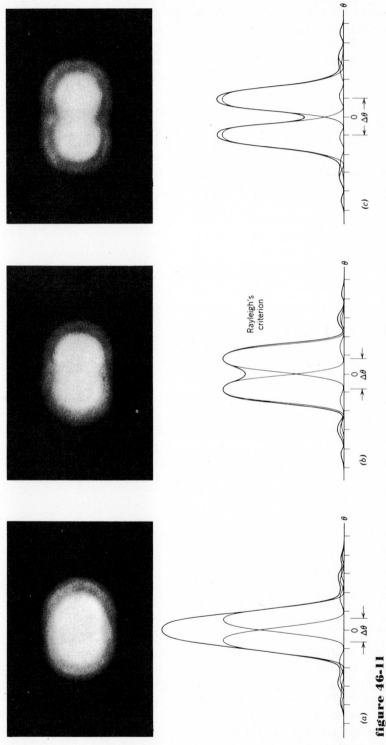

figure 46-11

The images of two distant point objects (simulated stars) are formed by a converging lens whose diameter (= 10 cm) is 200,000 times the effective wavelength (= 500 nm). Sketches of the images as they appear in the focal plane of the telescope objective lens are shown with the corresponding intensity plots below them. (a) The angular separation of the objects (see vertical ticks) is so small that the images are not resolved. (b) The objects are further apart and the images meet Rayleigh's criterion for resolution. (c) The objects are still farther apart and the images are well resolved.

disk but because the larger lens collects more light. Thus fainter objects, for example, distant galaxies, can be seen.

To reduce diffraction effects in *microscopes* we often use ultraviolet light, which, because of its shorter wavelength, permits finer detail to be examined than would be possible for the same microscope operated with visible light. We shall see in Chapter 50 that beams of electrons behave like waves under some circumstances. In the *electron microscope* such beams may have an effective wavelength of 4×10^{-3} nm, of the order of 10^5 times shorter than visible light ($\lambda \cong 500$ nm or 5000 Å). This permits the detailed examination of tiny objects like viruses. If a virus were examined with an optical microscope, its structure would be hopelessly concealed by diffraction.

46-6 DOUBLE SLIT INTERFERENCE AND DIFFRACTION COMBINED

In Young's double-slit experiment (Section 45-1) we assumed that the slits were arbitrarily narrow (that is, $a \ll \lambda$), which means that the central part of the diffusing screen was uniformly illuminated by the diffracted waves from each slit. When such waves interfere, they produce fringes of uniform intensity, as in Fig. 45-9. This idealized situation cannot occur with actual slits because the condition $a \ll \lambda$ cannot usually be met. Waves from the two actual slits combining at different points of the screen will have intensities that are *not* uniform but are governed by the diffraction pattern of a single slit. The effect of relaxing the assumption that $a \ll \lambda$ in Young's experiment is to leave the fringes relatively unchanged in location but to alter their intensities.

The interference pattern for infinitesimally narrow slits is given by Eq. 45-11b and 45-11c, or, with a small change in nomenclature,

$$I_{\theta,\text{int}} = I_{m,\text{int}} \cos^2 \beta, \tag{46-12}$$

where

$$\beta = \frac{\pi d}{\lambda} \sin \theta \tag{46-13}$$

in which d is the distance between the center-lines of the slits.

The intensity for the diffracted wave from either slit is given by Eqs. 46-8b and 46-8c, or, with a small change in nomenclature,

$$I_{\theta,\text{dif}} = I_{m,\text{dif}} \left(\frac{\sin \alpha}{\alpha}\right)^2, \tag{46-14}$$

where

$$\alpha = \frac{\pi a}{\lambda} \sin \theta. \tag{46-15}$$

We find the combined effect by regarding $I_{m,\text{int}}$ in Eq. 46-12 as a variable amplitude, given in fact by $I_{\theta,\text{dif}}$ of Eq. 46-14. This assumption, for the combined pattern, leads to

$$I_\theta = I_m \, (\cos \beta)^2 \left(\frac{\sin \alpha}{\alpha}\right)^2, \tag{46-16}$$

in which we have dropped all subscripts referring separately to interference and diffraction.

Let us express this result in words. At any point on the screen the available light intensity from each slit, considered separately, is given by the diffraction pattern of that slit (Eq. 46-14). The diffraction patterns for the two slits, again considered separately, coincide because parallel rays in Fraunhofer diffraction are focused at the same spot (see Fig. 46-5). Because the two diffracted waves are coherent, they will interfere.

The effect of interference is to redistribute the available energy over

figure 46-12
Interference fringes for a double slit with slit separation $d = 50\lambda$. Three different slit widths, described by $a/\lambda = 1$, 5, and 10 are shown.

the screen, producing a set of fringes. In Section 45-1, where we assumed $a \ll \lambda$, the available energy was virtually the same at all points on the screen so that the interference fringes had virtually the same intensities (see Fig. 45-9). If we relax the assumption $a \ll \lambda$, the available energy is *not* uniform over the screen but is given by the diffraction pattern of a slit of width a. In this case the interference fringes will have intensities that are determined by the intensity of the diffraction pattern at the location of a particular fringe. Equation 46-16 is the mathematical expression of this argument.

Figure 46-12 is a plot of Eq. 46-16 for $d = 50\lambda$ and for three values of a/λ. It shows clearly that for narrow slits $(a = \lambda)$ the fringes are nearly uniform in intensity. As the slits are widened, the intensities of the fringes are markedly modulated by the "diffraction factor" in Eq. 46-16, that is, by the factor $(\sin \alpha/\alpha)^2$.

Equation 46-16 shows that the fringe envelopes of Fig. 46-12 are precisely the single-slit diffraction patterns of Fig. 46-9. This is especially clear in Fig. 46-13, which shows, for the curve of Fig. 46-12b, (a) the "interference factor" in Eq. 46-16 (that is, the factor $\cos^2 \beta$), (b) the "diffraction factor" $(\sin \alpha/\alpha)^2$, and (c) their product.

figure 46-14
(a) Interference fringes for a double-slit system in which the slit width is *not* negligible in comparison with the wavelength. The fringes are modulated in intensity by the diffraction pattern of a single slit. (b) If one of the slits is covered up, the interference fringes disappear and we see the single slit diffraction pattern. (Courtesy G. H. Carragan, Rensselaer Polytechnic Institute.)

If we put $a = 0$ in Eq. 46-16, then (see Eq. 46-15) $\alpha = 0$ and $\lim_{\alpha \to 0} \sin \alpha / \alpha = \alpha/\alpha = 1$. Thus this equation reduces, as it must, to the intensity equation for a pair of vanishingly narrow slits (Eq. 46-12). If we put $d = 0$ in Eq. 46-16, the two slits coalesce into a single slit of width a, as Fig. 46-17 shows; $d = 0$ implies $\beta = 0$ (see Eq. 46-13) and $\cos^2 \beta = 1$. Thus Eq. 46-16 reduces, as it must, to the diffraction equation for a single slit (Eq. 46-14).

Figure 46-14 shows some actual double-slit interference photographs. The uniformly spaced interference fringes and their intensity modulation by the diffraction pattern of a single slit are clear. If one slit is covered up, as in Fig. 46-14b, the interference fringes disappear and we see the diffraction pattern of a single slit.

Starting from the curve of Fig. 46-12b, what is the effect of (a) increasing the slit width, (b) increasing the slit separation, and (c) increasing the wavelength?

(a) If we increase the slit width a, the envelope of the fringe pattern changes so that its central peak is sharper (compare Fig. 46-12c). The fringe spacing, which depends on d/λ, does not change.

EXAMPLE 6

(b) If we increase d, the fringes become closer together, the envelope of the pattern remaining unchanged.

(c) If we increase λ, the envelope becomes broader and the fringes move further apart. Increasing λ is equivalent to decreasing both of the ratios a/λ and d/λ. The general relationship of the envelope to the fringes, which depends only on d/a, does not change with wavelength.

EXAMPLE 7

In double-slit Fraunhofer diffraction what is the fringe spacing on a screen 50 cm away from the slits if they are illuminated with blue light (λ = 480 nm = 4800 Å), if $d = 0.10$ mm, and if the slit width $a = 0.02$ mm? What is the linear distance from the central maximum to the first minimum of the fringe envelope?

The intensity pattern is given by Eq. 46-16, the fringe spacing being determined by the interference factor $\cos^2 \beta$. From Example 2, Chapter 45, we have

$$\Delta y = \frac{\lambda D}{d},$$

where D is the distance of the screen from the slits. Substituting yields

$$\Delta y = \frac{(480 \times 10^{-9} \text{ m})(50 \times 10^{-2} \text{ m})}{0.10 \times 10^{-3} \text{ m}} = 2.4 \times 10^{-3} \text{ m} = 2.4 \text{ mm}.$$

The distance to the first minimum of the envelope is determined by the diffraction factor $(\sin \alpha/\alpha)^2$ in Eq. 46-16. The first minimum in this factor occurs for $\alpha = \pi$.

From Eq. 46-15,

$$\sin \theta = \frac{\alpha \lambda}{\pi a} = \frac{\lambda}{a} = \frac{480 \times 10^{-9} \text{ m}}{0.02 \times 10^{-3} \text{ m}} = 0.024.$$

This is so small that we can assume that $\theta \cong \sin \theta \cong \tan \theta$, or

$$y = D \tan \theta \cong D \sin \theta = (50 \text{ cm})(0.024) = 1.2 \text{ cm}.$$

There are about ten fringes in the central peak of the fringe envelope.

EXAMPLE 8

What requirements must be met for the central maximum of the envelope of the double-slit Fraunhofer pattern to contain exactly eleven fringes?

The required condition will be met if the sixth minimum of the interference factor $(\cos^2 \beta)$ in Eq. 46-16 coincides with the first minimum of the diffraction factor $(\sin \alpha/\alpha)^2$.

The sixth minimum of the interference factor occurs when

$$\beta = \tfrac{11}{2}\pi$$

in Eq. 46-12.

The first minimum in the diffraction term occurs for

$$\alpha = \pi.$$

Dividing (see Eqs. 46-13 and 46-15) yields

$$\frac{\beta}{\alpha} = \frac{d}{a} = \frac{11}{2}.$$

This condition depends only on the slit geometry and not on the wavelength. For long waves the pattern will be broader than for short waves, but there will always be eleven fringes in the central peak of the envelope.

The double-slit problem as illustrated in Fig. 46-12 combines interference and diffraction in an intimate way. At root both are superposition effects and depend on adding wave disturbances at a given point, taking phase differences properly into account. If the waves to be combined originate from a *finite* (and usually small) number of elementary coherent radiators, as in Young's double-slit experiment, we call the effect *interference*. If the waves to be combined originate by subdividing a wave into *infinitesimal* coherent radiators, as in our treatment of a single slit (Fig. 46-6), we call the effect *diffraction*. This distinction between interference and diffraction is convenient and useful. However, it should not cause us to lose sight of the fact that both are superposition effects and that often both are present simultaneously, as in Young's experiment.*

questions

1. Why is the diffraction of sound waves more evident in daily experience than that of light waves?

2. Why do radio waves diffract around buildings, although light waves do not?

3. A person holds a single narrow vertical slit in front of the pupil of his eye and looks at a distant light source in the form of a long heated filament. Is the diffraction pattern that he sees a Fresnel or a Fraunhofer pattern?

4. Do diffraction effects occur for virtual images as well as for real images? Explain.

5. Do diffraction effects occur for images formed by (*a*) plane mirrors and (*b*) spherical mirrors? Explain.

6. Comment on this statement: "Diffraction occurs in all regions of the electromagnetic spectrum." Consider the X-ray region and the micro-wave region for example, and give your arguments for believing it to be true or false.

7. Figure 46-1 shows the general case of diffraction from an aperture of arbitrary shape in an otherwise opaque screen. Discuss in general terms the inverse case, following the arguments of Section 46-1, in which diffraction occurs around an opaque *object*, such as a key or a paper clip, no screen B being present.

8. Distinguish between Fresnel and Fraunhofer diffraction. Do different physical principles underlie them? If so, what are they? If the same broad principle underlies them, what is it?

9. Most of us have noticed that the reception quality on a car radio deteriorates when we drive under a high-voltage transmission line. Does this have anything to do with the current in the line; with diffraction effects; with interference effects? Discuss.

10. *Poisson's Bright Spot:* If a coin or a ball bearing (or even a bowling ball) is suspended between a point source of light and a photographic film a bright spot, called the Poisson bright spot appears at the center of the geometrical shadow; see Fig. 46-15. Diffraction rings appear, both within the shadow and out of it. One might think that the center of the geometrical shadow, being most shielded from the light source, would be dark but just the reverse is true. Can you see the qualitative possibility that this experimental result can be consistent with Fresnel's diffraction theory?

* In this chapter we have not discussed *holography*, a technique in which a *three-dimensional* image of an object (rather than a two-dimensional image as in ordinary photography) can be produced. Diffraction plays a major role. To follow this up see "Holography, 1948–1971" by Dennis Gabor, *Science*, July 1972. Gabor won the Nobel Prize in 1971 for this discovery (made in 1948) and this article is an adaptation of his lecture on receiving the prize. See also "An Elementary Introduction to Practical Holography," by Alan G. Porter and S. George, *American Journal of Physics*, November 1975.

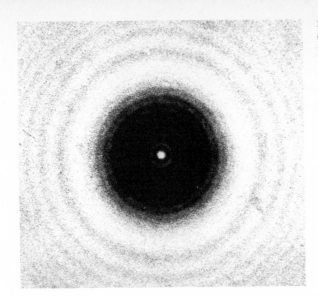

figure 46-15
Question 10
Fresnel's Bright Spot.

(Fresnel, who was trained as a civil engineer, submitted an essay on the diffraction of light to the French Academy of Sciences in response to a prize competition. He was then 30 years old. Siméon Poisson, a member of the prize committee and a distinguished mathematical physicist, was a confirmed believer in Newton's particle theory of light and totally disbelieved in the wave theory of Fresnel and others. He pointed out (as Fresnel had not) that Fresnel's theory predicted the existence of the bright spot that now bears Fresnel's name. Laboratory confirmation quickly followed. Although Fresnel won the prize, Poisson remained unconvinced of the wave theory of light and at his death, 22 years later, he still held to Newton's ideas).

11. A loud-speaker horn has a rectangular aperture 4 ft high and 1 ft wide. Will the pattern of sound intensity be broader in the horizontal plane or in the vertical?

12. We have claimed (correctly) that Maxwell's equations predict all the classical optical phenomena. Yet in Chapter 45 (Interference) and in this chapter (Diffraction) there is no mention of Maxwell's equations. Is there an inconsistency here? Where is the impact of Maxwell's equations felt? Discuss.

13. A radar antenna is designed to give accurate measurements of the height of an aircraft but only reasonably good measurements of its direction in a horizontal plane. Must the height-to-width ratio of the radar antenna be less than, equal to, or greater than unity?

14. In a single-slit Fraunhofer diffraction, what is the effect of increasing (a) the wavelength and (b) the slit width?

15. Sunlight falls on a single slit of width 10^3 nm. Describe qualitatively what the resulting diffraction pattern looks like.

16. In Fig. 46-5 rays r_1 and r_3 are in phase; so are r_2 and r_4. Why isn't there a *maximum* intensity at P_2 rather than a minimum?

17. Describe what happens to a Fraunhofer single-slit diffraction pattern if the whole apparatus is immersed in water.

18. When we speak of diffraction by a single "slit" we imply that the width of the slit must be much less than its height. Suppose that, in fact, the height was equal to twice the width. Make a rough guess at what the diffraction pattern would look like? An exact solution is of course not asked for.

19. In Fig. 46-3 we stated, correctly, that the optical path lengths from the slit to point P_0 are all the same. Why? Also, why are the optical path lengths of the rays in Fig. 46-5 from the slit to point P_2 *not* the same? What is the difference between these two figures?

20. In Fig. 46-7d why is E_θ, which represents the first maximum beyond the central maximum, not vertical? (Hint: Consider the effects of a slight winding or unwinding of the coil of vectors in this figure.)

21. Distinguish clearly between θ, α, and ϕ in Eq. 46-8c.

22. If we were to redo our analysis of the properties of lenses in Section 44-5 by the methods of geometrical optics but *without* restricting our considerations to paraxial rays and to "thin" lenses, would diffraction phenomena, such as that of Fig. 46-10, emerge from the analysis? Discuss.

23. Give at least two reasons why the usefulness of large reflecting telescopes increases as we increase the mirror diameter.

24. We have seen that diffraction limits the resolving power of optical telescopes (see Fig. 46-11). Does it also do so for radio telescopes, such as that of Fig. 41-2?

25. Distinguish carefully between interference and diffraction in Young's double-slit experiment.

26. In what way are interference and diffraction similar? In what way are they different?

27. In double-slit interference patterns such as that of Fig. 46-14a we said that the interference fringes were modulated in intensity by the diffraction pattern of a single slit. Could we reverse this statement and say that the diffraction pattern of a single slit is intensity-modulated by the interference fringes? Discuss.

SECTION 46-1

problems

1. *Babinet's principle:* A monochromatic beam of parallel light is incident on a "collimating" hole of diameter $x \gg \lambda$. Point P lies in the geometrical shadow region on a distance screen, as shown in Fig. 46-16a. Two obstacles, shown in Fig. 46-16b, are placed in turn over the collimating hole. A is an opaque circle with a hole in it and B is the "photographic negative" of A. Using superposition concepts, show that the intensity at P is identical for each of the two diffracting objects A and B.*

(a)   (b)

figure 46-16
Problem 1

SECTION 46-2

2. If the distance between the first and fifth minima of a single slit pattern is 0.35 mm with the screen 40 cm away from the slit and using light having a wavelength of 550 nm, what is the width of the slit?

3. In a single-slit diffraction pattern the distance between the first minimum on the right and the first minimum on the left is 5.2 mm. The screen on which the pattern is displayed is 80 cm from the slit and the wavelength is 5460 Å. Calculate the slit width.　　　*Answer:* 0.17 mm.

4. For a single slit the first minimum occurs at $\theta = 90°$, thus filling the forward hemisphere beyond the slit with light. What must be the ratio of the slit width to the wavelength be for this to take place?

5. A slit 1.0 mm wide is illuminated by light of wavelength 589 nm = 5890 Å. We see a diffraction pattern on a screen 3.0 m away. What is the distance

*In this connection, it can be shown that the diffraction pattern of a wire is that of a slit of equal width. See "Measuring the Diameter of a Hair by Diffraction" by S. M. Curry and A. L. Schawlow, *American Journal of Physics*, May 1974.

between the first two diffraction minima on either side of the central diffraction maximum?   *Answer:* 1.8 mm.

6. A plane wave ($\lambda = 590$ nm) falls on a slit with $a = 0.40$ mm. A converging lens ($f = +70$ cm) is placed behind the slit and focuses the light on a screen. What is the linear distance on the screen from the center of the pattern to (a) the first minimum and (b) the second minimum?

SECTION 46-4

7. What is the half-width of a diffracted beam for a slit whose width is (a) 1, (b) 5, and (c) 10 wavelengths?   *Answer:* (a) 53°, (b) 10°, (c) 5.1°.

8. A single slit is illuminated by light whose wavelengths are λ_a and λ_b, so chosen that the first diffraction minimum of λ_a coincides with the second minimum of λ_b. (a) What relationship exists between the two wavelengths? (b) Do any other minima in the two patterns coincide?

9. In Fig. 46-7d, calculate the angle E_θ makes with the vertical, assuming the slit to be divided into infinitesimal strips of width dx. See Question 20.   *Answer:* 12°.

10. (a) Show that the values of α at which intensity maxima for single-slit diffraction occur can be found exactly by differentiating Eq. 46-8b with respect to α and equating to zero, obtaining the condition

$$\tan \alpha = \alpha.$$

(b) Find the values of α satisfying this relation by plotting graphically the curve $y = \tan \alpha$ and the straight line $y = \alpha$ and finding their intersections; alternatively, use a pocket calculator. (c) Find the (nonintegral) values of m corresponding to successive maxima in the single-slit pattern. Note that the secondary maxima do not lie exactly halfway between minima.

11. In Example 4 solve the transcendental equation

$$\frac{1}{2} = \left(\frac{\sin \alpha_x}{\alpha_x}\right)^2$$

graphically for α_x, to an accuracy of three significant figures.
    Answer: 79.7°.

12. If you double the width of a single slit the intensity of the central maximum of the diffraction pattern increases by a factor of four, even though the energy passing through the slit only doubles. Can you explain this quantitatively?

SECTION 46-5

13. The two headlights of an approaching automobile are 140 cm apart. At what maximum distance will the eye resolve them? Assume a pupil diameter of 5.0 mm and $\lambda = 550$ nm. Assume also that diffraction effects alone limit the resolution.   *Answer:* 10 km (= 6.3 mi).

14. An astronaut in a satellite claims he can just barely resolve two point sources on the earth, 160 km below him. What is their separation, assuming ideal conditions? Take $\lambda = 550$ nm, and the pupil diameter to be 5.0 mm.

15. How closely may two small objects be located if they are to be resolved when viewed through the telescope of a transit having a 3.0-cm objective lens if the transit is 370 m from the objects? Take the wavelength to be 550 nm.   *Answer:* 8.3 mm.

16. (a) How small is the angular separation of two stars if their images are barely resolved by the Thaw refracting telescope at the Allegheny Observatory in Pittsburgh? The lens diameter is 30 in. and its focal length is 46 ft. Assume $\lambda = 550$ nm. (b) Find the distance between these barely resolved stars if each of them is 10 light years distant from the earth. (c) For the image of a single star in this telescope, find the diameter of the first dark ring in the diffraction pattern, as measured on a photographic plate placed at the focal plane.

Assume that the star image structure is associated entirely with diffraction at the lens aperture and not with (small) lens "errors," etc.

17. The wall of a large room is covered with acoustic tile in which small holes are drilled 5.0 mm from center to center. How far can a person be from such a tile and still distinguish the individual holes, assuming ideal conditions? Assume the diameter of the pupil to be 4.0 mm and λ to be 550 nm.
Answer: 37 m.

18. Find the separation of two points on the moon's surface that can just be resolved by the 200-in. telescope at Mount Palomar, assuming that this distance is determined by diffraction effects. The distance from the earth to the moon is 3.8×10^5 km. Assume $\lambda = 550$ nm $(= 5500 \text{ Å})$.

19. Under ideal conditions, estimate the linear separation of two objects on the planet Mars which can just be resolved by an observer on earth (a) using the naked eye, and (b) using the 200-in. Mt. Palomar telescope. Use the following data: distance to Mars $= 8.0 \times 10^7$ km; diameter of pupil $= 5.0$ mm; wavelength of light $= 550$ nm. *Answer:* (a) 1.1×10^4 km. (b) 11 km.

20. A "spy in the sky" satellite orbiting at, say, 100 mi above the earth's surface, has a lens with a focal length of 8.0 ft. Its resolving power for objects on the ground is 1.2 ft, that is, it could easily detect an automobile. What is the effective lens diameter, determined by diffraction considerations alone? Assume $\lambda = 550$ nm. Even more effective satellites are reported to be in operation today. See "Reconnaissance and Arms Control" by Ted Greenwood, *Scientific American* February 1973.

21. (a) A circular diaphragm 0.60 m in diameter oscillates at a frequency of 25 kHz in an underwater source of sound for submarine detection. Far from the source the sound intensity is distributed as a Fraunhofer diffraction pattern for a circular hole whose diameter equals that of the diaphragm. Take the speed of sound in water to be 1450 m/s and find the angle between the normal to the diaphragm and the direction of the first minimum. (b) Repeat for a source having an (audible) frequency of 1.0 kHz.
Answer: (a) 6.7°. (b) Because $1.22\lambda > d$, there is no answer for 1.0 kHz; explain.

22. It can be shown that, except for $\theta = 0$, a circular obstacle produces the same diffraction pattern as a circular hole of the same diameter. Furthermore, if there are many such obstacles, water droplets, say, located randomly, then the interference effects vanish leaving only the diffraction associated with a single obstacle. (a) Explain why one sees a "ring" around the moon on a foggy night. Usually the ring is reddish in color: explain. (b) Calculate the size of the water droplet in the air if the ring around the moon appears to have a diameter 1.5 times that of the moon. (c) At what distance from the moon might a bluish ring be seen? Sometimes the rings are white: explain. (d) The color arrangement is opposite to that in a rainbow: why should this be so?

23. Suppose that, as in Example 8, the envelope of the central peak contains eleven fringes. How many fringes lie between the first and second minima of the envelope? *Answer:* 5.

24. For $d = 2a$ in Fig. 46-17, how many interference fringes lie in the central diffraction envelope?

25. If we put $d = a$ in Fig. 46-17, the two slits coalesce into a single slit of width $2a$. Show that Eq. 46-16 reduces to the diffraction pattern for such a slit.

26. Construct qualitative vector diagrams like those of Fig. 46-7 for the double slit interference pattern. For simplicity consider d = 2a (see Fig. 46-17). Can you interpret the main features of the intensity pattern this way?

27. (a) How many complete fringes appear between the first minima of the fringe envelope to either side of the central maximum for a double slit pattern if $\lambda = 550$ nm, $d = 0.15$ mm and $a = 0.030$ mm? (b) What is the ratio of the intensity of the third fringe to the side of the center to that of the central fringe? *Answer:* (a) 9. (b) 0.25.

figure 46-17
Problems 24, 25, and 26

28. (*a*) Design a double-slit system in which the fourth fringe, not counting the central maximum, is missing. (*b*) What other fringes, if any, are also missing?

figure 46-18
Problem 29

29. An acoustic double-slit system (slit separation d, slit width a) is driven by two loudspeakers as shown in Fig. 46-18. By use of a variable delay line, the phase of one of the speakers may be varied. Describe in detail what changes occur in the intensity pattern at large distances as this phase difference is varied from zero to 2π. Take both the interference and diffraction effects into account.

47
gratings and spectra

In connection with Young's experiment (Sections 45-1 and 45-3) we discussed the interference of two coherent waves formed by diffraction at two elementary radiators (pinholes or slits). In our first treatment we assumed that the slit width was much less than the wavelength, so that light diffracted from each slit illuminated the observation screen essentially uniformly. Later, in Section 46-6, we took the slit width into account and showed that the intensity pattern of the interference fringes is modulated by a "diffraction factor" $(\sin \alpha/\alpha)^2$ (see Eq. 46-16).

Here we extend our treatment to cases in which the number N of radiators or diffracting centers is larger—and usually much larger—than two. We consider two situations:

1. An array of N parallel equidistant slits, called a *diffraction grating.*
2. A three-dimensional array of periodically arranged radiators—the atoms in a crystalline solid such as NaCl, for example. In this case the average spacing between the elementary radiators is so small that interference effects must be sought at wavelengths much smaller than those of visible light. We speak of *X-ray diffraction.*

In each case we distinguish carefully between the diffracting properties of a single radiator (slit or atom) and the interference of the waves diffracted, coherently, from the assembly of radiators.

A logical extension of Young's double-slit interference experiment is to increase the number of slits from two to a larger number N. An arrangement like that of Fig. 47-1, usually involving many more slits (as many as 10^4 slits/cm is not uncommon), is called a *diffraction grating.* As for

47-1
INTRODUCTION

47-2
MULTIPLE SLITS

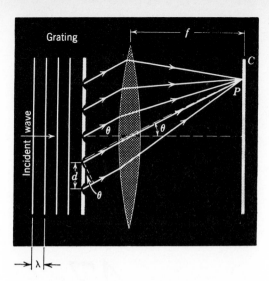

<figure>
figure 47-1
An idealized diffraction grating containing five slits. The slit width a is shown for convenience to be considerably smaller than λ, although this condition is not realized in practice. The figure is also distorted in that f is much greater than d in practice.
</figure>

a double slit, the intensity pattern that results when monochromatic light of wavelength λ falls on a grating consists of a series of interference fringes. The *angular separations* of these fringes are determined by the ratio λ/d, where d is the spacing between the centers of adjacent slits. The relative *intensities* of these fringes are determined by the diffraction pattern of a single grating slit, which depends on the ratio λ/a, where a is the slit width.

Figure 47-2, which compares the intensity patterns for $N = 2$ and $N = 5$, shows clearly that the "interference" fringes are modulated in intensity by a "diffraction" envelope, as in Fig. 46-14. Figure 47-3 presents a theoretical calculation of the intensity patterns for three fringes near the centers of the patterns of Fig. 47-2. These two figures show that increasing N (a) does not change the spacing between the (principal) interference fringe maxima, provided d and λ remain unchanged, (b) sharpens the (principal) maxima, and (c) introduces small secondary maxima between the principal maxima. Three such secondaries are present (but not readily visible) between each pair of adjacent principal maxima in Fig. 47-2b.

A principal maximum in Fig. 47-1 will occur when the path difference between rays from adjacent slits ($=d \sin \theta$) is given by

$$d \sin \theta = m\lambda \qquad m = 0, 1, 2, \ldots \qquad \text{(principal maxima),} \quad (47\text{-}1)$$

where m is called the *order number*. This equation is identical with Eq. 45-1, which locates the intensity maxima for a double slit. The loca-

<figure>
figure 47-2
Intensity patterns for "gratings" with (a) $N = 2$ and (b) $N = 5$ for the same value of d and λ. Note how the intensities of the fringes are modulated by a diffraction envelope as in Fig. 46-14; thus the assumption $a \ll \lambda$ is not realized in these actual "gratings." For $N = 5$ three very faint secondary maxima, not visible in this photograph, appear between each pair of adjacent primary maxima.
</figure>

<figure>
figure 47-3
Calculated intensity patterns for (a) a two-slit and (b) a five-slit grating for the same value of d and λ. This figure shows the sharpening of the principal maxima and the appearance of faint secondary maxima for $N > 2$. The letters on the five-slit pattern refer to Fig. 47-5. Three fringes only, centered around $\theta = 0$ in Fig. 47-1, are shown; compare Fig. 47-2. We may thus assume essentially equal intensities for each of the principal maxima shown.
</figure>

tions of the (principal) maxima are thus determined only by the ratio λ/d and are independent of N. As for the double slit, the ratio a/λ determines the relative *intensities* of the principal maxima but does not alter their locations appreciably.

The sharpening of the principal maxima as N is increased can be understood by a graphical argument, using phasors. Figures 47-4a and 47-4b show conditions at any of the principal maxima for a two-slit and a nine-slit grating. The small arrows represent the amplitudes of the wave disturbances arriving at the screen at the position of each principal maximum. For simplicity we consider the central principal maximum only, for which $m = 0$, and thus $\theta = 0$, in Eq. 47-1.

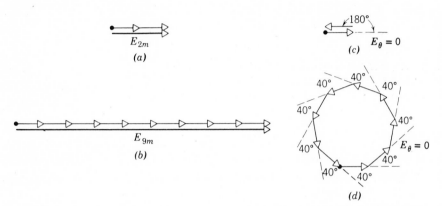

figure 47-4
Drawings (a) and (b) show conditions at the central principal maximum for a two-slit and a nine-slit grating, respectively. Drawings (c) and (d) show conditions at the minimum of zero intensity that lies on either side of this central principal maximum. In going from (a) to (c) the phase shift between waves from adjacent slits changes by 180° ($\Delta\phi = 2\pi/2$); in going from (b) to (d) it changes by 40° ($\Delta\phi = 2\pi/9$).

Consider the angle $\Delta\theta_0$ corresponding to the position of zero intensity that lies on either side of the central principal maximum. Figures 47-4c and 47-4d show the phasors at this point. The phase difference between waves from adjacent slits, which is zero at the central principal maximum, must increase by an amount $\Delta\phi$ chosen so that the array of phasors just closes on itself, yielding zero resultant intensity. For $N = 2$, $\Delta\phi = 2\pi/2$ (=180°); for $N = 9$, $\Delta\phi = 2\pi/9$ (=40°). In the general case it is given by

$$\Delta\phi = \frac{2\pi}{N}.$$

This increase in phase difference for adjacent waves corresponds to an increase in the path difference Δl given by

$$\frac{\text{phase difference}}{2\pi} = \frac{\text{path difference}}{\lambda},$$

or

$$\Delta l = \left(\frac{\lambda}{2\pi}\right)\Delta\phi = \left(\frac{\lambda}{2\pi}\right)\left(\frac{2\pi}{N}\right) = \frac{\lambda}{N}.$$

From Fig. 47-1, however, the path difference Δl at the first minimum is also given by $d \sin \Delta\theta_0$, so that we can write

$$d \sin \Delta\theta_0 = \frac{\lambda}{N},$$

or

$$\sin \Delta\theta_0 = \frac{\lambda}{Nd}.$$

Since $N \gg 1$ for actual gratings, $\sin \Delta\theta_0$, will ordinarily be quite small (that is, the lines will be sharp), and we may replace it by $\Delta\theta_0$, expressed in radians, to good approximation, or

$$\Delta\theta_0 = \frac{\lambda}{Nd} \qquad \text{(central principal maximum).} \qquad (47\text{-}2)$$

This equation shows specifically that if we increase N for a given λ and d, $\Delta\theta_0$ will decrease, which means that the central principal maximum becomes sharper.

We state without proof,* and for later use, that for principal maxima other than the central one (that is, for $m \neq 0$) the angular distance between the position θ_m of the principal maximum of order m and the minimum that lies on either side is given by

$$\Delta\theta_m = \frac{\lambda}{Nd \cos \theta_m} \qquad \text{(any principal maximum).} \qquad (47\text{-}3)$$

For the central principal maximum we have $m = 0$, $\theta_m = 0$, and $\Delta\theta_m = \Delta\theta_0$, so that Eq. 47-3 reduces, as it must, to Eq. 47-2.

The origin of the secondary maxima that appear for $N > 2$ can also be understood using the phasor method. Figure 47-5a shows conditions at the central principal maximum for a five-slit grating. The vectors are in phase. As we depart from the central maximum, θ in Fig. 47-1 increases from zero and the phase difference between adjacent vectors increases from zero to $\Delta\phi = \frac{2\pi}{\lambda} d \sin \theta$. Successive figures show how the resultant wave amplitude E_θ varies with $\Delta\phi$. Verify by graphical construction that a given figure represents conditions for both $\Delta\phi$ and $2\pi - \Delta\phi$. Thus we start at $\Delta\phi = 0$, proceed to $\Delta\phi = 180°$, and then trace backward through the sequence, following the phase differences shown in parentheses, until we reach $\Delta\phi = 360°$. This sequence corresponds to traversing the

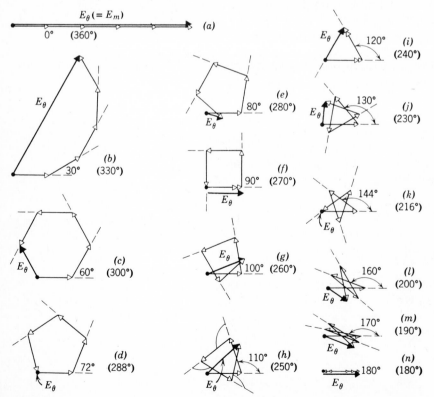

figure 47-5
The figures taken in sequence from (a) to (n) and then from (n) to (a) show conditions as the intensity pattern of a five-slit grating is traversed from the central principal maximum to an adjacent principal maximum. Phase differences between waves from adjacent slits are shown directly or, when going from (n) to (a), in parentheses. Principal maxima occur at (a), secondary maxima at, or near, (h) and (n), and points of zero intensity at (d) and (k). Compare Fig. 47-3b.

* See Problem 10.

intensity pattern from the central principal maximum to an adjacent one. Figure 47-5, which should be compared with Fig. 47-3*b*, shows that for $N = 5$ there are three secondary maxima, corresponding to $\Delta\phi = 110°$, 180°, and 250°. Make a similar analysis for $N = 3$ and show that only one secondary maximum occurs. In actual gratings, which commonly contain 10,000 to 50,000 "slits," the secondary maxima lie so close to the principal maxima or are so reduced in intensity that they cannot be distinguished from them experimentally.

The *grating spacing d* for a typical grating that contains 12,000 "slits" distributed over a 1-in. width is 2.54 cm/12,000, or 2100 nm. Gratings are often used to measure wavelengths and to study the structure and intensity of spectrum lines. Few devices have contributed more to our knowledge of modern physics.

Gratings are made by ruling equally spaced parallel grooves on a glass or a metal* plate, using a diamond cutting point whose motion is automatically controlled by an elaborate ruling engine. Once such a master grating has been prepared, replicas can be formed by pouring a collodion solution on the grating, allowing it to harden, and stripping it off. The stripped collodion, fastened to a flat piece of glass or other backing, forms a good grating.

Figure 47-6 shows a cross section of a common type of grating ruled on glass. In the rudimentary grating of Fig. 47-1 open slits were separated by opaque strips; the *amplitude* of the wave disturbance varied in a periodic way as light passed through the grating, dropping to zero on the opaque strips. The grating of Fig. 47-6 is transparent everywhere, so that there is little periodic change in amplitude as the grating is crossed. The effect of the rulings is to change the *optical thickness* of the grating in a periodic way, rays traversing the grating between the rulings (*b* in Fig. 47-6) containing more wavelengths than rays traversing the grating in the center of the rulings (*a* in Fig. 47-6). This results in a periodic change of *phase* as light crosses the grating at right angles to the rulings. Reflection gratings also depend for their operation on a periodic change in phase of the reflected wave as one crosses the grating, the change in amplitude under these conditions being negligible. The principal maxima for *phase gratings*, assuming that the incident light falls on the grating at right angles, is given by the same formula derived earlier for idealized amplitude or slit gratings, namely,

$$d \sin \theta = m\lambda \qquad m = 0, 1, 2\ldots,$$

where d is the distance between the rulings and the integer m is called the *order* of the particular principal maximum. Essentially all gratings used in the visible spectrum, whether of the transmission type, as in Fig. 47-6, or the reflection type, are phase gratings.

Figure 47-7 shows a simple grating spectroscope, used for viewing the spectrum of a light source, assumed to emit a number of discrete wavelengths, or *spectrum lines*. The light from source S is focused by lens L_1 on a slit S_1 placed in the focal plane of lens L_2. The parallel light emerging from collimator C falls on grating G. Parallel rays associated with a particular interference maximum occurring at angle θ fall on lens L_3, being brought to a focus in plane F-F'. The image formed in this

47-3
DIFFRACTION GRATINGS

figure 47-6
An enlarged cross section of a diffraction grating ruled on glass. Such gratings, in which the phase of the emerging wave changes as one crosses the grating, are called *phase gratings*.

---

* Gratings ruled on metal are called *reflection gratings* because the interference effects are viewed in reflected rather than in transmitted light. For an insight into the highly complex technique of preparing gratings, see "Ruling Engines," by A. G. Ingolls, *Scientific American*, June, 1952, and "Diffraction Gratings," by E. W. Palmer and J. F. Verrill, *Contemporary Physics*, May, 1968.

plane is examined, using a magnifying lens arrangement *E*, called an eyepiece. A symmetrical interference pattern is formed on the other side of the central position, as shown by the dotted lines. The entire spectrum can be viewed by rotating telescope *T* through various angles. Instruments used for scientific research or in industry are more complex than the simple arrangement of Fig. 47-7. They invariably employ photographic or photoelectric recording and are called *spectrographs.* Figure 49-12 shows a small portion of the spectrum of iron, produced by examining the light produced in an arc struck between iron electrodes, using a research type spectrograph with photographic recording. Each line in the figure represents a different wavelength that is emitted from the source.

Grating instruments can be used to make absolute measurements of wavelength, since the grating spacing *d* in Eq. 47-1 can be measured accurately with a traveling microscope. Several spectra are normally produced in such instruments, corresponding to $m = \pm 1, \pm 2$, etc., in Eq. 47-1 (see Fig. 47-8). This may cause some confusion if the spectra overlap. Further, this multiplicity of spectra reduces the recorded intensity of any given spectrum line because the available energy is divided among a number of spectra.

This disadvantage of the grating instrument can be overcome by shaping the profile of the grating grooves so that a large fraction of the light is thrown into a particular order on a particular side (for a given wavelength). This technique, called *blazing,* so alters the diffracting properties of the individual grooves (by controlling their profiles) that the light of wavelength λ diffracted by a single groove has a sharp peak of maximum intensity at a selected angle $\theta\ (\neq 0)$.

Gratings can, of course, separate wavelengths that are distributed continuously and not as sharp spectral lines. The emission from a red hot poker is an example. At the Flandrau Planetarium at the University of Arizona a "live" (m = 1) spectrum* of the visible radiation

* See "Eight Feet of the Solar Spectrum," O. Richard Norton, *Sky and Telescope,* September 1977. Even here sharp spectral lines are present. They appear as *absorption* lines (rather than *emission* lines) as the sun's radiation passes through its atmosphere. These are the so-called (dark) Fraunhofer lines. The element helium (from a Greek word meaning the sun) was discovered from an analysis of these lines. This is the only one of the elements not first discovered on earth.

from the sun is displayed to visitors; it is 8 ft long and about 5.5 in. high.

Light can also be analyzed into its component wavelengths if the grating in Fig. 47-7 is replaced by a prism. In a *prism spectrograph* each wavelength in the incident beam is deflected through a definite angle θ, determined by the index of refraction of the prism material for that wavelength. Curves such as Fig. 43-2, which gives the index of refraction of fused quartz as a function of wavelength, show that the shorter the wavelength, the larger the angle of deflection θ. Such curves vary from substance to substance and must be found by measurement. Prism instruments are not adequate for accurate *absolute* measurements of wavelength because the index of refraction of the prism material at the wavelength in question is usually not known precisely enough. Both prism and grating instruments make accurate *comparisons* of wavelength, using a suitable comparison spectrum such as that shown in Fig. 49-12, in which careful absolute determinations have been made of the wavelengths of the spectrum lines. The prism instrument has an advantage over a grating instrument in that its light energy is concentrated into a single spectrum so that brighter lines may be produced.

EXAMPLE 1

A grating with 8000 rulings/in. is illuminated with white light at perpendicular incidence. Describe the diffraction pattern. Assume that the wavelength of the light extends from 4000 to 7000 Å (= 400 to 700 nm).

The grating spacing d is 2.54 cm/8000, or 3170 nm. The central or zero-order maximum corresponds to $m = 0$ in Eq. 47-1. All wavelengths present in the incident light are superimposed at $\theta = 0$, as Fig. 47-8 shows.

The first-order diffraction pattern corresponds to $m = 1$ in Eq. 47-1. The 400 nm line occurs at an angle given by

$$\theta = \sin^{-1}\frac{m\lambda}{d} = \sin^{-1}\frac{(1)(400 \text{ nm})}{3170 \text{ nm}} = \sin^{-1} 0.126 = 7.3°.$$

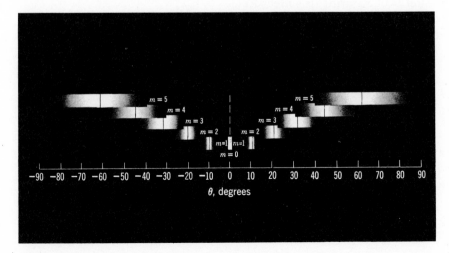

figure 47-8
Example 1. The spectrum of white light as viewed in a grating instrument like that of Fig. 47-7. The different orders, identified by the order number m, are shown separated vertically for clarity. As actually viewed, they would not be so displaced. The central line in each order corresponds to $\lambda = 550$ nm (= 5500 Å).

In the same way the angle for the 7000-A line is found to be 12.8°, and the entire pattern of Fig. 47-8 can be calculated. Note that the *first-order spectrum* ($m = 1$) is isolated but that the second-, third-, and fourth-order spectra overlap.

EXAMPLE 2

A diffraction grating has 10^4 rulings uniformly spaced over 1 in. It is illuminated at normal incidence by yellow light from a sodium vapor lamp. This light contains two closely spaced lines (the well-known *sodium doublet*) of wavelengths 5890.0 and 5895.9 Å (= 589.00 and 589.59 nm). (a) At what angle will the first-order maximum occur for the first of these wavelengths?

The grating spacing d is 10^{-4} in., or 2540 nm. The first-order maximum corresponds to $m = 1$ in Eq. 47-1. We thus have

$$\theta = \sin^{-1}\frac{m\lambda}{d} = \sin^{-1}\frac{(1)(589 \text{ nm})}{2540 \text{ nm}} = \sin^{-1}0.232 = 13.3°.$$

(b) What is the angular separation between the first-order maxima for these lines?

The straightforward way to find this separation is to repeat this calculation for $\lambda = 589.59$ nm and to subtract the two angles. A difficulty, which can best be appreciated by carrying out the calculation, is that we must carry a large number of significant figures to obtain a meaningful value for the difference between the angles. To calculate the difference in angular positions *directly*, let us write down Eq. 47-1, solved for $\sin \theta$, and differentiate it, treating θ and λ as variables:

$$\sin \theta = \frac{m\lambda}{d}$$

$$\cos \theta \, d\theta = \frac{m}{d} \, d\lambda.$$

If the wavelengths are close enough together, as in this case, $d\lambda$ can be replaced by $\Delta\lambda$, the actual wavelength difference; $d\theta$ then becomes $\Delta\theta$, the quantity we seek. This gives

$$\Delta\theta = \frac{m \, \Delta\lambda}{d \cos \theta} = \frac{(1)(0.59 \text{ nm})}{(2540 \text{ nm})(\cos 13.3°)} = 2.4 \times 10^{-4} \text{ rad} = 0.014°.$$

Note that although the wavelengths involve five significant figures, our calculation, done this way, involves only two or three, with consequent reduction in numerical manipulation.

The quantity $d\theta/d\lambda$, called the *dispersion D* of a grating, is a measure of the angular separation produced between two incident monochromatic waves whose wavelengths differ by a small wavelength interval. From this example we see that

$$D = \frac{d\theta}{d\lambda} = \frac{m}{d \cos \theta}. \tag{47-4}$$

To distinguish light waves whose wavelengths are close together, the maxima of these wavelengths formed by the grating should be as narrow as possible. Expressed otherwise, the grating should have a high *resolving power R*, defined from

47-4
RESOLVING POWER OF A GRATING

$$R = \frac{\lambda}{\Delta\lambda}. \tag{47-5}$$

Here λ is the mean wavelength of two spectrum lines that can barely be recognized as separate and $\Delta\lambda$ is the wavelength difference between them. The smaller $\Delta\lambda$ is, the closer the lines can be and still be resolved; hence the greater the resolving power R of the grating. It is to achieve a high resolving power that gratings with many rulings are constructed.

The resolving power of a grating is usually determined by the same consideration (that is, the Rayleigh criterion) that we used in Section 46-5 to determine the resolving power of a lens. If two principal maxima are to be barely resolved, they must, according to this criterion, have an angular separation $\Delta\theta$ such that the maximum of one line coincides with the first minimum of the other; see Fig. 46-11. If we apply this criterion, we can show that

$$R = Nm, \qquad (47\text{-}6)$$

where N is the total number of rulings in the grating and m is the order. As expected, the resolving power is zero for the central principal maximum ($m = 0$), all wavelengths being undeflected in this order.

Let us derive Eq. 47-6. The angular separation between two principal maxima whose wavelengths differ by $\Delta\lambda$ is found from Eq. 47-4, which we recast as

$$\Delta\theta = \frac{m \, \Delta\lambda}{d \cos\theta}. \qquad (47\text{-}4)$$

The Rayleigh criterion (Section 46-5) requires that this be equal to the angular separation between a principal maximum and its adjacent minimum. This is given from Eq. 47-3, dropping the subscript m in $\cos\theta_m$, as

$$\Delta\theta_m = \frac{\lambda}{Nd \cos\theta}. \qquad (47\text{-}3)$$

Equating Eqs. 47-4 and 47-3 leads to

$$R \; (=\lambda/\Delta\lambda) = Nm,$$

which is the desired relation.

EXAMPLE 3

In Example 2 how many rulings must a grating have if it is barely to resolve the sodium doublet in the third order?

From Eq. 47-5 the required resolving power is

$$R = \frac{\lambda}{\Delta\lambda} = \frac{589 \text{ nm}}{(589.59 - 589.00)\text{nm}} = 1000.$$

From Eq. 47-6 the number of rulings needed is

$$N = \frac{R}{m} = \frac{1000}{3} = 330.$$

This is a modest requirement.

The resolving power of a grating must not be confused with its dispersion. Table 47-1 shows the characteristics of three gratings, each illuminated with light of $\lambda = 589$ nm ($= 5890$ Å), the diffracted light being viewed in the first order ($m = 1$ in Eq. 47-1).

Table 47-1
Some characteristics of three gratings
($\lambda = 589$ nm, $m = 1$)

Grating	N	d, nm	θ	R	D 10^{-2} degrees/nm
A	10,000	2540	13.3°	10,000	2.32
B	20,000	2540	13.3°	20,000	2.32
C	10,000	1370	25.5°	10,000	4.64

figure 47-9
The intensity patterns for light of wavelengths λ_1 and λ_2 near 589 nm (= 5890 Å), incident on the gratings of Table 47-1. Grating B has the highest resolving power and grating C the highest dispersion.

You should verify that the values of D and R given in the table can be calculated from Eqs. 47-4 and 47-6, respectively.

For the conditions of use noted in Table 47-1, gratings A and B have the same *dispersion* and A and C have the same *resolving power*. Figure 47-9 shows the intensity patterns that would be produced by these gratings for two incident waves of wavelengths λ_1 and λ_2, in the vicinity of $\lambda = 589$ nm. Grating B, which has high resolving power, has narrow intensity maxima and is inherently capable of distinguishing lines that are much closer together in wavelength than those of Fig. 47-9. Grating C, which has high dispersion, produces twice the angular separation between rays λ_1 and λ_2 that grating B does.

EXAMPLE 4

The grating of Example 1 has 8000 lines illuminated by light from a mercury vapor discharge. (*a*) What is the expected dispersion, in the third order, in the vicinity of the intense green line ($\lambda = 546$ nm $= 5460$ Å)?

Noting that $d = 3170$ nm, we have, from Eq. 47-1,

$$\theta = \sin^{-1} \frac{m\lambda}{d} = \sin^{-1} \frac{(3)(546 \text{ nm})}{3170 \text{ nm}} = \sin^{-1} 0.517 = 31.1°.$$

From Eq. 47-4 we have

$$D = \frac{m}{d \cos \theta} = \frac{3}{(3170 \text{ nm})(\cos 31.1°)} = 1.1 \times 10^{-3} \text{ rad/nm} = 6.3 \times 10^{-2} \text{ deg/nm}.$$

(*b*) What is the expected resolving power of this grating in the fifth order? Equation 47-6 gives

$$R = Nm = (8000)(5) = 40,000.$$

Thus near $\lambda = 546$ nm a wavelength difference $\Delta\lambda$ given by Eq. 47-5, or

$$\Delta\lambda = \frac{\lambda}{R} = \frac{546 \text{ nm}}{40,000} = 0.014 \text{ nm},$$

can be distinguished.

47-5
X-RAY DIFFRACTION

Figure 47-10 shows how X-rays are produced when electrons from a heated filament F are accelerated by a potential difference V and strike a metal target T. X-rays are electromagnetic radiation with wavelengths of the order of 0.1 nm (or 1 Å). This value is to be compared with 550 nm

figure 47-10

X-rays are generated when electrons from heated filament F, accelerated by potential difference V, are brought to rest on striking metallic target T. W is a "window"—transparent to X-rays—in the evacuated metal container C.

(or 5500 Å) for the center of the visible spectrum. For such small wavelengths a standard optical diffraction grating, as normally employed, cannot be used. For $\lambda = 0.10$ nm and $d = 3000$ nm, for example, Eq. 47-1 shows that the first-order maximum occurs at

$$\theta = \sin^{-1}\frac{m\lambda}{d} = \sin^{-1}\frac{(1)(0.10 \text{ nm})}{3 \times 10^3 \text{ nm}} = \sin^{-1} 0.33 \times 10^{-4} = 0.002°.$$

This is too close to the central maximum to be practical. A grating with $d \cong \lambda$ is desirable, but, because X-ray wavelengths are about equal to atomic diameters, such gratings cannot be constructed mechanically.

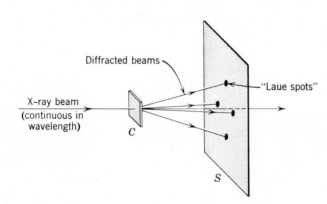

Diffracted beams

X-ray beam (continuous in wavelength)

"Laue spots"

C

S

figure 47-11

A nonmonochromatic beam of X-rays falls on a crystal C, which may be NaCl. Strong diffracted beams appear in certain directions, forming a so-called Laue pattern on a photographic film S.

In 1912 it occurred to physicist Max von Laue that a crystalline solid, consisting as it does of a regular array of atoms, might form a natural three-dimensional "diffraction grating" for X-rays. Figure 47-11 shows that if a collimated beam of X-rays, continuously distributed in wavelength, is allowed to fall on a crystal, such as sodium chloride, intense beams corresponding to constructive interference from the many diffracting centers of which the crystal is made up appear in certain sharply defined directions. If these beams fall on a photographic film, they form an assembly of "Laue spots." Figure 47-12, which is an actual example of these spots, shows that the hypothesis of Laue is indeed correct. The atomic arrangements in the crystal can be deduced from a careful study of the positions and intensities of the Laue spots* in much the same way that we might deduce the structure of an optical grating (that is, the detailed profile of its slits) by a study of the positions and intensities of the lines in the interference pattern.

* Other experimental arrangements have supplanted the Laue technique to a considerable extent today; the principle remains unchanged, however (see Question 21).

figure 47-12
Laue X-ray diffraction pattern from sodium chloride. A crystal of ordinary table salt was used in making this plate. (Courtesy of W. Arrington and J. L. Katz, X-ray Laboratory, Rensselaer Polytechnic Institute.)

Figure 47-13 shows how sodium and chlorine atoms (strictly, Na^+ and Cl^- ions) are stacked to form a crystal of sodium chloride. This pattern, which has *cubic* symmetry, is one of the many atomic arrangements exhibited by solids. The model represents the *unit cell* for sodium chloride. This is the smallest unit from which the crystal may be built up by repetition in three dimensions. You should verify that no smaller assembly of atoms possesses this property. For sodium chloride the length of the cube edge of the unit cell is 0.562737 nm.

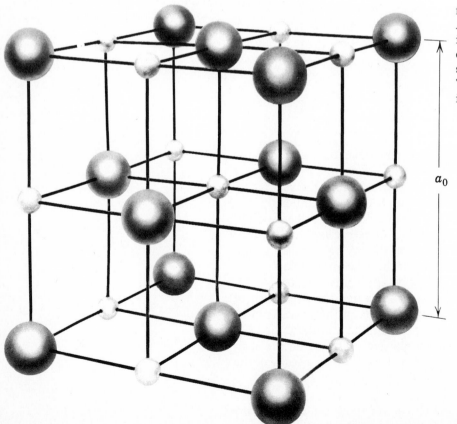

figure 47-13
A model showing how Na^+ and Cl^- ions are stacked to form a unit cell of NaCl. The small spheres represent sodium ions, the large ones chlorine. The edge a_0 of the (cubical) unit cell is 0.562737 nm.

a_0

Each unit cell in sodium chloride has four sodium ions and four chlorine ions associated with it. In Fig. 47-13 the sodium ion in the center belongs entirely to the cell shown. Each of the other twelve sodium ions shown is shared with three adjacent unit cells so that each contributes one-fourth of an ion to the cell under consideration. The total number of sodium ions is then $1 + \frac{1}{4}(12) = 4$. By similar reasoning you can show that although there are fourteen chlorine ions in Fig. 47-13, only four are associated with the unit cell shown.

The unit cell is the fundamental repetitive diffracting unit in the crystal, corresponding to the slit (and its adjacent opaque strip) in the optical diffraction grating of Fig. 47-1. Figure 47-14a shows a particular plane in a sodium chloride crystal. If each unit cell intersected by this plane is represented by a small cube, Fig. 47-14b results. You may imagine each of these figures extended indefinitely in three dimensions.

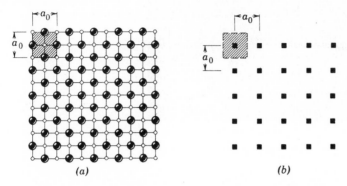

(a)   (b)

figure 47-14
(a) A section through a crystal of sodium chloride, showing the sodium and chlorine ions. (b) The corresponding unit cells in this section, each cell being represented by a small black square.

Let us treat each small cube in Fig. 47-14b as an elementary diffracting center, corresponding to a slit in an optical grating. The *directions* (but not the intensities) of all the diffracted X-ray beams that can emerge from a sodium chloride crystal (for a given X-ray wavelength and a given orientation of the incident beam) are determined by the geometry of this three-dimensional lattice of diffracting centers. In exactly the same way the *directions* (but not the intensities) of all the diffracted beams that can emerge from a particular optical grating (for a given wavelength and orientation of the incident beam) are determined only by the geometry of the grating, that is, by the grating spacing d. Representing the unit cell by what is essentially a point, as in Fig. 47-14b, corresponds to representing the slits in a diffraction grating by lines, as we did in discussing Young's experiment in Section 45-1.

The *intensities* of the lines from an optical diffraction grating are determined by the diffracting characteristics of a single slit, as Fig. 46-14 shows. In the idealized case of Fig. 47-1 these characteristics depend on the slit width a. In practical optical gratings these characteristics depend on the detailed shape of the profile of the grating rulings.

In exactly the same way the *intensities* of the diffracted beams emerging from a crystal depend on the diffracting characteristics of the unit cell.* Fundamentally the X-rays are diffracted by electrons, diffraction by nuclei being negligible in most cases. Thus the diffracting characteristics of a unit cell depend on how the electrons are distributed

---

* For some directions in which a beam might be expected to emerge, from interference considerations, no beam will be found because the diffracting characteristics of the unit cell are such that no energy is diffracted in that direction. Similarly, in optical gratings some lines, permitted by interference considerations, may not appear if their predicted positions coincide with a null in the single-slit diffraction pattern (see Fig. 46-12).

(a)

(b)

figure 47-15

(a) A photograph of an oscilloscope screen arranged to display projected electron density contours for phthalocyanine ($C_{32}H_{18}N_8$). Such plots, constructed electronically from X-ray diffraction data by an analog computer, provide a vivid picture of the structure of molecules. (Courtesy of Ray Pepinsky.) (b) A structural representation of the molecule phthalocyanine. The student should make a detailed comparison with (a), locating the various atoms identified in (b). Note that hydrogen atoms, which contain only a single electron, are not prominent in (a).

throughout the volume of the cell. By studying the *directions* of diffracted X-ray beams, we can learn the basic symmetry of the crystal. By also studying the *intensities* we can learn how electrons are distributed in the unit cell. Figure 47-15 shows the average density of electrons projected onto a particular plane through a unit cell of a crystal of phthalocyanine. This remarkable figure suggests something of the full power of X-ray methods for studying the structure of solids.

Bragg's law predicts the conditions under which diffracted X-ray beams from a crystal are possible. In deriving it, we ignore the structure of the unit cell, which is related only to the intensities of these beams. The dashed sloping lines in Fig. 47-16a represent the intersection with the plane of the figure of an arbitrary set of planes passing through the ele-

47-6
BRAGG'S LAW

mentary diffracting centers. The perpendicular distance between adjacent planes is d. Many other such families of planes, with different *interplanar spacings*, can be defined.

Figure 47-16b shows a plane wave that lies in the plane of the figure falling on one member of the family of planes defined in Fig. 47-16a, the incident rays making an angle θ with the plane.* Consider a family of diffracted rays lying in the plane of Fig. 47-16b and making an angle β with the plane containing the elementary diffracting centers. The diffracted rays will combine to produce maximum intensity if the path difference between adjacent rays is an integral number of wavelengths or

$$ae - bd = h(\cos \beta - \cos \theta) = l\lambda \qquad l = 0, 1, 2, \ldots. \qquad (47\text{-}7)$$

For $l = 0$ this leads to $\qquad \beta = \theta$,

and the plane of atoms acts like a mirror for the incident wave, no matter what the value of θ.

For other values of l, β does not equal θ, but the diffracted beam can always be regarded as being "reflected" from a *different* set of planes than that shown in Fig. 47-16a with a different interplanar spacing d. Since we wish to describe each diffracted beam as a "reflection" from a particular set of planes and since we are dealing in the present argument only with the particular set of planes shown in Fig. 47-16a, we ignore all values of l other than $l = 0$ in Eq. 47-7. It can also be shown that a plane of diffracting centers acts like a mirror (that is, $\beta = \theta$) whether or not the incident wave lies in the plane of Fig. 47-16b.

Figure 47-16c shows an incident wave striking the *family* of planes, a single member of which was considered in Fig. 47-16b. For a single plane, mirrorlike "reflection" occurs for *any* value of θ, as we have seen. To have a constructive interference in the beam diffracted from the entire family of planes in the direction θ, the rays from the separate planes must reinforce each other. This means that the path difference for rays from adjacent planes (*abc* in Fig. 47-16c) must be an integral number of wavelengths or

$$2d \sin \theta = m\lambda \qquad m = 1, 2, 3, \ldots. \qquad (47\text{-}8)$$

This relation is called *Bragg's law* after W. L. Bragg who first derived it. The quantity d in this equation (the interplanar spacing) is the perpendicular distance between the planes. For the planes of Fig. 47-16a analysis shows that d is related to the unit cell dimension a_0 by

$$d = \frac{a_0}{\sqrt{5}}. \qquad (47\text{-}9)$$

If an incident *monochromatic* X-ray beam falls at an *arbitrary* angle θ on a particular set of atomic planes, a diffracted beam will *not* result because Eq. 47-8 will not, in general, be satisfied. If the incident X-rays are *continuous* in wavelength, diffracted beams will result when wavelengths given by

$$\lambda = \frac{2d \sin \theta}{m} \qquad m = 1, 2, 3, \ldots$$

are present in the incident beam (see Eq. 47-8).

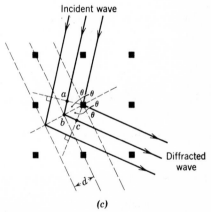

figure 47-16
(a) A section through the NaCl unit cell lattice of Fig. 45-14b. The dashed sloping lines represent an arbitrary family of planes, with interplanar spacing d. (b) An incident wave falls, at grazing angle θ, on one of the planes, xx', shown in (a). (c) An incident wave falls on the entire family of planes shown in (a). A strong diffracted wave is formed.

---

* In X-ray diffraction it is customary to specify the direction of a wave by giving the angle between the ray and the plane (the *glancing angle*) rather than the angle between the ray and the normal.

EXAMPLE 5

At what angles must an X.ray beam with $\lambda = 0.110$ nm fall on the family of planes represented in Fig. 47-16c if a diffracted beam is to exist? Assume the material to be sodium chloride.

The interplanar spacing d for these planes is given by Eq. 47-9 or

$$d = \frac{a_0}{\sqrt{5}} = \frac{0.563 \text{ nm}}{2.24} = 0.252 \text{ nm}.$$

Equation 47-8 gives

$$\sin \theta = \frac{m\lambda}{2d} = \frac{(m)(0.110 \text{ nm})}{(2)(0.252 \text{ nm})} = 0.218 \, m.$$

Diffracted beams are possible at $\theta = 12.6°$ $(m = 1)$, $\theta = 25.9°$ $(m = 2)$, $\theta = 40.9°$ $(m = 3)$, and $\theta = 60.7°$ $(m = 4)$. Higher-order beams cannot exist because they require $\sin \theta$ to exceed unity. Actually, the odd-order beams $(m = 1, 3)$ prove to have zero intensity because the unit cell in cubic crystals such as NaCl has diffracting properties such that the intensity of the light scattered in these orders is zero.

X-ray diffraction is a powerful tool for studying the arrangements of atoms in crystals.[*] To do so quantitatively requires that the wavelength of the X-rays be known. In one of several approaches to this problem the unit cell dimension for NaCl[†] is determined by a method that does not involve X-rays. X-ray diffraction measurements on NaCl can then be used to determine the wavelength of the X-ray beam which in turn can be used to determine the structures of solids other than NaCl.

If ρ is the measured density of NaCl, we have, for the unit cell of Fig. 47-13,[‡] recalling that each unit cell contains four NaCl "molecules,"

$$\rho = \frac{m}{V} = \frac{4m_{NaCl}}{a_0^3}.$$

Here m_{NaCl}, the mass of a NaCl molecule, is given by

$$m_{NaCl} = \frac{M}{N_0},$$

where M is the molecular weight of NaCl and N_0 is Avogadro's number. Combining these two equations and solving for a_0 yields

$$a_0 = \left(\frac{4M}{N_0\rho}\right)^{1/3}$$

which permits us to calculate a_0. Once a_0 is known, the wavelengths of monochromatic X-ray beams can be found, using Bragg's law (Eq. 47-8).

questions

1. Discuss this statement: "A diffraction grating can just as well be called an interference grating."
2. How would the spectrum of an enclosed source that is formed by a diffraction grating on a screen change (if at all) when the source, grating and screen are all submerged in water?

---

[*] See "X-ray Crystallography" by Sir Lawrence Bragg, *Scientific American*, July 1968.
[†] In practice, calcite $(CaCO_3)$ proves to be more useful as a standard crystal for a number of technical reasons.
[‡] This relation cannot be written down unless it is known that the structure of NaCl is cubic. This can be determined, however, by inspection of the symmetry of the spots in Fig. 47-12; the wavelength of the X-rays need not be known.

3. Assume that the limits of the visible spectrum are 4300 and 6800 Å. Is it possible to design a grating, assuming that the incident light falls normally on it, such that the first-order spectrum *barely overlaps* the second-order spectrum?

4. In a grating spectrograph, several lines having different wavelengths and formed in different orders might appear near a certain angle. How could you distinguish between their orders?

5. For the simple spectroscope of Fig. 47-7, show (a) that θ increases with λ for a grating and (b) that θ decreases with λ for a prism.

6. You are given a photograph of a spectrum on which the angular positions and the wavelengths of the spectrum lines are marked. (a) How can you tell whether the spectrum was taken with a prism or a grating instrument? (b) What information could you gather about either the prism or the grating from studying such a spectrum?

7. Explain in your own words why increasing the number of slits, N, in a diffraction grating sharpens the maxima. Why does decreasing the wavelength do so? Why does increasing the slit spacing, d, do so?

8. (a) What is a "phase grating"? (b) Can you make a diffraction grating out of parallel rows of fine wires strung closely together?

9. How much information can you discover about the structure of a diffraction grating by analyzing carefully the spectrum it forms of a monochromatic light source. Let = 589 nm, for an example.

10. (a) Why does a diffraction grating have closely spaced rulings? (b) Why does it have a large number of rulings?

11. Two nearly equal wavelengths are incident on a grating of N slits and are just not resolvable. However, they become resolved if the number of slits is increased. Formulas aside, is the explanation of this that: (a) more light can get through the grating? (b) the principal maxima become more intense and hence resolvable? (c) the diffraction pattern is spread more and hence the wavelengths become resolved? (d) there are a larger number of orders? or (e) the principal maxima become narrower and hence resolvable?

12. The relation $R = Nm$ suggests that the resolving power of a given grating can be made as large as desired by choosing an arbitrarily high order of diffraction. Discuss.

13. Show that at a given wavelength and a given angle of diffraction the resolving power of a grating depends only on its width $W (= Nd)$.

14. According to Eq. 47-3 the principal maxima become wider (that is, $\Delta\theta_m$ increases) the higher the order m (that is, the larger θ_m becomes). According to Eq. 47-6 the resolving power becomes greater the higher the order m. Explain this apparent paradox.

15. How would you *measure* (a) the dispersion D and (b) the resolving power R for either a prism or a grating spectrograph.

16. Is the pattern of Fig. 47-12 more properly described as a diffraction pattern or as an interference pattern?

17. For a given family of planes in a crystal, can the wavelength of incident X-rays be (a) too large or (b) too small to form a diffracted beam?

18. Why cannot a simple cube of edge $a_0/2$ in Fig. 47-13 be used as a unit cell for sodium chloride?

19. In some respects Bragg reflection *differs from* plane grating diffraction. Of the following statements which one is *true for Bragg reflection but not true for grating diffraction?* (a) two different wavelengths may be superposed. (b) radiation of a given wavelength may be sent in more than one direction. (c) long waves are deviated more than short waves. (d) there is only one grating spacing. (e) diffraction maxima of a given wavelength occur only for particular angles of incidence.

20. If a parallel beam of X-rays of wavelength λ is allowed to fall on a randomly oriented crystal of any material, generally no intense diffracted beams will

occur. Such beams appear if (a) the X-ray beam consists of a continuous distribution of wavelengths rather than a single wavelength or (b) the specimen is not a single crystal but a finely divided powder. Explain.

21. In Fig. 47-17(a) we show schematically the Debye-Scherrer experimental arrangement and in Fig. 47-17(b) a corresponding x-ray diffraction pattern. (a) Keeping in mind that the Laue method uses a large single crystal and an x-ray beam continuously distributed in wavelength, explain the origin of the spots. (Hint: each spot corresponds to the direction of scattering from a family of planes.)

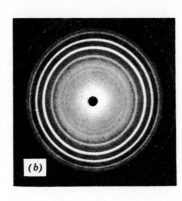

figure 47-17
Question 21 (a) and (b)
(a) A monochromatic beam of X-rays falls on an aggregate P of very small crystals oriented at random (called a microcyrstalline powder). Strong diffracted beams form rings surrounding a central spot, the so-called Debye-Scherrer rings, on a photographic film S.
(b) Debye-Scherrer X-ray diffraction pattern from zirconium oxide crystals. (From U. Fano and L. Fano, *Basic Physics of Atoms and Molecules*, John Wiley and Sons, 1959)

(b) Keeping in mind that the Debye-Scherrer method uses a large number of small single crystals randomly oriented and a monochromatic beam of x-rays, explain the origin of the rings. (Hint: because the small crystals are randomly oriented, all possible angles of incidence are obtained.)

SECTION 47-2

problems

1. Show that in a grating with alternately transparent and opaque strips of *equal* width, all the even orders (except $m = 0$) are absent.

2. The central intensity maximum formed by a grating, along with its subsidiary secondary maxima, can be viewed as the diffraction pattern of a single "slit" whose width is that of the entire grating. Treating the grating as a single wide slit, assuming that $m = 0$, and using the methods of Section 46-4, show that Eq. 47-2 can be derived.

3. A diffraction grating is made up of slits of width 300 nm with a 900 nm separation between their centers. The grating is illuminated by monochromatic plane waves, $\lambda = 600$ nm, the angle of incidence being zero. (a) How many diffracted lines are there? (b) What is the angular width of the spectral lines observed, if the grating has 1000 slits? *Angular width* is defined to be the angle between the two positions of zero intensity on either side of the maximum.   *Answer:* (a) 3, corresponding to $m = 0, \pm 1$. (b) 1.8×10^{-3} rad.

4. Assume that light is incident on a grating at an angle ψ as shown in Fig. 47-18. Show that the condition for a diffraction maximum is

$$d(\sin \psi + \sin \theta) = m\lambda \qquad m = 0, 1, 2, \ldots.$$

Only the special case $\psi = 0$ has been treated in this chapter (compare Eq. 47-1).

5. A transmission grating with $d = 1.50 \times 10^{-4}$ cm is illuminated at various angles of incidence by light of wavelength 600 nm. Plot as a function of angle of incidence (0 to 90°) the angular deviation of the first-order diffracted beam from the incident direction.

6. Derive this expression for the intensity pattern for a three-slit "grating":

figure 47-18
Problem 4

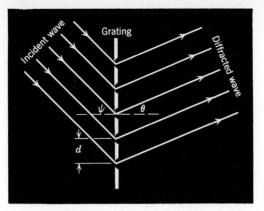

$$I_\theta = \tfrac{1}{9} I_m (1 + 4 \cos \phi + 4 \cos^2 \phi),$$

where
$$\phi = \frac{2\pi d \sin \theta}{\lambda}.$$

Assume that $a \ll \lambda$ and be guided by the derivation of the corresponding double-slit formula (Eq. 45-9).

7. (a) Using the result of Problem 6, show that the half-width of the fringes for a three-slit diffraction pattern, assuming θ small enough so that $\sin \theta \cong \theta$, is

$$\Delta\theta \cong \frac{\lambda}{3.2d}.$$

(b) Compare this with the expression derived for the two-slit pattern in Problem 16, Chapter 45. (c) Do these results support the conclusion that for a fixed slit spacing the interference maxima become sharper as the number of slits is increased?

8. (a) Using the result of Problem 6, show that a three-slit "grating" has only one secondary maximum. Find (b) its location and (c) its relative intensity.

9. A three-slit grating has separation d between adjacent slits. If the middle slit is covered up, will the half-width of the intensity maxima become broader or narrower? See Problem 7.
   Answer: Narrower by the factor 0.78.

10. Derive Eq. 47-3, that is, the expression for $\Delta\theta_m$, the angular distance between a principal maximum of order m and either adjacent minimum.

11. A diffraction grating has a large number N of slits, each of width d. Let I_{max} denote the intensity at some principal maximum, and let I_k denote the intensity of the kth adjacent secondary maximum. (a) If $k \ll N$, show from the phasor diagram that, approximately, $I_k/I_{max} = 1/(k + \tfrac{1}{2})^2 \pi^2$. (Compare this with the single-slit formula.) (b) For those secondary maxima which lie roughly midway between two adjacent principal maxima, show that roughly $I_k/I_{max} = 1/N^2$. (c) Consider the central principal maximum and those adjacent secondary maxima for which $k \ll N$. Show that this part of the diffraction pattern quantitatively resembles that for one single slit of width Nd.

SECTION 47-3
12. A diffraction grating 2.0 cm wide has 6000 rulings. At what angles will maximum-intensity beams occur if the incident radiation has a wavelength of 589 nm?

13. With light from a gaseous discharge tube incident normally on a grating with a distance 1.73×10^{-4} cm between adjacent slit centers, a green line appears with sharp maxima at measured transmission angles $\theta = \pm 17.6°$, $37.3°$, $-37.1°$, $65.2°$, and $-65.0°$. Compute the wavelength of the green line that best fits the data.   *Answer:* 524 nm.

14. A diffraction grating has 5000 rulings/in., and a strong diffracted beam is noted at $\theta = 30°$. (a) What are the possible wavelengths of the incident light? (b) What colors are they (see Figure 42-1)?

15. A grating has 3150 rulings/cm. For what wavelengths in the visible spectrum can fifth-order diffraction be observed?
    Answer: All wavelengths shorter than 635 nm.

16. Given a grating with 4000 lines/cm, how many orders of the entire visible spectrum (400–700 nm) can be produced?

17. A diffraction grating 3.00 cm wide produces a deviation of 30.0° in the second order with light of wavelength 600 nm. What is the total number of lines on the grating?   *Answer:* 12,500.

18. Light of wavelength 600 nm is incident normally on a diffraction grating. Two *adjacent* principal maxima occur at $\sin\theta = 0.2$ and $\sin\theta = 0.3$, respectively. The fourth order is a missing order. (a) What is the separation between adjacent slits? (b) What is the smallest possible individual slit width? (c) Name *all* orders actually appearing on the screen with the values chosen in (a) and (b).

19. Assume that the limits of the visible spectrum are arbitrarily chosen as 430 and 680 nm (= 4300 and 6800 Å). Design a grating that will spread the first-order spectrum through an angular range of 20°.
    Answer: 11,000 lines/cm.

20. Light containing a mixture of two wavelengths, 500 nm and 600 nm, is incident normally on a diffraction grating. It is desired (1) that the first and second principal maxima for each wavelength appear at $\theta \le 30°$, (2) that the dispersion be as high as possible, *and* (3) that the third order for 600 nm be a missing order. (a) What is the separation between adjacent slits? (b) What is the smallest possible individual slit width? (c) Name *all* orders for 600 nm that actually appear on the screen with the values chosen in (a) and (b).

21. A narrow beam of monochromatic light strikes a grating at normal incidence and produces sharp maxima at the following angles from the normal: 6°40′, 13°30′, 20°20′, 35°40′. No other maxima appear at any angle between 0° and 35°40′. The separation between adjacent slit centers in the grating is 5.04×10^{-4} cm. (a) Compute the wavelength of the light used. (b) Make the most complete *quantitative* statement that can be inferred from the above data concerning the width of each slit.
    Answer: (a) 586 nm. (b) The slit width lies between 1.01 and 1.68 μm.

22. A grating designed for use in the infrared region of the electromagnetic spectrum is "blazed" to concentrate all its intensity in the first order $(m = 1)$ for $\lambda = 8,000$ nm. If visible light (400 nm $< \lambda <$ 700 nm) were allowed to fall on this grating, what visual appearance would the diffracted beams present?

23. Show that the dispersion of a grating can be written as

$$D = \frac{\tan\theta}{\lambda}.$$

24. A grating has 3500 rulings/cm and is illuminated at normal incidence by white light. A spectrum is formed on a screen 30 cm from the grating. If a 1.0-cm square hole is cut in the screen, its inner edge being 5.0 cm from the central maximum, what range of wavelengths passes through the hole?

25. Two spectral lines have wavelengths λ and $\lambda + \Delta\lambda$, respectively, where $\Delta\lambda \ll \lambda$. Show that their angular separation $\Delta\theta$ in a grating spectrometer is given approximately by $\Delta\theta = \Delta\lambda/\sqrt{(d/m)^2 - \lambda^2}$, where d is the separation of adjacent slit centers and m is the order at which lines are observed. Notice that the angular separation is greater in the higher orders.

26. An optical grating with a spacing $d = 15,000$ Å is used to analyze soft X-rays of wavelength $\lambda = 5.0$ Å. The angle of incidence θ is $90° - \gamma$, where γ is a *small* angle. The first-order maximum is found at an angle $\theta = 90° - 2\beta$. Find the value of β.

SECTION 47-4

27. A grating has 6000 rulings/cm and is 6.0 cm wide. (*a*) What is the smallest wavelength interval that can be resolved in the third order at $\lambda = 500$ nm? (*b*) How many higher orders can be seen? Assume normal incidence of light on the grating throughout. *Answer:* (*a*) 4.6×10^{-3} nm. (*b*) None.

28. A grating has 40,000 rulings spread over 3.0 in. (*a*) What is its expected dispersion D for sodium light ($\lambda = 589$ nm $= 5890$ Å) in the first three orders? (*b*) What is its resolving power in these orders?

29. A source containing a mixture of hydrogen and deuterium atoms emits a red doublet at $\lambda = 656.3$ nm ($= 6563$ Å) whose separation is 0.18 nm ($= 1.8$ Å). Find the minimum number of lines needed in a diffraction grating which can resolve these lines in the first order. *Answer:* 3650.

30. In a particular grating the sodium doublet (see Example 2) is viewed in third order at 10° to the normal and is barely resolved. Find (*a*) the grating spacing and (*b*) the total width of the rulings.

31. A diffraction grating has a resolving power $R = \lambda/\Delta\lambda = Nm$. (*a*) Show that the corresponding frequency range $\Delta\nu$ that can just be resolved is given by $\Delta\nu = c/Nm\lambda$. (*b*) From Fig. 47-1, show that the "times of flight" of the two extreme rays differ by an amount $\Delta t = Nd \sin \theta/c$. (*c*) Show that $(\Delta\nu)(\Delta t) = 1$, this relation being independent of the various grating parameters. Assume $N \gg 1$.

32. A diffraction grating is made up of slits of width 300 nm with a 900 nm separation between centers. The grating is illuminated by monochromatic plane waves, $\lambda = 600$ nm, the angle of incidence being zero. (*a*) How many diffracted lines are there? (*b*) What is the angular width of the spectral lines observed, if the grating has 1000 slits? (*c*) How is the angular width of the spectral lines related to the resolving power of the grating?

SECTION 47-5

33. Consider an infinite two-dimensional square lattice as in Fig. 47-14b. One interplanar spacing is obviously a_0 itself. (*a*) Calculate the next five smaller interplanar spacings by sketching figures similar to Figure 47-16a. (*b*) Show that your answers obey the general formula

$$d = a_0/\sqrt{h^2 + k^2}$$

where h, k are both relatively prime integers (no common factor other than unity).

SECTION 47-6

34. Monochromatic high energy x-rays are incident on a crystal. If first order reflection is observed at Bragg angle 3.4°, at what angle would second order reflection be expected?

35. Monochromatic X-rays are incident on a NaCl crystal whose lattice spacing is 0.3 nm. When the beam is rotated 60° from the normal, first-order Bragg reflection is observed. What is the wavelength of the x-rays?
Answer: 0.30 nm.

36. In comparing the wavelengths of two monochromatic X-ray lines, it is noted that line *A* gives a first-order reflection maximum at a glancing angle of 30° to the smooth face of a crystal. Line *B*, known to have a wavelength of 0.097 nm, gives a third-order reflection maximum at an angle of 60° from the same face of the same crystal. Find the wavelength of line *A*.

37. Monochromatic X-rays ($\lambda = 1.25$ Å) fall on a crystal of sodium chloride, making an angle of 45° with the reference line shown in Fig. 47-19. The planes shown are those of Fig. 47-16a, for which $d = 2.52$ Å. Through what angles must the crystal be turned to give a diffracted beam associated with the planes shown? Assume that the crystal is turned about an axis that is perpendicular to the plane of the page. Ignore the possibility (see Problem 40) that some of these beams may be of zero intensity.
Answer: 30.6°, 15.3°, (clockwise); 3.1°, 37.8°, (counterclockwise).

figure 47-19
Problems 37, 38

38. Assume that the incident X-ray beam in Fig. 47-19 is not monochromatic but contains wavelengths in a band from 0.095 to 0.130 nm. Will diffracted beams, associated with the planes shown, occur? Assume $d_0 = 0.275$ nm.

39. Prove that it is impossible to determine both wavelength of radiation and spacing of Bragg reflecting planes in a crystal by measuring the angles for Bragg reflection in several orders.

40. *Missing orders in X-ray diffraction.* In Example 5 the $m = 1$ beam, permitted by interference considerations, has zero intensity because of the diffracting properties of the unit cell for this geometry of beams and crystal. Prove this. (*Hint:* Show that the "reflection" from an atomic plane through the top of a layer of unit cells is canceled by a "reflection" from a plane through the middle of this layer of cells. All odd-order beams prove to have zero intensity.)

48
polarization

Light, like all electromagnetic radiation, is predicted by electromagnetic theory to be a *transverse wave*, the directions of the vibrating electric and magnetic vectors being at right angles to the direction of propagation instead of parallel to it, as in a longitudinal wave. The transverse waves of Figs. 48-1 and 41-13 have the additional characteristic that they are *plane-polarized*. This means that the vibrations of the **E** vector are parallel to each other for all points in the wave. At any such point the vibrating **E** vector and the direction of propagation form a plane, called the *plane of vibration*; in a plane-polarized wave all such planes are parallel.

The transverse nature of light waves cannot be deduced from the interference or diffraction experiments so far described because longitudinal waves such as sound waves also show these effects. Thomas Young in 1817 provided an experimental basis for believing that light waves are transverse. Two of his contemporaries, Dominique-François Arago (1786–1853) and Augustin Jean Fresnel (1788–1827), were able, by allowing a light beam to fall on a crystal of calcite, to produce two separate

48-1
POLARIZATION

figure 48-1
An instantaneous "snapshot" of a plane-polarized wave showing the vectors **E** and **B** along a particular ray. The wave is moving to the right with speed *c*. The plane containing the vibrating **E** vector and the direction of propagation is a *plane of vibration.*

beams (see Section 48-4). Astonishingly, these beams, although co-herent, produced no interference fringes but only a uniform illumina-tion. Young deduced from this that light must be a transverse wave and that the planes of vibration in the two beams must be at right angles to each other. Wave disturbances that act at right angles to each other can-not show interference effects; you are asked to prove this in Problem 1. Young's words to Arago were these:

I have been reflecting on the possibility of giving an imperfect explanation of the affection of light which constitutes polarization without departing from the genuine doctrine of undulations. It is a principle in this theory that all undula-tions are simply propagated through homogeneous mediums in concentric spherical surfaces like the undulations of sound, consisting simply in the direct and retrograde motions of the particles in the direction of the radius with their concomitant condensation and rarefactions [that is, longitudinal waves]. And yet, it is possible to explain in this theory a transverse vibration, propagated also in the direction of the radius, and with equal velocity, the motions of the particles being in a certain constant direction with respect to that radius; and this is a *polarization*.

Note how Young presents the possibility of a transverse vibration as a novel idea, light having been generally — but incorrectly — assumed to be a longitudinal vibration.

In a plane-polarized transverse wave it is necessary to specify two directions, that of the wave disturbance (**E,** say) and that of propagation. In a longitudinal wave these directions are identical. In plane-polarized transverse waves, but not in longitudinal waves, we may thus expect a lack of symmetry about the direction of propagation. Electromagnetic waves in the radio and microwave range exhibit this lack of symmetry readily. Such a wave, generated by the surging of charge up and down in the dipole that forms the transmitting antenna of Fig. 48-2, has (at large distances from the dipole and at right angles to it) an electric field vector parallel to the dipole axis. When this plane-polarized wave falls on a second dipole connected to a microwave detector, the alternating elec-tric component of the wave will cause electrons to surge back and forth in the receiving antenna, producing a reading on the detector. If we turn the receiving antenna through 90° about the direction of propagation, the detector reading drops to zero. In this orientation the electric field vector is not able to cause charge to move along the dipole axis because it points at right angles to this axis. We can reproduce the experiment of Fig. 48-2 by turning the receiving antenna of a television set (assumed an electric dipole type) through 90° about an axis that points toward the transmitting station.

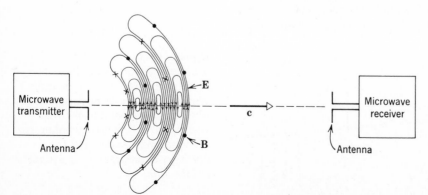

figure 48-2
The vectors **E** in the transmitted wave are parallel to the axis of the receiving antenna so that the wave will be detected. If the receiving antenna is rotated through 90° about the direction of propagation, essentially no signal will be detected.

Common sources of visible light differ from radio and microwave sources in that the elementary radiators, that is, the atoms and molecules, act independently. The light propagated in a given direction consists of independent wavetrains whose planes of vibration are randomly oriented about the direction of propagation, as in Fig. 48-3b. Such light, though still transverse, is *unpolarized*. The random orientation of the planes of vibration produces symmetry about the propagation direction, which, on casual study, conceals the true transverse nature of the waves. To study this transverse nature, we must find a way to unsort the different planes of vibration.

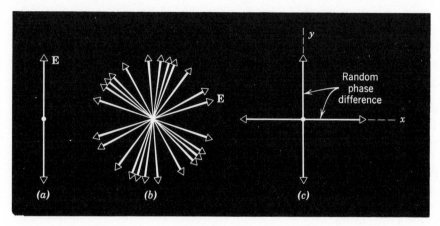

figure 48-3
(a) A plane-polarized transverse wave moving toward the reader, showing only the electric vector. (b) An unpolarized transverse wave viewed as a random superposition of many plane-polarized wavetrains. (c) A second, completely equivalent, description of an unpolarized transverse wave; here the unpolarized wave is viewed as two plane-polarized waves with a random phase difference. The orientation of the x and y axes about the propagation direction is completely arbitrary.

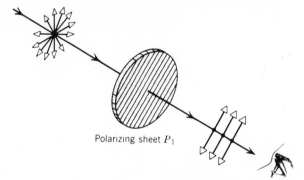

Polarizing sheet P_1

figure 48-4
A polarizing sheet produces plane-polarized light from unpolarized light. The parallel lines, which are not actually visible on the sheet, suggest the characteristic polarizing direction of the sheet.

48-2
POLARIZING SHEETS

Figure 48-4 shows unpolarized light falling on a sheet of commercial polarizing material called *Polaroid*.* There exists in the sheet a certain characteristic polarizing direction, shown by the parallel lines. The sheet will transmit only those wave-train components whose electric vectors vibrate parallel to this direction and will absorb those that vibrate at right angles to this direction. The emerging light will be plane-polarized. This polarizing direction is established during the manufacturing process by embedding certain long-chain molecules in a flexible plastic sheet and then stretching the sheet so that the molecules are aligned parallel to each other. Polarizing sheets more than 2 ft wide and 100 ft long may be produced.

In Fig. 48-5 the polarizing sheet or *polarizer* lies in the plane of the

---

* There are other ways of producing polarized light without using this well-known commercial product. We mention some of them below. Also see "The Amateur Scientist" by Jearl Walker, *Scientific American*, December 1977 for ways of making polarizing sheets, quarter-wave and half-wave plates and doing various experiments with them.

page and the direction of propagation is into the page. The arrow **E** shows the plane of vibration of a randomly selected wavetrain falling on the sheet. Two vector components, E_x (of magnitude $E \sin \theta$) and E_y (of magnitude $E \cos \theta$), can replace **E,** one parallel to the polarizing direction and one at right angles to it. Only the former will be transmitted; the latter is absorbed within the sheet.

Let us place a second polarizing sheet P_2 (usually called, when so used, an *analyzer*) as in Fig. 48-6. If P_2 is rotated about the direction of propagation, there are two positions, 180° apart, at which the transmitted light intensity is almost zero; these are the positions in which the polarizing directions of P_1 and P_2 are at right angles.

If the amplitude of the plane-polarized light falling on P_2 is E_m, the amplitude of the light that emerges is $E_m \cos \theta$, where θ is the angle between the polarizing directions of P_1 and P_2. Recalling that the intensity of the light beam is proportional to the square of the amplitude, we see that the transmitted intensity I varies with θ according to

$$I = I_m \cos^2 \theta, \qquad (48\text{-}1)$$

in which I_m is the maximum value of the transmitted intensity. It occurs when the polarizing directions of P_1 and P_2 are parallel, that is, when $\theta = 0$ or 180°. Figure 48-7a, in which two overlapping polarizing sheets are in the parallel position ($\theta = 0$ or 180° in Eq. 48-1) shows that the light transmitted through the region of overlap has its maximum value. In Fig. 48-7b one or the other of the sheets has been rotated through 90° so that θ in Eq. 48-1 has the value 90 or 270°; the light transmitted through the region of overlap is now a minimum.

figure 48-5
A wavetrain **E** is equivalent to two component wavetrains E_y and E_x. Only E_y is transmitted by the polarizer.

figure 48-6
Unpolarized light is not transmitted by crossed polarizing sheets.

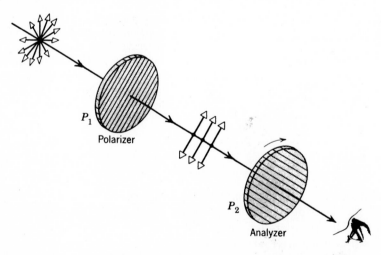

P₁
Polarizer

P₂
Analyzer

Equation 48-1, called the law of Malus, was discovered by Étienne Louis Malus (1775–1812) experimentally in 1809, using polarizing techniques other than those so far described (see Section 48-3). Equation 48-1 describes precisely the lack of symmetry about the propagation direction that must be exhibited by plane-polarized transverse waves. Longitudinal waves could not possibly show such effects. Interestingly enough the human eye, under certain conditions, can detect polarized light.*

* The so-called *Haidinger's brushes*; the interested student is referred to *Concepts of Classical Optics* by John Strong, W. H. Freeman & Co., 1958.

figure 48-7
Two square sheets of Polaroid are laid over a book. In (a) the axes of polarization of the two sheets are parallel and light passes through both sheets. In (b) one sheet has been rotated 90° and no light passes through. The book is opened to an illustration of the Luxembourg Palace in Paris. Malus discovered the phenomenon of polarization by reflection while looking at sunlight reflected off the palace windows through a calcite crystal.

EXAMPLE 1

Two polarizing sheets have their polarizing directions parallel so that the intensity I_m of the transmitted light is a maximum. Through what angle must either sheet be turned if the intensity is to drop by one-half?

From Eq. 48-1, since $I = \frac{1}{2}I_m$, we have

$$\frac{1}{2}I_m = I_m \cos^2 \theta$$

or

$$\theta = \cos^{-1} \pm \frac{1}{\sqrt{2}} = \pm 45°, \pm 135°.$$

The same effect is obtained no matter which sheet is rotated or in which direction.

Historically polarization studies were made to investigate the nature of light. Today we reverse the procedure and deduce something about the nature of an object from the polarization state of the light emitted by, or scattered from, that object. It has been possible to deduce, from studies of the polarization of light reflected from them, that the grains of cosmic dust present in our galaxy have been oriented in the weak galactic magnetic field ($\sim 5 \times 10^{-10}$ T) so that their long dimension is parallel to this field. Polarization studies have shown that Saturn's rings consist of ice crystals. The size and shape of virus particles can be determined by the polarization of ultraviolet light scattered from them. Much useful information about the structure of atoms and nuclei is gained from polarization studies of their emitted radiations in all parts of the electromagnetic spectrum. Thus we have a useful research technique for structures ranging in size from a galaxy ($\sim 10^{+22}$ m) to a nucleus ($\sim 10^{-15}$ m). Polarized light also has many practical applications in industry and in engineering science.

48-3
POLARIZATION BY REFLECTION

Malus discovered in 1809 that light can be partially or completely polarized by reflection. Anyone who has watched the sun's reflection in water, while wearing a pair of sunglasses made of polarizing sheet, has probably noticed the effect. It is necessary only to tilt the head from side to side, thus rotating the polarizing sheets, to observe that the intensity of the reflected sunlight passes through a minimum.

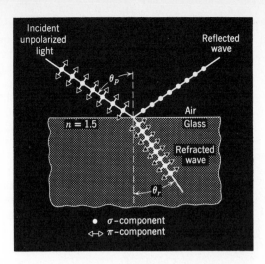

figure 48-8

For a particular angle of incidence θ_p, the reflected light is completely polarized, as shown. The transmitted light is partially polarized.

Figure 48-8 shows an unpolarized beam falling on a glass surface. The **E** vector for each wavetrain in the beam can be resolved into two components, one perpendicular to the plane of incidence—which is the plane of Fig. 48-8—and one lying in this plane. The first component, represented by the dots, is called the *σ-component,* from the German *senkrecht,* meaning perpendicular. The second component, represented by the arrows, is called the *π-component* (for *parallel*). On the average, for completely unpolarized incident light, these two components are of equal amplitude.

Experimentally, for glass or other dielectric materials, there is a particular angle of incidence, called the *polarizing angle* θ_p, at which the reflection coefficient for the π-component is zero. This means that the beam reflected from the glass, although of low intensity, is plane-polarized, with its plane of vibration at right angles to the plane of incidence. This polarization of the reflected beam can easily be verified by analyzing it with a polarizing sheet.

The π-component at the polarizing angle is entirely refracted; the

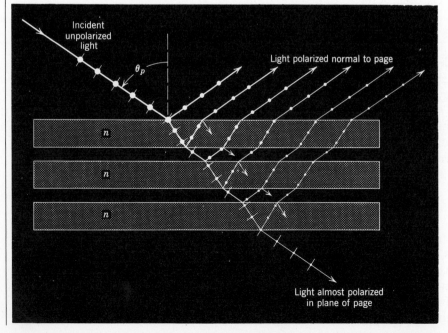

figure 48-9

Polarization of light by a stack of glass plates. Unpolarized light is incident on a stack of glass plates at Brewster's angle θ_p. (Polarization in the plane of the page is shown by the short lines and polarization normal to the page by the dots.) All light reflected out of the original ray is polarized normal to the page. After passing through several reflecting interfaces, the transmitted light no longer contains any appreciable component polarized normal to the page.

σ-component is only partially refracted. Thus the transmitted beam, which is of high intensity, is only partially polarized. By using a stack of glass plates rather than a single plate, we obtain reflections from successive surfaces and we can increase the intensity of the emerging reflected (σ-component) beam (see Fig. 48-9). By the same token, the σ-components are progressively removed from the transmitted beam, making it more completely π-polarized.

At the polarizing angle it is found experimentally that the reflected and the refracted beams are at right angles, or (Fig. 48-8)

$$\theta_p + \theta_r = 90°.$$

From Snell's law, $n_1 \sin \theta_p = n_2 \sin \theta_r.$

Combining these equations leads to

$$n_1 \sin \theta_p = n_2 \sin (90° - \theta_p) = n_2 \cos \theta_p$$

or $$\tan \theta_p = \frac{n_2}{n_1}, \tag{48-2}$$

where the incident ray is in medium 1 and the refracted ray in medium 2. This can be written as

$$\tan \theta_p = n, \tag{48-3}$$

where n $(= n_2/n_1)$ is the index of refraction of medium 2 with respect to medium 1. Equation 48-3 is known as *Brewster's law* after Sir David Brewster (1781–1868), who deduced it empirically in 1812. It is possible to prove this law rigorously from Maxwell's equations (see also Question 6).

We wish to use a plate of glass $(n = 1.50)$ as a polarizer. What is the polarizing angle? What is the angle of refraction?

From Eq. 48-3,

$$\theta_p = \tan^{-1} 1.50 = 56.3°.$$

The angle of refraction follows from Snell's law:

$$(1) \ \sin \theta_p = n \sin \theta_r$$

or $$\sin \theta_r = \frac{\sin 56.3°}{1.50} = 0.555 \qquad \theta_r = 33.7°.$$

EXAMPLE 2

In earlier chapters we assumed that the speed of light, and thus the index of refraction, is independent of the direction of propagation in the medium and of the state of polarization of the light. Liquids, amorphous solids such as glass, and crystalline solids having cubic symmetry normally show this behavior and are said to be *optically isotropic*. Many other crystalline solids are optically *anisotropic* (that is, not isotropic).*

Solids may be anisotropic in many properties. Mica cleaves readily in

48-4
DOUBLE REFRACTION

---

* Many transparent amorphous solids such as glasses and plastics become optically anisotropic when they are mechanically stressed. This fact is useful in engineering design studies in that strains in gears, bridge structures, etc., can be studied quantitatively by building plastic models, stressing them appropriately, and examining the optical anisotropy that results, using polarization techniques. The interested reader should consult "Photoelasticity," a chapter by H. T. Jessop in Vol. 6 of the *Encyclopedia of Physics*, edited by H. Flugge (1958), Springer Verlag, Berlin.

figure 48-10
Two images, one polarized 90°
relative to the other, are formed by a
calcite crystal. The book on which
the crystal is lying is Huygens' *Opera
Reliqua,* wherein the phenomenon
of double refraction is discussed.

one plane only; a cube of crystalline graphite does not have the same electric resistance between all pairs of opposite faces; a cube of crystalline nickel magnetizes more readily in certain directions than in others, etc. If a solid is a mixture of a large number of tiny crystallites, it may appear to be isotropic because of the random orientations of the crystallites. Powdered mica, for example, compacted to a solid mass with a binder, does not exhibit the cleavage properties that characterize the crystallites making it up.

Figure 48-10, in which a polished crystal of calcite ($CaCO_3$) is laid over some printed letters, shows the optical anisotropy of this material; *the image appears double.* Figure 48-11 shows a beam of unpolarized light falling on a calcite crystal at right angles to one of its faces. The

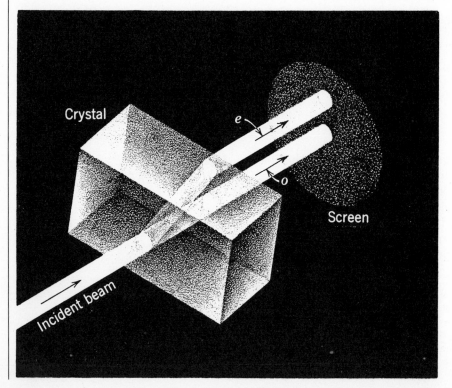

figure 48-11
A beam of unpolarized light falling
on a calcite crystal is split into two
beams which are polarized at right
angles to each other.

single beam splits into two at the crystal surface. The "double-bending" of a beam transmitted through calcite, exhibited in Figs. 48-10 and 48-11, is called *double refraction*.

If the two emerging beams in Fig. 48-11 are analyzed with a polarizing sheet, they are found to be plane-polarized with their planes of vibration at right angles to each other, a fact discovered by Huygens in 1678. Huygens used a second calcite crystal to investigate the polarization states of the beams labeled *o* and *e* in the figure.

If experiments are carried out at various angles of incidence, one of the beams in Fig. 48-11 (represented by the *ordinary ray*, or *o*-ray) will be found to obey Snell's law of refraction at the crystal surface, just like a ray passing from one isotropic medium into another. The second beam (represented by the *extraordinary ray*, or *e*-ray) will not. In Fig. 48-11, for example, the angle of incidence for the incident light is zero but the angle of refraction of the *e*-ray, contrary to the prediction of Snell's law, is not. In general, the *e*-ray does not even lie in the plane of incidence.

This difference between the waves represented by the *o*- and *e*-rays with respect to Snell's law can be explained in these terms:

1. The *o*-wave travels in the crystal with the same speed v_o in all directions. In other words, the crystal has, for this wave, a single index of refraction n_o, just like an isotropic solid.

2. The *e*-wave travels in the crystal with a speed that varies with direction from v_o to a larger value (for calcite) v_e. In other words, the index of refraction, defined as c/v, varies with direction from n_o to a smaller value (for calcite) n_e.

The quantities n_o and n_e are called the *principal indices of refraction* for the crystal. Problem 12 suggests how to measure them. Table 48-1 shows these indices for six doubly refracting crystals. For three of them the *e*-wave is slower; for the other three it is faster than the *o*-wave. Some doubly refracting crystals (mica, topaz, etc.) are more complex optically than calcite and require *three* principal indices of refraction for a complete description of their optical properties. Crystals whose basic crystal structure is cubic (see Fig. 47-13) are optically isotropic, requiring only *one* index of refraction.

Table 48-1
Principal indices of refraction of several doubly refracting crystals
(for sodium light, $\lambda = 589$ nm $= 5890$ Å)

Crystal	Formula	n_o	n_e	$n_e - n_o$
Ice	H_2O	1.309	1.313	+0.004
Quartz	SiO_2	1.541	1.553	+0.012
Wurzite	ZnS	2.356	2.378	+0.022
Calcite	$CaCO_3$	1.658	1.486	−0.172
Dolomite	$CaO \cdot MgO \cdot 2CO_2$	1.681	1.500	−0.181
Siderite	$FeO \cdot CO_2$	1.875	1.635	−0.240

The behavior for the speeds of the two waves traveling in calcite is summarized by Fig. 48-12, which shows two wave surfaces spreading out from an imaginary point light source S imbedded in the crystal. The *o*-wave surface is a sphere, as we would expect if the medium were isotropic. The *e*-wave surface is an ellipsoid of revolution about a characteristic direction in the crystal called the *optic axis*. The two wave

figure 48-12
Huygens' wave surfaces generated by a point source S imbedded in calcite. The polarization states for three o-rays and three e-rays are shown by the dots and lines, respectively. Note that in general (ray Sb) the bars representing the polarization direction are not perpendicular to the e-rays.

surfaces represent light having two different polarization states. If we consider for the present only rays lying in the plane of Fig. 48-12, then (a) the plane of polarization for the o-rays is perpendicular to the figure, as suggested by the dots, and (b) that for the e-rays coincides with the plane of the figure, as suggested by the short lines. We describe the polarization states more fully at the end of this section.

Figure 48-13, which shows a typical calcite crystal that may be obtained by cleavage from a naturally occurring crystal, shows how to locate the optic axis. The edges of calcite crystals may have any lengths but the angles at which the edges intersect always have one or another of two values, 78° 05′ or 101° 55′. The optic axis is found by erecting a line at either of the two corners where three obtuse angles meet (the "blunt" corners), making equal angles with the crystal edges. *Any line in the crystal parallel to this line is also an optic axis.*

We can use Huygens' principle to study the propagation of light waves in doubly refracting crystals. Figure 48-14a shows the special case in which unpolarized light falls at normal incidence on a calcite slab cut from a crystal in such a way that the optic axis is normal to the surface. Consider a wavefront that, at time $t = 0$, coincides with the crystal surface. Following Huygens, we may let any point on this surface serve as a radiating center for a double set of Huygens' wavelets, such as those in Fig. 48-12. The plane of tangency to these wavelets represents the new position of this wavefront at a later time t. The incident beam in Fig. 48-14a is propagated through the crystal without deviation at speed v_o. The beam emerging from the slab will have the same polarization character as the incident beam. The calcite slab, in these special circumstances only, behaves like an isotropic material, and no distinction can be made between the o- and the e-waves.

Figure 48-14b shows two views of another special case, namely, unpolarized incident light falling at right angles on a slab cut so that the optic axis is parallel to its surface. In this case too the incident beam is propagated without deviation. However, we can now identify o- and e-waves that travel through the crystal with different speeds, v_o and v_e, respectively. These waves are polarized at right angles to each other.

Some doubly refracting crystals have the interesting property, called *dichroism*, in which one of the polarization components is strongly absorbed within the crystal, the other being transmitted with little loss. Dichroism, illustrated in Fig. 48-15, is the basic operating principle of the commercial Polaroid sheet. The

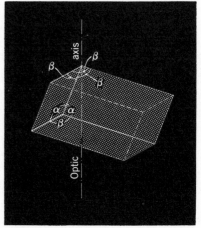

figure 48-13
A calcite crystal; α is 78° 05′; β is 101° 55′.

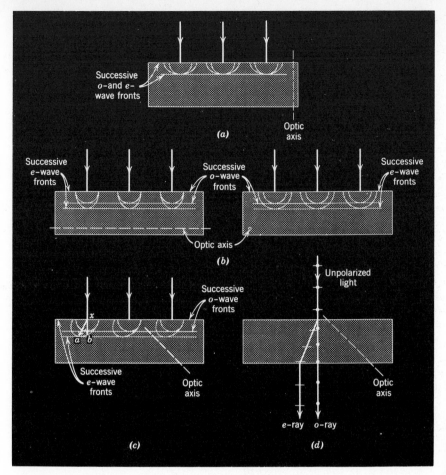

figure 48-14
Unpolarized light falls at normal incidence on slabs cut from a calcite crystal. The Huygens' wavelets are appropriate sections of the figure of revolution about the optic axis represented by Fig. 48-13. (a) No double refraction or speed difference occurs. (b) No double refraction occurs but there is a speed difference. (c) Both double refraction and a speed difference occur. (d) Same as (c) but showing the polarization states and the emerging rays.

many small crystallites, imbedded in a plastic sheet with their optic axes parallel, have a polarizing action equivalent to that of a single large crystal slab.

Figure 48-14c shows unpolarized light falling at normal incidence on a calcite slab cut so that its optic axis makes an arbitrary angle with the crystal surface. Two spatially separated beams are produced, as in Fig. 48-11. They travel through the crystal at different speeds, that for the o-wave being v_o and that for the e-wave being intermediate between v_o and v_e. Note that ray xa represents the shortest *optical* path for the transfer of light energy from point x to the e-wavefront. Energy transferred along any other ray, in particular along ray xb, would have a longer transit time, a consequence of the fact that the speed of e-waves varies with direction.* Figure 48-14d represents the same case as Fig. 48-14c. It shows the rays emerging from the slab, as in Fig. 48-11, and makes clear that the emerging beams are polarized at right angles to each other, that is, they are *cross-polarized*.

We now seek to understand, in terms of the atomic structure of optically anisotropic crystals, how cross-polarized light waves with different speeds can exist. Light is propagated through a crystal by the action of the vibrating **E** vectors of the wave on the electrons in the crystal. These electrons, which experience electrostatic restoring forces if they are moved from their equilibrium positions, are set into forced periodic oscillation about these positions and pass along the transverse wave disturbance that constitutes the light wave. The strength of the

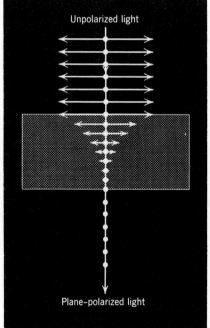

figure 48-15
Showing the absorption of one polarization component inside a dichroic crystal of the type used in Polaroid sheets.

---

* The reader who has not previously read Section 43-6 on Fermat's principle may wish to do so now.

restoring forces may be measured by a force constant k, as for the simple harmonic oscillator discussed in Chapter 15 (see Eq. 15-4).

In optically isotropic materials the force constant k is the same for all directions of displacement of the electrons from their equilibrium positions. In doubly refracting crystals, however, k varies with direction. For electron displacements that lie in a plane at right angles to the optic axis k has the constant value k_o, no matter how the displacement is oriented in this plane. For displacements parallel to the optic axis, k has the larger value (for calcite) k_e.* Note carefully that the speed of a wave in a crystal is determined by the direction in which the **E** vectors vibrate and *not* by the direction of propagation. It is the transverse **E**-vector vibrations that call the restoring forces into play and thus determine the wave speed. Note too that the stronger the restoring force, that is, the larger k, the faster the wave. For waves traveling along a stretched cord, for example, the restoring force for the transverse displacements is determined by the tension F in the cord. Equation 19-12 shows that an increase in F means an increase in the wave speed v.

Figure 48-16, a long weighted "tire chain" supported at its upper end, provides a one-dimensional mechanical analogy for double refraction. It applies specifically to o- and e-waves traveling at right angles to the optic axis, as in Fig. 48-14b. If the supporting block is oscillated, as in Fig. 48-16a, a transverse wave travels along the chain with a certain speed. If the block is oscillated lengthwise, as in Fig. 48-16b, another transverse wave is also propagated. The restoring force for the second wave is greater than for the first, the chain being more rigid in the plane of Fig. 48-16b than in the plane of Fig. 48-16a. Thus the second wave travels along the chain with a greater speed.

In the language of optics we would say that the speed of a transverse wave in the chain depends on the orientation of the plane of vibration of the wave. If we oscillate the top of the chain in a random way, the wave disturbance at a point along the chain can be described as the sum of two waves, polarized at right angles and traveling with different speeds. This corresponds exactly to the optical situation of Fig. 48-14b.

For waves traveling parallel to the optic axis, as in Fig. 48-14a, or for waves in optically isotropic materials, the appropriate mechanical analogy is a single weighted hanging chain. Here there is only one speed of propagation, no matter how the upper end is oscillated. The restoring forces are the same for all orientations of the plane of polarization of waves traveling along such a chain.

These considerations allow us to understand more clearly the polarization states of the light represented by the double-wave surface of Fig. 48-12. For the (spherical) o-wave surface, the **E**-vector vibrations must be everywhere at right angles to the optic axis. If this is so, the same force constant k_o will always hold, and the o-waves will travel with the same speed in all directions. More specifically, if we draw a ray in Fig. 48-12 from S to the o-wave surface, considered three-dimensionally (that is, as a sphere), the **E**-vector vibrations will always be at right angles to the plane defined by this ray and the optic axis. Thus these vibrations will always be at right angles to the optic axis.

For the (ellipsoidal) e-wave surface, the **E**-vector vibrations in general have a component parallel to the optic axis. For rays such as Sa in Fig. 48-12 or for the e-rays of Fig. 48-14b, the vibrations are completely parallel to this axis. Thus a relatively strong force constant (in calcite) k_e is operative, and the wave speed v_e will be relatively high. For e-rays, such as Sb in Fig. 48-12, the parallel component of the **E**-vector vibrations is less than 100%, so that the corresponding wave speed will be less than v_e. For ray Sc, in Fig. 48-12, the parallel component is zero, and the distinction between o- and e-rays disappears.

(a)   (b)

figure 48-16
Two views of a one-dimensional mechanical model for double refraction.

---

* For doubly refracting crystals with $n_e > n_o$ (see Table 48-1) k for displacements parallel to the optic axis is *smaller* than for those at right angles to it. Also, for crystals with three principal indices of refraction, there will be three principal force constants. Such crystals have two optic axes and are called *biaxial*. The crystals listed in Table 48-1 have only a single optic axis and are called *uniaxial*.

Let plane-polarized light of angular frequency ω ($=2\pi\nu$) fall at normal incidence on a slab of calcite cut so that the optic axis is parallel to the face of the slab, as in Fig. 48-17. The two waves that emerge will be plane-polarized at right angles to each other, and, if the incident plane of vibration is at 45° to the optic axis, they will have equal amplitudes. Since the waves travel through the crystal at different speeds, there will be a phase difference ϕ between them when they emerge from the crystal. If the crystal thickness is chosen so that (for a given frequency of light) $\phi = 90°$, the slab is called a *quarter-wave plate*. The emerging light is said to be *circularly polarized*.

In Section 15-7 we saw that the two emerging plane-polarized waves just described (vibrating at right angles with a 90° phase difference) can be represented as the projections on two perpendicular axes of a vector rotating with angular frequency ω about the propagation direction. These two descriptions of circularly polarized light are completely equivalent. Figure 48-18 clarifies the relationship between these two descriptions.

figure 48-17
Plane-polarized light falls on a doubly refracting slab of thickness x cut with its optic axis parallel to the surface. The plane of vibration of the incident light is oriented to make an angle of 45° with the optic axis.

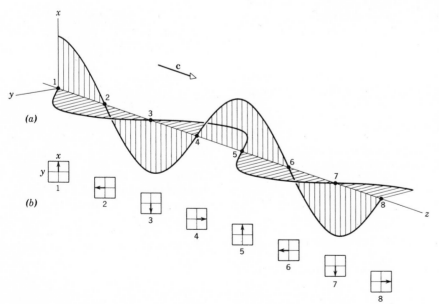

figure 48-18
(*a*) Two plane-polarized waves of equal amplitude and at right angles to each other are moving in the z direction. They differ in phase by 90°; where one wave has maximum values, the other is zero. (*b*) Views of the resultant amplitude of the approaching wave as seen by observers located at the positions shown on the z axis. Note that the resultant vector rotates clockwise with time.

A quartz quarter-wave plate is to be used with sodium light ($\lambda = 589$ nm = 5890 Å). What must its thickness be?

Two waves travel through the slab at speeds corresponding to the two principal indices of refraction given in Table 48-1 ($n_e = 1.553$ and $n_o = 1.541$). If the

EXAMPLE 3

crystal thickness is x, the number of wavelengths of the first wave contained in the crystal is

$$N_e = \frac{x}{\lambda_e} = \frac{xn_e}{\lambda},$$

where λ_e is the wavelength of the *e*-wave in the crystal and λ is the wavelength in air. For the second wave the number of wavelengths is

$$N_o = \frac{x}{\lambda_o} = \frac{xn_o}{\lambda},$$

where λ_o is the wavelength of the *o*-wave in the crystal. The difference $N_e - N_o$ must be one-fourth, or

$$\frac{1}{4} = \frac{x}{\lambda}(n_e - n_o).$$

This equation yields

$$x = \frac{\lambda}{4(n_e - n_o)} = \frac{589 \text{ nm}}{(4)(1.553 - 1.541)} = 0.012 \text{ mm}.$$

This plate is rather thin; most quarter-wave plates are made from mica, splitting the sheet to the correct thickness by trial and error.

A beam of circularly polarized light falls on a polarizing sheet. Describe the emerging beam.

   The circularly polarized light, as it enters the sheet, can be represented by

$$E_x = E_m \sin \omega t$$

and

$$E_y = E_m \cos \omega t,$$

where x and y represent arbitrary perpendicular axes. These equations correctly represent the fact that a circularly polarized wave is equivalent to two plane-polarized waves with equal amplitude and a 90° phase difference.

   The resultant amplitude in the incident circularly polarized wave is

$$E_{\text{cp}} = \sqrt{E_x^2 + E_y^2} = \sqrt{E_m^2 (\sin^2 \omega t + \cos^2 \omega t)} = E_m,$$

an expected result if the circularly polarized wave is represented as a rotating vector. The resultant intensity in the incident circularly polarized wave is proportional to E_m^2, or

$$I_{\text{cp}} \propto E_m^2. \qquad (48\text{-}4)$$

   Let the polarizing direction of the sheet make an arbitrary angle θ with the x axis as shown in Fig. 48-19. The instantaneous value of the plane-polarized wave transmitted by the sheet is

$$E = E_y \sin \theta + E_x \cos \theta$$

$$= E_m \cos \omega t \sin \theta + E_m \sin \omega t \cos \theta$$

$$= E_m \sin (\omega t + \theta).$$

The intensity of the wave transmitted by the sheet is proportional to E^2, or

$$I \propto E_m^2 \sin^2 (\omega t + \theta).$$

The eye and other measuring instruments respond only to the average intensity $\bar{I}$, which is found by replacing $\sin^2 (\omega t + \theta)$ by its average value over one or more cycles $(= \frac{1}{2})$, or

$$\bar{I} \propto \tfrac{1}{2} E_m^2.$$

Comparison with Eq. 48-4 shows that inserting the polarizing sheet reduces the

EXAMPLE 4

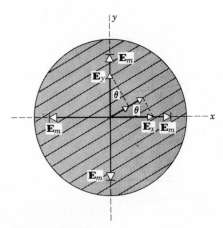

figure 48-19
Circularly polarized light falls on a polarizing sheet. $\mathbf{E}_x$ and $\mathbf{E}_y$ are instantaneous values of the two components, their maximum values being $\mathbf{E}_m$.

intensity by one-half. The orientation of the sheet makes no difference, since θ does not appear in this equation; this is to be expected if circularly polarized light is represented by a rotating vector, all azimuths about the propagation direction being equivalent. Inserting a polarizing sheet in an *unpolarized* beam has just the same effect, so that a simple polarizing sheet cannot be used to distinguish between unpolarized and circularly polarized light.

EXAMPLE 5

A beam of light is thought to be circularly polarized. How may this be verified?

Insert a quarter-wave plate. If the beam is circularly polarized, the two components will have a phase difference of 90° between them. The quarter-wave plate will introduce a further phase difference of ±90° so that the emerging light will have a phase difference of either zero or 180°. In either case the light will now be *plane-polarized* and can be made to suffer complete extinction by rotating a polarizer in its path.

Does the quarter-wave plate have to be oriented in any particular way to carry out this test?

EXAMPLE 6

A plane-polarized light wave of amplitude E_0 falls on a calcite quarter-wave plate with its plane of vibration at 45° to the optic axis of the plate, which is taken as the y axis; see Fig. 48-20. The emerging light will be circularly polarized. In what direction will the rotating electric vector appear to rotate? The direction of propagation is out of the page.

The wave component whose vibrations are parallel to the optic axis (the *e*-wave) can be represented as it emerges from the plate as

$$E_y = (E_0 \cos 45°) \sin \omega t = \frac{1}{\sqrt{2}} E_0 \sin \omega t = E_m \sin \omega t.$$

The wave component whose vibrations are at right angles to the optic axis (the *o*-wave) can be represented as

$$E_x = (E_0 \sin 45°) \sin (\omega t - 90°) = -\frac{1}{\sqrt{2}} E_0 \cos \omega t = -E_m \cos \omega t,$$

the 90° phase shift representing the action of the quarter-wave plate. Note that E_x reaches its maximum value one-fourth of a cycle *later* than E_y does, for, in calcite, wave E_x (the *o*-wave) travels *more slowly* than wave E_y (the *e*-wave).

To decide the direction of rotation, let us locate the tip of the rotating electric vector at two instants of time, (a) $t = 0$ and (b) a short time t_1 later chosen so that ωt_1 is a small angle. At $t = 0$ the coordinates of the tip of the rotating vector (see Fig. 48-20a) are

$$E_y = 0 \qquad \text{and} \qquad E_x = -E_m.$$

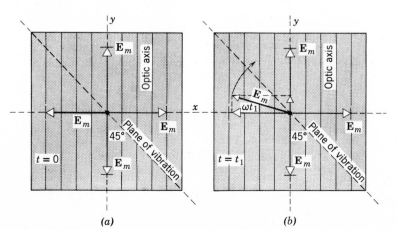

(a)   (b)

figure 48-20
Plane-polarized light falls from behind on a quarter-wave plate oriented so that the light emerging from the page is circularly polarized. In this case the electric vector E_m rotates clockwise as seen by an observer facing the light source.

At $t = t_1$ these coordinates become, approximately,

$$E_y = E_m \sin \omega t_1 \cong E_m(\omega t_1)$$

$$E_x = -E_m \cos \omega t_1 \cong -E_m.$$

Figure 48-20b shows that the vector which represents the emerging circular polarized light is rotating clockwise; by convention such light is called *right-circularly polarized*, the observer always being considered to face the light source.

If the plane of vibration of the incident light in Fig. 48-20 is rotated through $\pm 90°$, the emerging light will be *left-circularly polarized*.

48-6
ANGULAR MOMENTUM OF LIGHT

That light waves can deliver *linear momentum* to an absorbing screen or to a mirror is in accord with classical electromagnetism, with quantum physics, and with experiment. The facts of circular polarization suggest that light so polarized might also have *angular* momentum associated with it. This is indeed the case; once again the prediction is in accord with classical electromagnetism and with quantum physics. Experimental proof was provided in 1936 by Richard A. Beth, who showed that when circularly polarized light is produced in a doubly refracting slab, the slab experiences a reaction torque.

The angular momentum carried by light plays a vital role in understanding the emission of light from atoms and of γ-rays from nuclei. If light carries away angular momentum as it leaves the atom, the angular momentum of the residual atom must change by exactly the amount carried away; otherwise the angular momentum of the isolated system *atom plus light* will not be conserved.

Classical and quantum theory both predict that if a beam of circularly polarized light is completely absorbed by an object on which it falls, an angular momentum given by

$$L = \frac{U}{\omega} \tag{48-5}$$

is transferred to the object, where U is the amount of absorbed energy and ω the angular frequency of the light. You should verify that the dimensions in Eq. 48-5 are consistent.

48-7
SCATTERING OF LIGHT

A light wave, falling on a transparent solid, causes the electrons in the solid to oscillate periodically in response to the time-varying electric vector of the incident wave. The wave that travels through the medium is the resultant of the incident wave and of the radiations from the oscillating electrons. The resultant wave has a maximum intensity in the direction of the incident beam, falling off rapidly on either side. The lack of sideways scattering, which would be essentially complete in a large "perfect" crystal, comes about because the oscillating charges in the medium act cooperatively or coherently.

When light passes through a gas, we find much more sideways scattering. The oscillating electrons in this case, being separated by relatively large distances and not being bound together in a rigid structure, act independently rather than cooperatively. Thus the rigid cancellation of wave disturbances that are not in the forward direction is less likely to occur; there is more sideways scattering.

Light scattered sideways from a gas can be wholly or partially polarized, even though the incident light is unpolarized. Figure 48-21 shows an unpolarized beam moving upward on the page and striking a gas atom at *a*. The electrons at *a* will oscillate in response to the electric components of the incident wave, their motion being equivalent to two oscillating dipoles whose axes are represented by the arrow and the dot at *a*. An oscillating dipole does not radiate along its own line of action.

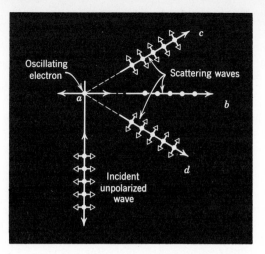

figure 48-21
Light is polarized either partially (c and d) or completely (b) by scattering from a gas molecule at a.

Thus an observer at b would receive no radiation from the dipole represented by the arrow at a. The radiation reaching him would come entirely from the dipole represented by the dot at a; thus this radiation would be plane-polarized, the plane of vibration passing through the line ab and being normal to the page.

Observers at c and d would detect partially polarized light, since the dipole represented by the arrow at a would radiate somewhat in these directions. Observers viewing the transmitted or the back-scattered light would not detect any polarization effects because both dipoles at a would radiate equally in these two directions.

A familiar example is the scattering of sunlight by the molecules of the earth's atmosphere. If the atmosphere were not present, the sky would appear black except when we looked directly at the sun. This has been verified by measurements made above the atmosphere in manned and unmanned space vehicles. We can easily check with a polarizer that the light from the cloudless sky is at least partially polarized. This fact is used in polar exploration in the so-called *solar compass*. In this device we establish direction by noting the nature of the polarization of the scattered sunlight. As is well known, magnetic compasses are not useful in these regions. It has been learned* that bees orient themselves in their flights between their hive and the pollen sources by means of polarization of the light from the sky; bees' eyes contain built-in polarization-sensing devices.

It still remains to be explained why the light scattered from the sky is predominantly blue and why the light received directly from the sun—particularly at sunset when the length of the atmosphere that it must traverse is greatest—is red. The cross section of an atom or molecule for light scattering depends on the wavelength, blue light being scattered more effectively than red light. Since the blue light is largely scattered, the transmitted light will have the color of normal sunlight with the blues largely removed; it is therefore more reddish in appearance.

The fact that the scattering cross section for blue light is higher than that for red light can be made reasonable. An electron in an atom or molecule is bound there by strong restoring forces. It has a definite natural frequency, like a small mass suspended in space by an assembly of springs. The natural frequency for

* See "Polarized Light and Animal Navigation" by Talbot H. Watermann in *Scientific American*, July 1955; *Bees: Their Vision, Chemical Sense, and Language* by K. von Frisch, Cornell University Press, 1950; and "Polarized-Light Navigation by Insects" by Rüdiger Wehner, *Scientific American*, July 1976.

electrons in atoms and molecules is usually in a region corresponding to violet or ultraviolet light.

When light is allowed to fall on such bound electrons, it sets up forced oscillations at the frequency of the incident light beam. In mechanical resonant systems it is possible to "drive" the system most effectively if we impress on it an external force whose frequency is as close as possible to that of the natural resonant frequency. In the case of light the blue is closer to the natural resonant frequency of the bound electron than is the red light. Therefore, we would expect the blue light to be more effective in causing the electron to oscillate, and thus it will be more effectively scattered.

48-8
DOUBLE SCATTERING

When X-rays were discovered in 1898, there was much speculation whether they were waves or particles. In 1906 they were established as transverse waves by Charles Glover Barkla (1877–1944) by means of a polarization experiment.

When the unpolarized X-rays strike scattering block S_1 in Fig. 48-22, they set the electrons into oscillatory motion. The considerations of the preceding section require that the X-rays scattered toward the second block be plane-polarized as shown in the figure. Let this wave be scattered from the second scattering block, and let us examine the radiation scattered from it by rotating a detector D in a plane at right angles to the line joining the blocks. The electrons will oscillate parallel to each other, and the positions of maximum and zero intensity will be as shown. A plot of detector reading as a function of the angle ϕ supports the hypothesis that X-rays are transverse waves. If the X-rays were a stream of particles or a longitudinal wave, these effects could by no means be so readily understood. Thus Barkla's important experiment established that X-rays are a part of the electromagnetic spectrum.

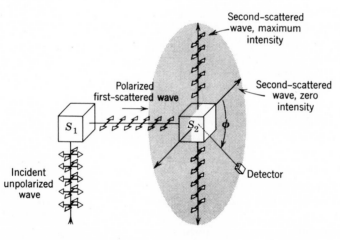

figure 48-22
A double-scattering experiment used by Barkla to show that X-rays are transverse waves.

In later studies you will learn that beams of particles such as electrons, protons, and pions can be viewed as waves. Scattering (including double scattering) techniques are often used to investigate the polarization characteristics of such beams.*

questions

1. It is said that light from ordinary sources is unpolarized. Can you think of any common sources that emit polarized light?
2. Why do sunglasses made of polarizing materials have a marked advantage over those that simply depend on absorption effects?
3. Unpolarized light falls on two polarizing sheets so oriented that no light is

*See "Polarized Accelerator Targets" by Gilbert Shapiro, *Scientific American*, July, 1976.

transmitted. If a third polarizing sheet is placed between them, can light be transmitted?

4. 3-D movies were very popular in the early 1950's. Viewers wore polarizing glasses and a polarizing sheet was placed in front of each of the *two* projectors needed. Explain how the system worked. Can you suggest any problems that may have led to the early abandonment of the system?

5. A wire grid, consisting of an array of wires arranged parallel to one another, can polarize an incident unpolarized beam of electromagnetic waves that pass through it. Explain the facts that (a) the diameter of the wires and the spacing between them must be much less than the incident wavelength to obtain effective polarization and (b) the transmitted component is the one whose electric vector oscillates in a direction perpendicular to the wires.

6. Brewster's law determines the polarizing angle on reflection from a dielectric. A plausible interpretation for zero reflection of the π-component at that angle is that the charges in the dielectric are caused to oscillate parallel to the reflected ray by this component and produce no radiation in this direction. Explain this and comment on the plausibility.

7. Can polarization by reflection occur if the light is incident on the interface from the side with the higher index of refraction (glass to air, for example)?

8. Devise a way to identify the polarizing direction of a sheet of Polaroid.

9. Is the optic axis of a doubly refracting crystal simply a line or a direction in space? Has it a direction sense, like an arrow? What about the characteristic direction of a polarizing sheet?

10. If ice is doubly refracting (see Table 48-1), why don't we see two images of objects viewed through an ice cube?

11. Is it possible to produce interference effects between the o-beam and the e-beam, which are separated by the calcite crystal from the incident unpolarized beam in Fig. 48-11, by recombining them? Explain your answer.

12. From Table 48-1, would you expect a quarter-wave plate made from calcite to be thicker than one made from quartz?

13. Does the *e*-wave in doubly refracting crystals always travel at a speed given by c/n_e?

14. In Fig. 48-14a and 48-14b describe qualitatively what happens if the incident beam falls on the crystal with an angle of incidence that is not zero. Assume in each case that the incident beam remains in the plane of the figure.

15. Devise a way to identify the direction of the optic axis in a quarter-wave plate.

16. If plane-polarized light falls on a quarter-wave plate with its plane of vibration making an angle of (a) 0° or (b) 90° with the axis of the plate, describe the transmitted light. (c) If this angle is arbitrarily chosen, the transmitted light is called *elliptically polarized;* describe such light.

17. You are given an object which may be (a) a disk of grey glass, (b) a polarizing sheet, (c) a quarter-wave plate, or (d) a half-wave plate (see Problem 14). How could you identify it?

18. Can a plane-polarized light beam be represented as a sum of two circularly polarized light beams of opposite rotation? What effect has changing the phase of one of the circular components on the resultant beam?

19. How can a right-circularly polarized light beam be transformed into a left-circularly polarized beam?

20. Could (a) a radar beam and (b) a sound wave in air be circularly polarized?

21. A beam of light is said to be unpolarized, plane-polarized, or circularly polarized. How could you choose among them experimentally?

22. A parallel beam of light is absorbed by an object placed in its path. Under what circumstances will (a) linear momentum and (b) angular momentum be transferred to the object?

23. When observing a clear sky through a polarizing sheet, one finds that the intensity varies by a factor of two on rotating the sheet. This does not happen when one views a cloud through the sheet. Can you devise an explanation?

24. In 1949 it was discovered that light from distant stars in our galaxy is slightly plane-polarized, with the preferred plane of vibration being parallel to the plane of our galaxy. This is probably due to non-isotropic scattering of the starlight by elongated and slightly aligned interstellar grains. If the grains are oriented with their long axis parallel to the interstellar magnetic field lines, as discussed in the text, and they absorb and radiate electromagnetic waves like the oscillating electrons in a radio antenna, how must the magnetic field be oriented with respect to the galactic plane?

problems

SECTION 48-1

1. Prove that two plane-polarized light waves of equal amplitude, their planes of vibration being at right angles to each other, cannot produce interference effects. (*Hint:* Prove that the intensity of the resultant light wave, averaged over one or more cycles of oscillation, is the same no matter what phase difference exists between the two waves.)

SECTION 48-2

2. Unpolarized light falls on two polarizing sheets placed one on top of the other. What must be the angle between the characteristic directions of the sheets if the intensity of the transmitted light is (a) one-third the maximum intensity of the transmitted beam or (b) one-third the intensity of the incident beam? Assume that the polarizing sheet is ideal, that is, that it reduces the intensity of unpolarized light by exactly 50%.

3. A beam of plane polarized light strikes two polarizing sheets. The characteristic direction of the first sheet is at an angle θ with respect to the incident beam, the characteristic direction of the second sheet being at 90° with respect to the incident beam. Find the angle θ for a transmitted beam intensity that is 1/10 the incident beam intensity.   *Answer:* 20° or 70°.

4. An unpolarized beam of light is incident on a group of four polarizing sheets which are lined up so that the characteristic direction of each is rotated by 30° clockwise with respect to the preceding sheet. What fraction of the incident intensity is transmitted?

5. A beam of light is a mixture of plane polarized light and randomly polarized light. When it is sent through a Polaroid sheet, it is found that the transmitted intensity can be varied by a factor of five depending on the orientation of the Polaroid. Find the relative intensities of these two components of the incident beam.
Answer: Plane polarized $\frac{2}{3}$, randomly polarized $\frac{1}{3}$.

6. Partially polarized light (a mixture of unpolarized and plane-polarized beams) can be represented as two plane-polarized beams of unequal intensities, I along the x-axis and i along the y-axis say, with a random phase difference. The degree of polarization is defined as $p = (I - i)/(I + i)$.
  (a) Suppose a beam of partially polarized light passes through a Polaroid sheet with its characteristic direction at an angle θ with the x-axis; show that the transmitted intensity I_t is

$$I_t = I \frac{1 + p \cos 2\theta}{1 + p}.$$

  (b) Does this reduce to expected results for $p = 1$ and $p = 0$?

7. It is desired to rotate the plane of polarization of a beam of plane-polarized light by 90°. (a) How might this be done using only Polaroid sheets? (b) How many sheets are required in order that the total intensity loss is less than 5%? Assume that each Polaroid sheet is ideal.   *Answer:* 48.

SECTION 48-3

8. (a) At what angle of incidence will the light reflected from water be completely polarized? (b) Does this angle depend on the wavelength of the light?

9. Calculate the range of polarizing angles for white light incident on fused quartz. Assume that the wavelength limits are 400 and 700 nm (4000 and 7000 Å) and use the dispersion curve of Fig. 43-2.
   Answer: 55°30′ to 55°46′

SECTION 48-4

10. Plane-polarized light of wavelength 525 nm strikes, at normal incidence, a wurzite crystal, cut with its faces parallel to the optic axis. What is the smallest possible thickness of the crystal if the emergent o and e-rays combine to form plane-polarized light? See Table 48-1.

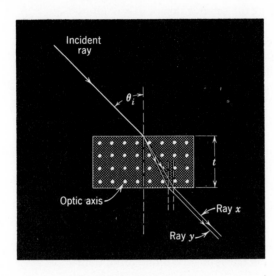

figure 48-23
Problem 11

11. A narrow beam of unpolarized light falls on a calcite crystal cut with its optic axis as shown in Fig. 48-23. (a) For $t = 1.0$ cm and for $\theta_i = 45°$, calculate the perpendicular distance between the two emerging rays x and y. (b) Which is the o-ray and which the e-ray? (c) What are the states of polarization of the emerging rays? (d) Describe what happens if a polarizer is placed in the incident beam and rotated. (*Hint:* Inside the crystal the **E**-vector vibrations for one ray are always perpendicular to the optic axis and for the other ray they are always parallel. The two rays are described by the indices n_o and n_e; *in this plane* each ray obeys Snell's law.)
   Answer: (a) 0.49 mm. (b) x is the e-ray, y is the o-ray. (c) **E** in ray y lies in the plane of the figure; **E** in ray x lies at right angles to the plane of the figure. (d) Every 90° one or the other beam, alternately, will be extinguished.

12. A prism is cut from calcite so that the optic axis is parallel to the prism edge as shown in Fig. 48-24. Describe how such a prism might be used to measure the two principal indices of refraction for calcite. (*Hint:* See hint in Problem 11; see also Example 3, Chapter 43.)

13. How thick must a sheet of mica be if it is to form a quarter-wave plate for yellow light ($\lambda = 589$ nm $= 5890$ Å)? Mica cleaves in such a way that the appropriate indices of refraction, for transmission at right angles to the cleavage plane, are 1.6049 and 1.6117.   *Answer:* 0.022 mm.

SECTION 48-5

14. What would be the action of a *half-wave plate* (that is, a plate twice as thick as a quarter-wave plate) on (a) plane-polarized light (assume the plane of vibration to be at 45° to the optic axis of the plate), (b) circularly polarized light, and (c) unpolarized light.

figure 48-24
Problem 12

15. Describe the state of polarization represented by these sets of equations:

(a) $E_x = E \sin(kz - \omega t)$

$E_y = E \cos(kz - \omega t)$,

(b) $E_x = E \cos(kz - \omega t)$

$E_y = E \cos\left(kz - \omega t + \dfrac{\pi}{4}\right)$,

(c) $E_x = E \sin(kz - \omega t)$

$E_y = -E \sin(kz - \omega t)$.

Answer: Assuming that a right-handed coordinate system is used, (a) circular, counterclockwise as seen facing the scource; (b) elliptical, counterclockwise as seen facing the scource, major axis of ellipse along $y = x$; (c) plane, along $y = -x$.

16. A sheet of Polaroid and a quarter-wave plate are glued together in such a way that, if the combination is placed with face A against a shiny coin, the face of the coin can be seen when illuminated with light of appropriate wavelength. When the combination is placed with face A away from the coin, the coin cannot be seen. (a) Which component is on face A and (b) what is the relative orientation of the components?

SECTION 48-6

17. Show that in a parallel beam of circularly polarized light the angular momentum per unit volume L_r is given by

$$L_r = \frac{P}{\omega c},$$

where P is the power per unit area of the beam. Start from Eq. 48-5.

18. Assume that a parallel beam of circularly polarized light whose intensity is 100 W is absorbed by an object, (a) At what rate is angular momentum transferred to the object? (b) If the object is a flat disk of diameter 5.0 mm and mass 1.0×10^{-2} g, after how long a time (assuming it is free to rotate about its axis) would it attain an angular speed of 1.0 rev/s? Assume a wavelength of 500 nm.

49
light and
quantum physics

Thus far we have studied radiation — and by that we include not only visible light but the entire electromagnetic spectrum — under the headings of propagation, reflection, refraction, interference, diffraction, and polarization. We have been able to understand all these phenomena by treating radiation as an (electromagnetic) *wave*, governed by Maxwell's equations. The evidence for this assumption is, in fact, overwhelming.

We now move off in an entirely new direction and consider experiments that can be understood only by making a quite different assumption about radiation, namely, that it behaves like a *particle*.

You may well ask: "Well, which is it, wave or particle?" These concepts are so different that it is hard to see how radiation can model itself after both at the same time. We shall face this question squarely in Section 50-12. Meanwhile, we ask you not to worry about this puzzle for the time being and simply to look, with us, at the strong experimental evidence that radiation is particlelike. The path we are taking will open the door to the world of quantum physics, one of the bedrock foundations on which our modern view of the universe rests.

We see most objects by the light that is reflected from them. At high enough temperatures, however, bodies become self-luminous and we can see them glow in the dark. Incandescent lamp filaments and bonfires (see Fig. 49-1) are familiar examples. Although we see such objects by the visible light that they emit, we do not have to linger too long near a bonfire to believe that it also emits copiously in the infrared region of the spectrum. It is a curious fact that quantum physics, which domi-

49-1
A NEW DIRECTION

49-2
THERMAL RADIATION
AND THE QUANTUM

figure 49-1
Students contemplating thermal radiation. The study of such radiation, under controlled laboratory conditions, laid the foundation for modern quantum physics. (Courtesy Alice Halliday.)

nates our modern view of the world around us, arose from the study — under controlled laboratory conditions — of the radiations emitted by hot objects.

Radiation given off by a body because of its temperature is called *thermal radiation.* All bodies not only emit such radiation but also absorb it from their surroundings. If a body is hotter than its surroundings it will emit more radiation than it absorbs and it will tend to cool off. Normally, it will come to thermal equilibrium with its surroundings, a condition in which its rate of absorption and emission of radiation are equal.

The spectrum of the thermal radiation from a hot solid body is continuous, its details depending strongly on the temperature. If we were to steadily raise the temperature of such a body, we would notice two things: (1) The higher the temperature, the more thermal radiation is emitted — at first the body appears dim, then it glows brightly; and (2) the higher the temperature, the shorter is the wavelength of that part of the spectrum radiating most intensely — the predominant color of the hot body shifts from dull red through bright yellow-orange to bluish "white heat." Since the characteristics of its spectrum depend on the temperature, we can estimate the temperature of a hot body, such as a glowing steel ingot or a star, from the radiation it emits. The eye sees chiefly the color corresponding to the most intense emission in the visible range.

The radiation emitted by a hot body depends not only on the temperature but also on the material of which the body is made, its shape, and the nature of its surface. For example, at 2000 K a polished flat tungsten surface emits radiation at a rate of 23.5 W/cm²; for molybdenum, however, the corresponding rate is 19.2 W/cm². In each case the rate increases somewhat if the surface is roughened. Other differences appear if we measure the distribution in wavelength of the emitted radiation. Such details make it hard to understand thermal radiation in terms of simpler physical ideas; it reminds us of the complications that come up in trying to understand the properties of real gases in terms of a simple atomic model. The "gas problem" was managed by introducing the notion of an ideal gas. In much the same spirit, the "radiation problem" can be made manageable by introducing an "ideal radiator" for which the spectrum of the emitted thermal radiation depends *only* on the temperature of the radiator and not on the material, the nature of the surface, or other factors.

We can make such an ideal radiator, in fact, by forming a cavity within a body, the walls of the cavity being held at a uniform temperature. We must pierce a small hole through the wall so that a sample of the radiation inside the cavity can escape into the laboratory to be examined. It turns out that such thermal radiation, called *cavity radiation*, has a very simple spectrum whose nature is indeed determined only by the temperature of the walls and not in any way by the material of the cavity, its shape, or its size. Cavity radiation (radiation in a box) helps us to understand the nature of thermal radiation, just as the ideal gas (matter in a box) helped us to understand matter in its gaseous form.

Figure 49-2 shows a cavity radiator made of a thin-walled cylindrical tungsten tube about 1 mm in diameter and heated to incandescence by passing a current through it. A small hole has been drilled in its wall. It is clear from the figure that the radiation emerging from this hole is much more intense than that from the outer wall of the cavity, even though the temperatures of the outer and inner walls are more or less equal.

There are three interrelated properties of cavity radiation—all well verified in the laboratory—that any theory of cavity radiation must explain.

figure 49-2
An incandescent tungsten tube with a small hole drilled in its wall. The radiation emerging from the hole is cavity radiation.

1. *The Stefan-Boltzmann law.* The total radiated power per unit area of the cavity aperture, summed over all wavelengths, is called its *radiancy $R(T)$* and is related to the temperature by

$$R = \sigma T^4 \qquad (49\text{-}1a)$$

in which $\sigma \,(= 5.670 \times 10^{-8}\,\text{W/m}^2 \cdot \text{K}^4)$ is a universal constant, called the *Stefan-Boltzmann constant.*

Ordinary hot objects always radiate less efficiently than do cavity radiators. We express this by generalizing Eq. 49-1a to

$$R = \epsilon \sigma T^4 \qquad (49\text{-}1b)$$

in which ϵ, a dimensionless quantity, is called the *emissivity* of the surface material. For a cavity radiator, $\epsilon = 1$, but for the surfaces of ordinary objects, the emissivity is always less than unity and is almost always a function of temperature.

2. *The spectral radiancy.* The *spectral radiancy $\mathcal{R}(\lambda,T)$* tells us how the intensity of the cavity radiation varies with wavelength for a fixed temperature. It is defined so that the product $\mathcal{R}(\lambda)\,d\lambda$ gives the radiated power per unit area that lies in the wavelength band that extends from λ to $\lambda + d\lambda$. We can find the radiancy $R(T)$ for any temperature by adding up (that is, by integrating) the spectral radiancy over the complete range of wavelengths. Thus

$$R(T) = \int_0^\infty \mathcal{R}(\lambda)\,d\lambda \qquad (T = \text{a constant}). \qquad (49\text{-}2)$$

figure 49-3
Spectral radiancy curves for cavity
radiation at four selected temperatures.
Note that as the temperature increases,
the wavelength of maximum spectral
radiancy shifts to lower values.

Figure 49-3 shows the spectral radiancy for cavity radiation at four selected temperatures. A glance at Eq. 49-2 shows that we can interpret the radiancy $R(T)$ as the area under the appropriate spectral radiancy curve. We see from the figure that, as the temperature increases, so does this area — and thus the radiancy — as Eq. 49-1a predicts.

3. *The Wien displacement law.* We can see from the spectral radiancy curves of Fig. 49-3 that λ_{max}, the wavelength at which the spectral radiancy is a maximum, decreases as the temperature increases. Wilhelm Wien was able to show from classical theory that, in fact, the product $\lambda_{max}T\ (=w)$ is a universal constant. Its measured value is

$$\lambda_{max}T\ (=w) = 2898\ \mu\text{m} \cdot \text{K} \qquad (49\text{-}3)$$

This relationship is often called the *Wien displacement law* and w the *Wien constant.*

How hot is a star? The "surfaces" of stars are not sharp boundaries like the surface of the earth. Most of the radiation that a star emits is in good thermal equilibrium with the hot gases that make up the star's outer layers. Without too much error, then, we can treat starlight as cavity radiation. Here are the wavelengths at which the spectral radiancies of three stars have their maximum values:

EXAMPLE 1

Star	λ_{max}	Appearance
Sirius	240 nm	blue-white
The sun	500 nm	yellow
Betelgeuse	850 nm	red

(a) What are the surface temperatures of these stars?
From Eq. 49-3 we find, for Sirius,

$$T = \frac{w}{\lambda_{max}}$$

$$= \left(\frac{2898 \ \mu m \cdot K}{240 \ nm}\right)\left(\frac{1000 \ nm}{1 \ \mu m}\right) = 12,000 \ K.$$

The temperatures for the sun and for Betelgeuse work out in the same way to be 5800 K and 3400 K, respectively. At 5800 K the sun's surface is near the temperature at which most of its radiation lies within the visible region of the spectrum. This suggests that over ages of human evolution our eyes have adapted to the sun to become most sensitive to those wavelengths that it radiates most intensely.

(b) What are the radiancies of these three stars?

For Sirius we have, from the Stefan-Boltzmann law (Eq. 49-1a)

$$R = \sigma T^4 = (5.67 \times 10^{-8} \ W/m^2 \cdot K^4)(12,000 \ K)^4$$

$$= 1.2 \times 10^9 \ W/m^2 = 120 \ kW/cm^2.$$

The radiancies for the sun and for Betelgeuse work out to be 6.4 kW/cm² and 0.767 kW/cm², respectively.

(c) The radius of the sun is 7.0×10^8 m and that of Betelgeuse is over 500 times larger, or 4.0×10^{11} m. What is the total radiated power output (that is, the *luminosity L*) of these stars?

We find the luminosity of a star by multiplying its radiancy by its surface area. Thus, for the sun,

$$L = R(4\pi r^2) = (6.4 \ kW/cm^2)(4\pi)(700 \times 10^8 \ cm)^2$$

$$= 3.9 \times 10^{26} \ W.$$

For Betelgeuse the luminosity works out to be 1.5×10^{31} W, about 38,000 times larger. The enormous size of Betelgeuse, which is classified as a "red giant," much more than makes up for the relatively low radiancy associated with its low surface temperature.

Interestingly enough, the colors of stars are not strikingly apparent to the average observer because the retinal cones, which are responsible for color vision, do not function well in dim light. If this were not so, the night sky would be spangled with color.

Is there a simple formula, derivable from basic principles, that fits the experimental radiancy curves of Fig. 49-3? In September 1900 there were two such formulas.

The first, due originally to Lord Rayleigh but later derived independently by Einstein and modified by James Jeans, was developed rigorously from its classical base. Alas, it fails utterly to fit the curves, not even passing through a maximum. However, the Rayleigh-Jeans formula, as it is called, *does* fit the curves quite well in the limit of very long wavelengths. Figure 49-4 shows the spectral radiancy curve for cavity radiation at 2000 K, along with the Rayleigh-Jeans prediction. The good fit we speak of occurs for wavelengths $\gg 50 \ \mu m$, far beyond the scale of that figure. The Rayleigh-Jeans formula, unsatisfactory though it may be, is the best that classical physics has to offer.

Wilhelm Wien also derived a theoretical expression for the spectral radiancy. His formula (see also Fig. 49-4) is much better. It fits the curves quite well at short wavelengths, passes through a maximum, but departs noticeably at the long wavelength end of the scale. However, Wien's formula was based on a conjecture — it has been called a "guess" — that there is an analogy between the spectral radiancy curves and the Maxwell speed distribution curves for the molecules of an ideal gas. For this reason its classical base is somewhat uncertain.

49-3
PLANCK'S RADIATION LAW

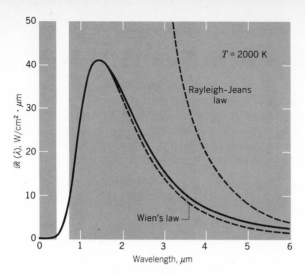

figure 49-4
The solid curve shows the
experimental spectral radiancy for
radiation from a cavity at 2000 K. The
classical Rayleigh-Jeans law and the
Wien law are shown as dashed lines.
The shaded bar represents the range of
visible wavelengths.

Thus we have two formulas, one agreeing with experiment at long
wavelengths and the other at short wavelengths. Max Planck, seeking to
reconcile these two radiation laws, made an inspired interpolation between them that turned out to fit the data at *all* wavelengths. Planck's
radiation formula, announced to the Berlin Physical Society on October
19, 1900, is

$$\mathcal{R}(\lambda,T) = \frac{a}{\lambda^5} \frac{1}{e^{b/\lambda T} - 1}, \tag{49-4}$$

in which a and b are empirical constants, chosen to give the best fit of
Eq. 49-4 to the experimental data. Figure 49-5 shows how good the
agreement is. Planck's formula, though interesting and (as it turned out)
correct, was still empirical at this stage and did not constitute a true
theory.

Planck set to work at once to derive his formula from simple assumptions and, in two months, he succeeded. In the process he recast his
formula slightly, presenting the two arbitrary constants it contained in
a different form. In this new notation, Planck's radiation law becomes

$$\mathcal{R}(\lambda,T) = \frac{2\pi c^2 h}{\lambda^5} \frac{1}{e^{hc/\lambda kT} - 1}. \tag{49-5}$$

figure 49-5
Planck's radiation law fitted to the
experimental data for a cavity radiator
at 1595 K.

The two adjustable constants a and b in Eq. 49-4 are here replaced by quantities involving two new constants, the *Boltzmann constant k* and a totally new constant, now called the *Planck constant h*; the quantity c is the speed of light.

By fitting Eq. 49-5 to the experimental data, Planck could find values for k and h. They proved to be within a percent or so of their presently accepted values, which are

$$k = 1.381 \times 10^{-23} \text{ J/K} \qquad \text{Boltzmann constant,}$$

and

$$h = 6.626 \times 10^{-34} \text{ J} \cdot \text{s} \qquad \text{Planck constant.}$$

49-4
THE QUANTIZATION OF ENERGY

We turn now to the assumptions made by Planck in deriving his radiation law and to the significance of the constant h that appears in it. These assumptions and their consequences were not immediately clear to Planck's contemporaries or, for that matter (as he confirmed later) to Planck himself. In what follows we describe the situation as it appeared some six or seven years after Planck first advanced his theory. It seems to be true that the basic premise underlying Planck's radiation law — the quantization of energy — was not understood at any earlier date.

Planck derived his radiation law by analyzing the interplay between the radiation in the cavity volume and the atoms that make up the cavity walls. He assumed that these atoms behave like tiny oscillators, each with a characteristic frequency of oscillation. These oscillators radiate energy into the cavity and absorb energy from it. It should be possible to deduce the characteristics of the cavity radiation from the characteristics of the oscillators that generate it.

Classically, the energy of these tiny oscillators is a smoothly continuous variable. We certainly assume this for large-scale oscillators such as pendulums or mass-spring systems. It turns out, however, that in order to derive Planck's radiation law it is necessary to make a radical assumption, namely: *Atomic oscillators may not have any energy E but only energies chosen from a discrete set, defined by*

$$E = nh\nu \qquad n = 1, 2, 3, \ldots, \tag{49-6}$$

in which ν is the oscillator frequency. Here h — the Planck constant — is introduced into physics for the first time. We say that the energy of an atomic oscillator is *quantized* and that the integer n is a *quantum number.* Equation 49-6 tells us that the oscillator energy levels are evenly spaced, the interval between adjacent levels being $h\nu$; see Fig. 49-6.

The assumption of energy quantization is indeed a radical one, and Planck himself resisted accepting it for many years. In his words, "My futile attempts to fit the elementary quantum of action [that is, h] somehow into the classical theory continued for a number of years, and they cost me a great deal of effort." Max von Laue, the 1914 Nobel laureate in physics, has written, "After 1900 Planck strove for many years to bridge, if not to close, the gap between the older and the quantum physics. The effort failed, but it had value in that it provided the most convincing proof that the two could not be joined."

Let us look at energy quantization in the context of a large-scale oscillator such as a swinging pendulum. It would be a stoutly defended belief that a pendulum can oscillate with *any* reasonable total energy

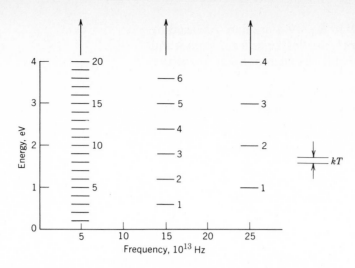

figure 49-6
The energy levels for atomic oscillators at three selected frequencies. The quantum numbers of some of the levels are indicated. On the right is shown kT, the average energy for a classical oscillator at 2000 K.

and not only with certain selected energies. As friction causes the pendulum amplitude to decay, it would seem that the energy should be dissipated in a perfectly continuous way and not in "jumps" or "quanta." However, because the Planck constant is so small, there is no basis in such everyday experience to dismiss energy quantization as a violation of "common sense." The "jumps" are there; they are just far too small for us to detect.

We would not apply quantum theory to a pendulum because classical theory works perfectly well in that case. We now see classical theory as a limiting case of quantum theory, the two being connected by the *correspondence principle*, which states that: *Quantum theory must agree with classical theory in the limit in which classical theory is known to agree with experiment.* It is in this way that the swinging pendulum and the oscillating atom relate to each other. We shall explore this principle further in the following two examples.

An oscillating mass-spring system. A 300-g body, connected to a spring whose spring constant k is 3 N/m, is oscillating with an amplitude A of 10 cm. Treat this system as a quantum oscillator and find (a) the energy interval between adjacent energy levels and (b) the quantum number that describes the oscillations.

EXAMPLE 2

(a) The frequency of oscillation is found from

$$v = \frac{1}{2\pi} \sqrt{\frac{k}{m}} = \frac{1}{2\pi} \sqrt{\frac{3 \text{ N/m}}{0.3 \text{ kg}}} = 0.50 \text{ s}^{-1}.$$

The total mechanical energy E of the oscillating system is

$$E = \frac{1}{2} kA^2 = \frac{1}{2} (3 \text{ N/m})(0.1 \text{ m})^2 = 0.015 \text{ J}.$$

As the amplitude of the oscillations dies away by friction, quantum theory predicts that the energy E will fall in "jumps" whose size is

$$\Delta E = hv = (6.63 \times 10^{-34} \text{ J} \cdot \text{s})(0.50 \text{ s}^{-1})$$
$$= 3.3 \times 10^{-34} \text{ J}.$$

Thus

$$\frac{\Delta E}{E} = \frac{3.3 \times 10^{-34} \text{ J}}{0.015 \text{ J}} = 2 \times 10^{-32}.$$

Energy measurements of such precision simply cannot be made. The quantum jumps of this oscillator are too small to be detected, and we are quite safe in treating the problem by the methods of classical physics.

(b) From the quantization relation, Eq. 49-6, we have

$$n = \frac{E}{h\nu} = \frac{0.015\ \text{J}}{3.3 \times 10^{-34}\ \text{J}} = 5 \times 10^{31},$$

an enormous number! We are justified in restating the correspondence principle in the form: *Quantum theory must agree with classical theory in the limit of large quantum numbers.*

The quantization of energy simply does not show up for large-scale oscillators. The smallness of the Planck constant h makes the graininess in the energy too fine to detect. In much the same way we cannot tell by waving our hand through it that air is made up of molecules.

The Planck constant might as well be zero as far as classical systems are concerned and, indeed, one way to reduce quantum formulas to their limiting classical counterparts is to let $h \rightarrow 0$ in those formulas. In much the same way we reduce relativistic formulas to their limiting classical counterparts by letting $c \rightarrow \infty$, where c is the speed of light.

EXAMPLE 3

Planck's radiation law — the classical limit. Figure 49-4 suggests that Planck's radiation law (Eq. 49-5) approaches the classical Rayleigh-Jeans law at long wavelengths. To what expression does Planck's law reduce as $\lambda \rightarrow \infty$?

For algebraic convenience, we can write Eq. 49-5 in the form

$$\mathscr{R}(\lambda) = \frac{2\pi c^2 h}{\lambda^5} \frac{1}{e^x - 1} \qquad (T = \text{a constant}),$$

in which $x = hc/\lambda kT$. As $\lambda \rightarrow \infty$ we see that $x \rightarrow 0$. Recalling that

$$e^x = 1 + x + \frac{x^2}{2!} + \frac{x^3}{3!} \cdots$$

(see Appendix I) allows us to make the approximation,

$$e^x - 1 \cong x.$$

Thus we have

$$\mathscr{R}(\lambda) \cong \frac{2\pi c^2 h}{\lambda^5} \frac{1}{x} = \frac{2\pi c^2 h}{\lambda^5} \left(\frac{\lambda kT}{hc} \right) = \frac{2\pi c kT}{\lambda^4}.$$

Note that the Planck constant h, a sure identifier of a quantum formula, conveniently cancels out as we approach the classical long-wave limit. The above result, in fact, is precisely the classical Rayleigh-Jeans expression for the spectral radiancy.

This leaves the question: "Why should letting the wavelength increase mean that we are approaching a realm in which classical physics holds?" The answer is that as the wavelength increases, the frequency decreases and thus the basic energy quantum ($= h\nu$) becomes smaller. To tell whether we are in a classical or a quantum situation we must compare this quantity with kT, which is a (classical) measure of the mean translational energy of a particle at temperature T. If $h\nu \ll kT$, the "graininess" of the energy of the atomic oscillators (which is measured by $h\nu$) will not be noticed and we are in the classical realm. To sum up then, we approach classical situations as $\nu \to 0$, as $\lambda \to \infty$, or (for that matter) as $T \to \infty$; all three lead to the condition that $h\nu \ll kT$.

Energy quantization was slow to be accepted, not an unusual fate for a radically new idea. It is not hard to see why. The systems whose energies were first quantized were the hypothetical "oscillators" that Planck assumed to form the walls of a cavity radiator. In fact, there are no such simple, one-dimensional, harmonic oscillators. The atoms that make up the wall are far more complex.

Energy quantization started to come into its own only after 1907, when Einstein showed that the same ideas that had worked so well for the cavity radiation problem could be used to solve another problem, that of the heat capacities of solids. Here, as we shall see, the systems whose energies are to be quantized are real and familiar atoms.

If you add heat Q to a solid of mass m and if a temperature rise ΔT results, the specific heat capacity is defined from

$$c_v = \frac{Q/m}{\Delta T} \qquad \text{(constant volume).} \qquad (49\text{-}7)$$

We have chosen to add heat under conditions of constant volume, so that the distances between atoms remain constant and any added energy appears entirely as energy of oscillation of the atoms about their fixed lattice sites. We also take the unit of mass to be the mole, so that comparisons from element to element can be made on the basis of the same number of atoms. When this is done, the reported quantity is called the *molar heat capacity.*

Table 49-1 shows the molar heat capacities of some elemental solids at or near room temperature. A glance at the table shows a regularity known as the Dulong and Petit rule, after the investigators who first pointed it out in 1819. It asserts simply that, with a few exceptions, all solids have the same molar heat capacity, namely, about 25 J/mol · K. Values that were substantially less than this were called "anomalous" in those early days.

Figure 49-7, which shows the molar heat capacity of lead, aluminum, and beryllium as a function of temperature, clarifies the situation. We see that c_v for all three elements approaches the same limiting value at high temperatures. That beryllium appears "anomalous" in Table 49-1 simply reflects the fact that, for this element, room temperature is not a very high temperature.

What does classical physics predict for the molar heat capacity of a solid?

The atoms in a solid are arranged in a three-dimensional lattice. Each atom, bound to its lattice site by electromagnetic forces, oscillates about that site with an amplitude that increases as the temperature increases. Each atom behaves like a tiny mass-spring oscillator with three independent degrees of freedom, corresponding to three independent directional axes along which the atom is free to move.

The classical equipartition of energy theorem associates a total energy (including both kinetic and potential forms) of kT with each directional degree of freedom. The energy per mole of a solid is then

$$u = (3kT)(N_A) = 3RT, \qquad (49\text{-}8)$$

49-5
ENERGY QUANTIZATION AND THE HEAT CAPACITY OF SOLIDS

Table 49-1
The molar heat capacities of some solids*

Solid	c_v J/mol · K	
Aluminum	23	
Beryllium		11
Bismuth	25	
Boron		13
Cadmium	25	
Carbon (diamond)		6
Copper	24	
Gold	25	
Lead	25	
Platinum	25	
Silver	24	
Tungsten	24	

* All measurements were made at room temperature; three "anomalous" values have been offset for emphasis.

figure 49-7
The molar heat capacities of three solids as a function of temperature.

1101

ENERGY QUANTIZATION AND THE HEAT CAPACITY OF SOLIDS   SEC. 49-5

in which N_A is the Avogadro constant and R is the universal gas constant.

If the solid is held at a constant volume — which we assume — we can replace Q/m in Eq. 49-7 (the heat added per mole) by Δu, the change in internal energy per mole. Doing so yields $c_v = \Delta u/\Delta T$, which becomes

$$c_v = \frac{du}{dT} \qquad (49\text{-}9)$$

in the differential limit. Substituting from Eq. 49-8 yields finally

$$c_v = \frac{d}{dT}\,(3RT) = 3R. \qquad (49\text{-}10)$$

Thus, classical theory predicts the molar heat capacity to be constant, the same for all substances, and independent of temperature. Substituting the value of R (= 8.31 J/mol · K) yields $c_v = 24.9$ J/mol · K. This agrees very well with the limiting value of c_v at high temperatures, as a glance at Fig. 49-7 and Table 49-1 shows. However, there is no suggestion of the variation at lower temperatures that is the outstanding feature of Fig. 49-7.

We turn now to the prediction of quantum theory. Einstein assumed that the energies of the atomic oscillators in the solid were quantized according to Eq. 49-6, and he assigned to each oscillator an average energy per directional degree of freedom, not of kT as in the classical case, but of

$$E = \frac{h\nu}{e^{h\nu/kT} - 1}. \qquad (49\text{-}11)$$

This is the same expression used by Planck for the average energy of the oscillators in the cavity radiation problem. In Eq. 49-11, ν is the assumed natural frequency of the oscillating atom, which Einstein left as an adjustable constant.

Multiplying the above by three to take account of the three degrees of freedom and also by the Avogadro constant gives the energy per mole. Differentiating, as in Eq. 49-9, gives eventually

$$c_v(T) = 3R(h\nu/kT)^2 \,\frac{e^{h\nu/kT}}{(e^{h\nu/kT} - 1)^2} \qquad (49\text{-}12)$$

figure 49-8
Showing the excellent agreement of the Debye quantum theory of heat capacity with experiment. The horizontal scale is the dimensionless ratio of the temperature T to the Debye temperature T_D, the latter having a characteristic value for each substance.

as Einstein's prediction for the molar heat capacity. Example 4 shows that, at high temperatures, this expression reduces to the classical prediction of $3R$. There is only one adjustable parameter in Eq. 49-12, the oscillator frequency v. Commonly, this frequency v is subsumed into a characteristic *Einstein temperature* T_E $(= hv/k)$. This temperature can be chosen so that Einstein's equation fits the data rather well, although there are small deviations at low temperatures, deviations that had not yet been experimentally established when Einstein proposed his theory.

The failure to agree with experiment at low temperatures can be traced to the fact that Einstein — perhaps deliberately — made an overly simple assumption, namely, that the oscillations of a particular atom are not influenced by those of its neighbors. In 1912 the Dutch physicist Peter Debye refined Einstein's theory by taking the interaction of the atomic oscillators with neighboring atoms into account. Figure 49-8 shows the excellent agreement of the Debye theory with experiment for a number of solids. The temperature scale in that figure is dimensionless, T_D being a constant that has a different value for each material. When these characteristic *Debye temperatures*, as they are called, are properly assigned, we see how nicely all of the experimental points fall on the same theoretical curve. A major triumph for quantum theory!

EXAMPLE 4

Einstein's heat capacity equation—the classical limit. Show that Einstein's quantum theory expression for the molar heat capacity (Eq. 49-12) reduces to the classical expression (Eq. 49-10) as $T \rightarrow \infty$.

For convenience, we can write Einstein's equation in the form

$$c_v(T) = 3Rx^2 \frac{e^x}{(e^x - 1)^2}, \qquad (v \text{ constant})$$

in which

$$x = hv/kT.$$

As $T \rightarrow \infty$, we see that $x \rightarrow 0$ so that we can make the approximations (see Appendix I)

$$e^x = 1 + x + \frac{x^2}{2!} \cdots \cong 1 \qquad \text{and} \qquad e^x - 1 \cong x.$$

Thus

$$c_v(T) \cong (3R)x^2 \frac{1}{x^2} = 3R.$$

This is indeed the classical prediction. Notice again that the Planck constant h has conveniently vanished somewhere along the line in our derivation of this limiting classical expression.

Why should Einstein's expression reduce to the classical expression at high temperatures? It is for the same reasons that we discussed at the end of Example 3. In a solid, the frequency v of the atomic oscillators was assumed by Einstein to have a single constant value, characteristic of the substance. Thus, as $T \to \infty$, we approach the condition in which $hv \ll kT$. This, as we have seen, means that the energy interval between adjacent levels for the atomic oscillators ($= hv$) is much less than the mean translational energy of the atoms (measured classically by kT). Under these conditions, the energy quantization of the atomic oscillators is not apparent, and classical conditions hold.

We were led to the idea of energy quantization by looking at the interplay between matter and radiation at the walls of a cavity radiator. Here is another example of a radiation–matter interaction, the *photoelectric effect*. It involves the Planck constant in a central way and extends the idea of quantization to the very nature of radiation itself.

Figure 49-9 shows a typical apparatus. Light of frequency v falls on a metal surface (emitter E) and, if the frequency is high enough, the light will kick electrons out of the surface. If we set up a suitable potential difference V between E and collector C, we can sweep up these *photoelectrons*, as we call them, and measure them as a photoelectric current i.

The potential difference V that acts between the emitter and the collector is not the same as the potential difference V_{ext} supplied by the external battery and read on the voltmeter in Fig. 49-9. There is also a second emf — a hidden battery, if you will — associated with the fact that the emitter and the collector are almost always made of different materials. If suitable precautions are taken, this *contact potential difference* V_{cpd} remains constant throughout the experiment. The potential V that the electrons "see" is the algebraic sum of these two quantities, or

$$V = V_{ext} + V_{cpd}. \qquad (49\text{-}13)$$

In all that follows we shall assume that this contact potential difference has been measured and taken into account and we shall express all our results in terms of V as defined by Eq. 49-13.

Figure 49-10 (curve a) shows the photoelectric current as a function of the potential difference V. We see that if V is positive and large enough, the photoelectric current reaches a constant saturation value, at which *all* photoelectrons ejected from E are collected by C.

If we reduce V to zero and then reverse it, the photoelectric current does not immediately drop to zero because the electrons emerge from emitter E with a definite speed. Some will reach the collector even though the potential difference opposes their motion. However, if we make the reversed potential difference large enough, we reach a value V_0 — called the *stopping potential* — at which the photoelectric current does indeed drop to zero. This potential difference, multiplied by the electronic charge e, gives us the kinetic energy K_{max} of the most energetic of the emitted photoelectrons. Thus

$$K_{max} = eV_0. \qquad (49\text{-}14)$$

The stopping potential V_0 — and thus K_{max} — is independent of the intensity of the incident light. Curve b in Fig. 49-10, in which the light

49-6
THE PHOTOELECTRIC EFFECT

figure 49-9
An apparatus used to study the photoelectric effect. The collected photoelectrons move in a direction opposite to that indicated by the conventional current arrows.

intensity has been doubled, shows this. Although the saturation current is also doubled, the stopping potential remains unchanged.

Figure 49-11 is a plot of the stopping potential as a function of the frequency of the incident light. We see by extrapolation that there is a sharp *cutoff frequency* v_0 corresponding to a stopping potential of zero. For light of a lower frequency than this, no photoelectrons at all are emitted. There simply is no photoelectric effect.

Three major features of the photoelectric effect cannot be explained in terms of the classical wave theory of light. As for the cavity radiation and the heat capacity problems, the failure of classical wave theory in this case is not a matter of a small numerical disagreement. The failure is total and indisputable. Here are the three problems:

1. *The intensity problem.* Wave theory requires that the oscillating electric vector **E** of the light wave increases in amplitude as the intensity of the light beam is increased. Since the force applied to the electron is $e\mathbf{E}$, this suggests that the *kinetic energy* of the photoelectrons should also increase as the light beam is made more intense. However, Fig. 49-10 shows that K_{max} ($= eV_0$) *is independent of the light intensity*; this has been tested over a range of intensities of 10^7.

2. *The frequency problem.* According to the wave theory, the photoelectric effect should occur for any frequency of the light, provided only that the light is intense enough to provide the energy needed to eject the photoelectrons. However, Fig. 49-11 shows that there exists, for each surface, a characteristic cutoff frequency v_0. *For frequencies less than v_0, the photoelectric effect does not occur, no matter how intense the illumination.*

3. *The time delay problem.* If the energy acquired by a photoelectron is absorbed directly from the wave incident on the metal plate, the "effective target area" for an electron in the metal is limited and probably not much more than that of a circle of the order of an atomic diameter. In the classical theory the light energy is uniformly distributed over the wave front. Thus, if the light is feeble enough, there should be a measurable time lag, which we shall estimate in Example 5, between the impinging of the light on the surface and the ejection of the photoelectron. During this interval the electron should be absorbing energy from the beam until it had accumulated enough to escape. However, *no detectable time lag has ever been measured.*

figure 49-10
A plot (not to scale) of data taken with the apparatus of Fig. 49-9. The intensity of the incident light is twice as great for curve *b* as for curve *a*.

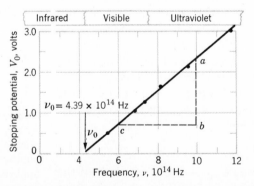

figure 49-11
The stopping potential (measured in volts) as a function of frequency for sodium. Data reported by R. A. Millikan in 1916.

EXAMPLE 5

1105   EINSTEIN'S PHOTON THEORY   SEC. 49-7

Wave theory and the time delay problem. A potassium foil is placed a distance r (= 3.5 m) from a light source whose output power P_0 is 1.0 W. How long would it take for the foil to soak up enough energy (= 1.8 eV) from the beam to eject an electron? Assume that the ejected photoelectron collected its energy from a circular area of the foil whose radius is one Bohr radius (= 5.3×10^{-11} m).

If the source radiates uniformly in all directions, the intensity of the light at a distance r is given by

$$I = \frac{P_0}{4\pi r^2} = \frac{1.0 \text{ W}}{4\pi(3.5 \text{ m})^2} = 6.5 \times 10^{-3} \text{ W/m}^2.$$

The target area A is $\pi(5.3 \times 10^{-11} \text{ m})^2$ or 8.8×10^{-21} m^2 so that the rate at which energy falls on the target is given by

$$P = IA = (6.5 \times 10^{-3} \text{ W/m}^2)(8.8 \times 10^{-21} \text{ m}^2)$$
$$= 5.7 \times 10^{-23} \text{ J/s}.$$

If all this incoming energy is absorbed, the time required to accumulate enough energy for the electron to escape is

$$t = \left(\frac{1.8 \text{ eV}}{5.7 \times 10^{-23} \text{ J/s}}\right)\left(\frac{1.6 \times 10^{-19} \text{ J}}{1 \text{ eV}}\right) = 1.4 \text{ h}.$$

Our selection of a radius for the effective target area was somewhat arbitrary, but no matter what reasonable area we choose, we would still calculate a "soak-up time" within the range of easy measurement. The fact is that no time delay has ever been observed under any circumstances, the early experiments setting an upper limit of about 10^{-9} s for such delays.

49-7 EINSTEIN'S PHOTON THEORY

In 1905 Einstein made a remarkable assumption about the nature of light, namely that, under some circumstances, it behaves as if its energy was concentrated into localized bundles, later called *photons*. The energy E of a single photon is given by

$$E = h\nu \qquad (49\text{-}15)$$

where ν is the frequency of the light. This notion that a light beam behaves like a stream of particles is in sharp contrast to the notion that it behaves like a wave. In the wave theory of light the energy is *not* concentrated into bundles but is spread out uniformly over the wave fronts.

When Planck, in 1900, derived his radiation law and first introduced the quantity h into physics he made use of the relation $E = h\nu$. He applied it, however, not to the radiation within the cavity but to the atomic oscillators that made up its walls. Planck treated the cavity radiation on the basis of wave theory. Einstein, however, was later able to derive Planck's radiation law on the basis of his photon concept. His method was both clear and simple and avoided many of the special assumptions that Planck had found it necessary to make in his pioneering effort.

If we apply Einstein's photon concept to the photoelectric effect, we can write

$$h\nu = \phi + K_{\text{max}} \qquad (49\text{-}16)$$

where $h\nu$ is the energy of the photon. Equation 49-16 says that a photon carries an energy $h\nu$ into the surface where it is absorbed by a single electron. Part of this energy (ϕ, called the *work function* of the emitting surface) is used in causing the electron to escape from the metal surface.

The excess energy $(hv - \phi)$ becomes the electron kinetic energy; if the electron does not lose energy by internal collisions as it escapes from the metal, it will still have this much kinetic energy after it emerges. Thus K_{max} represents the maximum kinetic energy that the photoelectron can have outside the surface.

Consider how Einstein's photon hypothesis meets the three objections raised against the wave-theory interpretation of the photoelectric effect. As for the first objection ("the intensity problem"), there is complete agreement of the photon theory with experiment. If we double the light intensity, we double the number of photons and thus double the photoelectric current; we do not change the energy of the individual photons or the nature of the individual photoelectric processes described by Eq. 49-16.

The second objection ("the frequency problem") is met by Eq. 49-16. If K_{max} equals zero, we have

$$hv_0 = \phi,$$

which asserts that the photon has just enough energy to eject the photoelectrons and none extra to appear as kinetic energy. If v is reduced below v_0, the individual photons, no matter how many of them there are (that is, no matter how intense the illumination), will not have enough energy to eject photoelectrons.

The third objection ("the time delay problem") follows from the photon theory because the required energy is supplied in a concentrated bundle. It is not spread uniformly over the beam cross-section as in the wave theory.

Let us rewrite Einstein's photoelectric equation (Eq. 49-16) by substituting for K_{max} from Eq. 49-14. This yields, after rearrangement,

$$V_0 = (h/e)v - (\phi/e). \qquad (49\text{-}17)$$

Thus Einstein's theory predicts a linear relationship between V_0 and v, in complete agreement with experiment; see Fig. 49-11. The slope of the experimental curve in this figure should be h/e, or

$$\frac{h}{e} = \frac{ab}{bc} = \frac{2.30 \text{ V} - 0.68 \text{ V}}{(10 \times 10^{14} - 6.0 \times 10^{14}) \text{ Hz}} = 4.1 \times 10^{-15} \text{ V} \cdot \text{s}.$$

We can find h by multiplying this ratio by the electron charge e,

$$h = (4.1 \times 10^{-15} \text{ V} \cdot \text{s})(1.6 \times 10^{-19} \text{ C}) = 6.6 \times 10^{-34} \text{ J} \cdot \text{s}.$$

From a more careful analysis of these and other data, including data taken with lithium surfaces, Millikan found the value $h = 6.57 \times 10^{-34}$ J · s with an accuracy of about 0.5%. This agreement with the value of h derived from Planck's radiation formula is a striking confirmation of Einstein's photon concept.

For the 26-year-old Einstein 1905 was a banner year. During that year he published, in sequence and in the same prestigious journal: (1) his theory of photons, (2) a theory of Brownian motion that contributed greatly to further understanding of the atomic nature of matter, (3) his special theory of relativity, and (4) the germ of his now-famous $E = mc^2$ relationship, derived from this theory. During all this time (and for four additional years), he was employed as a minor official in the Swiss patent office, working on physics privately and in his spare time!

When Einstein first advanced his photon theory of light, the facts of photoelectricity were not nearly as well established experimentally as we have described. Precise photoelectric measurements are difficult, and it was not until 1916 that the American physicist R. A. Millikan successfully subjected Einstein's photoelectric equation to rigorous experimental test. Although Millikan showed that this equation agreed with experiment in every detail, he himself remained unconvinced that Einstein's light particles were real. He wrote of Einstein's ". . . bold, not to say reckless, hypothesis . . ." and wrote further that Einstein's photon concept ". . . seems at present to be wholly untenable."

Planck, the very originator of the constant h, did not at once accept Einstein's photons either. In recommending Einstein for membership in the Royal Prussian Academy of Sciences in 1913 he wrote: "that he may sometimes have missed the target in his speculations, as for example in his theory of light quanta, cannot really be held against him." Far from being unusual, it is almost commonplace that truly novel ideas are accepted only slowly, even by such men of genius as Millikan and Planck. It was, incidentally, for his photon theory that Einstein received the Nobel Prize in physics for 1921.

EXAMPLE 6

The work function of sodium.   Find the work function of sodium from the data plotted in Fig. 49-11.

The intercept of the straight line in Fig. 49-11 on the frequency axis is the cutoff frequency v_0. Putting $V_0 = 0$ and $v = v_0$ in Eq. 49-17 yields

$$\phi = hv_0 = (6.63 \times 10^{-34} \text{ J} \cdot \text{s})(4.4 \times 10^{14} \text{ Hz})$$

$$= 2.9 \times 10^{-19} \text{ J} = 1.8 \text{ eV}.$$

We note from Eq. 49-17 that a determination of the Planck constant h involves only the slope of the straight line in Fig. 49-11 and a determination of the work function ϕ involves only the intercept. Convince yourself that in the first case you need not take the contact potential difference V_{cpd} into account but in the second case you must do so.

EXAMPLE 7

A rain of photons.   At what rate per unit area R do photons strike the metal plate in Example 5? Assume a wavelength of 589 nm (yellow sodium light).

The intensity of the light falling on the plate is, from Example 5,

$$I = (6.5 \times 10^{-3} \text{ J/m}^2 \cdot \text{s})(1 \text{ eV}/1.6 \times 10^{-19} \text{ J})$$

$$= 4.1 \times 10^{16} \text{ eV/m}^2 \cdot \text{s}.$$

Each photon has an energy given by

$$E = hv = \frac{hc}{\lambda} = \frac{(6.63 \times 10^{-34} \text{ J} \cdot \text{s})(3.00 \times 10^8 \text{ m/s})}{(5.89 \times 10^{-7} \text{ m})}$$

$$= (3.4 \times 10^{-19} \text{ J}) \left(\frac{1 \text{ eV}}{1.6 \times 10^{-19} \text{ J}} \right) = 2.1 \text{ eV}.$$

The rate per unit area at which photons strike the plate is then

$$R = \frac{I}{E} = \frac{4.1 \times 10^{16} \text{ eV/m}^2 \cdot \text{s}}{2.1 \text{ eV/photon}} = 1.9 \times 10^{14} \text{ photons/m}^2 \cdot \text{s}$$

Even at this low light intensity ($\sim 1 \ \mu\text{W/cm}^2$) the photon rate is very great, $\sim 10^{10}$ photons falling on 1 mm² each second. It is small wonder that we do not ordinarily notice the granularity of light. In the same way, because of the smallness of atoms, we do not ordinarily notice the granularity of matter.

49-8
THE COMPTON EFFECT

Cavity radiation, which involved largely the infrared part of the spectrum, was our first example of the interaction of radiation with matter. The photoelectric effect, our second example, involved visible and ultraviolet light. Here we describe the *Compton effect,* in which the key experiments occur in the X-ray and the gamma-ray regions of the electromagnetic spectrum.

The Compton effect can be readily understood in terms of billiard-ball-like collisions between photons and electrons. In the explanation we must take into account not only the energy of the photons but also their linear momentum, a property that we have not needed to introduce so far. We have seen that Einstein's analysis of the heat capacity of a solid in quantum terms went far to convince people to accept the notion of energy quantization. In the same way, Compton's analysis of the effect that bears his name went far to convince people of the reality of photons.

In 1923 Arthur Holly Compton, at Washington University in St. Louis, allowed a beam of X rays with sharply defined wavelength λ to fall on a graphite scatterer S, as in Fig. 49-12. He measured, for various angles of scattering, the intensity of the scattered X rays as a function of their wavelength. Figure 49-13 shows his experimental results. We see that although the incident beam consists essentially of a single wavelength λ, the scattered X rays have intensity peaks at two wavelengths; one of them is the same as the incident wavelength, the other, λ', being larger by an amount $\Delta\lambda$. This so-called *Compton shift* $\Delta\lambda$ varies with the angle at which the scattered X rays are observed.

The presence of a scattered wave of wavelength λ' cannot be understood if the incident X rays are regarded as an electromagnetic wave. On this picture the incident wave of frequency ν causes electrons in the scattering block to oscillate at that same frequency. These oscillating electrons, like charges surging back and forth in a small radio transmitting antenna, radiate electromagnetic waves that again have this same frequency ν. Thus, on the wave picture the scattered wave should have the same frequency — and the same wavelength — as the incident wave.

Compton explained his experimental results by postulating that the incoming X-ray beam was not a wave but an assembly of photons of energy E $(= h\nu)$ and that these photons experienced billiard-ball-like collisions with the free electrons in the scattering block. The "recoil" photons emerging from the block constitute, on this view, the scattered radiation. Since the incident photon transfers some of its energy to the electron with which it collides, the scattered photon must have a lower energy E'; it must therefore have a lower frequency ν' $(= E'/h)$, which implies a larger wavelength λ' $(= c/\nu')$. This point of view accounts, at least qualitatively, for the wavelength shift $\Delta\lambda$. Notice how different this particle model of X-ray scattering is from that based on the wave picture. Now let us analyze a single photon-electron collision quantitatively.

Figure 49-14 shows a collision between a photon and an electron, the electron assumed to be initially at rest and essentially free, that is, not bound to the atoms of the scatterer. Let us apply the law of conservation of energy to this collision. Since the recoil electrons may have a speed v that is comparable with that of light, we must use the relativistic expression for the kinetic energy of the electron. From the conservation of energy principle we may write

$$h\nu = h\nu' + (m - m_0)c^2,$$

in which the second term on the right (see Eq. 8-21) is the relativistic expression for the kinetic energy of the recoiling electron, m being the relativistic mass and m_0 the rest mass of that particle. Substituting c/λ

figure 49-12
Compton's experimental setup.

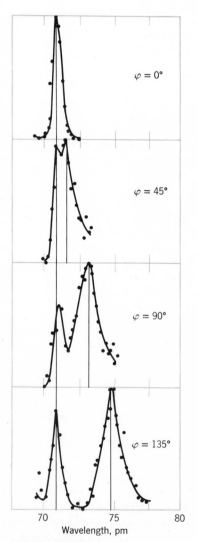

figure 49-13
Compton's experimental results for four values of the scattering angle φ. (1 pm = 1 picometer = 10^{-12} m.)

for v (and c/λ' for v') and using Eq. 8-20 to eliminate the relativistic mass m leads us to

$$\frac{hc}{\lambda} = \frac{hc}{\lambda'} + m_0 c^2 \left(\frac{1}{\sqrt{1 - (v/c)^2}} - 1 \right). \qquad (49\text{-}18)$$

Now let us apply the (vector) law of conservation of linear momentum to the collision of Fig. 49-14. We first need an expression for the momentum of a photon. In Section 42-2 we saw that if an object completely absorbs an energy U from a parallel light beam that falls on it, the light beam, according to the wave theory of light, will simultaneously transfer to the object a linear momentum given by U/c. On the photon picture we imagine this momentum to be carried along by the individual photons, each photon transporting linear momentum in amount $p = h\nu/c$, where $h\nu$ is the photon energy. Thus, if we substitute λ for c/ν, we can write

$$p = \frac{E}{c} = \frac{h\nu}{c} = \frac{h}{\lambda}. \qquad (49\text{-}19)$$

For the electron, the relativistic expression for the linear momentum is given by Eq. 9-13, or

$$\mathbf{p} = \frac{m_0 \mathbf{v}}{\sqrt{1 - (v/c)^2}}.$$

We can then write for the conservation of the x-component of linear momentum

$$\frac{h}{\lambda} = \frac{h}{\lambda'} \cos \varphi + \frac{m_0 v}{\sqrt{1 - (v/c)^2}} \cos \theta \qquad (49\text{-}20)$$

and for the y-component

$$0 = \frac{h}{\lambda'} \sin \varphi - \frac{m_0 v}{\sqrt{1 - (v/c)^2}} \sin \theta. \qquad (49\text{-}21)$$

Our aim is to find $\Delta\lambda$ ($= \lambda' - \lambda$), the wavelength shift of the scattered photons, so that we may compare it with the experimental results of Fig. 49-13. Compton's experiment did not involve observations of the recoil electron in the scattering block. Of the five collision variables (λ, λ', v, φ, and θ) that appear in the three equations (49-18, 49-20, 49-21) we may eliminate two. We chose to eliminate v and θ, which deal only with the electron, thereby reducing the three equations to a single relation among the variables.

Carrying out the necessary algebraic steps (see Problem 57) leads to this simple result:

$$\Delta\lambda = \lambda' - \lambda = \frac{h}{m_0 c}(1 - \cos \varphi) \qquad \text{(the Compton shift)}. \quad (49\text{-}22)$$

Thus the Compton shift $\Delta\lambda$ depends only on the scattering angle φ and not on the initial wavelength λ. Equation 49-22 predicts within experimental error the experimentally observed Compton shifts of Fig. 49-13. Note from the equation that $\Delta\lambda$ varies from zero (for $\varphi = 0$, corresponding to a "grazing" collision in Fig. 49-14, the incident photon being scarcely deflected) to $2h/m_0 c$ (for $\varphi = 180°$, corresponding to a "head-on" collision, the incident photon being reversed in direction).

Remember that the Compton shift $\Delta\lambda$ is a purely quantum effect, not expected to occur on the basis of classical physics. Equation 49-22 sup-

figure 49-14
A photon of wavelength λ strikes an electron at rest. The photon is scattered at an angle φ with an increased wavelength λ'. The electron moves off with a speed v at an angle θ.

ports this contention by showing that $\Delta\lambda \rightarrow 0$ as $h \rightarrow 0$. The device of letting the Planck constant approach zero is a formal way of testing quantum equations to see whether they predict what would happen if the laws of classical physics applied not only to large objects but also to atoms and electrons.

It remains to explain the presence of the peak in Fig. 49-13 for which the wavelength does not change on scattering. This peak results from collisions between photons and electrons which, instead of being free, are tightly bound in an ionic core in the scattering block. During photon collisions the bound electrons behave like very heavy free electrons. This is because the ionic core as a whole recoils during the collision. Thus the effective mass M for a carbon scatterer is approximately the mass of a carbon nucleus. Since this nucleus contains six protons and six neutrons, we have approximately, $M = 12 \times 1840 m_0 = 22{,}000 m_0$. If we replace m_0 by M in Eq. 49-22, we see that the Compton shift for collisions with tightly bound electrons is immeasurably small.

As in the cavity radiation problem and the photoelectric effect, the Planck constant h is centrally involved in the Compton effect. The quantity h is the central constant of quantum physics. In a universe in which $h = 0$ there would be no quantum physics and classical physics would be valid in the subatomic domain. In particular, as Eq. 49-22 shows, there would be no Compton efect (that is, $\Delta\lambda = 0$) in such a universe.

In recent years a phenomenon based on the Compton effect has received a lot of public attention. It is the so-called *electromagnetic pulse* or EMP, which describes one of the effects of the explosion of a thermonuclear device high in the atmosphere. The X rays and gamma rays generated in the explosion have Compton collisions with electrons in the upper atmosphere, knocking them sharply forward. This sudden and enormous forward surge of charge sets up electromagnetic fields that can play havoc with unshielded electric circuits on earth. The effect was first noticed when power and communication circuits in Hawaii failed at the time of a thermonuclear test explosion in Bikini, some 2500 miles away.

EXAMPLE 8

A Compton collision analyzed. X rays with $\lambda = 100$ pm are scattered from a carbon block. The scattered radiation is viewed at 90° to the incident beam. (a) What is the Compton shift $\Delta\lambda$? (b) What kinetic energy is imparted to the recoiling electron?

(a) Putting $\varphi = 90°$ in Eq. 49-22, we have, for the Compton shift,

$$\Delta\lambda = \frac{h}{m_0 c}(1 - \cos\varphi)$$

$$= \frac{6.63 \times 10^{-34}\,\text{J}\cdot\text{s}}{(9.11 \times 10^{-31}\,\text{kg})(3.00 \times 10^8\,\text{m/s})}(1 - \cos 90°)$$

$$= 2.43 \times 10^{-12}\,\text{m} = 2.43\,\text{pm}.$$

(b) If we put K for the kinetic energy of the electron, we can write Eq. 49-18 as

$$\frac{hc}{\lambda} = \frac{hc}{\lambda'} + K.$$

Since $\lambda' = \lambda + \Delta\lambda$, we obtain

$$\frac{hc}{\lambda} = \frac{hc}{\lambda + \Delta\lambda} + K,$$

which reduces to

$$K = \frac{hc\,\Delta\lambda}{\lambda(\lambda + \Delta\lambda)}$$

$$= \frac{(6.63 \times 10^{-34}\,\text{J}\cdot\text{s})(3.00 \times 10^{8}\,\text{m/s})(2.43 \times 10^{-12}\,\text{m})}{(1.00 \times 10^{-10}\,\text{m})(100 + 2.43) \times 10^{-12}\,\text{m}}$$

$$= 4.72 \times 10^{-17}\,\text{J} = 295\ \text{eV}.$$

You can show that the initial photon energy E in this case $(= h\nu = hc/\lambda)$ is 12.4 keV so that the photon lost about 2.4% of its energy in this collision. A photon whose energy was ten times as large $(= 124\ \text{keV})$ can be shown to lose 20% of its energy in a similar collision. This is consistent with the fact that $\Delta\lambda$ does not depend on the initial wavelength. More energetic X rays, which have smaller wavelengths, will experience a larger percent increase in wavelength and thus a larger percent loss in energy.

As Fig. 49-3 shows, the radiation emitted by a heated solid is continuously distributed in wavelength. We say that such radiation forms a *continuous spectrum.* By contrast, Fig. 49-15 shows some *line spectra,*

49-9
LINE SPECTRA

figure 49-15
Selected lines from various regions of the electromagnetic spectrum; see Table 49-2.

Table 49-2
Some selected line spectra; see Figure 49-15

Figure	Entity	Wavelength m	Spectrum Region	Mode
a	Hg-198 nucleus	$\sim 3 \times 10^{-12}$	Gamma ray	Emission
b	Mo atom	$\sim 6 \times 10^{-11}$	X ray	Emission
c	Fe atom	$\sim 3 \times 10^{-7}$	Ultraviolet	Emission
d	HCl molecule	$\sim 3 \times 10^{-6}$	Infrared	Absorption
e	NH_3 molecule	$\sim 1 \times 10^{-2}$	Microwave	Absorption
f	H_2 molecule	~ 40	Radio	Absorption

associated with the emission (or absorption) of radiation by isolated atoms, molecules, and nuclei. These curves, which involve wavelengths from widely separate regions of the electromagnetic spectrum (see Table 49-2), only begin to suggest the bewildering variety of line spectra that can be measured in the laboratory.

The line spectra are the characteristic identifying signatures of the atoms, or other entities, that emit them. We shall see that, by studying such spectra, we can learn something about how atoms are put together. We start by analyzing the spectrum of the hydrogen atom. It is the simplest atom and, not surprisingly, it has the simplest spectrum.

49-10
NIELS BOHR AND THE HYDROGEN ATOM

The interaction between radiation and matter is a theme that runs through this chapter. We study it here in its simplest form: A hydrogen atom (matter) emits or absorbs a photon (radiation). The wavelengths of the lines in the spectrum of atomic hydrogen (see Fig. 49-16) have been known with precision for many years and stand as a testing ground for any theory of the hydrogen atom that may be put forward.

Classically, we can imagine that the electron in the hydrogen atom revolves about the central nucleus in an orbit of radius r; see Fig. 49-17. We can then hypothesize that the frequency of the radiation that the atom emits is equal to the frequency at which the electron circulates in its orbit. Classical theory predicts that such an accelerating electron will indeed radiate *and* at its frequency of revolution. However, this theory has a fatal flaw. The orbiting electron will thereby radiate its store of energy completely away, emitting a continuous spectrum of radiation as it spirals into the nucleus and disappears.

Thus the great classical theories of Newton and Maxwell stand helpless before the spectrum of the simplest atom. They cannot even account for the existence of the spectrum lines, let alone predict their wavelengths.

In 1913, just two years after Rutherford had put forward the idea that

figure 49-16
The emission spectrum of atomic hydrogen in the visible and the near ultraviolet region. The series of lines shown is called the Balmer series. The arrow marks the series limit.

the atom has a nucleus, the great Danish physicist Niels Bohr (see Fig. 49-18) proposed a model for the hydrogen atom that not only accounted for the presence of the spectrum lines but — with no adjustable parameters whatever — predicted their wavelengths to an accuracy of about 0.02%. Although Bohr's theory was wonderfully successful for hydrogen and for other one-electron atoms such as He^+, it proved less useful for more complex atoms. We now regard Bohr's theory as an inspired and ingenious first step toward the more comprehensive quantum theory that followed it. Let us retrace Bohr's footsteps.

The wavelengths of the lines of the spectrum of atomic hydrogen can be represented with great accuracy by an empirical formula, first proposed in 1885 by Johannes Balmer, a Swiss high school teacher and a part-time instructor at the University of Basel. As later modified by Rydberg and generalized, the formula is

$$\frac{1}{\lambda} = R\left(\frac{1}{m^2} - \frac{1}{n^2}\right) \quad \text{(Balmer-Rydberg formula).} \quad (49\text{-}23)$$

Here R, called the *Rydberg constant*, has the (measured) value of

$$R = 1.097 \times 10^7 \text{ m}^{-1}.$$

The quantities m and n in Eq. 49-23 are positive integers. If we put $m = 2$ and allow n to have the values 3, 4, 5, . . . , the formula generates the wavelengths of the lines of the so-called Balmer series of the hydrogen spectrum, shown in Fig. 49-16. Other values of m generate other series of lines in this spectrum, lines not shown in that figure and not known at the time that Bohr developed his theory. Bohr's challenge: Find a theoretical basis for Eq. 49-23, the successful but empirical Balmer-Rydberg formula.

figure 49-17
A classical model of the one-electron atom. An electron of charge $-e$ and mass m_e moves about a nucleus of charge Ze and mass M, in a circular orbit. We assume that $M \gg m_e$.

figure 49-18
Niels Bohr with Aage Bohr, one of his five sons. Both earned Nobel prizes in physics, Niels in 1922 and Aage in 1975. (American Institute of Physics, Neils Bohr Library, Margarethe Bohr Collection.)

Bohr, realizing that classical physics had come to a dead end on the hydrogen atom problem, put forward two bold postulates. Both turned out to be enduring features that carry over in full force to our modern point of view. Moreover, both turned out to be quite general, applying not only to the hydrogen atom but to atomic, molecular, and nuclear systems of all kinds. We discuss each postulate in turn.

1. *The postulate of stationary states.* Bohr assumed that the hydrogen atom can exist *without radiating* in any one of a number of *stationary states* of well-defined energy. This assumption flies in the face of classical theory but Bohr's attitude was, "Let's assume it anyway and see what happens." Note that this postulate says nothing at all about what these states look like. There is, for example, no mention of orbits.

2. *The frequency postulate.* Bohr assumed that the hydrogen atom can emit or absorb radiation *only* when the atom changes from one of its stationary states to another. The energy of the emitted (or absorbed) photon is equal to the difference in energy between these two states. Thus, if an atom changes from an initial state of energy E_n to a final state of (lower) energy E_m, the energy of the emitted photon is given by

$$h\nu_{nm} = E_n - E_m \qquad \text{(Bohr's frequency postulate).} \qquad (49\text{-}24)$$

This postulate ties together two new ideas (the photon hypothesis and energy quantization) with one familiar old idea (the conservation of energy).

Bohr now sought to interpret the empirical Balmer-Rydberg formula in terms of his postulates. We start by recasting this formula (Eq. 49-23) in the general format of Bohr's frequency postulate (Eq. 49-24). If we multiply each side of Eq. 49-23 by hc and if we replace c/λ by its equal, ν_{nm}, we can write

$$h\nu_{nm} = \left(-\frac{hcR}{n^2}\right) - \left(-\frac{hcR}{m^2}\right).$$

Comparing, term by term, with Eq. 49-24 allows us to infer

$$E_n = -\frac{hcR}{n^2} \qquad n = 1, 2, 3, \ldots \qquad (49\text{-}25)$$

for the energies of the stationary states of the hydrogen atom. The fact that the energy is negative recognizes the fact that the atom is in a bound state, that is, work must be done by some external agent to pull it apart. In the same way, the earth – sun system is a bound state; work must be done by an external agent to tear *this* system apart against the gravitational force that holds it together.

Figure 49-19 shows an energy level diagram for the hydrogen atom, the energies being calculated from Eq. 49-25. Each level is marked with its quantum number n. A downward-pointing arrow connecting two levels represents the emission of a photon, in accord with Bohr's frequency postulate (Eq. 49-24). Table 49-3 displays the wavelengths of some of the lines shown in this figure.

Equation 49-25 is not yet the end of the path that Bohr followed. The value of the Rydberg constant in that formula — at this stage — can be found only from experiment. What is needed is a way of expressing his

figure 49-19
Energy levels and transitions in the spectrum of atomic hydrogen. When Bohr
proposed his theory only the Balmer series (see Fig. 49-16) was known. Bohr
predicted the existence of a state with $n = 1$ before the Lyman series, based on that
state, was discovered.

Table 49-3
The hydrogen spectrum
(Some selected lines)

Name of Series	Quantum Number		Wavelength, nm
	n (Lower State)	m (Upper State)	
Lyman	1	2	121.6
	1	3	102.6
	1	4	97.0
	1	5	94.9
	1	6	94.0
	1	∞ Series limit	91.2
Balmer	2	3	656.3
	2	4	486.1
	2	5	434.1
	2	6	410.2
	2	7	397.0
	2	∞ Series limit	364.6
Paschen	3	4	1875.1
	3	5	1281.8
	3	6	1093.8
	3	7	1005.0
	3	8	954.6
	3	∞ Series limit	822.0

constant in terms of other, known physical constants. This, as we shall see, is precisely what Bohr did next.

EXAMPLE 9

The binding energy of the hydrogen atom. Calculate the binding energy of the hydrogen atom, that is, the energy that must be added to the atom to remove its electron.

The energy of the atom when the electron has been removed from it, found by letting $n \to \infty$ in Eq. 49-25, is zero. The binding energy E_b is therefore numerically equal to the energy of the atom in its lowest energy state, found by putting $n = 1$ in Eq. 49-25. That is

$$E_b = -E_1 = -\left(-\frac{hcR}{1^2}\right)$$

$$= (6.63 \times 10^{-34} \text{ J} \cdot \text{s})(3.00 \times 10^8 \text{ m/s})(1.097 \times 10^7 \text{ m}^{-1})$$

$$= 2.18 \times 10^{-18} \text{ J} = 13.6 \text{ eV}.$$

This calculated value agrees with the experimentally observed binding energy for the hydrogen atom.

EXAMPLE 10

Calculating wavelengths. (a) What is the wavelength of the least energetic photon in the Balmer spectrum?

We identify the Balmer series (see Fig. 49-19) by putting $m = 2$ in Eq. 49-23. From the relation $E = h\nu$, the least energetic photon has the smallest frequency and thus the greatest wavelength. This means that we must put $n = 3$ (the smallest possible value) in Eq. 49-23; any higher value of n would yield a smaller wavelength. With these substitutions we have

$$\frac{1}{\lambda} = R\left(\frac{1}{m^2} - \frac{1}{n^2}\right)$$

$$= (1.097 \times 10^7 \text{ m}^{-1})\left(\frac{1}{2^2} - \frac{1}{3^2}\right) = 1.524 \times 10^6 \text{ m}^{-1}$$

or

$$\lambda = (1.524 \times 10^6 \text{ m}^{-1})^{-1} = 6.563 \times 10^{-7} \text{ m} = 656.3 \text{ nm}.$$

(b) What is the wavelength of the series limit for the Balmer series?

Again we put $m = 2$ in Eq. 49-23. To find the series limit (see Fig. 49-19) we let $n \to \infty$. Equation 49-23 then becomes

$$\frac{1}{\lambda} = (1.097 \times 10^6 \text{ m}^{-1})\left(\frac{1}{2^2} - \frac{1}{\infty}\right) = 2.743 \times 10^6 \text{ m}^{-1}$$

which yields $\lambda = 3.646 \times 10^{-7}$ m $= 364.6$ nm. Note that both of these numerical results appear in Table 49-3.

49-11
BOHR PUSHES ON—THE CORRESPONDENCE PRINCIPLE

Bohr's next task was to derive an expression for the Rydberg constant R in terms of known quantities. He did this by invoking the *correspondence principle*, which states: *Quantum theory must agree with classical theory in the limit in which classical theory is known to agree with experiment.*

Specifically, Bohr wrote down expressions for the frequency at which an atom radiates (1) according to classical laws and (2) according to quantum laws. Although these two expressions do not agree in general, the correspondence principle tells us that they *must* agree if we apply them to systems much larger than a normal atom. Let us see how Bohr took advantage of his fact.

Here we adopt the classical atom model of Fig. 49-17, assuming a central charge of Ze for generality, Z being the atomic number of the nucleus. This model corresponds to the hydrogen atom if we set $Z = 1$. It would apply to a singly ionized helium atom for $Z = 2$, and to a doubly ionized lithium atom for $Z = 3$.

Although the model of Fig. 49-17 fails for systems of atomic dimensions it does give correct results for macroscopic systems, that is, for $r \rightarrow \infty$, where r is the orbit radius. In seeking a theoretical underpinning for the Rydberg constant, Bohr was careful to use the classical model of Fig. 49-17 *only in his limiting case.*

In Example 11 we shall show that, using Newtonian mechanics, a number of mechanical properties of the system of Fig. 49-17 can be expressed in terms of the orbit radius r. Two of them concern us now, the frequency of revolution v of the electron in its orbit and the total energy E. The expressions are:

$$v(r) = \left(\frac{Ze^2}{16\pi^3\epsilon_0 m_e r^3}\right)^{1/2} \quad \text{and} \quad E(r) = -\frac{Ze^2}{8\pi\epsilon_0 r}. \quad (49\text{-}26)$$

There is, of course, no sign of quantization in these classical expressions. They merely tell us that, given the orbit radius, we can easily calculate the angular frequency of the electron and the total energy of the system.

Eliminating the radius r between the two expressions above yields a relation between the angular frequency and the energy, namely,

$$v_{\text{cm}} = \left(-\frac{32 Z\epsilon_0^2 E^3}{m_e e^4}\right)^{1/2}$$

in which we have added the subscript cm to remind us that this expression is derived on the basis of classical mechanics.

Substituting for E from Eq. 49-25 gives, as our final classical expression for the angular frequency,

$$v_{\text{cm}} = \left(\frac{32\epsilon_0^2 h^3 c^3 R^3}{Z^2 m_e e^4}\right)^{1/2} \frac{1}{n^3} \quad (49\text{-}27)$$

In classical physics, this angular frequency of revolution is also the frequency of the emitted radiation.

We turn now to the quantum point of view. In quantum terms, the frequency v_{qm} that corresponds to the classical frequency we have just calculated is associated with a transition from a state with quantum number n to the next lower state, whose quantum number is $n - 1$. Putting $m = n - 1$ in Eq. 49-23 gives us

$$v_{\text{qm}} = \frac{c}{\lambda} = cR\left(\frac{1}{(n-1)^2} - \frac{1}{n^2}\right) = cR\frac{(2n-1)}{(n-1)^2 n^2}.$$

This expression should agree with the classical expression in the limit of large quantum numbers. Letting $n \rightarrow \infty$ in the above yields

$$v_{\text{qm}} \cong \frac{2cR}{n^3} \quad \text{for } n \gg 1, \quad (49\text{-}28)$$

which is the relationship we seek.

We are ready at last to apply the correspondence principle. This

Table 49-4
The correspondence principle and the hydrogen atom

Quantum Number, n	Frequency of Revolution in Orbit (Eq. 49-27) Hz	Frequency of Transition to Next Lowest State (Eq. 49-28), Hz	Difference, %
2	8.22×10^{14}	24.7×10^{14}	67
5	5.26×10^{13}	7.40×10^{13}	29
10	6.58×10^{12}	7.72×10^{12}	15
50	5.26×10^{10}	5.43×10^{10}	3.1
100	6.580×10^{9}	6.680×10^{9}	1.5
1,000	6.5797×10^{6}	6.5896×10^{6}	0.15
10,000	6.5797×10^{3}	6.5807×10^{3}	0.015
25,000	4.2110×10^{2}	4.2113×10^{2}	0.007
100,000	6.5798	6.5799	0.0007

principle tells us that, in the limit of large quantum numbers, the frequency ν_{qm} calculated from Eq. 49-28 (a quantum expression) must equal the frequency ν_{cm} calculated from Eq. 49-27 (a classical expression). Table 49-4 shows this principle in action.

Equating Eqs. 49-27 and 49-28 and solving for the Rydberg constant R gives

$$R = \frac{m_e Z^2 e^4}{8\epsilon_0^2 h^3 c},\qquad(49\text{-}29)$$

which is what we set out to find.

Here we have a *theoretically predicted value* for the Rydberg constant in terms of other fundamental constants—the charge e and the mass m_e of the electron, the speed c of light, and the Planck constant h. Bohr, using data available in his time for these constants, obtained good agreement with the experimentally determined value of R, the agreement today being within extremely narrow limits of experimental error.

We can now regard the constant R as theoretically determined and, by substituting the above expression for R into Eq. 49-25, obtain

$$E_n = -\frac{m_e Z^2 e^4}{8\epsilon_0^2 h^2}\frac{1}{n^2},\qquad(49\text{-}30)$$

a purely quantum expression for the energies of the stationary states of the hydrogen atom. This expression is Bohr's triumph. Everything that he has done so far, including the postulate of stationary states, the frequency postulate, the correspondence principle, and Eq. 49-30, the expression for the energy of the hydrogen atom states, carries over unchanged into modern quantum mechanics.

A Bohr orbit—some classical results. Let us apply Newton's second law ($F = m_e a$) to the motion of the electron in the classical orbit shown in Fig. 49-17. The results, though classically derived, will be valid in the limit of large quantum numbers. For generality, we take the central charge to be Ze rather than e, where Z is the atomic number, $Z = 1$ identifying hydrogen. We assume further that $M \gg m_e$, where M is the nuclear mass and m_e is the mass of the orbiting electron.

EXAMPLE 11

Combining Coulomb's force law with Newton's second law gives us

$$\frac{1}{4\pi\epsilon_0} \frac{(Ze)(e)}{r^2} = m_e \frac{v^2}{r},$$

in which v is the speed of the electron in its orbit. Solving for v yields

$$v(r) = \sqrt{\frac{Ze^2}{4\pi\epsilon_0 m_e r}},$$

which tells us what the orbital speed is if we know the orbit radius.

From this result we can write an expression for the frequency of revolution of the electron in its orbit, or

$$v(r) = \frac{v}{2\pi r} = \sqrt{\frac{Ze^2}{16\pi^3\epsilon_0 m_e r^3}}. \qquad \text{[see Eq. 49-26]}$$

The kinetic energy follows from

$$K(r) = \frac{1}{2} m_e v^2 = \frac{Ze^2}{8\pi\epsilon_0 r}.$$

The potential energy is given by

$$U(r) = -\frac{Ze^2}{4\pi\epsilon_0 r}$$

so that the total mechanical energy $E(r)$ follows from

$$E(r) = K(r) + U(r) = -\frac{Ze^2}{8\pi\epsilon_0 r}. \qquad \text{[see Eq. 49-26]}$$

Finally, we can show (using the expression given earlier for v) that the angular momentum is given by

$$L(r) = m_e v r = \sqrt{\frac{Ze^2 m_e r}{4\pi\epsilon_0}}. \qquad (49\text{-}31)$$

Thus, if we knew the orbit radius, we could find the orbital linear speed, the frequency of revolution, the kinetic energy, the potential energy, the total mechanical energy, and the angular momentum. We see from their interconnections that if any one of these quantities turns out to be quantized, all of them will be. There is, however, no hint of the quantization of anything in these purely classical calculations.

Every theory in science should be tested by pushing it to the limit of its validity. Bohr did this for his theory by asking, in effect, "What will the atom model of Fig. 49-17 predict if I assume that it holds, not only for large quantum numbers, but for all quantum numbers?" We discuss three of his results.

1. *The size of atoms.* By eliminating the energy E between the (classical) Eq. 49-26 and the (quantum) 49-30, we can find an expression for the radii of the quantized Bohr orbits. It is

$$r_n = \left(\frac{\epsilon_0 h^2}{Ze^2 \pi m_e}\right) n^2 = a_0 n^2 \qquad n = 1, 2, 3, \ldots \qquad (49\text{-}32)$$

The quantity a_0, called the *Bohr radius*, has the value

$$a_0 = 5.292 \times 10^{-11} \text{ m} = 52.92 \text{ pm}.$$

In a formal sense, a_0 is the radius of the Bohr orbit corresponding to $n = 1$, which defines the ground state of the hydrogen atom in Bohr's semiclassical planetary model of the one-electron atom, where we visu-

49-12
PUSHING BOHR THEORY TO THE LIMIT

alize the electron moving in planetary orbits. Today we do not believe in such orbits but, based on experiment, we do have some notion of the size of atoms. They are all roughly of the order of magnitude of the Bohr radius! It is amazing that, although Bohr puts nothing into his theory that had anything to do with the size of atoms, it nevertheless generated a number that is just about the right size. Today we use the Bohr radius as a convenient unit in which to measure lengths on the scale of atomic dimensions.

2. *The quantization of angular momentum.* The fact that the energy (see Eq. 49-30) and the radius (see Eq. 49-32) of the Bohr semiclassical atom are quantized means that other mechanical properties are also quantized in his planetary model. The quantization of the angular momentum of the orbiting electron turns out to be particularly simple. It is (see Problem 73)

$$L = n \left(\frac{h}{2\pi} \right) = n\hbar \qquad n = 1, 2, 3, \ldots, \qquad (49\text{-}33)$$

in which we have written $\hbar$ (pronounced "ħ-bar") as a convenient short-hand for $h/2\pi$.

Here we begin to see departures from modern quantum theory. In that theory the quantization of angular momentum plays a truly central role; the quantization relation, however, differs somewhat from Eq. 49-33. For the ground state of the hydrogen atom, for example, this

figure 49-20
A poster commemorating the centenary of Bohr's birth, showing Bohr as atom builder. (American Institute of Physics, Neils Bohr Library.)

equation predicts $L = \hbar$; modern quantum theory, on the other hand, predicts $L = 0$, in agreement with experiment. It is to Bohr's credit that he foresaw the crucial importance of angular momentum quantization and, indeed, proposed Eq. 49-33 as an alternative basic hypothesis from which his theory could be developed; see Problem 79.

In 1924, eleven years after Bohr had presented his theory, Louis de Broglie gave a satisfying physical interpretation of the Bohr quantization rule for angular momentum (see Section 50-2), which also gave some insight into why there is no radiation from stationary states of the atom.

3. *Beyond the hydrogen atom.* Bohr quite properly pushed his theory for all it was worth, taking as his goal the development of a theoretical basis for the periodic table of the elements. Figure 49-20, a poster designed to celebrate the centenary of his birth, shows Bohr as atom builder. However, beyond hydrogen and hydrogen-like atoms such as He$^+$, Bohr had limited success. His "great and shining moment" was no doubt his prediction that element 72, then a blank space in the periodic table, should have chemical properties like those of zirconium. This led to the discovery of element 72 in zirconium ores and the naming of the new element *hafnium*, after an early name for Copenhagen, Bohr's hometown. Word of this discovery came dramatically to Bohr in Stockholm, just hours before he was scheduled to receive the Nobel prize.

EPILOG

We have seen in this chapter the breakdown of classical ideas and the emerging of quantum concepts. In our studies of cavity radiation, the heat capacities of solids, and the hydrogen atom, we discovered the quantization of energy. In the photoelectric effect and the Compton effect, we found evidence that radiation has a particle aspect to its nature, giving rise to the quantization of radiation itself in the form of the photon. Throughout, we saw the central role of the Planck constant as a measure of the "graininess" or quantization of the energy, the angular momentum, and the particulate nature of both matter and radiation. Throughout, the correspondence principle has taught us to look at classical physics as a limiting case of the more general quantum physics. Guided by this principle, we have tested all our quantum formulas to make sure that they reduce to familiar classical ones in situations in which they ought to do so.

In the next chapter we shall encounter still more new ideas. Together with those discussed here, they will provide the evidence and the necessary foundation for an understanding of modern quantum theory.

questions

1. "Pockets" formed by the charcoal briquettes in a barbeque grill seem brighter than the briquettes themselves. Is the temperature in such pockets appreciably higher than the surface temperature of an exposed glowing briquette?

2. The relation $R = \sigma T^4$ (Eq. 49-1a) is exact for true cavities and holds for all temperatures. Why don't we use this relation as the basis of a definition of temperature at, say, 100°C?

3. Do all incandescent solids obey the fourth-power law of temperature, as Eq. 49-1b seems to suggest?

4. A hole in the wall of a cavity radiator is sometimes called a *black body*. Why?

5. If you look into a cavity whose walls are maintained at a constant temperature, no details of the interior are visible. Does this seem reasonable?

6. By simply looking at the sky at night, can you pick out stars that are hotter than the sun? Cooler than the sun? What do you look for? Is the star's brightness a clue?

7. Betelgeuse, the prominent red star in the constellation Orion, has a surface temperature that is much lower than that of the sun, yet it radiates energy into space at a much higher rate than the sun does. How can that be?

8. Less than a few percent of the energy supplied to a 100-W lamp appears in the form of visible light. What happens to the rest of it? What could be done to increase this percentage? Why hasn't it already been done?

9. Suppose that your skin temperature is about 300 K. In what region of the electromagnetic spectrum do you emit thermal radiation most intensely?

10. Spectral radiancy curves for cavity radiators at different temperatures never intersect; see Fig. 49-3. Suppose, however, that they did. Can you show that this would violate the second law of thermodynamics?

11. We claim that all objects radiate energy by virtue of their temperature and yet we cannot see all objects in the dark. Why not?

12. Are there quantized quantities in classical physics? If so, give examples. Is energy quantized in classical physics?

13. Show that Planck's constant has the dimensions of angular momentum.

14. For you to be able to detect energy quantization by watching, say, a swinging pendulum, what order of magnitude would the Planck constant have to be? (*Hint:* See Example 2.)

15. For quantum effects to be "everyday" phenomena in our lives, what order of magnitude value would h need to have? [See G. Gamow, *Mr. Tompkins in Wonderland* (Cambridge University Press, Cambridge, 1957), for a delightful popularization of a world in which the physical constants c, G, and h make themselves obvious.]

16. Explain in your own words just why assuming that the energy of the atomic oscillators in a solid is quantized leads to a reduction in the heat capacity at low temperatures.

17. The classical law of equipartition of energy (see Section 23-8) leads to the Rayleigh-Jeans radiation law when applied to cavity radiation and to the Dulong-Petit law when applied to the heat capacities of solids. In both cases there is serious disagreement with experiment. Can you relate these two failures of the equipartition law and explain why energy quantization leads, in each case, to theories that *do* agree with experiment?

18. "For the cavity radiation problem and the heat capacity of solids problem, the disagreements between experiment and classical theory in certain ranges of the variables are not small but are total, and beyond all dispute." Can you identify, in each case, the specific disagreements to which this statement refers?

19. In Fig. 49-10, why doesn't the photoelectric current rise vertically to its maximum (saturation) value when the applied potential difference is slightly more positive than V_0?

20. In the photoelectric effect, why does the existence of a cutoff frequency speak in favor of the photon theory and against the wave theory?

21. Why are photoelectric measurements so sensitive to the nature of the photoelectric surface?

22. An insulated metal plate yields photoelectrons when first illuminated by ultraviolet light but then doesn't give up any more. Explain.

23. Why is it that even for incident radiation that is monochromatic the photoelectrons are emitted with a spread of velocities?

24. How can a *photon* energy be given by $E = h\nu$ (Eq. 49-15) when the very presence of the frequency ν in the formula implies that light is a *wave?*

25. Does Einstein's theory of photoelectricity, in which light is postulated to be a photon, invalidate Young's interference experiment?

26. Explain the statement that one's eyes could not detect faint starlight if light were not corpuscular.

27. Distinguish between the Planck relation $E = nh\nu$ (Eq. 49-6) and the Einstein relation $E = h\nu$ (Eq. 49-15).

28. A photon that reaches us from a distant galaxy has a wavelength that is shifted toward the red (longer waves) because the galaxy is receding from us as the universe expands. What happened to this "missing" energy? [*Hint:* See discussion in *The Physics Teacher* (December 1983).]

29. What is the direction of a Compton scattered electron with maximum kinetic energy compared with the direction of the incident monochromatic photon beam?

30. Why, in the Compton scattering picture (Fig. 49-13), would you expect $\Delta\lambda$ to be independent of the materials of which the scatterer is composed?

31. Why don't we observe a Compton effect with visible light?

32. Light from distant stars is Compton scattered many times by free electrons in outer space before reaching us. This shifts the light toward the red. How can this shift be distinguished from the Doppler red shift due to the motion of receding stars?

33. In both the photoelectric effect and the Compton effect there is an incident photon and an ejected electron. What is the difference between these two effects?

34. Why is the Compton effect more supportive of the photon theory of light than is the photoelectric effect?

35. Why was the Balmer series, rather than the Lyman or Paschen series, the first to be detected and analyzed in the hydrogen spectrum?

36. Only a relatively small number of Balmer lines can be observed from laboratory discharge tubes, whereas a large number are observed in stellar spectra. Explain this in terms of the small density, high temperature, and large volume of gases in stellar atmospheres.

37. Discuss the analogy between the Kepler-Newton relationship in the development of Newton's law of gravitation and the Balmer-Bohr relationship in developing the Bohr theory of atomic structure.

38. In Bohr's theory for the hydrogen atom orbits, what is the implication of the fact that the potential energy is negative and is greater in magnitude than the kinetic energy?

39. Consider a hydrogen-like atom in which a positron (a positively charged electron) circulates about a (negatively charged) antiproton. In what way, if any, would the emission spectrum of this "antimatter atom" differ from the spectrum of a normal hydrogen atom?

40. (a) Can a hydrogen atom absorb a photon whose energy exceeds its binding energy (13.6 eV)? (b) What minimum energy must a photon have to initiate the photoelectric effect in hydrogen gas?

41. On emitting a photon, an isolated hydrogen atom recoils to conserve momentum. Explain the fact that the energy of the emitted photon is slightly less than the energy difference between the energy levels involved in the emission process.

42. Why are some lines in the hydrogen spectrum brighter than others?

43. According to classical mechanics, an electron moving in an orbit should be able to do so with any angular momentum whatever. According to Bohr's theory of the hydrogen atom, however, the angular momentum is quantized. Reconcile these two statements, using the correspondence principle.

44. Describe several methods that can be used to determine experimentally the value of Planck's constant h.

45. The correspondence principle is expressed by taking n to infinity *or* by

taking h to zero. Explain the relation between these two techniques for finding the classical limit.

46. Can you use the device of letting $h \rightarrow 0$ to obtain classical results from quantum results in the case of the photoelectric effect? Explain.

problems

SECTION 49-2

1. A cavity whose walls are held at 2000 K has a small hole, 1.00 mm in diameter, drilled in its wall. At what rate does energy escape through this hole from the cavity interior?   *Answer:* 713 mW.

2. Estimate the thermal power radiated from a fireplace assuming an emissivity of 0.90, an effective radiating surface of 0.50 m², and a radiating temperature of 500°C. Does your answer seem reasonable?

3. What is the power radiated from a nichrome wire at 800°C? The wire is 1.0 m long and has a diameter of 1.5 mm. The emissivity of nichrome is 0.92.   *Answer:* 326 W.

4. A 100-W incandescent lamp has a coiled tungsten filament whose diameter is 0.40 mm and whose extended length is 30 cm. The effective emissivity under operating conditions is 0.21. What is the operating temperature of the filament?

5. Let us assume that the earth, whose mean surface temperature may be taken to be 280 K, is in thermal equilibrium and radiates energy into space at the same rate that it receives energy from the sun. If the earth were removed to an orbit of twice its present radius, what would be its expected mean surface temperature?   *Answer:* 198 K (= −103°F).

6. The average rate of solar radiation incident per unit area on the earth is 335 W/m². (a) Explain the consistency of this number with the *solar constant* (the solar energy falling per unit time at normal incidence on unit area of the earth's surface) whose value is 1340 W/m². (b) Consider the earth to behave like a cavity radiating energy into space at this same rate. What surface temperature would the earth have under these circumstances?

7. An oven with an inside temperature $T = 227°C$ is in a room with a temperature of $T_r = 27°C$. There is a small opening of area 5.0 cm² in one side of the oven. How much net power is transferred from the oven to the room? (*Hint:* Consider both oven and room as cavities.)   *Answer:* 1.5 W.

8. A *thermograph* is a medical instrument used to measure radiation from the skin. For example, normal skin radiates at a temperature of about 34°C and the skin over a tumor radiates at a slightly higher temperature. (a) Derive an approximate expression for the fractional difference in the radiancy between adjacent areas of the skin that are at slightly different temperatures. (b) Evaluate this expression for a temperature difference of 1°C. Assume that the skin radiates with a constant emissivity.

9. (a) Assuming the surface temperature of the sun to be 5800 K, use the Stefan-Boltzmann law to determine the rate at which the sun loses rest mass by radiation. Take the sun's diameter to be 1.4×10^9 m and assume an emissivity of unity. (b) What fraction of the sun's rest mass is lost each year in this way? Take the sun's rest mass to be 2.0×10^{30} kg. *Answer:* (a) 4.4×10^9 kg/s. (b) 6.9×10^{-14}.

10. *Some temperatures and some wavelengths.* Calculate the wavelength of maximum spectral radiancy and identify the region of the electromagnetic spectrum to which it belongs for each of the following: (a) The 3-K cosmic background radiation, a remnant of the primordial fireball. (b) Your body, assuming a skin temperature of 34°C. (c) A tungsten lamp filament at 1800 K. (d) The sun, at an assumed surface temperature of 5800 K. (e) An exploding thermonuclear device, at an assumed fireball temperature of

10^7 K. (f) The universe immediately after the Big Bang, at an assumed temperature of 10^{38} K. Assume cavity radiation conditions throughout.

11. *Birth of a planetary system?* In 1983 the Infrared Astronomical Satellite (IRAS) detected a cloud of solid particles surrounding the star Vega, radiating maximally at a wavelength of 32 μm. What is the temperature of this cloud of particles? Assume an emissivity of unity.   *Answer: 91 K.*

12. A black body (cavity radiator) has its maximum spectral radiancy at a wavelength of 25 μm, in the infrared region of the spectrum. The temperature of the body is now increased so that the radiancy $R(T)$ of the body is doubled. (a) What is this new temperature? (b) At what wavelength will the spectral radiancy now have its maximum value?

13. Low-temperature physicists would not consider a temperature of 2.0 mK (= 0.0020 K) to be particularly low. At what wavelength would a cavity whose walls were at this temperature radiate most copiously? To what region of the electromagnetic spectrum would this radiation belong? What are some of the practical difficulties of operating a cavity radiator at such a low temperature?   *Answer: 1.5 m; short radio wave.*

SECTION 49-3

14. A cavity whose walls are held at 4000 K has a circular aperture 5.0 mm in diameter. (a) At what rate does energy in the visible range (defined to extend from 0.40 to 0.70 μm) escape from this hole? (b) What fraction of the total radiation escaping from the cavity does this represent? Solve either analytically or graphically.

15. (a) Prove the Stefan-Boltzmann law by showing directly from Planck's formula (Eq. 49-5) that

$$\int_0^\infty \mathcal{R}(\lambda)\, d\lambda = AT^4$$

where A is a constant. (b) Evaluate A and show that it is identical with the Stefan-Boltzmann constant σ (= 5.670×10^{-8} W/m$^2 \cdot$ K^4).

SECTION 49-5

16. The specific heat capacities at constant pressure and at room temperature for boron and for gold are 0.136 cal/g $\cdot$ F$°$ and 0.0171 cal/g $\cdot$ F$°$, respectively. For each, calculate the molar heat capacity (in J/mol $\cdot$ K) and state whether the substance obeys the Dulong and Petit rule.

17. Supply the missing steps in the derivation of Eq. 49-12, Einstein's expression for the molar heat capacity of a solid.

18. The Einstein temperatures of lead, aluminum, and beryllium may be taken as 68 K, 290 K, and 690 K, respectively. For each of these elements, find (a) the resonant frequency v of its atomic oscillators, (b) the spacing ΔE between adjacent oscillator levels, and (c) the effective spring constant κ. Interpret your results in terms of the heat-capacity curves displayed in Fig. 49-7.

19. *The molar internal energy.* (a) Show that the molar internal energy of a solid can be written, according to Einstein's theory of heat capacities, as

$$u(T) = 3RT\left(\frac{x}{e^x - 1}\right),$$

in which $x = T_E/T$, where T_E is the *Einstein temperature* hv/k. (b) Verify that $u(T)$ approaches its classical value of $3RT$ as $T \to \infty$.

20. The Einstein temperature of aluminum may be taken as 290 K. According to Einstein's theory of heat capacity, what are (a) its oscillator frequency; (b) its molar internal energy (see Problem 19) at 150 K; (c) its molar heat capacity, under constant-volume conditions, at 150 K?

21. A 12.0-g block of aluminum is heated from 80 K up to 180 K, under presumed constant-volume conditions. How much thermal energy is required according to (a) the classical theory of heat capacity and (b) Einstein's quantum theory of heat capacity? The Einstein temperature for aluminum may be taken to be 290 K. (*Hint:* Use the expression for the molar internal energy, displayed in Problem 19.)   *Answer:* (a)1110 J. (b) 714 J.

22. *Conditions at the Einstein temperature.* In terms of Einstein's theory of heat capacity, (a) what is the molar heat capacity at constant volume of a solid at its Einstein temperature? Express your answer as a percentage of its classical value of 3R. (b) What is the molar internal energy (see Problem 19) at the Einstein temperature? Express your answer as a percentage of its classical value of 3RT.

23. In terms of Einstein's theory of heat capacity, (a) at what temperature will the molar internal energy of a solid (see Problem 19) achieve one-half its classical value of 3RT? (b) At what temperature will the molar heat capacity at constant volume of a solid achieve one-half its classical value of 3R? Express each answer in terms of the Einstein temperature of the solid. (c) Evaluate for aluminum, for which $T_E = 290$ K.
    Answer: (a) $0.80T_E$. (b) $0.34T_E$. (c) 230 K; 100 K.

SECTION 49-6

24. In Example 5 suppose that the "target" is a single gas atom of 0.10 nm radius and that the intensity of the light source is reduced to 1.0×10^{-5} W. If the binding energy of the most loosely bound electron in the atom is 2.0 eV, what time lag for the photoelectric effect is expected on the basis of the wave theory of light?

SECTION 49-7

25. Show that the energy E of a photon (in eV) is related to its wavelength λ (in nm) by

$$E = \frac{1240}{\lambda}.$$

    This result can be useful in solving many of the problems in this section.

26. (a) A spectral emission line, important in radioastronomy, has a wavelength of 21 cm. What is its corresponding photon energy? (b) At one time the meter was defined as 1,650,763.73 wavelengths of the orange light emitted by a light source containing krypton-86 atoms. What is the corresponding photon energy of this radiation?

27. An atom absorbs a photon having a wavelength of 375 nm and immediately emits another photon having a wavelength of 580 nm. What was the net energy absorbed by the atom in this process?   *Answer:* 1.17 eV.

28. What are (a) the frequency, (b) the wavelength, and (c) the momentum of a photon whose energy equals the rest energy of the electron?

29. *Focusing laser light.* The emerging beam from a 1.0-W argon laser ($\lambda = 515$ nm) has a diameter d of 3.0 mm. (a) At what rate per unit area do photons pass through any cross section of the incident laser beam? (b) The beam is focused by a lens system (assumed ideal) whose effective focal length f is 2.5 mm. The focused beam forms a circular diffraction pattern whose central disk has a radius R given by 1.22 $f\lambda/d$. (See Section 46-5). It can be shown that 84% of the incident power strikes this central disk, the rest falling in the fainter, concentric diffraction rings that surround the central disk. At what rate per unit area do photons fall on the central disk of the diffraction pattern?
    Answer: (a) 3.7×10^{23} photons/m^2 · s. (b) 2.5×10^{30} photons/m^2 · s.

30. Solar radiation is falling on the earth at a rate of 1340 W/m^2, on a surface normal to the incoming rays. At what rate per unit area do solar photons fall on such a surface? Assume an average solar photon wavelength of 550 nm.

31. Under ideal conditions the normal human eye will record a visual sensation at 550 nm if incident photons are absorbed at a rate as low as 100 s^{-1}. To what power level does this correspond?   *Answer:* 3.6×10^{-17} W.

32. An ultraviolet light bulb, emitting at 400 nm, and an infrared light bulb, emitting at 700 nm, each are rated at 400 W. (*a*) Which bulb radiates photons at the greater rate? (*b*) How many more photons does it generate per second than does the other bulb?

33. A satellite in earth orbit maintains a panel of solar cells of 1.0-m^2 area at right angles to the direction of the sun's rays. (*a*) At what rate does solar energy strike the panel? (*b*) At what rate do solar photons strike the panel? Assume that the solar radiation is monochromatic with a wavelength of 550 nm. (*c*) How long would it take for a "mole of photons" to strike the panel? The solar constant (see Problem 6) is 1390 W/m^2. *Answer:* (*a*) 1.4 kW. (*b*) 3.8×10^{21} photon/s. (*c*) 160 s.

34. Light of a wavelength 200 nm falls on an aluminum surface. In aluminum 4.2 eV are required to remove an electron. What is the kinetic energy of (*a*) the fastest and (*b*) the slowest emitted photoelectrons? (*c*) What is the stopping potential? (*d*) What is the cutoff wavelength for aluminum?

35. (*a*) If the work function for a metal is 1.8 eV, what would be the stopping potential for light having a wavelength of 400 nm? (*b*) What would be the maximum velocity of the emitted photoelectrons at the metal's surface? *Answer:* (*a*) 1.3 V. (*b*) 6.8×10^5 m/s.

36. Incident photons strike a sodium surface having a work function of 2.2 eV, causing photoelectric emission. When a stopping potential of 5.0 V is imposed, there is no photocurrent. What is the wavelength of the incident photons?

37. You wish to pick a substance for a photocell operable with visible light. Which of the following will do (work function in parentheses): tantalum (4.2 eV); tungsten (4.5 eV); aluminum (4.2 eV); barium (2.5 eV); lithium (2.3 eV)?   *Answer:* Barium and lithium.

38. (*a*) The energy needed to remove an electron from metallic sodium is 2.28 eV. Does sodium show a photoelectric effect for red light, with $\lambda = 680$ nm? (*b*) What is the cutoff wavelength for photoelectric emission from sodium and to what color does this wavelength correspond?

39. The stopping potential for photoelectrons emitted from a surface illuminated by light of wavelength 491 nm is 0.71 V. When the incident wavelength is changed to a new value, the stopping potential is found to be 1.43 V. (*a*) What is this new wavelength? (*b*) What is the work function for the surface? (*c*) Can you identify the material, given that it is an elemental substance?   *Answer:* (*a*) 382 nm. (*b*) 1.82 eV. (*c*) Cesium.

40. In a photoelectric experiment in which a sodium surface is used, one finds a stopping potential of 1.85 V for a wavelength of 300 nm and a stopping potential of 0.82 V for a wavelength of 400 nm. From these data find (*a*) a value for the Planck constant, (*b*) the work function for sodium and (*c*) the cutoff wavelength for sodium.

41. Consider monochromatic light falling on a photographic film. The incident photons will be recorded if they have enough energy to dissociate a AgBr molecule in the film. The minimum energy required to do this is about 0.6 eV. Find the cutoff wavelength greater than which the light will not be recorded. In what region of the spectrum does this wavelength fall?   *Answer:* 2.1 μm; infrared.

42. Millikan's photoelectric data for lithium are:

Wavelength, nm	433.9	404.7	365.0	312.5	253.5
Stopping potential, V	0.55	0.73	1.09	1.67	2.57

Make a plot like Fig. 49-11, which is for sodium, and find (*a*) the Planck constant and (*b*) the work functon for lithium.

43. Photosensitive surfaces are not necessarily very efficient. Suppose the fractional efficiency of a cesium surface (work function 1.8 eV) is 1.0×10^{-16}, that is, one photoelectron is produced for every 10^{16} photons striking the surface. What would be the photocurrent from such a cesium surface if it is illuminated with 600-nm light from a 2.0-mW laser and all of the photoelectrons produced take part in charge flow?   *Answer:* 9.7×10^{-14} A.

44. X rays with a wavelength of 71 pm eject photoelectrons from a gold foil, the electrons originating from deep within the gold atoms. The ejected electrons move in circular paths of radius r in a region of uniform magnetic field **B**. Experiment shows that $rB = 1.88 \times 10^{-4}$ T · m. Find (a) the maximum kinetic energy of the photoelectrons and (b) the work done in removing the electrons from the gold atoms that make up the foil.

45. To remove an inner, most tightly bound, electron from an atom of molybdenum requires an energy of 20 keV. If this is to be done by allowing a photon to strike the atom, (a) what must be the associated wavelength of the photon? (b) In what region of the spectrum does the photon lie? (c) Could this process be called a photoelectric effect? Discuss your answers. *Answer:* (a) 0.062 nm, or 62 pm. (b) X-ray region.

SECTION 49-8

46. Photons of wavelength 2.4 pm are incident on free electrons. (a) Find the wavelength of a photon that is scattered 30° from the incident direction. (b) Do the same if the scattering angle is 120°.

47. Calculate the percent change in photon energy for a Compton collision with φ in Fig. 49-14 equal to 90° for radiation in (a) the microwave range, with $\lambda = 3.0$ cm, (b) the visible range, with $\lambda = 500$ nm, (c) the X-ray range, with $\lambda = 0.10$ nm, and (d) the gamma-ray range, the energy of the gamma-ray photons being 1.0 MeV. What are your conclusions about the importance of the Compton effect in these various regions of the electromagnetic spectrum, judged solely by the criterion of energy loss in a single Compton encounter?
Answer: (a) 8.1×10^{-9} %. (b) 4.8×10^{-4} %. (c) 2.4%. (d) 68%.

48. An X-ray photon of wavelength $\lambda = 10$ pm strikes an electron head on ($\varphi = 180°$). Determine (a) the change in wavelength of the photon, (b) the change in energy of the photon, and (c) the final kinetic energy of the electron.

49. A 6.2-keV X-ray photon falling on a carbon block is scattered by a Compton collision and its frequency is shifted by 0.010%. (a) Through what angle is the photon scattered? (b) What kinetic energy is imparted to the electron? *Answer:* (a) 7.4°. (b) 0.62 eV.

50. A 0.510-MeV gamma-ray photon is Compton-scattered from a free electron in an aluminum block. (a) What is the wavelength of the incident photon? (b) What is the wavelength of the scattered photon? (c) What is the energy of the scattered photon? Assume a scattering angle of 90°.

51. What is the maximum kinetic energy of the Compton-scattered electrons knocked out of a thin copper foil by an incident beam of 17.5-keV X rays? *Answer:* 1.12 keV.

52. Photons of wavelength 2.40 pm are incident on a target containing free electrons. (a) Find the wavelength of a photon that is scattered at 30° from the incident direction and also the kinetic energy imparted to the recoil electron. (b) Do the same if the scattering angle is 120°. (*Hint:* See Example 8.)

53. What fractional increase in wavelength leads to a 75% loss of photon energy in a Compton collision with a free electron?   *Answer:* 300%.

54. Through what angle must a 200-keV photon be scattered by a free electron so that it loses 10% of its energy?

55. (a) Show that $\Delta E/E$, the fractional loss of energy of a photon during a Comp-

ton collision, is given by $(h\nu'/m_0c^2)(1 - \cos \varphi)$. (b) Plot $\Delta E/E$ versus φ and interpret the curve physically.

56. The quantity h/m_0c in Eq. 49-22 is often called the Compton wavelength, λ_c, of the scattering particle and that equation is written as

$$\Delta\lambda = \lambda_c(1 - \cos \varphi).$$

(a) What is the Compton wavelength of an electron? of a proton? (b) What is the energy of a photon whose wavelength is equal to the Compton wavelength of the electron? of the proton? (c) Show that in general the energy of a photon whose wavelength is equal to the Compton wavelength of a particle is just the rest energy of that particle.

57. Carry out the necessary algebra to eliminate v and θ from Eqs. 49-18, 49-20, and 49-21 to obtain the Compton shift relation (Eq. 49-22).

58. Find the maximum wavelength shift for a Compton collision between a photon and a free proton.

SECTION 49-10

59. Light with a wavelength of 486.3 nm is emitted by a hydrogen atom. (a) What transition of the hydrogen atom is responsible for this radiation? (b) To what series does this radiation belong?
Answer: (a) $n = 4$ to $n = 2$. (b) The Balmer series.

60. Show on an energy-level diagram for hydrogen the quantum numbers corresponding to a transition in which the wavelength of the emitted photon is 121.6 nm.

61. What are (a) the energy, (b) momentum, and (c) wavelength of the photon that is emitted when a hydrogen atom undergoes a transition from the state $n = 3$ to $n = 1$? *Answer:* (a) 12 eV. (b) 6.5×10^{-27} kg · m/s. (c) 103 nm.

62. Using Bohr's formula for the frequencies of the lines of the hydrogen spectrum, calculate the three longest wavelengths of the Balmer series.

63. (a) What are the wavelength intervals over which the Lyman, the Balmer, and the Paschen series extend? (The interval extends from the longest wavelength to the series limit.) (b) What are the corresponding frequency intervals?
Answer: (a) 30.5 nm; 291 nm; 1050 nm. (b) 8.25×10^{14} Hz; 3.65×10^{14} Hz; 2.06×10^{14} Hz.

64. A hydrogen atom is excited from a state with $n = 1$ to one with $n = 4$. (a) Calculate the energy that must be absorbed by the atom. (b) Calculate and display on an energy-level diagram the different photon energies that may be emitted if the atom returns to the $n = 1$ state. (c) Calculate the recoil speed of the hydrogen atom, assumed initially at rest, if it makes the transition from $n = 4$ to $n = 1$ in a single quantum jump.

65. A hydrogen atom in a state having a *binding energy* (the energy required to remove an electron) of 0.85 eV makes a transition to a state with an *excitation energy* (the difference in energy between the state and the ground state) of 10.2 eV. (a) Find the energy of the emitted photon. (b) Show this transition on an energy-level diagram for hydrogen, labeling the appropriate quantum numbers. *Answer:* (a) 2.6 eV. (b) $n = 4$ to $n = 2$.

66. From the energy-level diagram for hydrogen, explain the observation that the frequency of the second Lyman-series line is the sum of the frequencies of the first Lyman-series line and the first Balmer-series line. This is an example of the empirically discovered *Ritz combination principle*. Use the diagram to find some other valid combination.

67. (a) How much energy is required to remove the electron from a hydrogen atom in a state with $n = 8$? with $n = 25$? (b) From a He^+ ion in its ground state? (c) From a Li^{++} ion in a state with $n = 3$?
Answer: (a) 0.213 eV; 0.0218 eV. (b) 54.4 eV. (c) 13.6 eV.

68. Use Bohr's theory to compare the spectrum of singly ionized helium with the spectrum of hydrogen.

69. In stars the Pickering series is found in the He⁺ spectrum. It is emitted when the electron in He⁺ jumps from higher levels to the level with $n = 4$. (a) Show that the wavelengths in this series are given by

$$\lambda = \frac{1}{R_\infty} \frac{4n^2}{n^2 - 16}$$

in which $n = 5, 6, 7, \ldots$. Assume that $m_e \ll M_{\text{He}}$, where M_{He} is the mass of the helium nucleus. (b) Calculate the wavelength of the first line of this series and of the series limit. (c) In what region of the spectrum does this series occur?   *Answer:* (b) $1.013\ \mu\text{m}$; $0.365\ \mu\text{m}$. (c) Infrared and visible.

SECTION 49-11

70. If an electron is revolving in an orbit at frequency ν_0, classical electromagnetism predicts that it will radiate energy not only at this frequency but also at $2\nu_0$, $3\nu_0$, $4\nu_0$, and so on. Show that this is also predicted by Bohr's theory of the hydrogen atom in the limiting case of large quantum numbers.

71. In Table 49-4 show that the quantity in the last column is given by

$$\frac{100(\nu - \nu_0)}{\nu} \cong \frac{150}{n}$$

for large quantum numbers.

SECTION 49-12

72. (a) If the angular momentum of the earth due to its motion around the sun were quantized according to Bohr's relation $L = nh/2\pi$, what would the quantum number be? (b) Could such quantization be detected if it existed?

73. Combining Eq. 49-32 with the results of Example 11, show that in the semiclassical planetary one-electron model of the atom (a) the orbital speeds are quantized as $v_n = (Ze^2/2\epsilon_0 h)(1/n)$ and (b) the orbital angular momenta are quantized as $L_n = (h/2\pi)n$.

74. In the ground state of the hydrogen atom, according to Bohr's semiclassical planetary model, what are (a) the quantum number, (b) the orbit radius, (c) the angular momentum, (d) the linear momentum, (e) the angular velocity, (f) the linear speed, (g) the force on the electron, (h) the acceleration of the electron, (i) the kinetic energy, (j) the potential energy, and (k) the total energy?

75. How do the quantities (b) to (k) in Problem 74 vary with the quantum number?
Answer: (b) n^2. (c) n. (d) $1/n$. (e) $1/n^3$. (f) $1/n$. (g) $1/n^4$. (h) $1/n^4$. (i) $1/n^2$. (j) $1/n^2$. (k) $1/n^2$.

76. A diatomic gas molecule consists of two atoms of mass m separated by a fixed distance d rotating about an axis as indicated in Fig. 49-21. Assuming that its angular momentum is quantized as in the Bohr atom, determine (a) the possible angular velocities and (b) the possible quantized rotational energies. (c) Show these on an energy-level diagram.

77. *Muonic atoms.* Apply Bohr's model to a muonic atom, which consists of a nucleus of charge Ze with a negative muon (an elementary particle with a charge of $-e$ and a mass m that is 207 times as large as the electron mass) circulating about it. Calculate (a) the muon-nucleus separation in the first Bohr orbit, (b) the ionization energy, and (c) the wavelength of the most energetic photon that can be emitted. Assume that the muon is circulating about a hydrogen nucleus $(Z = 1)$. See "The Muonium Atom" by Vernon W. Hughes, *Scientific American*, April 1966.
Answer: (a) 2.6×10^{-13} m. (b) 2.8 keV. (c) 0.49 nm.

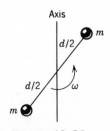

figure 49-21
Problem 76.

78. *Positronium.* Apply Bohr's model to the positronium atom. This consists of a positive and a negative electron revolving around their center of mass, which lies halfway between them. (*a*) What relationship exists between this spectrum and the hydrogen spectrum? (*b*) What is the radius of the ground state orbit? (*Hint:* It will be necessary to analyze this problem from first principles because this "atom" has no nucleus: both particles revolve about a point halfway between them.) See "Exotic Atoms" by E. H. S. Burhop, *Contemporary Physics*, July 1970.

79. Bohr proposed that, as an alternative to the correspondence principle, the quantization expression for the angular momentum (Eq. 49-33) could be taken as a basic postulate. Starting from this point, and using only classical results such as those derived in Example 11, derive Bohr's expression for the quantized energies of the stationary states of the hydrogen atom (Eq. 49-30).

50
the wave nature of matter

Physicists have rarely gone wrong in banking on the underlying symmetry of nature. For example, when you learn that a changing magnetic field produces an electric field it is a good bet to guess — and it turns out to be true — that a changing electric field will produce a magnetic field. For another example: The electron has a so-called antiparticle, that is, a particle of the same mass but of opposite charge. Is it not reasonable to expect that the proton should also have an antiparticle? A 5-GeV proton accelerator was built at the University of California at Berkeley to search for the antiproton. It was found.

In 1924 Louis de Broglie, a physicist and a member of a distinguished French aristocratic family, puzzled over the fact that radiation seemed to have a dual wave–particle aspect but matter — at that time — seemed to be entirely particlelike. This in the face of the fact that both radiation and matter are forms of energy, that each can be transformed into the other, and that they are both governed by the spacetime symmetries of the theory of relativity. He then began to think that matter should also have a dual character and that particles such as electrons should have wavelike properties.

For a wave, the wavelength λ and the frequency ν are the important things to know. For a particle, the momentum p and the energy E are what count. For radiation, these quantities are related by Eq. 49-19 ($p = h/\lambda$) and by Eq. 49-15 ($E = h\nu$). After considerable analysis, de Broglie made the bold suggestion that these same relations apply to matter as well as to radiation. If we want to describe a moving particle by a *matter wave*, we must be able to give the wavelength and the frequency of this wave. Of the two, the wavelength is far easier to measure, using interference and diffraction methods, so we focus our attention on it. The *de Broglie wavelength* (as we call it) of a matter wave is given by Eq. 49-19. Thus

$$\lambda = \frac{h}{p} \qquad \text{(de Broglie wavelength).} \qquad (50\text{-}1)$$

Note that the Planck constant h seems to be the connecting link between the wave–particle nature of both matter and radiation.

A question naturally arises. The tracks in Fig. 50-1 are strings of bubbles left in the wake of charged elementary particles as they pass through a bubble chamber. Can these tracks possibly have anything to do with waves? It seems very unlikely! However, we ask you to withhold judgment until the end of this chapter, where we will address the question. Meanwhile consider, with us, the strong experimental evidence for the wave nature of matter.

EXAMPLE 1

Calculating the de Broglie wavelength. What wavelength does Eq. 50-1 predict for (*a*) a 150-g baseball traveling at 40 m/s and (*b*) an electron whose kinetic energy is 150 eV?

(*a*) From Eq. 50-1 we have

$$\lambda = \frac{h}{p} = \frac{h}{mv} = \frac{6.63 \times 10^{-34}\,\text{J}\cdot\text{s}}{(0.15\,\text{kg})(40\,\text{m/s})} = 1.1 \times 10^{-34}\,\text{m}.$$

There is no hope of measuring so small a length and thus verifying the wave nature of a moving baseball. The number 10^{-34} is about equal to the ratio of the size of a virus particle to that of the known universe.

(*b*) For the electron we have

$$\lambda = \frac{h}{p} = \frac{h}{\sqrt{2mK}}$$

$$= \frac{6.63 \times 10^{-34}\,\text{J}\cdot\text{s}}{\sqrt{2(9.11 \times 10^{-31}\,\text{kg})(150\,\text{eV})(1.60 \times 10^{-19}\,\text{J/eV})}}$$

$$= 1.0 \times 10^{-10}\,\text{m} = 100\,\text{pm}.$$

On the atomic scale, this is *not* a small distance. It is, in fact, about the size of an atom or the distance between atomic planes in a crystalline solid. We can hope to verify the wave nature of electrons in the laboratory.

Although we have not mentioned it before, we should in general use the relativistic expression for momentum in Eq. 50-1. In part (*a*), however, the speed of the baseball is so much less than the speed of light that we were totally justified in using the classical momentum expression there. In part (*b*) the electron speed is also much less than that of light. We can tell this because the electron's kinetic energy K (= 150 eV) is so much less than its rest energy m_0c^2 (= 511 keV or 511,000 eV). Thus we were able to use the classical momentum expression there also.

De Broglie showed cleverly how his notion of matter waves meshed with Bohr's model of the hydrogen atom. If we are to view the electron orbiting the central proton as a wave, that wave must join onto itself in the same phase after every revolution. If it did not, the wave would destroy itself by destructive interference. Put another way, the wavelength of the orbiting electron must fit around the circumference of the orbit an integral number of times. As Fig. 50-2 suggests, this leads to

$$n\lambda = 2\pi r \qquad n = 1, 2, 3, \ldots .$$

Substituting h/p for λ allows us to predict that only orbits whose angular momentum L has the values

$$L = pr = n(h/2\pi) = n\hbar \qquad n = 1, 2, 3, \ldots \qquad (49\text{-}33)$$

can exist. This is exactly Bohr's result for the quantization of angular momentum! Although we no longer look at the hydrogen atom in terms of classical orbits, this neat result foreshadows a solid conclusion: atoms exist in stationary states in which both their energy and their angular momentum are quantized.

50-2
DE BROGLIE WAVES AND BOHR ORBITS

figure 50-2
Bohr orbits represented as de Broglie waves.

If you want to prove that you are dealing with a wave, a convincing thing to do is to measure the wavelength. In 1801, for example, Thomas Young made a strong case for the wave nature of visible light when he measured its wavelength, using the double slit inteference method that he invented for that purpose; see Fig. 45-1.

From this example you can see that to measure a wavelength you need two or more diffracting centers (slits) separated by a distance that is of the order of magnitude of the wavelength itself. Example 1a shows at once that it is hopeless to try to measure the wavelength of a pitched baseball. You would need a pair of "slits" spaced 10^{-34} m apart! That is why our daily experiences with large moving objects gives no clues to the wave nature of matter. Example 1b, however, suggests that we *should* be able to measure the wavelength of a moving electron. We now describe three ways to do so.

1. *The Davisson-Germer experiment.* Example 1b suggests that we should be able to use a crystal as a three-dimensional grating to measure the de Broglie wavelengths of electrons with kinetic energies of a few hundred electron volts. Figure 50-3 shows the apparatus used for this purpose by C. J. Davisson and L. H. Germer of what is now the AT&T Bell Laboratories. In 1937 Davisson shared the Nobel prize for this work, the first of seven such prizes awarded (as of 1986) to scientists from this remarkable laboratory.

In the Davisson-Germer apparatus of Fig. 50-3, electrons are accelerated from heated filament F by an adjustable potential difference V. The beam from this "electron gun" G, made up of electrons whose kinetic energy is eV, is then allowed to fall on a crystal C, which, in their experiment, was of nickel. They set detector D at an arbitrary angle ϕ and read the current I for various values of the potential difference V. Figure 50-4 shows the results of five such runs. We see that there is a strong diffracted beam for $\phi = 50°$ and $V = 54$ V. If either the angle or the accelerating potential are changed, the intensity of the diffracted beam drops.

Figure 50-5 is a simplified representation of the nickel crystal C of Fig. 50-3. The diffracted beam is formed by Bragg reflection of the electron matter wave from a particular family of atomic planes within the crystal. However, the beam bends at the surface when it emerges from the crystal so that its angle of reflection within the crystal is not the same as angle ϕ of Fig. 50-5, which is measured outside the crystal. When this surface bending is taken into account, it can be shown* that

50-3
TESTING DE BROGLIE'S HYPOTHESIS

figure 50-3
The apparatus of Davisson and Germer. Electrons from filament F are accelerated by a variable potential difference V. After reflection from crystal C they are collected by detector D and read as a current I.

*See, for example, *Basic Concepts in Relativity and Early Quantum Theory*, 2nd ed., Robert Resnick and David Halliday (John Wiley & Sons, New York, 1985), Supplementary Topic H.

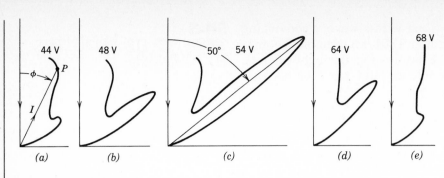

figure 50-4
The results of five runs with the apparatus of Fig. 50-3. A strong diffraction peak appears in run (c).

the crystal behaves like a two-dimensional diffraction grating, its grating lines being the parallel lines of atoms lying on the crystal surface and its grating spacing being the interval marked D in Fig. 50-5. The principal maxima for such a grating must satisfy Eq. 47-1, or

$$m\lambda = D \sin \phi \qquad m = 1, 2, 3, \ldots . \qquad (50\text{-}2)$$

For their crystal Davisson and Germer knew that $D = 215$ pm. For $m = 1$, which corresponds to a first-order diffraction peak, Eq. 50-2 leads to

$$\lambda = D \sin \phi = (215 \text{ pm})(\sin 50°) = 165 \text{ pm}.$$

The expected de Broglie wavelength for a 54-eV electron, calculated as in Example 1b, is 165 pm, in agreement with the measured value. De Broglie's prediction is exactly confirmed.

   2. **G. P. Thomson's experiment.** In 1927 George P. Thomson, working at the University of Aberdeen in Scotland, independently confirmed de Broglie's equation, using a somewhat different method. As Fig. 50-6a shows, he directed a monoenergetic beam of 15-keV electrons through a thin metal target foil. The target was specifically *not* a single crystal (as in the Davisson-Germer experiment) but was made up of a large number of tiny, randomly oriented crystallites. With this arrangement there will always, by chance, be a certain number of crystallites oriented at the proper angle to produce a diffracted beam.

   If a photographic film is placed parallel to the target, as shown in Fig. 50-6a, the central beam spot will be surounded by diffraction rings.

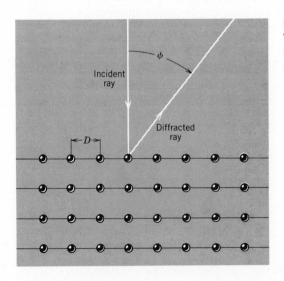

figure 50-5
The crystal surface acts like a diffraction grating, with grating spacing D.

(a)

(b)

(c)

figure 50-6

1137 TESTING DE BROGLIE'S HYPOTHESIS   SEC. 50-3

(a) An arrangement for producing a diffraction pattern that is characteristic of a powdered or polycrystalline target. (b) The pattern for an incident X-ray beam. (c) The pattern for an incident electron beam of the same wavelength.

Figure 50-6b shows this pattern for an X-ray beam incident on an aluminum target. Figure 50-6c shows the pattern for aluminum when an electron beam of the same wavelength replaces the X-ray beam. A simple glance at these two diffraction patterns leaves no doubt that both originated in the same way. Numerical analysis of the patterns confirms de Broglie's hypothesis in every detail.

Thomson shared the 1937 Nobel prize with Davisson for his electron diffraction experiments. George P. Thomson was the son of J. J. Thomson, who won the Nobel prize in 1906 for his discovery of the electron and for his measurement of its charge-to-mass ratio. It has been said that "... one may feel inclined to say that Thomson, the father, was awarded the Nobel prize for having shown that the electron is a particle, and Thomson, the son, for having shown that the electron is a wave."

3. *Diffraction at a straight edge.* Figure 50-7a shows an arrangement in which a beam—of either visible light or electrons—is allowed to strike a photographic plate P after passing an interposed straight edge E. Diffraction takes place at the edge, and the patterns of maxima and minima shown in Figs. 50-7b and 50-7c are formed on the plate. Without reading the caption, can you tell which of these two figures was made with an incident light beam and which with an incident electron beam? They are certainly similar, and it is hard to doubt that we are dealing in each case with a wave phenomenon.

Today the wave nature of matter is taken for granted and diffraction studies by beams of electrons or neutrons are used routinely to study the atomic structures of solids or liquids. Matter waves serve as a valuable supplement to X rays as tools for structure analysis. Electrons, for example, are less penetrating than X rays and so are particularly useful for studying surfaces. Again, X rays interact largely with the electrons in a target, and for that reason it is not easy to use them to locate light atoms—particularly hydrogen—which have few electrons. Neutrons,

figure 50-7
(a) An arrangement for showing a diffraction pattern for a straight edge E. (b) The pattern on screen P if the incident beam is visible light . . . (c) if it is an electron beam.

on the other hand, interact largely with the nucleus of the atom and can be used to fill this gap. Figure 50-8, for example, shows the structure of solid benzene as deduced from neutron diffraction studies. The six carbon atoms (solid lines) that form the familiar benzene ring are clearly there to see, as are the six hydrogen atoms (dashed lines) that couple to them.

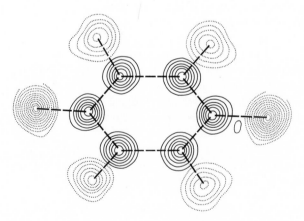

figure 50-8
The atomic structure of solid benzene as deduced by neutron diffraction methods. The familiar benzene ring is clearly present.

EXAMPLE 2

Neutron diffraction. A nuclear reactor is a very efficient "neutron factory." As in Fig. 50-9, let the shield wall of such a reactor be pierced by a solid graphite cylinder, one end of the cylinder being exposed to the reactor core and the other sticking out into the laboratory. The neutrons that emerge from this collimating cylinder will have made many collisions with the carbon atoms of the cylinder and will be in thermal equilibrium with them. Such neutrons, called

figure 50-9
A neutron diffraction arrangement, viewed from the top. The neutrons emerging from the graphite thermalizing column have a distribution of energies. After Bragg reflection from the crystal, the neutron beam is monoenergetic.

thermal neutrons, have a continuous distribution of speeds and thus of de Broglie wavelengths. Analysis shows that the de Broglie wavelength at which the emerging neutron beam is the most intense is given by

$$\lambda_{\text{max}} = \frac{h}{\sqrt{5mkT}}$$

in which m (= 1.68×10^{-27} kg) is the mass of the neutron, k is the Boltzmann constant, and T (= 300 K, say) is the temperature at which the neutrons are in thermal equilibrium.

(*a*) Find λ_{max}.

Substituting in the above equation yields

$$\lambda_{\text{max}} = \frac{6.63 \times 10^{-34}\,\text{J} \cdot \text{s}}{\sqrt{5(1.68 \times 10^{-27}\,\text{kg})(1.38 \times 10^{-23}\,\text{J/K})(300\,\text{K})}}$$

$$= 113\,\text{pm}.$$

(*b*) The emerging neutron beam is allowed to fall on a crystal of calcite, cut so that atomic planes separated by 303 pm are parallel to its surface; see Fig. 50-9. For what angle θ with the crystal surface will there be a first-order reflected beam whose wavelength is that calculated in (*a*)? Note that, although the beam that falls on the crystal has a distribution of de Broglie wavelengths, the reflected first-order beam has only a single de Broglie wavelength.

A reflected beam will occur when Bragg's law (see Section 47-6),

$$n\lambda = 2d \sin \theta \qquad n = 1, 2, 3, \ldots,$$

is satisfied. Thus

$$\theta = \sin^{-1} \frac{n\lambda}{2d} = \sin^{-1} \frac{(1)(113\,\text{pm})}{(2)(303\,\text{pm})}$$

$$= \sin^{-1} 0.187 = 10.8°.$$

Note that, no matter to what value of θ the crystal in Fig. 50-9 is rotated, a monoenergetic reflected neutron beam will appear if the appropriate wavelength is present in the neutron beam that falls on the crystal. If the crystal is set so that $\theta = 10.8°$, however, this reflected beam will have its maximum intensity.

The reflected neutron beam of Fig. 50-9 is available to use as a tool in the analysis of solid-state structure problems. The structure of solid benzene, displayed in Fig. 50-8, was worked out in this way.

As we have just seen, the evidence that matter is wavelike is very strong. Still, we cannot forget that the evidence that matter is particlelike is just as strong. The basic difference between these two points of view is that

50-4
WAVES, WAVE PACKETS, AND PARTICLES

figure 50-10
(a) A harmonic wave viewed at $t = 0$.
(b) The wave has a unique wave
number $\kappa_0 \; (= 1/\lambda_0)$.

a particle is localized in both space and time but a wave is not, being spread out in both of these dimensions. Let us begin to reconcile these two approaches by seeing whether we can put together an assembly of waves in such a way that we end up with something that reminds us of a particle. What we will have to say will hold for all kinds of waves, whether they be water waves, sound waves, electromagnetic waves, or matter waves. We discuss, in sequence, localizing a wave in space and in time.

1. *Localizing a wave in space.* Figure 50-10a is a "snap shot" of a generic wave taken at an arbitrary time, say $t = 0$. The wave extends from $x = -\infty$ to $x = +\infty$ and has a sharply defined wavelength λ_0 and a correspondingly sharply defined wave number $\kappa_0 \; (= 1/\lambda_0)$, as Fig. 50-10b shows.* However, there is nothing about this wave that suggests the localization in space that we associate with the word "particle." Put another way, if there is a particle hiding in Fig. 50-10a, the uncertainty Δx of its position along the x-axis is pretty complete.

We recall that it is possible mathematically (see Section 19-4) to manufacture almost any wave shape we want by piling up sine waves with properly chosen wave numbers, amplitudes and phases. Figure 50-11a shows a *wave packet* that can be put together in this way. The collection of infinite waves that makes it up reproduces the wave of Fig. 50-10a over a certain region but — by destructive interference — adds up to zero everywhere else. We now have a localization in space, measured by Δx, the length of the packet. The price we have paid is the sacrifice of the "purity" of our original wave because our packet now no longer contains a single wave number κ_0 but rather a spread of wave numbers, centered about κ_0; see Fig. 50-11b.

Let $\Delta \kappa$ in Fig. 50-11b be a rough measure of the spread of wave numbers that forms the packet of Fig. 50-11a. It is reasonable to believe that the sharper (that is, the more particlelike) we wish the wave packet to be, the broader the range of wave numbers that we must use to build it up. In Fig. 50-10a, for example, the "packet" was not sharp at all $(\Delta x \rightarrow \infty)$ but, on the other hand, we needed only a single wave number to "build it up" $(\Delta \kappa = 0.)$ At the other extreme, we could build a very sharp wave packet $(\Delta x \rightarrow 0)$, but we would need to pile up waves with a

---

* The wave number $\kappa \; (= 1/\lambda)$ must not be confused with the angular wave number k $(= 2\pi/\lambda)$; see Section 19-3.

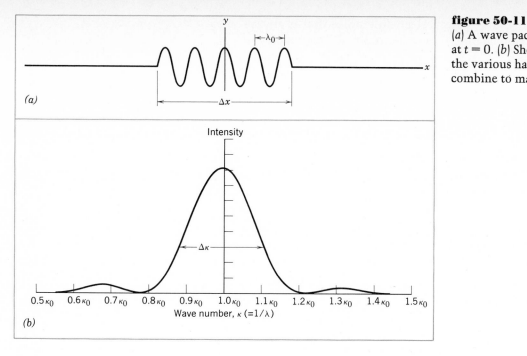

very large spread in wave number $(\Delta\kappa \to \infty)$ to do so. In general, as Δx decreases, $\Delta\kappa$ increases, and conversely. The relationship between them proves to be very simple, namely

$$\Delta x \cdot \Delta\kappa \cong 1. \qquad (50\text{-}3)$$

The approximation sign is there because we have deliberately chosen not to define $\Delta\kappa$ very precisely. We are much more interested in the general idea behind Eq. 50-3 than in a precise quantitative formula.

2. *Localizing a wave in time.* A particle is localized in time as well as in space. If we replaced the space variable x in Fig. 50-10a by the time variable t (and the wavelength λ_0 by the period T_0), that figure would then show how our generic wave would vary with time as it passes a particular fixed point, say x = 0. As before, there is nothing at all about this wave that suggests the localization in time that we associate with the word "particle."

We can build up a wave packet in time as well as in space. Figure 50-11a can illustrate this, provided we replace the space variables by the corresponding time variables, as above, and also replace the wave number κ_0 by the frequency ν_0. By analogy with Eq. 50-3, the duration Δt of our new wave packet is related to the spread $\Delta\nu$ of frequencies needed to make up the wave packet by

$$\Delta t \cdot \Delta\nu \cong 1. \qquad (50\text{-}4)$$

This equation has many practical applications. For example, much of our information in today's society, whether it involves telephony, radar, computer data storage, or whatever, is sent from point to point in the form of pulses. The electronic amplifiers through which these pulses pass should be sensitive over the full range of frequencies included in the pulses they are designed to handle. Equation 50-4 tells us that the shorter the time duration of the pulse, the greater must be the frequency band width (as it is called) of the amplifier, and conversely. On the other hand, the band width of the amplifier should not be larger than necessary. Otherwise the random electronic noise that is always present in the transmission channel — at essentially all frequencies — will be uselessly amplified.

EXAMPLE 3

Designing a radar receiver. A radar transmitter emits pulses 0.15 μs long at a wavelength of 1.2 cm. (*a*) To what central frequency should the radar receiver be tuned? (*b*) What should be its bandwidth? That is, to what range of frequencies should it be able to respond?

(*a*) The central frequency is given by

$$\nu_0 = \frac{c}{\lambda_0} = \frac{3 \times 10^8 \text{ m/s}}{0.012 \text{ m}} = 2.5 \times 10^{10} \text{ Hz} = 25 \text{ GHz.}$$

The wave number corresponding to this wavelength is simply $1/\lambda_0 = 1/(0.012 \text{ m}) = 83.3 \text{ m}^{-1}$.

(*b*) The receiver's bandwidth is given approximately by Eq. 50-4, or

$$\Delta\nu \cong \frac{1}{\Delta t} = \frac{1}{0.15 \times 10^{-6} \text{ s}} = 6.7 \times 10^6 \text{ Hz} = 6.7 \text{ MHz.}$$

If the receiver cannot respond to frequencies throughout this range, it will not be able to reproduce faithfully the shape of the transmitted radar pulse. On the other hand, if the receiver's bandwidth is substantially greater than this, random "noise" frequencies will be amplified and it will be more difficult to detect weak signals.

50-5
HEISENBERG'S UNCERTAINTY PRINCIPLE

Equation 50-3 applies to all kinds of waves. Let us apply it to matter waves. We write, for the quantity $\Delta\kappa$ that appears in that equation,

$$\Delta\kappa = \Delta\left(\frac{1}{\lambda}\right) = \Delta\left(\frac{p_x}{h}\right) = \left(\frac{1}{h}\right)\Delta p_x.$$

Here we have identified λ with the de Broglie wavelength of the particle and substituted h/p_x for it. The subscript on the momentum reminds us that we are dealing with motion along the x-axis only. Substituting this result into Eq. 50-3 leads to

$$\Delta\kappa \cdot \Delta x = \left(\frac{1}{h}\right)\Delta p_x \cdot \Delta x \cong 1$$

or

$$\Delta p_x \cdot \Delta x \cong h.$$

Taking into account the fact that momentum is a vector, we can generalize this relationship to (see Fig. 50-12)

$$\Delta p_x \cdot \Delta x \cong h$$

$$\Delta p_y \cdot \Delta y \cong h \qquad\qquad (50\text{-}5)$$

$$\Delta p_z \cdot \Delta z \cong h.$$

Equations 50-5 are a statement of Heinsenberg's uncertainty principle, first derived by Werner Heisenberg in 1927. Here is what it says: *It is not possible to measure both the position and the momentum of a particle simultaneously with unlimited precision.*

You can, in principle, measure the position of a particle as precisely as you wish. However, the farther you go in this direction, the less precisely will you be able to measure its momentum, and conversely.

figure 50-12
Physics is not all serious!

Put another way, corresponding uncertainties such as Δx and Δp_x cannot be zero at the same time. If you choose to tinker with your measuring equipment to make one of them smaller, the other will automatically grow larger, their product remaining constant. This constant product is — we note — nothing other than the Planck constant h, that certain indicator of the quantum realm.

These limitations have nothing whatever to do with the practical problems of measurement. Equations 50-5, in fact, *assume* ideal instruments. In practice, you will always do worse.*

When we use the word "particle" to describe objects like electrons, it conjures up in our mind the image of a tiny dot moving along a path, its position and velocity being well-defined at every moment. This way of thinking is a natural extension of familiar experiences with objects like baseballs and pebbles that we can see and touch. We must, however, accept the fact that this picture simply does not hold up experimentally beyond the limits set by the uncertainty principle. The quantum world is a world beyond our direct experience, and we must be prepared for new ways of thinking.

In Example 5 we shall see that the uncertainty principle does *not* limit our precision of measurement when we are dealing with large objects like golf balls. Here ordinary instrumental errors swamp the fundamental limits set by this principle. When we deal with electrons and other elementary particles, however, the situation is quite different, as Example 4 shows.

EXAMPLE 4

An electron. A free 10-eV electron moves in the x-direction with a speed of 1.88×10^6 m/s. Assume that you can measure this speed to a precision of 1%. With what precision can you simultaneously measure its position?

The electron's momentum is

$$p_x = mv_x = (9.11 \times 10^{-31} \text{ kg})(1.88 \times 10^6 \text{ m/s})$$

$$= 1.71 \times 10^{-24} \text{ kg} \cdot \text{m/s}.$$

The uncertainty Δp_x in momentum is 1% of this or 1.71×10^{-26} kg $\cdot$ m/s. The uncertainty in position is then, from Eq. 50-5,

$$\Delta x \cong \frac{h}{\Delta p_x} = \frac{6.63 \times 10^{-34} \text{ J} \cdot \text{s}}{1.71 \times 10^{-26} \text{ kg} \cdot \text{m/s}} = 40 \text{ nm},$$

which is about 200 atomic diameters. Given your measurement of the electron's momentum, there is simply no way that you can simultaneously pin down its position to a better precision than this.

EXAMPLE 5

A golf ball. A golf ball has a mass of 45 g and a speed, which you can measure with a precision of 1%, of 40 m/s. What limits does the uncertainty principle place on your ability to measure its position?

This example is exactly like Example 4 except that the golf ball is much more massive and much slower than the electron. The same calculation yields, in this case, $\Delta x \cong 4 \times 10^{-32}$ m.

---

* The "$\cong$" sign in Eqs. 50-5 is often replaced by a "$\gtrsim$" sign to recognize this fact. Also, in some formulations $h/2\pi$ or $h/4\pi$ may appear in place of h. The startling new fact, however, is that the product of the uncertainties in Eqs. 50-5 is *not zero*, as it would be if classical physics prevailed. It is enough here for us to know that this product is of the *order of magnitude* of the Planck constant; finer details do not concern us.

This is a very small distance, about 10^{17} times smaller than the diameter of a typical atomic nucleus. Where large objects are concerned the uncertainty principle sets no meaningful limit to the precision of measurement. You could never have discovered this principle by studying flying golf balls! All of this is in accord with the correspondence principle, which tells us that, in situations in which classical physics is known to give correct answers (that is, golf balls), the predictions of quantum physics merge with those of classical physics.

Here we learn more about the uncertainty principle by seeing how it works in a particular case. Consider a beam of electrons of speed v_0, moving upward as in Fig. 50-13. We set ourselves this task: Measure simultaneously the horizontal position x and the velocity component v_x for an electron in this beam. If you can dream up a scheme to measure each of these quantities simultaneously with unlimited precision, you will have beaten the uncertainty principle. As we shall see, this is easier said than done.

To measure x let us block the beam with a screen A in which we put a slit of width Δx. If an electron passes through this slit, we can claim to know its horizontal position to this precision. By narowing the slit we can improve the precision of this measurement as much as we wish.

So far so good. However, something happens that perhaps we hadn't counted on. The electron beam — being wavelike — flares out by diffraction as it passes through the slit. If we put a suitably sensitive screen at B in Fig. 50-13, a typical single-slit diffraction pattern shows up. Electrons that form the left half of this pattern must have been moving to the left (some faster, some slower) as they emerged from the slit. Those that form the right half must have been moving to the right. Even though — as the symmetry of the arrangement requires — the *average* value of v_x for the emerging electrons is zero, individual electrons have non-zero values.

There is a particular value of v_x that will cause the electron to land at the first minimum of the diffraction pattern, identified by the angle θ_1 in Fig. 50-13. We take this value of v_x — somewhat arbitrarily — as a

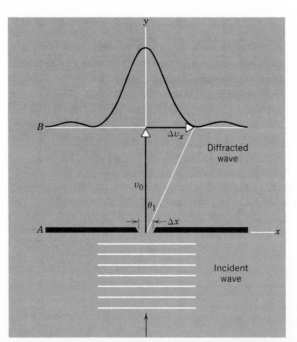

figure 50-13
An incident beam of electrons is diffracted at the slit in screen A. If the slit is made narrower, the diffraction pattern becomes wider, and conversely.

measure of the uncertainty of our knowledge of v_x and we call it Δv_x. A rough measure of this uncertainty is all that we need. Twice the value we have chosen would seem too large and half that value too small; our choice is reasonable.

The first minimum of the diffraction pattern is given by Eq. 46-1 (sin $\theta_1 = \lambda/\Delta x$). If we assume that θ_1 is small enough, we can replace sin θ_1 by θ_1, obtaining

$$\theta_1 \cong \lambda/\Delta x.$$

To reach the first minimum it must be true that

$$\theta_1 \cong \Delta v_x/v_0.$$

Combining these two relations leads to

$$\Delta v_x \cdot \Delta x \cong \lambda v_0.$$

Now λ, the de Broglie wavelength of the electron, is equal to h/p or h/mv_0; putting this into the above and rearranging yields

$$\Delta p_x \cdot \Delta x \cong h.$$

This is precisely the uncertainty relation for the x-component of the electron's motion! The agreement turns out to be *exact* only because of the way we have chosen to define Δp_x; see the footnote on page 1143.

See how the uncertainty principle operates in this case. If we want to pin down the horizontal position of the electron, we must narrow the slit. This, however, broadens the diffraction pattern so that Δp_x increases. On the other hand, if we want to pin down the horizontal momentum component of the electron, we must somehow reduce the angular width of the diffraction pattern. The only way to do this is to widen the slit but that, in turn, means that we no longer know the horizontal position of the electron as precisely as we did. Nobody has been able to think up a scheme to get around these limitations. If you can think of one, buy a ticket to Stockholm; your Nobel prize is assured.

50-7 HEISENBERG'S UNCERTAINTY PRINCIPLE—AN ALTERNATIVE FORM

So far we have considered only the *wavelengths* of matter waves and have said nothing about their *frequencies*. de Broglie proposed that the frequency v of a matter wave is related to the energy E of the corresponding particle by $E = hv$, the same relation that Einstein put forward for radiation when he introduced the photon concept. We can combine this relation with Eq. 50-4 to cast the uncertainty principle in a new form.

From Einstein's photon equation, the uncertainty in the frequency of a matter wave is related to the uncertainty in the energy E of the corresponding particle by $\Delta v = \Delta E/h$. Substituting this into Eq. 50-4 yields $(\Delta E/h)(\Delta t) \cong 1$, or

$$\Delta E \cdot \Delta t \cong h, \tag{50-6}$$

which is the uncertainty principle expressed in terms of different parameters. In words, it says: *If you wish to measure the energy E of a system and you have a time Δt in which to do so, your measurement will be uncertain by an amount ΔE given by $h/\Delta t$.*

Thus, all energy measurements carry an inherent uncertainty unless you have an infinite time available for the measurement. In an atom, for example, the lowest energy state (the so-called ground state) has a well-defined energy because the atom normally exists indefinitely in that state. The energies of all states of higher energy (the so-called excited states) are less precisely defined because the atom — sooner or later — will move spontaneously to a state of lower energy. On average, you have only a certain time Δt available so that your energy measurement will be uncertain by an amount ΔE given by $h/\Delta t$.

EXAMPLE 6

A new particle. In 1974 an important new particle, more than three times as massive as the proton, was discovered simultaneously and independently by two groups of physicists, using the giant accelerators at the Brookhaven National Laboratory and at Stanford University. The measured rest energy of this particle turned out to be 3097 MeV, the uncertainty of the measurement being only 0.063 MeV. Such a massive particle was expected to decay extremely rapidly into particles of smaller mass. What is the mean time interval between production and decay for these short-lived particles?

The answer follows from the uncertainty principle, in the form of Eq. 50-6. Solving for Δt yields

$$\Delta t \cong \frac{h}{\Delta E} = \frac{6.63 \times 10^{-34} \text{ J} \cdot \text{s}}{(0.063 \text{ MeV})(1.60 \times 10^{-13} \text{ J/MeV})} \cong 7 \times 10^{-20} \text{ s}.$$

By ordinary standards this seems to be a short time interval but, in fact, the experimenters were astonished by how *long* it was or — equivalently — by how sharply the rest energy of the particle was defined. Theory had predicted that the decay of this massive particle (called the J/ψ) would be very much faster. One observer remarked that the observed slowing down of the decay was as if Cleopatra, floating on the Nile in her Royal Barge, had dropped a pebble over the side and the falling pebble, as of today, had not yet hit the water! This new particle proved to be so significant that the leaders of the two groups, Burton Richter (who called it the ψ particle) and Samuel Ting, (who called it the J-particle), were awarded the Nobel prize for 1976 for its discovery.

The uncertainty principle is used in this way to measure the lifetimes of the excited states of molecules, atoms, and unstable fundamental particles of all kinds.

50-8
THE WAVE FUNCTION AND SCHRÖDINGER'S EQUATION

By this time you should be comfortable with the fact that a moving particle can be viewed as a wave and you should know how to measure its wavelength. It remains to ask: "What is the quantity whose variation in time and space makes up this wave?" To put it loosely: "What is waving?"

For a wave in a string we can represent the wave disturbance by the transverse displacement y. For sound waves we can choose the differential pressure p and for electromagnetic waves the electric field vector $\mathbf{E}$. For matter waves we introduce the so-called *wave function* Ψ (spelled "psi" and pronounced "sigh"). The problem at hand may be that of a proton traveling along the axis of an evacuated tube in a particle accelerator, a conduction electron moving through a copper wire, or an electron moving about the nucleus of a hydrogen atom. Whatever the case may be, if we know the wave function $\Psi(x,y,z,t)$ for every point of space and for every instant of time, we know all that can be known about that particle.

Before we look into the physical meaning of the wave function, let us consider a problem that involves radiation rather than matter: a plane electromagnetic wave traveling through free space. We can think of such a wave (following Maxwell) as an arrangement of electric and magnetic fields that varies in space and time or (following Einstein) as a beam of photons, each moving with the speed of light. On the first picture, the rate per unit area at which energy is transported by the wave (see Example 6, Section 41-9) is proportional to E^2, in which E is the amplitude of the electric field vector. On the second picture, this rate is proportional to the average number of photons per unit volume of the beam, each photon having an energy $h\nu$. We see here a connection between the wave and particle picture of radiation, namely, the notion — first advanced by Einstein — that the square of the electric field intensity is a direct measure of the average density of photons.

Max Born proposed that the wave function Ψ for a beam of particles be interpreted in this same way, namely, that its square be a direct measure of the average density of particles in the beam. In many problems, however, such as the structure of the hydrogen atom, there is only a single electron present. What can it then mean to speak of the "average density of particles"? Born proposed that in such cases we should interpret the square of the wave function at any point as the probability (per unit volume) that the particle will be at that point.* Specifically, if dV is a volume element located at a point whose coordinates are x, y, z, then the probability that the particle will be found in that volume element at time t is proportional to $\Psi^2\,dV$. Perhaps by analogy with ordinary mass density (a mass per unit volume), we call the square of the wave function a *probability density*, that is, a probability per unit volume.

Note that the relationship between the wave function and its associated particle is *statistical*, involving only the *probability* that the particle will find itself within a specified volume element. In classical physics we also deal with particles on a statistical basis (see Chapters 23 and 24), but in those cases statistical methods are just a handy way of dealing with large numbers of particles. In quantum mechanics, however, the statistical nature is inherent and is dictated by the uncertainty principle, which, as we have seen, sets limits to the meaning that we can attach to the word "particle."

The probability that our particle will be *somewhere* must be equal to unity (which corresponds to a certainty) so that we have

$$\int \Psi^2\,dV = 1 \qquad \text{(normalization condition)}, \qquad (50\text{-}7)$$

the integration being taken over all space. To *normalize* a wave function means to multiply it by a constant factor, chosen so that Eq. 50-7 is satisfied.

We save for last an obvious question: In any given problem, how do we know what the wave function is? Waves on strings and sound waves are governed by Newton's laws of mechanics. Electromagnetic waves are predicted and described by Maxwell's equations. Where do the wave functions come from?

In 1926 Erwin Schrödinger, inspired by de Broglie's matter–wave concept, thought along these lines: *Geometrical optics deals with rays and with the motion of light in a straight line; it turned out to be a special case of a much more general wave optics. Newtonian mechanics also has "rays" (trajectories) and straight-line motion (of free particles). Could it turn out to be a special case of a much more general — but as yet undiscovered — wave mechanics?*

* The wave function Ψ is usually a complex quantity, that is, it involves $\sqrt{-1}$, which we represent by the symbol i. By Ψ^2 (more properly written as $|\Psi|^2$) we mean the square of the absolute value of the wave function. This is always a real quantity. We give a physical interpretation only to the square of the wave function, not to the wave function itself.

Schrödinger derived a remarkably successful theory based on this analogy. Its central feature is a differential equation, now known as *Schrödinger's equation*, which governs the variation in space and time of the wave function Ψ for a wide range of problems. We obtain solutions to problems in classical mechanics by manipulating Newton's laws of motion; we obtain solutions to problems in electromagnetism by manipulating Maxwell's equations; in exactly the same spirit, we obtain solutions to atomic problems by manipulating Schrödinger's equation.*

In the next section we shall study an important problem from the point of view of wave mechanics, that of a particle trapped in a well from which it can never escape (or can it?).

50-9
A PARTICLE TRAPPED IN A WELL

Consider a particle — which we take to be an electron — trapped by external forces within a limited region of space. Atoms are such electron traps. Most of the electrons in the atoms that make up our planet and the living things that inhabit it have been so trapped since before the solar system was formed.

Let us consider trapping in one dimension only and let us use the setup of Fig. 50-14a as a guide. Although this is clearly a large-scale device, it is possible with modern technology to build up — atom layer by atom layer — solid "quantum well structures" in which the distance corresponding to *l* in Fig. 50-14a can be as small as a few atomic diameters. Such structures have many practical applications, in light wave communication devices and as optical logic gates, for two examples.

Figure 50-14b shows the potential energy $U(x)$ of an electron in the arrangement of Fig. 50-14a. If the electron finds itself in the central section of the arrangement, no force acts on it. However, as it approaches either end, a strong force appears, tending to turn it back.

(a)   (b)

figure 50-14
(a) An arrangement that can be used to confine an electron to a length *l* of the x-axis. (b) The potential energy $U(x)$ for various positions of the electron. In a practical device the corners of this curve would be somewhat rounded and the "vertical" walls would slope outward slightly.

---

* Schrödinger's equation is a differential equation for a particle of mass m in a force field given by a potential energy function U. For one dimensional motion the wave function $\psi(x,t)$ must satisfy this equation:

$$-\frac{\hbar^2}{2m}\frac{\partial^2\psi(x,t)}{\partial x^2} + U(x,t)\psi(x,t) = i\hbar\frac{\partial\psi(x,t)}{\partial t}.$$

Newton's law of motion can be derived from the Schrödinger equation in the classical limit of macroscopic systems. See Chapters 5 and 6 of 'Quantum Physics of Atoms, Molecules, Solids, Nuclei and Particles', 2nd Edition, by Robert Eisberg and Robert Resnick (John Wiley and Sons, New York, 1985) for details.

(Recall that the force acting on the electron is given by $-dU/dx$; see Eq. 8-7). If the total energy E of the electron is less than U_0, there is no way, classically, that the electron can escape from the central section. Our electron is, as we say, trapped in a potential well.

Our job is to describe the motion of the electron in its potential well in the language of wave mechanics. If we can do this, we will be well on our way to solving the problem of the hydrogen atom. Here too an electron is trapped in a potential well, although admittedly of different form than that of Fig. 50-14b.

We note that, as far as the space variables are concerned, the wave function for the electron in Fig. 50-14 depends only on x, being independent of y and z. Further, we make an important assumption, namely, that the wave function can be written in the form*

$$\Psi(x,t) = \psi(x)F(t) \qquad (50\text{-}8)$$

in which $\psi(x)$ is the spatial part of the wave function and $F(t)$ is the time-dependent part. In all problems in this book we will be interested only in how the wave function varies with x, that is, in $\psi(x)$.

We look first at the special case of Fig. 50-14b, that for which $U_0 \to \infty$, an infinitely deep potential well. It turns out that the Schrödinger equation for an electron trapped in an infinite well of width l is, mathematically, exactly the same as the classical equation (derived from Newton's laws) governing the motion of a string of length l stretched between rigid supports. All we have to do is to solve this familiar classical problem and then replace $y(x)$ by $\psi(x)$ where $y(x)$ is the amplitude of oscillations for the vibrating string and $\psi(x)$ is the wave function we seek.

We saw in Section 19-3 that, if a stretched string is infinitely long, waves of *any* wavelength can travel along it. However, as we learned in Section 19-9, if the string has a finite length l and is clamped between rigid supports, this freedom is removed. The act of clamping the string means that it cannot move at its endpoints and this, in turn, means that only standing waves of selected discrete wavelengths can exist in the string; see Fig. 19-17. Note carefully how localizing the motion of the string leads directly to the (classical) quantization of that motion.

The allowed wavelengths for a stretched string of length l are given by

$$\lambda_n = \frac{2l}{n} \qquad n = 1, 2, 3, \ldots \qquad (50\text{-}9)$$

Only if this equation is satisfied can there be a point of no motion (a node) at each end of the string. The amplitude of oscillation at any point along the string is given by (see Eq. 19-18b)

$$y_n(x) = y_{max} \sin k_n x \qquad 0 < x < l,$$

where k_n is the wave number, defined from $k_n = 2\pi/\lambda_n$, and y_{max}, the "amplitude of amplitudes," is the greatest amplitude of oscillation experienced by any point along the string. For $x < 0$ or for $x > l$, the wave amplitude $y(x)$ is zero.

Combining the above two equations yields

$$y_n(x) = y_{max} \sin \frac{n\pi x}{l} \qquad n = 1, 2, 3, \ldots \qquad (50\text{-}10)$$

---

* This assumption turns out to be true for almost all problems in quantum mechanics. $F(t)$ emerges as a complex quantity, the square of its absolute value being unity. Thus, for all problems that we shall meet, Ψ^2 and ψ^2 are identical.

You can check that $y_n(x)$ is indeed zero at $x = 0$ and at $x = l$, no matter what the value of n. Figure 50-15 shows plots of $y_n(x)$ for the three lowest modes of oscillation and for one higher mode.

Now let's return to our original problem, an electron trapped in an infinitely deep well. We can use Eq. 50-1 to write, for the momentum of the trapped electron,

$$p_n = \frac{h}{\lambda_n} = \frac{nh}{2l} \qquad n = 1, 2, 3, \ldots ,$$

in which we have identified the wavelength of Eq. 50-9 with the de Broglie wavelength of the electron. The energy of the electron is entirely kinetic and is given by

$$E_n = \frac{p_n^2}{2m_e} = n^2 \frac{h^2}{8m_e l^2} \qquad n = 1, 2, 3, \ldots , \qquad (50\text{-}11)$$

in which m_e is the mass of the electron and n now emerges as a quantum number. Equation 50-11 tells us that the energy of the trapped electron is quantized.

We turn now to the wave function of the trapped electron, which we find, as promised, by simply substituting $\psi_n(x)$ for $y_n(x)$ in Eq. 50-10. Thus the curves of Fig. 50-15 serve equally well to describe the allowed oscillation modes of a clamped string or the allowed stationary states of a trapped electron. Our interest, however, lies in the square of the wave function because this, rather than the wave function itself, tells us where the electron is likely to be found.

From Eq. 50-10 we have, for the square of the wave function (or the probability density $P_n(x)$, as we have called it)

$$P_n(x) = \psi_n^2(x) = A_n \sin^2 (n\pi x/l), \qquad (50\text{-}12)$$

in which A_n is a constant. Figure 50-16 shows $P_n(x)$ for the four allowed states illustrated in Fig. 50-15.

Once we have found the energies (Eq. 50-11) and the probability densities (Eq. 50-12) for an electron trapped in an infinite well, we have essentially solved the problem. Let us, however, take a closer look at what we have learned, under three headings.

1. *The electron cannot be at rest in its trap.* The state of lowest energy for the trapped electron, called its ground state, corresponds to $n = 1$ in Eq. 50-11. *This lowest energy is not zero.* This surprising result — which is not at all what we expect classically — follows directly from the uncertainty principle. To see this, let us estimate the uncertainties in the position and in the momentum for the trapped electron and see where we are led.

As for the position of the electron, we assume that its uncertainty is approximately equal to l, the width of the well. Thus $\Delta x \cong l$.

What about the uncertainty in momentum? You may think that because the ground state is precisely known (see Eq. 50-11), the ground state momentum is also precisely known so that the uncertainty in the momentum must be zero. If you think this you have forgotten that momentum is a vector. Although its magnitude is known precisely, we do not know anything about its direction. If we measure the momentum of the electron as it bounces back and forth between the walls of its trap, we are just as likely to find $+p_x$ as $-p_x$. We can take, then, as a rough measure of the uncertainty in momentum, $\Delta p_x \cong p_x - (-p_x) = 2p_x$.

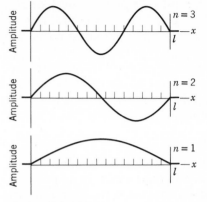

figure 50-15
This figure represents equally well the vibrational amplitudes of a stretched string (Eq. 50-10) or the wave function of a trapped electron. The three lowest modes and one higher mode are shown.

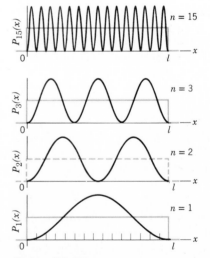

figure 50-16
The solid curves show the probability density $P(x)$ for four values of the quantum number n. Compare Fig. 50-15, which shows the wave functions for the same four cases. The dashed lines shows the classical expectations for the probability density.

The uncertainty relation (Eq. 50-5) then gives us

$$\Delta p_x \cdot \Delta x \cong h \qquad \text{or} \qquad (2p_x) \cdot (l) \cong h$$

so that

$$p_x \cong h/2l.$$

The energy E of that electron inside the trap is entirely kinetic so that we can write, introducing the above equation,

$$E = \frac{p_x^2}{2m_e} = \frac{h^2}{8m_e l^2}.$$

This is precisely Eq. 50-11 for the case of $n = 1$! That the agreement happens to be exact is a consequence of the way we have chosen to define Δx and Δp_x; see the footnote on page 1143. Thus we see how this *zero-point energy*, as it is called, follows directly from the uncertainty principle.

2. *The electron spends more time in certain parts of the trap than in others.* For this one-dimensional problem a "volume" element becomes a line element so that the product $P_n(x)\, dx$ is the probability that the electron will be found in the interval x to $x + dx$. A glance at the probability density curves of Fig. 50-16 shows that in the ground state ($n = 1$), the electron is much more likely to be found near the center of the trap than near its ends. Once again we have a prediction in sharp contrast with the prediction of classical mechanics. According to classical theory all positions between the walls of the trap are equally likely, as the horizontal lines in Fig. 50-16 suggest. For both the quantum and the classical curves, the area under the curve is unity, as the normalization condition (Eq. 50-7) requires.

For states of higher quantum number — and thus of higher energy — the distribution of the electron across the trap becomes more uniform and the quantum prediction merges with the classical prediction, just as the correspondence principle requires.

It may be disturbing that the quantum prediction, even at high quantum numbers, always contains points at which the probability density is zero. How can an electron pass through such a point? Note that no such difficulty arises if we stick to the wave picture for the electron; such points represent vibrational nodes in standing waves, as in Fig. 19-17. If we wish to look at the electron as a particle, we can do so only within the limitations of the uncertainty principle. It turns out (see Problem 35) that the uncertainty in the position of the electron in any given state is about equal to the distance between adjacent nodes in that state. Thus, the notion of "particle" is not well enough defined for the question that we have raised to pose any problem.

3. *The electron can escape from its trap.* So far we have dealt with a well of infinite depth. A major quantum surprise awaits us if we relax this requirement and deal with the more realistic case of a well of finite depth. In Fig. 50-17 we compare two wells of the same width ($= 2 \times 10^{-10}$ m, about the size of a large atom), one of the wells being infinitely deep and the other having a depth of only 20 eV. To find the allowed energies and the corresponding probability densities for a *finite* well, we need the full power of the Schrödinger equation. Here we will simply give the results, without proof.* We consider the ground state only.

---

* See, for example, Robert Eisberg and Robert Resnick, *Quantum Physics* (John Wiley & Sons, New York, 1985, 2nd ed.), Appendices G and H.

(a) ·   (b)

figure 50-17

A potential well of infinite depth and one of finite depth (= 20 eV) are compared. Both wells have the same width l (= 2×10^{-10} m). The energies E_1 and the probability densities $P(x)$ for the ground state are compared.

Figure 50-17 shows that the ground state energy for the finite well (= 4.45 eV) is substantially less than the ground state energy for the infinite well (= 9.41 eV, calculated from Eq. 50-11). We can tell that this should be the case simply by inspecting the probability density curves of Fig. 50-17. For the infinite well, half a de Broglie wavelength fits neatly and exactly between the rigid walls of the well. For the finite well, however, the de Broglie wavelength is too large to fit in this way; it spills over beyond the walls. Now if the de Broglie wavelength is *larger* for the finite well, the momentum (= h/λ) must be *smaller*, which means the energy must also be smaller, just as we observe. Thus the reduced energy for the finite well is consistent with the form of its probability density curve.

The exponential tails of the probability density curve (Fig. 50-17*b*) spilling over beyond the walls means that there is a finite probability that the electron will find itself outside the well! This curve has been normalized (see Eq. 50-7) so that the area under the curve is unity. The area under the two exponential tails is 0.074, which means that, if you measured the position of the trapped electron, you would find it outside the well 7.4% of the time.

How can an electron whose energy is only 4.45 eV escape from a well that is 20 eV deep? It is clearly a classical impossibility. It is as if you put a jelly bean into a closed box and sometimes (but not always) the jelly bean materialized outside. This is so unlikely for jelly beans that we can safely use the word "impossible." However, electrons are not jelly beans. They are governed by quantum laws and not at all by classical Newtonian laws. How are we to understand this behavior of the electron?

Once again the uncertainty principle rises to comfort us. Recall that, when we applied the uncertainty principle to an electron trapped in an infinite well, we assumed that $\Delta x \cong l$, the width of the well, and $\Delta p_x \cong 2p_x$. For the finite well the electron's momentum — we have just seen — is *smaller* than it is for the infinite well. Therefore, for our finite well, the uncertainty in position in the ground state must be *larger* than it is for the infinite well. Thus, for the finite well, *the uncertainty in position is larger than the width of the well!* We should not, then, be surprised to find the electron outside the well from time to time.

EXAMPLE 7

A trapped electron. Consider an electron confined by electrical forces to an infinitely deep potential well whose length l is 100 pm, which is roughly one atomic diameter. What are the energies of its three lowest allowed states and of the state with $n = 15$?

From Eq. 50-11, with $n = 1$, we have

$$E_n = n^2 \frac{h^2}{8m_e l^2} = (1)^2 \frac{(6.63 \times 10^{-34} \text{ J} \cdot \text{s})^2}{(8)(9.11 \times 10^{-31} \text{ kg})(100 \times 10^{-12} \text{ m})^2}$$

$$= 6.03 \times 10^{-18} \text{ J} = 37.7 \text{ eV}.$$

The energies of the remaining states ($n = 2$, 3 and 15) are $2^2 \times 37.7$ eV, $3^2 \times 37.7$ eV, and $15^2 \times 151$ eV or 339 eV, and 8480 eV, respectively.

EXAMPLE 8

A trapped dust particle. Consider a 1-μg speck of dust moving back and forth between two rigid walls separated by 0.1 mm. It moves so slowly that it takes 100 s for the particle to cross this gap. What quantum number describes this motion?

The energy of the particle is

$$E \ (= K) = \tfrac{1}{2} mv^2 = \tfrac{1}{2} (1 \times 10^{-9} \text{ kg})(1 \times 10^{-6} \text{ m/s})^2$$

$$= 5 \times 10^{-22} \text{ J}.$$

Solving Eq. 50-11 for n yields

$$n = \sqrt{8mE} \left(\frac{l}{h} \right)$$

$$= \sqrt{(8)(10^{-9} \text{ kg})(5 \times 10^{-22} \text{ J})} \ [(1 \times 10^{-4} \text{ m})/(6.63 \times 10^{-34} \text{ J} \cdot \text{s})]$$

$$\cong 3 \times 10^{14}.$$

This is a very large number. It is experimentally impossible to distinguish between $n = 3 \times 10^{14}$ and $n = 3 \times 10^{14} + 1$, so that the quantized nature of this motion would never reveal itself. If you compare this example with the previous one, you will see that, although its mass and its kinetic energy are both extremely small, our speck of dust is still a gross macroscopic object when compared to an electron. Quantum mechanics gives the correct answers but, since these answers coincide in this case with the answers given by classical physics, the complications of the quantum calculations are not needed. We are in a region governed by the correspondence principle.

EXAMPLE 9

Normalizing the probability density. Evaluate the normalization constant A_n in Eq. 50-12, which gives the probability density for a particle trapped in an infinitely deep well of width l.

For this one-dimensional problem, the "volume" element is a length element and the normalization equation (Eq. 50-7) becomes

$$\int_0^l P_n(x)\, dx = 1,$$

in which l is the width of the well. Substituting for $P_n(x)$ from Eq. 50-12 yields

$$A_n \int_0^l \sin^2\left(\frac{n\pi}{l}\right) x\, dx = 1. \tag{50-13}$$

This integral is carried out most easily by introducing a new variable θ, defined from

$$\theta = \left(\frac{n\pi}{l}\right) x.$$

With this change, the above integral becomes

$$A_n\left(\frac{1}{n\pi}\right)\int_0^{n\pi}\sin^2\theta\, d\theta = A_n\left(\frac{1}{n\pi}\right)\left|\left(\frac{\theta}{2}\right)-\left(\frac{1}{4}\right)\sin 2\theta\right|_0^{n\pi} = 1.$$

Evaluating and solving for A_n leads eventually to

$$A_n = \frac{2}{l}.$$

Note that the normalization constant A_n does not involve the quantum number n and is thus the same for all states of the system.

EXAMPLE 10

Where the electron spends its time. An electron is trapped in an infinitely deep well of width l. If the electron is in its ground state, what fraction of its time does it spend in the central third of the well?

In the preceding example we showed that the normalization constant A_n that appears in Eq. 50-12 is $2/l$, so that the probability density for the ground state, which corresponds to $n = 1$, is given by

$$P_1(x) = \left(\frac{2}{l}\right)\sin^2\left(\frac{\pi}{l}\right)x.$$

The integral of this quantity over the entire well is unity, and the fraction that we seek is given by

$$f = \int_{l/3}^{2l/3} P_1(x)\, dx = \left(\frac{2}{l}\right)\int_{l/3}^{2l/3}\sin^2\left(\frac{\pi}{l}\right)x\, dx.$$

Evaluating this integral as in the previous example leads to

$$f = 0.61.$$

Thus the electron — assuming that it is in its ground state — spends 61% of its time in the central third of its trap, and about 20% in each of the outer two thirds $(0.20 + 0.61 + 0.20 \cong 1)$. If the electron obeyed the laws of classical physics, it would spend exactly one third of its time in each of these regions of its trap. The probability density curve for the ground state, displayed in Fig. 50-17, supports graphically the calculation that we have made in this example.

50-10
THE HYDROGEN ATOM

We defined the problem of a particle trapped in a well by writing down the potential energy functions $U(x)$ shown in Fig. 50-17a (infinite well) and Fig. 50-17b (finite well). To define the problem of the hydrogen atom, we introduce a potential function that is readily derived from Coulomb's law. This law gives the force of attraction between the electron and the central proton in the hydrogen atom. From Eq. 29-13 we have

$$U(r) = -\frac{1}{4\pi\epsilon_0}\frac{e^2}{r}, \qquad (50\text{-}14)$$

in which r is the distance between the electron and the central proton.

If we insert this potential energy function into Schrödinger's equation and carry out the needed mathematical manipulations, we are able to derive an expression for the energies of the allowed stationary states of the atom and also for the wave functions that describe those states.

The expression for the energies of the stationary states turns out to be exactly Eq. 49-30, the expression derived from Bohr theory. We focus our attention here on the ground state of the hydrogen atom, that is, on the state of lowest energy.

The wave function for the ground state also emerges from the Schrödinger equation and turns out to depend only on the single variable r. It is given by

$$\psi(r) = \frac{1}{\sqrt{\pi a_0^3}}e^{-r/a_0}, \qquad (50\text{-}15)$$

in which a_0 is the Bohr radius.

This wave function has spherical symmetry, by which we mean that it depends only on the magnitude of the position vector r and not on its direction. This is perhaps not surprising. The potential energy function (Eq. 50-14) was also spherically symmetric so that the atom has no built-in preferred direction. Like a billiard ball, the atom looks the same in all directions.

The square of the wave function, which we have called the *probability density*, has the property that $\psi^2(r)\ dV$ gives the probability of finding the electron in a volume element dV located at a position defined by the position vector r. From Eq. 50-15 we have

$$\psi^2(r) = \frac{1}{\pi a_0^3}e^{-2r/a_0}. \qquad (50\text{-}16)$$

Figure 50-18*a* is a "dot plot" representation of Eq. 50-16, the density of the dots suggesting the probabilistic nature of the electron's location. A circle of one Bohr radius has been superscribed to show the scale.

A more useful quantity for representing the electron's position is the *radial probability density* $P_r(r)$. This is defined so that $P_r(r)\ dr$ gives the probability that, regardless of direction, the electron will be found to lie between two spherical shells whose radii are r and $r + dr$. The volume

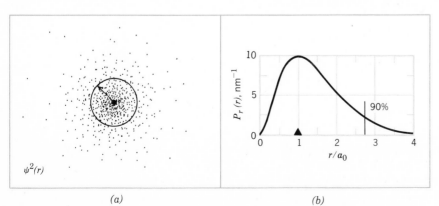

(a)   (b)

figure 50-18
(*a*) A "dot plot" representation of the probability density for the ground state of the hydrogen atom; see Eq. 50-16. (*b*) The radial probability density; see Eq. 50-17. The filled triangle marks the maximum, at $r = a_0$. The line marked "90%" marks the radius of the 90% probability sphere.

between those shells is $(4\pi r^2)(dr)$ so that we can write

$$P_r(r)\ dr = \psi^2(r)\ dV = \psi^2(r)(4\pi r^2)(dr),$$

which, combined with Eq. 50-16, leads to

$$P_r(r) = \psi^2(r)(4\pi r^2) = \frac{4}{a_0^3}\ r^2 e^{-2r/a_0}. \tag{50-17}$$

Figure 50-18b shows a plot of Eq. 50-17. We note (see Example 11) that the maximum value of $P(r)$ occurs at $r = a_0$. In wave mechanics we do not say that the electron in the ground state of the hydrogen atom moves in an orbit of one Bohr radius. We say instead that the electron is more likely to be found at this distance from the central nucleus than at any other distance, either larger or smaller. The so-called 90% radius is also indicated in the figure. It defines a sphere such that the probability that the electron, on measurement, will be found inside is 90%. The probability that it will be found outside is, of course, 10%.

We see that the answer to the question: "How big is the hydrogen atom?" is not so simple. You can say that it is one Bohr radius, but (see Problem 46) 68% of the times that you measure it, you will find that the electron is farther away than this. A more reasonable answer is to give the radius of the 90% probability sphere, which turns out to be 2.7 Bohr radii. The inherent quantum fuzziness of the atom simply does not permit us to answer the question any more precisely.

Wave mechanics allows us to solve another puzzle that blocked classical attempts to understand the atom. How can an atom exist in a stable stationary state without radiating away its energy? Recall that, in Bohr's theory, it was necessary to put in "by hand," as it were, the blind assumption that such stable nonradiating states exist. On the wave picture, however, we replace the Bohr orbit by a three-dimensional probability cloud that represents the electron in a statistical way. This cloud can be viewed as a smeared-out distribution of charge, and it can be shown that the radiations from its various moving elements cancel at all external points by destructive interference. In this way does wave mechanics give a ready explanation of the stability of stationary states.

EXAMPLE 11

Where is the electron most likely to be?   Verify that the maximum of the radial probability density curve of Fig. 50-18b falls at $r = a_0$, where a_0 is the Bohr radius.

It will simplify matters if we replace r by the quantity $a_0 x$ (where x is a new dimensionless variable) and correspondingly, dr by $a_0\ dx$. We can then recast Eq. 50-17 as

$$P_r(r)\ dr = P_r(x)\ dx = 4x^2 e^{-2x}\ dx$$

Differentiating $P_r(x)$, using the rule for products, yields

$$dP_r/dx = 4x^2 e^{-2x}(-2) + 4e^{-2x}(2x)$$
$$= 8x(1-x)e^{-2x}.$$

At the maximum of the curve we must have $dP_r/dx = 0$ and, as inspection of the above shows, this does indeed occur at $x = 1$, which is equivalent to $r = a_0$. Note that we also have $dP_r/dx = 0$ at $x = 0$ and as $x \to \infty$. However, these conditions are quite consistent with Fig. 50-18b and do not correspond to maxima.

In Section 50-9 we saw that an electron trapped in a well from which — classically — it could not escape has, nevertheless, a finite probability of being found outside the well. We pointed out that it was as if you put a jelly bean in a closed box and found it outside the box a certain fraction of the times that you checked. Things like this don't happen to massive objects like jelly beans, but they do happen to electrons and to other light particles.

Here we discuss a related quantum phenomenon, the penetration of classically impenetrable barriers. In this case it is as if you tossed a jelly bean at a window pane and — to your surprise — it materialized on the other side with the glass unbroken. Again, don't expect this to happen for jelly beans. *Barrier tunneling*, as it is called, certainly *does* happen for electrons and is — as we shall see — a phenomenon of great practical importance.

Figure 50-19a shows such a barrier, of height U_0 and thickness l. An electron of total energy E approaches the barrier from the left. Classically, because $E < U_0$, the electron would be reflected at the barrier and would move back in the direction from whence it came, its incident energy intact. In wave mechanics, however, there is a finite chance that the electron will penetrate the barrier and continue its motion to the right.

We can describe the situation by assigning a reflection coefficient R and a transmission coefficient T, the sum of these two quantities being unity. Thus, if $T = 0.05$, of every 100 electrons fired at the barrier 5, on average, will get through and 95 will be reflected.

Figure 50-19b shows the probability density for the situation. To the left of the barrier the reflected matter wave has a smaller amplitude than the incident wave so that there are no points at which the cancellation is total. Within the barrier the wave decays exponentially, just as it did outside the potential well in Fig. 50-17b. On the far side of the barrier we have a traveling matter wave of reduced amplitude.

From the Schrödinger equation we can show that the transmission coefficient T is given by*

$$T \cong e^{-2kl}, \qquad (50\text{-}18)$$

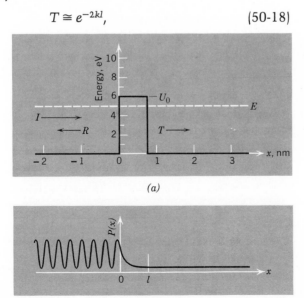

(a)

(b)

figure 50-19
(a) A particle of total energy E falls on a barrier of height U_0. I (of unit amplitude) represents the incident particle; R and T are the reflection coefficient and the transmission coefficient, respectively. (b) The probability density $P(x)$.

* See, for example Robert Eisberg and Robert Resnick, *Quantum Physics* (John Wiley & Sons, New York, 1985), 2nd ed., Section 6-5.

in which

$$k = \sqrt{\frac{8\pi^2 m(U_0 - E)}{h^2}}.$$

This formula is an approximation that holds only for barriers that are either high enough and/or thick enough so that the transmission coefficient T is small ($T \ll 1$). Nevertheless, Eq. 50-18 displays well enough the main features of the barrier-tunneling phenomenon.

The value of the transmission coefficient is very sensitive to the thickness of the barrier l and to the factor k, which, in turn, depends on the mass m of the particle and the height U_0 of the barrier. Equation 50-18 shows us that the transmission coefficient decreases if we increase either the thickness l or the height U_0 of the barrier. This is just what we expect from the correspondence principle. The transmission coefficient also decreases as the mass of the particle increases, becoming vanishingly small very rapidly indeed as we proceed from electrons to jelly beans. Again, this is just what we expect from the correspondence principle. Example 12 shows some numerical predictions of Eq. 50-18.

Barriers and Waves.   The penetration of barriers by waves of all kinds is not uncommon in classical physics. It is only when the wave is a matter wave and when, in addition, we choose to focus our attention on its associated particle, that nonclassical behavior presents itself.

Consider Fig. 50-20a, which represents an incident electromagnetic wave (visible light) falling on a glass–air interface at an angle of incidence such that total internal reflection occurs. When we treated this subject in Section 43-5, we assumed that there was no penetration of the incident ray into the air space beyond the interface. However, that treatment was based on geometrical optics which, as we know, is always an approximation, being a limiting case of the more general wave optics. Recall that, in much the same way, Newtonian mechanics (with its raylike trajectories) is a limiting case of the more general wave mechanics.

If we analyze total internal reflection from the wave optics point of view, we learn that there *is* a penetration of the wave, for a distance of the order of a few wavelengths, beyond the interface. Speaking very loosely, we can say that such a penetration is necessary because the incident wave must "feel out" the situation locally before it can "know for sure" that an interface is present.

In Fig. 50-20b, we place the face of a second glass prism parallel to the interface, the gap between them being no more than a few wavelengths. The incident wave can then "tunnel" through this narrow "barrier" and generate a transmitted wave T. The energy in the transmited wave

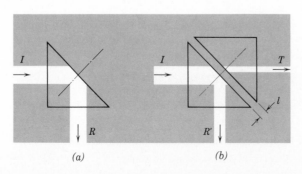

(a)   (b)

figure 50-20
(a) An incident light beam I undergoes total internal reflection at the glass–air interface. (b) The incident beam tunnels through the narrow air gap, generating a transmitted beam T. The intensity of the reflected beam R' is reduced in proportion.

comes at the expense of the reflected wave R, which is now reduced in intensity. The comparison with the barrier tunneling of a matter wave is pretty exact. In one case we deal with an electromagnetic wave (governed by Maxwell's equations) and in the other with a matter wave (governed by Schrödinger's equation.)

You can check out the phenomenon shown in Fig. 50-20b easily enough, using a glass of water. Look down into the glass at the side wall, at such an angle that the light entering your eye has been totally reflected from the wall. The wall will look silvery when this condition holds. Then press your (moistened) fingertip against the outside of the glass. You will be able to see the ridges of your fingerprint because, at those points, you have interfered with the total reflection process, as in Fig. 50-20b. The valleys between the ridges of your prints are still far enough away from the glass surface that the reflection here remains total and you see simply a silvery whorl.

It is also possible to demonstrate the phenomenon of Fig. 50-20b on a large scale by using incident microwaves and large paraffin prisms. In this case the wavelength may be a few centimeters so that the gap between the prisms can also be of this order of magnitude.

Barrier Tunneling — Some Examples.   The barrier tunneling of matter waves is an important phenomenon in the natural world and has many practical applications. For a simple example consider a bare copper wire that has been cut and the two ends twisted together. It still conducts electricity readily, in spite of the fact that the wires are coated with a thin layer of copper oxide, an insulating material. How do the electrons get through this (extremely thin) oxide layer? By barrier tunneling.

For a more exotic example, consider the core of the sun, where the sun's energy is being generated by thermonuclear fusion processes. Such processes involve the fusing together of light nuclei to form heavier ones, with the release of energy. Suppose that two protons are rushing together at high speed. They have to get extremely close before their strong attractive nuclear forces can grapple with each other and cause them to fuse. Meanwhile, they are slowed down by the repulsive Coulomb force that tends to drive them apart. They are, we can say, separated by a Coulomb barrier. The likelihood of fusion depends critically on the ability of the protons to tunnel through this mutual barrier. Without barrier tunneling the solar furnace would shut down and the sun would collapse into itself.

The emission of (positively charged) alpha particles by radioactive nuclei and the spontaneous fission of heavy nuclei into two large fragments are among other natural processes in which tunneling plays a role.

Among practical applications we may list the tunnel diode, in which the flow of electrons (by tunneling) through a device can be rapidly turned on or off by controlling the height of the barrier. This can be done with a very short response time (of the order of 10^{-11} s or 10 ps) so that the device is suitable for applications where speed of response is critical. The 1973 Nobel prize was shared by three "tunnelers," Leo Esaki (tunneling in semiconductors), Ivar Giaver (tunneling in superconductors), and Brian Josephson (the Josephson junction, a quantum switching device based on tunneling).

In a *scanning tunneling microscope*, a fine needle tip is scanned mechanically (in a TV-like raster pattern) over the surface of the sample

Scanning
needle tip

Electron cloud

Surface of sample

figure 50-21
A needle is scanned in a raster-like
pattern over the surface of a sample in
a scanning tunneling microscope. See
"The Scanning Tunneling
Microscope," Gerd Binnig and
Heinrich Rohrer, *Scientific American*,
August 1985. (Courtesy of the authors
and IBM, Zurich Research Laboratory)

being investigated, as in Fig. 50-21. Electrons from the sample tunnel
through the gap between the sample and the needle and are recorded as a
"tunnel current." Normally, this tunnel current would vary widely as
the gap between the sample and the needle changes during the scan.

However, a mechanism is provided that automatically moves the
needle up or down during the scan, so as to keep the tunnel current —
and thus the gap — constant. The needle's vertical position can then be
displayed on a screen as a function of its location, producing a three-dimensional plot of the surface.

Figure 50-22 shows the result of a scan over a silicon surface. The
"bumps" suggest individual silicon atoms. Features as small as 1/100 of
an atomic diameter can be resolved with this remarkable device, which
promises to be an invaluable tool in the study of surface phenomena.

figure 50-22
The surface of a silicon sample, as
revealed by a scanning tunneling
microscope. See "The Scanning
Tunneling Microscope," Gerd Binnig
and Heinrich Rohrer, *Scientific
American*, August 1985. (Courtesy of
the authors and IBM, Zurich Research
Laboratory)

Barrier tunneling.   Consider an electron whose total energy E is 5 eV approaching a barrier whose height U_0 is 6 eV, as in Fig. 50-19a. Let the barrier thickness
l be 0.70 nm.

(a) What is the de Broglie wavelength of the incident electron?

Before the electron reaches the barrier, its total energy E is entirely kinetic,
the potential energy in that region being zero. Proceeding as in Example 1b we
find $\lambda = 0.55$ nm. Thus the barrier is about 0.70 nm/0.55 nm or about 1.3 de
Broglie wavelengths thick.

(b) What transmission coefficient follows from Eq. 50-18?

EXAMPLE 12

We have

$$k = \sqrt{\frac{8\pi^2 m (U_0 - E)}{h^2}}$$

$$= \sqrt{\frac{8\pi^2 (9.11 \times 10^{-31}\ \text{kg})(6\ \text{eV} - 5\ \text{eV})(1.60 \times 10^{-19}\ \text{J/eV})}{(6.63 \times 10^{-34}\ \text{J} \cdot \text{s})^2}}$$

$$= 5.12 \times 10^9\ \text{m}^{-1}.$$

The quantity kl is then $(5.12 \times 10^9\ \text{m}^{-1})(700 \times 10^{-12}\ \text{m})$ or 3.58, and the transmission coefficient is

$$T = e^{-2kl} = e^{-2 \times 3.58} = 7.7 \times 10^{-4}.$$

Thus, of every 100,000 electrons striking the barrier, only 77 will tunnel through it.

(c) What would be the transmission coefficient if the barrier thickness were reduced to 0.30 nm?

Carrying out the calculation just as above except for this change yields $T = 0.10$. The thinner barrier is much easier to tunnel.

(d) What would be the transmission coefficient under the conditions of (b) above if the barrier height were increased to 7 eV?

Carrying out the calculation as in (b) except for this change yields $T = 5.9 \times 10^{-5}$. The taller barrier is harder to tunnel.

(e) What would be the transmission coefficient under the conditions of (b) above if the incident particle was a proton?

Carrying out the calculation once more but with the proton mass ($= 1.67 \times 10^{-27}$ kg) substituted for the electron mass yields $T = 10^{-130}$. The transmission coefficient has been very substantially reduced for this more massive particle. (Imagine how small it would be for a jelly bean!)

On several earlier occasions we have promised to address the question "How can an electron (or a photon) be wavelike under some circumstances and particlelike under others?" We now keep that promise. First we remind you, in Table 50-1, of the hard experimental evidence that both matter and radiation do indeed have this dual character.

Our mental images of "wave" and "particle" are drawn from our familiarity with large-scale objects such as ocean waves and tennis balls. In a way it is fortunate that we are able to extend these concepts into the atomic domain and to apply them to entities such as the electron, which we can neither see nor touch. We say at once, however, that no single concrete mental image, combining the features of *both* wave and particle, is possible in the quantum world. As Paul Davies, physicist and

50-12
WAVES AND PARTICLES

Table 50-1
Selected experiments showing the dual wave–particle nature of matter and of radiation

	Matter	Radiation
Wave nature	Davisson and Germer's electron diffraction experiments	Young's double-slit interference experiment
	—	—
	Section 50-3	Section 45-1
Particle nature	J. J. Thomson's measurement of e/m for the electron	The Compton effect
	—	—
	Section 33-8	Section 49-8

science writer, has written, "It is impossible to visualize a wave-particle, so don't try." What, then, are we to do?

Niels Bohr, who not only played a major role in the development of quantum mechanics but also served as its major philosopher and interpreter, has shown the way with his *principle of complementarity*, which states: *The wave and the particle aspects of a quantum entity are both necessary for a complete description. However, both aspects cannot be revealed simultaneously in a single experiment. The aspect that is revealed is determined by the nature of the experiment being done.*

Consider a beam of light, perhaps from a laser, that passes across a laboratory table. What is the nature of the light beam? Is it wave or a stream of particles?

You cannot answer this question unless you interact with the beam in some way. If you put a diffraction grating in the path of the beam you reveal it as a wave. If you interpose a photoelectric apparatus such as that of Fig. 49-9, you will need to regard the beam as a stream of particles (photons) if you are to interpret your measurements in a satisfactory way. Try as you will, there is no single experiment that you can carry out with the beam that will require you to interpret it as a wave *and* as a particle *at the same time.* You may not like having to switch back and forth between wave and particle descriptions, depending on the experiment you are doing, but there is at least no confusion caused by overlap of the two models.

Bohr thought so much of his principle of complementarity that he sought parallels to it in other fields, such as biology and ethics. His choice of a coat-of-arms (Fig. 50-23) suggests the importance he attached to the matter.

Complementarity — A Case Study. Let us see how complementarity works by trying to set up an experiment that will force nature to reveal both the wave and the particle aspects of electrons at the same time. In Fig. 50-24 a beam of electrons falls on a double-slit arrangement in

figure 50-23
Niels Bohr's coat-of-arms. The Latin words mean "Opposites complement each other." (American Institute of Physics, Neils Bohr Library, Margarethe Bohr Collection.)

figure 50-24
An electron beam falling on a double-slit in screen *A* produces interference fringes on screen *B*. Screen *B* can be replaced by an electron detector, which can be moved up and down, as shown.

screen *A* and sets up a pattern of interference fringes on screen *B*. This is convincing proof of the wave nature of the incident electron beam.

Suppose now that we replace screen *B* with a small electron detector, designed to generate and record a "click" every time an electron hits it. We find that such clicks do indeed occur. If we move the detector up and down in Fig. 50-24, we can, by plotting the click rate against the detector position, trace out the pattern of interference fringes. Have we not succeeded in demonstrating both wave and particle? We see the fringes (wave) and we hear the clicks (particle).

We have not. A mere "click" is not enough evidence that we are dealing with a particle. The concept of "particle" involves the concept of "trajectory" and a mental image of a dot following a path. As a minimum, we want to be able to know which of the two slits in screen *A* the electron passed through on its way to generating a click in the detector. Can we find out?

We can, in principle, by putting a very thin detector in front of each slit, designed so that, if an electron passes through it, it will generate an electronic signal. We can then try to correlate each click, or "screen arrival signal" with a "slit passage signal," thus identifying the path of the electron involved.

If we succeed in modifying the apparatus to do this, we find a surprising thing. The interference fringes have disappeared! In passing through the slit detectors the electrons were affected in ways that destroyed the interference pattern. Although we have now shown the particle nature of the electron, the evidence for its wave nature has vanished.

The converse to our thought experiment is also true. If we start with an experiment that shows that electrons are particles and if we tinker with it to bring out the wave aspect, we will always find that the evidence for particles has vanished. Also, our experiment would work in precisely the same way if we substituted a light beam for the incident electron beam in Fig. 50-24.

A Quantum Puzzle Resolved.   At the beginning of this chapter we asked how tracks such as those shown in Fig. 50-1, made up of tiny bubbles and so clearly suggesting the wake of a fast charged particle moving through the chamber, can be associated with waves.

As the beginnings of an answer we look again at the thought experiment of Fig. 50-24, in which the pattern of fringes on screen *B* is neatly accounted for by the alternating constructive and destructive interference of matter wavelets radiating from each of the two slits in screen *A*.

(a)

(b)

(c)

figure 50-25
Showing the buildup of interference fringes as electrons fall on screen *B* in Fig. 50-24. In (*a*) about 30 electrons have landed on the screen, in (*b*) about 1000, and in (*c*) about 10,000. The probability density of the matter wave at various points of the screen determines where the electrons will land. (A computer simulation.)

We can think of these as "guiding waves," their connection with the particle being that the square of their associated wave function at any point gives the probability (per unit volume) that the particle will be found at that point. Thus, on screen *B*, electrons will pile up at those places where this probability amplitude is large, and they will be found in lesser abundance at those places where it is small.

These considerations apply even if the incident beam is deliberately made so weak that, by calculation, there should be — on average — only a single electron in the apparatus at any given time. You might think that, because the single electron that chances to be in the apparatus must go through one slit or the other, the fringes must vanish; after all — you reason — the electron cannot interfere with itself and there is nothing else for it to interfere with. However, experiment shows that the fringes will *still* be formed, built up slowly as electron after electron falls on screen *B*. Even under these conditions the associated matter wave always passes through *both* slits and *it* is what determines where the electrons are likely to fall on screen *B*. Figure 50-25, a computer simulation, shows how the fringes build up with time for a weak incident beam.

With this background we are ready to answer the question posed by Fig. 50-1. Consider Fig. 50-26, which shows an electron being generated at point *I* and detected at point *F*. How does "it" traverse the empty space between them?

The quantum answer is that the associated matter wave explores all possible paths, as the figure suggests, assigning an equal probability to each. However, only for the straight line connecting the two points do the matter waves add constructively, yielding a high probability that the particle will be found there if sought. For points not on this straight line, the waves can be shown to cancel each other by destructive interference, the cancellation being more severe the more massive the particle. It is in this way that the trajectories of particles in Newtonian mechanics are related to their associated matter waves. It is on this basis that we can claim that the tracks in Fig. 50-1 can be given a wave interpretation.

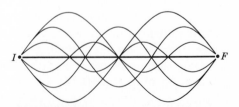

figure 50-26
A particle is generated at point *I* and detected at point *F*. The probability waves that describe its journey between these two points reinforce each other along the straight line connecting *I* and *F* and cancel each other by destructive interference over all other paths.

1. Why is the wave nature of matter not more apparent in our daily observations?

2. If, in the de Broglie formula $\lambda = h/mv$, we let $m \to \infty$, do we get the classical result for particles of matter?

3. An electron and a neutron each have the same kinetic energy. Which has the shorter deBroglie wavelength?

4. Can the de Broglie wavelength associated with a particle be smaller than a linear dimension of the particle? larger? Is there any relation necessarily between such quantities?

5. How could Davisson and Germer be sure that the "54-eV" peak of Fig. 50-4 was a first-order diffraction peak, that is, that $m = 1$ in Eq. 50-2.

6. In a repetition of Thomson's experiment for measuring e/m for the electron (see Section 33-8), a beam of electrons is collimated by passage through a slit. Why is the beamlike character of the emerging electrons not destroyed by diffraction of the electron wave at this slit?

7. Considering the wave behavior of electrons, we should expect to be able to construct an electron microscope. This, indeed, has been done. (a) How might an electron beam be focused? (b) What advantages might an electron microscope have over a light microscope? (c) Why not make a proton microscope? a neutron microscope?

8. How does the wave nature of matter enter into the design and operation of an electron microscope?

9. Do electron diffraction experiments give different information about crystals than can be obtained from X-ray diffraction experiments? from neutron diffraction experiments? Give examples.

10. In Fig. 50-6b (made with X rays) the diffraction circles are speckled, but in Fig. 50-6c (made with electrons) they are smooth. Can you explain why?

11. Electromagnetic waves will penetrate sea water to a certain extent if their frequency is low enough. This is the basis of one plan to communicate with submerged submarines. A difficulty with this plan is that the lower the frequency, the longer the time it takes to transmit a message (in Morse code pulses, say). Can you explain why this should be?

12. (a) Give examples of how the process of measurement disturbs the system being measured. (b) Can the disturbances be taken into account ahead of time by suitable calculations?

13. Why is the Heisenberg uncertainty principle not more readily apparent in our daily observations?

14. You measure the pressure in a tire, using a pressure gauge. The gauge, however, bleeds a little air from the tire in the process, so that the act of measuring changes the property that you are trying to measure. Is this an example of the Heisenberg uncertainty principle? Explain.

15. Why does the concept of Bohr orbits violate the uncertainty principle?

16. Several groups of experimenters are trying to detect gravity waves, perhaps coming from our galactic center, by measuring small distortions in a massive object through which the hypothesized waves pass. They seek to measure displacements as small as $\sim 10^{-21}$ m (the radius of a proton is $\sim 10^{-15}$ m, a million times larger!) Does the uncertainty principle put any restriction on the precision with which this measurement can be carried out?

17. Figure 50-16 shows that for $n = 3$ the probability function for a particle confined between rigid walls is zero at two points between the walls. How can the particle ever move across these positions?

18. "The energy of the ground state of an atomic system can be precisely known, but the energies of its excited states are always subject to some uncertainty." Can you explain this statement on the basis of the uncertainty principle?

19. "If an electron is localized in space, its momentum becomes uncertain. If it is localized in time, its energy becomes uncertain." Explain this statement

in terms of Heisenberg's uncertainty principle and describe some "thought experiments" that illustrate your argument.

20. Discuss the analogy between (a) physical optics → geometrical optics and (b) wave mechanics → classical mechanics.

21. The quantity $\psi(x)$, the amplitude of a matter wave, is called a *wave function*. What is the relationship between this quantity and the particles that form the matter wave?

22. Discuss similarities and differences between a matter wave and an electromagnetic wave.

23. A standing wave can be viewed as the superposition of two traveling waves. Can you apply this to the problem of a particle confined between rigid walls, giving an interpretation in terms of the motion of the electron?

24. The allowed energies for a particle confined between rigid walls are given by Eq. 50-11. First, convince yourself that, as n increases, the energy levels become farther apart. How can this possibly be? The correspondence principle would seem to require that they move closer together as n increases, approaching a continuum.

25. In the $n = 1$ mode, for a particle confined between rigid walls, what is the probability that the particle will be found in a small-volume element at the surface of either wall?

26. What are the dimensions of $P_n(x)$ in Fig. 50-16? What is the value of the classically expected probability density, represented by the horizontal lines? What value do the areas under the curves have? How does the area under any curve compare with the area under the horizontal line? All these questions can be answered by inspection of the figure.

27. In Fig. 50-16 what do you imagine the curve for $P_n(x)$ for $n = 100$ looks like? Convince yourself that these curves approach classical expectations as $n \to \infty$.

28. In Section 50-9 we solved the wave mechanical problem of a particle trapped in an infinitely deep well without ever using (or even writing down) Schrödinger's equation. How were we able to do that?

29. How can the predictions of wave mechanics be so exact if the only information we have about the positions of the electrons is statistical?

30. What is the relationship between the wave function $\psi(r)$ and the probability density function $P(r)$ as applied to the hydrogen atom? In what units are each of these quantities expressed? What value does each of these quantities have at the center of the atom in its ground state?

31. Figure 50-18 suggests that the square of the wave function for the electron in the hydrogen atom becomes very large at the center of the atom, but the radial probability density is zero at that point. Both quantities are related to the probability that the electron will be at a certain location. Why this (apparent) conflict between their predictions?

32. We have seen that barrier tunneling works for matter waves and for electromagnetic waves. Do you think that it also works for water waves? for sound waves?

33. Comment on the statement, "A particle can't be detected while tunneling through a barrier, so that it doesn't make sense to say that such a thing actually happens."

34. List examples of barrier tunneling occuring in nature and in man-made devices.

35. A proton and a deuteron, each having 3 MeV of energy, attempt to penetrate a rectangular potential barrier of height 10 MeV. Which particle has the higher probability of succeeding? Explain in qualitative terms.

36. Considering electrons and photons to be particles, how are they different from each other?

37. A laser projects a beam of light across a laboratory table. If you put a diffraction grating in the path of the beam and observe the spectrum, you

declare the beam to be a wave. If instead you put a clean metal surface in the path of the beam and observe the ejected photoelectrons, you declare this same beam to be a stream of particles (photons). What can you say about the beam if you don't put anything in its path?

38. State and discuss (a) the correspondence principle, (b) the uncertainty principle, and (c) the complementarity principle.

39. In Figure 50-24, why would you expect the electrons from each slit to arrive at the screen over a range of positions? Shouldn't they all arrive at the same place? How does your answer relate to the complementarity principle?

40. How many experiments can you recall whose explanation calls for the wave theory of light? for the particle theory of light? for the wave theory of matter? for the particle theory of matter?

problems

SECTION 50-1

1. Derive the commonly memorized (nonrelativistic) relation between the de Broglie wavelength λ of an electron (in Å) and the corresponding accelerating potential V (in volts): $\lambda = \sqrt{150/V}$.   (*Note:* 1 Å $= 10^{-10}$ m $= 100$ pm.)

2. Particles that are in thermal equilibrium with their surroundings have an average kinetic energy of $\frac{3}{2} kt$. What is the wavelength of a hydrogen atom moving with a speed corresponding to the average kinetic energy at 20°C? Express your answer in SI units and also in terms of the Bohr radius a_0.

3. If the de Broglie wavelength of a proton is 1.0×10^{-13} m, (a) what is the speed of the proton and (b) through what electric potential would the proton have to be accelerated to acquire this speed?
Answer: (a) 4.0×10^6 m/s. (b) 82 kV.

4. A bullet of mass 40 g travels at 1000 m/s. (a) What wavelength can we associate with it? (b) Why does the wave nature of the bullet not reveal itself through diffraction effects?

5. (a) Photons and electrons travel in free space with wavelengths of 1.0 nm. What are the energy of the photon and the kinetic energy of the electron? (b) Repeat for a wavelength of 1.0 fm.
Answer: (a) 1240 eV; 1.50 eV. (b) 1.24 GeV; 1.24 GeV.

6. (a) A photon in free space has an energy of 1.0 eV and an electron, also in free space, has a kinetic energy of that same amount. What are their wavelengths? (b) Repeat for an energy of 1.0 GeV.

7. The existence of the atomic nucleus was uncovered in 1911 by Ernest Rutherford, who properly interpreted some experiments in which a beam of alpha particles was scatered from a foil of atoms such as gold. (a) If the alpha particles had a kinetic energy of 7.5 MeV, what was their de Broglie wavelength? (b) Should the wave nature of the incident alpha particles have been taken into account in interpreting these experiments? The mass of an alpha particle is 4.00 u and its distance of closest approach to the nuclear center in these experiments was about 30 fm. (The wave nature of matter was not postulated until more than a decade after these crucial experiments were first performed.)   *Answer:* (a) 5.2 fm.

8. Neutrons that are in thermal equilibrium with their surroundings have an average kinetic energy $(3/2)kT$. (a) What is the average energy of such thermal neutrons for $T = 300$ K? (b) What is the corresponding de Broglie wavelength?

9. In a repetition of Thomson's experiment for measuring e/m for the electron, a beam of 10-keV electrons is collimated by passage through a slit of width 0.5 mm. Why is the beamlike character of the emergent electrons not destroyed by diffraction of the electron wave at this slit?

10. Consider a balloon filled with helium at room temperature and pressure. (a) How does the average de Broglie wavelength of the helium atoms compare with the average distance between atoms under these conditions? The aver-

age kinetic energy of an atom is equal to $(3/2)kT$. (b) Can the molecules be treated as small particles under these conditions?

11. The 20-GeV electron accelerator at Stanford (1 GeV $= 10^9$ eV) provides an electron beam of small wavelength, suitable for probing the fine details of nuclear structure by scattering experiments. What will this wavelength be and how does it compare with the size of an average nucleus? [At these energies it is sufficient to use the extreme relativistic relationship between momentum and energy, namely, $p = E/c$. This is the same relationship used for light (Eqs. 42-2) and is justified when the kinetic energy of a particle is much greater than its rest energy m_0c^2, as in this case. The radius of a middle-mass nucleus is about 5 fm or 5×10^{-15} m.]   *Answer:* 0.062 fm.

12. The highest achievable resolving power of a microscope is limited only by the wavelength used: that is, the smallest detail that can be separated is about equal to the wavelength. Suppose one wishes to "see" inside an atom. Assuming the atom to have a diameter of 100 pm, this means that we wish to resolve detail of separation about 10 pm. (a) If an electron microscope is used, what minimum energy of electrons is needed? (b) If a light microscope is used, what minimum energy of photons is needed? (c) Which microscope seems more practical for this purpose? Why?

13. What accelerating voltage would be required for electrons in an electron microscope to obtain the same ultimate resolving power as that which could be obtained from a gamma-ray microscope using 100-keV gamma rays?   *Answer:* 9.7 kV.

SECTION 50-2

14. According to the correspondence principle, as $n \to \infty$ we expect classical results in the Bohr atom. Hence, the de Broglie wavelength associated with the electron (a quantum result) should get smaller compared with the radius of the Bohr orbit as n increases. Indeed, we expect that $\lambda/r \to 0$ as $n \to \infty$. Show that this the case.

SECTION 50-3

15. In the experiment of Davisson and Germer, (a) at what angles would the second- and third-order diffracted beams corresponding to the strong maximum of Fig. 50-4c occur, provided they are present, and (b) at what angle would the first-order diffracted beam occur if the accelerating potential were changed from 54 to 60 V? The experimenter is free to rotate the crystal.   *Answer:* (a) The beams are not present. (b) 47°.

16. A potassium chloride crystal is cut so that the layers of atomic planes parallel to its surface have an interplanar spacing of 314 pm. An incident beam makes an angle θ with the surface and a first-order diffracted peak is generated. Find θ if the incident beam is (a) a 40-keV X-ray beam and if it is (b) a beam of electrons whose kinetic energy is 40 keV. (*Hint:* At these relatively high energies the bending of the electron beam at the crystal surface can be ignored and the electron diffraction treated in strict analogy with X-ray diffraction; see Section 47-5.)

17. A neutron crystal spectrometer utilizes crystal planes of spacing $d = 73.2$ pm in a beryllium crystal. What must be the Bragg angle θ so that only neutrons of energy $K = 4.0$ eV are reflected? Consider only first-order reflections.   *Answer:* 5.6°.

18. A beam of thermal neutrons from a nuclear reactor falls on a crystal of calcium fluoride, the beam direction making an angle θ with the surface of the crystal. The atomic planes parallel to the crystal surface have an interplanar spacing of 54.64 pm. The de Broglie wavelength corresponding to the maximum intensity of the incident beam is 11.00 pm. For what values of θ will the first three orders of Bragg-reflected neutron beams occur? (*Hint:* Neutrons, which carry no charge and are thus not subject to electrical forces, are not refracted as they pass through a crystal surface. Thus, neutron diffraction can be treated in strict analogy with X-ray diffraction; see Example 2 and Section 47-5.)

SECTION 50-4

19. Using a rotating shutter arrangement, you listen to a 540-Hz standard tuning fork for 0.1 s. What approximate spread of frequencies is contained in this acoustic pulse? *Answer:* 10 Hz.

20. The signal from a television station contains pulses of full width $\Delta t \simeq 10^{-8}$ s. Explain why it is not feasible to transmit television in the AM broadcasting band.

SECTION 50-5

21. A microscope using photons is employed to locate an electron in an atom to within a distance of 10 pm. What is the minimum uncertainty in the momentum of the electron located in this way? *Answer:* 3.6×10^7 m/s.

22. The uncertainty in the position of an electron is given as 50 pm, which is about the radius of the first Bohr orbit in hydrogen. What is the uncertainty in the momentum of the electron?

23. Suppose we wished to test the possibility that electrons in atoms move in orbits by "viewing" them with photons with sufficiently short wavelength, say 10 pm or less. (a) What would be the energy of such photons? (b) How much energy would such a photon transfer to a free electron in a head-on Compton collision? (c) What does this tell you about the possibility of confirming orbital motion by "viewing" an atomic electron at two or more points along its path? *Answer:* (a) 124 keV. (b) 40.5 keV.

24. Imagine playing baseball in a universe where Planck's constant was 0.6 J · s. What would be the uncertainty in the position of a 0.5-kg baseball moving at 20 m/s with an uncertainty of 1 m/s? Why should it be hard to catch such a ball?

25. Estimate the energy of a neutron, of mass about 10^{-27} kg, that is assumed to move freely through a nucleus of diameter about 10^{-14} m but that is confined to that nucleus. (The answer is close to the actual energy of a neutron in the lowest energy state of a typical nucleus.) *Answer:* About 8 MeV.

26. Show that if the uncertainty in the location of a particle is equal to its de Broglie wavelength, the uncertainty in its velocity is equal to its velocity.

SECTION 50-7

27. The lifetime of an electron in the state $n = 2$ in hydrogen is about 10^{-8} s before making a transition to the ground state. What is the uncertainty in the energy of the $n = 2$ state? Compare this with the energy of this state. *Answer:* 4.1×10^{-7} eV; $E_2 = -3.4$ eV.

28. A nucleus in an excited state will return to its ground state, emitting a gamma ray in the process. If its mean lifetime in a particular excited state is about 10^{-12} s, what is the uncertainty in the energy of the corresponding emitted gamma-ray photon?

29. An atom in an excited state has a lifetime of 1.2×10^{-8} s; in a second excited state the lifetime is 2.3×10^{-8} s. What is the uncertainty in energy for a photon emitted when an electron makes a transition between these two states? *Answer:* 5.2×10^{-7} eV.

SECTION 50-9

30. A particle is confined between rigid walls separated by a distance l. What is the probability that it will be found within a distance $l/3$ from one wall (a) for $n = 1$, (b) for $n = 2$, (c) for $n = 3$, and (d) under the assumption of classical physics?

31. (a) A proton or (b) an electron is trapped in a one-dimensional well 100 pm in length. What is the minimum energy these particles can have? *Answer:* (a) 0.021 eV. (b) 38 eV.

32. A particle is confined between rigid walls located at $x = 0$ and $x = l$. For the fourth allowed energy state, (a) sketch the probability density curve for the particle's location. (b) Calculate the approximate probabilities of finding the particle within a $\Delta x = 0.0003l$ when Δx is located at $l/8$ and at $x =$

3l/16. Refer to your figure to see whether or not your results seem reasonable. (*Hint:* No integration is necessary.)

33. Where are the points of maximum and minimum probability for a particle trapped in an infinitely deep well of length l if the particle is in the state n? *Answer:* $x = Nl/2n$, $N = 1,3,5 \ldots (n-1)$; $x = Nl/n$, $N = 0,1,2, \ldots n$.

34. Consider an electron trapped in an infinitely deep well whose length is 100 pm. If it is in a state with $n = 15$, as in Fig. 50-16, what are (a) its energy? (b) the uncertainty in its momentum? (c) the uncertainty in its position?

35. (a) Prove that, for a particle trapped between rigid walls, the distance s_n between adjacent minima in the probability density $P_n(x)$ is given by $s_n = l/n$. (b) Prove that, for such a trapped particle, the uncertainty principle gives for the uncertainty Δx in the position of the particle, $\Delta x \cong s_n$. (*Hint:* In applying this principle take Δp_x, the momentum uncertainty, to be $2p_x$, using the argument advanced in Section 50-9.) (c) Use the above results to show that, as n increases, the position of the trapped particle can be located with increasing precision but the points for which $P_n(x) = 0$ can never be resolved experimentally.

36. In Example 7, the electron's energy is determined *exactly* by the size of the box. How do you reconcile this with the fact that the uncertainty in the location of the electron cannot exceed 100 pm and, if the uncertainty principle is to be obeyed, the electron's momentum must be correspondingly uncertain?

37. (a) Find, approximately, the smallest allowed energy of an electron confined to an atomic nucleus (diameter about 1.4×10^{-14} m). (b) Compare this with the several MeV of energy binding protons and neutrons inside the nucleus; on this basis should we expect to find electrons inside nuclei? *Answer:* (a) 1900 MeV

38. (a) What is the separation in energy between the lowest two energy levels for a container 20 cm on a side containing argon atoms? (b) How does this compare with the thermal energy of the argon atoms at 300 K? (c) At what temperature does the thermal energy equal the spacing between these two energy levels? Assume, for simplicity, that the argon atoms are trapped in a one-dimensional well 20 cm long. The atomic weight of argon is 39.9 g/mol.

39. Calculate the fractional difference between two adjacent energy levels of a particle confined in a one-dimensional well of infinite depth. *Answer:* $(2n + 1)/n^2$.

SECTION 50-10

40. In the ground state of the hydrogen atom, evaluate the wave function $\psi(r)$, the square of this quantity, $\psi^2(r)$, and the radial probability density $P(r)$ for the positions (a) $r = 0$ and (b) $r = a_0$. Explain what these quantities mean.

41. The radial probability density for the hydrogen atom in its ground state can be written in the form $P(x) = bx^2e^{-2x}$, in which x is the dimensionless ratio r/a_0 and b is a constant. (a) What value of b follows from Eq. 50-17? (b) Show that, with this value of b, the above equation is properly normalized (see Example 9.) *Answer:* (a) $4/a_0$.

42. For the hydrogen atom in its ground state, find the radius of a sphere such that the probability of finding the electron inside it is 99%.

43. In atoms there is a finite, though very small, probability that any given orbital electron will actually be inside the nucleus. In fact, some unstable nuclei use this occasional appearance of the electron to decay by *electron capture.* Assuming that the proton itself is a sphere of radius 1.1×10^{-15} m and that the hydrogen atom electronic wave function holds all the way to

the proton's center, estimate the probability that the hydrogen atom electron is inside its nucleus. *Answer:* 1.2×10^{-14}.

44. Find the ratio of the probabilities of finding the electron in the hydrogen atom at the Bohr radius to that of finding it at twice that distance.

45. In the ground state of the hydrogen atom show that the probability $P(r)$ that the electron lies within a sphere of radius r is given by

$$P(r) = 1 - e^{-2x}(1 + 2x + 2x^2),$$

in which $x = r/a_0$, a dimensionless ratio. Does this yield expected values for (a) $r = 0$ and (b) $r = \infty$? (c) State clearly the difference in meaning between this expression and that given in Eq. 50-17.

46. According to wave mechanics, what is the probability that the electron in the ground state of the hydrogen atom will be found outside a sphere whose radius is the Bohr radius a_0?

SECTION 50-11

47. Consider a barrier such as that of Fig. 50-19a, whose height U_0 is 6 eV and whose thickness l is 0.7 nm. What is the energy of an incident electron such that its penetration probability is one in 1000? *Answer:* 5.07 eV.

48. A proton and a deuteron (which has the same charge as a proton but twice the mass) are incident on a barrier of thickness 10^{-14} m and height 10 MeV. If each particle has an energy of 3 MeV, find the penetration probabilities for them.

49. Consider the barrier tunneling situation defined by Example 12. What fractional change in the transmission coefficient occurs for a 1% increase in (a) the barrier height, (b) the barrier thickness, and (c) the incident energy of the electron? *Answer:* (a) −21%. (b) −7.2%. (c) +18%.

50. In Example 12, suppose that an incident beam of 5-eV protons fell on the barrier, at a rate equivalent to a current of 1000 A. How long would you have to wait — on average — for one proton to be transmitted? How thick is the barrier in terms of de Broglie wavelengths of the incident proton?

51

the structure of atoms

Ever since it has been known that matter is made up of atoms, the fundamental question has been: "What is an atom like?" It is our aim in this chapter to answer this question from the point of view of wave mechanics. Unless we understand the structure of individual atoms we will not be able to understand how atoms join together to form molecules and bulk solids. Both chemistry and solid-state physics — and all they imply for our technological society — rest on this foundation.

We start with hydrogen which, with its single electron, is the simplest atom. It is also an important atom, being a constituent of water and of all organic compounds. On the cosmic scale, this element makes up 90% of the material universe. Understanding hydrogen will lead to an understanding of multielectron atoms, particularly those — sodium is one example — in which all of the electrons but a single outlying valence electron form a core of one or more filled subshells.

Beyond this, it turns out that the quantum numbers that we use to describe the stationary states of the hydrogen atom can be used to label the states of a single electron in *any* atom — vanadium or plutonium, for example — no matter how many electrons that atom contains. This fact permits us to account completely for the structure of the periodic table — and thus for all of chemistry — from the principles of wave mechanics.

For the physicist, however, the most important reason to study the hydrogen atom is that this atom, because of its stark simplicity, has long been a meeting ground for theory and experiment in any proposed study of the nature of matter at the atomic level. A theory that cannot explain the hydrogen atom can certainly not hope to go further. From the early days of Bohr theory (1913), through wave mechanics (1926), and into

51-1
WHAT IS AN ATOM LIKE?

modern quantum electrodynamics (1948–49), the hydrogen atom has served as a laboratory for testing, in exquisite detail, the depth of our understanding of the nature of matter.

Figure 51-1 suggests symbolically how to solve any problem involving the motion of an electron, using the methods of wave mechanics. The "input" in that figure is the information that you must provide to define the problem.

In classical mechanics we define problems by giving the force acting on the particle. Wave mechanics, however, is set up to deal most easily with the potential energy. These two quantities are related and (as we learned in Section 8-3) if you know either one, you can find the other.

For the hydrogen atom the force is given by Coulomb's law, which gives the force exerted on the electron by the (positively charged and relatively massive) central proton. The potential energy function is easily derived from this law and is (see Eq. 29-13)

$$U(r) = -\frac{1}{4\pi\epsilon_0}\frac{e^2}{r},\qquad(51\text{-}1)$$

in which r is the distance of the electron from the central proton. It is Eq. 51-1 that must be inserted into the Schrödinger wave equation, in a place specifically provided for it. In Fig. 51-1, this function is our "Input."

Recall that the potential energy is related to the force by the relation $F(r) = -dU(r)/dr$. You can easily check that, if you take the derivative of $U(r)$ above and change its sign, you do indeed end up with Coulomb's law. It is interesting that Coulomb's law continues to hold for problems at the atomic level, even though Newtonian mechanics fails and must be replaced by wave mechanics. In fact, no failures of Coulomb's law have ever been found experimentally, no matter how small (or how large, for that matter) the distance between the interacting charges.

The hydrogen atom is a three-dimensional structure, and our treatment of it is made a lot easier because of its spherical symmetry. That is, the potential energy function depends only on the *magnitude* of the position vector **r** that points from the proton to the electron and not at all on the direction of this vector. Figure 51-2 shows a plot of Eq. 51-1. The energy is given in electron volts, an appropriate unit for atomic problems. The distance r is given in terms of another unit suitable for atomic problems, the *Bohr radius* a_0, where (see Eq. 49-32)

figure 51-1

Solving Schrödinger's wave equation. For one dimensional motion this equation is $\dfrac{-\hbar^2}{2m}\dfrac{\partial^2 \Psi(x,t)}{\partial x^2} + U(x,t)\,\Psi(x,t) = i\hbar\,\dfrac{\partial \Psi(x,t)}{\partial t}$. The device shown might be a computer, programmed to solve this equation for various problems. The nature of the problem is defined by the input to the device, the potential energy function, which you must supply. The output includes the wave functions, Ψ, and corresponding energies and quantum numbers.

$$a_0 = \epsilon_0 h^2 / e^2 \pi m_e = 5.292 \times 10^{-11} \text{ m} = 52.92 \text{ pm}.$$

Although the Bohr radius (see Section 49-12) happens to be equal to the radius of the first Bohr orbit in the hydrogen atom, we do not use it here with that significance. We use it instead simply as a convenient unit of measure at the scale of atomic dimensions.

Having properly introduced the potential into Schrödinger's wave equation, our next task is to solve that equation. It can be solved explicitly for the hydrogen atom, but the methods are beyond our scope here.* It is best to think of the "Schrödinger machine" shown in Fig. 51-1 as a computer, programmed to solve the Schrödinger equation once it has been supplied with the appropriate potential energy function. Indeed, this is how the vast majority of problems in wave mechanics are solved.

What happens, then, when you press the "RUN" button on the device of Fig. 51-1, thus activating the equation-solving program? For the case of the hydrogen atom the program generates a complete description of each of its stationary states. For each state the program provides (1) its energy, (2) its wave function, and (3) the quantum numbers (there is more than one!) that identify it.

For the electron trapped in an infinite well, there is only a single quantum number n; it appears in the expressions for the energy (Eq. 50-11) and for the probability density (Eq. 50-12). This seems appropriate because there is only one degree of freedom for this one-dimensional problem.

To locate an electron in a hydrogen atom, however, we must specify three position coordinates (for example, the magnitude of the position vector $\mathbf{r}$ and two angular coordinates to fix its direction) so we are not surprised to learn that three quantum numbers are needed to define its various stationary states. Table 51-1 displays these quantum numbers, gives their names, and tells us what feature of the hydrogen atom they are associated with.

The principal quantum number n gives the energies E_n of the stationary states and, indeed, the formula generated by solving the Schrödinger equation happens to be exactly the same as Bohr's formula, namely (see Eq. 49-30),

$$E_n = -\frac{me^4}{8\epsilon_0^2 h^2} \frac{1}{n^2} \qquad n = 1, 2, 3 \ldots \qquad (51\text{-}2)$$

We should perhaps not be surprised at this coincidence because Bohr's theory was in such excellent agreement with the measured wavelengths of the lines of the hydrogen spectrum.

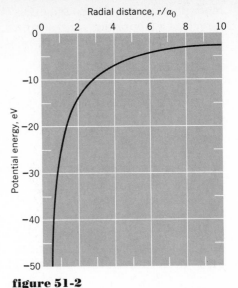

figure 51-2
The potential energy function $U(r)$ for the hydrogen atom. The radial distance r between the electron and the central nucleus (proton) is measured in terms of the Bohr radius a_0.

Table 51-1
The hydrogen atom quantum numbers*

Symbol	Name	Associated With
n	Principal quantum number	Energy
l	Orbital quantum number	Orbital angular momentum
m_l	Magnetic quantum number	

* A fourth quantum number, associated with the fact that the electron, quite apart from its orbital motion, has an intrinsic spinning motion, will crop up later.

* See Robert Eisberg and Robert Resnick, *Quantum Physics*, 2nd ed. Chapter 7 (John Wiley & Sons, New York, 1985) for the full treatment.

You may well ask: "If Bohr's theory gives the correct answer for the energies of the stationary states, why do we need wave mechanics, with all its complications?" The answer is that wave mechanics is a complete and coherent theory that applies to atomic systems of all kinds, explaining in principle the properties of not only the hydrogen atom, but also of more complex systems such as, say, the vanadium atom, the ammonia molecule, and solid copper. Bohr theory, on the other hand, works for the hydrogen atom (but only partially so — it cannot predict the intensities of its spectrum lines, for example) but fails pretty badly for other atoms, even for so simple an atom as helium, with its two electrons. Today we view Bohr theory as a remarkably ingenious first step toward the development of the complete quantum theory that followed.

51-3
ANGULAR MOMENTUM AND MAGNETISM

Kepler's three laws, which describe the motions of the planets as they orbit the sun, can be derived in classical mechanics by invoking the laws of conservation of energy and of angular momentum. These two conservation laws hold also in wave mechanics. As a matter of fact, the first questions one is likely to ask about an atom in a given stationary state are: "What is its energy?" and "What is its angular momentum?" We recall that Bohr recast his original model of the hydrogen atom in a form in which angular momentum was a central feature.

Consider a hydrogen atom in a state with a very high principal quantum number, say $n \approx 500,000$. In such an atom the correspondence principle applies and it is safe to think in terms of the semiclassical picture of a Bohr orbit of radius r. The circulating electron has an angular momentum associated with this orbital motion, its magnitude L being given classically by $m_e vr$, in which m_e is the mass of the electron and v is its orbital speed. Although this mental picture fades away at low quantum numbers, where wave mechanics governs, the hydrogen atom continues to have an associated angular momentum; we seek a proper wave mechanical expression for it.

It turns out that the electron, like most other elementary particles, has an *intrinsic* angular momentum, quite apart from any orbital angular momentum it may have. We shall discuss this *spin*, as it is called, in Section 51-6. For the present we shall focus our attention on angular momentum associated only with the orbital motion of the electron about the central nucleus.

The Bohr model also suggests that the orbiting electron — a tiny current loop — should have an (orbital) magnetic dipole moment associated with it. Both the angular momentum **L** and the magnetic dipole moment μ are vectors and share a common axis. Because the electron has a negative charge, however, these vectors point in opposite directions along this axis. In Example 2 of Chapter 37 we showed that these two vectors are related by

$$\mu = -(e/2m_e)\mathbf{L}, \qquad (51\text{-}3)$$

the minus sign suggesting the opposite directions of **L** and μ. Although Eq. 51-3 was derived on a semiclassical basis, it remains true in wave mechanics.

Consider a state in which the angular momentum is $h/2\pi$. Substituting this value for **L** in the above equation yields

$$\mu_B = eh/4\pi m_e$$
$$= 9.274 \times 10^{-24} \text{ J/T} = 5.788 \times 10^{-5} \text{ eV/T} \qquad (51\text{-}4)$$

for the magnitude of the magnetic dipole moment. We take μ_B, which we call the *Bohr magneton*, as a convenient unit in which to measure atomic magnetic moments, much as we took the Bohr radius a_0 as a convenient unit in which to measure atomic distances.

Bohr theory predicts that the magnetic dipole moment of the hydrogen atom in its ground state will be one Bohr magneton. The theory is simply not correct on this point. Experiment shows that, in accordance with the predictions of wave mechanics, both the (orbital) angular momentum and the (orbital) magnetic dipole moment of the hydrogen atom in its ground state are zero. This is, in fact, what we would expect from the spherical symmetry of the electron cloud that characterizes this state. This failure of Bohr theory, however, does not stop us from using the Bohr magneton as a convenient unit of measure.

The magnetic dipole moment of the atom — much like a compass needle — can respond to an external magnetic field. This gives the atom a convenient "handle" by means of which we can explore its inner workings by probing from the outside. Because the magnetic dipole moment is rigidly coupled to the angular momentum, keeping track of the former automatically keeps track of the latter as well.

We do not have to look far to find evidence that atoms can be the carriers of magnetism. An ordinary iron bar shows no external magnetic properties because its elementary atomic magnets are randomly arranged, their effects canceling at all external points. When these elementary atomic magnets are lined up, however, as they are in a bar magnet, their combined magnetic strength is there for all to see.

When the magnetic dipole moments of an assembly of atoms are lined up, their angular momenta — to which they are rigidly coupled — must also be lined up. In 1915 Einstein, working with W. J. de Haas (the son-in-law of the great Dutch physicist H. A. Lorentz), carried out an experiment to explore this phenomenon. If an iron bar is suddenly magnetized, perhaps by switching on a current in a solenoid as in Fig. 51-3, the angular momenta of its atoms suddenly become lined up. Because angular momentum must be conserved, the bar as a whole must start to rotate in the opposite sense. This *Einstein-de Haas effect*, as it is called, is small and the measurements are difficult. Bear in mind that in 1915, when this experiment was carried out, wave mechanics had not been discovered, Bohr theory was only two years old, and the intrinsic spin of the electron had yet to be discovered.

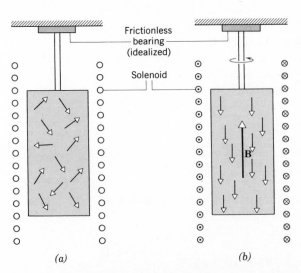

(a)   (b)

figure 51-3
The Einstein-de Haas effect. (a) The atomic angular momentum vectors in an iron cylinder are randomly oriented. (b) When an axial magnetic field is applied, the atomic angular momenta line up as shown and the cylinder as a whole starts to rotate in the opposite sense.

It turned out later that the Einstein-de Haas effect (and also, for that matter, the ferromagnetism of a bar magnet) are due largely to the intrinsic angular momentum (spin) of the electrons rather than to their orbital angular momentum. This, however, does not alter the fact that this experiment demonstrates, in a macroscopic way, that atoms can be the carriers of both magnetism and angular momentum.

EXAMPLE 1

Measuring the angular momentum. An unmagnetized iron cylinder, whose radius R is 5 mm, hangs from a frictionless bearing so that it can rotate freely about its axis; see Fig. 51-3. A magnetic field is suddenly applied parallel to this axis, causing the magnetic dipole moments of the atoms to align themselves parallel to the field. The atomic angular momentum vectors, which are coupled back-to-back with the magnetic dipole moment vectors, also become aligned and the cylinder will start to rotate in the opposite sense. Find T, the period of rotation of the cylinder. Assume that each iron atom has an angular momentum of $h/2\pi$. The atomic weight A of iron is 55.8 g/mol (= 0.0558 kg/mol).

The angular momentum of the rotating cylinder (= $\mathbf{L}_{cyl}$) must be equal in magnitude (though opposite in direction) to the angular momentum associated with the aligned atoms (= $\mathbf{L}_{atoms}$). If N is the number of atoms in the cylinder, N_A is the Avogadro constant, and M is the mass of the cylinder, we can write

$$L_{atoms} = N(h/2\pi) = (N_A M/A)(h/2\pi).$$

For the rotating cylinder we have

$$L_{cyl} = I\omega = (\tfrac{1}{2}MR^2)(2\pi/T),$$

in which I is the rotational inertia of the cylinder about its rotational axis and ω is its angular speed.

Setting these two expressions equal to each other and solving for T yields

$$T = \frac{2\pi^2 R^2 A}{N_A h}$$

$$= \frac{(2\pi^2)(5 \times 10^{-3}\,\text{m})^2(0.0558\,\text{kg/mol})}{(6.02 \times 10^{23}\,\text{mol}^{-1})(6.63 \times 10^{-34}\,\text{J} \cdot \text{s})}$$

$$= 6.90 \times 10^4\,\text{s} = 19.2\,\text{h}.$$

This is indeed a slowly rotating cylinder! Actually, Einstein and de Haas suspended their cylinder from a torsion fiber and used more refined techniques in their experiment dealing with this effect.

51-4
ANGULAR MOMENTUM IS QUANTIZED

The energy of a state is a scalar and, in the hydrogen atom, it is specified by a single quantum number n. The angular momentum of a state, however, is a vector and we are perhaps not surprised to learn from Table 51-1 that it takes two quantum numbers, l and m_l, to describe it. The angular momentum, like the energy, is not only quantized but doubly so — in both magnitude and direction. We discuss each in turn.

The Magnitude of L. The magnitude of the (orbital) angular momentum — so the Schrödinger wave equation tells us — is given by

$$L = \sqrt{l(l + 1)}\,\hbar \qquad (51\text{-}5)$$

in which l is the orbital quantum number. For convenience, we have again introduced the symbol $\hbar$ (pronounced "h-bar") as an abbreviation for $h/2\pi$.

The values that l can have in Eq. 51-5 depend on the value of the principal quantum number n and are given by

$$l = 0, 1, 2, \ldots (n-1). \tag{51-6}$$

Thus the ground state of the hydrogen atom, which has $n = 1$, must have $l = 0$ (and thus $\mathbf{L} = 0$), no other value being permitted by Eq. 51-6.

Equation 51-6 tells us further that a state with $n = 2$ can have $l = 0$ and $l = 1$ but no other values. These two states of the hydrogen atom, which share the same principal quantum number and thus have the same energy, are said to be *degenerate*.

Referring to two or more states as degenerate is not a moral judgment but simply a way of saying that the states in question have the same energy. This does not mean, however, that these states are identical and, indeed, they are not. Their wave functions are quite different and they represent quite different modes of motion for the electron. If we put the atom in a strong enough external magnetic field, these degenerate states will separate in energy and reveal themselves as the independent states that they are. We say then that the degeneracy has been "lifted." When the external magnetic field is removed, these states "collapse" or "degenerate" back into a single energy level.

The Direction of L. An isolated hydrogen atom has no built-in reference axis with respect to which directions can be measured. We must supply one from the outside. In our earth-bound existence we are accustomed to the reference axis provided by the earth's gravitational field. The sense of "up" and "down" that it provides is entrenched in our patterns of living, our language, and even our theology. The gravitational force, however, is far too weak to affect the internal arrangements of the hydrogen atom.

For our present purposes the external field that we supply need only be strong enough to establish a reference direction, which, quite arbitrarily, we shall call the z-axis. In fact, it is quite good enough to *imagine* that such a field has been provided.

It turns out that the angular momentum vector $\mathbf{L}$ cannot take up *any* position with respect to the z-axis, but only those positions that have a component along the z-axis given by

$$L_z = m_l \hbar, \tag{51-7}$$

in which m_l, the *magnetic quantum number*, may have only the values

$$m_l = 0, \pm 1, \pm 2, \ldots \pm l. \tag{51-8}$$

This restriction on the direction of $\mathbf{L}$ is called *space quantization*.

We see from the above equations that, for a hydrogen atom state with $l = 2$, the magnitude of $\mathbf{L}$ is $\sqrt{2(2+1)}\,\hbar$ or $2.45\,\hbar$. L_z, the component of $\mathbf{L}$ along the z-axis, may have the values $0, \pm 1\,\hbar$, and $\pm 2\,\hbar$, five components in all. *No other orientations of the angular momentum vector with respect to the z-axis are allowed.* Note that the maximum value of L_z (= $2\,\hbar$ in this case) is less than the magnitude of $\mathbf{L}$ (= $2.45\,\hbar$). This will always be the case; the angular momentum vector $\mathbf{L}$ can never be fully lined up with the z-axis.

Figure 51-4 shows the allowed values of L_z for $l = 1, 2$, and 10. In the latter case we begin to approach the classical situation, in which space quantization has faded away and *any* orientation of the angular momentum vector is allowed. Example 2 gives further details of how the correspondence principle operates in this case.

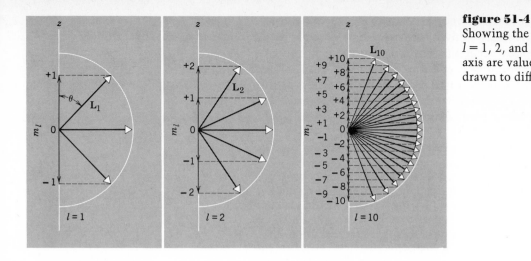

figure 51-4
Showing the allowed values of L_z for $l = 1, 2,$ and 10. The numbers on the z-axis are values of m_l. The figures are drawn to different scales.

The quantization of L_z means that the angle θ between **L** and the z-axis (see Fig. 51-4) is quantized, its values being restricted to (see Eqs. 51-5 and 51-7)

$$\theta = \cos^{-1} \frac{L_z}{L} = \cos^{-1} \frac{m_l}{\sqrt{l(l+1)}}. \tag{51-9}$$

The minimum value of θ occurs when m_l has its greatest value, which is l. For $l = 2$, for example, you can easily show that this minimum value is

$$\theta_{\min} = \cos^{-1} [2/\sqrt{2(2+1)}] = \cos^{-1} 0.817 = 35.3°.$$

For hydrogen atom states with $l = 2$ it is simply not possible for the (orbital) angular momentum vector **L** to make any smaller angle with the z-axis.

If the angular momentum is quantized, the magnetic dipole moment must also be quantized, and in the same fashion. Combining Eq. 51-3 (z-components only) and Eq. 51-7 allows us to write

$$\mu_z = -(e/2m_e)L_z = -(e/2m_e)(m_l\hbar) = -m_l(eh/4\pi m_e).$$

From Eq. 51-4, we see that the z-components of the magnetic dipole moment are given simply by

$$\mu_z = -m_l\mu_B, \tag{51-10}$$

in which μ_B is the Bohr magneton.

Figure 51-5 suggests a classical vector model that helps in visualizing the space quantization of both **L** and μ. It shows both vectors precessing about the z-direction, like a spinning top or a gyroscope precessing about a vertical axis in the earth's gravitational field. The projections L_z and μ_z remain constant as the motion proceeds. It is, in fact, these projections, and not the corresponding magnitudes, that are of direct experimental interest. When we quote a value for the (orbital) angular momentum of a state or for its (orbital) magnetic dipole moment, these values essentially always refer to the maximum allowed values of L_z and μ_z, and not to L and μ.

Heisenberg's uncertainty principle helps us to understand the space quantization of the angular momentum vector. In its angular form (compare Eq. 50-5) this principle is

$$\Delta L_z \cdot \Delta\phi \cong h \qquad \text{(z-component)} \tag{51-11}$$

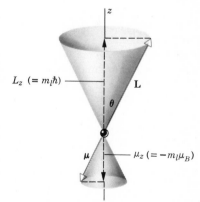

figure 51-5
A vector model design to represent the space quantization of the angular momentum vector **L** and the magnetic dipole moment vector μ.

in which ϕ is the angle of rotation about the z-axis in Fig. 51-5. Equation 51-7 tells us that L_z is precisely known, once we have specified the quantum number m_l. It follows that ΔL_z, the uncertainty in L_z, must be zero. Equation 51-11 then requires that $\Delta\phi \to \infty$, which means that we have no information at all about the angular position about the z-axis of the precessing angular momentum vector **L**. We know the magnitude of **L** and its projection L_z on the z-axis, and *nothing else.*

You may imagine from what we have said about space quantization that the z-axis is in some sense favored over the x- and y-axes. We could, however, just as easily have labeled the space quantization axis the x-axis (or the y-axis); it is simply convention to call it the z-axis. What *should* give you pause to think is the fact that the atom accepts without question whatever axis we impose on it from the outside. If we put the atom in a magnetic field, for example, and if we set up an experiment to measure the direction of its space quantization axis, we *always* find that this axis coincides with the field direction. This remains true no matter how arbitrarily we have chosen that direction. Once again the nature of quantum reality presents itself and we see the close connection between the act of measurement and the thing measured.

EXAMPLE 2

Space quantization and the correspondence principle. Find the minimum value of θ in Fig. 51-5 for $l = 1$, 10^2, 10^3, 10^4, and 10^9.

The minimum value of θ occurs when we put $m_l = l$ in Eq. 51-9. Doing so and rearranging leads to

$$\theta_{min} = \cos^{-1}\frac{l}{\sqrt{l(l+1)}} = \cos^{-1}\left[1 + \frac{1}{l}\right]^{-1/2}.$$

We see by inspection that if we let $l \to \infty$, then $\theta \to \cos^{-1} 1 = 0$. This is just what we expect from the correspondence principle.

Substituting for l in the above leads to these results:

l	θ_{min}
1	45.0°
10^2	5.7°
10^3	1.8°
10^4	0.57°
10^9	0.0018°

For a macroscopic object like a spinning top or a phonograph record, l would be enormously larger than 10^9 and θ_{min} would be so close to zero that the difference would be beyond the possibility of measurement. Thus, as the angular momentum of a spinning object gets larger and larger, the space quantization of wave mechanics merges gently into the continuous distribution of classical mechanics. We see once again that the correspondence principle really works.

Computational note: If you use the above formula to calculate θ_{min} for $l = 10^9$, your calculator will probably overflow. Take advantage of the fact that $(1/l) \ll 1$ and develop an approximate formula. You will need to use both the binomial expansion and the series expansion for $\cos\theta$; see Appendix I.

EXAMPLE 3

Space quantization. For $n = 4$, (a) what is the largest allowed value of l? (b) What is the magnitude of the corresponding angular momentum? (c) How many projections on the z-axis may this angular momentum vector have? (d)

What is the magnitude of the largest projected component? (e) What is the smallest angle that the angular momentum vector can make with the z-axis?

(a) From Eq. 51-6, the largest allowed value of l is $n - 1$ or $l = 3$.

(b) From Eq. 51-5 we have

$$L = \sqrt{l(l + 1)}\,(h/2\pi)$$
$$= \sqrt{3(3 + 1)}\,(6.63 \times 10^{-34}\ \text{J} \cdot \text{s})/(2\pi)$$
$$= 3.66 \times 10^{-34}\ \text{J} \cdot \text{s}.$$

In practice, atomic angular momenta are rarely reported in SI units. It would be more customary to report the magnitude of the angular momentum in this case as simply $\sqrt{12}\ \hbar$ or $3.46\ \hbar$. See (d) below.

(c) The number of projections that the angular momentum vector may have on the z-axis is equal to the number of allowed values of the magnetic quantum number m_l. From Eq. 51-8 this number is $2l + 1$ or $2 \times 3 + 1$ or 7.

(d) The largest projected component is found from Eq. 51-7, in which the magnetic quantum number m_l is given its largest possible value. From Eq. 51-8 this largest value is just l, so we have

$$L_z = l(h/2\pi)$$
$$= (3)(6.63 \times 10^{-34}\ \text{J} \cdot \text{s})/(2\pi)$$
$$= 3.17 \times 10^{-34}\ \text{J} \cdot \text{s}.$$

Note from (b) above that this is smaller than the magnitude of the angular momentum vector, as it must be. See also the remark in (b) above. In this case the maximum projected angular momentum component would be reported as simply $3\ \hbar$. When we refer to angular momentum, in fact, we almost always mean this maximum projected value. The magnitude of the angular momentum rarely enters quantum calculations and is rarely given.

(e) The smallest angle that the angular momentum vector can make with the z-axis follows from Eq. 51-9, with $m_l = l$. We have, then

$$\theta = \cos^{-1}[l/\sqrt{l(l + 1)}]$$
$$= \cos^{-1}[3/\sqrt{3(3 + 1)}]$$
$$= \cos^{-1} 0.866 = 30°.$$

The angular momentum vector can make no smaller angle with the z-axis than this.

51-5
THE STERN-GERLACH EXPERIMENT

Space quantization, that is, the notion that an atomic angular momentum vector **L** or an atomic magnetic dipole moment vector μ can have only a certain discrete set of projections on a selected axis, is not an easy concept for the classically oriented mind to accept. Nevertheless, it was predicted theoretically (by Wolfgang Pauli) and verified experimentally (in 1922, by Otto Stern and Walther Gerlach) several years before the development of wave mechanics.

Figure 51-6 shows the apparatus of Stern and Gerlach. Silver is vaporized in an electrically heated "oven" and silver atoms spray into the external vacuum of the apparatus from a small hole in the oven wall. The atoms (which are electrically neutral but which have a magnetic dipole moment) are formed into a narrow beam as they pass through a slit in an interposed screen. The collimated beam then passes between the poles of an electromagnet and, finally, deposits itself on a glass detector plate.

Often in laboratory experiments we want the magnetic field to be

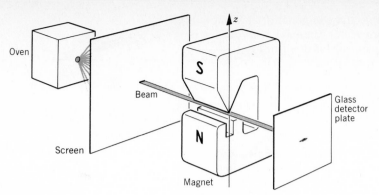

figure 51-6
The apparatus of Stern and Gerlach, used to demonstrate space quantization.

uniform, but in this case the pole faces are shaped to make the field as *nonuniform* as possible. The atomic beam passes very close to the sharp V-shaped ridge in the upper pole piece, where the nonuniformity of the magnetic field is greatest.

A Dipole in a Nonuniform Field. We now digress to ask what force acts on a magnetic dipole placed in a magnetic field, either uniform or nonuniform. Figure 51-7*a* shows a dipole of magnetic moment μ, making an angle θ with a uniform magnetic field. We can imagine the dipole to be a tiny bar magnet, with the magnetic dipole moment vector μ pointing (by convention) from its South pole to its North pole. We may imagine the forces to be concentrated at the poles as shown in the figure. We see that, for a uniform field, there is no net force on the dipole. The upward and downward forces on the poles are of the same magnitude and they cancel, no matter what the orientation of the dipole.

Figures 51-7*b,c* show the situation in a nonuniform field. Here the upward and downward forces do *not* have the same magnitude because the two poles are immersed in fields of different strengths. In this case there *is* a net force, both its magnitude and direction depending on the orientation of the dipole, that is, on the value of θ. In Fig. 51-7*b* this net force is up and in Fig. 51-7*c* it is down. Thus the silver atoms in the beam

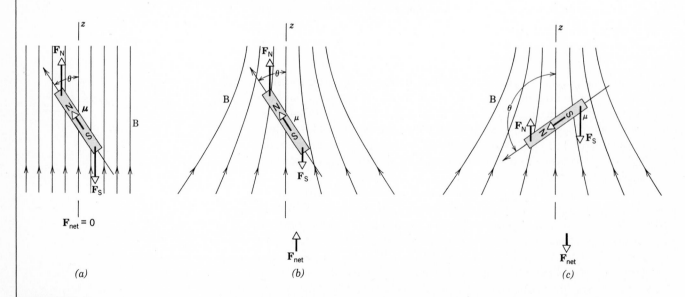

figure 51-7
A magnetic dipole, represented as a small bar magnet with two poles, in (*a*) a uniform magnetic field and (*b,c*) a nonuniform field. The net force $\mathbf{F}_{net}$ is zero in (*a*), points up in (*b*), and points down in (*c*).

of Fig. 51-6, as they pass through the electromagnet, will be deflected up or down, and in greater or lesser amounts, depending on the orientation of their magnetic moment dipole vectors with respect to the magnetic field.

Now let us calculate the deflecting force quantitatively. The magnetic potential energy of a dipole in a magnetic field **B** is given by Eq. 33-12, or

$$U(\theta) = -\mu \cdot \mathbf{B} = -(\mu \cos \theta)B.$$

In our mind's eye let us follow the silver atoms in the beam as they move through the electromagnet of Fig. 51-6 parallel to the knife edge. From symmetry (see also Fig. 51-7b,c), the magnetic field at this central position has no x- or y-components. Thus $B = B_z$. Because $\mu \cos \theta = \mu_z$, we can write the above equation as

$$U(\theta) = -\mu_z B_z. \tag{51-12}$$

From Eq. 8-7, the net force F_z on the atom is $-(dU/dz)$ or, from Eq. 51-12,

$$F_z = \mu_z(dB_z/dz). \tag{51-13}$$

Note that the deflecting force is determined by the gradient of the magnetic field and does not depend in any way on the magnitude of the field itself. In Fig. 51-7b,c, B_z increases as z increases so that the gradient is positive. Thus the sign of the deflecting force F_z in Eq. 51-13 depends on the sign of μ_z. If μ_z is positive (as in Fig. 51-7b) the atom will be deflected upward; if it is negative (as in Fig. 51-7c) the deflection will be downward.

One troublesome point remains. If the individual atoms in the beam behave like tiny bar magnets, why don't they all simply line up with the magnetic field? Why should any of them point, even partially, in the opposite direction? The answer is that the atoms not only have a magnetic dipole moment; they also have angular momentum. The result is that they precess around the field direction (see Fig. 51-5) rather than line up with it. In the same way a top that is not spinning will simply fall over if you place it at an angle with the earth's gravitational field. We saw in Section 13-2, however, that if the top is spinning, it will precess about this direction. It is the angular momentum that does it!

The Experimental Results. When the electromagnet in Fig. 51-6 is turned off (or is operating at very low power), there will be no deflections of the atoms and the beam will form a narrow line on the detecting plate.

When the electromagnet is turned on, however, strong deflecting forces come into play. Then there are two possibilities, depending upon whether space quantization exists or not. (Don't forget that the entire object of this experiment is to find out!) If there is no space quantization, the atomic magnetic dipole moment vectors will have a continuous distribution of angles with the direction of the magnetic field. The quantity μ_z in Eq. 51-13 will have a continuous distribution of values, some positive and some negative; the beam will simply broaden.

On the other hand, if space quantization *does* exist, there will be only a discrete set of values of μ_z. This means that there will be only a discrete set of values for the deflecting force F_z in Eq. 51-13 and the beam will split up into a number of discrete components.

Figure 51-8 shows what happens. The beam does not broaden but

(a)

(b)

figure 51-8
The results of the Stern-Gerlach experiment, showing the silver deposit on the glass detecting screen with the magnetic field (a) turned off and (b) turned on. The beam has been split into two subbeams by the action of the field. The vertical bar at the right in (b) represents 1 mm.

splits cleanly into two subbeams. Space quantization exists! Stern and Gerlach ended the published report of their work with the words, "We view these results as direct experimental verification of space quantization in a magnetic field." Physicists everywhere agree.

EXAMPLE 4

The Stern-Gerlach experiment. In an experiment of the Stern-Gerlach type, the magnetic field gradient dB_z/dz at the beam position was 1.4 T/mm and the length h of the beam path through the magnet was 3.5 cm. The temperature of the "oven" in which the silver was evaporated was adjusted so that the most probable speed v for the atoms in the beam was 750 m/s. Find the separation d between the two deflected subbeams as they emerge from the magnet; see Fig. 51-8b. The mass M of a silver atom is 1.8×10^{-25} kg and its magnetic moment μ_z is 1 Bohr magneton ($= 9.28 \times 10^{-24}$ J/T).

The acceleration of a silver atom as it passes through the electromagnet is given (see Eq. 51-13) by

$$a = \frac{F_z}{M} = \frac{\mu_z(dB_z/dz)}{M}.$$

The vertical deflection of either of the subbeams as it clears the magnet is

$$\tfrac{1}{2}d = \tfrac{1}{2}at^2 = \tfrac{1}{2}\frac{\mu_z(dB_z/dz)}{M}\frac{h^2}{v}$$

so that

$$d = \frac{\mu_z(dB_z/dz)h^2}{Mv^2}$$

$$= \frac{(9.28 \times 10^{-24}\text{ J/T})(1.4 \times 10^3\text{ T/m})(3.5 \times 10^{-2}\text{ m})^2}{(1.8 \times 10^{-25}\text{ kg})(750\text{ m/s})^2}$$

$$= 1.6 \times 10^{-4}\text{ m} = 0.16\text{ mm}.$$

This is the order of magnitude of the separation displayed in Fig. 51-8b; note the scale in that figure.

51-6
THE SPINNING ELECTRON

In earlier sections we dropped several hints that there is more to say about angular momentum than we have yet said. Consider a curious feature of the Stern-Gerlach experiment. As Fig. 51-8 shows, the beam of silver atoms is separated by the electromagnet into two subbeams, an *even* number. A glance at Fig. 51-4 shows that the number of allowed projections (that is, the number of values of m_l) is always *odd.* From Eq. 51-8 you can see that this number is just $2l + 1$, which is odd if l is an integer. To make this number come out to be two we would have to have $l = \tfrac{1}{2}$, which (as Eq. 51-6 shows) is not possible for the orbital angular momentum quantum number.

Question: "Does there lurk within the hydrogen atom another kind of angular momentum, which might possibly have a half-integral quantum number?"

Answer: "Yes. It is the intrinsic angular momentum of the electron itself, called its *spin.*"

The spin angular momentum of the electron can have components in the z-direction given by

$$S_z = m_s \hbar, \tag{51-14}$$

in which the *spin quantum number* m_s can have only the values $+\frac{1}{2}$ and $-\frac{1}{2}$. We have used the symbol "S" for angular momentum associated with spin, to distinguish it from "L," the angular momentum associated with orbital motion.

Now we can understand why there are two beams in the Stern-Gerlach experiment. The silver atom has 47 electrons, 46 of which pair off in such a way that their resultant angular momentum is zero. The remaining electron is in a state with $L = 0$ so that the entire angular momentum of the silver atom is contributed by the spin of this single valence electron. As Eq. 51-14 tells us, the angular momentum vector has only two possibilities for alignment, either with the magnetic field or opposed to it; thus the two subbeams.

There were yet other failures of agreement between theory and experiment at the time of the Stern-Gerlach experiment. The lines of the hydrogen spectrum, for example, when examined carefully with a spectroscope of high resolving power, are shown to be made up of a number of close components. This *fine structure*, as it is called, cannot be explained by Schrödinger's wave mechanics.

It eventually became clear to the perceptive physicist Wolfgang Pauli (see Fig. 51-9) that a fourth quantum number, in addition to n, l, and m_l, was needed to complete the hydrogen atom story, but the nature of this new quantum number remained a mystery. In 1925, however, two Dutch graduate students (George Uhlenbeck and Samuel Goudsmit), two years away from their doctorates at the University of Leiden, proposed the notion of electron spin as the physical interpretation of Pauli's proposed new quantum number.

figure 51-9
Wolfgang Pauli and Niels Bohr watching a "tippy top," a top that spins for a while and then turns upside down. They seem to be waiting for the "spin flip." See "The Tippy Top Revisited," Richard J. Cohen, *American Journal of Physics*, January 1977. (American Institute of Physics, Neils Bohr Library, Margrethe Bohr Collection.)

We turn now to the magnetic dipole moment vector μ_s that is associated with the spin of the electron. To make sure that we do not confuse this with the magnetic dipole moment vector associated with the electron's orbital motion, which we have so far called simply μ, we shall from now on label the latter as μ_l. A host of experimental data requires us to assume that the projections of μ_s on the z-axis are given by

$$\mu_{s,z} = -2m_s\mu_B, \qquad (51\text{-}15)$$

in which μ_B is the Bohr magneton. Compare this with the projections of the orbital magnetic dipole moment vector on the z-axis, which we rewrite for clarity as

$$\mu_{l,z} = -m_l\mu_B. \qquad (51\text{-}10)$$

Notice the factor of 2. This tells us that *spin angular momentum is twice as effective as orbital angular momentum in generating magnetism*. This factor of 2 was, in fact, conjectured by Uhlenbeck and Goudsmit when they first introduced the concept of electron spin.

Schrödinger's wave mechanics says nothing at all about electron spin. This is because Schrödinger's equation does not take into account the theory of relativity. When the English mathematical physicist P. A. M. Dirac developed a relativistic quantum theory of the electron, the spin concept and the spin quantum number arose in a natural way. To include spin, we must replace the Schrödinger machine of Fig. 51-1 by a Dirac machine!

The most complete theory of the electron and its interactions, called *quantum electrodynamics* (or QED) tells us that the factor 2 in Eq. 51-15 above should, in fact, be a little greater than 2, by about 0.1%. This prediction has been brilliantly confirmed by experiment. The results for "2" are

Predicted by theory: 2.002 319 304 76

Measured in the laboratory: 2.002 319 304 82

The uncertainties in both numbers are confined to the last two figures. "Magnificent" is not too strong a word to describe this agreement between theory and experiment. We seem to understand the electron and its interactions pretty well!

It is tempting to compare the spin and the orbital motion of an electron in the hydrogen atom to the rotation and the orbital motion of the earth as it goes around the sun. The analogy, though it may be useful, is superficial. For the spinning and the orbiting earth the two angular momenta are of the same nature; they are simply measured with respect to different origins. The spin and the orbital motion of the electron, however, are intrinsically different, as the factor 2 above demonstrates.

The spin of the electron, like its charge and its rest mass, is an intrinsic property of the electron, no matter whether the electron is part of an atom or is a free particle. The neutron and the proton (and, indeed, the quarks that combine to make them up) also have a spin of $\frac{1}{2}$, measured in units of $\hbar$. So does the neutrino, that elusive particle that is streaming at the rate of billions per second through the book you are now reading. In fact, spin comes close to being about the only non-zero property that the neutrino has!

Most nuclei have spin, either integral or half-integral. Even the photon has an associated spin (of unity, again in $\hbar$ units). When a beam of light is circularly polarized, we may think of it as a beam of photons with all of its spins aligned. When such a beam falls on an absorbing screen, it can, in fact, exert a measurable torque on that screen. The concept of spin is powerfully useful in atomic, nuclear, and solid-state physics and is a major organizing principle in the study of elementary particles. In fact, one of the first questions asked about a new elementary particle is: "What is its spin?"

51-7
SCIENCE, TECHNOLOGY, AND SPIN

Most discoveries in pure science turn out, sooner or later, to have applications in technology. On November 8, 1895, for example, William Röntgen, working in his physics laboratory at the University of Wurzburg in Germany, discovered X rays. Less than three months later a skater, having fallen on the ice of the Connecticut River, had his broken arm X-rayed at Dartmouth College — the first medical application of X rays in this country.

It took longer for spin to make the journey from pure science to technology, but it has now done so. Consider a proton, perhaps residing in a drop of water. The proton, like the electron, has a spin of $\frac{1}{2}$ in units of $\hbar$. Its magnetic dipole moment vector μ_p can occupy either of two quantized orientations with respect to a magnetic field $\mathbf{B}$, as Fig. 51-10a shows.* These two positions differ in energy by $2\mu_p B$, this being the work required to turn a magnetic dipole of moment μ_p end for end in a magnetic field.

If a drop of water containing our proton is subjected to an electromagnetic field that is alternating with a frequency ν, transitions between the two orientations of μ_p — called *spin flips* — may be induced. For this to happen the equation

$$h\nu = 2\mu_p B \qquad (51\text{-}16)$$

must be satisfied. That is, the energy $h\nu$ of the photons associated with the alternating electromagnetic field must be just equal to the energy difference between the two spin orientations. (Recall that, in Section 37-7, we gave a classical analysis of such spin flips, based on an analogy to a spinning top.)

The spin flip may go either way (that is, from up to down or from down to up) with equal probability. However, if the water drop is in thermal equilibrium, there will be more proton spins in the lower energy state than in the higher energy state, as the dots in Fig. 51-10b suggest. This means that there will be a net absorption of energy from the electromagnetic field. Figure 37-22 shows such an absorption peak, taken under conditions of relatively low resolution.

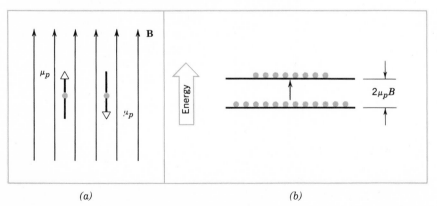

(a) (b)

figure 51-10
(a) A proton, whose spin is $\frac{1}{2}$ in units of $\hbar$, can occupy either of two quantized orientations in an external magnetic field. (b) If Eq. 51-16 is satisfied, the protons in the sample can flip from one orientation to the other.

* The spin angular momentum $(= \frac{1}{2}\hbar)$ and the magnetic dipole moment $(= \mu_p)$ refer, as usual, to the maximum projections of these quantities on the z-axis and not to their magnitudes.

The usefulness of magnetic resonance studies, carried out under conditions of high resolution, lies in the fact that the factor B in Eq. 51-16 is *not* the imposed external magnetic field B_{ext} but rather it is that field as modified by the small local internal magnetic field B_{local} due to the electrons and nuclei in the molecule of which our proton is a part. Thus we can rewrite Eq. 51-16 as

$$hv = 2\mu_p(B_{local} + B_{ext}). \qquad (51\text{-}17)$$

In practice, it is customary to leave the frequency v of the electromagnetic oscillator fixed and to vary B_{ext} until Eq. 51-17 is satisfied and an absorption peak is recorded.

Figure 51-11 shows a *nuclear magnetic resonance spectrum*, as it is called, for ethanol, whose formula we may write as $CH_3\text{—}CH_2\text{—}OH$. The various resonance peaks all represent spin flips of protons. They occur at different values of B_{ext}, however, because the local environments of the protons within the ethanol molecule are different. The spectrum of Fig. 51-11 is a unique signature for ethanol, and it should be believable that the nuclear magnetic resonance method is of great value as an analytical tool in organic chemistry. Nuclear magnetic resonance apparatus is commercially available and is to be found in laboratories all over the world.

Spin technology has also been applied to medical diagnostic imaging. The protons in the various tissues and organs of the human body also find themselves in different local magnetic environments. When the body, or part of it, is immersed in a strong external magnetic field, these environmental differences can be detected by spin flip techniques and translated by computer processing into an image similar to one produced by X rays. Figure 51-12, for example, shows a cross section of a human head taken by this method. The method supplements X-ray imaging in many important ways. All this from a suggestion by two Dutch graduate students, more than 50 years ago!

figure 51-11
A nuclear magnetic resonance spectrum for ethanol. All of the absorption lines are due to the spin flips of protons. The three groups of lines correspond, as indicated, to protons in the OH group, the CH_2 group, and the CH_3 group within the molecule. The entire horizontal scale in the figure is considerably less than 10^{-4} T $(= 1$ gauss$)$.

figure 51-12
A cross section of a human head, taken by spin-flip techniques. It shows anatomical and functional detail not visible on X-ray images. (Courtesy of General Electric Medical Systems, Inc.)

EXAMPLE 5

Flipping the proton. A drop of water is suspended in a 1.8-T magnetic field and an alternating electromagnetic field is applied, its frequency chosen so as to produce spin flips for the protons in the sample. What is the required frequency of the oscillator that supplies this alternating field? What is the wavelength? The magnetic dipole moment of the proton is 1.41×10^{-26} J/T. Ignore local magnetic fields within the sample.

From Eq. 51-16 we have,

$$\nu = \frac{2\mu_p B}{h}$$

$$= \frac{2(1.41 \times 10^{-26} \text{ J/T})(1.8 \text{ T})}{6.63 \times 10^{-34} \text{ J} \cdot \text{s}}$$

$$= 7.66 \times 10^7 \text{ Hz} = 76.6 \text{ MHz}.$$

This frequency is in the short radio wave band. The corresponding wavelength is

$$\lambda = \frac{c}{\nu} = \frac{3 \times 10^8 \text{ m/s}}{7.66 \times 10^7 \text{ Hz}} = 3.92 \text{ m}.$$

Note that the magnetic dipole moment of the proton ($= 1.41 \times 10^{-26}$ J/T) is about 650 times smaller than the magnetic dipole moment of the electron ($= 9.27 \times 10^{-24}$ J/T). It is generally true that nuclear magnetic dipole moments are smaller than atomic magnetic dipole moments by something like this factor.

51-8
COUNTING THE HYDROGEN ATOM STATES

We have now described the four quantum numbers that define the stationary states of the hydrogen atom, and we have shown how each of them can be interpreted physically. Although we did not prove it, it is nevertheless true that these facts emerge lock, stock, and barrel from Schrödinger's wave equation, with an important assist from Dirac's equation in the case of the electron spin. If you take a full course in quantum mechanics, you will get a chance to solve the Schrödinger wave equation yourself and you will have the pleasure of seeing in detail just how the quantum numbers and the other information come tumbling out in a natural way.

Our next task is to see whether we can arrange the hydrogen atom states in some orderly fashion. Consider first the principal quantum number n. All states with the same value of n have the same energy, and we say that the assembly of such states forms a *shell*. Equation 51-4 tells us that the number of different values of l that are possible for a given value of n is just equal to n. Thus, for $n = 3$ we can have three values of l ($= 0$, 1, and 2).

The shells can be further subdivided. Within a given shell, all states with the same value of l have the same angular momentum and are said to form a *subshell*. Thus, the shell defined by $n = 3$ contains three subshells, each with the same energy but a different angular momentum.

The subshells are themselves degenerate because of the different ways the angular momentum vector can be oriented. For a given l, Eq. 51-6 tells us that there are $2l + 1$ values of m_l. In our example, then, the subshell with $l = 2$ contains 5 ($= 2 \times 2 + 1$) and those with $l = 1$ and 0 contain 3 states and 1 state, respectively. This adds up to a total of 9 ($= 5 + 3 + 1$) states in the shell with $n = 3$.

The effect of the spin quantum number m_s is simply to double the number of states. Each combination of n, l, and m_l that we have identified can now be associated with either $m_s = +\frac{1}{2}$ or $m_s = -\frac{1}{2}$, thus producing two states where one existed before. To continue with our example, there are not 9 but 18 states in the shell with $n = 3$.

Table 51-2 summarizes this classification of hydrogen atom states into shells and subshells.

Do the shell population numbers (that is: 2, 8, and 18) in the bottom row of Table 51-2 seem familiar? They are the lengths of the horizontal periods in the periodic table of the elements! As Appendix E shows, period 1 contains 2 elements, periods 2 and 3 contain 8 elements each, and periods 4 and 5 contain 18. In Chapter 52 we shall see in detail just how the order of the elements in the periodic table arises from wave mechanical principles.

Atoms Beyond Hydrogen. At the beginning of this chapter we described how the hydrogen atom potential energy function (Eq. 51-1) is inserted into the Schrödinger equation and how that equation can be solved to yield quantum numbers and wave functions. All of our results so far, then, apply to the hydrogen atom only. What about other atoms?

For atoms with substantially more than one electron the potential energy function that we must substitute into the Schrödinger equation

Table 51-2
States of the hydrogen atom*

n	1	2		3		
l	0	0	1	0	1	2
m_l	0	0	$0, \pm 1$	0	$0, \pm 1$	$0, \pm 1, \pm 2$
m_s	$\pm\frac{1}{2}$	$\pm\frac{1}{2}$	$\pm\frac{1}{2}$	$\pm\frac{1}{2}$	$\pm\frac{1}{2}$	$\pm\frac{1}{2}$
Populations of the subshells	2	2	6	2	6	10
Populations of the shells	2	8		18		

* Complete to $n = 3$ only.

involves the Coulomb interaction between many pairs of particles and can rapidly become hopelessly complicated. In neon, for example, it can be shown that, including the nucleus and the ten electrons, there are no fewer than about 2×10^7 independent pairs of charges whose Coulomb interactions must be taken into account if the Schrödinger equation is to be solved rigorously. At this level of complexity rigorous solutions are not possible, and we must rely on methods that are both approximate and numerical.

Our work on the hydrogen atom, however, has by no means gone to waste. *It turns out that the states of an electron in a many-electron atom are described by exactly the same set of quantum numbers that we developed for the hydrogen atom.* The energies and the wave functions are different, but the labeling scheme and the classification of states into shells and subshells remain unchanged, a tremendously useful state of affairs.

The hydrogen atom contains only one electron so that of all the states listed in Table 51-2 (and also of the many with $n > 3$ that are not listed) only one can be occupied. The rest are vacant. In a many-electron atom, however, we can fill as many states as there are available electrons.

In sprinkling the electrons among the various states in such atoms, we are guided by the *Pauli exclusion principle*, which asserts that no state can contain more than one electron. If it were not for this rule, all of the electrons in many-electron atoms would fall into the $n = 1$ shell and the world would not be what it is. We shall discuss this important one-state-one-electron principle more fully in Section 52-4.

Consider an atom of neon, with its ten electrons. Two of them occupy the shell with $n = 1$, filling it completely. The remaining eight occupy the shell with $n = 2$. This shell contains two subshells, corresponding to $l = 0$ and to $l = 1$. The eight available electrons are just enough to fill both of these subshells. Thus the neon atom in its lowest energy state has its electrons arranged in three filled subshells.

In a filled subshell all allowable projections of both the orbital and the spin angular momentum vectors are present in the same atom and thus cancel out for the atom as a whole. Such a filled subshell has zero net angular momentum and, correspondingly, zero net magnetic dipole moment. As we know, the neon atom (with its filled subshells) is chemically inert.

Beyond neon comes sodium, with its eleven electrons. Ten of them form a neon-like core, leaving the remaining electron in a state with $n = 3$ and $l = 0$. To a first approximation we can think of the sodium nucleus (whose charge is $+11e$) as partially screened by the neon-like core (charge $-10e$), leaving a reduced net central charge to govern the motion of this outer electron. The entire angular momentum and magnetic dipole moment of the sodium atom is due to this single, relatively loosely bound, valence electron.

EXAMPLE 6

Counting states for $n = 4$. A shell is defined by a principal quantum number n of 4. (a) How many subshells does this shell contain? (b) What is the population of each of these subshells? (c) What is the population of the shell?

(a) If $n = 4$, we know from Eq. 51-6 that the allowed values of l are 0, 1, 2, and 3. This is a total of 4 values, in agreement with the fact that the number of allowed values of l for a given n is just equal to n. Each value of n defines a shell and each value of l defines a subshell within that shell. Thus there are four subshells in the $n = 4$ shell.

(b) The population of a subshell is given by $2(2l + 1)$, the factor of 2 coming from the two allowed values of the spin magnetic quantum number. For the populations of the various subshells in the $n = 4$ shell we then have

l	$2(2l + 1)$
0	2
1	6
2	10
3	14

(c) The population of the $n = 4$ shell is found by adding up the populations of the subshells that it contains. From the table above we have $2 + 6 + 10 + 14$ or 32. Note with interest that 32 is the number of elements in horizontal row 6 of the periodic table of the elements. See Appendix E.

Can you prove that, in general, the population of a shell defined by principal quantum number n is given by $2n^2$? That works out in this case because $2 \times 4^2 = 32$, as we found by explicit counting.

We have already discussed this state of the hydrogen atom in Section 50-10. We summarize our findings in Table 51-3.

We said nothing in Section 50-10 about the role of the spin of the electron in the ground state of the hydrogen atom. As Table 51-2 reminds us, the shell corresponding to $n = 1$ in the hydrogen atom contains two states, corresponding to the two allowed values $(= \pm\frac{1}{2})$ of the spin quantum number m_s. These two states, however, have exactly the same energy and, for an isolated atom, there is no way to tell them apart experimentally. It is best to think of the atom in its ground state as spending — on average — half its time in the state with $m_s = +\frac{1}{2}$ and half in the state with $m_s = -\frac{1}{2}$. The energy of the atom and the probability density curve of Fig. 50-18b are the same for each of these states.

If you *really* want to distinguish between these two spin orientation states you can do so by putting the atom in an external magnetic field. This not only provides a natural reference axis with respect to which the electron's spin angular momentum vector (and its spin magnetic dipole moment vector) can orient themselves but it also separates the two states in energy. This, in fact, is exactly what was done in the Stern-Gerlach experiment.

51-9
THE HYDROGEN ATOM — THE GROUND STATE

Table 51-3
The ground state of the hydrogen atom*

Quantity	Symbol	Function	Equation Reference
Wave function	$\psi(r)$	$\dfrac{1}{\sqrt{\pi a_0^3}}\, e^{-r/a_0}$	50-15
Probability density	$\psi^2(r)$	$\dfrac{1}{\pi a_0^3}\, e^{-2r/a_0}$	50-16
Radial probability density	$P_r(r)$	$\dfrac{4}{a_0^3}\, r^2 e^{-2r/a_0}$	50-17

* See Section 50-10 for full details.

The state next highest in energy above the ground state is called the first excited state. Its energy $(= -3.40 \text{ eV})$ is found by putting $n = 2$ in the energy equation (Eq. 51-2). As Table 51-2 reminds us, the $n = 2$ shell contains two subshells, corresponding to $l = 0$ and to $l = 1$. We deal with each in turn.

The $n = 2$, $l = 0$ subshell. The wave function for this state proves to be

$$\psi_{200}(r) = \frac{1}{4\sqrt{2\pi a_0^3}} (2 - r/a_0)e^{-r/2a_0}, \qquad (51\text{-}18)$$

in which, again, the subscripts represent the quantum number sequence $n = 2$, $l = 0$, and $m_l = 0$. This state, just like the ground state, has spherical symmetry in that it is a function of r only and involves no angles as variables.

The probability density $\psi^2(r)$ and the radial probability density $P_r(r)$ are given by

$$\psi^2(r) = \frac{1}{32\pi a_0^3} (2 - r/a_0)^2 e^{-r/a_0} \qquad (51\text{-}19)$$

and

$$P_r(r) = \psi^2(r)(4\pi r^2)$$
$$= \left(\frac{1}{8a_0}\right)\left(\frac{r}{a_0}\right)^2 \left(2 - \frac{r}{a_0}\right)^2 e^{-r/a_0}, \qquad (51\text{-}20)$$

respectively. Figure 51-13a is a "dot plot" of Eq. 51-19, the probability density $\psi^2(r)$. Figure 51-13b is a plot of Eq. 51-20, the radial probability density. Note that the latter curve has two maxima and goes to zero at $r = 2a_0$, as simple inspection of Eq. 51-20 makes clear.

The remarks about spin at the end of Section 51-9 apply with equal force here. This state also has complete spherical symmetry. Its angular momentum is $\frac{1}{2}$, in units of $\hbar$, due (as before) entirely to the spin of the electron.

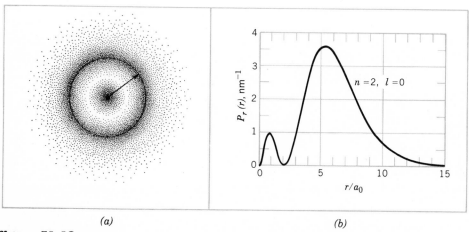

(a)   (b)

figure 51-13
(a) The *probability density* $\psi^2(r)$ for the state of the hydrogen atom with $n = 2$ and $l = 0$. The radius of the circle is 4 Bohr radii. (b) The *radial probability density* $P_r(r)$ for that same state.

* An extended coverage section; may be omitted.

The $n = 2$, $l = 1$ Subshell. The states that comprise this subshell *do* have orbital angular momentum, its z-axis projections being given by $m_l\hbar$, where m_l can take on the values of 0 or ± 1. The wave functions for these three states are *not* spherically symmetric, being functions not only of r but also of the co-latitude angle θ, defined in Fig. 51-14a.

Figure 51-14a shows "dot plots" of the three probability densities, $\psi_{11-1}^2(r,\theta)$, $\psi_{110}^2(r,\theta)$, and $\psi_{11+1}^2(r,\theta)$. As usual, the subscripts represent the quantum number sequence n, l, and m_l. All three plots have rotational symmetry about the z-axis, the plots for $m_l = -1$ and for $m_l = +1$ being identical.

You are entitled to be a little puzzled about the lack of spherical symmetry shown in Fig. 51-14a. After all, the potential energy function that we inserted into Schrödinger's equation depended only on r. Does the electron in a state with $n = 2$, $l = 1$, and $m_l = 0$ *really* like to cluster about the z-axis, avoiding the equatorial plane? Who decides in what direction this axis should point anyway?

The answer to this puzzle comes when we realize that the three states in question have the same energy and, in the absence of an external magnetic field, there is no way to isolate them experimentally. We assume that—on average—the atom spends one-third of its time in each of the three states shown in Fig. 51-14a, and on this basis we can calculate a *weighted average probability density* for the subshell as a whole. The result is

$$\psi_{21}^2(r) = (\tfrac{1}{3})\psi_{21-1}^2(r,\theta)$$
$$+ (\tfrac{1}{3})\psi_{210}^2(r,\theta) + (\tfrac{1}{3})\psi_{21+1}^2(r,\theta)$$
$$= \frac{1}{96\pi a_0^5}\,r^2 e^{-r/a_0}. \tag{51-21}$$

The subscript on the probability density gives the values of $n\ (= 2)$ and of $l\ (= 1)$. Note that the angular variable θ has disappeared from the final result! The probability density for the subshell as a whole depends only on r and has the spherical symmetry that we expect it to have. This means that if you superimpose the three cylindrically symmetrical dot plots of Fig. 51-14a, the resulting dot plot (imagined in three dimensions) will be spherically symmetric.

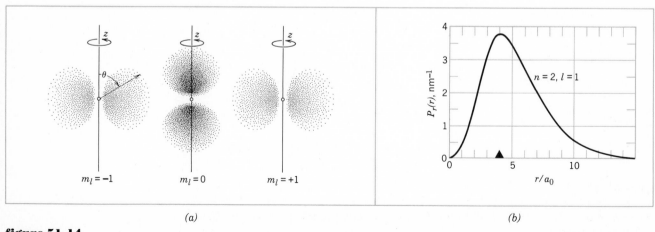

(a)   (b)

figure 51-14
(a) The probability densities of the hydrogen atom in a state with $n = 2$, $l = 1$, and m_l as indicated. All three figures have rotational symmetry about the z-axis. (b) The radial probability density $P_r(r)$ for the subshell of the hydrogen atom defined by $n = 2$ and $l = 1$. Note the maximum at 4 Bohr radii.

We now find the radial probability density for this subshell, proceeding as we did in Section 51-9 for the ground state, namely

$$P_r(r) = \psi_{21}^2(r)(4\pi r^2)$$

$$= \frac{1}{24a_0^5}\, r^4 e^{-r/a_0}. \qquad (51\text{-}22)$$

Figure 51-14b shows a plot of this radial probability density. Note that the maximum of the distribution occurs at $r = 4a_0$, which (see Eq. 49-31) is just the radius of the second Bohr orbit.

More about spin. In a hydrogen atom state with $l = 1$, both orbital and spin angular momentum are present. How do we combine them to find the total angular momentum of the atom? The quantum number l ($= 1$) describes the orbital angular momentum **L** and s ($= \frac{1}{2}$) describes the spin angular momentum **S**. For the total angular momentum **J** we introduce the *total quantum number j*, the relationship between them being (compare Eq. 51-5)

$$J = \sqrt{j(j+1)}\,\hbar.$$

The two possible values of j in this case are given by

$$j = l + s = 1 + \tfrac{1}{2} = \tfrac{3}{2} \qquad \text{(parallel)}$$

and

$$j = l - s = 1 - \tfrac{1}{2} = \tfrac{1}{2} \qquad \text{(antiparallel)}.$$

It is this total angular momentum that takes up space quantized positions in a magnetic field, the number of such positions being equal to $2j + 1$. Thus, states with $j = \frac{3}{2}$ will have $2 \times (\frac{3}{2}) + 1$ or four positions. Those with $j = \frac{1}{2}$ will have $2 \times (\frac{1}{2}) + 1$ or two positions. Table 51-2 shows that the $l = 0$ subshell is sixfold degenerate so everything works out nicely. Four of those states have $j = \frac{3}{2}$ and the other two have $j = \frac{1}{2}$.

The two values of j correspond to parallel and antiparallel combinations of the orbital and spin angular momentum vectors. You cannot combine angular momentum vectors without simultaneously combining their associated magnetic dipole moment vectors. We note that different geometric arrangements of two bar magnets will have different energies. This means that the four states with $j = \frac{3}{2}$ will have a somewhat different energy than the two states with $j = \frac{1}{2}$. Convince yourself that, because $j = \frac{3}{2}$ corresponds to two tiny magnets aligned parallel to each other, this level will have the higher energy.

In this way the degeneracy of the $l = 1$ subshell is partially lifted, its six states now being distributed between two energy levels. In the hydrogen atom, these two $l = 1$ levels, though very close together, have been explored experimentally in the finest detail. For the valence electron in sodium, these two states account for the fact that the yellow light that characterizes a sodium flame is made up of two closely lying wavelengths, as Fig. 51-15 shows. You may well have peered at this familiar sodium doublet through a spectroscope in your physics laboratory.

Michael Faraday, one of the greatest experimental scientists of all time, was convinced that if you put a light source in a strong magnetic field, something would happen to the spectrum lines of its emitted light. Faraday's intuition was seldom wrong and, on March 12, 1862, at the age of 70, he did the experiment. Alas, his notebook tells us: "not the slightest trace of effect . . . was observed." It was, however, his apparatus and not his intuition that had failed him in this difficult experiment.

About 30 years later, in 1896, the Dutch physicist Pieter Zeeman reasoned that if the great Faraday thought the experiment worthwhile,

* Extended coverage material; may be omitted.

figure 51-15
Showing the three energy levels that account for the two lines of the familiar sodium doublet. The energy difference between the two upper levels is only 1/1000 of the energy difference between those levels and the lower level. To the right of each level we see the total quantum number j and also the degree of degeneracy ($= 2j + 1$).

PROBING THE ATOM WITH A MAGNETIC FIELD

Magnet

OFF

ON

OFF

figure 51-16
The Zeeman effect in rhodium. The top and bottom portions show a small portion of
the spectrum, with the magnet turned OFF. The central portion shows the splitting of
the lines when the magnet is turned ON.

it was probably worth repeating, with more sensitive apparatus. He did
so and discovered that Faraday was right. Figure 51-16, taken with
modern equipment and in a very strong magnetic field, shows how each
spectrum line splits into several components when the magnetic field is
turned on.

The atomic spectra alone posed enough problems for the early physicists as they tried to learn how atoms are put together. Now the *Zeeman
effect*, as it is called, revealed hidden and unsuspected properties of the
atom and posed new challenges. Studies of this effect played a major
role in unraveling the puzzles of atomic structure and in providing data
for the testing of wave mechanics. It ranks high among the triumphs of
that theory that the number of Zeeman components into which a given
line splits, their intensities and relative spacings in wavelength, and
even the nature of the polarization of the light emitted by the individual
components, all agree with the predictions of wave mechanics.

The Zeeman effect is now so well understood that astronomers work
it backwards, starting from the observed Zeeman splittings and deducing the strengths of the magnetic fields present on the solar disk, on the
surfaces of stars, and in the interstellar spaces of our galaxy. Figure
51-17, for example, shows a typical magnetogram of the sun, its surface
magnetism revealed by measuring the Zeeman splittings of the light it
emits.

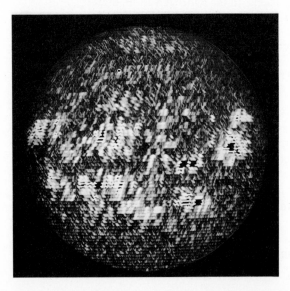

figure 51-17
A magnetogram of the sun.
The plot is based on Zeeman effect
measurements. (Nasa.)

Zeeman, together with H. A. Lorentz, who gave a deeply perceptive classical interpretation of the effect, shared the 1902 Nobel prize — only the second such prize to be awarded. As for Faraday, his biographer, L. Pearce Williams, writes, "It is not denigrating Zeeman's achievement to suggest that the Zeeman effect was Faraday's last discovery."

You cannot tell by looking at a spectrum line how many Zeeman components it will have in a magnetic field. In Fig. 51-16, for example, the four spectrum lines at the top and the bottom of the figure look pretty much alike. Yet the numbers of their Zeeman components, scanning from right to left in the figure, are 4, 8, 4, and 10. What determines these numbers?

Every spectrum line represents a transition between two levels. If the line splits into wavelength components by the action of the magnetic field, it must mean that one or both of its associated levels has been split into energy components.

Consider, for example, the component of the sodium doublet whose wavelength is 588.995 nm, shown at the left in Fig. 51-15. If a sodium flame is put into a magnetic field, this single line splits into six closely lying wavelength components. Figure 51-18 shows how this comes about.

When the sodium atom is in the state represented by the upper level in Fig. 51-18a, it has a total angular momentum quantum number j of $\frac{3}{2}$ and (as always) an associated magnetic dipole moment. In a magnetic field this magnetic dipole moment vector can take up four ($= 2j + 1$) space-quantized orientations so that the single level splits into four levels, each with its identifying total magnetic quantum number m_j, as

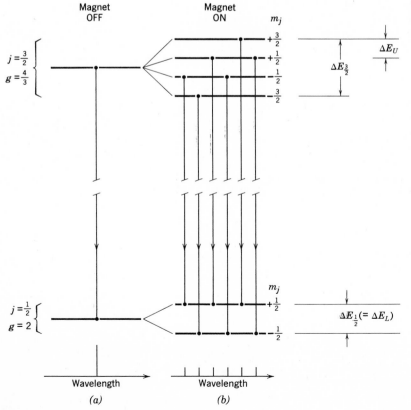

figure 51-18

(a) The two levels responsible for one of the lines of the sodium doublet. (b) When a magnetic field is applied, each of these levels splits, revealing its degeneracy. The spectrum line separates into six closely spaced components.

Fig. 51-18b shows. In just the same way, the lower level (with $j = \frac{1}{2}$) splits into two $(= 2j + 1)$ components.

If we drew all possible transitions between the upper and the lower sets of levels in Fig. 51-18b, we would end up with eight Zeeman components. However, wave mechanics predicts that two of these transitions (can you find them?) will not occur, and indeed they do not. The number of observed Zeeman components is six, just as predicted.

Calculating the level splittings. To find the wavelengths of the Zeeman components we must first find the energy differences between the levels that give rise to those transitions. How do we do this?

Let us see how to calculate quantities such as $\Delta E_{3/2}$ and $\Delta E_{1/2}$, shown in Fig. 51-18b. Such energy differences, which we generalize as ΔE_j, represent the energy required to turn a magnetic dipole end for end in a magnetic field of strength B. Thus

$$\Delta E_j = 2\mu_j B \qquad (51\text{-}23)$$

in which μ_j is the magnetic dipole moment.

Now we need to know the value of the magnetic dipole moment μ_j. Our task is complicated by the fact that orbital motion and spin motion differ (by a factor of 2) in their effectiveness in generating magnetism. In the general case, when both orbital and spin motion are present, we allow for this complication by writing

$$\mu_j = gj\mu_B, \qquad (51\text{-}24)$$

in which μ_B is the Bohr magneton and g (called simply the *g-factor*) is a number whose value depends on the nature of the angular momentum we are dealing with.

If the angular momentum could be entirely orbital, then j would be the same as l and we would have $g = 1$, as implicitly shown in Eq. 51-10. If, on the other hand, the angular momentum is entirely associated with spin, then j is the same as s and we have $g = 2$, as in Eq. 51-15. In the general case, in which the angular momentum is a mixture of spin and oribtal motions, the *g*-factor must be computed from an appropriate wave mechanical formula; see Problem 35.

We state without proof that the *g*-factors for the upper and the lower states in Fig. 51-18b are $\frac{4}{3}$ and 2, respectively. The value of the *g*-factor for the lower state should not be surprising because this state resembles the ground state of the hydrogen atom in that it has $l = 0$. The angular momentum of the sodium atom in this state is associated entirely with the spin of its valence electron; thus $g = 2$ is appropriate.

Combining Eqs. 51-23 and 51-24 gives

$$\Delta E_j = 2gj\mu_B B \qquad (51\text{-}25)$$

for the quantity we seek. We can easily find the energy interval ΔE_{adj} between *adjacent* levels by dividing this quantity by $2j$, which is the number of such intervals. This yields

$$\Delta E_{adj} = g\mu_B B. \qquad (51\text{-}26)$$

We see at once that this equation yields a correct result $(= 2\mu_B B)$ for the lower level in Fig. 51-18b, for which the angular momentum is entirely due to spin so that $g = 2$.

The Zeeman level splitting. A sodium vapor lamp is placed in a magnetic field of 2.5 T. What are the energy intervals between adjacent levels for (a) the upper and (b) the lower groups of levels in Fig. 51-18b?

EXAMPLE 7

(a) For the upper levels (see the figure) the g-factor is $\frac{4}{3}$ and Eq. 51-26 gives us

$$\Delta E_U = g\mu_B B$$

$$= (\tfrac{4}{3})(5.79 \times 10^{-5} \text{ eV/T})(2.5 \text{ T})$$

$$= 1.93 \times 10^{-4} \text{ eV},$$

in which the subscript "U" refers to the upper set of levels in Fig. 51-18b.

(b) A similar calculation for the energy interval between the lower two levels in Fig. 51-18b (for which $g = 2$) yields

$$\Delta E_L = 2.89 \times 10^{-4} \text{ eV}.$$

By comparison, the energy difference between the levels in Fig. 51-18a that are responsible for the 589-nm emission line is

$$\Delta E = h\nu = h(c/\lambda)$$

$$= \frac{(4.14 \times 10^{-15} \text{ eV} \cdot \text{s})(3 \times 10^8 \text{ m/s})}{(589 \times 10^{-9} \text{ m})}$$

$$= 2.11 \text{ eV}.$$

Thus the splitting of the levels in a 2.5-T magnetic field is about 1 part in 10,000 of the separation between levels. The Zeeman effect is not trivially observed! Zeeman himself, in fact, first observed a mere broadening of the line rather than a break-up into components.

questions

1. If Bohr theory and wave mechanics predict the same result for the energies of the hydrogen atom states (see Eq. 51-2), then why do we need wave mechanics, with all its complexity?

2. Compare Bohr theory and wave mechanics. In what respects do they agree? In what respects do they differ?

3. How would you show in the laboratory that an atom has angular momentum? that it has a magnetic dipole moment?

4. Justify the statement that, in the Einstein-de Haas effect, the angular momentum of the iron bar as a whole must be conserved when the bar is suddenly magnetized.

5. In the Einstein-de Haas experiment (see Example 1), can you justify the fact that the predicted period of rotation of the cylinder depends only on the cylinder radius and not, for example, on its height. What assumptions were made in deriving the expression for the period of rotation?

6. Convince yourself that the directions of the arrows in Fig. 51-3b representing the current in the solenoid, the magnetic field, the atomic angular momenta, and the direction of rotation of the cylinder are consistent with each other.

7. Does the Einstein-de Haas effect provide any evidence that angular momentum is quantized?

8. Why don't we observe space quantization for a spinning top?

9. The angular momentum of the electron in the hydrogen atom is quantized. Why isn't the linear momentum also quantized? (Hint: Consider the implications of the uncertainty principle.)

10. Angular momentum is a vector and you might expect that it would take three quantum numbers to describe it, corresponding to the three space components of a vector. Instead, in an atom, only two quantum numbers characterize the angular momentum. Explain why.

11. A beam of neutral silver atoms is used in a Stern-Gerlach experiment. What is the origin of both the force and the torque that act on the atom? How is the atom affected by each?

12. In a Stern-Gerlach apparatus, is it possible to have a magnetic field configuration in which the magnetic field itself is zero along the beam path but the field gradient is not? If your answer is yes, can you design an electromagnet that will produce such a field configuration?

13. What determines the number of subbeams into which a beam of neutral atoms is split in a Stern-Gerlach experiment?

14. The silver atoms in the Stern-Gerlach experiment of Example 4 are uncharged. Suppose that a silver atom in the apparatus were suddenly to lose an electron, becoming a silver ion. What would be the nature and the relative magnitude of the forces acting on it (a) before and (b) after this event?

15. How do we arrive at the conclusion that the spin magnetic quantum number m_s can have only the values $\pm\frac{1}{2}$? What kinds of experiments support this conclusion?

16. Why is the magnetic moment of the spinning electron directed opposite to its spin angular momentum?

17. Discuss how good an analogy the rotating earth revolving about the sun is to a spinning electron moving about a proton in the hydrogen atom.

18. In Section 51-6 we said that a beam of circularly polarized light, viewed as a beam of photons whose spins are aligned, can exert a torque on an absorbing screen. Develop the analogy to the Einstein-de Haas experiment.

19. Compare and contrast X-ray imaging methods with nuclear magnetic resonance ones.

20. "If the angular momentum of electrons in atoms were not quantized, the periodic table of the elements would not be what it is." Discuss this statement.

21. How would the properties of helium differ if the electron had no spin, that is, if the only operative quantum numbers were n, l, and m_l?

22. We assert that the number of quantum numbers needed for a complete description of the motion of the electron in the hydrogen atom is equal to the number of degrees of freedom that that electron possesses. What is this number? How can you justify it?

23. Define and distinguish among the terms wave function, probability density, and radial probability density.

24. What are the dimensions and the SI units of a wave function, a probability density, and a radial probability density? Are the dimensions what you expect?

25. In the hydrogen atom state with $l = 1$, the spin and the orbital angular momentum vectors can be aligned either parallel or antiparallel. Which arrangement has the larger energy and why?

26. How can you account for the fact that in the state of the hydrogen atom with $n = 2$ and $l = 0$, the probability density is a maximum at $r = 0$ but the radial probability density is zero there? See Fig. 51-13.

27. Figure 51-14a shows the three probability densities for the hydrogen atom states with $n = 2$ and $l = 1$. What determines the direction in space that we choose for the z-axis?

28. Consider the three probability density "dot plots" of Fig. 51-14, each of which is a figure of revolution about the z-axis. Do you see any connection between these figures and the semiclassical vector model of the atom (Fig. 51-5) for he case of $l = 1$?

29. Use Heisenberg's uncertainty principle to show that the probability densities in an $l = 2$ state have cylindrical symmetry about the z-axis.

30. Explain how the interaction between the spin and the orbital motions of the valence electron in sodium lead to the splitting of the spectral lines of sodium, producing the familiar sodium doublet. See Fig. 51-15.

31. In Fig. 51-18 why do the upper and the lower sets of levels have different energy splittings when an external magnetic field is applied?

32. We have noted that, experimentally, spin angular momentum is twice as effective as orbital angular momentum in generating magnetism. How is this fact related to the number of Zeeman components induced in the sodium doublet by an external magnetic field? See Fig. 51-18.

33. What would the Zeeman effect look like if space quantization did not exist? if electron spin did not exist?

problems

SECTION 51-2

1. An electron is one Bohr radius from a proton and has a total energy E of $+5.0$ eV. What are (a) its potential energy U and (b) its kinetic energy K? (c) What kinetic energy would this electron have at very large distances from the proton, assuming always that its total energy E remains unchanged? (d) Describe this state of the hydrogen atom. Can a quantum number be assigned to it?
   Answer: (a) -27.2 eV. (b) 32.2 eV. (c) 5.0 eV. (d) Unbound; No.

SECTION 51-3

2. Verify that $\mu_B = 9.274 \times 10^{-24}$ J/T $= 5.788 \times 10^{-5}$ eV/T, as reported in Eq. 51-4.

3. Show that the magnetic moments of the electrons in the various Bohr orbits are given by

$$\mu = n\mu_B$$

in which μ_B is the Bohr magneton and $n = 1, 2, 3, \ldots$.

SECTION 51-4

4. Calculate and tabulate, for a hydrogen atom in a state with $l = 3$, the allowed values of L_z, μ_z, and θ. Find also the magnitudes of $\mathbf{L}$ and μ.

5. Consider the relation

$$\cos \theta_{\min} = \left(1 + \frac{1}{l}\right)^{-1/2}$$

in Example 2 for the limiting case of large l. (a) By expanding both sides of this equation in series form (see Appendix I), show that, to a good approximation for large l,

$$\theta_{\min} \cong \frac{1}{\sqrt{l}},$$

where $\theta_{\min}$ is to be expressed in radian measure. (b) Test the validity of this approximate formula for the five values of l given in Example 2. (c) Enlarge the table in Example 2 by finding $\theta_{\min}$ for $l = 10^5$, 10^6, and 10^7. (d) What computational difficulties do you encounter if you do *not* use the above approximate formula to calculate $\theta_{\min}$ for large values of l?
   Answer: (c) $0.18°$; $0.057°$; $0.018°$.

6. Of the three scalar components of $\mathbf{L}$, one, L_z, is quantized, according to Eq. 51-7. In view of the restrictions imposed by Eq. 51-5 and 51-6, taken together, show that the most that can be said about the other two components of $\mathbf{L}$ is

$$(L_x^2 + L_y^2)^{1/2} = [l(l + 1) - m_l^2]^{1/2}\, \hbar$$

Note that these two components are not separately quantized. Show also that

$$l\hbar \le (L_x^2 + L_y^2)^{1/2} \le [l(l + 1)]^{1/2}\, \hbar.$$

Correlate these results with Fig. 51-5.

7. Show, by reanalyzing Problem 76 of Chapter 49 for a diatomic molecule with the angular momentum quantized by Eqs. 51-5 and 51-6, that the energy levels can be written as

$$E_l = \frac{h^2 l(l+1)}{4\pi^2 m d^2}, \qquad l = 1, 2, 3 \ldots .$$

Calculate the energies of the first three levels of the O_2 molecule, for which the two atoms are spaced 0.20 nm apart. The mass of an oxygen atom is 16 u.   *Answer:* 1.3×10^{-4} eV; 3.9×10^{-4} eV; 7.8×10^{-4} eV.

8. Estimate (a) the quantum number l for the orbital motion of the earth around the sun and (b) the number of quantized orientations of the plane of the earth's orbit, according to the rules of space quantization. (c) Also find θ_{min}, the half-angle of the smallest cone that can be swept out by a perpendicular to the earth's orbit as the earth revolves around the sun. Discuss from the point of view of the correspondence principle. The earth's mass is $\sim 6 \times 10^{24}$ kg and the mean earth–sun distance is $\sim 1.5 \times 10^{11}$ m.

SECTION 51-5

9. What is the acceleration of the silver atom as it passes through the deflecting magnet in the Stern-Gerlach experiment of Example 4?
Answer: 7.3×10^4 m/s^2.

10. Suppose a hydrogen atom (in its ground state) moves 80 cm in a direction perpendicular to a magnetic field that has a gradient, in the vertical direction, of 1.6×10^{-2} T/m. (a) What is the force on the atom due to the magnetic moment of the electron, which we take to be 1 Bohr magneton? (b) What is its vertical displacement if its speed is 1.2×10^5 m/s?

11. Assume that in the Stern-Gerlach experiment described for neutral silver atoms the magnetic field **B** has a magnitude of 0.500 T. (a) What is the energy difference between the orientations of the silver atoms in the two subbeams? (b) What is the frequency of the radiation that would induce a transition between these two states? (c) What is its wavelength, and to what part of the electromagnetic spectrum does it belong? The magnetic moment of a neutral silver atom is 1 Bohr magneton.
Answer: (a) 5.8×10^{-5} eV. (b) 1.4×10^{10} Hz. (c) 2.1 cm; short radio wave region.

SECTION 51-6

12. What is the wavelength of a photon that will induce a transition of an electron spin from parallel to antiparallel orientation in a magnetic field of magnitude 0.20 T? Assume that $l = 0$.

13. The proton as well as the electron has spin $\frac{1}{2}$. In the hydrogen atom in its ground state, with $n = 1$ and $l = 0$, there are two energy levels, depending on whether the electron and the proton spins are in the same direction or in opposite directions. The state with the spins in the opposite direction has the higher energy. If an atom is in this state and one of the spins "flips over," the small energy difference is released as a photon of wavelength 21 cm. This spontaneous spin-flip process is very slow, the mean life for the process being about 10^7 y. However, radio astronomers observe this 21-cm radiation in deep space, where the density of hydrogen is so small that an atom can flip before being disturbed by collisions with other atoms. What is the effective magnetic field (due to the magnetic dipole moment of the proton) experienced by the electron in the emission of this 21-cm radiation?   *Answer:* 5.1×10^{-2} T.

14. Write down the quantum numbers for all the hydrogen atom states belonging to the subshell for which $n = 4$ and $l = 3$.

15. How many hydrogen atom states are there with $n = 5$? How are they distributed among the subshells?
Answer: The five subshells have 1, 3, 5, 7, 9 electrons, respectively, for a total of 25. Consideration of spin would double these numbers.

16. A hydrogen atom state is known to have the quantum number $l = 3$. What are the possible n, m_l, and m_s quantum numbers?

17. A hydrogen atom state has a maximum m_l value of $+4$. What can you say about the rest of its quantum numbers? *Answer: $l = 4$, $n \geq 5$.*

18. Label as true or false these statements involving the quantum numbers n, l, m_l. (a) One of these subshells cannot exist: $n = 2$, $l = 1$; $n = 4$, $l = 3$; $n = 3$, $l = 2$; $n = 1$, $l = 1$. (b) The number of values of m_l that are allowed depends only on l and not on n. (c) The $n = 4$ shell contains four subshells. (d) The smallest value of n that can go with a given l is $l + 1$. (e) All states with $l = 0$ also have $m_l = 0$, regardless of the value of n. (f) Every shell contains n subshells.

19. In connection with Table 51-2, show that the number of states in any shell is given by $2n^2$.

20. What are the quantum numbers n, l, m_l, m_s for the two electrons of the helium atom in its ground state?

SECTION 51-9

21. For a hydrogen atom in its ground state, what are the values, at $r = a_0$, of (a) the wave function $\psi(r)$, (b) the probability density $\psi^2(r)$, and (c) the radial probability density $P(r)$?
Answer: (a) 5.40×10^{14} m$^{-3/2}$. (b) 2.91×10^{29} m^{-3}. (c) 1.02×10^{10} m^{-1}.

22. A small sphere of radius $0.1a_0$ is located a distance a_0 from the nucleus of a hydrogen atom in its ground state. What is the probability that the electron will be found inside this sphere?

23. For a hydrogen atom in its ground state, what is the probability of finding the electron between two spheres of radii $r = 1.00a_0$ and $r = 1.01a_0$?
Answer: 5.4×10^{-3}.

SECTION 51-10

24. For a hydrogen atom in a state with $n = 2$ and $l = 0$, what are the values, at $r = 5a_0$, of (a) the wave function $\psi(r)$, (b) the probability density $\psi^2(r)$, and (c) the radial probability density $P(r)$?

25. For a hydrogen atom in a state with $n = 2$ and $l = 0$, what is the probability of finding the electron between two spheres of radii $r = 5.00a_0$ and $r = 5.01a_0$?
Answer: 1.9×10^{-3}.

26. Locate the two maxima for the radial probability density curve of Fig. 51-13b.

27. For a hydrogen atom in a state with $n = 2$ and $l = 0$, what is the probability of finding the electron somewhere within the smaller of the two maxima of its radial probability density function? See Fig. 51-13b.
Answer: 2.25×10^{-2}.

28. Using Eq. 51-22, show that, for the hydrogen atom state with $n = 2$ and $l = 1$,

$$\int_0^\infty P_r(r)\, dr = 1.$$

What is the physical interpretation of this result?

SECTION 51-11

29. The wavelengths of the lines of the sodium doublet (see Fig. 51-15) are 588.995 nm and 589.592 nm. (a) What is the difference in energy between the two upper levels in that figure? (b) This energy difference comes about because the electron's spin magnetic dipole moment can be oriented either parallel or antiparallel to the internal magnetic field associated with the electron's orbital motion. Use the result you have just calculated to find the strength of this internal magnetic field. The electron's spin magnetic dipole moment has a magnitude of 1 Bohr magneton.
Answer: (a) 2.13×10^{-3} eV. (b) 18 T.

30. Potassium ($Z = 19$), like sodium ($Z = 11$), is an alkali metal, its single valence electron moving around a filled 18-electron argon-like core. As in sodium, there is a potassium doublet, its wavelengths being 764.5 nm and 769.9 nm. The quantum numbers of the levels that give rise to these lines are just the same as for sodium (see Fig. 51-15) except that $n = 4$. Calculate (a) the energy splitting between the two upper states and (b) the approximate energy difference between the two upper states (either of them) and the ground state.

31. Figure 51-18 describes the Zeeman splitting of the left-hand member of the sodium doublet shown in Fig. 51-15. (a) Draw a figure, along the lines of Fig. 51-18, showing the Zeeman splitting of the right-hand member of this doublet. (b) Solve the problem posed in Example 7 for this member of the doublet.   Answer: (b) 9.65×10^{-5} eV; 2.89×10^{-4} eV.

32. In Fig. 51-18b, suppose that the energy difference between adjacent levels was the same for both the upper and the lower set of levels. How many Zeeman components would there then be for this spectrum line of Fig. 51-18a when the sodium light source is placed in a strong magnetic field?

33. What is the ratio of the level splitting for the upper group of levels in Fig. 51-18b to that for the lower group?   Answer: 0.67

34. A source emitting the potassium doublet (see Problem 30) is placed in a magnetic field with $B = 1.2$ T. (a) Into how many levels is each of the three states involved split? (b) For each state, what is the energy difference between adjacent levels when the magnetic field is imposed?

35. The g-factor of Eq. 51-24 is related to the quantum numbers by

$$g = 1 + \frac{j(j + 1) + s(s + 1) - l(l + 1)}{2j(j + 1)}$$

in which s, the spin quantum number, always has the value $\frac{1}{2}$. (a) Show that if $l = 0$, then $g = 2$, as Eq. 51-15 requires. (b) Show that if $s = 0$, then $g = 1$, as Eq. 51-10 requires. (Note: Although s cannot actually be zero for a one-electron atom, we can at least contemplate a spinless electron.) (c) Show that $g = \frac{4}{3}$ for the upper level and $g = 2$ for the lower level in Fig. 51-18a.

36. A hydrogen atom is in a state with $n = 3$ and $l = 2$. (a) List the possible values of the total angular momentum quantum number j and (b) the g-factor associated with each of these values. Use the g-factor formula displayed in Problem 35. (c) Into how many components will each of these levels split if the atom is placed in a strong magnetic field?

52
atomic physics: three selected topics

In the preceding three chapters we have developed the foundations of wave mechanics and, using it, we have been able to understand something about how atoms are put together. In this chapter and the next we want to suggest that wave mechanics and the understanding of atomic structure that we have gained from it are broadly useful, not only in physics, but also in chemistry, engineering, and technology.

We do that in this chapter by taking a close look at three selected topics, (1) X rays, (2) the building up of the periodic table, and (3) the laser. In the following chapter we will focus our attention on selected topics in solid-state physics, with special emphasis on practical devices such as the transistor and the light-emitting diode.

So far we have dealt with the behavior of single electrons in atoms, either the lone electron of hydrogen or the single valence electron of sodium. We now shift our attention to the behavior of electrons deep within the atom. We move from a region of relatively low binding energy (~ 5 eV for the work required to remove the valence electron from sodium, for example) to a region of higher energy (~ 70 keV for the work required to remove an innermost electron from tungsten, for example). The radiations we deal with, though of course still part of the electromagnetic spectrum, differ drastically in wavelength, from $\sim 6 \times 10^{-7}$ m for the sodium doublet lines, say, to $\sim 2 \times 10^{-11}$ m for one of the tungsten characteristic radiations, say, a ratio of $\sim 30{,}000$. We are now speaking of X rays.

The usefulness of these penetrating radiations in medical and dental diagnostics and in therapy is well known, as are their many industrial

52-1
WAVE MECHANICS—THE PAYOFF

52-2
X RAYS AND THE NUMBERING OF THE ELEMENTS

applications, such as examining welded joints in pipe lines. In Section 47-5 we described how X rays can be used to deduce the atomic structures of crystalline materials. The structures of such complex substances as insulin and DNA have been worked out by these methods. In astronomy, X-ray satellites have shown us an entirely new view of our universe, that provided by the X rays emitted by various objects in it.

Our concern in this section, however, is what X rays can teach us about the structure of the atoms that emit or absorb them. We focus on the work of the British physicist H. G. J. Moseley, who, by X-ray studies, developed the concept of atomic number and gave physical meaning in terms of atomic structure to the ordering of the elements in the periodic table.

We saw in Section 47-5 that X rays are produced when energetic electrons strike a solid target and are brought to rest in it. Figure 52-1 shows the wavelength spectrum of the X rays produced when 35-keV electrons strike a molybdenum target.

The Continuous X-ray Spectrum. We first examine the continuous background spectrum of Fig. 52-1, ignoring — for the time being — the two prominent peaks that rise from it. Consider an electron of kinetic energy K that happens to pass close to the nucleus of one of the molybdenum atoms in the target, as in Fig. 52-2. The electron may well lose a certain part ΔK of its kinetic energy, which will appear as the energy $h\nu$ of an X-ray photon that radiates away from the site of the encounter. Such *bremsstrahlung* processes (German: "braking radiation") account for the continuous X-ray spectrum.

If the target on which the incident electrons fall is sufficiently thick, electrons in the kinetic energy range $0 < K < eV$ will be present, V being the accelerating potential and eV the initial kinetic energy. All electrons in this energy range can undergo bremsstrahlung processes, thus contributing X-ray photons to the continuous X-ray spectrum.

A prominent feature of the continuous spectrum of Fig. 52-1 is the sharply defined cutoff wavelength λ_{min} below which the continuous spectrum does not exist. This minimum wavelength corresponds to a decelerating event in which one of the incident electrons (with initial kinetic energy eV) loses *all* of this energy in a single encounter, radiating it away as a single photon. Thus,

$$eV = h\nu_{max} = \frac{hc}{\lambda_{min}},$$

or

$$\lambda_{min} = \frac{hc}{eV}. \tag{52-1}$$

Equation 52-1 shows that if $h \to 0$, then $\lambda_{min} \to 0$, which is the prediction of classical theory. The existence of a minimum wavelength is a quantum phenomenon.

Note that as you change the target material, perhaps from molybdenum to copper, the general shape and intensity of the continuous spectrum will change but the cutoff wavelength will not change. This wavelength depends only on the kinetic energy of the electrons that bombard the target and not at all on the target material.

The Characteristic X-ray Spectrum. We now turn our attention to the two peaks of Fig. 52-1, labeled K_α and K_β. These peaks are characteristic of the target material and, together with other peaks that appear at

figure 52-1
The wavelength spectrum of the X rays produced when 35-keV electrons strike a molybdenum target. (1 pm = 10^{-12} m.)

figure 52-2
An electron passing near the nucleus of a target atom is accelerated and radiates a photon, losing part of its kinetic energy in the process.

longer wavelengths, form the *characteristic X-ray spectrum* of the element in question.

Here is how these X-ray photons arise: (1) An energetic incoming electron strikes an atom in the target and knocks out one of its deeply-lying electrons. If the electron is in the shell with $n = 1$ (called, for historical reasons, the K shell) there remains a vacancy, or a "hole" as we shall call it, in this shell. (2) One of the outer electrons moves in to fill this hole and, in the process, the atom emits a characteristic X-ray photon. If the electron falls from the shell with $n = 2$ (called the L shell), we have the K_α line of Fig. 52-1; if it falls from the next outermost shell (called the M shell) we have the K_β line, and so on. Of course, such a transition will leave a hole in either the L or the M shell, but this will be filled in by an electron from still further out in the atom, the atom, in the process, emitting still another characteristic X-ray spectrum line.

Figure 52-3 shows an X-ray atomic energy-level diagram for molybdenum, the element to which Fig. 52-1 refers. The base line ($E = 0$) represents the energy of a neutral molybdenum atom in its ground state. The level marked K ($E = 20$ keV) represents the energy of a molybdenum atom with a hole in its K shell. Similarly, the level marked L ($E = 2.7$ keV) represents the energy of an atom with a hole in its L shell, and so on. Note that in representing the energy levels for the hydrogen atom (see Fig. 49-19), we chose a different base line. Rather than the atom in its ground state, we there selected the atom with its electron removed to infinity as our $E = 0$ configuration. Actually, the atomic configuration for which we choose to put $E = 0$ does not matter. Only differences in energy are physically significant, and these are the same no matter what our choice of an $E = 0$ base line is.

The transitions K_α and K_β in Fig. 52-3 show the origin of the two peaks in Fig. 52-1. The K_α line, for example, originates when an electron from the L shell of molybdenum — moving *upward* on the energy-level diagram — fills the hole in the K shell. This is the same as saying that a hole — moving *downward* on the diagram — moves from the K shell to the L shell. It is easier to keep track of a single hole than of the 41 electrons in ionized molybdenum that are potentially available to fill it. We have drawn the arrows in Fig. 52-3 from the point of view of hole transitions.

In his investigation of the atomic number concept, Moseley generated characteristic X rays by using as many elements as he could find — he found 38 — as targets for electron bombardment in a special evacuated X-ray tube of his own design. He measured the wavelengths of a number of the lines of the characteristic X-ray spectrum by the crystal diffraction method described in Section 47-5. He then sought, and readily found, regularities in the spectra as he moved from element to element in the periodic table. In particular, he noted that if, for a given spectrum line such as K_α, he plotted the square root of the line frequency ($= \sqrt{v} = \sqrt{c/\lambda}$) against the position of the associated element in the periodic table, a straight line resulted. Figure 52-4 shows a portion of his data. We shall see below why it is logical to plot the data in this way and why a straight line is to be expected. Moseley's conclusion from the full body of his data was

We have here a proof that there is in the atom a fundamental quantity, which increases by regular steps as we pass from one element to the next. This quantity can only be the charge on the central atomic nucleus.

Moseley's achievement can be appreciated all the more when we realize the status of understanding of atomic structure at that time

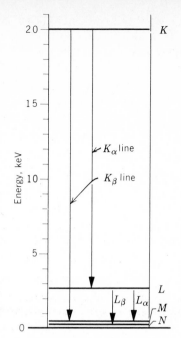

figure 52-3
An atomic energy level diagram for molybdenum, showing the transitions that give rise to the characteristic X rays of that element. (All levels, except the K level, contain a number of closely-lying components, not shown in the figure.)

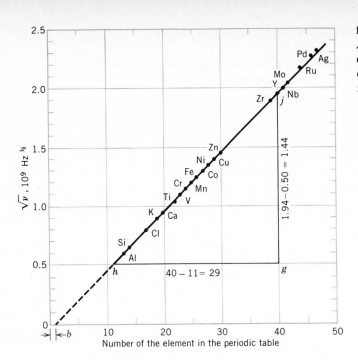

figure 52-4
A Moseley plot of the K_α line of the characteristic X-ray spectra of 21 elements. The frequency ν is calculated from the measured wavelength.

(1913). The nuclear model of the atom had been proposed by Rutherford only two years earlier. Little was known about the magnitude of the nuclear charge or the arrangement of the atomic electrons; Bohr published his first paper on atomic structure only in that same year. An element's place in the periodic table was at that time assigned by atomic *weight*, although there were several cases in which it was necessary to invert this order to fit the demands of the chemical evidence. The table had several empty squares, and a surprisingly large number of claims for the discovery of new elements had been advanced; the rare earth elements, because of the problems caused by their similar chemical properties, had not yet been properly sorted out.

Due to Moseley's work, the characteristic X-ray spectrum became the universally accepted signature of an element. Through such studies it became possible to string the elements in a line, so to speak, and to assign consecutive numbers to them, all without the slightest need to know anything about their chemical properties.

It is not hard to see why the characteristic X-ray spectrum shows such impressive regularities from element to element and the optical spectrum does not. The key to the identity of an element is the charge on its nucleus. This determines the number of its atomic electrons and thus its chemical and physical properties. Gold, for example, is what it is because its atoms have a nuclear charge of $+79e$. If it had just one more unit of charge, it would not be gold but mercury; if it had one fewer, it would be platinum. The K electrons, which play such a large role in the production of the characteristic X-ray spectrum, lie very close to the nucleus and are sensitive probes of its charge. The optical spectrum, on the other hand, is associated with transitions of the outermost or valence electrons. Such outermost electrons are heavily screened from the nucleus by the remaining $Z - 1$ electrons of the atom; they are not sensitive probes of the nuclear charge.

Bohr Theory and the Moseley Plot. It is possible to account for the Moseley plot of Fig. 52-4 in terms of Bohr's theory of atomic structure. It may seem surprising that this theory, which works so well for hydrogen but fails even for helium, the next element in sequence, should provide an excellent first approximation when applied to the inner electrons of heavy atoms. This proves to be the case, however, although its success seems to be, in part at least, a matter of the fortuitous cancellation of errors.

Consider an electron in the L shell of an atom that is about to move into a hole in the K shell, emitting a K_α X-ray photon in the process. This electron "sees" the charge Ze of the atomic nucleus, but it sees it partially screened by the charge $-e$ of the single remaining K electron. In part because of this screening and in part because of readjustments that take place in the electron cloud as a whole, the effective atomic number for the transition turns out to be $Z - b$, where $b \approx 1$.

Bohr's formula for the frequency of the radiation corresponding to a transition between any two atomic levels in hydrogen-like atoms is

$$v = \frac{m_e e^4 Z^2}{8\epsilon_0^2 h^3}\left(\frac{1}{m^2} - \frac{1}{n^2}\right). \qquad [49\text{-}23,29]$$

For the K_α transition of Fig. 52-3, it is appropriate to replace Z by $Z - b$ and to substitute 1 for m and 2 for n. Doing so yields

$$v = \frac{m_e e^4}{8\epsilon_0^2 h^3}\left(\frac{1}{1^2} - \frac{1}{2^2}\right)(Z-b)^2.$$

Taking the square root of each side leads to

$$\sqrt{v} = \left(\frac{3m_e e^4}{32\epsilon_0^2 h^3}\right)^{1/2}(Z-b), \qquad (52\text{-}2)$$

which we can write in the form

$$\sqrt{v} = aZ - ab, \qquad (52\text{-}3)$$

where a is the indicated constant and $b \approx 1$.

Equation 52-3 represents a straight line, in full agreement with the experimental data of Fig. 52-4. If this plot is extended to higher atomic numbers, however, it turns out to be not quite straight but somewhat concave upwards. Nevertheless, the quantitative agreement with Bohr theory is surprisingly good, as Example 2 shows.

EXAMPLE 1

The cutoff wavelength. Calculate the wavelength λ_{min} for the continuous spectrum of X rays emitted when 35-keV electrons fall on a molybdenum target, as in Fig. 52-1.

From Eq. 52-1, we have

$$\lambda_{min} = \frac{hc}{eV} = \frac{(4.14 \times 10^{-15} \text{ eV} \cdot \text{s})(3.00 \times 10^8 \text{ m/s})}{(35.0 \times 10^3 \text{ eV})}$$

$$= 3.54 \times 10^{-11} \text{ m} = 35.4 \text{ pm}.$$

This is in agreement with the experimental result shown by the vertical arrow in Fig. 52-1. Note that Eq. 52-1 contains no reference to the target material. For a given accelerating potential all targets, no matter what they are made of, exhibit the same cutoff wavelength.

EXAMPLE 2

The Moseley plot. Calculate the quantity a in Eq. 52-3 and compare it with the measured slope of the straight line in Fig. 52-4.

Comparing Eqs. 52-2 and 52-3 allows us to write

$$a = \left(\frac{3m_e e^4}{32\epsilon_0^2 h^3}\right)^{1/2}$$

$$= \frac{\sqrt{3}\,(9.11 \times 10^{-31}\,\text{kg})^{1/2}(1.60 \times 10^{-19}\,\text{C})^2}{4\sqrt{2}\,(8.85 \times 10^{-12}\,\text{F/m})(6.63 \times 10^{-34}\,\text{J}\cdot\text{s})^{3/2}}$$

$$= 4.95 \times 10^7\,\text{Hz}^{1/2}.$$

Careful measurement of Fig. 52-4, using the triangle hgj, yields

$$a = \frac{gj}{hg} = \frac{(1.94 - 0.50) \times 10^9\,\text{Hz}^{1/2}}{(40 - 11)} = 4.96 \times 10^7\,\text{Hz}^{1/2},$$

which is in agreement with the value predicted by Bohr theory within the uncertainty of the graphical measurement. Note also that the intercept in Fig. 52-4 is in fact ≈ 1, as expected from our shielding argument above.

The agreement with Bohr theory is not nearly as good for other lines in the X-ray spectrum, corresponding to the transitions of electrons farther from the nucleus; here one must rely on calculations based on wave mechanics.

EXAMPLE 3

Tracking down an unknown. A cobalt target is bombarded with electrons, and the wavelengths of its characteristic spectrum are measured. A second, fainter, characteristic spectrum is also found, because of an impurity in the target. The wavelengths of the K lines are 178.9 pm (cobalt) and 143.5 pm (impurity). What is the impurity?

Let us apply Eq. 52-3 both to cobalt and to the impurity. Putting c/λ for ν (and assuming $b = 1$), we obtain

$$\sqrt{\frac{c}{\lambda_{\text{co}}}} = aZ_{\text{co}} - a \quad \text{and} \quad \sqrt{\frac{c}{\lambda_{\text{x}}}} = aZ_{\text{x}} - a.$$

Dividing yields

$$\sqrt{\frac{\lambda_{\text{co}}}{\lambda_{\text{x}}}} = \frac{Z_{\text{x}} - 1}{Z_{\text{co}} - 1}.$$

Substituting gives us

$$\sqrt{\frac{178.9\,\text{pm}}{143.5\,\text{pm}}} = \frac{Z_{\text{x}} - 1}{27 - 1}.$$

Solving for the unknown, we find $Z_{\text{x}} = 30.0$; a glance at the periodic table identifies the impurity as zinc.

52-3
PUTTING ATOMS TOGETHER

In the preceding section we saw how, by measuring the wavelengths of the characteristic X-ray spectrum of an element, we could assign an *atomic number Z* to each element and thus string them in a line according to a logical principle.

Here we go a step further. We try to see how far the principles of wave mechanics can take us in breaking up this line into a series of segments, corresponding to the horizontal periods of the periodic table of the elements.

The attempt meets with essentially total success. Every detail of the table (see Appendix E) can be accounted for, including (1) the numbers of elements in the seven horizontal periods into which the table is divided, (2) the similarity of the chemical properties of the elements in the various vertical columns — the alkali metals and the noble gases, for example — and (3) the existence of the rare earth, or lanthanide, series of elements, all crammed into one square of the table. In short, wave mechanics, supplemented by certain guiding principles that we discuss below, accounts for every feature of this table and thus, essentially, for all of chemistry.

Let us imagine that — in Tinker Toy fashion — we are going to construct a typical atom for each of the 100+ elements that make up the periodic table. Our starting materials will be a supply of nuclei, each characterized by a charge $+Ze$ and by an appropriate mass, with Z ranging by integers from 1 to 100+. We also need an ample supply of electrons. Our plan is to add Z electrons to each nucleus in such a way as to produce a neutral atom in its ground state.

Success follows only if we observe these three principles of atom building:

1. *The quantum number principle.* The electron in a hydrogen atom may — to mention one possibility — be in a state described by the quantum numbers $n = 2$, $l = 1$, $m_l = +1$, and $m_s = -\frac{1}{2}$. It turns out that a particular electron in an atom of, say, zirconium, may also be fully indentified by this same set of quantum numbers. That is not to say that the electrons in these two cases will move in the same way, because they will not. Put another way, although these two electrons may share the same set of quantum numbers, the potentials in which they move — and thus their wave functions — will be quite different. Specifically, the quantum number principle asserts: *The hydrogen atom quantum numbers can be used to describe electron states, and to assign electrons to shells and subshells, in any atom, no matter how many electrons it contains. Furthermore, the restrictions among the quantum numbers discussed in Section 51-8 remain in force.*

2. *The Pauli exclusion principle.* This powerful principle was put forward by the Austrian-born physicist Wolfgang Pauli in 1925. Speaking generally, it tells us that no two electrons can be in the same state of motion at the same time. More specifically, it asserts that: *In a multielectron atom there can never be more than one electron in any given quantum state.*

    If this principle did not hold, all of the electrons in an atom would pile up in its K shell, and chemistry as we know it would not exist. You would not be here to read this sentence, and we would not have been here to have written it. Pauli's exclusion principle is no trivial assertion!

3. *The minimum energy principle.* As we fill subshells with electrons in the course of atom building, the question arises: In what order shall we fill them? The answer is: When one subshell is filled, put the next electron in whichever vacant subshell will lead to an atom lowest in energy. To do otherwise would be to depart from our stated aim of building atoms in their ground states.

    The lowest-energy subshell can be identified with the help of the following rule, which we first state and then try to make reasonable: *For a given principal quantum number n in a multielectron atom, the order of increasing energy of the subshells is the order of increasing l.*

Table 52-1
Energy levels for electrons with $n = 4$, in three different atoms

Orbital Quantum Number l	Energy, eV		
	Hydrogen* $Z = 1$	"Lead"** $Z = 82$	Lead*** $Z = 82$
0	−0.85	−5720	−890
1	−0.85	−5720	−710
2	−0.85	−5720	−420
3	−0.85	−5720	−140

* A neutral hydrogen atom; see Eq. 49-30.
** A hypothetical one-electron atom with $Z = 82$; see Eq. 49-30.
*** An actual neutral lead atom ($Z = 82$); data from experiment.

Table 52-1 helps to clarify this rule. Consider first a hydrogen atom whose single electron is in a state with $n = 4$. There are four allowed values of l, namely, 0, 1, 2, and 3. For electrons in true one-electron atoms — such as hydrogen — the energy does not depend on l at all but only on n, being given by

$$E = -\frac{m_e e^4}{8\epsilon_0 h^2}\frac{Z^2}{n^2}, \qquad n = 1, 2, 3, \ldots \qquad [49\text{-}30]$$

Recall that this relation is predicted not only by Bohr theory but also by wave mechanics. Putting $Z = 1$ and $n = 4$ in this relation yields, for hydrogen, $E = -0.85$ eV, as Table 52-1 shows.

Consider now a lead nucleus ($Z = 82$) around which only a single electron circulates, again in a state with $n = 4$. Equation 49-30 also applies to this (admittedly rather unlikely) one-electron atom. For $Z = 82$ and $n = 4$, the table shows that we have $E = -5720$ eV, once more independent of l. It is not surprising that the single electron in this "atom" lies much lower in energy (that is, has a much higher binding energy) than the electron in hydrogen. It moves in the field of a nucleus with a charge of $+82e$; furthermore, it is drawn in very close to this nucleus, the equivalent Bohr orbit radius (see Eq. 49-32) being 82 times smaller than for hydrogen.

Finally, let us construct a normal, neutral lead atom by "sprinkling in" the missing 81 electrons. The outermost or valence electrons in lead have $n = 6$, so that an electron with $n = 4$ would lie somewhere in the middle of the smeared-out electron cloud surrounding the lead nucleus.

Equation 49-30 no longer holds for this multielectron atom, but we can find the energies of the four $n = 4$ subshells experimentally from X-ray studies. Their approximate values are shown in the last column of Table 52-1. We see at once that they lie higher in energy (that is, the binding energies are smaller) than for our hypothetical one-electron lead "atom" and that they vary with l just as the minimum energy rule predicts.

That the electrons in lead become more loosely bound when the entire electron cloud is present follows because some substantial fraction of this cloud screens the nucleus electrically, following the principles of Gauss's law (see Example 2, Chapter 28). A typical $n = 4$ electron no longer "sees" the full positive nuclear charge, but rather sees this charge reduced by the negative charge of that part of the electron cloud that lies between the nucleus and the effective radius of the electron in question.

As for the variation of energy with l, let us ask ourselves what an $l = 0$ orbit would have to look like under the mechanical constraints of the Bohr picture. Truly to have no angular momentum, the electron would have to oscillate back and forth on a straight line segment passing directly through the nucleus. This does not happen, of course. The equivalent wave mechanical statement is that an electron with $l = 0$ must spend a larger fraction of its time near the nucleus than do electrons with higher values of l. Such electrons would then, on the average, "see" a higher effective nuclear charge and would be more tightly bound; they would lie lower in energy, just as the minimum energy principle and Table 52-1 predict. It is interesting to reexamine Fig. 51-13, which shows the $n = 2$, $l = 0$ state of hydrogen. In this state there is indeed a marked tendency for the electron to cluster near the nucleus — note the close-in secondary maximum — just as our qualitative argument suggests that it should.

52-4 PUTTING THE PERIODIC TABLE TOGETHER

Figure 52-5 shows how the periodic table is put together, using the three rules for atom building that we have described in the previous section. Energy increases upward in this figure. States with the same value of l have been displaced to the left for clarity and grouped into columns according to their l value.

Before we look more closely at this table, we need to introduce a notation (which, so far, we have not used) for the angular momentum quantum number l. For historical reasons the values of l have been given letter equivalents, according to this scheme:

l	1	2	3	4	5	6. . .
Symbol	s	p	d	f	g	h. . .

In this notation, which is universally used, a state with $n = 1$ and $l = 0$ is called a "1s" state. Similarly, a state with $n = 4$ and $l = 3$ is called a "4d" state, and so on.

The dependence of energy on l is a dominant feature of Fig. 52-5. Look, for example, at the sequence of states 4s, 4p, 4d, and 4f. They lie in the figure in the order of increasing energy, just as the minimum energy rule requires. In fact, the 4f states lie so high that they are above the 5s and the 5p states in the next shell.

By a *shell* we mean a group of states, lying close together in energy, that has a particular stability when those states are fully occupied. When dealing with the hydrogen atom (for which the energy depends only on the principal quantum number n), we identified the shells by giving the value of that quantum number. We now see that, in many-electron atoms, the principal quantum number alone is no longer a good indicator. As Fig. 52-5 shows, the shell we have labeled "6," corresponding to the sixth horizontal period of the periodic table, does indeed contain all of the 6s and the 6p states, corresponding to $n = 6$. However, it also contains all of the 4f and 5d states. Moreover, the 6d states do not lie in this shell at all but in the shell above.

By starting with hydrogen in Fig. 52-5 and following the dashed line, we can see how the seven horizontal periods of the periodic table are built up, each starting with an alkali metal and ending with a noble gas. Consider again the long sixth period, which starts with the alkali metal cesium $(Z = 55)$ and ends with the noble gas radon $(Z = 86)$. The order in which the subshells are filled, as the dashed line indicates, is 6s, 4f, 5d, and 3p.

figure 52-5
If you trace the dashed line, starting at the bottom with hydrogen, you will pass in sequence through the seven horizontal periods of the periodic table. Each period starts with an alkali metal and ends with a noble gas.

This sixth period contains a run of 15 elements, listed separately at the bottom of the periodic table of the elements in Appendix E and called the *lanthanide elements* or the *rare earths*. Their chemical properties are so similar that all are grouped into a single square of the table. This similarity arises because, while the 4f subshell is being filled deep within the electron cloud, an outer screen of one or two *6s* valence electrons remains in place. It is these outermost electrons that determine the chemical properties of an atom.

Electron configurations. (a) From Fig. 52-5, write down the subshells to which the electrons in the xenon atom belong. This element ($Z = 54$) is a noble gas.

Starting from hydrogen ($Z = 1$) and following the dashed line in Fig. 52-5 to xenon allows us to write:

EXAMPLE 4

$$1s^2 2s^2 2p^6 3s^2 3p^6 4s^2 3d^{10} 4p^6 5s^2 4d^{10} 5p^6.$$

Such a listing is called the *electron configuration* of the element in question. The superscripts indicate the number of electrons in the corresponding subshell. Note that the superscripts add up to 54, the atomic number of xenon.

An *s* state ($l = 0$) can accommodate two electrons, a *p* state ($l = 1$) six, and a *d* state ($l = 2$) ten; see Table 51-2. Thus all the subshells in xenon are filled, as Fig. 52-5 verifies. The ionization potential of xenon, that is, the energy needed to pull an electron from its filled 5*p* subshell (= 12.2 eV), is relatively large.

(b) What is the electron configuration for cesium ($Z = 55$), the next element in the periodic table above xenon?

Proceeding as above allows us to write

$$[Xe]6s^1$$

in which [Xe] is simply an abbreviation for the xenon electron configuration, given above. Thus cesium, an alkali metal, has a single *6s* electron outside a xenon-like core. Because of this single, loosely bound valence electron, cesium is chemically active and readily combines with many other elements. Its ionization potential (= 3.89 eV) is relatively low. However, the energy required to pull a *second* electron from cesium (= 25.1 eV) is quite large; this electron would have to be torn out of the tightly bound xenon core.

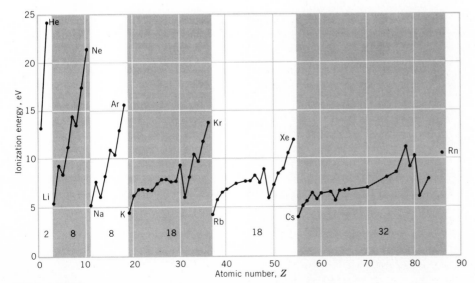

figure 52-6
The ionization energy of the elements as a function of their atomic number. The numbers in the alternate shaded strips are the numbers of elements in the first six horizontal periods of the periodic table.

Figure 52-6 shows the ionization potentials of the elements. The systematic variation through the horizontal periods of the periodic table is clear. Note the values for xenon and for cesium.

Note: Writing down electron configurations from Fig. 52-5, as we have done above, gives correct answers for all but a handful of elements. The energy levels in that figure are not fixed but have values that depend on how many electrons have previously been added to the atom. In a few cases these small shifts reverse the energy sequence of two adjacent subshells.

52-5 LASERS AND LASER LIGHT

In the late 1940s and again in the early 1960s quantum physics made two enormous contributions to technology, the transistor and the laser. The first stimulated the growth of *microelectronics*, which deals with the interaction (at the quantum level) between electrons and bulk matter. The laser is leading the way in a new field — sometimes called *photonics* — which deals with the interaction (again at the quantum level) between photons and bulk matter.

To see the importance of lasers, let us look at some of the characteristics of laser light (see Fig. 52-7). We shall compare it as we go along with the light emitted by such sources as a tungsten filament lamp (continuous spectrum) or a neon gas discharge tube (line spectrum). We shall see that referring to laser light as "the light fantastic" goes far beyond whimsy.

1. *Laser light is highly monochromatic.* Tungsten light, spread over a continuous spectrum, gives us no basis for comparison. The light from selected lines in a gas discharge tube, however, can have wavelengths in the visible region that are precise to about 1 part in 10^6. The sharpness of definition of laser light can easily be a thousand times greater, or 1 part in 10^9.

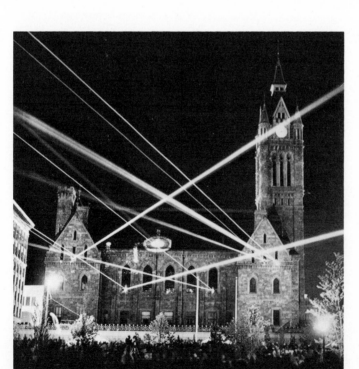

figure 52-7
Laser beams in the sky at Holyoke, Massachusetts. (Dave Roback/Wide World Photos.)

2. *Laser light is highly coherent.* Wavetrains for laser light (see Fig. 50-11a) may be several hundred kilometers long. Interference fringes can be set up by combining two beams that have followed separate paths whose lengths differ by as much as this amount. The corresponding coherence length for light from a tungsten filament lamp or a gas discharge tube is typically considerably less than 1 m.

3. *Laser light is highly directional.* A laser beam departs from strict parallelism only because of diffraction effects, determined (see Section 46-5) by the wavelength and the diameter of the exit aperture. Light from other sources can be made into an approximately parallel beam by a lens or a mirror, but the beam divergence is much greater than for laser light. Each point on, say, a tungsten filament source forms its own separate beam, the angular divergence of the overall composite beam being determined — not by diffraction — but by the spatial extent of the filament.

4. *Laser light can be sharply focused.* This property is related to the parallelism of the laser beam. As for star light, the size of the focused spot for a laser beam is limited only by diffraction effects and not by the size of the source. Flux densities for focused laser light of $\sim 10^{15}$ W/cm^2 are readily achieved. An oxyacetylene flame, by contrast, has a flux density of only $\sim 10^3$ W/cm^2.

The smallest lasers, used for telephone communication over optical fibers, have as their active medium a semiconducting gallium arsenide crystal about the size of a pinhead. The largest lasers, used for laser fusion research (see Fig. 52-8), fill a large building. They can generate pulses of laser light of $\sim 10^{-10}$ s duration which, during the pulse, have a power level of $\sim 10^{13}$ W. This is about ten times the total power-generating capacity of all the electric power stations on earth.

Other laser uses include spot-welding detached retinas, drilling tiny

figure 52-8
A view in the NOVA laser room at the Lawrence Livermore National Laboratory. The laser, used for research in thermonuclear fusion (see Section 55-10), generates power (in nanosecond pulses) at a peak level of $\sim 10^{14}$ W, about 200 times the present power generating capacity of the United States. (Roger Ressmeyer/ Wheeler Pictures.)

figure 52-9
A Michelson interferometer arrangement at the National Bureau of Standards, used to measure the x-, y-, z- coordinates of a point in space with extreme precision. Note the many laser beam segments. (National Bureau of Standards.)

holes in diamonds for drawing fine wires, cutting cloth (50 layers at a time, with no frayed edges) in the garment industry, precision surveying, precise length measurements by interferometry, precise fluid-flow velocity measurements using the Doppler effect, and the generation of holograms, which are stabilized two-dimensional interference "gratings" that make it possible to reconstruct the images of objects in true three-dimensional perspective.

Figure 52-9 shows another example of laser technology, namely, a facility at the National Bureau of Standards used to measure the x-, y-, z-coordinates of a point, by laser interference techniques, with a precision of $\pm 2 \times 10^{-8}$ m $(= \pm 20$ nm$)$. It is used for measuring the dimensions of special three-dimensional gauges which, in turn, are used in industry to check the dimensional accuracy of complicated machined parts. A number of laser beams, visible by scattered light, appear in the figure.

52-6
EINSTEIN AND THE LASER

In 1917 Einstein introduced into physics a new concept, that of *stimulated emission*, which we shall define and discuss below. Even though the world had to wait until 1960 to see its first operating laser, the groundwork for its invention was put in place just at that moment. The word "laser" is an acronym for **L**ight **A**mplification by the **S**timulated **E**mission of **R**adiation, and we are thus not surprised to learn that stimulated emission is the key to its operation.

What was Einstein working on when the concept of stimulated emission occurred to him? Nothing other than the cavity radiation problem which, in the hands of Planck and others, established the new science of quantum mechanics. In 1917 Einstein succeeded in deriving the Planck radiation law in terms of beautifully simple assumptions and in a way that made quite clear the role of energy quantization and the photon concept.*

It is interesting that Einstein was also thinking deeply about this same fundamental cavity radiation problem when, in 1905, he first proposed the concept of the photon and realized that the photoelectric effect could be explained with its use. We learn from both of these examples that practical devices of major importance can flow from a concern over problems that seem to have no relevance to technology. When you next see a photoelectric elevator door opener or listen to a laser-disc stereo system, think of Einstein.

Now let us take a look at three processes that involve the interaction between matter and radiation. Two of them, absorption and spontaneous emission, have long been familiar; the third is stimulated emission.

1. *Absorption.* Figure 52-10a suggests an atomic system in the lower of two possible states, of energies E_1 and E_2. A continuous spectrum of radiation is present. Let a photon from this radiation field approach the two-level atom and interact with it and let the associated frequency v of the photon be such that

$$hv = E_2 - E_1. \tag{52-4}$$

The result is that the photon vanishes and the atomic system moves to its upper energy state. We call this familiar process *absorption.*

2. *Spontaneous emission.* In Fig. 52-10b the atomic system is in its upper state and there is no radiation nearby. After a mean time τ, this (isolated) atomic system moves of its own accord to the state of lower energy, emitting a photon of energy hv in the process. We call this familiar process *spontaneous emission,* in that no outside influence triggered the emission.

   Normally the mean life τ for spontaneous emission by excited atoms is $\sim 10^{-8}$ s. However, there are some states for which τ is much longer, perhaps $\sim 10^{-3}$ s. We call such states *metastable;* they play an essential role in laser operation.

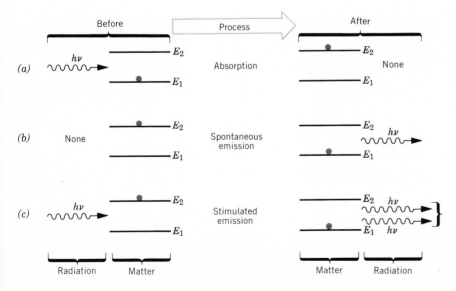

figure 52-10
The interaction of matter and radiation for the processes of (a) absorption, (b) spontaneous emission, and (c) stimulated emission.

The light from a glowing lamp filament is generated by spontaneous emission. Photons produced in this way are totally independent of each other. In particular, they have different directions and phases. Put another way, the light they produce has a low degree of coherence.

3. *Stimulated emission.* In Fig. 52-10c the atomic system is again in its upper state, but this time radiation of frequency given by Eq. 52-4 is present. As in absorption, a photon of energy $h\nu$ interacts with the system. The result is that the system is driven down to its lower state and there are now *two* photons where only one existed before.

The emitted photon in Fig. 52-10c is in every way identical with the "triggering" or "stimulating" photon. It has the same energy, direction, phase, and state of polarization. This is how laser light acquires its characteristics. We call the process of Fig. 52-10c *stimulated emission.* We can see how a chain reaction of similar processes could be triggered by one such event. This is the "*a*mplification" of the laser acronym.

Figure 52-10 refers to the interaction with radiation of a single atom. In the usual case, however, we find ourselves dealing with a large number of atoms. For the two-level system of Fig. 52-10, how many of these atoms will be in level E_1 and how many in level E_2? Ludwig Boltzmann has shown that, in thermal equilibrium, the number n_x of atoms in any level whose energy is E_x is given by

$$n_x = Ce^{-E_x/kT}, \qquad (52\text{-}5)$$

in which C is a constant. This seems reasonable.* The quantity kT is the mean energy of agitation of an atom at temperaure T and we see that, the higher the temperature, the more atoms — on long-term average — will be "bumped up" by thermal agitation to the level E_x.

If we apply Eq. 52-5 to the two levels of Fig. 52-10 and divide, the constant C cancels out and we find, for the ratio of the number of atoms in the upper level to the number in the lower level

$$n_2/n_1 = e^{-(E_2 - E_1)/kT}. \qquad (52\text{-}6)$$

Figure 52-11a illustrates this situation. Because $E_2 > E_1$ the ratio n_2/n_1 will always be less than unity, which means that there will always be fewer atoms in the higher energy level than in the lower. Again, this is what we would expect if the level populations are determined only by the action of thermal agitation.

If we expose a system like that of Fig. 52-11a to radiation, the dominant process — by sheer weight of numbers — will be absorption. However, if the level populations were inverted, as in Fig. 52-11b, the dominant process in the presence of radiation would be stimulated emission, and with it the generation of laser light. A *population inversion* like that of Fig. 52-11b is a nonequilibrium situation; we must use clever tricks to bring it about.

Figure 52-12 shows schematically how a population inversion can be achieved so that laser action — or "lasing" as it is called — can occur. Atoms from ground state E_1 are "pumped" up to an excited state E_3 by some mechanism or other. One possibility, called *optical pumping,* is

* See Supplementary Topic IV.

figure 52-11
(a) The normal thermal equilibrium distribution of atomic systems between two states. (b) An inverted population, obtained by special techniques.

52-7
HOW A LASER WORKS

the absorption of light energy from an intense, continuous-spectrum source that is placed so that it surrounds the lasing material.

From E_3 the atoms decay rapidly to a state of energy E_2. For lasing to occur this state must be metastable, that is, it must have a relatively long mean life against decay by spontaneous emission. If conditions are right, state E_2 can then become more heavily populated than state E_1, thus providing the needed population inversion. A stray photon of the right energy can then trigger an avalanche of stimulated emission events, resulting in the production of laser light. A number of lasers using crystalline solids (such as ruby) as a lasing material operate in this three-level mode.

Figure 52-13 shows the elements of a type of laser that is often found in student laboratories. The glass discharge tube is filled with an 80%–20% mixture of the noble gases helium and neon, the helium being the "pumping" medium and the neon the "lasing" medium. Figure 52-14 is a simplified version of the level structures for these two atoms. Note that four levels, labeled E_0, E_1, E_2, and E_3, are involved in this lasing scheme, rather than three levels as in Fig. 52-12.

Pumping is accomplished by setting up an electrically induced gas discharge in the helium-neon mixture. Electrons and ions in this discharge, occasionally enough, collide with helium atoms, raising them to level E_3 in Fig. 52-14. This level is metastable, spontaneous emission to the ground state (level E_0) being very infrequent. Level E_3 in helium (= 20.61 eV) is, by chance, very close to level E_2 in neon (= 20.66 eV), so that, during collisions between helium and neon atoms, the excitation energy of the former can readily be transferred to the latter. In this way level E_2 in Fig. 52-14 can become more highly populated than level E_1 in that figure. This population inversion is maintained because (1) the metastability of level E_3 ensures a ready supply of neon atoms in level E_2 and (2) level E_1 decays rapidly (through intermediate stages not shown) to the neon ground state, E_0. Stimulated emission from level E_2 to level E_1 predominates and red laser light, of wavelength 632.8 nm, is generated.

Most stimulated emission photons initially produced in the discharge tube of Fig. 52-13 will not happen to be parallel to the tube axis, and will be quickly stopped at the walls. Stimulated emission photons that *are* parallel to the axis, however, can move back and forth through the discharge tube many times by successive reflections from mirrors M_1 and M_2. A chain reaction thus builds up rapidly in this direction and the inherent parallelism of the laser light finds its origin.

Rather than thinking in terms of the photons bouncing back and forth between the mirrors, it is perhaps more useful to think of the

figure 52-12
The basic three-level scheme for laser operation. Metastable state E_2 is more heavily populated than the ground state E_1.

figure 52-13
The elements of a helium-neon gas laser.

entire arrangement of Fig. 52-13 as an optical resonant cavity that, like an organ pipe for sound waves, can be tuned to be sharply resonant at one (or more) wavelengths.

The mirrors M_1 and M_2 are concave, with their focal points nearly coinciding at the center of the tube. Mirror M_1 is coated with a dielectric film whose thickness is carefully adjusted to make the mirror as close as possible to totally reflective at the wavelength of the laser light; see Section 45-5. Mirror M_2, on the other hand, is coated so as to be slightly "leaky," so that a small fraction of the laser light can escape at each reflection to form the useful beam.

The windows W, W in Fig. 52-13, which close the ends of the discharge tube, are slanted so that their normals make an angle θ_p, the Brewster angle, with the tube axis, where

$$\tan \theta_p = n, \qquad [48\text{-}3]$$

n being the index of refraction of the glass for the laser light wavelength. In Section 48-3 we showed that such windows transmit light without loss by reflection, provided only that the light is polarized with its plane of polarization in the plane of Fig. 52-13. If the windows were square to the tube ends, beam loss by reflection ($\sim 4\%$ from each surface of each window) would make laser operation impossible.

EXAMPLE 5

Level populations. A three-level laser of the type shown in Fig. 52-12 emits laser light at a wavelength of 550 nm, near the center of the visible band.

(a) If the optical pumping mechanism is shut off, what will be the ratio of the population of the upper level (energy E_2) to that of the lower level (energy E_1)? Assume that $T = 300$ K.

From the Bohr frequency condition, the energy difference between the two levels is given by

$$E_2 - E_1 = h\nu = \frac{hc}{\lambda}$$

$$= \frac{(6.63 \times 10^{-34} \text{ J} \cdot \text{s})(3 \times 10^8 \text{ m/s})}{(550 \times 10^{-9} \text{ m})(1.60 \times 10^{-19} \text{ J/eV})}$$

$$= 2.26 \text{ eV}.$$

The mean energy of thermal agitation is equal to

$$kT = (8.62 \times 10^{-5} \text{ eV/K})(300 \text{ K}) = 0.0259 \text{ eV}.$$

From Eq. 52-6 we have then, for the desired ratio,

$$n_2/n_1 = e^{-(E_2 - E_1)/kT} = e^{-(2.26 \text{ eV})/(0.0259 \text{ eV})}$$

$$= e^{-87.3} \cong 1.3 \times 10^{-38}!$$

This is an incredibly small number. It is not unreasonable, however. An atom whose mean thermal agitation energy is only 0.0259 eV will not often impart an energy of 2.26 eV (87 times as great) to another atom in a collision.

(b) At what temperature for the conditions of (a) above would the ratio n_2/n_1 be $\frac{1}{2}$?

Making this substitution in Eq. 52-6, taking the natural logarithm of each side, and solving for T yields

$$T = \frac{E_2 - E_1}{k(\ln 2)} = \frac{2.26 \text{ eV}}{(8.62 \times 10^{-5} \text{ eV/K})(0.693)}$$

$$= 37,800 \text{ K}.$$

This is much hotter than the surface of the sun. It is clear that, if we are to invert the populations of these two levels, some rather special pumping or other mechanism is needed. Without population inversion, lasing is not possible.

A ruby laser. A pulsed ruby laser has as its active element a synthetic ruby crystal in the form of a cylinder 6 cm long and 1 cm in diameter. Ruby consists of Al_2O_3 in which — in this case — one aluminum ion in every 3500 has been replaced by a chromium ion, Cr^{3+}. It is in fact the optical absorption properties of this small chromium "impurity" that account for the characteristic color of ruby. These same ions also account for the lasing ability of ruby, which occurs — by the three-level mechanism of Fig. 52-12 — at a wavelength of 694.4 nm.

Suppose that *all* of the Cr^{3+} ions are in a metastable state corresponding to state E_2 of Fig. 52-12 and that *none* are in the ground state E_1. How much energy is available for release in a single pulse of laser light if *all* of these ions revert to the ground state in a single stimulated emission chain reaction episode? Our answer will be an upper limit only because the conditions postulated cannot be realized in practice. The density ρ of Al_2O_3 is 3700 kg/m³, and its molecular weight A is 0.102 kg/mol.

The number of Al^{3+} ions is

$$N_{Al} = \frac{2N_A M}{A} = \frac{2N_A \rho V}{A}.$$

where M is the mass of the ruby cylinder and the factor 2 accounts for there being two aluminum ions in each "molecule" of Al_2O_3.

The volume V $(= \pi d^2 h/4)$ is given by

$$V = (\pi/4)(1.0 \times 10^{-2} \text{ m})^2(6.0 \times 10^{-2} \text{ m})$$

$$= 4.7 \times 10^{-6} \text{ m}^3.$$

EXAMPLE 6

Thus

$$N_{Al} = \frac{(2)(6.0 \times 10^{23}/\text{mol})(3.7 \times 10^3 \text{ kg/m}^3)(4.7 \times 10^{-6} \text{ m}^3)}{0.102 \text{ kg/mol}}$$

$$= 2.1 \times 10^{23}.$$

The number of Cr^{3+} ions is then

$$N_{Cr} = \frac{N_{Al}}{3500} = 6.0 \times 10^{19}.$$

The energy of the stimulated emission photon is

$$E = h\nu = \frac{hc}{\lambda} = \frac{(4.1 \times 10^{-15} \text{ eV} \cdot \text{s})(3.0 \times 10^8 \text{ m/s})}{(694 \times 10^{-9} \text{ m})} = 1.8 \text{ eV},$$

and the total available energy per laser pulse is

$$U = N_{Cr}E = (6.0 \times 10^{19})(1.8 \text{ eV})(1.6 \times 10^{-19} \text{ J/eV}) = 17 \text{ J}.$$

Such large pulse energies have indeed been achieved, but only by much more elaborate laser arrangements than that described here.

In this example we have postulated an ideal circumstance, namely, a total population inversion, in which the ground state remains virtually unpopulated. The actual population inversion in a working ruby laser will be very much less than total. For this and other reasons the pulse energy in practice will be very much less than the upper limit calculated above.

questions

1. What is the origin of the cutoff wavelength λ_{min} of Fig. 52-1? Why is it an important clue to the photon nature of X rays?

2. What are the characteristic X rays of an element? How can they be used to determine the atomic number of the element?

3. Compare Figs. 52-1 and 52-3. How can you be sure that the two prominent peaks in the former figure do indeed correspond numerically with the two transitions similarly labeled in the latter figure?

4. Can atomic hydrogen be caused to emit X rays? If so, describe how. If not, why not?

5. How does the X-ray energy-level diagram of Fig. 52-3 differ from the energy-level diagram for hydrogen, displayed in Fig. 49-19? In what respects are the two diagrams similar?

6. When extended to higher atomic numbers, the Moseley plot of Fig. 52-4 is not a straight line but is concave upwards. Does this affect the ability to assign atomic numbers to the elements?

7. Why is it that Bohr theory, which does not work very well even for helium $(Z = 2)$, gives such a good account of the characteristic X-ray spectra of the elements, or at least of that portion that originates deep within the atom?

8. Why does the characteristic X-ray spectrum vary in a systematic way from element to element but the spectrum in the visible range does not?

9. Why do you expect the wavelengths of radiations generated by transitions deep within the atom to be shorter than those generated by transitions occurring in the outer fringes of the atom?

10. Given the characteristic X-ray spectrum of a certain element, containing a number of lines, how would you go about identifying and labeling them?

11. How does the structure of the periodic table support the need for a fourth quantum number, corresponding to electron spin?

12. If there were only three quantum numbers (that is, if the electron did not spin), how would the chemical properties of helium be different?

13. Why does it take more energy to remove an electron from neon $(Z = 10)$ than from sodium $(Z = 11)$?

14. What can Fig. 52-5 tell you about why the noble gases are so chemically inert?

15. Does it make any sense to assign quantum numbers to a vacancy in an otherwise filled subshell?

16. Why do the lanthanide elements (see Appendix E) have such similar chemical properties? How can we justify putting them all into a single square of the periodic table? Why is it that, in spite of their similar chemical properties, they can be so easily sorted out by measuring their characteristic X-ray spectra?

17. In your own words, state the minimum energy principle for atom building and give a physical argument in support of it.

18. Figure 52–5 shows that the $2s$ state is lower in energy than the $2p$ state. Can you explain why this should be so, basing your argument on the radial probability densities of the two states (see Figs. 51-13b and 51-14b)?

19. If you start with a bare nucleus and fill in the electrons to form an atom in its ground state, the energies of the unfilled levels change as you proceed. Why do they change? Do they increase or decrease in energy as electrons are added?

20. Why is focused laser light inherently better than focused light from a tiny incandescent lamp filament for delicate surgical jobs such as spot-welding detached retinas?

21. Laser light forms an almost parallel beam. Does the intensity of such light fall off as the inverse square of the distance from the source?

22. In what ways are laser light and star light similar? In what ways are they different?

23. Arthur Schawlow, one of the laser pioneers, invented a typewriter eraser, based on focusing laser light on the unwanted character. Can you imagine what its principle of operation is?

24. In what ways do spontaneous emission and stimulated emission differ?

25. We have spontaneous emission and stimulated emission. From symmetry, why don't we also have spontaneous and stimulated absorption? Discuss in terms of Fig. 52-10.

26. Why is a population inversion necessary between two atomic levels for laser action to occur?

27. What is a metastable state? What role do such states play in the operation of a laser?

28. Comment on this statement: "Other things being equal, a four-level laser scheme such as that of Fig. 52-12 is preferable to a three-level scheme such as that of Fig. 52-14 because, in the latter scheme, one-half of the population of atoms in level E_1 must be moved to state E_2 before a population inversion can even begin to occur."

29. Comment on this statement: "In the laser of Fig. 52-13, only light whose plane of polarization lies in the plane of that figure is transmitted through the right-hand window. Therefore half of the energy potentially available to the output beam is lost." (*Hint:* Is this second statement really true? Consider what happens to photons whose effective plane of polarization is at right angles to the plane of Fig. 52-13? Do such photons participate fully in the stimulated emission amplification process?)

SECTION 52-2

1. Show that the short-wavelength cutoff in the continuous X-ray spectrum is given by

$$\lambda_{\min} = 1240 \text{ pm}/V,$$

problems

where V is the applied potential difference in kilovolts.

2. Determine Planck's constant from the fact that the mininum X-ray wavelength produced by 40.0-keV electrons is 31.1 pm.

3. Show that a moving electron cannot spontaneously change into an X-ray photon in a free space. A third body (atom or nucleus) must be present. Why is it needed?

4. X rays are produced in an X-ray tube by a target potential of 50 kV. If an electron makes three collisions in the target before coming to rest and loses one-half of its remaining kinetic energy on each of the first two collisions, determine the wavelengths of the resulting photons. Neglect the recoil of the heavy target atoms.

5. Electrons bombard a molybdenum target, producing both continuous and characteristic X rays as in Fig. 52-1. In that figure the energy of the incident electrons is 35.0 keV. If the accelerating potential applied to the X-ray tube is increased to 50.0 kV, what new values of (a) λ_{min}, (b) λ_{K_α}, and (c) λ_{K_β} result?   *Answer: (a) 24.8 pm; (b) and (c) remain unchanged.*

6. What is the minimum potential difference across an X-ray tube that will produce X rays with a wavelength of 0.1 nm (= 1Å)?

7. In Fig. 52-1, the X rays shown are produced when 35.0-keV electrons fall on a molybdenum target. If the accelerating potential is maintained at 35.0 kV but a silver target ($Z = 47$) is substituted for the molybdenum target, what values of (a) λ_{min}, (b) λ_{K_β}, and (c) λ_{K_α} result? The K, L, and M atomic X-ray levels for silver (compare Fig. 52-3) are 25.51, ~3.56, and ~0.53 keV. *Answer: (a) 35.4 pm, as for molybdenum; (b) 50 pm; (c) 57 pm.*

8. What is the ratio of the wavelength of the sodium yellow doublet line (mean value) to the wavelength of the molybdenum K_α characteristic X-ray line?

9. The wavelength of the K_α line from iron is 194 pm. What is the energy difference between the two states of the iron atom (see Fig. 52-3) that give rise to this transition? What is the corresponding energy difference for the hydrogen atom? Why is the difference so much greater for iron than for hydrogen? (*Hint:* In the hydrogen atom the K shell corresponds to $n = 1$ and the L shell to $n = 2$.)   *Answer: 6.39 keV; 10.2 eV.*

10. Find the K_α wavelength for an atom with $Z = 50$ but with a single muon (mass = 207 electron masses) moving around its nucleus.

11. Here are the K_α wavelengths of a few elements:

Ti	27.5 pm	Co	17.9 pm
V	25.0	Ni	16.6
Cr	22.9	Cu	15.4
Mn	21.0	Zn	14.3
Fe	19.3	Ga	13.4

Make a Moseley plot (see Fig. 52-4) and verify that its slope agrees with the value calculated in Example 2.

12. A tungsten target ($Z = 74$) is bombarded by electrons in an X-ray tube. (a) What is the minimum value of the accelerating potential that will permit the production of the characteristic K_β and K_α lines of tungsten? (b) For this same accelerating potential, what is λ_{min}? (c) What are λ_{K_β} and λ_{K_α}? The K, L, and M atomic X-ray levels for tungsten (see Fig. 52-3) are 69.5, ~11.3, and ~2.3 keV, respectively).

13. A molybdenum target ($Z = 42$) is bombarded with 35.0-keV electrons and the X-ray spectrum of Fig. 52-1 results. Here $\lambda_{K_\beta} = 63$ pm and $\lambda_{K_\alpha} = 71$ pm. (a) What are the corresponding quantum energies? (b) It is desired to filter these radiations through a material that will absorb the K_β line much more strongly than it will absorb the K_α line. What substance would you use? The

K ionization energies for molybdenum and for four neighboring elements are as follows:

Z	40	41	42	43	44
Element	Zr	Nb	Mo	Tc	Ru
E_K (keV)	18.00	18.99	20.00	21.04	22.12

(*Hint:* A substance will selectively absorb one of two X radiations more strongly if the photons of one have enough energy to eject a *K* electron from the atoms of the substance but the photons of the other do not.)
Answer: (*a*) 19.7 keV; 17.5 keV; (*b*) Zr or Nb; Zr better.

14. The binding energy of *K*-shell and *L*-shell electrons in copper are 8.979 keV and 0.951 keV, respectively. If a K_α X ray from copper is incident on a sodium chloride crystal and gives a first order Bragg reflection at 15.9° when reflected from the alternating planes of the sodium atoms, what is the spacing between these planes?

15. From Fig. 52-1, calculate approximately the energy difference $E_L - E_M$ for the X-ray atomic energy levels of molybdenum. Compare with the result that may be found from Fig. 52-3.   *Answer:* ~2.2 keV.

16. Using Bohr's theory, estimate the ratio of energies of photons from the K_α transitions from two atoms whose atomic numbers are Z and Z'. How much more energetic is a K_α X ray from uranium expected to be than from aluminum? From lithium?

17. Determine how close the theoretical K_α X-ray energies, obtained from Eq. 52-2, are to the measured energies in the light elements from lithium to magnesium. To do this you must first (*a*) determine the constant a^2h to five significant figures, using data in Appendix B. Next, (*b*) calculate the percentage deviation of the theoretical from the measured energies. (*c*) Plot the deviation and comment on the trend. Put *b* = 1. The measured values of the K_α X rays are:

Li	54.3 eV	O	524.9
Be	108.5	F	676.8
B	183.3	Ne	848.6
C	277	Na	1041
N	392.4	Mg	1254

(There is actually more than one K_α ray because of splitting of the *L* energy level, but that effect is negligible in the elements considered.)
Answer: (*a*) 24%; 15%; 11%; 7.9%; 6.5%; 4.7%; 3.5%; 2.5%; 2.0%; 1.5%.

SECTION 52-3

18. If a uranium nucleus (Z = 92) had only a single electron, what would be the radius of its ground state orbit, according to Bohr's theory?

19. In the alkali metals there is one electron outside a closed shell. (*a*) Using Eq. 49-30, calculate the energy of this electron for each of the alkali metals through cesium assuming, first, that Z = 1 in every case, and second, that Z has its actual value. (*b*) What logical support does each calculation have? (*c*) Which argument do the actual measured energies, being the negative of the ionization energies found in Fig. 52-6, seem to favor? The relevant quantum numbers are found in Fig. 52-5.
Answer: (*a*) H: −13.6, −13.6; Li: −3.4, −30.6; Na: −1.54, −183; K: −0.85, −307; Rb: −0.544, −745; Cs: −0.378, −1140; all energies in eV.

20. Four identical, noninteracting particles are placed in a one-dimensional box of length 0.5 nm. Find the lowest energy level of the system and list the quantum states (*a*) if the Pauli exclusion principle does not apply and (*b*) taking into account the exclusion principle.

21. If the electron had no spin, and if the Pauli exclusion principle still held, how would the periodic table be affected? In particular, which of the present elements would be noble gases?
   Answer: Only argon would remain a noble gas.

22. Suppose there are two electrons in the same system, both of which have $n = 2$ and $l = 1$. (a) If the exclusion principle did not apply, how many combinations of states are conceivably possible? (b) How many states does the exclusion principle forbid? Which ones are they?

23. Show formally that for a filled subshell, both the resultant angular momentum and the resultant magnetic moment are zero.

24. What elements do you think would be noble gases (a) If the energy of a state depended only on the principal quantum number n and not at all on the orbital quantum number l? (b) If the electron had no spin?

SECTION 52-4

25. What are the quantum numbers of the three electrons in lithium according to the n, l, m_l, m_s scheme? Assume the atom to be in its ground state.
   Answer: $(1, 0, 0, \pm\frac{1}{2})$ and $(2, 0, 0, \pm\frac{1}{2})$.

26. Two electrons in lithium $(Z = 3)$ have as their quantum numbers n, l, m_l, m_s, the values $1, 0, 0, \pm\frac{1}{2}$. (a) What quantum numbers can the third electron have if the atom is to be in its ground state? (b) If the atom is to be in its first excited state?

27. Aluminum $(Z = 13)$ in its ground state has, according to Fig. 52-5, filled $1s$, $2s$, $2p$, and $3s$ subshells and has a single electron in the $3p$ subshell. (a) What values of the total angular momentum quantum number j might this electron have? (b) How would you use a Stern-Gerlach type of experiment to decide between the possibilities?
   Answer: (a) $n = 3; l = 1, j = \frac{3}{2}$ or $\frac{1}{2}$, whichever is the lower in energy. (b) If $j = \frac{3}{2}$ is lower, there will be four sub-beams in a Stern-Gerlach experiment; if $j = \frac{1}{2}$ is lower, there will be two. Experiment would reveal that $j = \frac{1}{2}$ is lower.

28. By inspection of Fig. 52-5, what do you think might be the atomic number of the next higher noble gas above radon $(Z = 86)$?

29. A chlorine atom $(Z = 17)$ in its ground state has, according to Fig. 52-5, filled $1s$, $2s$, $2p$, and $3s$ subshells and has five $3p$ electrons, leaving one "hole" in that subshell. (a) What quantum numbers $n, l,$ and j can you assign to the hole? Make use of the fact that in a given subshell states with lower j are filled first. What are (b) the angular momentum (maximum projected value) and (c) the magnetic moment (maximum projected value) of the entire chlorine atom?
   Answer: (a) $n = 3, l = 1, j = \frac{3}{2}$; (b) $\frac{3}{2}\hbar$; (c) $2\mu_B$.

SECTION 52-5

30. A He-Ne laser emits radiation at a wavelength of 632.8 nm and has an output power of 2 mW. How many photons per second does this correspond to?

31. A ruby laser emits light at wavelength 694.4 nm. If a laser pulse is emitted for 10^{-11} s and the energy release per pulse is 0.15 J, (a) what is the spatial length of the pulse and (b) how many photons are in each pulse?
   Answer: (a) 3.0 mm; (b) 5.2×10^{17}.

32. Lasers have become very small as well as very large. The active volume of a laser constructed of the semiconductor GaAlAs measures only $5 \times 200 \times 0.2 \ \mu m^3$ and yet it can continuously deliver 5 mW of power at 0.8-μm wavelength. What is the production rate of photons?

33. A high-powered laser beam $(\lambda = 600$ nm) with a beam diameter of 12 cm is aimed at the moon, 3.8×10^5 km distant. The spreading of the beam is caused only by diffraction effects. The angular location of the edge of the central diffraction disc (see Fig. 46-10) is given by

$$\sin \theta = \frac{1.22\lambda}{d},$$

where d is the diameter of the beam aperture. What is the diameter of the central diffraction disk at the moon's surface?   *Answer: 4.6 km.*

34. The beam from an argon laser ($\lambda = 515$ nm) has a diameter d of 3.0 mm and a continuous-wave power output of 5.0 W. The beam is focused onto a diffuse surface by a lens (assumed ideal) whose focal length f is 3.5 cm. A diffraction pattern such as that of Fig. 46-10 is formed, the radius of the central disk being given by

$$R = \frac{1.22\,f\lambda}{d}.$$ [see Eq. 46-11]

The central disk can be shown to contain 84% of the incident power. Find (a) the radius R of the central disc, (b) the average power flux density in the incident beam, and (c) in the central disk.

35. The use of lasers for missile defense has been proposed as part of the Strategic Defense Initiative (Star Wars). A beam of intensity 10^8 W/m² would probably burn into and destroy a hardened missile in 1 s. If the laser has 5-MW power, 3-μm wavelength and 4.0-m beam diameter (a very powerful laser indeed), would it destroy a missile at a distance of 3000 km? If the wavelength could be changed, what maximum value would work? Use the equation for the central disk given in Problem 34 and take the focal length to be the distance to the target.   *Answer: No; 0.14 μm.*

36. It is entirely possible that techniques for modulating the frequency or amplitude of a laser beam will be developed so that such a beam can serve as a carrier for television signals, much as microwave beams do now. Assume also that laser systems will be available whose wavelengths can be precisely "tuned" to anywhere in the visible range, that is, in the range 450 nm < λ < 650 nm. If a television channel occupies a bandwidth of 10 MHz, how many channels could be accommodated with this laser technology? Comment on the intrinsic superiority of visible light to microwaves as carriers of information.

SECTION 52-6

37. Consider two single atomic levels with energies E_2 and E_1, where $E_2 > E_1$. If a system of such atoms is in thermal equilibrium at temperature T, the ratio of the populations of these two states is given (see Eq. 52-6) by

$$\frac{n_2}{n_1} = e^{-(E_2 - E_1)/kT},$$

in which k (= 8.63×10^{-5} eV/K) is Boltzmann's constant. In a particular atomic system, $E_2 - E_1 = 1.0$ eV. (a) Calculate n_2/n_1 at room temperature ($T = 300$ K). The smallness of your answer shows the small role played by thermal agitation in populating the upper state. (b) At what temperature would 0.01% of the atoms be in the state with the higher energy? *Answer: (a) 10^{-17}; (b) 1260 K.*

SECTION 52-7

38. The mirrors in the laser of Fig. 52-13 form a cavity in which standing waves of laser light are set up. In the vicinity of 533 nm, how far apart in wavelength are the adjacent allowed operating modes? The mirrors are 8 cm apart.

39. A population inversion for two levels is often described by assigning a negative Kelvin temperature to the system. Show that such a negative temperature would indeed correspond to an inversion. What negative temperature would describe the system of Example 5 if the population of the upper level exceeds that of the lower by 10%?   *Answer: -2.75×10^5 K.*

40. The NOVA laser at Lawrence Livermore National Laboratory is used in

thermonuclear fusion experiments (See Fig. 52-8 and Section 55-10). Its beam is designed to deliver between 0.1 and 1 MJ of energy in nanoseconds to a target of solid deuterium and tritium (isotopes of hydrogen) to heat the material and increase its density. (a) Assume that two 150-kJ laser pulses simultaneously impact from opposite directions on a sphere of 1.0-mm radius containing 4.0 μg of fuel. If the pulses last 3 ns and are completely absorbed, estimate the maximum pressure on the sphere. [For simplicity, assume that each pulse is uniformly spread over an area $\pi (1.0 \text{ mm})^2$ and use Eq. 42-2a.] (b) If 1% of the energy is effective in raising the temperature, how hot will the fuel get? Assume the fuel is entirely deuterium with an atomic weight of 2.

41. The active medium in a particular ruby laser ($\lambda = 694.4$ nm) is a synthetic ruby crystal 6.0 cm long and 1.0 cm in diameter. The crystal is silvered at one end and—to permit the formation of an external beam—only partially silvered at the other. (a) Treat the crystal as an optical resonant cavity in analogy to a closed organ pipe and calculate the number of standing-wave nodes there are along the crystal axis. (b) By what amount $\Delta \nu$ would the beam frequency have to shift to increase this number by one? Show that $\Delta \nu$ is just the inverse of the travel time of light for one round trip back and forth along the crystal axis. (c) What is the corresponding fractional frequency shift $\Delta \nu / \nu$? The appropriate index of refraction is 1.75.

Answer: (a) 3.0×10^5; (b) 1430 MHz; (c) 3.3×10^{-6}; this is just the inverse of (a).

53
electrical
conduction
in solids

We have seen how well wave mechanics works when we apply it to individual atoms. In this chapter we hope to show, by a single broad example, that this powerful theory works just as well when we apply it to aggregates of atoms in the form of solids.

Every solid has an enormous range of properties that we can choose to examine. Is it transparent? Can you hammer it out into a flat sheet? What kinds of waves travel through it and at what speeds? Does it have interesting magnetic properties? Is it a good heat conductor? What is its crystal structure? Does it have special surface properties? . . . The list goes on.

We choose here to focus on a single question: "What are the mechanisms by which a solid conducts — or does not conduct — electricity? From this point of view we classify solids as conductors, semiconductors, and insulators. Among the elements, copper, silicon, and sulfur are, respectively, examples in each class.

Figure 53-1 suggests the striking progress that has been made during the last 250 years in understanding how solids conduct electricity. The upper figure is not a parlor stunt but a demonstration by Stephen Gray (1666 – 1736) that the human body is a conductor of electricity. It was Stephen Gray, in fact, who first recognized the difference between conductors and insulators and realized that solids could be classified on that basis. Figure 53-1*b* shows a megabit chip developed at the AT&T Bell Laboratories. Although smaller than a dime, it can store over a million bits of information. It is hard to imagine that this progress would have been possible without wave mechanics.

53-1
THE PROPERTIES OF SOLIDS

(a)

(b)

figure 53-1
(a) An experiment by Stephen Gray, showing that the human body is a conductor of electricity. (b) A computer memory chip that can store over one million digital bits of information. (AT&T Bell Laboratories.)

An isolated copper atom has 29 electrons. In solid copper, 28 of these are held close to their lattice sites by electromagnetic forces and are thus not free to move throughout the volume of the solid. The remaining electron *is* free to so move and, if we apply an emf between the ends of a copper wire, it is these *conduction electrons* (one per atom) that constitute the current that is set up in the wire.

In Section 31-4 we looked at this problem from the point of view of classical physics, comparing the conduction electrons in a metal cube to the atoms of a gas confined to a cubical box. Using this (classical) *free electron gas* model, we derived an expression for the resistivity of the metal. It is (see Eq. 31-14).

$$\rho = \frac{m}{ne^2\tau} \qquad (53\text{-}1)$$

53-2
CONDUCTION ELECTRONS IN A METAL

in which m is the mass and e the charge of the electron, n is the number of conduction electrons per unit volume, and τ is the average time between collisions of the electrons that make up the classical free electron gas.

We showed in Section 31-4 that τ is essentially constant, independent of whether or not an electric field has been set up inside the cube by an externally applied emf. Thus the resistivity ρ is independent of the applied electric field, which is another way of saying that metals obey Ohm's law.

Although this derivation of the form of Ohm's law is a fine achievement for classical physics, it is no simple matter to go much further. Also, there is one problem—the heat capacities of metals—about which this classical theory *does* have something to say but, alas, its predictions simply do not agree with experiment. Looking beyond this level of concern, it is hard to imagine how we would explain something as complicated as a transistor on the basis of the classical free electron gas model. We had better see what wave mechanics has to offer.

The first step in solving any wave mechanical problem is to specify the potential energy of the particle—which we take to be a single conduction electron—as a function of its position. As Fig. 51-1 reminds us, we need this information to substitute into the Schrödinger equation. We start with the simplest reasonable assumption, namely, that the potential energy is zero for all points within the cubical metal sample and that it approaches an infinitely great value for all points outside. We are still dealing with a free electron gas, but it is now one that is governed by quantum—rather than classical—rules.

This potential variation reminds us of the problem of an electron trapped in an infinite well that we solved in Section 50-9. Note, however, two differences: the present problem is three-dimensional, and it involves a well of macroscopic—rather than atomic—dimensions.

We represent a single conduction electron trapped in its metal cube by a (standing) matter wave $\psi(\mathbf{r})$, in which $\mathbf{r}$ is a position vector, and we impose the condition that the probability density $\psi^2(\mathbf{r})$ be zero both at the surface of the cube and at all outside points. This is our way of recognizing that the electron is truly trapped inside the metal cube. Figures 50-15 and 50-16 remind us that we proceeded in just the same way in the one-dimensional case.

If we impose these boundary conditions on the wave function, the Schrödinger equation tells us that the total energy E of the electron will be quantized, just as it was for an electron trapped in a one-dimensional well. There is a big difference, however. Because our metal cube is so very large on the scale of atomic dimensions, the number of standing matter waves that we can fit into the volume of the cube and still satisfy the boundary requirements is enormous and the allowed electron energies are extremely close together. Example 1 will show us that, for a cube 1 cm on edge, there are no fewer than $\sim 10^{22}$ quantized states that lie between $E = 5$ eV and $E = 6$ eV! Compare this with the limited array of well-spaced levels shown, for example, for the hydrogen atom in Fig. 49-19.

We cannot possibly deal with this vast number of states one at a time; we must use statistical methods. Instead of asking, "What is the energy of this state?" we must ask, "How many states have energies that lie in the range E to $E + dE$?"

We have met situations like this before. For example, in describing the speeds of the molecules of an ideal gas in Section 24-2, we saw that

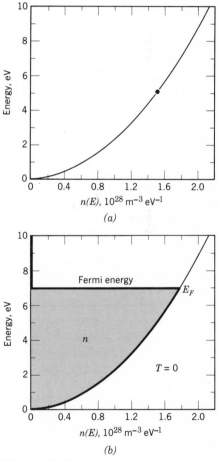

figure 53-2

(*a*) The density of states $n(E)$ as a function of the energy E. The dot shows the point calculated in Example 1. (*b*) For copper, the available conduction electrons fill up the states (at $T = 0$) to an energy E_F (= 7.0 eV), called the Fermi energy.

the only way to proceed was to pose the question: "How many molecules have speeds that lie in the range v to $v + dv$?"

For the conduction electrons, the number of states (per unit volume of the solid) whose energies lie in the range E to $E + dE$ can be written as $n(E)dE$, where $n(E)$ is a function called the *density of states*. For our (quantum) free electron gas it can be shown to be*

$$n(E) = \frac{8\sqrt{2}\pi m^{3/2}}{h^3} E^{1/2}. \qquad (53\text{-}2)$$

We remind ourselves that — at this stage — we are simply counting the states that are available to a single conduction electron. In the following section we shall see how to go about filling those states.

Figure 53-2a is a plot of Eq. 53-2. Note that there is nothing in this figure or this equation that depends on the material of which our sample is made. Fitting patterns of standing waves into a cubical box is a purely geometrical problem.

You may well ask, "If the energies of the allowed states are so close together, why don't we just forget about the quantization and assume a continuous distribution in energy?" The answer, as we shall see in the next section, rests on the fact that the Pauli exclusion principle applies to electrons wherever we find them, whether as orbital electrons in atoms or as conduction electrons in metals. Even though there are a lot of states in our problem there are also a lot of conduction electrons available to occupy them and Pauli's principle allows us to put only one electron in each of these states. Thus, even though we cannot easily detect directly the quantized nature of the energies of the conduction electrons, the fact of quantization remains an absolutely central feature and has important consequences.

EXAMPLE 1

Counting the allowed states. A cube of copper is 1 cm on edge. How many states are available for its conduction electrons in the energy interval between $E = 5.00$ eV and 5.01 eV? Assume that the conduction electrons behave like a (quantum) free electron gas.

These energy limits are so close together that we can safely say that the answer, on a per unit volume basis, is $n(E)\Delta E$, where $E = 5$ eV and $\Delta E = 0.01$ eV. From Eq. 53-2 we have

$$n(E) = \frac{8\sqrt{2}\pi m^{3/2}}{h^3} E^{1/2}$$

$$= \frac{(8\sqrt{2}\pi)(9.11 \times 10^{-31} \text{ kg})^{3/2}(5 \text{ eV})^{1/2}(1.6 \times 10^{-19} \text{ J/eV})^{1/2}}{(6.63 \times 10^{-34} \text{ J} \cdot \text{s})^3}$$

$$= 9.48 \times 10^{46} \text{ m}^{-3}\text{J}^{-1} = 1.52 \times 10^{28} \text{ m}^{-3}\text{eV}^{-1}.$$

This result appears as a dot on the curve of Fig. 53-2a. Note that we must express the energy E in joules before substituting into Eq. 53-2, even though we wish our final result to be given in terms of electron volts.

The actual number N of states that lie in the range from $E = 5.00$ eV to $E = 5.01$ eV in our cube is, if a is the length of the cube edge,

$$N = n(E)\,\Delta E\,a^3$$

$$= (1.52 \times 10^{28} \text{ m}^{-3}\text{eV}^{-1})(0.01 \text{ eV})(1 \times 10^{-2} \text{ m})^3$$

$$= 1.52 \times 10^{20}.$$

* See Robert Eisberg and Robert Resnick, *Quantum Physics* (John Wiley & Sons, New York, 1985) 2nd ed., Section 11-11.

The average energy interval ΔE_{adj} between adjacent levels near $E = 5$ eV follows readily from

$$\Delta E_{adj} = \frac{\Delta E}{N} = \frac{0.01 \text{ eV}}{1.52 \times 10^{20}} \cong 7 \times 10^{-23} \text{ eV}.$$

We conclude that, even in this narrow energy band, there are a lot of states and they lie exceedingly close together in energy.

Our conclusions are completely independent of the material of the sample. Nor is it important that the sample be cubical; any other shape enclosing the same volume would give the same final result. What we *have* assumed to be true is that the conduction electrons behave like a (quantum) free electron gas. That is, we have assumed that their potential energy is constant — we have taken it to be zero — for all points within the sample. For actual metals this assumption is never strictly true. Nevertheless, our central conclusion holds: many states are available to a conduction electron in a metal and they lie very close together in energy.

Now that we have seen how many states there are, we are ready to start filling them with electrons. We went through this process in Section 52-3 in connection with building up the periodic table of the elements. There we saw the central importance of Pauli's exclusion principle, which tells us that we can allocate only one electron to a given state. This powerful principle is just as important for our present problem.

Let us first consider conditions at the absolute zero of temperature. This represents the lowest energy state of our sample and we achieve it by sprinkling the conduction electrons into the unfilled states that lie lowest in energy. Figure 53-2b suggests the process. The highest occupied state under these conditions is called the *Fermi level*, and its energy, marked E_F in Fig. 53-2b, is called the *Fermi energy*. At the absolute zero of temperature, all states with energies below this value are filled and all of those above it are vacant. On the assumption that the conduction electrons in copper behave like a free electron gas, the Fermi energy for this metal is 7.0 eV.

We can find the Fermi energy for a metal by adding up (integrating) the number of occupied states in Fig. 53-2b between the limits of $E = 0$ and $E = E_F$. The result must equal n, the number of conduction electrons per unit volume for the metal. In equation form we have

$$n = \int_0^{E_F} n(E) \, dE. \tag{53-3}$$

Note that n is represented by the shaded area in Fig. 53-2b.

A glance at Fig. 53-2b should be enough to shatter the popular misconception that all motion ceases at the absolute zero of temperature. We see that, entirely because of Pauli's exclusion principle, the electrons are stacked up in energy from zero up to the Fermi energy. The *average* energy for the conditions of Fig. 53-2b turns out to be about 4.2 eV. By comparison, the average translational kinetic energy of a molecule of an ideal gas at room temperature is only 0.025 eV. The conduction electrons in a metal have plenty of energy at the absolute zero!

It seems clear that the molecules of a gas at ordinary temperatures and the conduction electrons in a metal behave in quite different ways. Formally, we say that the gas molecules obey the (classical) Maxwell-Boltzmann statistics and that the conduction electrons obey the (quantum) Fermi-Dirac statistics. The word "statistics" here refers to the formal rules for counting particles. In Maxwell-Boltzmann statistics, for example, we assume that we can tell identical particles apart, but in Fermi-Dirac statistics we assume that we cannot. Again, in Maxwell-Boltzmann statistics, Pauli's exclusion principle plays no role, but in Fermi-Dirac statistics its role — as we have seen — is vital.

What happens to the electron distribution of Fig. 53-2b as we raise the temperature? The short answer is that not very much happens, although the little that does is very important. Figure 53-3c shows what the distribution of Fig. 53-2b would look like at $T = 1000$ K, a temperature at which a metal sample would glow brightly in a dark room.

The striking feature of *this* figure is how little it differs from Fig. 53-2b, the distribution at the absolute zero. A few electrons have increased their energies from slightly below the Fermi energy (region 1 in Fig. 53-3c) to slightly above the Fermi energy (region 2 in Fig. 53-3c). Thus the average energy is a *little* larger than it was at the absolute zero but it is not much larger. This is again in striking contrast with the behavior of an ideal gas, for which the average kinetic energy of the molecules is proportional to the temperature.

Another impression that we get by comparing conditions at $T = 0$ and at $T = 1000$ K is that the "action" all takes place for conduction electrons whose energies are close to the Fermi energy. The great bulk of the electrons remain unchanged as the temperature is raised, their large store of energy being effectively locked up.

Let us see why this is so. Figure 53-3c displays the magnitude of kT, a measure of the energy available from thermal agitation; its value at 1000 K is only 0.086 eV. No electron can hope to have its energy changed by more than a few times this relatively small amount by thermal agitation alone. Because of the exclusion principle, only electrons whose energies are near the Fermi energy have vacant states close enough to them for such thermal transitions to occur. An electron with an energy of, say 2 eV, can neither gain nor lose energy because all states close enough to it in energy are already filled; it simply has nowhere to go. It has been said — accurately enough and somewhat poetically — that thermal agitation normally causes only ripples on the surface of the Fermi sea; the vast depths of that sea lie undisturbed.

Now let us see, in a more formal way, how the distribution of Fig. 53-3c comes about. We start with the density of states function, $n(E)$, shown again in Fig. 53-3a. This tells us how the *unoccupied* states are distributed in energy for a (quantum) free electron gas. We must now weight each energy by a factor $p(E)$ that gives the probability that a state with that energy will be occupied. The value of $p(E)$ must range between unity, corresponding to certain occupancy, and zero, corresponding to certain vacancy. The product of these two factors gives us $n_o(E)$, the density of *occupied* states. Thus

$$n_o(E) = n(E)\, p(E). \qquad (53\text{-}4)$$

We see by comparing Figs. 53-2a and 53-2b that, at the absolute zero, the function $p(E)$ must equal unity if $E < E_F$ and zero if $E > E_F$. This is borne out by the rectangular plot shown in Fig. 53-3b, marked $T = 0$.

In general, $p(E)$, which is called the *Fermi-Dirac probability function*, can be shown to be

$$p(E) = \frac{1}{e^{(E - E_F)/kT} + 1}, \qquad (53\text{-}5)$$

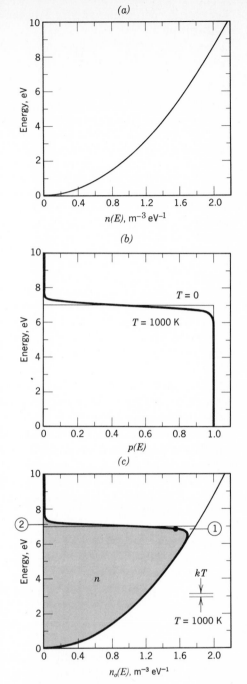

(a)

(b)

(c)

figure 53-3

(a) The density of states $n(E)$ as a function of the energy E. (b) The Fermi-Dirac probability function (Eq. 53-5), for $T = 0$ and for $T = 1000$ K. (c) The density of occupied states $n_0(E)$ as a function of energy E, for $T = 1000$ K. The dot shows the point calculated in Example 4c.

in which E_F is the Fermi energy. This equation is derived from the basic assumptions of the Fermi-Dirac statistics, to which we referred briefly above. Figure 53-3b shows a plot of $p(E)$ for $T = 1000$ K.

Equation 53-5 is worth a little study. We point out first that this equation does indeed yield the rectangular plot of Fig. 53-3b for $T = 0$. As $T \to 0$, the exponent $(E - E_F)/kT$ in Eq. 53-5 approaches $-\infty$ if $E < E_F$ and $+\infty$ if $E > E_F$. In the first case we have $p(E) = 1$ and in the second case $p(E) = 0$, just as required.

Equation 53-5 also shows us that the important quantity is not the energy E but rather $E - E_F$, the energy interval between E and the Fermi energy. We see further that, because of the exponential nature of the term in the denominator of Eq. 53-5, $p(E)$ is very sensitive to small changes in E. This confirms our assertion that electrons whose energies are close to the Fermi energy are the only ones that play an active role. As we shall see, one's first question in dealing with the electrons in a solid — be it a conductor, a semiconductor, or an insulator — is likely to be: "On an energy scale, where is the Fermi level?"

EXAMPLE 2

Calculating the Fermi energy. Calculate the Fermi energy for copper, given that the number of conduction electrons per unit volume (see Example 2, Chapter 31) is 8.4×10^{28} m^{-3}.

Combining Eqs. 53-2 and 53-3 leads to

$$n = \int_0^{E_F} n(E)\, dE = \frac{8\sqrt{2}\pi m^{3/2}}{h^3} \int_0^{E_F} E^{1/2}\, dE$$

$$= \left(\frac{8\sqrt{2}\pi m^{3/2}}{h^3}\right)\left(\frac{2}{3} E_F^{3/2}\right).$$

Solving for E_F leads eventually to

$$E_F = \frac{h^2}{8m}\left(\frac{3n}{\pi}\right)^{2/3} \tag{53-6}$$

$$= \frac{(6.63 \times 10^{-34}\text{ J}\cdot\text{s})^2}{(8)(9.11 \times 10^{-31}\text{ kg})}\left[\frac{(3)(8.4 \times 10^{28}\text{ m}^{-3})}{\pi}\right]^{2/3}$$

$$= 1.12 \times 10^{-18}\text{ J} = 7.0\text{ eV}.$$

EXAMPLE 3

The Fermi-Dirac probability function. What is the probability of occupancy for a state whose energy is (a) 0.1 eV above the Fermi energy, (b) 0.1 eV below the Fermi energy, and (c) equal to the Fermi energy? Assume a temperature of 800 K.

(a) The (dimensionless) exponent in Eq. 53-5 is

$$\frac{E - E_F}{kT} = \frac{0.1\text{ eV}}{(8.62 \times 10^{-5}\text{ eV/K})(800\text{ K})} = 1.45.$$

Inserting this exponent into Eq. 53-5 yields

$$p(E_+) = \frac{1}{e^{1.45} + 1} = 0.19.$$

Thus the occupancy probability for this state is 19%.

(b) For an energy 0.1 eV *below* the Fermi energy, the exponent in Eq. 53-5 has the same numerical value as above but is negative. Thus, from this equation

$$p(E_-) = \frac{1}{e^{-1.45} + 1} = 0.81.$$

The occupancy probability for this state is 81%.

For states whose energies lie below the Fermi energy, we are often more interested in the probability that the state is *not* occupied. This is, of course, just $1 - p(E)$, or 19% in the present case. An unfilled state in an energy range in which most of the states are filled is called a *hole*. We shall see later that this is a very useful concept.

(c) For $E = E_F$ the exponent in Eq. 53-5 is zero and that equation becomes

$$p(E_F) = \frac{1}{e^0 + 1} = \frac{1}{1 + 1} = 0.50.$$

Note that this result does not depend on the temperature. Note also that none of these three results depends on the actual value of the Fermi energy, only on the energy interval between the Fermi energy and the energy of the state in question.

EXAMPLE 4

The energy distribution of the conduction electrons.

(a) For copper at 1000 K, find the energy at which the probability $p(E)$ that a conduction electron state will be occupied is 90%. (Assume that the conduction electrons in copper behave like a free electron gas, with a Fermi energy of 7.0 eV.)

Substitution into Eq. 53-5 yields

$$p(E) = \frac{1}{e^{\Delta E/kT} + 1} = 0.9,$$

in which $\Delta E = E - E_F$. A little algebra leads to $\Delta E/kT = -2.20$ so that

$$\Delta E = -2.20 \, kT = -(2.20)(8.62 \times 10^{-5} \text{ eV/K})(1000 \text{ K})$$

$$= -0.19 \text{ eV}.$$

For copper, assuming that $E_F = 7.0$ eV, we have

$$E = E_F + \Delta E = 7.0 \text{ eV} - 0.19 \text{ eV} = 6.8 \text{ eV}.$$

(b) For this energy, what is $n(E)$, the distribution in energy of the unoccupied states?

Carrying out a calculation just like that of Example 1 for $E = 6.8$ eV yields $n(E) = 1.8 \times 10^{28}$ m^{-3} eV^{-1}.

(c) For this same energy, what is $n_o(E)$, the distribution in energy of the occupied states?

From Eq. 53-5 we have, again for $E = 6.8$ eV,

$$n_o(E) = n(E) \, p(E)$$

$$= (1.8 \times 10^{28} \text{ m}^{-3} \text{ eV}^{-1})(0.90)$$

$$= 1.6 \times 10^{28} \text{ m}^{-3} \text{ eV}^{-1}.$$

This result appears as a dot on the curve of Fig. 53-3c.

53-4
THE RESISTIVITY OF METALS

The (quantum) free electron gas model leads to the same expression for the electrical resistivity of a metal (Eq. 53-1) that we derived in Section 31-4 using the (classical) free electron gas model. The mechanism of conduction, however, is now a lot clearer. Although all of the conduction electrons take part in the conduction process, we now see that only those electrons whose energies are near the Fermi energy are able to take part in the collisions that account for the resistivity. Only those electrons have empty states, near in energy, into which they can scatter.

(These same electrons, incidentally, are also responsible for the transfer of energy to the lattice that we have earlier called Joule heating.) All such electrons move with approximately the same speed v_F, which we call the *Fermi speed*.

In Example 5 we show that, for copper at room temperature, the Fermi speed — which is the average speed of the conduction electrons between collisions — is 1.6×10^6 m/s, a not insignificant fraction of the speed of light. The mean time τ between collisions proves to be 2.5×10^{-14} s and the mean free path ~41 nm, which is about 150 nearest-neighbor distances in the copper lattice.

You may be surprised that, at room temperature, a conduction electron can move so far through a copper lattice without hitting an ion core. At lower temperatures — where the resistivity is lower — it can move even much further. It is, in fact, a perhaps unexpected prediction of wave mechanics that a perfectly periodic lattice at the absolute zero of temperature would be totally transparent to conduction electrons. There would never be any collisions!

There are, however, no perfectly periodic lattices. Vacant lattice sites and impurity atoms are always present, no matter how hard we try to eliminate them. Furthermore, at temperatures above the absolute zero the lattice is vibrating, and these motions also spoil the periodicity of the lattice. At room temperature the "collisions" of which we have spoken are largely interactions between the conduction electrons and the vibrations of the lattice.

EXAMPLE 5

A conduction electron in copper. Take the Fermi energy of copper to be 7.0 eV. (a) What is the speed of a conduction electron with this amount of kinetic energy?

Throughout this section we have assumed that the conduction electrons are moving in a region in which their potential energy is zero. Thus their total energy E is all kinetic and we can write, if $E = E_F$,

$$E_F = \tfrac{1}{2}mv_F^2,$$

in which v_F is the Fermi speed. Solving for v_F yields

$$v_F = \left(\frac{2E_F}{m}\right)^{1/2} = \left[\frac{(2)(7.0 \text{ eV})(1.6 \times 10^{-19} \text{ J/eV})}{9.11 \times 10^{-31} \text{ kg}}\right]^{1/2}$$

$$= 1.6 \times 10^6 \text{ m/s.}$$

You should not confuse this speed with the *drift speed* of the conduction electrons, which is of the order of 10^{-4} m/s and is thus smaller by about a factor of 10^{10}. As we explained more fully in Section 31-1, the drift speed is the average speed at which electrons actually drift through a conductor when an electric field is applied; the Fermi speed is their average speed between collisions.

(b) The resistivity of copper at room temperature is $1.7 \times 10^{-8} \; \Omega \cdot$ m. From Eq. 53-1, what is the average time τ between collisions?

Solving Eq. 53-1 for τ yields

$$\tau = \frac{m}{ne^2\rho}$$

$$= \frac{9.11 \times 10^{-31} \text{ kg}}{(8.5 \times 10^{28} \text{ m}^{-3})(1.60 \times 10^{-19} \text{ C})^2(1.7 \times 10^{-8} \; \Omega \cdot \text{m})}$$

$$= 2.5 \times 10^{-14} \text{ s.}$$

(c) What mean free path λ may be calculated from the results of (a) and (b) above?

We have

$$\lambda = v_F \tau = (1.6 \times 10^6 \text{ m/s})(2.5 \times 10^{-14} \text{ s})$$

$$= 4.1 \times 10^{-8} \text{ m} = 41 \text{ nm}.$$

In the copper lattice the centers of neighboring ion cores are 0.26 nm apart. Thus a typical conduction electron can move a substantial distance through a copper lattice at room temperature without making a collision.

Figure 53-4a suggests the potential energy variation that we have been using to describe a conduction electron in a metal. There are problems with this model. It tells us, for example, that, because of the infinite potential barrier, an electron could never escape from inside the sample through its surface. We know that this isn't true. Electrons can be "boiled out" of a metal by raising its temperature, as in the heated filament of a vacuum tube (thermionic emission.) They can also be "kicked out" if we shine light of high enough frequency on the metal surface (photoelectric effect.)

Figure 53-4b shows that we can take care of this difficulty easily enough by making the potential energy jump at the surface finite. We made the same realistic adjustment (see Fig. 50-17) for the electron trapped in an atom-sized, one-dimensional well. The quantity ϕ in Fig. 53-4b is the *work function* of the metal, defined so that $e\phi$ is the least amount of energy that must be supplied to an electron to remove it from the sample.

We take the opportunity of this change to bring our energy scale into agreement with that used for the hydrogen atom. That is, we choose the $E = 0$ configuration to represent an electron at rest for outside the sample. It is possible to make this change because the potential energy always contains an arbitrary additive constant and we are more concerned, in any case, with changes in the total energy E than with E itself. On our new energy scale, the total energies of electrons trapped in the sample are negative, just as they are for the hydrogen atom.

By far the largest defect remaining in the potential variation of Fig. 53-4b is that it still assumes that the potential energy of a conduction electron is constant throughout the volume of the sample. This ignores the fact that the conduction electrons move about among an array of positively charged ion cores. It is rather remarkable, in fact, that we have been able to learn as much as we have done about the resistivity of metals without taking into account the potential variations caused by the ion cores of the lattice. We have not, however, been able to answer such questions as: "Why is copper a conductor and diamond a nonconductor?" If we take the lattice periodicity into account, we shall be able to answer this question and to go far beyond.

Figure 53-4c shows a potential energy curve that takes the lattice cores into account. Substituting this variation or — much more likely — some approximation to it into the Schrödinger equation brings out an interesting new phenomenon. As Fig. 53-4c shows, *the allowed states are now grouped into bands, with energy gaps between them in which no states exist.* Note that electrons just below the Fermi level are free to move throughout the lattice, but electrons with lower energies, the *core electrons*, are not. Let us see if we can understand these bands and gaps in physical terms.

The distance between nearest neighbors in a copper lattice is

53-5
BANDS AND GAPS

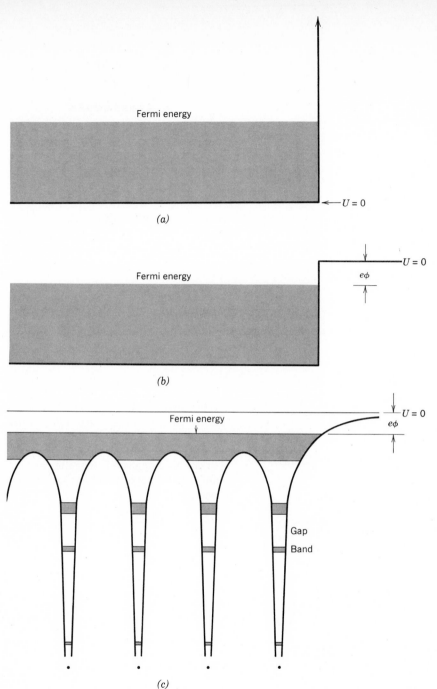

figure 53-4
(a) The potential energy variation
assumed for a metal in the free
electron gas model. (b) A more realistic
variation, showing a finite change in
potential at the surface of the sample.
(c) A still more realistic variation,
taking the lattice of ion cores into
account. This curve is a one-
dimensional cut along a line of ion cores.

0.27 nm. Consider, however, two copper atoms separated by a much
greater distance, say 50 nm, so that we may describe them as "isolated";
see Fig. 53-5a. In each atom the 29 electrons are assigned to the levels
shown in Fig. 53-5b.

Now let us bring the two atoms closer together so that an outer
electron in either atom can be influenced — however slightly — by
forces exerted on it by the other atom. In the language of wave mechanics we say that their wave functions begin to overlap. We state
without proof that the overlapping wave functions can be combined in
two independent ways, describing two states having (slightly) different
energies.

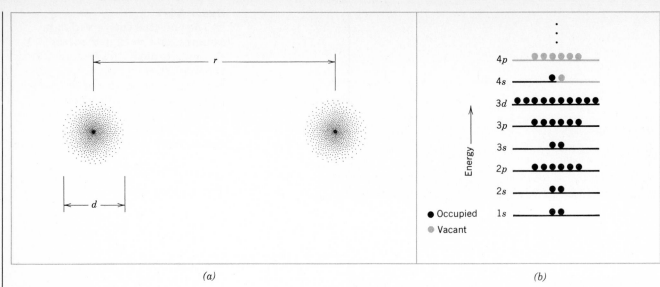

(a)

(b)

figure 53-5

(a) Two neutral copper atoms of diameter d, separated by a distance r, with $r \gg d$. (b) The atoms are independent systems and each, in its ground state, has the same quantum number assignments for its electrons, as shown. The energy scale is symbolic only.

figure 53-6

As atoms are brought together to form a lattice, the levels of the isolated atoms split, eventually forming bands of closely lying levels. For the case shown, the upper two bands overlap in energy.

Number of atoms ⟶

Put directly, when the two atoms begin to "touch," each of their energy levels splits into two levels. Because the overlap is greater for the outer electrons, the energy splitting will be greater for them than for the inner electrons.

By extension, if we bring N copper atoms together to form a copper lattice, each level of the isolated atom splits into N levels of the solid. Thus, the $1s$ level of the atom becomes the $1s$ band of the solid and so on. Figure 53-6 suggests the process.

From this point of view the forbidden gaps are not so hard to understand, being familiar from the level structure of the isolated atom. Indeed, we can say that Niels Bohr, even before wave mechanics, "invented" energy gaps when he said, in effect: "I assume that atoms can exist without radiating in a discrete set of stationary states of definite energy, *states of intermediate energy being forbidden.*"

53-6
CONDUCTORS, INSULATORS, AND SEMICONDUCTORS

Figure 53-7a represents the band structure of a conductor, such as copper. Its central feature is that the most energetic band that contains any electrons at all is only partially filled. There are vacant states above the Fermi level so that, if you apply an electric field **E**, every electron in this band is able to increase its momentum in the $-\mathbf{E}$ direction and there will be a current. The lower energy bands are completely filled and cannot contribute to the conduction process.

Figure 53-7b represents an insulator. Its central feature is that the most energetic band that contains any electrons at all is completely filled and the forbidden energy gap lying immediately above it, marked E_g in the figure, is substantial. By "substantial" we mean that $E_g \gg kT$, so that the probability that an electron will be lifted, by thermal agitation, into the empty band that lies above the gap is negligible. If you set up an electric field within an insulator, there is no way for any of the electrons to respond to it so that there will be no current.

Carbon in its diamond form is an excellent insulator, its energy gap being 5.5 eV, more than 200 times the value of kT at room temperature.

Figure 53-7c represents a semiconductor. It differs from an insulator in that its energy gap is small enough so that thermal excitation of electrons across it can occur to a useful extent at room temperature. This puts some electrons into the (nearly empty) band labeled *conduction band* in the figure and leaves an equal number of vacant states, or *holes*, in the (nearly filled) *valence band.*

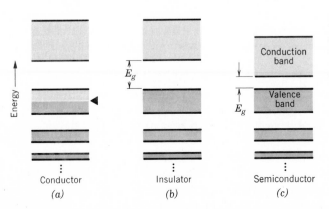

figure 53-7
An idealized representation of the energy bands for (a) a conductor, (b) an insulator, and (c) a semiconductor. The black triangle marks the Fermi level for the conductor.

Silicon is our prototype semiconductor. It has the same crystal structure as diamond but its gap width (= 1.1 eV) is considerably smaller. At the absolute zero of temperature, where thermal agitation is absent, all semiconductors are insulators. At any higher temperature the probability that an electron will be raised across the gap is very sensitive to the gap width. Thus the distinction between an insulator and a semiconductor is somewhat arbitrary. There is no hesitation, however, in calling diamond (E_g = 5.5 eV) an insulator and silicon (E_g = 1.1 eV) a semiconductor.

The revolution in microelectronics that has so influenced our lives is based on semiconductors, so we would do well to learn more about them. Table 53-1 compares some electrical properties of silicon, our prototype semiconductor, and copper, our prototype conductor. Let us look carefully at this table, one row at a time.

53-7
MORE ABOUT SEMICONDUCTORS

Table 53-1
Some electrical properties of copper and of silicon*

	Copper	Silicon
Mode of conduction	Conductor	Semiconductor
Density of charge carriers** n, m^{-3}	9×10^{28}	1×10^{16}
Resistivity ρ, $\Omega \cdot$ m	2×10^{-8}	3×10^3
Temperature coefficient of resistivity α, K^{-1}	$+4 \times 10^{-3}$	-70×10^{-3}

* All values refer to room temperature.
** Includes, for silicon, both electrons and holes.

1. *The density of charge carriers, n.* Copper has many more charge carriers than does silicon, by a factor of about 10^{13}. For copper the carriers are the conduction electrons, present in the number of one per atom. Figure 53-7c shows that, at the absolute zero of temperature, silicon would have no charge carriers at all. At room temperature, to which Table 53-1 refers, charge carriers arise only because, at thermal equilibrium, thermal agitation has caused a certain (very small) number of electrons to be raised to the conduction band, leaving an equal number of vacant states (holes) in the valence band.

The holes in the valence band of a semiconductor also serve effectively as charge carriers because they permit a certain freedom of movement to the electrons in that band. If an electric field is set up in a semiconductor, the electrons in the valence band, being negatively charged, will drift in the direction of $-\mathbf{E}$. This causes the holes to drift in the direction of $\mathbf{E}$. That is, the holes behave like particles carrying a charge $+e$ and, in all that follows, that is exactly how we shall regard them. Conduction by holes is an important fact of life for semiconductors.

If the concept of a migrating hole causes a problem, think of a bubble of air drifting slowly upwards in a pond. Does it not behave in every respect like a water droplet of negative gravitational mass? Is it not easier to analyze the motion of the bubble as it rises than that of the water as it streams downward around it?

2. *The resistivity, ρ.* At room temperature the resistivity of silicon is considerably higher than that of copper, by a factor of $\sim 10^{11}$. For both elements, the resistivity is determined by Eq. 53-1. As that equation shows, the resistivity increases as n, the density of charge carriers,

decreases. The vast difference in resistivity between copper and silicon can be accounted for by the vast difference in n. (The mean collision time τ will also be different for copper and for silicon, but the effect of this on the resistivity is swamped by the enormous difference in the density of charge carriers.)

For completeness, we mention that the resistivity of a good insulator (fused quartz or diamond, for example) may be as high as $10^{20}\ \Omega \cdot m$, about 10^{28} times higher than that of copper at room temperature. Perhaps no other physical property has as wide a range of measurable values as the electrical resistivity.

3. *The temperature coefficient of resistivity, α.* This quantity (see Eq. 31-10) is the fractional change in the resistivity ρ per unit change of temperature, or

$$\alpha = \frac{1}{\rho}\frac{d\rho}{dT}.$$

The resistivity of copper and other metals *increases* with temperature $(d\rho/dT > 0)$. This happens because collisions occur more frequently the higher the temperature, thus reducing τ in Eq. 53-1. For metals, the density of charge carriers n in that equation is independent of temperature.

On the other hand the resistivity of silicon (and other semiconductors) *decreases* with temperature $(d\rho/dT < 0)$. This happens because the density of charge carriers n in Eq. 53-1 increases rapidly with temperature. The decrease in τ mentioned above for metals also occurs for semiconductors, but its effect on the resistivity is swamped by the very rapid increase of the density of charge carriers.

In the laboratory, you can identify a semiconductor by its large resistivity ρ and, especially, by its large — and negative — temperature coefficient of resistivity α, both quantities being compared to their values for a typical metal.

53-8
DOPING

The versatility of semiconductors can be marvelously improved by introducing a small number of suitable replacement atoms (it seems pejorative to call them impurities) into the semiconductor lattice, a process called *doping*. We describe the semiconductor that results as *extrinsic*, to distinguish it from the pure undoped or *intrinsic* material. Essentially all semiconducting devices today are based on extrinsic material.

Figure 53-8a is a two-dimensional representation of a lattice of pure silicon. Each silicon atom forms a two-electron bond with each of its four nearest neighbors, the electrons involved in the bonding making up the valence band of the sample.

In Fig. 53-8b one of the silicon atoms (valence = 4) has been replaced by an atom of phosphorus (valence = 5). The "extra" electron is loosely bound to the phosphorus ion core, as Fig. 53-8b suggests. It is far easier for *this* electron to be thermally excited into the conduction band than it is for one of the silicon valence electrons to be so excited.

The phosphorus atom is called a *donor* atom because it so readily *donates* an electron to the conduction band. The "extra" electron in Fig. 53-8b can be said to lie in a localized donor level, as Fig. 53-9a shows. This level is separated from the bottom of the conduction band by an energy gap E_d, where $E_d \ll E_g$. By controlling the concentration of donor atoms it is possible to increase greatly the density of electrons in the conduction band.

Semiconductors doped with donor atoms are called *n-type* semiconductors, the "*n*" standing for "negative" because the negative charge carriers greatly outnumber the positive charge carriers. The former, called the *majority carriers*, are the electrons in the conduction band. The latter, called the *minority carriers*, are the holes in the valence band.

Figure 53-8*c* shows a silicon lattice in which a silicon atom (valence = 4) has been replaced by an aluminum atom (valence = 3). In this case there is a "missing" electron, and it is easy for the aluminum ion core to "steal" a valence electron from a nearby silicon atom, thus creating a hole in the valence band.

The aluminum atom is called an *acceptor* atom because it so readily *accepts* an electron from the valence band. The electron so accepted moves into a localized acceptor level, as Fig. 53-9*b* shows. This level is separated from the top of the valence band by an energy gap E_a, for which $E_a \ll E_g$. By controlling the concentration of acceptor atoms it is possible to greatly increase the number of holes in the valence band.

Semiconductors doped with acceptor atoms are called *p-type* semiconductors, the "*p*" standing for "positive" because the positive charge carriers in this case greatly outnumber the negative carriers. In *p*-type semiconductors the majority carriers are the holes in the valence band and the minority carriers are the electrons in the conduction band.

Table 53-2 summarizes the properties of a typical *n*-type and a typical *p*-type semiconductor. Note particularly that the donor and accep-

figure 53-9
(*a*) An *n*-type semiconductor, showing donor levels that have contributed electrons (majority carriers) to the conduction band. The small number of holes (minority carriers) in the valence band is also shown. (*b*) A *p*-type semiconductor, showing acceptor levels that have contributed holes (majority carriers) to the valence band. The small number of electrons (minority carriers) in the conduction band is also shown.

Table 53-2
Two typical extrinsic semiconductors

Matrix Material	Silicon	Silicon
Dopant	Phosphorus	Aluminum
Type of dopant	Donor	Acceptor
Type of semiconductor	*n*-type	*p*-type
Dopant valence*	5 (= 4 + 1)	3 (= 4 − 1)
Dopant energy gap**	45 meV	57 meV
Majority carriers	Electrons	Holes
Minority carriers	Holes	Electrons
Dopant ion core charge***	$+e$	$-e$

* The valence of silicon is 4.
** Recall that 1 meV = 10^{-3} eV. The energy gap in pure silicon is 1100 meV (= 1.1 eV).
*** The dopant ion cores are fixed in the lattice and are not carriers.

(a)

(b)

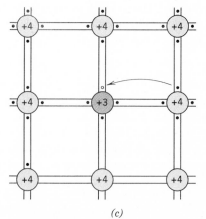

(c)

figure 53-8
(*a*) A two-dimensional representation of the silicon lattice. Each silicon ion core (charge = $+4e$) is bonded to each of its four nearest neighbors by a shared two-electron bond. The dots show these valence electrons. (*b*) A phosphorus atom (valence = $+5e$) is substituted for a silicon atom, creating a donor site. (*c*) An aluminum atom (valence = $+3e$) is substituted for a silicon atom, creating an acceptor site.

tor ion cores, although they are charged, are not charge carriers because, at normal temperatures, they remain fixed in their lattice sites.

EXAMPLE 6

Doping a silicon lattice. The density of conduction electrons in pure silicon at room temperature is about 10^{16} m^{-3}. Assume that, by doping the lattice with phosphorus, you want to increase this number by a factor of a million (10^6). What fraction of the silicon atoms must you replace by phosphorus atoms? (Assume that, at room temperature, the thermal agitation is effective enough so that essentially every phosphorus atom donates its "extra" electron to the conduction band.)

The density of the phosphorus atoms must be about $(10^{16}$ m$^{-3})(10^6)$ or $\sim 10^{22}$ m^{-3}. The density of silicon atoms in a pure silicon lattice may be found from

$$n_{Si} = \frac{N_A d}{A},$$

in which N_A is the Avogadro constant, d ($= 2330$ kg/m^3) is the density of silicon, and A ($= 28.1$ g/mol) is the atomic weight of silicon. Substituting yields

$$n_{Si} = \frac{(6.02 \times 10^{23} \text{ mol}^{-1})(2330 \text{ kg/m}^3)}{0.0281 \text{ kg/mol}} \cong 5 \times 10^{28} \text{ m}^{-3}.$$

The ratio of these two number densities is the quantity we are looking for. Thus

$$\frac{s_{Si}}{n_p} = \frac{5 \times 10^{28} \text{ m}^{-3}}{10^{22} \text{ m}^{-3}} \cong 5 \times 10^6.$$

We see that if only *one silicon atom in five million* is replaced by a phosphorus atom, the number of electrons in the conduction band will be increased by a factor of 10^6.

How can such a tiny admixture of phosphorus atoms have such a big effect? The answer is that, for pure silicon at room temperature, there were not many conduction electrons there to start with! The density of conduction electrons was $\sim 10^{16}$ m^{-3} before doping and $\sim 10^{22}$ m^{-3} after doping. For copper, however, the conduction electron density (see Table 53-1) is $\sim 10^{29}$ m^{-3}. Thus, even *after* doping, the conduction electron density of silicon remains much less than that of a typical metal such as copper.

EXAMPLE 7

A closer look at a donor atom. Assume that the "extra" electron in a phosphorus donor atom moves in a Bohr orbit around the central phosphorus ion core, as in Fig. 53-8b. Calculate (a) the binding energy and (b) the orbit radius for this electron.

(a) The Bohr theory expression for the binding energy is

$$E_b\ (= -E) = \frac{mZ^2 e^4}{8\epsilon_0^2 h^2}\frac{1}{n^2}. \qquad [49\text{-}30]$$

Here we put $n = 1$ because we are interested in the ground state of the donor atom and $Z = 1$ because the orbiting electron "sees" a net central charge of $+e$.

We derived Eq. 49-30 above by considering a hydrogen-like atom, its orbiting electron moving in a vacuum. In this case, however, the electron moves through a silicon lattice. One effect of this is to reduce the electrostatic force by a factor of κ, the dielectric constant of silicon. To realize this force reduction quantitatively, we must replace ϵ_0 in Coulomb's law by $\kappa\epsilon_0$. Making the same replacement in Eq. 49-30 above leads to

$$E_b = \frac{1}{\kappa^2}\left(\frac{me^4}{8\epsilon_0^2 h^2}\right),$$

in which the factor in parentheses is just 13.6 eV, the binding energy of the hydrogen atom. For silicon we have $\kappa = 12$, so that

$$E_b = \frac{13.6 \text{ eV}}{(12)^2} \cong 0.09 \text{ eV} = 90 \text{ meV}.$$

This result is in rough order-of-magnitude agreement with the value of 45 meV listed in Table 53-2.

(b) The orbit radius follows from Eq. 49-32. Substituting as above leads to

$$r = \kappa \left(\frac{\epsilon_0 h^2}{\pi m e^2} \right).$$

The factor in parentheses is just the Bohr radius (= 52.9 pm). Thus

$$r = (12)(52.9 \text{ pm}) \cong 600 \text{ pm}.$$

This is about ten atomic radii, large enough to justify our treatment of the silicon lattice as a continuous medium.

Note:. We should also replace the electron mass m in the above two equations by an effective electron mass m_{eff}, to take partially into account the periodic nature of the silicon lattice potential. Doing so reduces the estimated binding energy and increases the estimated orbit radius, both changes being in the direction of improving the agreement with experiment.

53-9
THE pn JUNCTION

In the next few sections we describe some commonly used semiconducting devices, such as diode rectifiers, light-emitting diodes, and transistors. There is no end to the number of such devices that we could have chosen to describe. With today's technology, in fact, it is possible to tailor-make complex semiconducting devices to meet specific needs, a specialty known as *band-gap engineering*.

Essentially all semiconducting devices involve one or more so-called *pn junctions*. Before we can understand how even the simplest of such devices work, we must first be very clear about what a *pn* junction is and what its properties are.

What is a *pn* junction? Pass a hypothetical plane across a rod of a pure crystalline semiconducting material such as silicon. Dope the rod on one side of the plane with donor atoms (thus creating *n*-type material) and on the other side with acceptor atoms (thus creating *p*-type material.) You have just made a *pn* junction.*

Figure 53-10*a* represents a *pn* junction at the imagined moment of its creation, showing the majority carriers only. We note the expected abundance of electrons in the *n*-type material and of holes in the *p*-type material. Let us focus our attention on the electrons, ignoring the holes for the time being.

Those electrons close to the junction plane will tend to diffuse across it, for much the same reason that gas molecules will diffuse through a

---

* In common practice, to make a *pn* junction one starts with, say, *n*-type material, made by adding donor atoms to the molten silicon from which the solid silicon crystal is drawn. Acceptor atoms are then diffused into the solid sample at high temperature in a special furnace, overcompensating the donor atoms to a certain (controllable) depth below the surface. The junction that we analyze here is idealized in that we assume that the *n*-type and the *p*-type regions are separated by a well-defined plane; in practice, these regions blend into each other gradually.

permeable membrane into a vacuum beyond it. The diffusing electrons, which move from right to left in Fig. 53-10*a*, will readily combine with the holes that they find on the other side of the junction plane.

For every such diffusion–recombination event, the portion of the bar on the right side of this plane acquires a positive charge and the portion on the left side a negative charge. These charges* cause a potential difference to build up across the junction, as Fig. 53-10*c* shows. Related to the potential difference (by the familiar equation $E = -dV/dx$) is an internal electric field E_0 that appears across the junction plane, pointing as shown in Fig. 53-10*c*. This field exerts a force on an electron that opposes its motion of diffusion. Put another way, for an electron to succeed in diffusing from right to left in Fig. 53-10*b*, it must be energetic enough to overcome the potential barrier represented by Fig. 53-10*c*.

While the electrons are diffusing from right to left in Fig. 53-10*a*, the holes are diffusing from left to right. We leave it to you to convince yourself that both of these motions have exactly the same effect, that is, they both operate to build up the space charges on either side of the junction plane. In so doing, they both contribute to the potential difference that builds up across the junction plane and to the related electric field E_0. Both motions constitute a current whose direction, in the usual conventional sense, is from left to right in Fig. 53-10*a*. We call this current the *diffusion current* and label it i_{diff}.

It is, of course, not possible to have an isolated silicon rod resting on a shelf with a current flowing indefinitely along its length. Something must happen to stop, or to compensate, this current. To find out what it is we turn our attention to the minority carriers.

As Fig. 53-9 and Table 53-2 show, although the majority carriers in *n*-type material are electrons, there are nevertheless also a few holes, the minority carriers. Likewise in *p*-type material, although the majority carriers are holes, there are also a few electrons.

Although the electric field in Fig. 53-10*c* acts to retard the motions of the majority carriers — being a barrier for them — it is a downhill trip for the minority carriers, be they electrons or holes. When, by thermal agitation, an electron close to the junction plane is raised from the valence band to the conduction band of the *p*-type material in Fig. 53-10*a*, it drifts steadily from left to right across the junction plane, swept along by the electric field E_0. Similarly, if a hole is created in the *n*-type material, it, too, drifts across to the other side. The space-charge region shown in Fig. 53-10*b* is effectively swept free of charge carriers by this process and, for that reason, we call it the *depletion zone*. The current represented by the motions of the minority carriers, called the *drift current* i_{drft}, is in the opposite direction to the diffusion current and just compensates it at equilibrium, as Fig. 53-10*d* shows.

Thus, at equilibrium, a *pn* junction resting on a shelf develops a contact potential difference V_0 between its ends. The diffusion current i_{diff} that moves through the junction plane from the direction *p* to *n* is just balanced by a drift current i_{drft} that moves in the opposite direction. An electric field E_0 acts across the depletion layer, whose width is d_0.

---

* The fixed charges, which are close to — and separated by — the junction plane, are those of the donor and the acceptor ion cores which, we recall, are not mobile. Normally the charges of these ion cores are compensated by the (opposite) charges of the mobile charge carriers. However, when charge carriers cross the junction plane, the ion cores are no longer fully compensated and are, so to speak, uncovered.

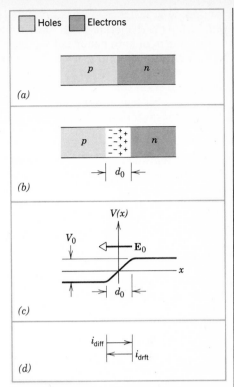

figure 53-10

(*a*) A *pn* junction at the imagined moment of its creation. Only the majority carriers are shown. (*b*) Diffusion of majority carriers across the junction plane causes a space charge to appear. (*c*) The space charge causes a contact potential difference V_0 and a related electric field E_0 to develop across the junction plane. (*d*) The diffusion of majority carriers across the junction plane is just balanced by the drift of minority carriers in the opposite direction.

Exploring the depletion zone. A silicon-based *pn* junction has an equal concentration n_0 of both donor and acceptor atoms. Its depletion zone, of width d, is symmetrical about the junction plane, as Fig. 53-11a shows.

EXAMPLE 8

(*a*) Derive an expression for E_{max}, the maximum intensity of the electric field in the depletion zone.

The electric field may be taken as zero in the *n*-type and the *p*-type material outside the depletion zone. The field points from right to left within the depletion zone and, from symmetry, has its maximum value in the center of this zone; see Fig. 53-11b.

Let us apply Gauss's law to the closed "box" (Gaussian surface) shown in Fig. 53-11a. This law is

$$\epsilon_0 \oint \kappa E \cdot dS = q, \qquad [30\text{-}17]$$

in which $\kappa \ (= 12)$ is the dielectric constant of silicon and $q \ [= n_0 eA(d/2)]$ is the free charge contained within the box. The integral is to be taken over the surface of the box.

The only contribution to the integral comes from the face of the box that lies in the junction plane so that the integral has the value $\kappa E_{max} A$. Making these substitutions in Eq. 30-17 and solving for E_{max} yields

$$E_{max} = \frac{n_0 e d}{2 \kappa \epsilon_0}, \qquad (53\text{-}7)$$

the relationship we seek.

(*b*) Derive an expression for V_0, the potential difference that exists across the depletion zone; see Fig. 53-10c.

As Fig. 53-11b shows, the electric field drops linearly from its central value of E_{max} to zero at each edge of the depletion zone. Its *average* value throughout the zone is thus $\frac{1}{2}E_{max}$. The potential difference V_0 is equal to the work per unit charge required to carry a test charge q_0 from one face of the depletion zone to the other. Thus, if F is the average force acting on the test charge,

$$V_0 = \frac{W}{q_0} = \frac{Fd}{q_0} = \frac{(\frac{1}{2}E_{max}q_0)(d)}{q_0} = \frac{1}{2}dE_{max}.$$

Substituting for E_{max} from Eq. 53-7 above leads to

$$V_0 = \frac{n_0 e d^2}{4 \kappa \epsilon_0}. \qquad (53\text{-}8)$$

(*c*) Assume that $n_0 = 3 \times 10^{22} \ \text{m}^{-3}$ and that V_0 is measured to be 0.6 V. Calculate the width of the depletion zone from Eq. 53-8 above.

Solving Eq. 53-8 for d and substituting the given values leads to

$$d = \left(\frac{4 \kappa \epsilon_0 V_0}{e n_0} \right)^{1/2}$$

$$= \left[\frac{(4)(12)(8.85 \times 10^{-12} \ \text{F/m})(0.60 \ \text{V})}{(1.60 \times 10^{-19} \ \text{C})(3 \times 10^{22} \ \text{m}^{-3})} \right]^{1/2}$$

$$= 2.3 \times 10^{-7} \ \text{m} = 230 \ \text{nm}.$$

(*d*) Using this value of d, calculate the value of E_{max} from Eq. 53-7. Substituting in this equation leads to

$$E_{max} = \frac{(3 \times 10^{22} \ \text{m}^{-3})(1.60 \times 10^{-19} \ \text{C})(2.3 \times 10^{-7} \ \text{m})}{(2)(12)(8.85 \times 10^{-12} \ \text{F/m})}$$

$$= 5.2 \times 10^6 \ \text{V/m}.$$

What assumptions were made in this problem that might lead to different values of the calculated quantities under practical laboratory conditions?

(*a*)

(*b*)

figure 53-11
(*a*) The depletion zone of a *pn* junction. The rectangle marks a Gaussian surface, the area of its end caps being A. (*b*) Showing how the electric field E varies throughout the depletion zone.

Although a *pn* junction can be used in many ways, it is basically a rectifier. That is, if you connect it across the terminals of a battery, the current (a few picoamperes) in the circuit will be very much smaller for one polarity of the battery connection than for the other. Figure 53-12 shows that, for a typical silicon-based *pn* junction diode, the current for the so-called back-bias connection is negligible by comparison with the current for the forward-bias connection.

Figure 53-13 shows one of many possible applications of a diode rectifier. A sine wave input potential generates a half-wave output potential, the diode rectifier acting as essentially a short circuit for one polarity of the input potential and as essentially an open circuit for the other. An ideal diode rectifier, in fact, has only these two modes of operation. It is either ON (zero resistance) or OFF (infinite resistance.)

Figure 53-13 displays the conventional symbol for a diode rectifier. The arrow head corresponds to the *p*-type terminal of the device and points in the direction of "easy" conventional current flow. That is, the diode is ON when the terminal with the arrow head is (sufficiently) positive with respect to the other terminal.

Figure 53-14 shows details of the two connections. In Fig. 53-14*a* — the back-bias arrangement — the battery emf simply adds to the contact potential difference, thus increasing the height of the barrier that the majority carriers must surmount. Fewer of them can do so and, as a result, the diffusion current decreases markedly.

The drift current, however, senses no barrier and thus is independent of the magnitude or direction of the external potential. The nice current balance that existed at zero bias (see Fig. 53-10*d*) is thus upset and, as shown in Fig. 53-14*a*, a current — but a very small one — appears in the circuit.

Another effect of back-bias is to widen the depletion zone, as a comparison of Figs. 53-10*b* and 53-14*a* shows. This seems reasonable because the positive battery terminal, connected to the *n*-type end of the junction, tends to pull electrons out of the depletion zone back into the *n*-type material and to repel holes back into the *p*-type material. Because the depletion zone contains very few charge carriers, it is a region of high resistivity. Thus its substantially increased width means a substantially increased resistance, consistent with the small value of the back-bias current.

Figure 53-14*b* shows the forward-bias connection, the positive terminal of the battery being connected to the *p*-type end of the *pn* junction. Here the applied emf *subtracts* from the contact potential, the diffusion current *rises* substantially, and a relatively *large* net forward current results. The depletion zone *narrows*, its low resistance being consistent with the large forward current.

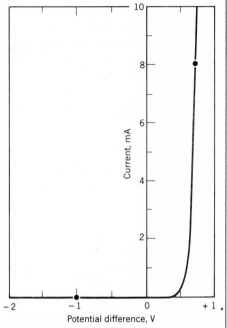

figure 53-12
The current–voltage curve for a typical silicon-based *pn* junction. It clearly conducts much more readily for one polarity of the applied potential difference than for the other. The dots refer to Problem 42.

figure 53-13 A *pn* junction diode connected as a so-called half-wave rectifier.

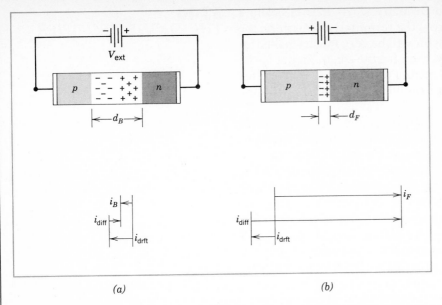

(a)   (b)

figure 53-14
(a) The back-bias connection of a *pn* junction, showing the wide depletion zone and the small back current i_B. (b) The forward-bias connection, showing the narrow depletion zone and the large forward current i_F. Note that the drift current i_{drft} is the same in each case.

We are all familiar with the brightly colored numbers that flash and glow at us from cash registers, gasoline pumps, and pocket calculators. In nearly all cases this light is emitted from an assembly of *pn* junctions operating as *light-emitting diodes* (LEDs).

Figure 53-15a shows the familiar seven-segment display from which the numbers are formed. Figure 53-15b shows that each element of this display is the end of a flat plastic lens, at the other end of which is a small LED, possibly about 1 mm² in area. Figure 53-15c shows a typical circuit, in which the LED is forward-biased; note the customary symbol.

How can a *pn* junction emit light? When an electron at the bottom of the conduction band of a semiconductor falls into a hole at the top of the valence band, an energy E_g is released, where E_g is the gap width. What happens to this energy? There are at least two possibilities. It might be transformed into thermal energy of the vibrating lattice and, with high probability, that is exactly what happens in a silicon-based semiconductor.

In some semiconducting materials, however, conditions are such that the emitted energy can also appear as electromagnetic radiation, the wavelength being given by

$$\lambda = \frac{c}{v} = \frac{c}{E_g/h} = \frac{hc}{E_g}. \tag{53-9}$$

53-11
THE LIGHT-EMITTING DIODE

(a)   (c)   (b)

figure 53-15
(a) The familiar seven-segment number display, activated to show the number "7." (b) One segment from such a display. (c) An LED connected to an external emf.

One segment

Plastic lens

LED

Commercial LEDs designed for the visible region are usually based on a semiconducting material that is a suitably chosen gallium-aresnic-phosphorus compound.* By adjusting the ratio of phosphorus to arsenic the gap width — and thus the wavelength of the emitted light — can be tailored to suit.

A question arises. If light is emitted when an electron falls from the conduction band to the valence band, will not light of that same wavelength be absorbed when an electron moves in the other direction, that is, from the valence band to the conduction band? It will indeed. To avoid having all of the emitted photons absorbed, it is necessary to have a great surplus of both electrons and holes present in the material, in much greater numbers than would be generated by thermal agitation in the intrinsic semiconducting material. These are precisely the conditions that result when majority carriers — be they electrons or holes — are injected across the central plane of a *pn* junction by the action of an external potential difference. That is why a simple intrinsic semiconductor will not serve as an LED. You need a *pn* junction! To provide lots of majority carriers — and thus lots of photons — it should be heavily doped and strongly forward-biased.

Apart from their use in visual displays, LEDs operating in the infrared are much used in optical communication systems, using optical fibers. The infrared region is chosen because the absorption per unit length of such fibers has well-defined minima at two different wavelengths in this region. In a development of the LED, the ends of a suitable *pn* junction crystal are polished so that a slice of the crystal across the junction plane serves as a laser. Such a device is called a *laser diode*. Figure 53-16 suggests the tiny scale of such devices.

figure 53-16
A laser diode developed at the AT&T Bell Laboratories. It is about the size of a grain of sand. The laser element is a groove at the bottom of the chip, only 5 μm wide. (Ivan Kaminov/AT&T Bell Laboratories.)

Refinements of this device are now routinely used in laser-disk audio systems and for the transmission of telephone messages and other signals over an intercity network of optical fibers. A transatlantic optical cable, designed to handle up to 40,000 simultaneous telephone conversations, is planned for service in 1988.

* Note that intrinsic semiconductors need not be simple elements, such as silicon.

EXAMPLE 9

A light emitting diode. An LED is constructed from a *pn* junction based on a certain Ga-As-P semiconducting material, whose energy gap is 1.9 eV. What is the wavelength of its emitted light?

From Eq. 53-9 we have

$$\lambda = \frac{hc}{E_g} = \frac{(6.63 \times 10^{-34} \text{ J} \cdot \text{s})(3.00 \times 10^8 \text{ m/s})}{(1.9 \text{ eV})(1.60 \times 10^{-19} \text{ J/eV})}$$

$$= 6.53 \times 10^{-7} \text{ m} = 653 \text{ nm}.$$

Light of this wavelength is red.

53-12
THE TRANSISTOR

The devices we have discussed so far have been diodes, that is, two-terminal devices. Here we introduce a three-terminal device, a *transistor*. As Fig. 53-17 suggests, the function of a transistor is to control a current flowing through the device from terminal *D* (the *drain*) to terminal *S* (the *source*) by varying the potential of terminal *G* (the *gate*).

For many applications, particularly those involving computers, we only need to be able to turn the drain–source current ON (gate open) or OFF (gate closed.) One of these conditions corresponds to a "0" and the other to a "1" in the binary arithmetic on which the computer logic is based. We wish further that the gate terminal draw essentially no current from the circuit to which it is attached, thus interfering with its operation as little as possible. In more formal language, we say that we wish the transistor to have a high input impedance.

The central question proves to be: "How can we control the current in a conductor without making direct electric contact with it?" The perhaps surprising answer is that, by using the variable gate potential that is at our disposal, we can change the effective cross-sectional area of the conductor, going even so far as to reduce it to zero.

Figure 53-14 gives a clue. Here we see that, by changing the bias potential of the *pn* junction, we can control the width of the depletion zone. Simply by changing a potential we can effectively transform a conductor (*n*-type or *p*-type material) into a nonconductor (the depletion zone material.) We can use the same trick in the transistor, but with a different geometrical arrangement.

Of the several types of transistor that are in common use, we choose to describe the MOSFET (**M**etal-**O**xide-**S**emiconductor **F**ield-**E**ffect **T**ransistor).* Figure 53-18 shows its essential features.

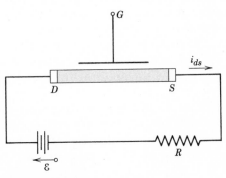

figure 53-17
A schematic representation of a transistor, showing a current i_{ds} moving through the device from the drain *D* to the source *S*. The magnitude of the current is controlled by the potential applied to *G*, the gate terminal.

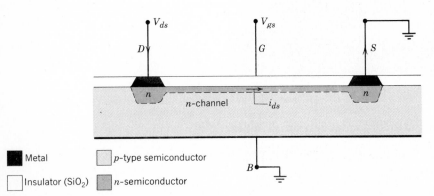

figure 53-18
Showing the construction of a MOSFET. The source *S* and the base *B* are grounded and a potential V_{ds} is applied to the drain terminal *D*. The magnitude of the current is controlled by the gate potential V_{gs}.

Metal
Insulator (SiO₂)
p-type semiconductor
n-semiconductor

* Sometimes also known as the IGFET (Insulated-Gate Field-Effect Transistor) or the MOST (Metal-Oxide-Semiconductor Transistor).

A lightly doped p-type substrate has imbedded in it two "islands" of heavily doped n-type material, forming the drain D and the source S. These terminals are connected by a thin channel of n-type material, called the n-channel. An insulating layer of silicon dioxide (hence Oxide in the acronym) is deposited on the substrate and penetrated by two metallic contacts (hence Metal) at D and S, so that electrical contact can be made with the drain and the source. A thin metallic layer—the gate G—is deposited opposite to the n-channel. Note that the gate makes no ohmic contact with the transistor proper, being separated from it by the insulating oxide layer. Thus a MOSFET has the desired high input impedance, perhaps as high as 10^{15} Ω.

Consider first the situation with the source and the substrate grounded, the gate "floating" (that is, not connected electrically to a source of emf), and a positive potential V_{ds} applied to the drain. A drain–source current i_{ds} will be set up, as shown.

The potential difference across the boundary between the n-channel and the p-type substrate will vary from zero at the source end of the channel to V_{ds} at the drain end. The polarity is such (compare Fig. 53-14) that the pn junction that exists at this boundary is back-biased for essentially its full length. A depletion zone will exist at this boundary, increasing in thickness from the source end of the channel to the drain end. For these conditions, the n-channel will not have the same cross-sectional area along its length, being invaded by the depletion zone to a greater and greater extent as one proceeds along the channel from the source toward the drain.

The thickness of the depletion zone along its length can be influenced by the potential that we choose to apply to the gate. If we make the gate negative with respect to the source, electrons will be repelled from the n-channel into the substrate, thus widening the depletion zone, constricting the channel and decreasing the drain–source current. Alternatively, a positive gate potential will attract electrons into the n-channel, narrow the depletion zone, widen the conducting channel, and increase the drain–source current. In this way a small change in the gate potential can generate a substantial change in the drain–source current, much as a valve controls the flow of water through a pipe.

Figure 53-19 shows a MOSFET (note the descriptive symbol) connected into a circuit as an amplifier. The input signal is applied to the gate and the output appears as a varying potential difference across the load resistor R. MOSFETs are available commercially as individual packaged units and are also readily adaptable to incorporation in integrated circuits such as that shown in Fig. 53-1b. The fabrication of such circuits is an incredible technological achievement.

figure 53-19
A MOSFET connected as an amplifier. A (time-varying) input signal V_{in} generates an amplified output signal, V_{out}. The letters G, D, B, and S have the same meaning as in Fig. 53-18. Note that, in this application, the gate G is connected directly to the source S.

questions

1. The conduction electrons in a metallic sphere occupy states of quantized energy. On the free electron gas model, does the average energy interval between adjacent states depend on (a) the material of which the sphere is made, (b) the radius of the sphere, (c) the energy of the state, or (d) the temperature of the sphere?

2. What role does the Pauli exclusion principle play in accounting for the electrical conductivity of a metal?

3. In what ways do the free electron gas model (see Section 31-4) and the Fermi-Dirac model (see Fig. 53-3) for the electrical conductivity of a metal differ?

4. If we compare the conduction electrons of a metal with the atoms of an

ideal gas, we are surprised to note (see Fig. 53-2b) that so much kinetic energy is locked into the conduction electron system at the absolute zero. Would it be better to compare the conduction electrons, not with the atoms of a gas, but with the inner electrons of a heavy atom? After all, a lot of kinetic energy is also locked up in this case, and we don't seem to find that surprising. Discuss.

5. What features of Fig. 53-3 make it specific for copper? What features are independent of the identity of the metal?

6. Why do the curves in Fig. 53-2b and 53-3c differ so little from each other?

7. Distinguish carefully among (a) the density of states function $n(E)$, (b) the density of occupied states function $n_o(E)$, and (c) the Fermi-Dirac probability function $p(E)$, all of which appear in Eq. 53-4.

8. Does the Fermi energy for a given metal depend on the volume of the sample? If, for example, you compare a sample whose volume is 1 cm³ with one whose volume is twice that, the latter sample has just twice as many available conduction electrons; it might seem that you would have to go to higher energies to fill its available levels. Do you?

9. In Section 23-8 we showed that the (molar) heat capacity of an ideal monatomic gas is $\frac{3}{2}R$. If the conduction electrons in a metal behaved like such a gas, we would expect them to make a contribution of about this amount to the measured specific heat of a metal. However, this measured specific heat can be accounted for quite well in terms of energy absorbed by the vibrations of the ion cores that form the metallic lattice. The electrons do not seem to absorb much energy as the temperature of the specimen is increased. How does Fig. 53-3 provide an explanation of this prequantum-days puzzle?

10. Distinguish between the drift speed and the Fermi speed of the conduction electrons in a metal.

11. To generate a current in a conductor, is it enough for every electron in the conduction band to increase its energy? The magnitude of its momentum? What is the critical requirement?

12. At room temperature a given applied electric field will generate a drift speed for the conduction electrons of silicon that is about 40 times as great as that for the conduction electrons of copper. Why isn't silicon a better conductor of electricity than copper?

13. Give a physical argument to account qualitatively for the existence of allowed and forbidden energy bands in solids. Why is it that no hint of bands appears in Fig. 53-2b?

14. Why is it that, in a solid, the allowed bands become wider as one proceeds from the inner to the outer atomic electrons?

15. Is the existence of a forbidden energy gap in an insulator any harder to accept than the existence of forbidden energies for an electron in, say, the hydrogen atom?

16. What can band theory tell us about solids that the free electron gas model (see Section 31-4) cannot?

17. On the band theory picture, what are the essential requirements for a solid to be (a) a conductor, (b) an insulator, or (c) a semiconductor?

18. Discuss this sentence: "The distinction between a conductor and a semiconductor is sharp and clear cut, but that between a semiconductor and an insulator is not."

19. How do you account for the fact that the resistivity of metals increases with temperature but that of semiconductors decreases (see Table 53-1)?

20. Consider these two statements: (a) At low enough temperatures silicon ceases to be a semiconductor and becomes a rather good insulator. (b) At high enough temperatures silicon ceases to become a semiconductor and becomes a rather good conductor. Discuss the extent to which each statement is either true or not true.

21. The energy gaps for the semiconductors silicon and germanium are 1.14 eV and 0.67 eV, respectively. Which substance do you expect would have the higher density of charge carriers at room temperature? At the absolute zero of temperature?

22. In a semiconductor we have seen that a hole behaves like a particle of charge $+e$. Will an electron and a hole experience the same magnitude of force in response to an applied electric field? Will they achieve the same equilibrium drift speed?

23. A semiconductor contains equal numbers of donor and acceptor impurities. Do they cancel each other in their electrical effects? If so, what is the mechanism? If not, why not?

24. Why does an n-type semiconductor have so many more electrons than holes? Why does a p-type semiconductor have so many more holes than electrons? Explain in your own words.

25. What elements other than phosphorus are good candidates to use as donor impurities in silicon? What elements other than aluminum are good candidates to use as acceptor impurities? Consult the periodic table.

26. Does one distinguish between majority and minority carriers for an intrinsic semiconductor such as silicon or germanium? If not, why not? If so, what criterion do you use?

27. In preparing n-type or p-type semiconductors by doping, why is it extremely important to avoid contamination of the sample with even very small concentrations of unwanted impurities?

28. Do (intrinsic) semiconductors obey Ohm's law?

29. Would you expect doping to change the resistivity of silicon by very much?

30. Germanium and silicon are similar semiconducting materials whose principal distinction is that the gap width E_g (see Fig. 53-9) is 0.67 eV for the former and 1.14 eV for the latter. If you wished to construct a pn junction (see Fig. 53-10) in which the back current is to be kept as small as possible, which material would you choose and why?

31. In a pn junction (see Fig. 53-10) we have seen that electrons and holes may diffuse, in opposite directions, through the junction region. What is the eventual fate of each such particle as it diffuses into the material on the opposite side of the junction?

32. Consider two possible techniques for fabricating a pn junction (see Fig. 53-10). (a) Prepare separately an n-type and a p-type sample and join them together, making sure that their abutting surfaces are plane and highly polished. (b) Prepare a single n-type sample and diffuse an excess acceptor impurity into it from one face, at high temperature. Which method is preferable and why?

33. The pn junction shown in Fig. 53-11a has equal dopant concentrations on each side of its junction plane. Suppose, however, that the donor concentration were significantly greater than the acceptor concentration. Would the depletion zone still be symmetrically located about the junction plane? If not, would the central plane of the zone move toward the n-type or toward the p-type face of the junction? Give your reason.

34. Why can't you measure the contact potential difference generated at a pn junction by simply connecting a voltmeter across it?

35. Explain in your own words how the thickness of the depletion zone of a pn junction can be decreased by (a) increasing the forward-bias voltage and (b) increasing the dopant concentration.

36. In Fig. 53-10b, why does the depletion zone build up close to the junction plane? Why does it not spread out throughout the volume of the sample?

37. If you increase the temperature of a back-biased pn junction, what happens to the current (see Fig. 53-14a)? Is the effect larger for silicon or for germanium? (The intrinsic energy gap E_g for silicon is larger than that for germanium.)

38. Does the diode rectifier whose characteristics are shown in Fig. 53-12 obey Ohm's law? What is your criterion for deciding?

39. We have seen that a simple intrinsic semiconductor cannot be used as a light emitting diode. Why not? Would a heavily doped n-type or p-type semiconductor work?

40. Explain in your own words how the MOSFET device of Fig. 53-18 works.

problems

SECTION 53-2

1. Calculate and compare the number of particles per unit volume for (a) the molecules of oxygen gas at standard temperature and pressure and (b) the conduction electrons in copper. (c) What is the ratio of these numbers? (d) What is the average distance between particles in each case? (Assume that this distance is the edge of a cube whose volume is equal to the volume per particle.)
   Answer: (a) 2.7×10^{25} m^{-3}; (b) 8.4×10^{28} m^{-3}; (c) 3100; (d) 3.3 nm for oxygen, 0.23 nm for the electrons.

2. At what pressure would an ideal gas have a density of molecules equal to that of the density of the conduction electrons in copper (= 8.5×10^{28} m^{-3})? Assume that $T = 300$ K.

3. Gold is a monovalent metal with an atomic weight of 197 g/mol and a density of 1.93×10^4 kg/m^3. What is the density of charge carriers?
   Answer: 5.90×10^{28} m^{-3}.

4. Show that Eq. 53-2 can be written as

$$n(E) = CE^{1/2},$$

   where $C = 6.78 \times 10^{27}$ m^{-3} eV$^{-3/2}$. Use this relation to verify a calculation of Example 1, namely that for $E = 5.00$ eV, $n(E) = 1.52 \times 10^{28}$ m^{-3} eV^{-1}.

5. Calculate the density $n(E)$ of conduction electron states in a metal for $E = 8.0$ eV and show that your result is consistent with the curve of Fig. 53-2a.   *Answer:* 1.98×10^{28} m$^{-3} \cdot$ eV^{-1}.

SECTION 53-3

6. Verify that at $T = 0$, the Fermi-Dirac probability function $p(E)$ is 1 for energies less than the Fermi energy, $\frac{1}{2}$ when they are equal, and 0 when they are greater.

7. Prove that the occupancy probabilities for two states whose energies are equally spaced above and below the Fermi energy add up to 100%.

8. Show that Eq. 53-6 can be written as

$$E_F = An^{2/3},$$

   where the constant A has the value 3.65×10^{-19} m^2 eV.

9. Show that at the absolute zero of temperature, the average energy of the conduction electrons in a metal is equal to $(\frac{3}{5})E_F$, where E_F is the Fermi energy. Evaluate for copper and sketch in this line in Fig. 53-3c.
   Answer: 4.21 eV.

10. Using the result of Problem 9 estimate how much energy would be released by the conduction electrons in a penny (assumed all copper; mass = 3.1 g) if we could suddenly turn off the Pauli exclusion principle. For how long would this amount of energy light a 100-W lamp? Note: There is no known way to turn off the Pauli principle!

11. The Fermi energy of aluminum is 11.7 eV; its density is 2.70 g/cm^3 and its atomic weight is 27.0 g/mol. From these data determine the number of conduction electrons per atom of aluminum.   *Answer:* 3.

12. The *Fermi temperature* is defined as a temperature such that $kT_F = E_F$. (a) What is the Fermi temperature for copper? (b) Describe qualitatively what

the distribution in energy of the conduction electrons would look like at temperatures near the Fermi temperature.

13. Show that the density of states function given by Eq. 53-2 can be written in the form

$$n(E) = 1.5nE_F^{-3/2}E^{1/2}.$$

Explain how it can be that $n(E)$ is independent of material when the Fermi energy E_F (= 7.0 eV for copper, 9.4 eV for zinc, etc.) appears explicitly in this expression.

14. A neutron star can be analyzed by techniques similar to those used for ordinary metals. In this case the neutrons (rather then electrons) obey the distribution function, Eq. 53-5. Consider a neutron star of 2 solar masses with a radius of 10 km. Calculate the Fermi energy of the neutrons.

15. White dwarf stars represent a late stage in the evolution of stars like the sun. They become dense enough and hot enough that we can analyze their structure as a solid in which all Z electrons per atom are free. For a white dwarf with a mass equal to that of the sun and a radius equal to that of the earth, calculate the Fermi energy of the electrons. Assume the atomic structure to be represented by iron atoms. *Answer:* 0.234 MeV.

16. Figure 53-3c shows the energy distribution $n_o(E)$ of the conduction electrons in copper at 1000 K. Calculate $n_o(E)$ for $E = 4.00$, 6.75, 7.00, and 9.00 eV and verify that the curve shown in the figure is correct at those energies. The Fermi energy of copper is 7.00 eV.

17. Carry out the calculation of Example 4 for the case of $p(E) = 0.10$. Compare your result carefully with Fig. 53-3c.
Answer: (a) 7.19 eV; (b) 1.82×10^{28} m^{-3} · eV^{-1}; (c) 1.82×10^{27} m^{-3} · eV^{-1}.

18. In Example 7 of Chapter 50 we considered the problem of an electron trapped in an infinitely deep well whose width was 100 pm. Consider that 100 electrons are placed in the well, two to a level, with opposite spins. What is the Fermi energy of this system? (*Note:* The Fermi energy is the energy of the highest occupied level at the absolute zero of temperature.)

19. Convince yourself, from an inspection of Figs. 53-3b and 53-3c, that a rough but reasonable estimate of the fraction of the conduction electrons in a metal that are in excited states at any temperature T is

$$\frac{(kT)\, n(E_F)}{n},$$

where n is the total number of conduction electrons per unit volume. This expression should be valid to within a factor of 2 or so. Evaluate for $T = $ (a) zero, (b) 300 K, and (c) 1000 K. For copper, $E_F = 7.0$ eV and $n = 8.5 \times 10^{28}$ m^{-3}. *Answer:* (a) Zero; (b) 0.5%; (c) 1.8%.

20. Compare how (a) the molecular speeds of an ideal gas, (b) the spectral radiancies for cavity radiation, and (c) the energies of the conduction electrons in a metal are described. Stress the formal similarities and draw rough sketches of the distribution functions.

21. Equation 24-2, which, with a small change of notation, is

$$n(v) = 4\pi n_0 \left(\frac{m}{2\pi kT}\right)^{3/2} v^2 e^{-mv^2/2kT},$$

gives the speed distribution for the molecules of an ideal gas at temperature T. Show that this can be rewritten in the form of an energy distribution, as

$$n(E) = \frac{1.13 n_0}{(kT)^{3/2}} E^{1/2} e^{-E/kT},$$

in which E (= $\frac{1}{2}mv^2$) is the kinetic energy. [*Hint:* Consider a group of particles whose speeds lie in a selected speed range dv, or in the corresponding kinetic energy range dE. The number of particles in this group can be written either as $n(v)\ dv$ or as $n(E)\ dE$, so that these two expressions can be set equal to each other.]

22. It can be shown that the conduction electrons in a metal do behave like an ideal gas of the ordinary kind if the temperature is high enough. In particular, we must have a temperature such that $kT \gg E_F$, the Fermi energy. What temperatures are required for copper ($E_F = 7.0$ eV) for this to be true? Study Fig. 53-3c in this connection and note that we have $kT \ll E_F$ for the conditions of that figure. This is just the reverse of the requirement cited above. Note also that copper boils at 2870 K!

23. Show that if $E \gg E_F$, the distribution in energy of the occupied states $n_o(E)$ can be written as

$$n_o(E) \cong CE^{1/2}e^{-E/kT},$$

in which C is a constant. Compare this result with that calculated for the Maxwell speed distribution in Problem 21. What do you conclude?

SECTION 53-4

24. Zinc is a bivalent metal with an atomic weight of 65.4 g/mol and a density of 7.1×10^3 kg/m³. Calculate (a) the number of conduction electrons per unit volume, (b) the Fermi energy E_F, (c) the Fermi speed v_F, and (d) the de Broglie wavelength corresponding to this speed.

25. Silver is a monovalent metal with an atomic weight of 107.9 g/mol and a density of 1.05×10^4 kg/m³. Calculate (a) the number of conduction electrons per unit volume, (b) the Fermi energy E_F, (c) the Fermi speed v_F, and (d) the de Broglie wavelength corresponding to this speed.
Answer: (a) 5.85×10^{28} m⁻³; (b) 5.48 eV; (c) 1.39×10^6 m/s; (d) 5.24 Å.

26. The density and atomic weight of sodium are 971 kg/m³ and 23 g/mol, respectively; the radius of the ion Na⁺ is 0.98×10^{-10} m. (a) What fraction of the volume of metallic sodium is available to its conduction electrons? (b) Carry out the same calculation for copper. Its density, atomic weight, and ionic radius are, respectively, 8900 kg/m³, 63.6 g/mol, and 1.35×10^{-10} m. (c) For which of these two metals do you think the conduction electrons behave more like a free electron gas?

27. Consider the conduction electrons in copper as a free electron gas. (a) Assume first that this gas behaves like an ideal gas of the familiar kind and calculate the root-mean-square speed v_{rms}. (b) Assume now that this gas follows the quantum rules exemplified by Fig. 53-3 and calculate the Fermi speed v_F. (c) What is the ratio of these two speeds? Note that assumption (b) is consistent with experiment, but assumption (a) is not.
Answer: (a) 1.2×10^5 m/s; (b) 1.6×10^6 m/s; (c) 13.

SECTION 53-6

28. Consider a solid consisting of roughly 1 cm³ of material (about 10^{22} atoms). Estimate the number of conduction electrons at 300 K if the material is (a) a metal; (b) an intrinsic semiconductor ($E_g \cong 1$ eV); (c) an insulator ($E_g \cong 6$ eV). (d) How would your estimates differ at 1000 K? at 4 K?

29. In a semiconductor electrons can be lifted by the action of light from the valence band to the conduction band. What is the maximum wavelength that will produce photoconduction in diamond, which has a band gap of 5.5 eV?
Answer: 226 nm.

SECTION 53-8

30. Identify the following as p-type or n-type semiconductors: (a) Sb in Si; (b) In in Ge; (c) Al in Ge; (d) P in Si.

31. A silicon crystal is doped with arsenic. Is the semiconductor that results n-type or p-type? What about a germanium crystal doped with indium?
Answer: n; p.

32. What mass of phosphorus would be required to dope a one-gram silicon sample to the extent described in Example 6?

33. A silicon crystal is doped with phosphorus to a concentration of 10^{22} phosphorus atoms per cubic meter. On average, how far apart are these atoms? See Example 6. *Answer:* 46 nm.

34. In Fig. 53-8*b* we represent the "extra" electrons as being in a Bohr orbit about the dopant ion core, whose net charge is $+e$. In Fig. 53-8*c* we can choose to represent the "missing" electron as a hole circulating in a Bohr orbit about the dopant ion core, whose net charge is now $-e$. Carry out and interpret a calculation like that of Example 7 based on this model.

35. Pure silicon at room temperature has an electron density in the conduction band of approximately $1 \times 10^{16}/m^3$ and an equal density of holes in the valence band. Suppose that one of every 10^7 silicon atoms is replaced by a phosphorus atom. (*a*) Which type will this doped semiconductor be, *n* or *p*? (*b*) What charge carrier density will the phosphorus add? (The density of silicon is $2.33 \times 10^3 \, kg/m^3$ and its atomic weight is 28.1 g/mol.) (*c*) What is the ratio of the charge carrier density in the doped silicon to that for the pure silicon? *Answer:* (*a*) *n*; (*b*) $5 \times 10^{21} \, m^{-3}$; (*c*) 2.5×10^5.

36. In Fig. 53-20 two energy bands of a hypothetical solid are represented. The bands are filled to level E_x, which may be in either band 1 or band 2. There may be an impurity level at E_i. Indicate whether the solid is a conductor, insulator, intrinsic semiconductor, or extrinsic semiconductor. The impurity type may be donor, acceptor, or none, and extrinsic semiconductors may be either *p*-type or *n*-type. Complete the table.

E_x	E_i (eV)	E_b	Solid	Impurity	Extrinsic Semiconductor
3.00	—	9.00			
3.00	4.06	4.10			
3.00	—	4.10			
1.49	—	9.00			
4.40	—	4.10			
3.00	3.04	4.10			

figure 53-20
Problem 36.

37. We defined the Fermi level for a metal (see Fig. 53-2*b*) as the highest occupied level at the absolute zero of temperature. Let us now redefine it as an energy located so that the occupancy probability of a state above it, added to the occupancy probability of a state an equal interval below it, is 100%. Show that: (*a*) For a metal, the two definitions are equivalent. (*b*) For an intrinsic semiconductor, the new definition places the Fermi level in the middle of the energy gap. (*c*) For an *n*-type semiconductor, the new definition places the Fermi level in the gap but close to the bottom of the conduction band. (*d*) For a *p*-type semiconductor, the new definition places the Fermi level in the gap but close to the top of the valence band. *Note:* The new definition is the one actually used.

38. In Problem 37 we saw that the Fermi level for an intrinsic semiconductor is in the center of the energy gap and for an extrinsic semiconductor doped with donor atoms (*n*-type) it is in the gap but near the bottom of the conduction band. Consider silicon, with an energy gap of 1.11 eV. Let it be doped with donor atoms, each of which has a donor level 0.15 eV below the bottom of the conduction band. Further, suppose that the concentration of dopant atoms is such that the Fermi level is 0.11 eV below the bottom of this band; see Fig. 53-21. What is the probability that a state at the bottom of the conduction band is occupied, for (*a*) pure silicon and (*b*) doped silicon? (*c*) What is the probability that a donor state in the doped material is occupied? Assume $T = 300$ K.

39. A silicon sample is doped with 5.6×10^{16} donor atoms/m³. Each atom contributes one state, with energy 0.11 eV below the bottom of the con-

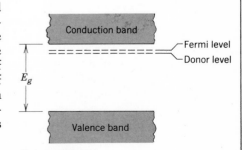

figure 53-21
Problem 38.

duction band. (a) If each of these states is occupied with probability 5.0×10^{-5} at temperature 300 K, where is the Fermi level relative to the bottom of the conduction band? (b) What then is the probability that a state at the bottom of the conduction band is occupied? See Problems 37 and 38. *Answer: (a) 0.74 eV; (b) 7.2×10^{-7}.*

SECTION 53-9

40. (a) Discuss the motions of the majority carriers (both electrons and holes) in a forward-biased pn junction. (b) Discuss the motions of the minority carriers in this same junction.

41. When a photon enters the depletion region of a pn junction, electron–hole pairs can be created as electrons absorb part of the photon's energy and are excited from the valence band to the conduction band. These junctions are thus often used as detectors for photons, especially for X rays and nuclear gamma rays. When a 662-keV gamma-ray photon is totally absorbed by a semiconductor with an energy gap of 1.1 eV, on the average how many electron–hole pairs are created?   *Answer: 6.02×10^5.*

SECTION 53-10

42. Calculate and compare the resistances of the diode rectifier for the two points shown on the characteristic curve of Fig. 53-12. The current for the left-hand dot (too small to show in the figure) is 5×10^{-11} A.

43. For an ideal pn junction diode, with a sharp boundary between the two semiconducting materials, the current i can be shown to be related to the potential difference V across the diode by

$$i = A(e^{eV/kT} - 1),$$

where A depends on the materials but not on the current or potential difference. V is positive if the junction is forward-biased and negative if it is back-biased. Verify that this expression predicts the behavior expected of a diode by (a) sketching i as a function of V over the range -0.12 V $< V <$ $+0.12$ V. Take $T = 300$ K and $A = 5 \times 10^{-9}$ A. (b) For the same temperature calculate the ratio of the current for a 0.50 V forward-bias to the current for a 0.50 V back-bias.   *Answer: (b) 2.49×10^8.*

44. A drop of lead (work function = 3.4 eV) is in close contact with a sheet of copper (work function = 4.5 eV). What is the contact potential difference that appears across the lead–copper interface? How might you measure it? Draw an energy diagram, showing (in the style of Fig. 53-4b) the relative Fermi levels both before and after the two metals are joined together. Can such a junction serve as a diode rectifier?

SECTION 53-11

45. Fill in the seven-segment display shown in Fig. 53-15a to show how all ten numbers may be generated. (b) If the numbers are displayed randomly, in what fraction of the displays will each of the seven segments be used?

46. You wish to make an LED that will emit green light. (a) What must the energy gap be? (b) Germanium has an energy gap of 0.67 eV. If it were otherwise suitable for use as an LED (which it is not), what wavelength would it emit?

47. Section 53-11 deals with the mode of operation of a light-emitting diode, in which light is emitted when charge carriers are injected across the central plane of a pn junction by an external potential. The reverse device, a *photodiode*, is also a possibility. That is, you can shine light on a pn junction and a current will develop across the junction plane. Discuss how such a device might operate. Would it be best to operate it in a forward- or a back-biased mode?

54
nuclear physics: an introduction

In the first years of this century not much was known about the structure of atoms beyond the fact that they contained electrons. This particle had only been discovered (by J. J. Thomson) in 1897 and its mass was unknown in those early days. Thus it was not possible even to say just how many electrons a given atom contained. Atoms were electrically neutral so they must also contain some positive charge, but nobody knew what form this compensating positive charge took. How the electrons moved within the atom and how the mass of the atom was divided between the electrons and the positive charge were also very open questions.

In 1911, Ernest Rutherford, interpreting some experiments carried out in his laboratory, was led to propose that the positive charge of the atom was densely concentrated at the center of the atom and that, furthermore, it was responsible for most of the mass of the atom. He had discovered the atomic nucleus!

Until this step had been taken, all attempts to understand the motions of the electrons within the atom were doomed to failure. It was only two years later that Bohr, seizing on this proposal, developed the semiclassical theory of atom structure that we have described in Chapter 49, and the Rutherford-Bohr atom was on its way.

How did Rutherford come to make this proposal? It was not an idle conjecture but was based firmly on the results of an experiment suggested by him and carried out by his collaborators, Hans Geiger (of Geiger counter fame) and Ernest Marsden, a 20-year-old student who had not yet earned his Bachelor's degree.

Rutherford's idea was to probe the forces acting within an atom by firing energetic α-particles through a thin target foil and measuring the

54-1
DISCOVERING THE NUCLEUS

extent to which they were deflected as they passed through the foil. α-Particles, which are about 7300 times more massive than electrons, carry a charge of $+2e$ and are emitted spontaneously (with energies of a few MeV) by many radioactive materials. We now know that these useful projectiles are the nuclei of the atoms of ordinary helium. Figure 54-1 shows the experimental arrangement of Geiger and Marsden. The experiment consists in counting the number of α-particles deflected through various scattering angles ϕ.

Figure 54-2 shows their results. Note especially that the vertical scale is logarithmic. We see that most of the α-particles are scattered through rather small angles, but — and this was the big surprise — a very small fraction of them are scattered through very large angles, approaching 180°. In Rutherford's words: "It was quite the most incredible event that ever happened to me in my life. It was almost as incredible as if you had fired a 15-inch shell at a piece of tissue paper and it came back and hit you."

Why was Rutherford so surprised? At the time of these experiments, most physicists believed in the so-called plum pudding model of the atom, which had been advanced by J. J. Thomson. In this view the positive charge of the atom was thought to be spread out through the entire volume of the atom. The electrons (the "plums") were thought to vibrate about fixed centers within this sphere of charge (the "pudding").

The maximum deflecting force acting on the α-particle as it passes through such a positive sphere of charge proves to be far too small to deflect the α-particle by even as much as one degree. The electrons in the atom would also have very little effect on the massive, energetic α-particle. They would, in fact, be themselves strongly deflected, much as a swarm of gnats would be brushed aside by a stone thrown through them. As for deflecting an α-particle backwards, in the direction from which it came, there is simply no mechanism in Thomson's atom model to account for this.

Rutherford saw that to produce such a large deflection there must be a large force, which could be provided if the positive charge, instead of being spread throughout the atom, was concentrated tightly at its center. On this model the incoming α-particle can get very close to the

figure 54-1

An arrangement (top view) used in Rutherford's laboratory to study the scattering of α-particles by thin metal foils. The detector can be rotated to various values of the scattering angle ϕ.

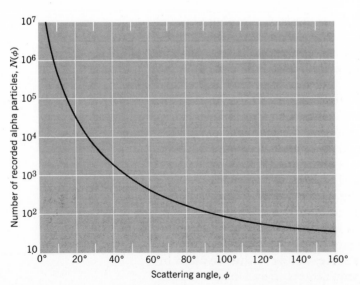

figure 54-2

A scattering curve for a gold foil, obtained by Geiger and Marsden using the apparatus of Fig. 54-1. Note that the vertical axis is marked off in powers of ten.

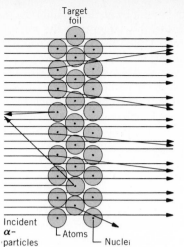

center of the positive charge without penetrating it, resulting in a large deflecting force; see Problem 1.

Figure 54-3 shows the paths taken by typical α-particles as they pass through the atoms of the target foil. As we see, most are only slightly deflected, but a few (those whose extended incoming paths pass, by chance, close to a nucleus) are deflected through large angles. From an analysis of the data, Rutherford concluded that the dimensions of the nucleus must be smaller than the diameter of an atom by a factor of about 10^4. The atom is mostly empty space! It is not often that the piercing insight of a gifted scientist, buttressed by a few simple calculations,* leads to results of such importance.

EXAMPLE 1

A close encounter. A 5.30-MeV α-particle happens, by chance, to be headed directly toward the nucleus of an atom of gold ($Z = 79$). How close does it get before it comes momentarily to rest and reverses its course? Neglect the recoil of the (relatively massive) gold nucleus.

Initially the total mechanical energy of the two interacting particles is just equal to K_α (= 5.30 MeV), the initial kinetic energy of the α-particle. At the moment the α-particle comes to rest, the total energy is the electrostatic potential energy of the system of two particles. Because energy must be conserved, these two quantities must be equal, or

$$K_\alpha = \frac{1}{4\pi\epsilon_0} \frac{Q_\alpha Q_{\text{Au}}}{d},$$

in which Q_α (= $2e$) is the charge of the α-particle, Q_{Au} (= $79e$) is the charge of the gold nucleus, and d is the distance between the centers of the two particles.

Substituting for the charges and solving for d yields

$$d = \frac{(2 \times 79)e^2}{4\pi\epsilon_0 K_\alpha}$$

$$= \frac{(158)(1.60 \times 10^{-19})^2}{(4\pi)(8.85 \times 10^{-12} \text{ F/m})(5.30 \text{ MeV})(1.60 \times 10^{-13} \text{ J/MeV})}$$

$$= 4.29 \times 10^{-14} \text{ m} = 42.9 \text{ fm}.$$

* For a mathematical analysis of this scattering experiment, see Robert Resnick and David Halliday, *Basic Concepts in Relativity and Early Quantum Theory* (John Wiley & Sons, New York, 1985, 2nd ed.), Supplementary Topic I.

This is a small distance by atomic standards but not by nuclear standards. As we shall see in the following section, it is considerably larger than the sum of the radii of the gold nucleus and the α-particle. Thus the α-particle reverses its course without ever "touching" the gold nucleus.

If the positive charge associated with the gold atom had been spread uniformly throughout the volume of the atom, the maximum retarding force acting on the α-particle would have occurred at the moment the α-particle began to touch the surface of the atom. This force (see Problem 1) would have been far too weak to have had much effect on the motion of the α-particle, which would have gone barreling right through such a "spongy" atom.

54-2
SOME NUCLEAR PROPERTIES

The nucleus, tiny as it may be, has a structure that is every bit as complex as that of the atom. Nuclei are made up of protons and neutrons. These particles (unlike the electron) are not true elementary particles, being made up of other particles, called *quarks*. However, nuclear physics — the subject of this chapter — is defined to be those studies of the nucleus into which the structure of the protons and neutrons themselves does not enter. The fundamental nature of these two particles is a topic in the field of elementary particle physics.

Table 54-1 shows some properties of a few selected nuclear species, or *nuclides*, as they are called.

Table 54-1
Some properties of selected nuclides*

Nuclide	Z	N	A	Stability**	Atomic Mass, u	Radius, fm	Binding Energy, MeV/nucleon	Spin	Magnetic Moment, μ_N
^{7}Li	3	4	7	92.5%	7.01604	2.1	5.60	$\frac{3}{2}$	+3.26
^{31}P	15	16	31	100%	30.97376	3.36	8.48	$\frac{1}{2}$	+1.13
^{81}Br	35	46	81	49.3%	80.91629	4.63	8.69	$\frac{3}{2}$	+2.27
^{120}Sn	50	70	120	32.4%	119.90221	5.28	8.50	0	0
^{157}Gd	64	93	157	15.7%	156.92397	5.77	8.21	$\frac{3}{2}$	−0.339
^{197}Au	79	118	197	100%	196.96655	6.23	7.91	$\frac{3}{2}$	+0.145
^{227}Ac	89	138	227	21.8 y	227.02777	6.53	7.65	$\frac{3}{2}$	+1.10
^{239}Pu	94	145	239	24,100 y	239.05217	6.64	7.56	$\frac{1}{2}$	+0.200

* The nuclides were chosen to lie along the stability line of Fig. 54-4.
** For stable nuclides the isotopic abundance is given; for radionuclides, the half-life.

The Nuclear Force. The force that controls the electronic structure and properties of the atom is the familiar Coulomb force. To bind the nucleus together, however, there must be a strong attractive force of a totally new kind, acting between the neutrons and the protons that make up the nucleus. This force must be strong enough to overcome the repulsive Coulomb force between the (positively charged) protons and to bind both neutrons and protons into the tiny nuclear volume.

Experiment suggests that this *strong force*, as it is simply called, has the same character between any pair of nuclear constituents, be they neutrons or protons. Because of this we refer to neutrons and protons collectively as *nucleons*, except when we specifically need to distinguish between them.

The "strong force" has a short range. This means that the attractive force between pairs of nucleons drops rapidly to zero for nucleon separations greater than a certain critical value. This in turn means that, except in the smallest nuclei, a given nucleon cannot interact with all the other nucleons in the nucleus but only with a few of its nearest neighbors. By contrast, the Coulomb force is not a short-range force. A given proton in a nucleus exerts a Coulomb repulsion on all the other protons, no matter how large their separation; see Problem 2.

Nuclear Systematics.   Nuclei are made up of protons and neutrons. The number of protons in the nucleus, as we have already seen, is called the *atomic number* and is represented by Z. The number of neutrons is called the *neutron number*, and we represent it by N. The total number of nucleons $(= Z + N)$ is called the *mass number*, and we represent it by A.

We use A, the total number of nucleons, as an identifying superscript in labeling nuclides. In ^{81}Br, for example there are 81 nucleons. The symbol "Br" tells us that we are dealing with bromine, for which $Z = 35$. The remaining 46 nucleons are neutrons, so that, for this nuclide, $Z = 35$, $N = 46$, and $A = 81$.

Figure 54-4 shows a chart of the nuclides. It is a plot of Z against N, each circle representing a nuclide. Filled black circles represent stable nuclides; filled gray circles represent radioactive nuclides — or radionuclides — that are relatively stable, having half-lives greater than an arbitrarily chosen value of 100 y; open circles represent radionuclides with half-lives less than this value.

Note that there is a reasonably well-defined linear zone of stability in Fig. 54-4, traced only by the black and the gray circles. Relatively unstable radionuclides tend to lie on either side of the stability zone.

Figure 54-4 also shows that the lightest stable nuclides tend to lie on or close to the line $Z = N$. The heavier stable nuclides lie well below this line and thus typically have many more neutrons than protons. The tendency to an excess of neutrons at large mass numbers is a Coulomb repulsion effect. Because a given nucleon interacts with only a small number of its neighbors through the strong force, the amount of energy tied up in strong-force bonds between nucleons increases just in proportion to A. The energy tied up in Coulomb-force bonds between protons increases more rapidly than this because each proton interacts with every other proton in the nucleus. Thus the Coulomb energy becomes increasingly important at high mass numbers.

Consider a nucleus with 238 nucleons. If it were to lie on the $Z = N$ line, it would have $Z = N = 119$. However, such a nucleus, if it could be assembled, would fly apart at once because of Coulomb repulsion. Relative stability is found only if we replace 27 of the protons by neutrons, thus greatly diluting the Coulomb repulsion effect. We then would have the nuclide ^{238}U, which has $Z = 92$ and $N = 146$, a neutron excess of 54.

Even in ^{238}U, Coulomb effects are evident in that (1) this nuclide is radioactive and emits α-particles and (2) it can easily break up (fission) into two fragments if it absorbs a neutron. Both of these processes reduce the Coulomb energy more than they do the energy in the strong-force bonds.

Nuclear Radii.   We have used the Bohr radius a_0 $(= 5.29 \times 10^{-11}$ m$)$ as a convenient unit for measuring the dimensions of atoms. Nuclei are smaller by a factor of about 10^4 and a convenient unit for measuring distances on *this* scale is the *femtometer* $(= 10^{-15}$ m$)$. This unit is often called the *fermi*, and shares the same abbreviation. Thus

$$1 \text{ fermi} = 1 \text{ femtometer} = 1 \text{ fm} = 10^{-15} \text{ m}.$$

figure 54-4

A plot of Z against N for all nuclides known as of 1983. Black circles represent stable nuclides. Gray circles represent "long-lived" radioactive nuclides (half-lives $>$ 100 y). Open circles represent "short-lived" radioactive nuclides (half-lives $<$ 100 y). Along the diagonal lines A is constant, equal to the labelled value.

We can learn about the size and structure of nuclei by doing a scattering experiment, much as suggested by Fig. 54-1, using an incident beam of high-energy electrons. The energy of the incident electrons must be high enough (~200 MeV) so that their de Broglie wavelength will be small enough for them to act as structure-sensitive nuclear probes.

The distribution of nuclear matter for a typical nucleus is shown in Fig. 54-5. We see that the nucleus does not have a sharply defined surface. It does, however, have a characteristic mean radius R. The nuclear density $\rho(r)$ has a constant value in the nuclear interior and drops to zero through the fuzzy surface zone. R is given by

$$R = R_0 A^{1/3}, \tag{54-1}$$

in which A is the mass number and R_0 is a constant with the value of 1.1 fm. For ^{63}Cu, for example,

$$R = (1.1 \text{ fm})(63)^{1/3} = 4.3 \text{ fm}.$$

By comparison, the mean radius of a copper ion in a lattice of solid copper is 1.8 Bohr radii, about 2×10^4 times larger.

Nuclear Masses and Binding Energies. Atomic masses can be measured with great precision using modern mass spectrometer and nuclear reaction techniques. We recall that such masses are measured in *unified atomic mass units* (abbr. u), chosen so that the atomic mass (*not* the nuclear mass) of ^{12}C is exactly 12 u. The relation of this unit to the SI mass standard is

$$1 \text{ u} = 1.66 \times 10^{-27} \text{ kg}.$$

Note that the mass number (symbol A) identifying a nuclide is so named because this number is equal to the atomic mass of the nuclide, rounded to the nearest integer. Thus the mass number of the nuclide ^{137}Cs is 137; this nuclide contains 55 protons and 82 neutrons, a total of 137 particles; its atomic mass is 136.90707 u, which rounds off numerically to 137.

In nuclear physics, as contrasted with atomic physics, the energy changes per event are commonly so great that Einstein's well-known mass-energy relation $E = \Delta mc^2$ is an indispensable work-a-day tool. See Fig. 54-6 for an example of the pervasiveness of this best-known of all equations in physics.

We shall often need to use the energy equivalent of 1 atomic mass unit, and we find it from

$$E = \Delta m \, c^2 = \frac{(1.66 \times 10^{-27} \text{ kg})(3.00 \times 10^8 \text{ m/s})^2}{1.60 \times 10^{-13} \text{ J/MeV}}$$

$$= 931 \text{ MeV}.$$

This means that we can write c^2 as 931 MeV/u and can thus easily find the energy equivalent (in MeV) of any mass or mass difference (in u), or conversely.

As an example, consider the deuteron, the nucleus of the heavy hydrogen atom. A deuteron consists of a proton and a neutron bound together by the strong interaction force. The energy E_B that we must add to the deuteron to tear it apart into its two constituent nucleons is called its *binding energy.* From the conservation of mass-energy we can write, for this pulling-apart process,

$$m_d c^2 + E_B = m_n c^2 + m_p c^2. \tag{54-2}$$

figure 54-5
The variation with radial distance of the density of nuclear matter for the nuclide ^{197}Au.

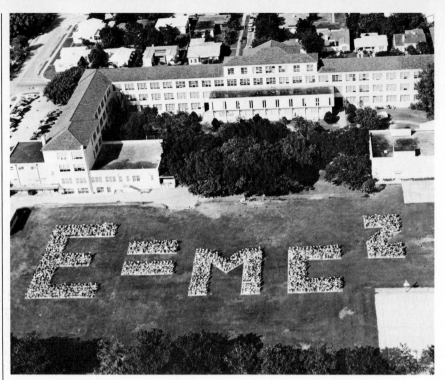

figure 54-6
Students of Shenandoah Junior High School, in Miami, Florida, honor Einstein in the year of the 100th anniversary of his birth by spelling out his famous formula with their bodies. (Courtesy of Rocky Raisen, their physics teacher; photo by Justo Alsonso.)

If we add $m_e c^2$, the energy equivalent of one electron mass, to each side of this equation, we have

$$(m_d + m_e)c^2 + E_B = m_n c^2 + (m_p + m_e)c^2,$$

or

$$m_{^2\mathrm{H}} c^2 + E_B = m_n c^2 + m_{^1\mathrm{H}} c^2. \qquad (54\text{-}3)$$

Here $m_{^2\mathrm{H}}$ and $m_{^1\mathrm{H}}$ are the masses of the neutral *heavy* hydrogen atom and the neutral *ordinary* hydrogen atom, respectively. They are atomic masses, not nuclear masses. Solving Eq. 54-3 for E_B yields

$$E_B = (m_n + m_{^1\mathrm{H}} - m_{^2\mathrm{H}})c^2 = \Delta m c^2, \qquad (54\text{-}4)$$

in which Δm is the change in atomic rest mass. In making calculations of this kind we always use atomic, rather than nuclear, masses, as this is what is normally tabulated. As in this example, the electron masses conveniently cancel.*

For the deuteron problem the needed atomic masses are

Symbol	Atomic Mass, u
$^1\mathrm{H}$	1.00783
n	1.00867
$^2\mathrm{H}$	2.01410

Substituting into Eq. 54-4 and replacing c^2 by its equal, 931 MeV/u, yields for the binding energy

$$E_B = (1.00867\ \mathrm{u} + 1.00783\ \mathrm{u} - 2.01410\ \mathrm{u})(931\ \mathrm{MeV/u})$$

$$= (0.00240\ \mathrm{u})(931\ \mathrm{MeV/u}) = 2.23\ \mathrm{MeV}.$$

* See, however, Problem 51, for an exception.

figure 54-7
The binding energy per nucleon as a
function of mass number. The nuclides
of Table 54-1 are identified, along
with a few other nuclides of interest.

Compare this with the binding energy of the hydrogen atom in its ground state, which is 13.6 eV, about five orders of magnitude smaller.

If we divide the binding energy of a nucleus by its mass number, we get the average binding energy per nucleon, a property we have chosen to list in Table 54-1. Figure 54-7 shows a plot of this quantity as a function of mass number. The fact that this *binding energy curve* "droops" at both high and low mass numbers has practical consequences of the greatest importance.*

The drooping of the binding energy at high mass numbers tells us that nucleons are more tightly bound when they are assembled into two middle-mass nuclei rather than into a single high-mass nucleus. In other words, energy can be released in the *nuclear fission* of a single massive nucleus into two smaller fragments.

The drooping of the binding-energy curve at low mass numbers, on the other hand, tells us that energy will be released if two nuclei of small mass number combine to form a single middle-mass nucleus. This process, the reverse of fission, is called *nuclear fusion*. It occurs inside our sun, and other stars, and is the mechanism by which the sun generates the energy it radiates to us.

Nuclear Spin and Magnetism.   Nuclei, like atoms, have an intrinsic angular momentum whose maximum projected value is given by $I\hbar$. Here I is a quantum number, which may be integral or half-integral, called the *nuclear spin*; some values for selected nuclides are shown in Table 54-1.

Again as for atoms, a nuclear angular momentum has a nuclear magnetic moment associated with it. Recall that, in atomic magnetism the *Bohr magneton* μ_B, defined as

$$\mu_B = \frac{eh}{4\pi m_e} = 5.79 \times 10^{-5} \text{ eV/T},$$

is a unit of convenience.

* *The Curve of Binding Energy* has even been adopted as the title of a book (by John McPhee) about the possibilities of nuclear terrorism!

In nuclear physics the corresponding unit of convenience is the *nuclear magneton*, defined as above except that the electron mass m_e is replaced by the proton mass m_p. That is,

$$\mu_N = \frac{eh}{4\pi m_p} = 3.15 \times 10^{-8} \text{ eV/T}.$$

Because the magnetic moment of the free electron is (very closely) one Bohr magneton, it might be supposed that the magnetic moment of the free proton would be (very closely) one nuclear magneton. It is not very close, however, the measured value being $+2.7929\ \mu_N$. Nuclear magnetic moments are not yet as well understood as are electronic and atomic magnetic moments.

EXAMPLE 2

What is the approximate density of the nuclear matter from which all nuclei are made?

We know that this number will be high, because virtually all of the mass of the atom resides in its tiny nucleus. The volume of the nucleus, approximated as a uniform sphere of radius R, is given by Eq. 54-1 as

$$V = \tfrac{4}{3}\pi R^3 = \tfrac{4}{3}\pi(R_0^3 A).$$

The density of nuclear matter, expressed in nucleons per unit volume, is then

$$\rho_0 = \frac{A}{V} = \frac{A}{(4\pi/3)R_0^3 A}$$

$$= \frac{1}{(4\pi/3)(1.1 \text{ fm})^3} = 0.18 \text{ nucleons/fm}^3.$$

The mass of a nucleon is 1.7×10^{-27} kg. In more familiar units, the density of nuclear matter is then

$$\rho_0 = (0.18 \text{ nucleons/fm}^3)\ (1.7 \times 10^{-27} \text{ kg/nucleon})(1 \text{ fm}/10^{-15} \text{ m})^3$$

$$= 3.1 \times 10^{17} \text{ kg/m}^3,$$

or 3.1×10^{14} times greater than the density of water! Unlike the orbital electrons, the nuclides have a density nearly independent of the number of their nucleons. To some extent nucleons are packed in like marbles in a bag.

EXAMPLE 3

Calculating the binding energy. Imagine that a typical middle-mass nucleus such as ^{120}Sn is picked apart into its constituent protons and neutrons. Find (a) the total energy required and (b) the energy per nucleon. The atomic mass of ^{120}Sn is 119.90221 u; see Table 54-1.

^{120}Sn contains 50 protons and $(120 - 50)$ or 70 neutrons. Their combined masses as free particles are, using the atomic mass data given above,

$$M = 50 \times 1.00783 \text{ u} + 70 \times 1.00867 \text{ u}$$

$$= 120.99840 \text{ u}.$$

This exceeds the atomic mass of ^{120}Sn by

$$\Delta m = 120.99840 \text{ u} - 119.90221 \text{ u}$$

$$\approx 1.10 \text{ u}.$$

Converting this to energy yields

$$E = \Delta mc^2 = (1.10 \text{ u})(931 \text{ MeV/u}) = 1020 \text{ MeV}.$$

Finally, the binding energy per nucleon is

$$E_B = \frac{E}{A} = \frac{1020 \text{ MeV}}{120} = 8.50 \text{ MeV/nucleon.}$$

This agrees with the value that may be read from the curve of Fig. 54-7.

54-3
RADIOACTIVE DECAY

As Fig. 54-4 shows, most of the nuclides that have been identified are radioactive. That is, they spontaneously emit a particle, transforming themselves in the process into a different nuclide. In this chapter we shall discuss the two most common situations, the emission of an α-particle (alpha decay) and the emission of an electron (beta decay).

No matter what the nature of the decay, its main feature is that it is statistical. Consider, for example, a one-gram sample of uranium metal. It contains 2.5×10^{21} atoms of the very long-lived alpha emitter ^{238}U. The nuclei of these atoms have existed without decaying since they were created (before the formation of our solar system) in the explosion of a supernova.

During any given second about 12,000 of the nuclei in our sample will decay, emitting an α-particle in the process. We have absolutely no way of predicting, however, whether any given nucleus in the sample will be among those that do so. Every single ^{238}U nucleus has exactly the same probability as any other to decay during any one-second observation period, namely, $12,000/(2.5 \times 10^{21})$, or one chance in 2×10^{17}.

In general, if a sample contains N radioactive nuclei, we can express the statistical character of the decay process by saying that the decay rate $R \; (= -dN/dt)$ is proportional to N, or

$$-\frac{dN}{dt} = \lambda N, \tag{54-5}$$

in which λ, the *disintegration constant*, has a characteristic value for every radioactive nuclide. We can rewrite Eq. 54-5 as

$$\frac{dN}{N} = -\lambda dt,$$

which integrates readily to

$$N = N_0 e^{-\lambda t}. \tag{54-6}$$

Here N_0 is the number of radioactive nuclei in the sample at $t = 0$. We see that the decrease of N with time follows a simple exponential law.

We are often more interested in the decay rate $R \; (= -dN/dt)$ of the sample than we are in N. Differentiating Eq. 54-6 yields

$$R = R_0 e^{-\lambda t}, \tag{54-7}$$

in which R_0 is the decay rate at $t = 0$. We assumed initially that R and N are proportional to each other, so we are not surprised to confirm that they both decrease with time according to the same exponential law.

A quantity of interest is the time $t_{1/2}$, called the *half-life*, after which both N and R are reduced to one-half of their initial values. Putting $R = \frac{1}{2}R_0$ in Eq. 54-7 gives

$$\tfrac{1}{2}R_0 = R_0 e^{-\lambda t_{1/2}}$$

which leads readily to

$$t_{1/2} = \frac{\ln 2}{\lambda} = \frac{0.693}{\lambda}, \qquad (54\text{-}8)$$

a relationship between the half-life and the disintegration constant.

The following two examples show how λ can be measured for decay processes with relatively short half-lives and also with relatively long half-lives.

EXAMPLE 4

Short-lived decays. We define short-lived decays as cases in which it is possible to measure directly the decrease in the decay rate R with time. The following table gives some data for a sample of ^{128}I, a radionuclide often used medically as a tracer to measure the iodine uptake rate of the thyroid gland. Find (a) the disintegration constant λ and (b) the half-life $t_{1/2}$ from these data.

Time, min	R, counts/s	Time, min	R, counts/s
4	392.2	132	10.9
36	161.4	164	4.56
68	65.5	196	1.86
100	26.8		

If we take the logarithm of each side of Eq. 54-7, we find that

$$\log R = \log R_0 - (\lambda \log e)t = \log R_0 + (-0.434\lambda)t. \qquad (54\text{-}9)$$

Thus, if we plot $\log R$ against t, we should obtain a straight line whose slope is -0.434λ. Figure 54-8 shows such a plot. Equating the slope of the line to -0.434λ yields

$$-0.434\lambda = -\frac{(2.64 - 0)}{(218 \text{ min} - 0)},$$

or

$$\lambda = 0.0279 \text{ min}^{-1}.$$

Equation 54-8 yields for $t_{1/2}$:

$$t_{1/2} = \frac{0.693}{\lambda} = \frac{0.693}{0.0279 \text{ min}^{-1}} = 24.8 \text{ min}.$$

figure 54-8
A plot of Eq. 54-9 for the radioactive decay of ^{128}I.

EXAMPLE 5

Long-lived decays. In the case of long-lived decays it is not possible to wait long enough to observe a measurable decrease in the decay rate R with time. We must find λ by measuring both N and $(-dN/dt)$ in Eq. 54-5.

A 1.00-g sample of pure KCl from the chemistry stock room is found to be radioactive and to decay at an absolute rate R of 1600 counts/s. The decays are traced to the element potassium and in particular to the isotope ^{40}K, which constitutes 1.18% of normal potassium. What is the half-life for this decay?

The molecular weight of KCl is 74.9 g/mol, so the number of potassium atoms in the sample is

$$N_K = \frac{(6.02 \times 10^{23} \text{ mol}^{-1})(1.00 \text{ g})}{(74.9 \text{ g/mol})} = 8.04 \times 10^{21}.$$

Of these the number of ^{40}K atoms is

$$N_{40} = (8.04 \times 10^{21})(1.18 \times 10^{-2}) = 9.49 \times 10^{19}.$$

From Eq. 54-5 we have

$$\lambda = \frac{-dN/dt}{N} = \frac{R}{N_{40}} = \frac{1600 \text{ s}^{-1}}{9.49 \times 10^{19}} = 1.69 \times 10^{-17} \text{ s}^{-1},$$

and the half-life, from Eq. 54-8, is

$$t_{1/2} = \frac{0.693}{\lambda} = \left(\frac{0.693}{1.69 \times 10^{-17} \text{ s}^{-1}}\right)\left(\frac{1 \text{ y}}{3.16 \times 10^7 \text{ s}}\right)$$

$$= 1.30 \times 10^9 \text{ y}.$$

This is of the order of magnitude of the age of the universe. No wonder we cannot measure the half-life of this nuclide by waiting around for its decay rate to decrease! (Interestingly, the potassium in our own bodies has its normal share of the ^{40}K isotope. We are all slightly radioactive.)

54-4
ALPHA DECAY

^{238}U, a typical alpha emitter, decays spontaneously according to the scheme

$$^{238}\text{U} \rightarrow {}^{234}\text{Th} + {}^{4}\text{He}, \tag{54-10}$$

the half-life of the decay being 4.47×10^9 y. In Example 6 we show that, in every such decay, an energy of 4.26 MeV is emitted, appearing as kinetic energy shared between the α-particle (^{4}He) and the recoiling residual nucleus (^{234}Th).

We now ask ourselves: "If energy is released in every such decay event, why did the ^{238}U nuclei not decay shortly after they were created?" The creation process presumably occurred in the violent explosions of ancestral stars (supernovae), predating the formation of our solar system. Why did these nuclei wait so very long before getting rid of their excess energy by emitting an α-particle? To answer this question, we must study the detailed mechanism of alpha decay.

We choose a model in which the α-particle is imagined to exist preformed inside the nucleus before it escapes. Figure 54-9 shows the approximate potential energy function $U(r)$ for the α-particle and the

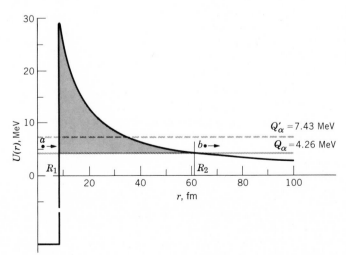

figure 54-9
A potential energy function describing the emission of α-particles by ^{2398}U. The shaded area represents a potential barrier that inhibits the decay process. The solid horizontal line shows the disintegration energy.

residual ^{234}Th nucleus as a function of their separation. It is a combination of a potential well associated with the (attractive) strong nuclear force that acts in the nuclear interior (region a) and a Coulomb potential associated with the (repulsive) electrostatic force that acts between the two particles after the decay has occurred. (region b)

The horizontal line marked $Q_\alpha = 4.26$ MeV shows the disintegration energy for the process, as calculated in Example 6. Note that this line intersects the potential energy curve at two points, R_1 and R_2. We now see why the α-particle is not immediately emitted from the ^{238}U nucleus! That nucleus is surrounded by an impressive potential barrier, shown by the shaded area in Fig. 54-9. Visualize this barrier as a spherical shell whose inner radius is R_1 and whose outer radius is R_2, its volume being forbidden to the α-particle under the laws of classical physics. If the α-particle found itself in that region, its potential energy U would exceed its total energy E, which would mean, classically, that its kinetic energy K $(= E - U)$ would be negative, an impossible situation.

Indeed, we now change our question and ask: "How can the ^{238}U nucleus *ever* emit an α-particle?" The α-particle seems permanently trapped inside the nucleus by the barrier.

The answer is that, as we learned in Section 50-11, in wave mechanics there is always a chance (described by Eq. 50-18) that a particle can tunnel through a barrier that is classically insurmountable. Interestingly enough, the explanation of alpha decay by wave mechanical barrier tunneling was one of the very first applications of the new quantum physics.

For the long-lived decay of ^{238}U the barrier is actually not very "leaky." We can show that the α-particle, presumed to be rattling back and forth within the nucleus must present itself at the inner surface of the barrier abut 10^{38} times before it succeeds in tunneling through. This is about 10^{20} times per second for about 10^9 years! We, of course, are waiting on the outside, taking note of only those α-particles that *do* manage to escape.

We can test this barrier tunneling explanation of alpha decay by looking at other alpha emitters, for which the barrier would be different. For an extreme contrast, consider the alpha decay of another uranium isotope, ^{228}U, which can be shown to have a disintegration energy Q'_α of 7.43 MeV. We show this as a dashed straight line in Fig. 54-9.

Note that the barrier in this case is both thinner and lower; if our barrier tunneling notions are correct, we would expect alpha decay to occur more readily for this nuclide than for ^{238}U. Indeed it does. As Table 54-2 shows, its half-life is only 550 s! We recall from Section 50-11 that the transmission coefficient of a barrier — because of the exponential nature of Eq. 50-18 — is very sensitive to small changes in the dimensions of the barrier. We see that an increase in Q_α by a factor of only 1.6 produces a decrease in half-life (that is, in the effectiveness of barrier tunneling) by a factor of 2.4×10^{14}. This is sensitivity indeed.

Table 54-2
The alpha decay of ^{238}U and ^{228}U

Nuclide	Q_α	Half-life
^{238}U	4.26 MeV	4.47×10^9 y
^{228}U	6.81 MeV	550 s

EXAMPLE 6

1277 BETA DECAY   SEC. 54-5

The energetics of alpha decay. (*a*) Find the energy released during the alpha decay of ^{238}U. (*b*) Show that this nuclide cannot spontaneously emit a proton or a neutron. The needed atomic masses are

^{238}U	238.05081 u	^{4}He	4.00260 u
^{234}Th	234.04363 u	^{1}H	1.00783 u
^{237}Pa	237.05121 u	n	1.00867 u
^{237}U	237.04874 u		

(*a*) In the alpha decay process of Eq. 54-10 the atomic mass of the decay products (= 238.04623 u) is less than the atomic mass of ^{238}U by $\Delta m = 0.00458$ u, whose energy equivalent is

$$Q_\alpha = \Delta m c^2 = (0.00458 \text{ u})(931 \text{ MeV/u}) = +4.26 \text{ MeV}.$$

This *disintegration energy* is available to share as kinetic energy between the α-particle and the recoiling ^{234}Th atom.

(*b*) If ^{238}U were to emit a proton, the decay process would be

$$^{238}\text{U} \rightarrow {}^{237}\text{Pa} + {}^1\text{H}.$$

In this case the mass of the decay products *exceeds* the mass of ^{238}U by $\Delta m = 0.00823$ u, the energy equivalent Q_p being -7.66 MeV. The minus sign tells us that ^{238}U is stable against spontaneous proton emission.

For neutron emission the process would be

$$^{238}\text{U} \rightarrow {}^{237}\text{U} + n,$$

and the energy release ($Q_n = -6.14$ MeV) again proves to be negative. Thus ^{238}U is also stable against spontaneous neutron emission.

54-5 BETA DECAY

A nucleus that decays spontaneously by emitting an electron (either positive or negative) is said to undergo *beta decay.** Here are two examples:

$$^{32}\text{P} \rightarrow {}^{32}\text{S} + e^- + v \qquad (t_{1/2} = 14.4 \text{ } d) \qquad (54\text{-}11)$$

and

$$^{64}\text{Cu} \rightarrow {}^{64}\text{Ni} + e^+ + v \qquad (t_{1/2} = 12.7 \text{ } h). \qquad (54\text{-}12)$$

The symbol "v" represents a *neutrino*, a neutral particle that is emitted from the nucleus along with the electron during the decay process. Neutrinos interact only very weakly with matter and — for that reason — are so extremely difficult to detect that, for many years, their presence went unnoticed.†

It may seem surprising that nuclei can emit electrons (and neutrinos) in view of the fact that we have said that nuclei are made up of neutrons and protons only. However, we saw earlier that atoms emit photons, and we certainly do not say that atoms "contain" photons. We say that the photons are created during the emission process.

So it is with the electrons and the neutrinos emitted from nuclei during beta decay. They are both created during the emission process, a

---

* Beta decay also includes electron capture, in which a nucleus decays by absorbing one of its orbital electrons. We do not consider that process here.
† The neutral particle emitted in Eq. 54-11 is actually an antineutrino, a distinction that we do not draw in this introductory treatment.

neutron transforming itself into a proton within the nucleus (or conversely) according to

$$n \rightarrow p + e^- + v \qquad \beta^-\text{-decay} \qquad (54\text{-}13)$$

or

$$p \rightarrow n + e^+ + v \qquad \beta^+\text{-decay} \qquad (54\text{-}14)$$

These are, in fact, the basic beta-decay processes.

In alpha decay the emitted α-particle has a sharply defined energy, which fits in well with our understanding that the same amount of energy is released in every individual alpha-decay process. In beta decay, however, the energy of the emitted electron is distributed in a continuous spectrum, from zero up to a certain maximum $K_{\max}$, as Fig. 54-10 shows for the beta decay of ^{64}Cu (see Eq. 54-12).

For many years, before the neutrino was identified, curves like that of Fig. 54-10 were a challenging puzzle. They seemed to suggest that not all decaying nuclei — though they are presumably identical — release the same amount of energy when they decay. Bohr went so far as to contemplate the possibility that the law of conservation of energy might hold only statistically, averaged over a large number of individual events.

The real answer is that the energy release displayed in Fig. 54-10 is only part of the story. In beta decay *two* particles are emitted, an electron and a neutrino, and the experimenters measured only that fraction of the disintegration energy that the electron carried. The neutrinos passed totally through their apparatus, leaving no trace of their presence.

We now know that, in every single beta-decay event, the same amount of energy — equal to $K_{\max}$ — is released. Sometimes the electron gets most of this energy, sometimes the neutrino does. In every case the sum of the electron's energy and the neutrino's energy is equal to $K_{\max}$ so that energy is conserved in each individual decay process. We only measure, of course, the electron's share of the released energy.

It was Pauli who (in 1927) suggested that some other particle must be emitted during beta decay, to account for the "missing" energy (and, incidentally, to solve an equally serious puzzle, involving "missing" angular momentum). It was Fermi who dubbed Pauli's proposed particle the *neutrino* ("tiny neutral one") and who (in 1934) worked up Pauli's suggestion into a successful formal theory.

The neutrino is a truly elusive particle. It has been calculated that the mean free path of an energetic neutrino in water is no less than several thousand light years! The sun emits neutrinos copiously from the nuclear furnace at its core and, at night, these messengers from the center of the sun come up at us from below, the earth being totally transparent to them.*

In spite of their elusive character, neutrinos have been detected in the laboratory, first (in 1953) by F. Reines and C. L. Cowan, the neutrinos they measured being generated in a high-power nuclear reactor. In spite of the difficulties of detection, experimental neutrino physics is now a well-developed branch of experimental physics, with avid practitioners at several major laboratories throughout the world. Neutrinos were generated in copious quantities during the formation of the universe

figure 54-10
The distribution in kinetic energy of the positive electrons emitted in the beta decay of ^{64}Cu. The maximum energy of the distribution ($K_{\max}$) is 0.653 MeV.

---

* For a poet's view of the neutrino, see John Updike's poem "Cosmic Gall."

and are thought today to be, by far, the most abundant of all of the known particles of physics.

Our study of alpha and beta decay permits us to look at the nuclidic chart of Fig. 54-4 in a new way. Let us construct a three-dimensional surface by plotting the mass of each nuclide in a direction at right angles to the *N-Z* plane of that figure. The surface so formed gives a graphic representation of nuclear stability. As Fig. 54-11 shows (for the light nuclides), it describes a "valley of the nuclides," the stability zone of Fig. 54-4 running along its bottom. Nuclides on the headwall of the valley (a region not displayed in Fig. 54-11) decay into it largely by chains of alpha decay and by spontaneous fission. Nuclides on the proton-rich side of the valley decay into it by emitting positive electrons and those on the neutron-rich side do so by emitting negative electrons.

EXAMPLE 7

Calculating the disintegration energy. Calculate the disintegration energy Q in the beta decay of ^{32}P, as described by Eq. 54-11. The needed atomic masses are 31.97391 u for ^{32}P and 31.97207 u for ^{32}S.

Because of the presence of the emitted electron, we must be especially careful to distinguish between nuclear and atomic masses. Let m'_P and m'_S represent the nuclear masses of ^{32}P and ^{32}S and let m_P and m_S be their atomic masses. We take the disintegration energy Q to be $\Delta m\, c^2$, where

$$\Delta m = m'_P - (m'_S + m_e),$$

m_e being the mass of the electron. If we add and subtract $15 m_e$ to the right-hand side, we have

$$\Delta m = (m'_P + 15 m_e) - (m'_S + 16 m_e).$$

The quantities in parentheses are the atomic masses. Thus we have

$$\Delta m = m_P - m_S.$$

If we subtract the atomic masses in this way, the mass of the emitted electron is automatically taken into account.*

---

* This is not the case for positron decay or for electron capture; see Problems 51 and 53. Note also that in this Example we neglect the (small) difference in the binding energies of the atomic electrons before and after the beta decay.

The disintegration energy for the ^{32}P decay is then

$$Q = \Delta m\, c^2 = (31.97391 \text{ u} - 31.97207 \text{ u})(931 \text{ MeV/u})$$

$$= 1.71 \text{ MeV}.$$

This is just equal to the measured value of K_{max}, the maximum energy of the emitted electrons. Thus, although 1.71 MeV is released every time a ^{32}P nucleus decays, in essentially every case the electron carries away less energy than this. The neutrino gets the rest, carrying it away from the laboratory undetected.

54-6 MEASURING IONIZING RADIATION

The effects of ionizing radiations on living tissue (particularly our own!) has become a matter of general public interest. Such radiations arise in nature from the cosmic rays and also from radioactive elements in the earth's crust. Man-made radiations also contribute, including diagnostic and therapeutic X rays and radiations from radionuclides used in medicine and in industry. The disposal of radioactive waste and the evaluation of the probabilities of nuclear accidents continue to be dealt with at the level of national policy.

It is not our task here to explore the various sources of ionizing radiations but simply to describe the units in which the properties and effects of these radiations are expressed. There are four such units, and they are often used loosely or incorrectly in popular reporting.

1. *The curie (abbr. Ci).* This is a measure of the *activity* of a radioactive source. It was originally defined as the activity of one gram of radium in equilibrium with its by-products, but it is now defined simply as

   1 curie = 1 Ci = 3.7×10^{10} disintegrations per second.

   This definition says nothing about the nature of the decays. Note also that this unit is not appropriate to describe the ionizing effects of X rays from, say, a medical X-ray machine. The radiations must be emitted from a radionuclide.

   An example of the proper use of the curie is the statement: "Plutonium-239 has a specific activity of 62 Ci/mg." The fact that ^{239}Pu is an alpha emitter does not enter.

2. *The roentgen (abbr. R).* This is a measure of *exposure*, that is, of the ability of a beam of X rays or gamma rays to produce ions in a particular substance. Specifically, one roentgen is defined as that exposure that would produce 1.6×10^{12} ion pairs per gram of air, the air being dry and at standard temperature and pressure. We might say, for example: "This dental X-ray beam provides an exposure of 300 mR/s." This says nothing about whether ions are actually produced or whether or not there is a patient in the chair.

3. *The rad.* This is an acronym for Radiation Absorbed Dose and is a measure, as its name suggests, of the dose actually delivered to a specific object, in terms of the energy transferred to it. An object, which might be a person (whole body) or a specific part of the body (the hands, say) is said to have received an absorbed dose of one rad when 10^{-5} J/g have been delivered to it by ionizing radiations. A typical statement to show the usage is: "A whole-body gamma-ray dose of 300 rad will cause death in 50% of the population exposed to it." By way of comfort we note that the present average exposure to

radiation from both natural and man-made sources is about 0.2 rad (= 200 mrad) per year.

4. *The rem.* This is an acronym for Roentgen Equivalent in Man and is a measure of *dose equivalent.* It takes account of the fact that, although different types of radiation (gamma rays and neutrons, say) may deliver the same energy per unit mass to the body, they do not have the same biological effect. The dose equivalent (in rems) is found by multiplying the absorbed dose (in rads) by a *quality factor* QF, which may be found tabulated in various reference sources. For X rays and electrons, QF = 1. For slow neutrons, QF = 5, and so on. Personnel monitoring devices such as film badges are designed to register the dose equivalent in rems.

An example of correct usage of the rem is: "The recommendation of the National Council on Radiation Protection is that no individual who is (nonoccupationally) exposed to radiations should receive a dose equivalent greater than 500 mrem (= 0.5 rem) in any one year." This includes radiations of all kinds, the appropriate quality factors being used.

EXAMPLE 8

Ionizing radiation. We have seen that a dose of 300 rad is lethal to 50% of the population that receives it. If the equivalent amount of energy were absorbed directly as heat, what temperature increase would result? Assume that c, the specific heat capacity of the human body, is the same as that of water, namely, 4180 J/kg · K.

An absorbed dose of 300 rad corresponds to an absorbed energy per unit mass of

$$(300 \text{ rad}) \left(\frac{10^{-2} \text{ J/kg}}{1 \text{ rad}} \right) = 3 \text{ J/kg}.$$

The temperature increase that would result from such an influx of thermal energy is found from

$$\Delta T = \frac{Q/m}{c} = \frac{3 \text{ J/kg}}{4180 \text{ J/kg} \cdot \text{K}} = 7.2 \times 10^{-4} \text{ K}.$$

We see from this tiny temperature increase that the damage done by ionizing radiation has very little to do with thermal heating. The harmful effects arise because the ionizing radiation succeeds in breaking molecular bonds and thus interfering with the normal functioning of the tissue in which it has been absorbed.

54-7
NUCLEAR REACTIONS

We can represent a nuclear reaction by

$$\text{X} + a \rightarrow \text{Y} + b \qquad (54\text{-}15a)$$

or, in more compact notation

$$\text{X}(a,b)\text{Y}. \qquad (54\text{-}15b)$$

We consider only those reactions in which the projectile (a) and the emerging particle (b) are considerably lighter than the target nuclide (X) and the residual nuclide (Y). The projectile will typically be raised to its desired energy in a Van de Graaff accelerator, such as that shown in Fig. 54-12.

The reaction energy Q is defined (see Example 7 of Chapter 10) as

$$Q = (m_X + m_a)c^2 - (m_Y + m_b)c^2 \qquad (54\text{-}16a)$$

or, equivalently, as

$$Q = (K_Y + K_b) - (K_X + K_a), \qquad (54\text{-}16b)$$

in which the K's are kinetic energies.

A typical reaction (see Example 7 of Chapter 10) is

$$^{19}\text{F}(p,\alpha)^{16}\text{O},$$

for which $Q = 8.13$ MeV. Equation 54-16 tells us that, in this reaction,
the system loses rest mass energy and gains kinetic energy, in amount
8.13 MeV per event. Reactions, like this one, for which $Q > 0$ are called
exothermic. Reactions for which $Q < 0$ are called *endothermic*; such

Atomic number, Z							
82	^{197}Pb 42 min	^{198}Pb 2.4 h	^{199}Pb 1.5 h	^{200}Pb 21.5 h	^{201}Pb 9.42 h	^{202}Pb 3×10^5 y	^{203}Pb 52.0 h
81	^{196}Tl 1.84 h	^{197}Tl 2.83 h	^{198}Tl 5.3 h	^{199}Tl 7.4 h	^{200}Tl 26.1 h	^{201}Tl 73.6 h	^{202}Tl 12.2 d
80	^{195}Hg 9.5 h	^{196}Hg 0.15%	^{197}Hg 64.1 h	^{198}Hg 10.0%	^{199}Hg 16.9%	^{200}Hg 23.1%	^{201}Hg 13.2%
79	^{194}Au 39.5 h	^{195}Au 183 d	^{196}Au 6.18 d	^{197}Au 100%	^{198}Au 2.70 d	^{199}Au 3.14 d	^{200}Au 48.4 min
78	^{193}Pt ~50 y	^{194}Pt 32.9%	^{195}Pt 33.8%	^{196}Pt 25.3%	^{197}Pt 18.3 h	^{198}Pt 7.2%	^{199}Pt 30.8 min
77	^{192}Ir 74.2 d	^{193}Ir 62.7%	^{194}Ir 19.2 h	^{195}Ir 2.5 h	^{196}Ir 52 s	^{197}Ir 9.85 min	^{198}Ir ~8 s
76	^{191}Os 15.4 d	^{192}Os 41.0%	^{193}Os 30.5 h	^{194}Os 6.0 h	^{195}Os 6.5 min	^{196}Os 35 min	—
	115	**116**	**117**	**118**	**119**	**120**	**121**

Neutron number, N

figure 54-13
A portion of the nuclidic chart of Fig.
54-4, expanded and with more
information added.

reactions will not "go" unless a certain minimum kinetic energy (the *threshhold energy*) is carried into the system by the projectile.

If a and b are identical particles, which requires that X and Y also be identical, we describe the reaction as *scattering*. If the kinetic energy of the system is the same both before and after the event (which means that $Q = 0$), we have *elastic scattering*. If these energies are different ($Q \neq 0$), we have *inelastic scattering*.

We can easily keep track of nuclear reactions by plotting them on a nuclidic chart like that of Fig. 54-4. Figure 54-13 shows an enlarged portion of such a chart, centered arbitrarily on the nuclide ^{197}Au. Stable nuclides are shaded, and their isotopic abundances are shown. The unshaded squares represent radionuclides, their half-lives being shown.

Figure 54-14 suggests a transparent overlay that we can place over a nuclidic chart such as that of Fig. 54-13. If the shaded central square of Fig. 54-14 overlays a particular target on the chart of Fig. 54-13, the residual nuclides resulting from the various reactions printed on the overlay are identified.

Thus, if we chose ^{197}Au as a target, a (p,α) reaction will produce (stable) ^{194}Pt, and either an (n,γ) or a (d,p) reaction will produce the radionuclide ^{198}Au, whose half-life is 2.70 d.

Nuclei, like atoms, have stationary states of definite energy, and reaction studies can be used to identify them. Consider, for example, the reaction

$$^{27}\text{Al}(d,p)^{28}\text{Al} \qquad Q = 5.49 \text{ MeV},$$

in which a thin aluminum target foil is bombarded with 2.10-MeV deuterons. In the laboratory the emerging protons are seen to come off with a number of well-defined discrete energies and are accompanied by gamma rays. Figure 54-15 shows this proton spectrum.

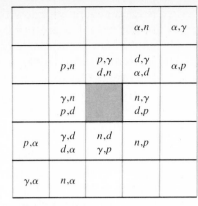

			α,n	α,γ
	p,n	p,γ d,n	d,γ α,d	α,p
	γ,n p,d		n,γ d,p	
p,α	γ,d d,α	n,d γ,p	n,p	
γ,α	n,α			

figure 54-14
An "overlay" to be placed over a nuclidic chart such as that of Fig. 54-12. The central square is placed over the target nuclide and the various possible residual nuclides are indicated.

figure 54-15
The distribution in energy of protons from the reaction ^{27}Al$(d,p)^{28}$Al. The energy of the incident deuteron is 2.10 MeV, and the protons are detected as they emerge from the target foil at right angles to the incident beam. (Adapted from Enge, Beuchner, and Sperduto, 1952.)

In every reaction event we know that an energy equal to the kinetic energy of the incident deuteron (= 2.10 MeV) plus the disintegration energy Q (= 5.49 MeV) is available to be shared between the two reaction products, that is, between the residual nucleus ^{28}Al and the emerging proton p. How is this total energy (2.10 MeV + 5.49 MeV = 7.59 MeV) to be shared between these two particles?

It all depends on whether the residual nucleus ^{28}Al is produced in its ground state or in one of its excited stationary states. In the former case, the emerging proton will have the maximum possible energy, corresponding to the peak on the extreme right of the proton spectrum in Fig. 54-15. If, however, the residual nucleus is formed in an excited state, that nucleus will retain more of the available energy and there will be less energy left for the emerging proton. The residual nucleus will not remain in its excited state very long but will get rid of its excess energy by emitting a gamma ray. Hence the spectrum of gamma rays that accompanies the spectrum of protons.

Thus we see that every proton peak in the spectrum of Fig. 54-15 corresponds to a stationary state of the residual nucleus ^{28}Al. Figure 54-16 shows the energy levels that may be deduced by analyzing this spectrum. We have seen that our understanding of the way atoms are put together rests on the measured energies of the hydrogen atom states as its firm foundation. In the same way, we can hope to learn how nuclei are put together by studying the energies and other properties of their stationary states.

54-8
NUCLEAR MODELS

To begin to understand atoms we had to make a model. That is, we had to think up a framework or — to put it another way — to develop a way of looking at atoms that allowed us to interpret as much of the experimental data as possible. We saw, for example, how Niels Bohr's semi-classical orbital model of the atom allowed him to account so very nicely for the measured wavelengths of the spectrum lines for the hydrogen atom. More refined models followed.

We follow the same path in trying to understand the nucleus, and we present here *two* models. Although based on assumptions that seem flatly to exclude each other, each accounts very well for a selected group of nuclear properties. After describing them separately, we shall see how these two models may be combined to form a single coherent picture of the atomic nucleus.

The Liquid Drop Model. In the liquid drop model, formulated by Niels Bohr, the nucleons are imagined to interact strongly with each other, like the molecules in a drop of liquid. A given nucleon collides frequently with the other nucleons in the nuclear interior, its mean free path as it moves about being substantially less than the nuclear diameter. This constant "jiggling around" reminds us of the thermal agitation of the molecules in a fluid. The liquid drop model has had impressive success in selected situations. It has been especially useful in explaining nuclear fission, as we shall make clear in Section 55-2.

Consider how the liquid drop model may be applied to nuclear reactions of the type

$$X + a \rightarrow C^* \rightarrow Y + b. \qquad (54\text{-}17)$$

Here C* represents an excited state of a so-called *compound nucleus* C. We imagine that projectile a enters target nucleus X, forming the compound nucleus C and conveying to it a certain excitation energy. The incoming particle *a* — let us say it is a neutron — is at once caught up in the random motions that characterize the nuclear interior. It quickly loses its identity, so to speak, and the excitation energy that it has carried with it comes quickly to be shared more or less equally by all of the nucleons.

The above quasi-stable state, represented by C* in Eq. 54-17, may endure for as long as, 10^{-16} s. This may seem to be a very short time, but it is a million times longer than the time it takes for a nucleon with a few MeV of kinetic energy to cross a nuclear diameter.

Eventually, by a statistical fluctuation, one of the nucleons in C* — perhaps a proton — will acquire enough energy to leave the compound nucleus, leaving as particle *b*.

The central feature of the compound nucleus model is that the formation of the compound nucleus and its eventual decay are totally independent events. A given compound nucleus may be formed in a variety of ways and may also decay in a variety of ways. At the time of its decay, we assume, the nucleus has "forgotten" how it was formed. We show here three of the possible ways in which a compound nucleus ^{20}Ne* might be formed and three different ways in which it might decay:

$$\left.\begin{matrix} ^{16}\text{O} + \alpha \\ ^{19}\text{F} + p \\ ^{20}\text{Ne} + \gamma \end{matrix}\right\} \rightarrow \underset{\substack{\text{An excited} \\ \text{compound} \\ \text{nucleus}}}{^{20}\text{Ne*}} \rightarrow \left\{\begin{matrix} ^{18}\text{F} + d \\ ^{19}\text{Ne} + n \\ ^{17}\text{O} + {}^{3}\text{He} \end{matrix}\right.$$

Three formation nodes   Three decay modes

These and similar sets of nuclear reactions have been studied quantitatively in the laboratory and support the liquid drop model (and the compound nucleus idea based on it) as useful and productive. There are, however, many other experimental results about which the liquid drop model simply has nothing to say.

The Independent Particle Model. In the liquid drop model, we assumed that the nucleons moved around at random and bumped into each other frequently. The independent particle model, however, is based on just the opposite assumption, namely, that each nucleon moves in a well-defined orbit within the nucleus and hardly makes any collisions at all! The nucleus — unlike the atom — has no fixed center of charge, and we assume in this model that each nucleon moves in a potential that is determined by the smeared-out motions of all the other nucleons.

A nucleon in a nucleus, like an electron in an atom, has a set of quantum numbers that defines its state of motion. Also nucleons, again like electrons, obey the Pauli exclusion principle. That is, no two nucleons may occupy the same state at the same time. In considering nucleon states, the neutrons and the protons are treated separately, each having its own array of available quantized states.

The fact that nucleons obey the Pauli principle helps us to understand the relative stability of the nucleon states. If two nucleons within the nucleus are to collide, the energy of each of them after the collision must correspond to the energy of an unoccupied stationary state. If these states (or even just one of them) are filled, the collision simply

figure 54-16
Energy levels of the nucleus ^{28}Al, deduced from data such as that of Fig. 54-15.

cannot occur. In time, any given nucleon will find it possible to collide, but meanwhile it will have made enough revolutions in its orbit to give meaning to the notion of a stationary nucleon state with a quantized energy.

In the atomic realm, the essence of the periodic table of the elements is that it is periodic. That is, certain properties of the elements repeat themselves in a regular fashion as one proceeds through the table. These repetitions are associated with the fact that the atomic electrons arrange themselves in shells and subshells that have a special stability when they are fully occupied. We can take the atomic numbers of the noble gases:

$$2, 10, 18, 36, 54, 86, \ldots$$

as *"magic electron numbers"* that mark the completion of such shells.

Nuclei also show such shell effects, associated with certain *magic nucleon numbers:*

$$2, 8, 20, 28, 50, 82, 126, \ldots .$$

Any nuclide whose proton number Z or neutron number N has one of these values turns out to have a special stability that may be made apparent in a variety of ways.

Examples of "magic" nuclides are ^{18}O $(Z = 8)$, ^{40}Ca $(Z = 20, N = 20)$, ^{92}Mo $(N = 50)$, and ^{208}Pb $(Z = 82, N = 126)$. Both ^{40}Ca and ^{208}Pb are said to be "doubly magic" because they contain filled shells of both protons *and* neutrons.

The magic number "2" shows up in the exceptional stability of the α-particle (^4He), which, with $Z = N = 2$, is doubly magic. For example, the binding energy per nucleon for this nuclide stands well above that of its neighbors on the binding energy curve of Fig. 54-7. The α-particle is so tightly bound, in fact, that it is impossible to add another particle to it; there is no stable nuclide with $A = 5$.

Figure 54-17 shows a small part of the evidence for $N = 20, 28, 50$, and 82 as magic neutron numbers. It displays the number of stable nuclides that occur in nature as a function of the neutron number N. The assumption is that, at the time the elements were formed, nuclides with especially stable neutron (or proton) configurations would have been formed in an especially large number of varieties. These two examples only hint at the overwhelming evidence for the stability of nuclides with magic neutron or proton numbers.

We have seen that wave mechanics can account beautifully for the magic electron numbers, that is, for the populations of the shell and the subshells into which the atomic electrons are grouped; see Table 51-2. It turns out that, by making certain reasonable assumptions, wave mechanics can account equally well for the magic nucleon numbers! The

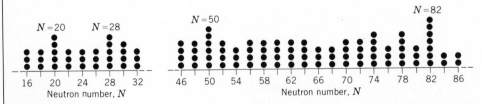

figure 54-17
The numbers of stable nuclides for various neutron numbers N. The "magic" neutron numbers are identified. For clarity the odd neutron numbers (for which the numbers of nuclides are small) are not shown.

1963 Nobel prize was, in fact, awarded to Maria Mayer and to Hans Jensen "for their discoveries concerning nuclear shell structure."

Table 54-3, which does for the neutrons and protons in the nucleus what Table 51-2 does for the electrons in the atom, summarizes the details. The table, which applies separately to neutrons and to protons, shows the nucleon levels on the left, each identified by its total angular momentum quantum number j. Note how the levels cluster in energy, forming themselves into shells.

Table 54-3
Nucleon levels in the shell model

The Levels	j	Level Population $(= 2j + 1)$	Shell Population	Cumulative Shell Population
$\vdots$				
——	$13/2$	14	44	126
——	$1/2$	2		
——	$5/2$	6		
——	$3/2$	4		
——	$9/2$	10		
——	$7/2$	8		
——	$11/2$	12	32	82
——	$3/2$	4		
——	$1/2$	2		
——	$7/2$	8		
——	$5/2$	6		
——	$9/2$	10	22	50
——	$1/2$	2		
——	$5/2$	6		
——	$3/2$	4		
——	$7/2$	8	8	28
——	$3/2$	4	12	20
——	$1/2$	2		
——	$5/2$	6		
——	$1/2$	2	6	8
——	$3/2$	4		
——	$1/2$	2	2	2

(left margin: Energy, with upward arrow)

A level with total angular momentum j, according to wave mechanics, can accommodate $2j + 1$ nucleons. These level populations are shown in the next column of Table 54-3. It is then a simple matter to add up the populations of the levels to find the populations of the shells. Finally, we can find the cumulative shell populations, adding up the shell populations, starting from the bottom of the table. These cumulative shell populations are exactly the magic nucleon numbers. Another triumph for wave mechanics!

The Collective Model. The collective model of nuclear structure, developed largely by Aage Bohr (a son of Niels Bohr) and Ben Mottleson, succeeds in combining the seemingly irreconcilable points of view of the liquid drop and the independent particle models.

Consider a nucleus with a small number of neutrons (or protons) outside a filled core that contains a "magic" number of neutrons (or protons). The "extra" nucleons moved in quantized orbits, in a potential established by the central core, thus preserving the central feature of the independent particle model. These extra nucleons can also interact with the core, deforming it and setting up vibrational and rotational "tidal wave" motions in it. These "liquid drop" motions of the core preserve the central feature of that model. The collective model has had very great success indeed in many areas of nuclear physics.

The compound nucleus and the uncertainty principle. Consider the neutron-capture reaction

$$^{109}\text{Ag} + n \rightarrow {}^{110}\text{Ag}^* \rightarrow {}^{110}\text{Ag} + \gamma. \qquad (54\text{-}18)$$

Figure 54-18 shows its cross section (see Section 10-7) as a function of the energy of the incident neutron. Analyze this figure in terms of the compound nucleus concept and the uncertainty principle.

The cross-section curve of Fig. 54-18 is sharply peaked, reaching a maximum cross section of 12,500 barns. This "resonance peak" suggests that we are dealing with a single excited level in the compound nucleus $^{110}\text{Ag}^*$. When the available energy just matches the energy of his level above the ^{110}Ag ground state, we have "resonance," and the reaction really "goes."

However, the resonance peak is not infinitely sharp. From the figure we can measure that it has an approximate width at half maximum (that is, at 6250 barns) of 0.20 eV. We account for this by saying that the excited ^{110}Ag level is not sharply defined in energy; it is "fuzzy," with an energy uncertainty ΔE of ~0.20 eV.

We can use the uncertainty principle, written in the form

$$\Delta E \cdot \Delta t \cong h \qquad (54\text{-}19)$$

to tell us something about any state of an atomic or nuclear system. We have seen that ΔE is a measure of the uncertainty of our knowledge of the energy of the state. The quantity Δt is interpreted as the time available to measure the energy of the state; it is in fact the mean life $\bar{t}$ of the state before it decays.

For the excited state $^{110}\text{Ag}^*$ we have, from Eq. 54-10,

$$\Delta t = \bar{t} \cong \frac{h}{\Delta E} = \frac{4.14 \times 10^{-15}\ \text{eV} \cdot \text{s}}{0.20\ \text{eV}} = 2.1 \times 10^{-14}\ \text{s}.$$

This is just the order of magnitude of the lifetime that we expect for a compound nucleus! Thus we have convincing evidence for the compound nucleus concept.

Testing the shell model. The nuclide ^{120}Sn ($Z = 50$) has a filled proton shell, 50 being one of the magic nucleon numbers. The nuclide ^{121}Sb ($Z = 51$) has an "extra" proton outside this shell. According to the shell concept, this extra proton should be easier to remove than a proton from the filled shell. Verify this by calculating the required energy in each case. Use the following mass data:

Nuclide	Z	N	Atomic Mass, u
^{121}Sb	$50 + 1$	70	120.9038
^{120}Sn	50	70	119.9022
^{119}In	$50 - 1$	70	118.9058

The proton atomic mass is 1.00783 u.

EXAMPLE 9

figure 54-18
The cross section for the reaction $^{109}\text{Ag}(n,\gamma)^{110}\text{Ag}$ for neutron energies near 5.2 eV. The so-called halfwidth of this resonance peak is 0.20 eV.

EXAMPLE 10

Removing the "extra" proton corresponds to the process

$$^{121}\mathrm{Sb} + E \rightarrow {}^{120}\mathrm{Sn} + p.$$

The required energy E follows from

$$E = (M_{Sn} + M_p - M_{Sb})c^2$$

$$= (119.9022\ \mathrm{u} + 1.00783\ \mathrm{u} - 120.9038\ \mathrm{u})(931\ \mathrm{MeV/u})$$

$$= 5.80\ \mathrm{MeV}.$$

Removing the proton from the filled shell corresponds to

$$^{120}\mathrm{Sn} + E \rightarrow {}^{119}\mathrm{In} + p.$$

The required energy follows from

$$E = (M_{In} + M_p - M_{Sn})c^2$$

$$= (118.9058\ \mathrm{u} + 1.00783\ \mathrm{u} - 119.9022\ \mathrm{u})(931\ \mathrm{MeV/u})$$

$$= 10.6\ \mathrm{MeV}.$$

This is considerably greater than the energy required to remove an "extra" proton (= 5.80 MeV), just as the shell model predicts. In much the same way the energy needed to remove an *electron* from a filled *electron shell* (= 22 eV for the filled shell of neon) is much greater than that needed to remove an "extra" electron from outside such a filled shell (= 5 eV for the "extra" electron from sodium).

questions

1. Explain why the Thomson model atom could not account for large angle scattering of α-particles from a thin foil whereas the Rutherford nuclear atom model could.

2. In what ways do the so-called strong force and the electrostatic or Coulomb force differ?

3. Why does the relative importance of the Coulomb force compared to the strong nuclear force increase at large mass numbers?

4. In your body, are there more neutrons than protons? More protons than electrons? Discuss.

5. Why do nuclei tend to have more neutrons than protons at high mass numbers?

6. Why do we use atomic rather than nuclear masses in analyzing nuclear decay and reaction processes?

7. How might the equality $1.000\ \mathrm{u} = 1.660 \times 10^{-27}\ \mathrm{kg}$ be arrived at in the laboratory?

8. Because the most stable nuclides have a mass number A near 60 (see Fig. 54-7), why don't *all* nuclides have mass numbers near 60?

9. If we neglect the very lightest nuclides, the binding energy per nucleon in Fig. 54-7 is roughly constant at 7 to 8 MeV/nucleon. Do you expect the mean electronic binding energy per electron in atoms also to be roughly constant throughout the periodic table?

10. Why is the binding-energy (Fig. 54-7) low at low mass numbers? At high mass numbers?

11. In the binding-energy curve of Fig. 54-7, what is special or notable about the nuclides $^2\mathrm{H}$, $^4\mathrm{He}$, $^{56}\mathrm{Fe}$, and $^{239}\mathrm{Pu}$?

12. The magnetic moment of the neutron is -1.9131 nuclear magnetons. What is a nuclear magneton and how does it differ from a Bohr magneton? What does the minus sign mean? How can the neutron, which carries no charge, have a magnetic moment in the first place?

13. A particular ^{238}U nucleus was created in a massive stellar explosion, perhaps 10^{10} y ago. It suddenly decays by α-emission while we are observing it. After all those years, why did it decide to decay at this particular moment?

14. Can you justify this statement: "In measuring half-lives by the method of Example 4, it is not necessary to measure the absolute decay rate R; any quantity proportional to it will suffice. However, in the method of Example 5 an absolute rate *is* needed."

15. You are running longevity tests on light bulbs. Do you expect their "decay" to be exponential? What is the essential diference between the decay of light bulbs and of radionuclides?

16. The half-life of ^{238}U is 4.47×10^9 y, about the age of the solar system. How can such a long half-life be measured?

17. In Fig. 54-9, make use of the relation $F = -dU/dr$ and show that the force in the nuclear interior is attractive whereas the force outside the nucleus is repulsive for α-particles.

18. Explain why, in alpha decay, short half-lives correspond to large disintegration energies, and conversely.

19. Can you give a justification, even a partial one, for the barrier tunneling phenomenon in terms of basic ideas about the wave nature of matter?

20. In beta decay the emitted electrons form a continuous spectrum, but in alpha decay they form a discrete spectrum. What difficulties did this cause in the explanation of beta decay, and how were these difficulties finally overcome?

21. How do neutrinos differ from photons? Each has zero charge and (presumably) zero rest mass and travels at the speed of light.

22. Make a list of the various sources of ionizing radiation encountered in our environment, whether natural or man-made.

23. Which of these conservation laws apply to all nuclear reactions? Conservation of (a) charge, (b) relativistic mass, (c) relativistic total energy, (d) rest mass, (e) kinetic energy, (f) linear momentum, (g) angular momentum, and (h) total numbers of nucleons.

24. In the development of our understanding of the atom, did we use atomic models as we now use nuclear models? Is Bohr theory such an atomic model? Are models now used in atomic physics? What is the difference between a model and a theory?

25. What are the basic assumptions of the liquid drop and the independent particle models of nuclear structure? How do they differ? How does the collective model reconcile these differences?

26. What is so special ("magic") about the magic nucleon numbers?

27. Why aren't the magic nucleon numbers and the magic electron numbers the same? What accounts for each?

28. The numbers of stable (or very long-lived) isotopes of the noble gases are

He	Ne	A	Kr	Xe	Rn
2	3	3	5	9	0

an average of 3.7. The average number of stable nuclides for the four magic neutron numbers shown in Fig. 54-17, however, is 5.8, considerably greater. If the noble gases are so stable, why were not more stable isotopes of them created when the elements were formed?

SECTION 54-1

1. (a) *The Thomson atom model.* At one time the positive charge of the atom was thought to be distributed uniformly throughout a sphere with a radius of about 1.0×10^{-10} m, that is, throughout the entire atom. Calculate the electric field at the surface of a gold atom on this (erroneous) assumption.

problems

Neglect the effect of the atomic electrons. (b) *The Rutherford, or nuclear, atom model.* Experiment shows that the positive charge of the atom is not spread uniformly throughout the volume of the atom, as (a) above suggests, but is concentrated in a small region (the nucleus) at the center of the atom. For gold the nuclear radius is 6.9×10^{-15} m. What is the electric field at the nuclear surface? Again, neglect effects associated with the atomic electrons.   *Answer:* (a) 1.1×10^{13} N/C; (b) 2.4×10^{21} N/C.

SECTION 54-2

2. (a) Show that the energy tied up in nuclear, or strong force, bonds is proportional to A, the mass number of the nucleus in question. (b) Show that the energy tied up in Coulomb force bonds between the protons is proportional to $Z(Z - 1)$. (c) Show that, as we move to larger and larger nuclei (see Fig. 54-4), the importance of (b) above increases more rapidly than does that of (a).

3. Locate the nuclides displayed in Table 54-1 on the nuclidic chart of Fig. 54-4. Verify that, as asserted, they lie along the stability line.

4. Using the nuclidic chart of Fig. 54-4, write the symbols for (a) all stable nuclides with $Z = 60$ (isotopes), (b) all radioactive nuclides with $N = 60$ (isotones), and (c) all nuclides with $A = 60$ (isobars).

5. Nuclear radii may be measured by scattering high-energy electrons from nuclei. (a) What is the de Broglie wavelength for 200-MeV electrons? (b) Are they suitable probes for this purpose?   *Answer:* (a) 6.2 fm; (b) yes.

6. Arrange the 25 nuclides given below in squares as a nuclidic chart similar to Fig. 54-4 but showing greater detail. Draw in and label (a) all isobaric (constant A) lines and (b) all lines of constant neutron excess, defined as $N - Z$. Nuclides: $^{118-122}$Te, $^{117-121}$Sb, $^{116-120}$Sn, $^{115-119}$In, and $^{114-118}$Cd.

7. Calculate and compare (a) the nuclear mass density ρ_m and (b) the nuclear charge density ρ_q for a fairly light nuclide such as ^{55}Mn and for a fairly heavy one such as ^{209}Bi. (c) Are the differences what you would expect?
Answer: (a) $\rho_m = 3.0 \times 10^{17}$ kg/m³ for each; (b) $\rho_q = 1.3 \times 10^{25}$ C/m³ for ^{55}Mn, and 1.1×10^{25} C/m³ for ^{209}Bi.

8. The atomic masses of ^{1}H, ^{12}C, and ^{238}U are 1.007825 u, 12.000000 u (by definition), and 238.050788 u, respectively. (a) What would these masses be if the mass unit were defined so that the mass of ^{1}H was (exactly) 1.000000 u? (b) Use your result to suggest why this perhaps obvious choice was not made.

9. Because the neutron has no charge, its mass must be found in some way other than by using a mass spectrometer. When a resting neutron and a proton meet, they combine and form a deuteron, emitting a γ-ray whose energy is 2.2233 MeV. The atomic masses of the proton and the deuteron are 1.007825035 u and 2.0141019 u, respectively. Find the mass of the neutron from these data, to as many significant figures as the data warrant. (A more precise value of the mass – energy conversion factor than that used in the text is 931.502 MeV/u.)   *Answer:* 1.0087 u.

10. Verify that the binding energy per particle given in Table 54-1 for ^{239}Pu is indeed 7.56 MeV/nucleon. The needed atomic masses are 239.05216 u (^{239}Pu), 1.00783 u (^{1}H), and 1.00867 u (neutron).

11. A penny has a mass of 3.0 g. Calculate the nuclear energy that would be required to separate all the neutrons and protons in this coin. Ignore the binding energy of the electrons. For simplicity assume that the penny is made entirely of ^{64}Cu atoms (mass = 63.92977 u). The atomic masses of the proton and the neutron are 1.00783 u and 1.00867 u, respectively.
Answer: 1.58×10^{25} MeV.

12. (a) Show that an approximate formula for the mass M of an atom is $M = AM_p$, where A is the mass number and M_p is the proton mass. (b) What percent error is committed in using this formula to calculate the masses of the atoms in Table 54-1? The proton mass is 1.007276 u.

13. You are asked to pick apart an α-particle (^{4}He) by removing, in sequence, a proton, a neutron, and a proton. Calculate (a) the work required for each step, (b) the total binding energy of the α-particle, and (c) the binding energy per particle. Some needed atomic masses are

^{4}He   4.00260 u, ^{2}H   2.01410 u,
^{3}H   3.01605 u, ^{1}H   1.00783 u,
            n   1.00867 u.

Answer: (a) 19.8 MeV, 6.26 MeV, 2.22 MeV; (b) 28.3 MeV; (c) 7.07 MeV.

14. The electrostatic potential energy of a uniform sphere of charge Q and radius R is

$$U = \frac{3Q^2}{20\pi\epsilon_0 R}.$$

(a) Find the electrostatic potential energy for the nuclide ^{239}Pu, assumed spherical; see Table 54-1. (b) Compare its electrostatic potential energy per particle with its binding energy per nucleon of 7.57 MeV. (c) What do you conclude?

15. A neutron star is a stellar object whose density is about that of nuclear matter, as calculated in Example 2. Suppose that the sun were to collapse into such a star without losing any of its present mass. What would be its expected radius?   *Answer:* 12 km.

16. *Mass excess.* To simplify calculations, atomic masses are sometimes tabulated, not as the actual atomic mass m, but as $(m - A)c^2$, where A is the mass number expressed in mass units. This quantity, usually reported in keV or MeV, is called the mass excess, symbol Δ. Using data from Example 3, find the mass excesses for the proton, the neutron, and ^{120}Sn.

17. *Mass excess* (See Problem 16). Show that the total binding energy of a nuclide can be written as

$$E = Z\Delta_p + N\Delta_n - \Delta.$$

where Δ_p, Δ_n, and Δ are the appropriate mass excesses. Using this method calculate the binding energy per particle for ^{197}Au. Compare your result with the value listed in Table 54-1. The needed mass excesses, rounded to three significant figures, are $\Delta_p = +7.29$ MeV, $\Delta_n = +8.07$ MeV, and $\Delta_{197} = -31.2$ MeV. Note the economy of calculation that results when mass excesses are used in place of the actual masses.
Answer: 7.92 MeV.

18. The strong neutron excess of heavy nuclei is illustrated by the fact that most heavy nuclides could never fission or break up into two stable nuclei without neutrons being left over. For example, consider the spontaneous fission of a ^{235}U nucleus. If the two daughter nuclei had atomic numbers 39 and 53, and were stable, by referring to Fig. 54-4 determine the daughter nuclides and the number of neutrons left over.

19. As Table 54-1 shows, the nuclide ^{197}Au has a nuclear spin of $\frac{3}{2}$. (a) If we regard this nucleus as a spinning rigid sphere with a radius given by Eq. 54-1, what rotational frequency results? (b) What rotational kinetic energy? (c) What is the acceleration for a point on the equator of the spinning nucleus? Note that this picture is overly mechanistic.
Answer: (a) $\sim 5 \times 10^{18}$ Hz; (b) ~ 15 keV; (c) $\sim 6 \times 10^{23}$ g.

20. How many particles make up the nuclide ^{7}Li on (a) the accepted neutron–proton theory of nuclear structure and (b) the long-since-abandoned proton–electron theory? (c) The nuclear spin of ^{7}Li is $\frac{3}{2}$. Can this fact be used to decide between these two theories? (d) The nuclear magnetic moment of ^{7}Li is $+3.26$ nuclear magnetons. Does this fact help to decide between the two theories?

21. The spin and the magnetic moment of ^{81}Br in its ground state (see Table 54-1) are $\frac{3}{2}$ and $+2.27$ nuclear magnetons, respectively. A free ^{81}Br nucleus is placed in a magnetic field of 2.0 T. (a) Into how many levels will the ground state split because of space quantization? (b) What is the energy difference between adjacent pairs of levels? (c) What is the wavelength that corresponds to a transition between such a pair of levels? (d) In what region of the electromagnetic spectrum does this wavelength lie?
    Answer: (a) Four; (b) 7.4×10^{-8} eV; (c) 17 m; (d) Radio region.

22. Because a nucleon is confined to a nucleus, we can take its uncertainty in position to be approximately the nuclear radius R. What does the uncertainty principle say about the kinetic energy of a nucleon in a nucleus with, say, $A = 100$? (*Hint:* Take the uncertainty in momentum Δp to be the actual momentum p.)

23. In the periodic table, the entry for magnesium is:

12
Mg
24.312

    There are three isotopes:

    ^{24}Mg, atomic mass $= 23.98504$ u.

    ^{25}Mg, atomic mass $= 24.98584$ u.

    ^{26}Mg, atomic mass $= 25.98259$ u.

    The abundance of ^{24}Mg is 78.99% by weight. What are the abundances of the other two isotopes?   *Answer:* ^{25}Mg: 9.30%; ^{26}Mg: 11.71%.

SECTION 54-3

24. From data presented in the first few paragraphs of Section 54-3, deduce (a) the disintegration constant λ and (b) the half-life of ^{238}U.

25. A 5.00-g charcoal sample from an ancient fire pit has a ^{14}C activity of 63.0 disintegrations/min. Carbon from a living tree has a specific activity of 15.3 disintegrations/g · min. The half-life of ^{14}C is 5730 y. How old is the charcoal sample?   *Answer:* 1600 y.

26. Uranium ^{238}U decays to lead ^{206}Pb with a half-life of 4.47×10^9 y. Although the series has many individual steps, the first has by far the longest half-life; therefore, one can often consider the decay to go directly to lead. That is,

    $$^{238}\text{U} \rightarrow {}^{206}\text{Pb} + \text{various decay products.}$$

    A rock is found to contain 4.2 mg of ^{238}U and 2.135 mg of ^{206}Pb. Assume that the rock contained no lead at formation, all of the lead now present arising from the decay of uranium. (a) How many atoms of ^{238}U and ^{206}Pb does the rock now contain? (b) How many atoms of ^{238}U did the rock contain at formation? (c) What is the age of the rock?

27. The radionuclide ^{64}Cu has a half-life of 12.7 h. What fraction of the ^{64}Cu nuclei initially present will decay between the 14th and the 16th hours?
    Answer: 0.048.

28. A radioactive isotope of mercury, ^{197}Hg, decays into gold, ^{197}Au, with a decay constant of 0.0108 h^{-1}. (a) What is its half-life? (b) What fraction of the original amount will remain after three half-lives? (c) After 10 days?

29. Many radionuclides have two or more competing modes of decay. For ^{212}Bi, for example, we have

    $$^{212}\text{Bi} \rightarrow {}^{208}\text{Tl} + \alpha \qquad \text{in 35.4\% of the cases}$$

    and

    $$^{212}\text{Bi} \rightarrow {}^{212}\text{Po} + e^- + \nu \qquad \text{in 64.6\% of the cases.}$$

The half-life for the ^{212}Bi decay is 60.6 min. What would the half-life be if only the alpha decay occurred? Although the answer is of interest, it must be realized that there is no known way to suppress the beta decay.
Answer: 171 min.

30. ^{238}U is an α-emitter with a half-life of 4.47×10^9 y. At what rate do ^{238}U alpha decays occur in a 1.00-g sample of uranium metal? The isotopic abundance of ^{238}U in naturally occurring uranium is 99.3%; the molecular weight of such uranium is 238 g/mol.

31. The radionuclide ^{32}P $(t_{1/2} = 14.28$ d$)$ is often used as a tracer to follow the course of biochemical reactions involving phosphorus. (*a*) If the counting rate in a particular experimental setup is 3050 counts/s, after what time will it fall to 170 counts/s? (*b*) A solution contaning ^{32}P is fed to the root system of an experimental tomato plant and the ^{32}P activity in a leaf is measured 3.48 days later. By what factor must this reading be multiplied to correct for the decay that has occurred since the experiment began?
Answer: (*a*) 59.5 d; (*b*) 1.18.

32. A 1.00-g sample of samarium emits α-particles at an absolute rate of 120 particles/s. ^{147}Sm, whose natural abundance in elemental samarium is 15.0%, is the responsible isotope. What is the half-life for the decay process?

33. After long effort, in 1902, Marie and Pierre Curie succeeded in separating from uranium ore the first substantial quantity of radium, one decigram of pure RaCl$_2$. The radium was the radioactive isotope ^{226}Ra, which has a decay half-life of 1600 y. (*a*) How many radium nuclei had they isolated? (*b*) What was the decay rate of their sample, in disintegrations/s? In curies? [The unit *curie* (abbr. Ci) was named in honor of the Curies, who received the 1903 Nobel prize in physics for their work on radiation phenomena. One curie equals 3.7×10^{10} disintegrations/s.]
Answer: (*a*) 2.0×10^{20}; (*b*) 2.8×10^9/s = 75 mCi.

34. A source contains two phosphorus radionuclides, ^{32}P $(t_{1/2} = 14.3$ d$)$ and ^{33}P $(t_{1/2} = 25.3$ d$)$. Initially 10% of the decays come from ^{33}P. How long must one wait until 90% do so?

35. As of this writing there is speculation that the proton may not actually be a stable particle but may be radioactive, with a half-life of about 10^{32} y. If this turns out to be true — and measurements continue — about how long would a person have to live to be sure that a single proton in his body had decayed? (*Hint:* Assume that you are made of water, which has ten protons per molecule. Only an approximate answer is called for, of course.)
Answer: 6200 y for a 70-kg person.

36. The plutonium isotope ^{239}Pu is produced as a by-product in nuclear reactors and hence is accumulating in our environment. It is radioactive, decaying by alpha decay with a half-life of 2.41×10^4 y. But plutonium is also one of the most toxic chemicals known; as little as 2 mg is lethal to a human. (*a*) How many nuclei constitute a chemically lethal dose? (*b*) What is the decay rate of this amount? If you were handling that quantity would you fear being poisoned or suffering radiation sickness?

37. One of the dangers of radioactive fallout from a nuclear bomb is ^{90}Sr, which beta decays with a 29-year half-life. Because it has chemical properties much like calcium, the strontium, if eaten by a cow, becomes concentrated in its milk and ends up in the bones of whoever drinks the milk. The energetic decay electrons damage the bone marrow and thus impair the production of red blood cells. A 1-megaton bomb produces approximately 400 g of ^{90}Sr. If the fallout was spread uniformly over a 2000-km^2 area, what area would have radioactivity equal to the allowed bone burden for one person of 0.002 mCi? (One curie = 1 Ci = 3.7×10^{10} disintegrations per second.)   *Answer:* 730 cm^2.

38. (*a*) Show that the *mean life* $\bar{t}$ of a radionuclide, that is, the average elapsed time before a nucleus decays, is $1/\lambda$. (*b*) Show that if it could be arranged —

and there is no known way to do so — for the sample to continue to decay at its initial rate, the radioactivity would drop to zero after one mean life. (c) After how many mean lives will the radioactivity drop to one-half of its initial value?

39. After a brief neutron irradiation of silver, two activities are present: ^{108}Ag ($t_{1/2} = 2.42$ min) with an initial decay rate of 3.1×10^5/s, and ^{110}Ag ($t_{1/2} = 24.6$ s) with an initial decay rate of 4.1×10^6/s. Make a semilog plot similar to Fig. 54-8 showing the total combined decay rate of the two isotopes as a function of time from $t = 0$ until $t = 10$ min. In Fig. 54-8, the extraction of the half-life for simple decays was illustrated. Given only the plot of total decay rate, can you suggest a way to analyze it in order to find the half-lives of both isotopes?

40. *Secular equilibrium.* A certain radionuclide is being manufactured, say, in a cyclotron, at a constant rate R. It is also decaying, with a disintegration constant λ. Let the production process continue for a time that is long compared to the half-life of the radionuclide. Convince yourself that the number of radioactive nuclei present at such times will be constant and will be given by $N = R/\lambda$. Convince yourself further that this result holds no matter how many of the radioactive nuclei were present initially. The nuclide is said to be in secular equilibrium with its source; in this state its decay rate is just equal to its production rate.

41. *Secular equilibrium* (see Problem 40). The radionuclide ^{56}Mn has a half-life of 2.58 h and is produced in a cyclotron by bombarding a manganese target with deuterons. The target contains only the stable manganese isotope ^{55}Mn and the reaction that produces ^{56}Mn is

$$^{55}\text{Mn} + d \rightarrow {}^{56}\text{Mn} + p.$$

After being bombarded for a time $\gg 2.58$ h, the activity of the target, due to ^{56}Mn, is 2.4 curie (Ci), where 1 Ci is 3.7×10^{10} disintegrations/s. (a) At what constant rate R are ^{56}Mn nuclei being produced in the cyclotron during the bombardment? (b) At what rate are they decaying (also during the bombardment)? (c) How many ^{56}Mn nuclei are present at the end of the bombardment? (d) What is their mass?
Answer: (a) 8.88×10^{10} s^{-1}; (b) $(8.88 \times 10^{10}$ s$^{-1})[1 - e^{-(7.46 \times 10^{-5})t}]$, t in seconds; (c) 1.19×10^{15}; (d) 0.11 μg.

42. *Secular equilibrium* (see Problems 40 and 41). A radium source contains 1.00 mg of ^{226}Ra, which decays with a half-life of 1600 y to produce ^{222}Rn, a noble gas. This radon isotope in turn decays by α-emission with a half-life of 3.82 d. (a) What is the rate of disintegration of ^{226}Ra in the source? (b) How long does it take for the radon to come to secular equilibrium with its radium parent? (c) At what rate is the radon then decaying? (d) How much radon is in equilibrium with its radium parent?

SECTION 54-4

43. A ^{238}U nucleus emits an α-particle of energy 4.196 MeV. What is the disintegration energy Q for this process, taking the recoil energy of the residual ^{234}Th nucleus into account? *Answer:* 4.269 MeV.

44. Consider a ^{238}U nucleus to be made up of an α-particle (^{4}He) and a residual nucleus (^{234}Th). Plot the electrostatic potential energy $U(r)$, where r is the distance between these particles. Cover the range ~ 10 fm $< r < \sim 100$ fm and compare your plot with that of Fig. 54-9.

45. Heavy radionuclides emit an α-particle rather than other combinations of nucleons because the α-particle is such a stable, tightly bound structure. To confirm his, calculate the disintegration energies for these hypothetical decay processes and discuss the meaning of your findings:

$$^{235}\text{U} \rightarrow {}^{232}\text{Th} + {}^{3}\text{He} + Q_3,$$
$$^{235}\text{U} \rightarrow {}^{231}\text{Th} + {}^{4}\text{He} + Q_4,$$
$$^{235}\text{U} \rightarrow {}^{230}\text{Th} + {}^{5}\text{He} + Q_5.$$

The needed atomic masses are

^{232}Th	232.0381 u	^{3}He	3.0160 u
^{231}Th	231.0363 u	^{4}He	4.0026 u
^{230}Th	230.0331 u	^{5}He	5.0122 u
	^{235}U	235.0439 u	

Answer: $Q_3 = -9.50$ MeV; $Q_4 = 4.66$ MeV; $Q_5 = -1.30$ MeV.

46. Consider that a ^{238}U nucleus emits (a) an α-particle or (b) a sequence of neutron, proton, neutron, proton. Calculate the energy released in each case. (c) Convince yourself both by reasoned argument and also by direct calculation that the difference between these two numbers is just the total binding energy of the α-particle. Find the binding energy. Some needed atomic masses are

^{238}U	238.05079 u	^{4}He	4.00260 u
^{237}U	237.04874 u	^{1}H	1.00783 u
^{236}Pa	236.04891 u	n	1.00867 u
^{235}Pa	235.04544 u		
^{234}Th	234.04363 u		

47. Under certain circumstances, a nucleus can decay by emitting a particle heavier than an α-particle. Such decays are very rare and have only recently been observed. Consider the decays

^{223}Ra $\rightarrow$ ^{209}Pb + ^{14}C and

^{223}Ra $\rightarrow$ ^{219}Rn + ^{4}He.

(a) Calculate the Q-values for these decays and determine that both are energetically possible. (b) The Coulomb barrier height for α-particles in this decay is 30 MeV. What is the barrier height for ^{14}C decay?

Masses:
^{223}Ra	223.01850 u	^{14}C	14.00324 u
^{209}Pb	208.98108 u	^{4}He	4.00260 u
^{219}Rn	219.01008 u		

Answer: (a) 5.98 MeV; 31.8 MeV; (b) 78 MeV.

48. Generally speaking, heavier nuclides tend to be more unstable to alpha decay. For example, the most stable isotope of uranium, ^{238}U has an alpha decay half-life of 4.5×10^9 y. The most stable isotope of plutonium is ^{244}Pu with a 8.2×10^7 y half-life, and for curium we have ^{248}Cm and 3.4×10^5 y. When half of an original sample of ^{238}U has decayed, what fractions of the original isotopes of plutonium and curium are left?

SECTION 54-5

49. A certain stable nuclide, after absorbing a neutron, emits a negative electron and then splits spontaneously into two α-particles. Identify the nuclide.   *Answer:* ^{7}Li.

50. *Recoil during beta decay.* The radionuclide ^{32}P decays to ^{32}S as described by Eq. 54-11. In a particular decay event, a 1.71-MeV electron is emitted, the maximum possible value. What is the kinetic energy of the recoiling ^{32}S atom in this event? (*Hint:* For the electron it is necessary to use the relativistic expressions for the kinetic energy and the linear momentum. Newtonian mechanics may safely be used for the relatively slow-moving ^{32}S atom.)

51. *Q for positron decay.* The radionuclide ^{11}C decays according to

$$^{11}\text{C} \rightarrow {}^{11}\text{B} + e^+ + \nu \qquad t_{1/2} = 20.3 \text{ min.}$$

The maximum energy of the positron spectrum is 0.960 MeV. (a) Show that the disintegration energy Q for this process is given by

$$Q = (m_C - m_B - 2m_e)c^2,$$

where m_C and m_B are the atomic mass of ^{11}C and ^{11}B, respectively and m_e is the electron (positron) mass. (b) Given that $m_C = 11.011434$ u, $m_B = 11.009305$ u, and $m_e = 0.0005486$ u, calculate Q and compare it with the maximum energy of the positron spectrum, given above. (Hint: Let m_C' and m_B' be the nuclear masses and proceed as in Example 7 for beta decay. Note that positron decay is an exception to the general rule that, if atomic masses are used in nuclear decay processes, the mass of the emitted electron is automatically taken care of.)   Answer: 0.961 MeV.

52. ^{137}Cs is present in the fallout from above-ground detonations of nuclear bombs. Because it beta decays with a 30.2-y half-life into ^{137}Ba and releases considerable energy in the process, it is an environmental concern. If the atomic masses of the parent and daughter are 136.9071 u and 136.9058 u, respectively, calculate the total energy released in each decay.

*53. *Electron capture.* Some radionuclides decay by capturing one of their own atomic electrons, a K-electron, say. An example is

$$^{49}V + e^- \rightarrow ^{49}Ti + v \qquad t_{1/2} = 331 \text{ d}.$$

Show that the disintegration energy Q for this process is given by

$$Q = (m_V - m_{Ti})c^2 - E_K,$$

where m_V and m_{Ti} are the atomic masses of ^{49}V and ^{49}Ti, respectively, and E_K is the binding energy of the vanadium K-electron. (Hint: Put m_V' and m_{Ti}' as the corresponding nuclear masses and proceed as in Example 7; see the footnotes in that example.)

54. *Electron capture.* Find the disintegration energy Q for the decay of ^{49}V by K-electron capture, as described in Problem 53. The needed data are $m_V = 48.94852$ u, $m_{Ti} = 48.94787$ u, and $E_K = 5.47$ keV.

55. A free neutron decays according to Eq. 54-13. If the neutron—hydrogen atom mass difference is 840 μu, what is the maximum energy of the β-spectrum?   Answer: 0.782 MeV.

56. An electron is emitted from a middle-mass nuclide ($A = 150$, say) with a kinetic energy of 1.0 MeV. (a) What is its de Broglie wavelength? (b) What is the radius of the emitting nucleus? (c) Can such an electron be confined as a standing wave in a "box" of such dimensions? (d) Can you use these numbers to disprove the argument (long since abandoned) that electrons actually exist in nuclei?

57. Two radioactive materials that are unstable to alpha decay, ^{238}U and ^{232}Th, and one that is unstable to beta decay, ^{40}K, are sufficiently abundant in granite to contribute significantly to the heating of the earth through the decay energy produced. The α-unstable isotopes give rise to decay chains that stop at stable lead isotopes. ^{40}K has a single beta decay. Decay information follows:

Parent Nuclide	Decay Mode	Half-life (y)	Stable Endpoint	Q (MeV)	f (ppm)
^{238}U	α	4.47×10^9	^{206}Pb	51.7	4
^{232}Th	α	1.41×10^{10}	^{208}Pb	42.7	13
^{40}K	β	1.28×10^9	^{40}Ca	1.31	4

Q is the total energy released in the decay of one parent nucleus to the final stable endpoint and f is the abundance of the isotope in kilograms per kilogram of granite; ppm means parts per million. (a) Show that these materials give rise to a total heat production of 9.8×10^{-10} W for each kilogram of granite. (b) Assuming that there is 2.7×10^{22} kg of granite in a 20-km-thick, spherical shell around the earth, estimate the power this will produce over the whole earth. (c) Compare this with the total solar power intercepted by the earth, 1.7×10^{17} W.   *Answer:* (b) 2.6×10^{13} W.

58. Heavy radionuclides, which may be either α- or β-emitters, belong to one of four decay chains, depending on whether their mass numbers A are of the form $4n$, $4n + 1$, $4n + 2$, or $4n + 3$, where n is a positive integer. (a) Justify this statement and show that if a nuclide belongs to one of these families, all its decay products will belong to the same family. (b) Classify these nuclides as to family: ^{235}U, ^{236}U, ^{238}U, ^{239}Pu, ^{240}Pu, ^{245}Cm, ^{246}Cm, ^{249}Cf, and ^{253}Fm.

SECTION 54-6

59. An airline pilot spends an average of 20 hours per week flying at 35,000 ft, at which altitude the equivalent dose due to cosmic radiation is 0.7 mrem/h. What is his annual equivalent dose from this source alone? Note that the maximum permitted yearly equivalent dose (from all sources) for the general population is 170 mrem and for radiation workers it is 5000 mrem. *Answer:* 0.73 rem.

60. A radioactive iodine isotope (^{125}I) much used in medicine costs $24/\mu$Ci. Its specific activity, in the form of free iodine, is 0.8 Ci/μmol. (a) What would a gram of this material cost? (b) At what rate would the value of a one-gram sample decrease as the sample decays? (c) At what rate would such a sample generate energy? The half-life of ^{125}I is 60 d; the energy release per disintegration is 176 keV.

61. Cancer cells are more vulnerable to X and γ radiation than are healthy cells. Though linear accelerators are now replacing it, in the past the standard source for radiation therapy has been radioactive ^{60}Co, which beta decays into an excited nuclear state of ^{60}Ni, which immediately drops into the ground state, emitting two γ-ray photons, each of approximate energy 1.2 MeV. The controlling beta decay half-life is 5.27 y. How many radioactive ^{60}Co nuclei are present in a 6000-Ci source used in a hospital? (1 Ci = 1 curie = 3.7×10^{10} disintegrations/s.)   *Answer:* 5.3×10^{22}.

SECTION 54-7

62. The radionuclide ^{60}Co ($t_{1/2} = 5.27$ y) is much used in cancer therapy. Tabulate possible reactions that might be used in preparing it. Limit the projectiles to neutrons, protons, and deuterons. Limit the targets to stable nuclides. The stable nuclides suitably close to ^{60}Co are ^{63}Cu, ^{60,61,62}Ni, ^{59}Co, and ^{57,58}Fe. [Commercially, ^{60}Co is made by bombarding elemental cobalt (^{59}Co; 100%) with neutrons in a reactor.]

63. A beam of deuterons falls on a copper target. Copper has two stable isotopes; ^{63}Cu(69.2%) and ^{65}Cu(30.8%). Tabulate the residual nuclides that can be produced by the reactions (d,n), (d,p), (d,α), and (d,γ). By inspection of Fig. 54-4, indicate which residual nuclides are stable and which are radioactive.
Answer: ^{64}Zn, ^{66}Zn, ^{64}Cu, ^{66}Cu, ^{61}Ni, ^{63}Ni, ^{65}Zn, ^{67}Zn; stable: ^{64}Zn, ^{66}Zn, ^{61}Ni, ^{67}Zn.

64. Consider the reactions:

$$^{19}\text{F}(p,\alpha)Y_1, \quad Q_1,$$

and

$$^{16}\text{O}(d,p)Y_2 \quad Q_2.$$

(a) Identify Y_1 and Y_2. (b) Identify a third reaction, of reaction energy Q_3, such that $Q_1 + Q_2 + Q_3 = 0$.

65. Consider the reaction $X(a,b)Y$, in which X is taken to be at rest in the laboratory reference frame. The initial kinetic energy in this frame is

$$K_{\text{lab}} = \left(\frac{1}{2}\right) m_a v_a^2.$$

(a) Show that the initial velocity of the center of mass of the system in the laboratory frame is

$$V = v_a \left(\frac{m_a}{m_X + m_a} \right).$$

Is this quantity changed by the reaction?

(b) Show that the initial kinetic energy, viewed now in a reference frame attached to the center of mass of the two particles, is given by

$$K_{cm} = K_{lab} \left(\frac{m_X}{m_X + m_a} \right).$$

Is this quantity changed by the reaction?

(c) In the reaction ^{27}Al$(d,p)^{28}$Al, the kinetic energy of the deuteron, measured in the laboratory frame, is 2.10 MeV. Find v_a $(= v_d)$, V, and K_{cm}. *Answer:* (c) 1.4×10^7 m/s; 9.8×10^5 m/s; 1.96 MeV.

66. In an endothermic reaction $(Q < 0)$, the interacting particles a and X must have a kinetic energy, measured in the center-of-mass reference frame, of at least $|Q|$ if the reaction is to "go." Show, using the result of Problem 65, that the so-called *threshold energy* for particle a, measured in the laboratory reference frame is

$$K_{th} = |Q| \frac{m_X + m_a}{m_X}.$$

Is it reasonable that K_{th} should be greater than $|Q|$?

67. *The characteristic nuclear time.* The characteristic nuclear time is a useful but loosely defined quantity, taken as equal to the time required for a nucleon with a few MeV of kinetic energy to travel a distance equal to the diameter of a middle-mass nuclide. What is the order of magnitude of this quantity? Consider 5-MeV neutrons traversing a nuclear diameter of ^{197}Au; see Table 54-1. *Answer:* 4×10^{-22} s.

68. Prepare an overlay like that of Fig. 54-14 in which that figure is extended to include reactions involving the light nuclides ^{3}H (tritium) and ^{3}He, considered both as projectiles and as emerging particles.

69. Prepare an overlay like that of Fig. 54-14 in which two nucleons or light nuclei may appear as emerging particles. The reaction ^{63}Cu$(\alpha,pn)^{65}$Zn is an example. Consider the combinations nn, np, and pd as possibilities.

70. Making mental use of the overlay of Fig. 54-14, write down the reactions by which the radionuclide ^{197}Pt $(t_{1/2} = 18.3$ h$)$ can be prepared, at least in principle. Except in special circumstances, only stable nuclides can serve as practical targets in cyclotron bombardment arrangements.

71. A platinum target is bombarded with cyclotron-accelerated deuterons for several hours and then the element iridium $(Z = 77)$ is separated chemically from it. What radioisotopes of iridium are present and by what reactions are they formed? (*Note:* ^{190}Pt and ^{192}Pt, not shown in Fig. 54-13, are stable platinum isotopes, but their isotopic abundances are so small that we may ignore their presence.) *Answer:* ^{192}Ir, ^{194}Pt(d,α), 74.2 d; ^{194}Ir, ^{196}Pt(d,α), 19.2 h; ^{196}Ir, ^{198}Pt(d,α), 52 s.

SECTION 54-8

72. An intermediate nucleus in a particular nuclear reaction decays within $\sim 10^{-22}$ s of its formation. Note that this is the characteristic nuclear time, defined and calculated in Problem 67. (a) What is the uncertainty ΔE in our knowledge of this intermediate state? (b) Can this state be called a compound nucleus?

73. A typical kinetic energy for a nucleon in a middle-mass nucleus may be taken as, say, 5 MeV. To what effective nuclear temperature does this correspond, using the assumptions of the liquid drop model of nuclear structure? (*Hint:* See Eq. 23-6.) *Answer:* 3.9×10^{10} K.

74. Consider the three formation modes shown for the compound nucleus ^{20}Ne* on p. 1285. What energy must (a) the α-particle, (b) the proton, and (c) the X-ray photon have to provide 25.0 MeV of excitation energy to the compound nucleus? Some needed atomic masses are

^{20}Ne	19.99244 u	α	4.00260 u
^{19}F	18.99840 u	p	1.00783 u
^{16}O	15.99491 u		

75. Consider the three decay modes shown for the compound nucleus ^{20}Ne* on p. 1285. If the compound nucleus is initially at rest and has an excitation energy of 25 MeV, what kinetic energy, measured in the laboratory, will (a) the deuteron, (b) the neutron, and (c) the ^{3}He nuclide have when the nucleus decays? Some needed atomic masses are

^{20}Ne	19.99244 u	d	2.01410 u
^{19}Ne	19.00188 u	n	1.00867 u
^{18}F	18.00094 u	^{3}He	3.01603 u
^{17}O	16.99913 u		

Answer: (a) 3.55 MeV; (b) 7.72 MeV; (c) 3.26 MeV.

76. From the following list of nuclides, identify (a) those with filled nucleon shells, (b) those with one nucleon outside a filled shell, and (c) those with one vacancy in an otherwise filled shell. Nuclides: ^{13}C, ^{18}O, ^{40}K, ^{49}Ti, ^{60}Ni, ^{91}Zr, ^{92}Mo, ^{121}Sb, ^{143}Nd, ^{144}Sm, ^{205}Tl, and ^{207}Pb.

77. In Fig. 54-4 note, for $Z \cong 85$, a region of relatively short-lived radionuclides ($t_{1/2} < 100$ y, by the criterion of that figure) followed at higher Z by a region of relatively long-lived radionuclides ($t_{1/2} > 100$ y). Account for this break in the zone of stability.

78. The nuclide ^{208}Pb is "doubly magic" in that both its proton number Z (= 82) and its neutron number N (= 126) represent filled nucleon shells. An additional proton would yield ^{209}Bi and an additional neutron ^{209}Pb. These "extra" nucleons should be easier to remove than a proton or a neutron from the filled shells of ^{208}Pb.

(a) Calculate the energy required to move the "extra" proton from ^{209}Bi and compare it with the energy required to remove a proton from the filled proton shell of ^{208}Pb.

(b) Calculate the energy required to remove the "extra" neutron from ^{209}Pb and compare it with the energy required to remove a neutron from the filled neutron shell of ^{208}Pb. Do your results agree with expectation? Use these atomic mass data:

Nuclide	Z	N	Atomic Mass, u
^{209}Bi	82 + 1	126	208.9804
^{208}Pb	82	126	207.9767
^{207}Tl	82 − 1	126	206.9774
^{209}Pb	82	126 + 1	208.9811
^{207}Pb	82	126 − 1	206.9759

The masses of the proton and the neutron are 1.00783 u and 1.00867 u, respectively.

79. The nucleus ^{91}Zr ($Z = 40$, $N = 51$) has a single neutron outside a filled 50-neutron core. Because 50 is a magic number, this neutron should perhaps be especially loosely bound. (a) What is its binding energy? (b) What is the binding energy of the next neutron, which must be extracted from the filled core? (c) What is the binding energy per particle for the entire nu-

cleus? Compare these three numbers and discuss. Some needed atomic masses are

^{91}Zr	90.90564 u	n	1.00867 u
^{90}Zr	89.90471 u	p	1.00783 u
	^{89}Zr	88.90890 u	

Answer: (a) 7.21 MeV; (b) 12.0 MeV; (c) 8.69 MeV.

80. (a) Extend the illustration in Fig. 54-17 of greater stability for nuclides having a magic number of neutrons to include *all* the magic numbers, showing several neighbors for each magic number. (b) Repeat the illustration for protons.

55

energy from the nucleus

When we get energy from coal by burning it in a furnace, we are tinkering with atoms of carbon and oxygen, rearranging their outer electrons in more stable combinations. When we get energy from uranium by consuming it in a nuclear reactor, we are tinkering with its nucleus, rearranging its nucleons in more stable combinations.

Electrons are held in atoms by the Coulomb force and it takes a few electron volts to pull one of them out. On the other hand, nucleons are held in nuclei by the strong nuclear force and it takes a few *million* electron volts to pull one of *them* out. This factor of several million is reflected in the fact that — after overcoming the greater technological difficulties — we can extract about that much more energy from a kilogram of uranium than we can from a kilogram of coal.

In comparing atoms and nuclei as sources of energy there are — from the point of view of the physics involved — two things to bear in mind:

1. In both cases the appearance of energy is accompanied by a decrease in rest mass, according to Einstein's $E = mc^2$ equation. The only difference between consuming uranium and burning coal is that, in the former case, a much larger fraction of the available mass (again, by a factor of several million) is converted to energy.

2. You must be clear about whether your concern is for energy or for power. In other words, are you concerned about the rate at which the energy is delivered? In the nuclear case will you burn your one kilogram of uranium slowly in a power reactor (Fig. 55-1*a*) or explosively in a bomb (Fig. 55-1*b*)? In the atomic case, are you thinking about exploding a stick of dynamite or digesting a jelly doughnut? (Sur-

55-1
THE NUCLEUS AND THE ATOM

1302

(a)

(b)

figure 55-1
(*a*) A drawing of the first nuclear reactor, assembled at the University of Chicago. It went critical on December 2, 1942. (Chicago Historical Society.) (*b*) The first nuclear bomb, exploded near Alamagordo, New Mexico, on July 16, 1945. (Courtesy Los Alamos National Laboratory.)

prisingly, the energy release is greater in the second case than in the first!)

Table 55-1 shows how much energy can be extracted from one kilogram of matter by doing various things to it. Instead of reporting the energy directly, we measure it by showing how long the extracted en-

Table 55-1
Obtaining energy from 1.0 kg of matter*

Form of Matter	Energy Extraction Process	Time
1. Water	A 50-m waterfall	5 s
2. Coal	Burning (2.7 kg of oxygen is also required)	8 h
3. UO_2 (enriched) to 3% ^{235}U)	Fission in a typical reactor	680 y
4. ^{235}U	Complete fission	3×10^4 y
5. Hot deuterium gas	Complete fusion to $^3He + n$	3×10^4 y
6. Matter and antimatter (0.5 kg each)	Total mutual annihilation	3×10^7 y

* Measured by the length of time the energy generated could operate a 100-W light bulb.

ergy could operate a 100-W light bulb. Only processes in the first three rows of the table have actually been carried out; the remaining three represent theoretical limits that may not be attainable in practice. Row 6, the total mutual annihilation of matter and antimatter, is the ultimate in extracting energy from matter. When you have used up all the available mass you can do no more.

Keep in mind that the comparisons of Table 55-1 are on a per-unit-mass basis. Pound for pound you get several millions times more energy from uranium than you do from coal or from falling water. On the other hand, there is a lot of coal in the earth's crust and there is a lot of water backed up behind the Bonneville dam in the Columbia river.

55-2
NUCLEAR FISSION: THE BASIC PROCESS

In 1932 the English physicist James Chadwick discovered the neutron. A few years later Enrico Fermi and his collaborators in Rome discovered that, if various elements are bombarded by these new projectiles, new radioactive elements are produced. Fermi had predicted that the neutron, being uncharged, would be a useful nuclear projectile; unlike the proton or the α-particle, it experiences no repulsive Coulomb force when it approaches a nuclear surface. Put in other words, there is no Coulomb barrier for it, so that the slowest neutron can interact with even the most massive, highly charged nucleus.

Thermal neutrons, which are neutrons in equilibrium with matter at room temperature, are convenient and effective bombarding particles. At 300 K, the mean kinetic energy of such neutrons is

$$E_{av} = \tfrac{3}{2}kT = \tfrac{3}{2}(8.6 \times 10^{-7} \text{ eV/K})(300 \text{ K}) \approx 0.04 \text{ eV} = 40 \text{ meV}.$$

In 1939 the German chemists Otto Hahn and Fritz Strassman, following up work initiated by Enrico Fermi and his collaborators, bombarded uranium with such thermal neutrons. They found by chemical analysis that after the bombardment a number of new radioactive elements were present, among them one whose chemical properties were remarkably similar to those of barium. Repeated tests finally convinced these able chemists that this "new" element was not new at all; it really *was* barium. How could this middle-mass element ($Z = 56$) be produced by bombarding uranium ($Z = 92$) with neutrons?

The riddle was solved within a few weeks by the physicists Lise Meitner and her nephew Otto Frisch. They showed that a uranium nucleus, having absorbed a neutron, could split, with the release of energy, into two roughly equal parts, one of which might well be barium. They named this process *nuclear fission.** Figure 55-2 shows the

figure 55-2
When a fast charged particle passes through a cloud chamber, it leaves a track of liquid droplets. The two back-to-back tracks represent fission fragments, produced by a fission event that took place in a thin vertical uranium foil in the center of the chamber. (Tracks representing irrelevant background events have been blacked out, for clarity.)

* See "The Discovery of Fission," by Otto Frisch and John Wheeler, *Physics Today,* November 1967, for a fascinating account of the early days of discovery.

tracks left in the gas of a cloud chamber by the two energetic fission fragments that result from a fission event occurring near the center of the chamber.

We can represent the fission of ^{235}U by thermal neutrons — a process of great practical importance — by

$$^{235}\text{U} + n \rightarrow {}^{236}\text{U}^* \rightarrow \text{X} + \text{Y} + bn, \qquad (55\text{-}1)$$

in which ^{236}U, as the asterisk indicates, is a compound nucleus. The factor b, which has the value 2.47 for fission events of this type, is the average number of neutrons released in such events.

Figure 55-3 shows the distribution by mass number of the fission fragments X and Y. We see that in only about 0.01% of the events will the fragments have equal mass. The most probable mass numbers, occurring in about 7% of the events, are $A = 140$ and $A = 95$. We can also tell from the difference in the length of the two fission fragment tracks in Fig. 55-2 that the two fragments in this particular fission event do not have the same mass.

^{236}U, which is the fissioning nucleus in Eq. 55-1, has 92 protons and $236 - 92$ or 144 neutrons, a neutron/proton ratio of ~ 1.6. The primary fragments formed immediately after fission will retain this same neutron/proton ratio. Study of the stability curve of Fig. 54-4, however, shows that stable nuclides in the middle-mass region ($75 < A < 150$) have a neutron/proton ratio of only 1.2 to 1.4. The primary fragments will thus be excessively neutron-rich and will "boil off" a small number of neutrons, 2.47 of them on the average for the reaction of Eq. 55-1. The fragments X and Y that remain are still too neutron-rich and will approach the stability line of Fig. 54-4 by a chain of successive β^- decays.

Equation 55-2 below is a specific instance of the generalized fission process of Eq. 55-1, in which X and Y are taken as ^{140}Xe and ^{94}Sr, respectively, and in which just two neutrons are emitted. Thus

$$^{235}\text{U} + n \rightarrow {}^{236}\text{U}^* \rightarrow {}^{140}\text{Xe} + {}^{94}\text{Sr} + 2n. \qquad (55\text{-}2)$$

As Table 55-2 shows, the fission fragments in Eq. 55-2 decay until each reaches a so-called daughter nuclide that is stable. The decays are β^- events, the half-lives being indicated at each stage. As expected, the mass numbers (140 and 94) remain unchanged as the decay continues.

The disintegration energy Q for fission is very much larger than for chemical processes. We can support this by a rough calculation. From the binding-energy curve of Fig. 54-7 we see that for heavy nuclides ($A = 240$, say) the mean binding energy per nucleon is about 7.6 MeV. In the intermediate range ($A = 120$, say), it is about 8.5 MeV. The difference in total binding energy between a single large nucleus ($A =$

figure 55-3
How the fission fragments X and Y (see Eq. 55-1) resulting from the fission of ^{235}U by thermal neutrons are distributed by mass number. The central dot corresponds to fission into fragments of equal mass. Note that the vertical scale is logarithmic.

Table 55-2
The radioactive decay of the fission fragments produced in the fission event of Eq. 55-2

$^{140}\text{Xe} \xrightarrow{\beta^-} {}^{140}\text{Cs} \xrightarrow{\beta^-} {}^{140}\text{Ba} \xrightarrow{\beta^-} {}^{140}\text{La} \xrightarrow{\beta^-} {}^{140}\text{Ce}$
14 s   65 s   13 d   40 h   Stable

$^{94}\text{Sr} \xrightarrow{\beta^-} {}^{94}\text{Y} \xrightarrow{\beta^-} {}^{94}\text{Zr}$
75 s   19 min   Stable

240) and two fragments (assumed equal) into which it may be split is then

$$Q = 2(8.5 \text{ MeV}) \left(\frac{A}{2}\right) - (7.6 \text{ MeV})A \approx 200 \text{ MeV}.$$

Example 1 shows a more careful calculation.

EXAMPLE 1

Analyzing a fission event. Calculate the disintegration energy Q for the fission event of Eq. 55-2, taking into account the decay of the fission fragments as outlined in Table 55-2. Needed atomic masses are

^{235}U	235.0439 u	^{140}Ce	139.9054 u
n	1.00866 u	^{94}Zr	93.9063 u.

If we combine Eq. 55-2 and Table 55-2, we see that the overall transformation is

$$^{235}\text{U} \rightarrow {}^{140}\text{Ce} + {}^{94}\text{Zr} + n.$$

The single neutron comes about because the (initiating) neutron on the left side of Eq. 55-2 cancels one of the two neutrons on the right side of that equation.

The mass difference for the above relation is

$$\Delta m = (235.0439 \text{ u}) - (139.9054 \text{ u} + 93.9063 \text{ u} + 1.00866 \text{ u})$$

$$= 0.224 \text{ u},$$

and the corresponding disintegration energy is

$$Q = \Delta m \, c^2 = (0.224 \text{ u})(931 \text{ MeV/u}) = 209 \text{ MeV},$$

in good agreement with the rough estimate made above.

In what form does this disintegration energy appear? Immediately after fission occurs, it is in the form of the kinetic energy of the (highly excited) primary fission fragments.

Eventually, if the fission event takes place in bulk solid, most of the disintegration energy appears as an increase in the internal energy of that body, revealing itself as a rise in temperature. Five percent or so of the disintegration energy, however, is associated with neutrinos that are emitted during the beta decay of the primary fission fragments. This energy is carried away into the depths of the cosmos and is lost to us forever.

55-3
NUCLEAR FISSION: THE MECHANISM

Soon after the discovery of fission, Niels Bohr and John Wheeler developed a theory, based on the analogy between a nucleus and a charged liquid drop, that explained its main features.

Figure 55-4 suggests how the fission process proceeds. When a heavy nucleus — let us say ^{235}U — absorbs a slow neutron, as in Fig. 55-4a, that neutron falls into the potential well associated with the strong nuclear forces that act in the nuclear interior. Its potential energy is then transformed into internal excitation energy, as Fig. 55-4b suggests. In much the same way the potential energy of a stone, dropped into a dry well, is transformed into (thermal) excitation energy at the bottom of the well.

The amount of excitation energy that a slow neutron carries into the nucleus that absorbs it is equal to the work required to pull the neutron back out of the nucleus, that is, to the binding energy E_n of the neutron. In much the same way, the amount of excitation energy delivered to a well when a stone is dropped into it is equal to the work required to pull

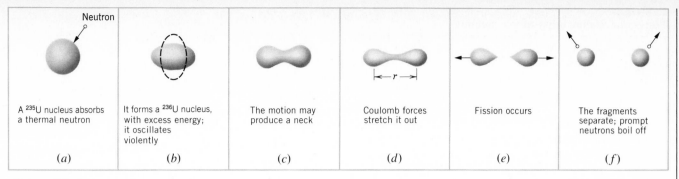

figure 55-4
Stages in a typical fission process.

the stone back out of the well, that is, to the "binding energy" E_s of the stone. In Example 2 we show that the binding energy E_n of a neutron in ^{236}U is 6.5 MeV.

Figure 55-4c shows that the nucleus, behaving like an energetically oscillating charged liquid drop, will sooner or later develop a short "neck" and will begin to separate into two charged "globs." If conditions are right, the electrostatic repulsion between these two globs will force them apart, breaking the neck. The two fragments, each still carrying some residual excitation energy, then fly apart. Fission has occurred.

So far this model gives a good qualitative picture of the fission process. It remains to be seen, however, whether it can answer a hard question: "Why are some heavy nuclides (^{235}U and ^{239}Pu, say) readily fissionable by slow neutrons but other, equally heavy, nuclides (^{238}U and ^{243}Am, say) are not?"

Bohr and Wheeler were able to answer this question. Figure 55-5 shows the potential energy curve for the fission process that they derived from their model. The horizontal axis displays the *distortion parameter r*, which is a rough measure of the extent to which the oscillating nucleus departs from a spherical shape. Figure 55-4d suggests how this parameter is defined before fission occurs. When the fragments are far apart, this parameter is simply the distance between their centers.

The energy interval between the initial state and the final state of the fissioning nucleus — that is, the disintegration energy Q — is displayed in Fig. 55-5. The central feature of that figure, however, is that the potential energy curve passes through a maximum at a certain value of r. There is a *potential barrier* of height E_b that must be surmounted (or tunneled!) before fission can occur. This reminds us of alpha decay (see Fig. 54-9), which is also a process that is inhibited by a potential barrier.

We see, then, that fission will occur *only* if the absorbed neutron provides an excitation energy E_n great enough to overcome the barrier. E_n need not be quite as great as the barrier height E_b because of the possibility of wave mechanical tunneling.

Table 55-3 shows a test of fissionability by thermal neutrons applied to four heavy nuclides, chosen from dozens of candidates that might have been considered. For each nuclide both the barrier height E_b and the excitation energy E_n are given. The former was calculated from the theory of Bohr and Wheeler; the latter was computed (as in Example 2) from the known masses.

For ^{235}U and ^{239}Pu we see that $E_n > E_b$. This means that fission by

figure 55-5
The potential energy at various stages in the fission process, showing the disintegration energy Q and the barrier height E_b.

Table 55-3
Test of the fissionability of four nuclides

Target Nuclide	Nuclide Being Fissioned	E_b, MeV	E_n, MeV	$E_n - E_b$, MeV	Fission Cross Section, Barns[a]
^{235}U	^{236}U	5.2	6.5	+1.3	500
^{238}U	^{239}U	5.7	4.8	−0.9	<0.5
^{239}Pu	^{240}Pu	4.8	6.4	+1.6	750
^{243}Am	^{244}Am	5.8	5.5	−0.3	<0.07

[a] The value given is for thermal neutrons. See Section 10-7 for the definition of cross section.

absorbing a thermal neutron is predicted to occur for these nuclides. This is confirmed by noting, in the table, the large measured cross sections for the process.

For the other two nuclides (^{238}U and ^{243}Am), we have $E_n < E_b$, so that there is not enough energy to surmount the barrier or to tunnel through it effectively. The excited nucleus (Fig. 55-4b) prefers to get rid of its excitation energy by emitting a gamma ray instead of by breaking into two large fragments. Table 55-3 shows, as we expect, that the cross sections for thermal neutron fission in these cases are exceedingly small. These nuclides *can* be made to fission, however, if they absorb a substantially energetic (rather than a thermal) neutron. For ^{238}U, for example, the absorbed neutron must have at least 1.3 MeV of energy for the fission process to "go" with meaningful probability.

EXAMPLE 2

Removing a neutron from a nucleus. Consider a nucleus ^{236}U in its ground state. How much energy is required to remove a neutron from it, leaving a ^{235}U nucleus behind? The needed atomic masses are

n	1.00867 u
^{235}U	235.04393 u
^{236}U	236.04557 u.

The increase in mass of the system as the neutron is pulled out is

$$\Delta m = (1.00867\ u + 235.04393\ u) - 236.04557\ u = 0.00703\ u.$$

This means that an energy equal to

$$E_n = \Delta m\ c^2 = (0.00703\ u)(931\ MeV/u) = 6.54\ MeV$$

must be expended. This, by definition, is just the binding energy of the neutron in the ^{236}U nucleus.

If a ^{235}U nucleus absorbs a thermal neutron, this energy (= 6.54 MeV) is just the amount of excitation energy that the thermal neutron will carry into the nucleus with it. The excited nucleus can get rid of this energy either by emitting a gamma ray (which leaves a ^{236}U nucleus in its ground state) or by fission (see Eq. 55-1). It turns out that fission is favored, by a ratio of six to one.

55-4
NUCLEAR REACTORS: THE BASIC PRINCIPLES

Energy releases in individual nuclear events such as α-emission are roughly a million times larger than those of chemical events, calculated on a per-atom basis. To make large-scale use of nuclear energy, however, we must arrange it so that one nuclear event triggers off another nearby until the process spreads throughout bulk matter like a flame through a burning log. The fact that more neutrons are generated in fission than are consumed (see Eq. 55-1) raises just this possibility; the neutrons that are produced can cause fission in nearby nuclei and in this way a chain of fission events can propagate itself. Such a process is called a *chain reaction*. It can either be rapid and uncontrolled as in a nuclear bomb or controlled as in a nuclear reactor.

Suppose that we wish to design a nuclear reactor based, as most present reactors are, on the fission of ^{235}U by slow neutrons. The fuel in such reactors is almost always artificially "enriched," so that ^{235}U makes up a few percent of the uranium nuclei rather than the 0.7% that occurs in natural uranium. Although on the average 2.47 neutrons are produced in ^{235}U fission for every thermal neutron consumed, there are serious difficulties in making a chain reaction "go." Here are three of the difficulties, together with their solutions:

1. *The neutron leakage problem.* A certain percentage of the neutrons produced will simply leak out of the reactor core and be lost to the chain reaction. If too many do so, the reactor will not work. Leakage is a surface effect, its magnitude being proportional to the square of a typical reactor core dimension ($= 4\pi r^2$ for a sphere). Neutron production, however, is a volume effect, proportional to the cube of a typical dimension ($= \frac{4}{3}\pi r^3$ for a sphere). The fraction of neutrons lost by leakage can be made as small as we wish by making the reactor core large enough, thereby decreasing its surface-to-volume ratio ($= 3/r$ for a sphere).

2. *The neutron energy problem.* The neutrons produced by fission are fast neutrons, with kinetic energies of ~ 2 MeV. However, fission is induced most effectively by slow neutrons. The fast neutrons can be slowed down by mixing the uranium fuel with a substance that has these properties: (a) It is effective in causing neutrons to lose kinetic energy by elastic collisions and (b) it does not absorb neutrons excessively, thereby removing them from the fission chain. Such a substance is called a *moderator*. Most power reactors in this country today use water as a moderator, the hydrogen nuclei (protons) being the effective moderating element; see Example 2 of Chapter 10.

3. *The neutron capture problem.* Neutrons may be captured by nuclei in ways that do not result in fission, capture with the emission of a γ-ray being the most common possibility. In particular, as the fast (~ 2 MeV) neutrons generated in the fission processes are slowed down in the moderator to thermal equilibrium (~ 0.04 eV), they must pass through an energy interval (~ 1 to ~ 100 eV) in which they are particularly susceptible to nonfission capture by ^{238}U.

   To minimize such *resonance capture*, as it is called, the uranium fuel and the moderator (water, say) are not intimately mixed but are "clumped," remaining in close contact with each other but occupying different regions of the reactor volume. The hope is that a fast fission neutron, produced in a uranium "clump" (which might be a fuel rod), will with high probability find itself in the moderator as it passes through the "dangerous" resonance energy range. Once it has reached thermal energies, it will very likely wander back into a clump of fuel and produce a fission event. It seems clear that finding the optimum geometric arrangement of fuel and moderator is no simple problem.

Figure 55-6 shows the neutron balance in a typical power reactor operating with a steady power output. We start at the top with 1000

figure 55-6
An accounting of one complete regeneration cycle for 1000 fast neutrons in a typical nuclear reactor, using natural uranium as a fuel.

typical fast neutrons produced by fission. Eight of them produce fission in the fuel before they have had a chance to interact with the moderator (fast fission). Another 292 are permanently lost to the chain reaction by leakage (192) and by resonance capture (100); these two effects can be minimized, as we indicated above, but they cannot be eliminated. This leaves 700 fast neutrons available to be turned into slow neutrons by the moderator.

Of the 700 slow neutrons, some 308 are also lost to the chain reaction, by leakage (75) and by nonfission capture in the fuel (208), in the materials of which the reactor is made (2), and in the accumulated fission products (23). This leaves 392 slow neutrons free to cause fission. Together with the original 8 fast neutrons, they produce a new generation of 2.5 (392 + 8) or 1000 fast neutrons, and the cycle starts all over again. Note that we have assumed that b in Eq. 55-1 is 2.5.

The *multiplication factor* k is an important reactor parameter. It is the ratio of the number of neutrons produced in each fission cycle to the number present at the beginning of the cycle. If $k = 1$, as in Fig. 55-6, the reactor operation is said to be critical. If $k > 1$ it is supercritical, and if $k < 1$ it is subcritical. Reactors are designed with the inherent ability to go supercritical. This condition cannot last very long, because the rate at which fission processes occur, and thus the rate of energy production in the fuel, would increase exponentially. Criticality is achieved, at any desired power level in the design range of the reactor, by suitably adjusting the *control rods*. These rods, made of a material such as cadmium, which absorbs slow neutrons readily, may be moved mechanically in and out of the reactor volume.

A vital factor in the control of a reactor is its response time. If, for example, one of the control rods is suddenly partially withdrawn, how rapidly would the reactor power level increase? This time is controlled by the extremely fortunate circumstance that a small fraction of the neutrons generated in the overall fission process are not boiled off promptly from the newly formed fission fragments but are emitted from these fragments along the β-decay chain (see Table 55-2 in which, however, neutron emission does not happen to occur; in other decay chains it does occur). Of the 1000 neutrons analyzed in Fig. 55-6, about 16 are delayed, being emitted from the fragments following beta decays whose half-lives range from 0.2 s to 55 s. These delayed neutrons are few in number, but they serve an important function in slowing down reactor response times to within the range of human reaction times.

Figure 55-7 shows the broad outlines of an electric power plant based on a so-called pressurized-water reactor (PWR), a type in common use in this country. In such a reactor, water is used both as the moderator and as the heat-transfer medium. In the primary loop, water at high temperature and pressure (possibly 600 K and 150 atm) circulates through the reactor vessel and transfers heat from the reactor core to the steam generator, which provides high-pressure steam to operate the turbine-generator combination. Low-pressure steam from the turbine is condensed to water and forced back into the steam generator by a pump, thus completing the secondary loop. To give some idea of scale, we note that a typical reactor pressure vessel for a 1000-MW (electric) plant may be 40 ft high and 15 ft in diameter, weighing 450 tons. Water flows through the primary loop at the rate of ~ 300,000 gal/min.

An unavoidable feature of nuclear power plants is the accumulation of radioactive wastes, including both fission products and heavy,

figure 55-7
A simplified layout of a nuclear power plant based on a pressurized water reactor.

figure 55-8
Thermal power released by the radioactive wastes generated by one year's operation of a typical large nuclear power plant. Note that both scales are logarithmic. (After B. L. Cohen.)

"transuranic" nuclides such as plutonium and americium. One measure of their radioactivity is the rate at which they release energy in thermal form. Figure 55-8 shows the variation with time of the thermal power generated by the radioactive wastes from one year's operation of a typical large power plant. The curve is the sum of contributions from the many radioactive nuclides that make up the waste. Note that both scales are logarithmic. The total activity of the waste 10 years after processing is $\sim 3 \times 10^7$ Ci, where 1 Ci is 3.7×10^{10} disintegrations/s.

EXAMPLE 3

Analyzing a nuclear power plant. A large electric power generating station operates on thermal energy derived from a pressurized-water nuclear reactor. Here are some of its characteristics:

Electric power output of the plant: 1100 MW

Thermal power in the reactor core: 3400 MW

Fuel charge: 86,000 kg of uranium in the form of 110 tons of uranium oxide, distributed among 57,000 fuel rods. The uranium is enriched to 3.0% ^{235}U.

(a) What is the efficiency of the plant?

$$e = \frac{\text{Useful output}}{\text{Energy input}} = \frac{1100 \text{ MW (electric)}}{3400 \text{ MW (thermal)}} = 32\%.$$

The efficiency is limited by the strictures of the second law of thermodynamics; see Section 25-5. It is necessary to discharge 3400 MW − 1100 MW = 2300 MW as thermal energy to the environment.

(b) At what rate R do fission events occur in the reactor core?

Let us assume conditions at start-up. If P (= 3400 MW) is the thermal power in the core and Q (= 200 MeV) is the energy released per fission event, we have

$$R = \frac{P}{Q} = \left(\frac{3.4 \times 10^9 \text{ W}}{200 \text{ MeV}}\right)\left(\frac{1 \text{ MeV}}{1.6 \times 10^{-13} \text{ J}}\right) = 1.1 \times 10^{20} \text{ s}^{-1}.$$

(c) At what initial rate (at start-up) is ^{235}U disappearing?

^{235}U disappears by fission at the rate calculated in (b). It also is consumed by (n,γ) reactions at a rate about one-fourth as large. The total loss rate is then $(1.25)(1.1 \times 10^{20} \text{ s}^{-1})$ or $1.4 \times 10^{20} \text{ s}^{-1}$. We recast this as a mass rate by putting

$$\frac{dM}{dt} = (1.4 \times 10^{20} \text{ s}^{-1})\left(\frac{235 \text{ g/mol}}{6.02 \times 10^{23} \text{ atoms/mol}}\right)\left(\frac{8.6 \times 10^4 \text{ s}}{1 \text{ d}}\right)$$

$$= 4.7 \text{ kg/d}.$$

The quantity in the second set of parentheses above is just the mass of a single ^{235}U atom.

(d) At this rate of fuel consumption, how long would the fuel supply last?

From the data given, we can calculate that, at start-up, $(0.030)(86,000 \text{ kg})$ or 2600 kg of ^{235}U were present. Thus a simplistic answer would be

$$T = \frac{2600 \text{ kg}}{4.7 \text{ kg/d}} = 550 \text{ d}.$$

However, complications associated with (1) the "poisoning" of the fuel by the build-up of fission products and (2) the build-up of (fissionable) ^{239}Pu render this simplistic result meaningless. In practice the fuel rods are replaced in batches—with their ^{235}U supply far from depleted—according to an efficiency-optimizing schedule.

(e) At what rate is mass being converted to energy in the reactor core?

From Einstein's $E = mc^2$ relationship, we can write

$$\frac{dM}{dt} = \frac{dE/dt}{c^2} = \frac{(3.4 \times 10^9 \text{ W})}{(3.0 \times 10^8 \text{ m/s})^2}$$

$$= 3.8 \times 10^{-8} \text{ kg/s} \quad \text{or} \quad 3.3 \text{ g/d}.$$

Note that this mass rate (conversion to energy) is quite a different physical quantity from the mass rate (loss of ^{235}U) calculated in part (c).

55-5
A NATURAL REACTOR

On December 2, 1942, when the reactor assembled by Enrico Fermi and his associates (see Fig. 55-1a) first went critical, they had every right to expect that they had put into operation the first fission reactor that had ever existed on this planet. About 30 years later it was discovered that, if they did in fact think that, they were wrong.

Some two billion years ago, in a uranium deposit now being mined in Gabon, West Africa, a natural fission reactor went into operation and ran for perhaps several hundred thousand years before shutting itself off.

The story of this discovery is totally fascinating at the level of the best detective thriller. More important, it provides a first-class example of the nature of the scientific evidence needed to back up what may seem at first to be an improbable claim. It sets a high standard for all who speculate about past events. We consider here only two points.*

1. *Was there enough fuel?* The fuel for a uranium-based fission reactor must be the easily fissionable isotope ^{235}U, which constitutes only 0.72% of natural uranium. This isotopic ratio has been measured not only for terrestrial samples but also in moon rocks and in meteorites; one always finds the same value. The initial clue to the discovery in Gabon was that the uranium from this deposit was deficient in this isotope, some samples having a ^{235}U abundance as low as 0.44%. Investigation led to the speculation that this deficit in ^{235}U could be accounted for if, at some time in the past, this isotope was partially consumed by the operation of a natural fission reactor.

The serious problem remains that, with an isotopic abundance of only 0.72%, a reactor can be assembled (as Fermi and his team learned) only with the greatest of difficulty. There seems no chance at all that it could have happened naturally.

However, things were different in the distant past. Both ^{235}U and ^{238}U are radioactive, with half-lives of 0.704×10^9 y and 4.47×10^9 y,

---

* See "A Natural Fission Reactor," by George A. Cowan, *Scientific American*, July 1976, for the complete story.

respectively. Thus the half-life of the readily fissionable ^{235}U is about 6.4 times shorter than that of ^{238}U. Because ^{235}U decays faster, there must have more of it, relative to ^{238}U, in the past. Two billion years ago, in fact, this abundance was not 0.72%, as it is now, but 3.8%. This abundance happens to be just about the abundance to which natural uranium is artificially enriched to serve as fuel in modern power reactors.

With this much readily fissionable fuel available in the distant past, the presence of a natural reactor (providing certain other conditions are met) is much less surprising. The fuel was there. Two billion years ago, incidentally, the highest order of life forms that had evolved were the blue-green algae.

2. *What is the evidence?* The mere depletion of ^{235}U in an ore deposit is not enough evidence on which to hang a claim for the existence of a natural fission reactor. One looks for more convincing proof.

If there was a reactor, there must also be fission products; see Fig. 55-3. Of the 30 or so elements whose stable isotopes are produced in this way, some must still remain. Study of their isotopic ratios could provide the convincing evidence we need.

Of the several elements investigated, the case of neodymium is spectacularly convincing. Figure 55-9a shows the isotopic abundances of the seven stable neodymium isotopes as they are normally found in nature. Figure 55-9b shows these abundances as they appear among the ultimate stable fission products of the fission of ^{235}U. The clear differences are not surprising, considering their totally different origins. The isotopes shown in Fig. 55-9a were formed in supernovae explosions that occurred before the formation of our solar system. The isotopes of Fig. 55-9b were cooked up in a reactor by totally different processes. Note particularly that ^{142}Nd, the dominant isotope in the natural element, is totally absent from the fission products.

The big question is, "What do the neodymium isotopes found in the uranium ore body in Gabon look like?" We must expect that, if a natural reactor operated there, isotopes from *both* sources (that is, natural isotopes as well as fission-produced isotopes) might be present. Figure 55-9c shows the results after this and other corrections have been made to the raw data. Comparison of Figs. 55-9b and 55-9c leaves little doubt that there was indeed a natural fission reactor at work!

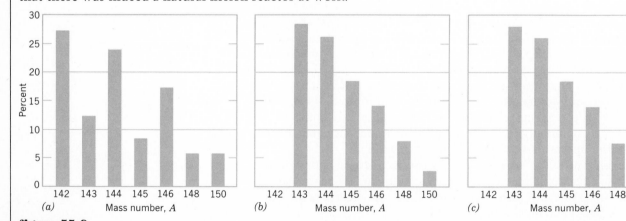

figure 55-9

The distribution by mass number of the isotopes of neodymium as they occur in (a) natural terrestrial deposits and (b) the spent fuel of a power reactor. (c) The distribution (after several corrections) found for neodymium from the uranium mine in Gabon, West Africa. Note that (b) and (c) are virtually identical. (From "A Natural Fission Reactor," by George A. Cowan. © Copyright *Scientific American*, July 1976, page 43.)

EXAMPLE 4

1315

THERMONUCLEAR FUSION: THE BASIC PROCESS   SEC. 55-6

Uranium in the distant past. The isotopic ratio of ^{235}U to ^{238}U in natural uranium deposits today is 0.0072. What was this ratio 2.0×10^9 y ago? The half-lives of the two isotopes are 0.704×10^9 y and 4.47×10^9 y, respectively.

Consider two samples that, at a time t in the past, contained $N_5(0)$ and $N_8(0)$ atoms of ^{235}U and ^{238}U, respectively. The numbers of atoms remaining at the present time are

$$N_5(t) = N_5(0)e\lambda_5 t \quad \text{and} \quad N_8(t) = N_8(0)e\lambda_8 t,$$

respectively, in which λ_5 and λ_8 are the corresponding disintegration constants. Dividing gives

$$\frac{N_5(t)}{N_8(t)} = \frac{N_5(0)}{N_8(0)} e^{-(\lambda_5 - \lambda_8)t}.$$

Expressed in terms of isotopic ratios, this becomes

$$R(0) = R(t)e^{(\lambda_5 - \lambda_8)t}$$

The disintegration constants are related to the half-lives by Eq. 54-8, or

$$\lambda_5 = \frac{0.693}{T_5} = \frac{0.693}{7.04 \times 10^8 \text{ y}} = 0.984 \times 10^{-9} \text{ y}^{-1}$$

and

$$\lambda_8 = \frac{0.693}{T_8} = \frac{0.693}{4.47 \times 10^9 \text{ y}} = 0.155 \times 10^{-9} \text{ y}^{-1}.$$

Substituting in the expression above for the isotopic ratio gives

$$R(0) = R(t)e^{(\lambda_5 - \lambda_8)t}$$

$$= (0.0072)e^{(0.984 - 0.155)(10^{-9} \text{ y}^{-1})(2.0 \times 10^9 \text{ y})}$$

$$= (0.0072)e^{1.65} = 0.0378 \text{ or } 3.78\%.$$

We see that, two billion years ago, the ratio of ^{235}U to ^{238}U in natural uranium deposits was much higher than it is today. When the earth was formed (4.5 billion years ago) this ratio was 30%.

55-6 THERMONUCLEAR FUSION: THE BASIC PROCESS

We pointed out in connection with the binding-energy curve of Fig. 54-7 that energy can be released if light nuclei are combined to form nuclei of somewhat larger mass number, a process called *nuclear fusion*. However, this process is hindered by the mutual Coulomb repulsion that tends to prevent two such (positively) charged particles from coming within range of each other's attractive nuclear forces and "fusing." This reminds us of the potential barrier that inhibits nuclear fission (see Fig. 55-5) and also of the barrier that inhibits alpha decay (see Fig. 54-9).

In the case of alpha decay, two charged particles — the α-particle and the residual nucleus — are initially *inside* their mutual potential barrier. For alpha decay to occur, the α-particle must leak through this barrier by the barrier-tunneling process and appear on the *outside*. In nuclear fusion the situation is just reversed. Here the two particles must penetrate their mutual barrier from the *outside* if a nuclear interaction is to occur. Example 5 gives some indication of the magnitudes of such Coulomb barriers for interacting light nuclei.

Although the calculation of Example 5 is somewhat simplistic, the result does indicate the order of magnitude of the Coulomb barrier for the deuteron–deuteron interaction. As we shall see later, this interaction is of particular importance in fusion research. The deuteron–

deuteron barrier, calculated below, is actually a fairly modest one, 200 keV. The corresponding barrier for two interacting ³He nuclei (charge = +2e), for example, is ~ 1 MeV. For more highly charged particles the barrier, of course, is correspondingly higher.

One way to arrange for light nuclei to penetrate their mutual Coulomb barriers is to use one light particle as a target and to accelerate the other by means of a cyclotron or some such device. To generate power in a useful way from the fusion process, however, we must have the interaction of matter in bulk, just as in the combustion of coal. The cyclotron technique holds no promise in this direction. The best hope for obtaining fusion in bulk matter in a controlled fashion is to raise the temperature of the material so that the particles have sufficient energy to penetrate the barrier due to their thermal motions alone. This process is called *thermonuclear fusion*.

The mean thermal kinetic energy of a particle in equilibrium at a temperature T is given, as we have seen, by

$$\overline{K} = \tfrac{3}{2}kT, \qquad (55\text{-}3)$$

where k (= 8.63×10^{-5} eV/K) is the Boltzmann constant. Substitution shows that at room temperature ($T \approx 300$ K) we have $K = 0.04$ eV, which is, of course, far too small for our purpose.

Even at the center of the sun, where $T \cong 1.5 \times 10^7$ K, the mean thermal kinetic energy calculated from Eq. 55-3 is only 1.9 keV. This still seems hopelessly small in view of the magnitude of the Coulomb barrier calculated in Example 5. Yet we know that thermonuclear fusion not only occurs in the solar interior but is the central and dominant feature of that body.

The puzzle is solved with the realization that (1) the energy calculated from Eq. 55-3 is a *mean* kinetic energy; particles with energies much greater than this mean value constitute the high-energy "tails" of the Maxwellian speed distribution curves (see Fig. 24-4). Also, (2) the barrier heights that we have quoted above are just that; they represent the *peaks* of the barriers. Barrier tunneling can occur to a significant extent at energies well below these peaks.

Figure 55-10 summarizes the situation by a quantitative example. The curve marked $n(K)$ in this figure is a Maxwell distribution curve drawn to correspond to the sun's central temperature, 1.5×10^7 K. Although the same curve holds no matter what particle is under consideration, we focus our attention on protons, bearing in mind that hydrogen forms about 35% of the mass of the sun's central core.

The distribution curve of Fig. 55-10 differs from the Maxwell distri-

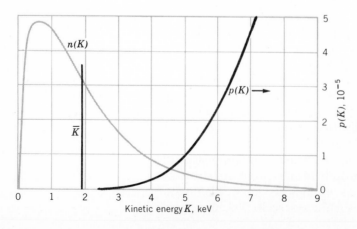

figure 55-10
The curve marked $n(K)$ gives the distribution in energy of particles (protons) at $T = 1.5 \times 10^7$ K, the temperature of the sun's interior. Its vertical scale is arbitrary. The curve marked $p(K)$ gives the probability of Coulomb barrier penetration for colliding protons at different energies. The scale is shown on the right.

bution curves of Fig. 24-4 in that it is drawn in terms of energy and not speed. Specifically, $n(K)\,dK$ gives the probability per unit volume that a proton will have a kinetic energy lying between K and $K + dK$. For $T = 1.5 \times 10^7\,K$, Eq. 55-3 yields $\overline{K} = 1.9$ keV, and this value is indicated by a vertical line in the figure. Note that there are many particles whose energies exceed this mean value.

The curve marked $p(K)$ in Fig. 55-10 is the probability of barrier penetration for two colliding protons. At $K = 6$ keV, for example, we have $p = 2.4 \times 10^{-5}$. This is the probability that two colliding protons, each with $K = 6$ keV, will succeed in penetrating their mutual Coulomb barrier and coming within range of each other's strong nuclear forces. Put another way, on the average, one of every $\sim 42{,}000$ such encounters will succeed.

It turns out that the most probable energy for proton–proton fusion events to occur—at the sun's central temperature—is ~ 6 keV. If the energy is much higher, the barrier is transparent enough but there are too few protons in the Maxwellian "tail." If the energy is much lower, there are plenty of protons but the barrier is now too formidable.

EXAMPLE 5

The Coulomb barrier for fusion. The deuteron (^{2}H) has a charge $+e$ and may be taken as a sphere whose effective radius R has been shown from scattering experiments to be 2.1 fm. Two such particles are fired at each other with the same initial kinetic energy K. What must K be if the particles are brought to rest by their mutual Coulomb repulsion when the two assumed spheres are just "touching"? We shall take K as a measure of the height of the Coulomb barrier.

Because the two deuterons are momentarily at rest when they "touch" each other, their kinetic energy has all been transformed into electrostatic potential energy associated with the Coulomb repulsion between them. If we treat them as charged mass points separated by a distance $2R$, we have

$$2K = \frac{1}{4\pi\epsilon_0}\frac{q_1 q_2}{r} = \frac{1}{4\pi\epsilon_0}\frac{e^2}{2R},$$

which yields

$$K = \frac{e^2}{16\pi\epsilon_0 R}$$

$$= \frac{(1.6 \times 10^{-19}\,\text{C})^2}{(16\pi)(8.9 \times 10^{-12}\,\text{C}^2/\text{J}\cdot\text{m})(2.1 \times 10^{-15}\,\text{m})}\left(\frac{1\,\text{keV}}{1.6 \times 10^{-16}\,\text{J}}\right)$$

$$\approx 200\,\text{keV}.$$

55-7
THERMONUCLEAR FUSION IN THE SUN AND OTHER STARS

Here we consider in more detail the thermonuclear fusion processes that take place in our sun and in other stars. In the sun's deep interior, where its mass is concentrated and where most of the energy production takes place, the (central) temperature, as we have already seen, is $\sim 1.5 \times 10^7$ K and the central density is $\sim 1.5 \times 10^5$ kg/m^3, about 13 times the density of lead. The central temperature is so high that, in spite of the high central pressure ($\sim 2 \times 10^{11}$ atm), the sun remains gaseous throughout.

The present composition of the sun's core is about 35% hydrogen by mass, about 65% helium, and about 1% other elements. At these temperatures the light elements are essentially totally ionized, so that our picture is one of an assembly of protons, electrons, and α-particles in random motion.

The sun radiates at the rate of 3.9×10^{26} W and has been doing so for as long as the solar system has existed, which is about 4.5×10^9 y. It has been known since the early 1930s that thermonuclear fusion processes in the sun's interior account for its prodigious energy output. Before analyzing this further, however, let us dispose of two other possibilities that had been put forward earlier. Consider first chemical reactions such as simple burning. If the sun, whose mass is 2.0×10^{30} kg, were made of coal and oxygen in just the right proportions for burning, it would last only $\sim 10^3$ y, which of course is far too short. The sun, as we shall see, does not burn coal but hydrogen, and in a nuclear furnace, not an atomic or chemical one.

Another possibility is that, as the core of the sun cools and the pressure there drops, the sun will shrink under the action of its own strong gravitational forces. By transferring gravitational potential energy to thermal energy (just as we do when we drop a stone onto the earth's surface), the temperature of the sun's core will rise so that radiation may continue. Calculation shows, however, that the sun could radiate from this cause for only $\sim 10^7$ y, too short by a factor of about 500. This leaves only thermonuclear fusion.

The sun's energy is generated by the thermonuclear "burning" (that is, "fusing") of hydrogen to form helium. Table 55-4 shows the *proton–proton cycle* by means of which this is accomplished. Note that each reaction shown is a fusion reaction, in that one of the products, be it ^{2}H, ^{3}He, or ^{4}He, has a higher mass number than any of the reacting particles that form it. As we expect for fusion reactions, the reaction energy Q, shown in the table, is positive. As we have seen, this characterizes an exothermic reaction, with the net production of energy.

The cycle is initiated by the thermal collision of two protons (^{1}H + ^{1}H) to form a deuteron (^{2}H), with the simultaneous creation of a positive electron (e^+) and a neutrino (ν). In Table 55-4 we follow the consequences of two such events, as indicated in the top row of the table. Such events are extremely rare. In fact, only once in $\sim 10^{26}$ proton–proton collisions is a deuteron formed; in the vast majority of cases the colliding protons simply scatter from each other. It is the slowness of this process that regulates the rate of energy production and keeps the sun from exploding. Interestingly, however, in spite of this slowness, there are so very many protons in the huge volume of the sun's core that deuterium is produced there in this way at the rate of $\sim 10^{12}$ kg/s!

Once a deuteron has been produced it quickly (within a few seconds) collides with another proton and forms a ^{3}He nucleus, as the second row of Table 55-4 shows. Two such ^{3}He nuclei may then eventually (within

Table 55-4
The proton–proton cycle

$$^1\text{H} + {}^1\text{H} \rightarrow {}^2\text{H} + e^+ + \nu \qquad Q_1 = 0.42 \text{ MeV} \qquad {}^1\text{H} + {}^1\text{H} \rightarrow {}^2\text{H} + e^+ + \nu$$

$$^2\text{H} + {}^1\text{H} \rightarrow {}^3\text{He} + \gamma \qquad Q_2 = 5.49 \text{ MeV} \qquad {}^2\text{H} + {}^1\text{H} \rightarrow {}^3\text{He} + \gamma$$

$$Q_3 = 12.86 \text{ MeV} \qquad {}^3\text{He} + {}^3\text{He} \rightarrow {}^4\text{He} + {}^1\text{H} + {}^1\text{H}$$

~10^5 y) collide, forming an α-particle (^{4}He) and two protons, as the third row of the table shows. There are other variations of the proton–proton cycle, involving other light elements, but we concentrate on the principal sequence as outlined above.

Taking an overall view of the proton–proton cycle, we see that it amounts to the combination of four protons to form an α-particle and two positive electrons with the release of energy Q' in the process. Note that although six protons are consumed in each cycle, two are generated, so that there is a net loss of four. Thus

$$Q' = (4m'_{1H} - m'_{4He} - 2m_e)c^2,$$

where the masses involved are here nuclear — not atomic — masses.

Consider now what happens to the two positive electrons that are produced in each cycle. Each will quickly encounter a negative electron and annihilate with it, turning their combined rest masses into γ-ray energy as described in Section 8-9. Thus the energy equivalent of the rest masses of four electrons (two positive and two negative) is also made available, or

$$Q'' = 4m_e c^2.$$

The total energy release for the cycle is then

$$\begin{aligned} Q &= Q' + Q'' \\ &= 4(m'_{1H} + m_e)c^2 - (m'_{4He} + 2m_e)c^2 \\ &= (4m_{1H} - m_{4He})c^2, \end{aligned}$$

in which the masses in the last line are now the customary atomic masses. Substitution yields

$$\begin{aligned} Q &= [(4)(1.00783 \text{ u}) - (4.00260 \text{ u})][931 \text{ MeV/u}] \\ &= 26.7 \text{ MeV.} \end{aligned}$$

Not quite all of this energy is available as thermal energy inside the sun. About 0.5 MeV is associated with the two neutrinos that are produced in each cycle. Neutrinos are so penetrating that in essentially all cases they escape from the sun, carrying this energy with them. Some are intercepted by the earth, bringing us our only direct information about the sun's interior.

Subtracting the neutrino energy leaves 26.2 MeV per cycle available within the sun. It is not hard to show (see Example 6) that this corresponds to a "heat of combustion," as chemists call it, for the nuclear burning of hydrogen into helium of 6.3×10^{14} J/kg of hydrogen consumed. By comparison, the heat of combustion of coal is about 3.3×10^7 J/kg, some 20 million times lower, reflecting roughly the general ratio of energies in nuclear and chemical processes.

The burning of hydrogen in the sun's core to form helium is alchemy on a grand scale in the sense that one element is turned into another. The medieval alchemists, however, were more interested in changing lead into gold than in changing hydrogen into helium. In a sense they were on the right track, except that their furnaces were too cool. Instead of being at ~600 K, they should have been at ~10^7 K, the temperature of stellar interiors!

We may ask how long the sun can continue to shine at its present rate before all the hydrogen in its core has been converted into helium. Hydrogen burning has been going on for about 4.5×10^9 y, and calculations show that there is enough available hydrogen left for about $5 \times$

10^9 y more. At that time major changes will begin to happen. The sun's core, which by then will be largely helium, will begin to collapse and to heat up while the outer envelope will expand greatly, perhaps so far as to encompass the earth's orbit. The sun will become what astronomers call a *red giant*.

If the core temperature heats up to $\sim 10^8$ K, energy can be produced once more by burning helium to make carbon. Helium does not burn readily, the only possible reaction being

$$^4\text{He} + {}^4\text{He} + {}^4\text{He} \rightarrow {}^{12}\text{C} + \gamma \qquad (Q = +7.3 \text{ MeV}).$$

Such a three-body collision of three α-particles must occur within $\sim 10^{-16}$ s if the reaction is to go. Nevertheless, if the density and temperature of the helium core are high enough, carbon will be manufactured by the burning of helium in this way.

As a star evolves and becomes still hotter, other elements can be formed by other fusion reactions. However, elements more massive than $A = 56$ (^{56}Fe, ^{56}Co, ^{56}Ni, etc.) cannot be manufactured by further fusion processes. $A = 56$ marks the peak of the binding-energy curve of Fig. 54-7, and fusion between nuclides beyond this point involves the consumption — and not the production — of energy.

If heavy elements cannot be produced by the fusion of middle-mass nuclides and if, as seems certain, they were not present in the primeval fireball that marked the creation of the universe, where then did they come from? The present view is that they were built up from medium-mass nuclides by the successive absorption of neutrons followed by β^- decay. A reaction such as

$$^{56}\text{Fe} + 3n \rightarrow {}^{59}\text{Fe},$$
$$^{59}\text{Fe} \rightarrow {}^{59}\text{Co} + e^- + \nu \qquad (t_{1/2} = 45 \text{ d}),$$

for example, produces ^{59}Co, a stable nuclide. Neutrons, being uncharged, experience no Coulomb barriers.

Conditions suitable for the build-up of the heaviest nuclides are thought to occur in *supernovae*, which are cataclysmic stellar explosions that occur in our galaxy, and also in other galaxies, with reasonable frequency. In such an event a very large amount of stellar matter is ejected violently into space where it mixes with, and becomes part of, the tenuous medium (~ 1 particle/cm^3) that fills the space between the stars. It is from this medium, continually enriched by debris from stellar explosions, that new stars form, by condensation under the influence of gravity.

The fact that elements heavier than hydrogen and helium are found on earth suggests to us that our solar system has condensed out of interstellar material that contained the remnants of such explosions. Thus all the elements around us — including those in our own bodies — were manufactured in the interiors of distant stars that have exploded long ago, returning their substance to the cosmic dust from which they evolved. As one scientist put it, "In truth we are the children of the Universe."

Burning hydrogen in the sun. At what rate is hydrogen being consumed in the core of the sun, assuming that all of the radiated energy is generated by the proton–proton cycle of Table 55-4?

EXAMPLE 6

We have seen that 26.2 MeV appear as thermal energy in the sun for every four protons consumed, a rate of 6.6 MeV/proton. We can express this as

$$6.6 \, \frac{\text{MeV}}{\text{proton}} \left(\frac{1 \text{ proton}}{1.67 \times 10^{-27} \text{ kg}} \right) \left(\frac{1.6 \times 10^{-13} \text{ J}}{1 \text{ MeV}} \right) = 6.3 \times 10^{14} \text{ J/kg},$$

which tells us that the sun radiates away 6.3×10^{14} J for every kilogram of protons consumed. The hydrogen consumption rate is then just the output power (= 3.9×10^{26} W) divided by the above quantity, or

$$\frac{dm}{dt} = \frac{3.9 \times 10^{26} \text{ W}}{6.3 \times 10^{14} \text{ J/kg}} = 6.2 \times 10^{11} \text{ kg/s}.$$

To keep this number in perspective, we point out that the sun's mass is 2.0×10^{30} kg.

55-8
CONTROLLED THERMONUCLEAR FUSION

Thermonuclear reactions have been going on in the universe since its creation in the presumed cosmic "big bang" of some 15 billion years ago. Such reactions have taken place on earth, however, only since November 1952, when the first fusion (or hydrogen) bomb was exploded. The high temperatures needed to sustain the thermonuclear reactions in this case are provided by using a fission bomb as a trigger.

A sustained and controllable thermonuclear power source — a fusion reactor — is proving much more difficult to achieve. The goal, however, is being vigorously pursued because many look to the fusion reactor as the ultimate power source of the future, at least as far as the generation of electricity is concerned.

The proton–proton interaction displayed in the top row of Table 55-4 is not suitable for use in a terrestrial fusion reactor because it is hopelessly slow. The reaction cross section is in fact so small that it cannot be measured in the laboratory. The reaction succeeds under the conditions that prevail in stellar interiors only because of the enormous number of protons available in the huge, high-density stellar cores.

The most attractive reactions for terrestrial use appear to be the deuteron–deuteron $(d–d)$ and the deuteron–triton $(d–t)$ reactions:

$$^{2}\text{H} + {}^{2}\text{H} \rightarrow {}^{3}\text{He} + n \quad (d–d), \quad Q = +3.27 \text{ MeV};$$
$$^{2}\text{H} + {}^{2}\text{H} \rightarrow {}^{3}\text{H} + {}^{1}\text{H} \quad (d–d), \quad Q = +4.03 \text{ MeV};$$

and

$$^{2}\text{H} + {}^{3}\text{H} \rightarrow {}^{4}\text{He} + n \quad (d–t), \quad Q = +17.59 \text{ MeV}.$$

As we see, *triton* is the name given to the nucleus of the hydrogen isotope ^{3}H. Note that each reaction above is indeed a fusion reaction and has a positive Q value. Deuterium, whose isotopic abundance in normal hydrogen compounds is 1/6500, is available in unlimited quantities as a component of sea water.

There are three basic requirements for the successful operation of a thermonuclear reactor.

1. *A high particle density n.* The number of interacting particles (deuterons, say) per unit volume must be great enough to ensure a sufficiently high deuteron–deuteron collision rate. At the high temperatures required, the deuterium gas would be completely ionized into a neutral *plasma* consisting of deuterons and electrons.

2. *A high plasma temperature T.* The plasma must be hot. Otherwise the colliding deuterons will not be energetic enough to penetrate the mutual Coulomb barrier that tends to keep them apart. In fusion research, temperatures are often reported by giving the corresponding value of kT (not $\frac{3}{2}kT$). A plasma temperature of 6.5 keV, corresponding to 7.5×10^7 K, has been achieved in the laboratory. This is higher than the sun's central temperature (1.3 keV, or 1.5×10^7 K).

3. *A long confinement time τ.* A major problem is containing the hot plasma long enough to ensure that its density and temperature remain sufficiently high. It is clear that no actual solid container can withstand the high temperatures necessarily involved, so that special techniques, to be described later, must be employed. By use of one such technique, confinement times as great as 800 ms have been achieved.

J. D. Lawson has shown that, for the successful operation of a thermonuclear reactor, it is necessary to have

$$n\tau \gtrsim 6 \times 10^{13} \text{ s} \cdot \text{cm}^{-3}, \tag{55-4}$$

a condition called *Lawson's criterion*. Equation 55-4 tells us, loosely speaking, that we have a choice between confining a lot of particles for a relatively short time or confining fewer particles for a somewhat longer time. Beyond meeting this criterion, it is also necessary that the plasma temperature be sufficiently high.

Let us see how Lawson's criterion comes about. To raise a plasma to a suitably high temperature and to maintain it there against losses, energy must be added to the plasma at a rate per unit volume P_h, where the subscript stands for "heating." The heating may be done by passing an electric current through the plasma, by firing a beam of energetic neutral particles into it, or in other ways. The denser the plasma, the greater the heating power required, in direct proportion, or

$$P_h = C_h n, \tag{55-5}$$

where C_h is a suitable constant.

If thermonuclear fusion occurs in the plasma, there will be a certain rate of energy generation per unit volume P_f, where the subscript now stands for "fusion." P_f is proportional to the confinement time τ. It is also proportional to n^2, where n is the particle density. To see this, suppose that we double the particle density. Not only will a given particle make twice as many collisions as it wanders through the plasma, but there will be twice as many wandering particles, an overall factor of four. Thus

$$P_f = C_f n^2 \tau. \tag{55-6}$$

To have a net production of power, we must have

$$P_f > P_h$$

or, from the above equations,

$$n\tau > \frac{C_h}{C_f},$$

which leads directly to Eq. 55-4 if the constants C_h and C_f are suitably evaluated. The condition in which $P_f = P_h$ is called *break-even*.

TOKAMAK, a Russian-language acronym for "toroidal magnetic chamber," refers to a thermonuclear fusion device, first developed in the USSR. It is thought by many to be the best bet for achieving (eventually) a self-sustaining thermonuclear fusion reaction in the laboratory. Large tokamaks have been built and operated in several countries, and several major new machines are in the design stage.

In a tokamak the charged particles that make up the hot plasma are confined by a particular magnetic field configuration to the geometry of a doughnut or torus. As Fig. 55-11 suggests, the confining magnetic field is a sheath of helical lines of force — only one of which is shown in the figure — that spiral around the plasma "doughnut." The magnetic forces acting on the moving charges of the plasma keep the hot plasma from touching the walls of the vacuum chamber.

The helical magnetic field is the resultant of two component fields, shown at the left in Fig. 55-11. (1) The *toroidal* field component B_t is generated by a current i_p in windings that wrap around the toroidal vacuum chamber just as for the toroid of Fig. 34-15; one such winding is shown at the left in Fig. 55-11. (2) The so-called *poloidal* field component B_p is generated by a toroidal current i_t (see figure) that is induced in the plasma itself by an alternating current in the windings of a transformer-magnet arrangement not shown. This induced toroidal current is also used to heat the plasma to the desired high temperature. Additional heating may be provided by firing one or more beams of energetic neutral particles into the plasma.

The Tokamak Fusion Test Reactor (TFTR) at Princeton is one of the leading tokamak devices operating in this country as of 1986. Figure 55-12 shows a worker inside the torus, whose major radius is 2.5 m. When in full operation this device is expected to reach a Lawson coefficient $n\tau$ of about 6×10^{13} s $\cdot$ cm^{-3} at a plasma temperature of about 3 keV.

This design Lawson coefficient is at least close to the Lawson threshold and the temperature considerably exceeds that at the core of the sun. Nevertheless, these conditions are still not good enough to ignite the plasma, that is, to cause a self-supporting thermonuclear fusion reaction such as occurs at the core of our sun. By supplying additional energy to the plasma, the TFTR workers hope to achieve the lesser — but highly significant — goal of "break-even." When this condition is reached, the energy produced in the plasma by thermonuclear events is just equal to the energy that must be supplied to the plasma to heat it.

Figure 55-13, a plot of the Lawson coefficient versus the plasma temperature, shows the steady progress being made in tokamak performance at various laboratories both in the United States and elsewhere. The "tokamak tree" is branching up toward the break-even region but has not yet reached it! It seems likely that the next generation

55-9
THE TOKAMAK

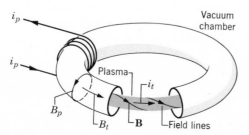

figure 55-11
The toroidal vacuum chamber that forms the heart of a tokamak. Note the plasma, the helical magnetic field that confines it, and the induced toroidal current i_t that heats it.

figure 55-12
A worker making adjustments inside
the torus of the Toroidal Fusion Test
Reactor at Princeton University.
(Courtesy Princeton University Plasma
Physics Laboratory.)

figure 55-13
The progress that has been made in
tokamak research in various
laboratories. Note that both scales are
logarithmic. The arrow at the left
represents the sun's central
temperature. The square box
represents a design goal.

of tokamaks can achieve true ignition. From there it is still a very long
step to a commercial thermonuclear power plant. This ultimate goal
seems to be no closer than the early decades of the next century.

EXAMPLE 7

Lawson's criterion. The Tokamak Fusion Test Reactor at Princeton has
achieved a confinement time of 400 ms. (a) What must be the density of particles in the plasma if Lawson's criterion is to be satisfied? (b) How does this
number compare with the particle density of the atoms of an ideal gas at stan-

dard conditions? (c) If a next-generation tokamak could achieve ignition, with a plasma temperature of 10 keV and a confinement time of 1 s, what would the particle density of its plasma have to be?

(a) We must have

$$n = \frac{6 \times 10^{13} \text{ s} \cdot \text{cm}^{-3}}{0.40 \text{ s}} = 1.5 \times 10^{14} \text{ cm}^{-3}.$$

(b) The number density of atoms in an ideal gas at standard conditions is given by $n' = N_A/V$, where N_A is the Avogadro constant and V (= 22,400 cm³/mol) is the so-called gram-molecular-volume. Thus we have

$$n' = \frac{N_A}{V} = \frac{6.02 \times 10^{23} \text{ mol}^{-1}}{22,400 \text{ cm}^3/\text{mol}} = 2.7 \times 10^{19} \text{ cm}^{-3}.$$

Thus the particle density of the plasma in (a) above is smaller than that of an ideal gas by a factor of about 2×10^5.

(c) From Fig. 55-13 we see that the 10-keV temperature line intersects the curve marked "ignition" at a value of the Lawson number of about 3×10^{14} s · cm⁻³. (In making this last estimate, bear in mind that the scale is logarithmic.) The necessary particle density is then

$$n = \frac{3 \times 10^{14} \text{ s} \cdot \text{cm}^{-3}}{1 \text{ s}} = 3 \times 10^{14} \text{ cm}^{-3}.$$

For comparison, this is a factor of 2 larger than the value calculated in (a) above. Also, note that no tokamak has yet achieved a plasma temperature as high as 10 keV.

55-10
LASER FUSION

A second technique for confining plasma so that thermonuclear fusion can take place is called *inertial confinement*. In terms of Lawson's criterion (Eq. 55-4), it involves working with extremely high particle densities n for extremely short confinement times τ. These times are arranged to be so short that the fusion episode is over before the particles of the plasma have time to move appreciably from the positions they occupy at the onset of fusion. The interacting particles are confined by their own inertia. The fusion bomb operates on the same principle; after ignition the bomb explodes before the interacting particles have time to drift away.

Laser fusion, which relies on the inertial-confinement principle, is being investigated in many laboratories, both here and abroad. At the Lawrence Livermore Laboratory, for example, in the NOVA laser fusion arrangement (see Fig. 52-8) deuterium–tritium fuel pellets, each smaller than a grain of sand (see Fig. 55-14), are to be "zapped" by 10 synchronized, high-powered laser pulses, symmetrically arranged around the pellet. The laser pulses are designed to deliver in total some 200 kJ of energy to each fuel pellet in less than a nanosecond. This is a delivered power of $\sim 2 \times 10^{14}$ W during the pulse, which far exceeds the total electrical power generating capacity of the United States!

The laser pulse energy serves to heat the fuel pellet, ionizing it to a plasma and — it is hoped — raising its temperature to $\sim 10^8$ K. As the surface layers of the pellet evaporate at these high thermal speeds, the reaction force of the escaping particles compresses the core of the pellet, increasing its density by a factor of $\sim 10^3$. If all these things happened, then conditions would be right for thermonuclear fusion to occur in the core of the highly compressed pellet of plasma, the fusion reaction being

$$^2\text{H} + {}^3\text{H} \rightarrow {}^4\text{He} + n, \qquad Q = +17.59 \text{ MeV}. \qquad (55\text{-}7)$$

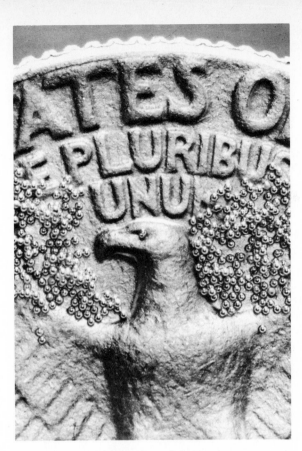

In an operating thermonuclear reactor of the laser fusion type, it is visualized that fuel pellets would be exploded, like miniature hydrogen bombs, at the rate of perhaps 10 to 100 per second. The energetic emerging particles of the fusion reaction (^{4}He, n) plus some of the reacting particles (^{2}H, ^{3}H) escaping at thermal ($\sim 10^8$ K) energies might be absorbed in a "blanket" consisting of a moving stream of molten lithium, heating it up. Thermal energy would then be extracted from the lithium stream at another location and used to generate steam, just as in a fission reactor or a fossil-fuel power plant. Lithium would be a suitable choice for a heat-transfer medium because the energetic neutron of Eq. 55-7 would, with high probability, deliver up its energy to the "blanket" by the reaction

$$^6\text{Li} + n \rightarrow {}^4\text{He} + {}^3\text{H};$$

the two charged particles would be readily brought to rest in the lithium matrix. The feasibility of laser fusion as the basis of a fusion reactor has not been conclusively demonstrated as of 1986; the matter is being pursued with vigor in many laboratories.

Suppose that a fuel pellet in a laser fusion device is made of a liquid deuterium–tritium mixture containing equal numbers of deuterium and tritium atoms. The density d ($= 200$ kg/m^3) of the pellet is increased by a factor of 10^3 by the action of the laser pulses. (*a*) How many particles per unit volume (either deuterons or tritons) does the pellet contain in its compressed state? (*b*) According to Lawson's criterion, for how long must the pellet maintain this particle density if break-even operation is to take place?

EXAMPLE 8

(a) We can write, for the density d^* of the compressed pellet,

$$d^* = 10^3\, d = m_d\left(\frac{n}{2}\right) + m_t\left(\frac{n}{2}\right)$$

in which n is the number of particles per unit volume (either deuterons or tritons) in the compressed pellet, m_d is the mass of a deuterium atom, and m_t is the mass of a tritium atom. These atomic masses are related to the Avogadro constant N_A and to the corresponding atomic weights (A_d and A_t) by

$$m_d = A_d/N_A \quad \text{and} \quad m_t = A_t/N_A.$$

Combining the above equations and solving for n leads to

$$n = \frac{2 \times 10^3\, d\, N_A}{A_d + A_t}$$

$$= \frac{(2 \times 10^3)(200 \text{ kg/m}^3)(6.02 \times 10^{23} \text{ mol}^{-1})}{2.0 \times 10^{-3} \text{ kg/mol} + 3.0 \times 10^{-3} \text{ kg/mol}}$$

$$= 4.8 \times 10^{31} \text{ m}^{-3} = 4.8 \times 10^{25} \text{ cm}^{-3}.$$

(b) From the above we have, where L is the Lawson parameter,

$$\tau = \frac{L}{n} = \frac{6 \times 10^{13} \text{ s} \cdot \text{cm}^{-3}}{4.8 \times 10^{25} \text{ cm}^{-3}} \cong 10^{-12} \text{ s}.$$

The pellet must remain compressed for at least this long if break-even operation is to occur. (It is also necessary for the effective temperature to be suitably high.)

A comparison with Example 7 shows that, unlike tokamak operation, laser fusion seeks to operate in the realm of very high particle densities and correspondingly very short confinement times.

questions

1. Can you say, from examining Table 55-1, that one source of energy, or of power, is better than another? If not, what other considerations enter?

2. To which of the processes in Table 55-1 does the relationship $E = mc^2$ apply? To all of them? To all except number 1? To all except numbers 1 and 2? And so on.

3. If it's so much harder to get a nucleon out of a nucleus than to get an electron out of an atom, why try?

4. Of the two fission fragment tracks shown in Fig. 55-2, which fragment has the larger (a) magnitude of momentum? (b) kinetic energy? (c) speed? (d) mass?

5. In Eq. 55-1, the generalized equation for the fission of ^{235}U by thermal neutrons, do you expect the Q of the reaction to depend on the identity of X and Y?

6. Is the fission fragment curve of Fig. 55-3 necessarily symmetrical about its central minimum?

7. In the chain decays of the primary fission fragments (see Table 55-2), why do no β^+ decays occur?

8. The half-life of ^{235}U is 7.0×10^8 y. Discuss the assertion that if it had turned out to be shorter by a factor of 10 or so, there would not be any atomic bombs today.

9. The uranium nuclides present in the earth today were originally built up and spewed into space during the explosion of a star, a so-called supernova event. This explosion, which occurred before the formation of our solar system, represented the collapse of the star under its own gravitational forces. Can you then say that energy derived from fission was once stored in a gravitational field? Does fission energy, then, in this limited sense,

have something in common with energy derived from hydroelectric sources?

10. From Table 55-3 we can see that ^{238}U is not fissionable by thermal neutrons. What minimum neutron energy do you think would be necessary to induce fission in this nuclide?

11. By bombardment of heavy nuclides in the laboratory it is possible to prepare other heavy nuclides that decay, at least in part, by *spontaneous fission*. That is, after a certain mean life they spontaneously break up into two major fragments. Can you explain this on the basis of the theory of Bohr and Wheeler?

12. The half-life for the decay of ^{235}U by α-emission is 7×10^8 y; by spontaneous fission, acting alone, it would be ~3×10^{17} y. Both are barrier-tunneling processes, as Figs. 54-9 and 55-5 reveal. Why this enormous difference in barrier-tunneling probability?

13. Compare fission with alpha decay in as many ways as possible.

14. How can a thermal neutron deliver several million electron volts of excitation energy to a nucleus that absorbs it, as in Figs. 55-4a and 55-4b? The neutron has essentially no energy to start with!

15. The binding energy curve of Fig. 54-7 tells us that any nucleus more massive than $A = 56$ can release energy by the fission process. Only very massive nuclides seem to do so, however. Why can't lead, for example, release energy by the fission process?

16. Compare a nuclear reactor with a coal fire. In what sense does a chain reaction occur in each? What is the energy-releasing mechanism in each case?

17. Not all neutrons produced in a reactor are destined to initiate a fission event. What happens to such neutrons?

18. Explain just what is meant by the statement that in a reactor core neutron leakage is a surface effect and neutron production is a volume effect.

19. Slow neutrons are more effective than fast ones in inducing fission. Can you make that plausible? (*Hint:* Consider how the de Broglie wavelength of a neutron might be related to its capture cross section in ^{235}U.)

20. Why is a moderator needed in a fission reactor?

21. Describe how to operate the control rods of a nuclear reactor (a) during initial start-up, (b) to reduce the power level, and (c) on a long-term basis, as fuel is consumed.

22. A reactor is operating at full power with its multiplication factor k adjusted to unity. If the reactor is now adjusted to operate stably at half power, what value must k now assume?

23. In Fig. 55-10, are you surprised that, as judged by the areas under the curve marked $n(K)$, the number of particles with $K > \overline{K}$ is smaller than the number with $K < \overline{K}$, where $\overline{K}$ is the average thermal energy?

24. The sun's energy is assumed to be generated by nuclear reactions such as the proton–proton cycle. What alternative ways of generating solar energy were proposed in the past, and why were they rejected?

25. Elements up to mass number 56 are created by thermonuclear fusion in the cores of stars. Why are heavier elements not created by this process?

26. Do you think that the thermonuclear fusion reaction controlled by the two curves plotted in Fig. 55-10 necessarily has its maximum effectiveness for the energy at which the two curves cross each other?

27. Why does it take so long (~10^6 y!) for γ-ray photons generated by nuclear reactions in the sun's central core to diffuse to the surface? What kinds of interactions do they have with the protons, α-particles, and electrons that make up the core?

28. The primordial matter of the early universe is thought to have been largely hydrogen. Where did all the silicon in the earth come from? all the gold?

problems

29. Do conditions at the core of the sun satisfy Lawson's criterion for a sustained thermonuclear fusion reaction?

30. To achieve ignition in a tokamak, why do you need a high plasma temperature? a high density of plasma particles? a long confinement time?

31. Which would generate more radioactive waste products, a fission reactor or a fusion reactor?

32. Does Lawson's criterion hold both for tokamaks and for laser fusion devices?

SECTION 55-2

1. (a) How much energy is produced by the complete fissioning of 1.0 kg of ^{235}U? (b) For how long would this energy light a 100-W lamp? Assume that $Q = 200$ MeV.   *Answer:* (a) 5.1×10^{26} MeV; (b) 2.6×10^4 y.

2. At what rate must ^{235}U nuclei undergo fission by thermal neutrons to generate 1.0 W? Assume that $Q = 200$ MeV.

3. Verify that, as reported in Table 55-1, the fission of the ^{235}U in 1.0 kg of UO_2 (enriched to 3.0% ^{235}U) could keep a 100-W lamp burning for 680 y.

4. Fill in the following table, which refers to the generalized fission reaction of Eq. 55-1.

X	Y	b
^{140}Xe		1
^{139}I		2
	^{100}Zr	2
^{141}Cs	^{92}Rb	

5. Show that, in Example 1, there is no need to take the masses of the electrons emitted during the beta decay of the primary fission fragments explicitly into account.

6. Calculate the disintegration energy Q for the fission of ^{52}Cr into two equal fragments. The needed masses are ^{52}Cr, 51.94051 u; and ^{26}Mg, 25.98260 u.

7. Calculate the disintegration energy Q for the fission of ^{98}Mo into two equal parts. The needed masses are ^{98}Mo, 97.90541 u; and ^{49}Sc, 48.95003 u. If Q turns out to be positive, discuss why this process does not occur spontaneously.   *Answer:* 5.0 MeV.

8. In a particular fission event of ^{235}U by slow neutrons, it happens that no neutron is emitted ($b = 0$ in Eq. 55-1) and that one of the primary fission fragments is ^{83}Ge. (a) What is the other fragment? (b) How is the disintegration energy Q split between the two fragments? (c) If $Q = 170$ MeV, what is the initial velocity of each fragment? Note that in this problem Q refers to the initial fission fragments as a final state, no account being taken of their subsequent decay.

9. Very occasionally a ^{233}U nucleus, having absorbed a neutron, breaks up into *three* fragments. If two of these fragments are identified chemically as isotopes of chromium and gallium and if no prompt neutrons are involved, what is at least one possibility for the identity of the fragments? Consult a nuclidic chart or table.

10. A million ^{235}U nuclei each absorb a thermal neutron, as in Eq. 55-1. (a) How many of the resulting fission fragments will have $A = 150$? What will be the mass number of the fission fragment that is formed along with an $A = 150$ fragment? (b) What will be the mass numbers of the most commonly occurring primary fission fragments and how many of each of them will there be?

11. Consider the fission of ^{238}U by fast neutrons. In one fission event no neutrons were emitted and the final stable end products, after the β decay of the primary fission fragments, were ^{140}Ce and ^{99}Ru. (a) How many β decay

events were there in the two β decay chains, considered together? (b) Calculate Q. The relevant atomic masses are

| ^{238}U | 238.05081 u | ^{140}Ce | 139.90549 u |
| n | 1.00867 u | ^{99}Ru | 98.90594 u. |

Answer: (a) 10; (b) 231 MeV.

SECTION 55-3

12. ^{238}Np has an activation energy for fission of 4.2 MeV. To remove a neutron from this nuclide requires an energy expenditure of 5.0 MeV. Is ^{237}Np fissionable by thermal neutrons?

13. A ^{236}U nucleus undergoes fission and breaks up into two middle-mass fragments, ^{140}Xe and ^{96}Sr. (a) By what percentage does the surface area of the ^{236}U nucleus change during this process? (b) By what percentage does its volume change? (c) By what percentage does its electrostatic potential energy change? The potential energy of a uniformly charged sphere of radius r and charge q is given by

$$U = \frac{3}{5}\left[\frac{q^2}{4\pi\epsilon_0 r}\right].$$

Answer: (a) +25%; (b) Zero; (c) −82%.

14. Assume that just after the fission of ^{236}U* according to Eq. 55-2, the resulting ^{140}Xe and ^{94}Sr nuclei are just touching at their surfaces. (a) Assuming the nuclei to be spherical, calculate the Coulomb potential energy of repulsion between the two fragments. (b) Compare this energy with the energy released in a typical fission process. In what form will this energy ultimately appear in the laboratory?

15. ^{235}U decays by α-emission with a half-life of 7.0×10^8 y. It also decays (rarely) by spontaneous fission, and if the α decay did not occur, its half-life due to this process alone would be $\sim 3 \times 10^{17}$ y. (a) At what rate do spontaneous fission decays occur in 1.0 g of ^{235}U? (b) How many α-decay events are there for every spontaneous fission event?
Answer: (a) $\sim$0.7 h^{-1}; (b) 4.3×10^8.

SECTION 55-4

16. A 200-MW fission reactor consumes half its fuel in 3 years. How much ^{235}U did it contain initially? Assume that all the energy generated arises from the fission of ^{235}U and that this nuclide is consumed *only* by the fission process.

17. You wish to produce 1 GJ of energy. Calculate and compare (a) the amount of coal needed if you obtain the energy by burning coal and (b) the amount of natural uranium needed if you obtain the energy by fission in a reactor. Assume that the combustion of 1 kg of coal releases 2.9×10^7 J and that the fission of 1 kg of uranium in a reactor releases 8.6×10^{13} J.
Answer: (a) 37 kg; (b) 11.6 mg.

18. In the United States, coal commonly contains about 3 parts per million (3 ppm) of fissionable uranium and thorium. Calculate and compare (a) the energy derived from burning 100 kg of coal and (b) the energy that could be derived from the fission of the fissionable impurities that remain in its ashes. Assume that the combustion of 1 kg of coal releases 2.9×10^7 J and that the fission of 1 kg of uranium or thorium in a reactor releases 8.6×10^{13} J.

19. Many fear that helping additional nations develop nuclear power reactor technology will increase the likelihood of nuclear war because reactors can be used not only to produce energy but, as a by-product through neutron capture with inexpensive ^{238}U, to make ^{239}Pu, which is a "fuel" for nuclear bombs. What simple series of reactions involving neutron capture and beta decay would yield this plutonium isotope?
Answer: $^{238}U + n \rightarrow {}^{239}U \rightarrow {}^{239}Np + e$; $^{239}N_p \rightarrow {}^{239}Pu + e$.

20. Neutrons in a reactor are in thermal equilibrium with the moderator at an effective temperature of 330 K. Find (a) the most probable energy and (b) the average energy. See Problem 33.

21. *Energy loss in a moderator.* (a) A neutron with initial kinetic energy K makes a head-on elastic collision with a resting atom of mass m. Show that the fractional energy loss of the neutron is given by

$$\frac{\Delta K}{K} = \frac{4m_n m}{(m + m_n)^2},$$

in which m_n is the neutron mass. (b) Find $\Delta K/K$ if the resting atom is hydrogen ($m = m_n$), deuterium ($m = 2m_n$), carbon ($m = 12m_n$), or lead ($m = 208m_n$). (c) If $K = 1.00$ MeV initially, how many successive head-on collisions would it take to reduce the neutron energy to thermal values (0.025 eV) if the material is deuterium, a commonly used moderator? (*Note:* In actual moderators most collisions are not "head-on.")
Answer: (b) 1.0; 0.89; 0.28; 0.019; (c) 8.

22. *The neutron generation time* t_{gen} in a reactor is the average time for a fast neutron emitted in a fission event to be slowed down to thermal energies by the moderator and to initiate another fission event. If $t_{gen} = 1.0$ ms and if the control rods are adjusted so that $k = 1.0005$, how many neutron generations would it take for the power output of the reactor to double? How much actual time?

23. The neutron generation time t_{gen} (see Problem 22) in a particular reactor is 1.0 ms. If the reactor is operating at a power level of 500 MW, about how many free neutrons are present in the reactor at any moment? Assume that all neutrons present will initiate a fission.   Answer: 1.6×10^{17}.

24. *Radionuclide power sources.* The thermal energy generated when radiations from radionuclides are absorbed in matter can be used as the basis for a small power source for use in satellites, remote weather stations, and so on. Such radionuclides are manufactured in abundance in nuclear power reactors and may be separated chemically from the spent fuel. One suitable radionuclide is ^{238}Pu ($t_{1/2} = 87.8$ y) which is an α-emitter with $Q = 5.50$ MeV. At what rate is thermal energy generated in 1.0 kg of this material?

25. *Radionuclide power sources* (see Problem 24). Among the many fission products that may be extracted chemically from the spent fuel of a nuclear power reactor is ^{90}Sr ($t_{1/2} = 29$ y). It is produced in a typical large reactor at the rate of about 18 kg/y. By its radioactivity it generates thermal energy at the rate of 0.93 W/g. (a) What is the effective disintegration energy Q_{eff} associated with the decay of a ^{90}Sr nucleus? (Q_{eff} includes contributions from the decay of the ^{90}Sr daughter products in its decay chain but not from neutrinos, which escape totally from the sample.) (b) It is desired to construct a power source generating 150 W (electric) to use in operating electronic equipment in an underwater acoustic beacon. If the source is based on the thermal energy generated by ^{90}Sr and if the efficiency of the thermal-electric conversion process is 5.0%, how much ^{90}Sr is needed?
Answer: (a) 1.15 MeV; (b) 3.2 kg, the 65-day output of one large power reactor.

SECTION 55-5

26. The natural fission reactor discussed in Section 55-5 is estimated to have generated 15 gigawatt-years of energy during its lifetime. (a) If the reactor lasted for 200,000 y, at what average power level did it operate? (b) How much ^{235}U did it consume during its lifetime?

27. Some uranium samples from the natural reactor site described in Section 55-5 were found to be slightly *enriched* in ^{235}U, rather than depleted. Account for this in terms of neutron absorption by the abundant isotope ^{238}U and the subsequent beta and alpha decay of its products.

SECTION 55-6

28. Calculate the height of the Coulomb barrier for the head-on collision of two protons. The effective radius of a proton may be taken to be 0.8 fm. See Example 5.

29. Estimate the Coulomb barrier height for two ^{7}Li nuclei, fired at each other with the same initial kinetic energy K. See Eq. 54-1 and Example 5. *Answer:* 1.5 MeV.

30. Methods other than heating the material have been suggested for overcoming the Coulomb barrier for fusion. For example, one might consider using particle accelerators. If you were to use two of them to accelerate two beams of deuterons directly toward each other so as to collide "head-on," (a) what voltage would each require to overcome the Coulomb barrier? See Example 5. (b) Would this voltage be difficult to achieve? (c) Why do you suppose this method is not presently used?

31. From information given in the text, collect and write down the approximate heights of the Coulomb barriers for (a) the alpha decay of ^{238}U, (b) the fission of ^{235}U by thermal neutrons, and (c) the head-on collision of two deuterons. *Answer:* (a) 30 MeV; (b) Zero; (c) 200 keV.

32. The equation of the curve $n(K)$ in Fig. 55-10 is (see Problem 21 of Chapter 53)

$$n(K) = 1.13\, n\, \frac{K^{1/2}}{(kT)^{3/2}}\, e^{-K/kT},$$

where n is the total density of particles, regardless of their energy. At the center of the sun the temperature is 1.5×10^7 K and the mean particle (proton) energy $\overline{K}$ is 1.94 keV. Find the ratio of the density of protons at 5.00 keV to that at the mean proton energy.

33. Expressions for the Maxwell speed and energy distributions for the molecules in a gas are given in Problem 21 of Chapter 53. (a) Show that the *most probable energy* is given by

$$K_p = \frac{1}{2}\, kT.$$

Verify this result from the energy distribution curve of Fig. 55-10, for which $T = 1.5 \times 10^7$ K.

(b) Show that the *most probable speed* is given by

$$v_p = (2kT/m)^{1/2}.$$

Find its value for protons at $T = 1.5 \times 10^7$ K.

(c) Show that *the energy corresponding to the most probable speed* (which is not the same as the most probable energy!) is

$$K_{p,v} = kT.$$

Locate this quantity on the energy-distribution curve of Fig. 55-10.

SECTION 55-7

34. We have seen that Q for the overall proton–proton cycle is 26.7 MeV. How can you relate this number to the Q-values for the three reactions that make up this cycle, as displayed in Table 55-4?

35. Calculate and compare the energy released by (a) the fusion of 1.0 kg of hydrogen deep within the sun and (b) the fission of 1.0 kg of ^{235}U in a fission reactor. *Answer:* (a) 4.0×10^{27} MeV; (b) 5.1×10^{26} MeV.

36. Verify the values of Q_1, Q_2, and Q_3 reported for the three reactions in Table 55-4. The needed atomic masses are

^{1}H	1.007825 u	^{3}He	3.016029 u
^{2}H	2.014102 u	^{4}He	4.002603 u
	$e^{\pm}$	0.0005486 u	

(*Hint:* Distinguish carefully between atomic and nuclear masses, and take the positrons properly into account.)

37. The sun has a mass of 2.0×10^{30} kg and radiates energy at the rate of 3.9×10^{26} W. (*a*) At what rate does the sun transfer its mass into energy? (*b*) What fraction of its original mass has the sun lost in this way since it began to burn hydrogen, about 4.5×10^{9} y ago?
    Answer: (*a*) 4.3×10^{9} kg/s; (*b*) 3.1×10^{-4}.

38. At the center of the sun the density is 1.5×10^{5} kg/m³ and the composition is essentially 35% hydrogen by mass and 65% helium. (*a*) What is the density of protons at the sun's center? (*b*) How much larger is this than the density of particles for an ideal gas at standard conditions of temperature and pressure?

39. Let us assume that the core of the sun includes half its mass and is compressed within a sphere whose radius is one-fourth of the solar radius. We assume further that the composition of the core is 35% hydrogen by mass and that essentially all of the sun's energy is generated there. If the sun continues to burn hydrogen at the rate calculated in Example 6, how long will it be before the hydrogen is entirely consumed? The sun's mass is 2.0×10^{30} kg.
    Answer: 1.8×10^{10} y; a more refined calculation yields about 5×10^{9} y.

40. Coal burns according to

$$C + O_2 \rightarrow CO_2.$$

The heat of combustion is 3.3×10^{7} J/kg of atomic carbon consumed. (*a*) Express this in terms of energy per carbon atom. (*b*) Express it in terms of energy per unit mass of the initial reactants, carbon and oxygen. (*c*) Suppose that the sun (mass $= 2.0 \times 10^{30}$ kg) were made of carbon and oxygen in combustible proportions and that it continued to radiate energy at its present rate of 3.9×10^{26} W. How long would it last?

41. In certain stars the so-called *carbon cycle* is more likely than the proton–proton cycle to be effective in generating energy. This cycle is

$$^{12}C + {}^{1}H \rightarrow {}^{13}N + \gamma, \qquad Q_1 = 1.95 \text{ MeV},$$
$$^{13}N \rightarrow {}^{13}C + \beta^{+} + \nu, \qquad Q_2 = 1.19,$$
$$^{13}C + {}^{1}H \rightarrow {}^{14}N + \gamma, \qquad Q_3 = 7.55,$$
$$^{14}N + {}^{1}H \rightarrow {}^{15}O + \gamma, \qquad Q_4 = 7.30,$$
$$^{15}O \rightarrow {}^{15}N + \beta^{+} + \nu, \qquad Q_5 = 1.73,$$
$$^{15}N + {}^{1}H \rightarrow {}^{12}C + {}^{4}He, \qquad Q_6 = 4.97.$$

(*a*) Show that this cycle of reactions is exactly equivalent in its overall effects to the proton–proton cycle of Table 55-4. (*b*) Verify that both cycles, as expected, have the same Q.

42. The effective Q for the proton–proton cycle of Table 55-4 is 26.2 MeV. (*a*) Express this as energy per unit mass of hydrogen consumed. (*b*) The luminosity of the sun is 3.9×10^{26} W. If its energy derives from the proton–proton cycle, at what rate is it losing hydrogen? (*c*) At what rate is it losing mass? Account for the difference in the results for (*b*) and (*c*). (*d*) The sun's mass is 2.0×10^{30} kg. If it loses mass at the constant rate calculated in (*c*), how long will it take before it loses 0.10% of its mass?

43. Show that the energy released when 3 alpha particles fuse to form ^{12}C is 7.3 MeV. The atomic mass of ^{4}He is 4.0026 u.

44. *Helium burning in stars.* After most of a star's hydrogen has been fused into helium, its core will collapse and heat up through the release of gravitational energy. When the core temperature reaches 10^{8} K, the helium itself will begin to fuse. By the "triple alpha process," it will produce stable ^{12}C with a considerable release of energy. (*a*) Except for possible gamma rays

produced, give the reaction involved. (b) Why would there be no neutrinos involved as there are with hydrogen fusion? (c) Why would the reaction $^4\text{He} + ^4\text{He} + ^4\text{He} + ^4\text{He} \rightarrow ^{16}\text{O} + \gamma$ not be significant?

45. *Nucleosynthesis.* The process of fusing light nuclei into heavier and heavier nuclei with consequent releases of energy inside stars must ultimately stop. (a) Referring to Fig. 54-7, determine the region of nuclides where this happens. (b) Explain why. (c) One way to continue the build-up of heavy nuclei is through a net addition of energy for each reaction. This is thought to occur in heavy stars that suffer gigantic supernova explosions. Explain how such explosions could account for the heavy nuclei.
Answer: (a) At mass number $A = 56$.

46. The gravitational potential energy of a uniform spherical mass of radius r and mass m is

$$U = -3\,Gm^2/5r$$

in which G is the universal gravitational constant.

(a) Use this to find the maximum energy that could be released by a uniform spherical mass, initially of infinite radius, in shrinking to the present size of the sun (use Appendix C for appropriate data). (b) Calculate the net energy the sun would radiate at the present rate (3.9×10^{26} W) for its lifetime (4.5×10^9 y). (c) What is the ratio of radiated to gravitational energy?

SECTION 55-8

47. Verify the Q values reported on p. 1321 for the two d–d reactions and the d–t reaction. The needed atomic masses are

^1H	1.007825 u	^3He	3.016029 u
^2H	2.014102 u	^4He	4.002603 u
^3H	3.016049 u	n	1.008665 u

48. Verify that, as reported in Table 55-1, the fusion of 1.0 kg of deuterium by the reaction

$$^2\text{H} + ^2\text{H} \rightarrow ^3\text{He} + n \qquad Q = +3.27 \text{ MeV},$$

could keep a 100-W light bulb burning for 2.5×10^4 y.

49. Ordinary water consists of roughly 0.015% of "heavy water," in which one of the two hydrogens is replaced with deuterium, ^2H. How much fusion power could be extracted if we "burned" all of the ^2H in one liter of water in one day through the reaction $^2\text{H} + ^2\text{H} \rightarrow ^3\text{He} + n$?   *Answer:* 14.4 kW.

50. Suppose we had a quantity of N deuterons (^2H nuclei). (a) Which of the following procedures for fusing these N nuclei releases more energy? (A) $N/2$ fusion reactions of the type $^2\text{H} + ^2\text{H} \rightarrow ^3\text{H} + p$, or (B) $N/3$ fusion reactions of the type $^2\text{H} + ^3\text{H} \rightarrow ^4\text{He} + n$, using $N/3$ nuclei of ^3H that are first made in $N/3$ reactions of type A. (b) List the ultimate product nuclei resulting from the two procedures and the quantity of each.

SECTION 55-9

51. By the summer of 1985 the TFTR tokamak at the Princeton Plasma Physics Laboratory could be run consistently with plasma number densities of 3×10^{13} cm^{-3}, confinement times of 400 ms, and ion temperatures of 3 keV. Under special experimental conditions these same parameters are 6×10^{12} cm^{-3}, 100 ms, and 10 keV. Plot the two points representing this data on Fig. 55-13. Realizing that this represents just one machine while the figure represents many, what is your impression of progress being made in this type of fusion research?

SECTION 55-10

52. In the deuteron–triton fusion reaction of Eq. 55-7, how is the reaction energy Q shared between the α-particle and the neutron? Neglect the relatively small kinetic energies of the two combining particles.

53. The (uncompressed) radius of the fuel pellet of Example 8 is 20 μm. Suppose that the compressed pellet "burns" with an efficiency of 10%. That is, only 10% of the deuterons and 10% of the tritons participate in the fusion reaction of Eq. 55-7. (a) How much energy is released in each such microexplosion of a pellet? (b) To how much TNT is each such pellet equivalent? The heat of combustion of TNT is 4.6×10^6 J/kg. (c) If a fusion reactor is constructed on the basis of 100 microexplosions per second, what power would be generated? (Note that part of this power must be used to operate the lasers.)   *Answer:* (a) 226 J; (b) 49 mg. (c) 22.6 kW.

54. Assume that a plasma temperature of 1×10^8 K is reached in a laser fusion device. (a) What is the most probable speed of a deuteron at this temperature? (See Example 3 of Chapter 24). (b) How far would such a deuteron move in the confinement time calculated in Example 8?

supplementary topics

In Section 11-6 we discussed the relations between the linear and angular kinematic variables for a particle moving in a plane but confined to move in a circle about an axis at right angles to the plane. Such a particle might be any particle in a rigid body rotating about a fixed axis. Here we relax the restriction and allow the particle to move freely in the plane. A planet moving in an elliptical orbit about the sun is an example.

We start from Eq. 11-11, $\mathbf{r} = \mathbf{u}_r r$, in which, however, we now take *both* $\mathbf{u}_r$ and r to be variables; the particle is no longer confined to a circle of constant radius. We find the velocity by differentiation, or

$$\mathbf{v} = \frac{d\mathbf{r}}{dt} = \mathbf{u}_r \frac{dr}{dt} + r \frac{d\mathbf{u}_r}{dt}.$$

Equation 11-13 shows us that $d\mathbf{u}_r/dt = \mathbf{u}_\theta \omega$. Thus we can write

$$\mathbf{v} = \mathbf{u}_r \frac{dr}{dt} + \mathbf{u}_\theta \omega r, \tag{I-1}$$

which shows that $\mathbf{v}$ has two components, a radial component $v_r = dr/dt$ and a component at right angles, $v_\theta = \omega r$. If we hold r constant, then $dr/dt = 0$ and Eq. I-1 reduces to Eq. 11.14a, as it must.

To find the acceleration we differentiate Eq. I-1, remembering that *all five* quantities on the right are variables. We obtain

$$\mathbf{a} = \frac{d\mathbf{v}}{dt} = \mathbf{u}_r \frac{d^2r}{dt^2} + \frac{dr}{dt}\frac{d\mathbf{u}_r}{dt} + (\mathbf{u}_\theta)\left(\omega \frac{dr}{dt} + r \frac{d\omega}{dt}\right) + (\omega r)\left(\frac{d\mathbf{u}_\theta}{dt}\right).$$

Now $d\mathbf{u}_r/dt = \mathbf{u}_\theta \omega$, $d\mathbf{u}_\theta/dt = -\mathbf{u}_r \omega$ (see Eq. 11-16), and $d\omega/dt = \alpha$. Substituting and rearranging leads us finally to

$$\mathbf{a} = \mathbf{u}_r\left(\frac{d^2r}{dt^2} - \omega^2 r\right) + \mathbf{u}_\theta\left(\alpha r + 2\omega \frac{dr}{dt}\right). \tag{I-2}$$

$$= \mathbf{u}_r(a_r - \omega^2 r) + \mathbf{u}_\theta(\alpha r + 2\omega v_r).$$

SUPPLEMENTARY TOPIC I
RELATION BETWEEN LINEAR AND ANGULAR KINEMATICS FOR A PARTICLE MOVING IN A PLANE

Once again, if $r =$ a constant, then $dr/dt = d^2r/dt^2 = 0$ and Eq. I-2 reduces to Eq. 11-17, which we derived especially for this case.

The two new terms in Eq. I-2, $\mathbf{u}_r d^2r/dt^2$ and $\mathbf{u}_\theta 2\omega\, dr/dt$, need a little explanation. We can understand the first of these terms by imagining that the particle moving in the plane is *not* rotating about the axis. If we put $\omega = \alpha = 0$ in Eq. I-2 this equation reduces to

$$\mathbf{a} = \mathbf{u}_r \frac{d^2r}{dt^2} = \mathbf{u}_r a_r,$$

which is just the familiar acceleration of a particle moving along a straight line. Hence this term in Eq. I-2 gives the radial acceleration due to the change in the *magnitude* of $\mathbf{r}$, the other radial acceleration term arising from the changing *direction* of $\mathbf{r}$ as the particle rotates.

There are also two θ-directed acceleration terms. The first one, $\mathbf{u}_\theta \alpha r$, arises simply from the angular acceleration α of a particle in circular motion ($r =$ constant) and is the tangential acceleration of Section 11-5. To understand the second term, $\mathbf{u}_\theta 2\omega\, dr/dt$, consider a man walking outward along a radial line painted on the floor of a merry-go-round. The merry-go-round is rotating with constant angular velocity ω so that its angular acceleration α is now zero. If the man were simply to stand still on the merry-go-round, ($d^2r/dt^2 = dr/dt = 0$, and $r =$ constant) his acceleration, as seen by an observer in a reference frame on the ground (see Eq. I-2), would be simply the familiar centripetal acceleration $-\mathbf{u}_r \omega^2 r$, directed radially inward. If he walks outward, however, $dr/dt \neq 0$ and then Eq. I-2 predicts that the ground observer would also measure a θ-directed acceleration given by $\mathbf{u}_\theta 2\omega v_r$, where $v_r = dr/dt$. This is called a *Coriolis acceleration*. It arises from the fact that even though the angular velocity of the man is constant his speed increases as r increases. Let us convince ourselves that this effect really exists.*

Figure I-1a shows the walking man (point P) as he appears to the ground observer at times t and $t + \Delta t$. We show at time t his radially directed velocity $\mathbf{v}_r (= \mathbf{u}_r\, dr/dt)$ and also a θ-directed velocity caused by the rotation of the merry-go-round and given by $\mathbf{v}_\theta (= \mathbf{u}_\theta \omega r)$. At a time Δt later each of these velocities has changed. The radial velocity has changed in direction, although its magnitude remains dr/dt. The θ-directed velocity has not only changed direction (we have learned to account for this as a centripetal acceleration), but, because the man has moved outward to a point at which the floor is moving faster, its *magnitude* has also changed, from ωr to $\omega(r + \Delta r)$.

Figure I-1b shows the change in velocity caused by the change in direction of the radial line along which the man is walking. If $\Delta\theta$ in the triangle shown is small enough, we have

$$\Delta v_r = v_r\, \Delta\theta.$$

Dividing by Δt and letting Δt approach zero yields

$$a' = \frac{dv_r}{dt} = v_r \frac{d\theta}{dt} = v_r \omega.$$

This is just half the term $2\omega v_r$ in Eq. I-2. However, we have considered only the change in the *radial* velocity; there is also a change in the *tangential* velocity.

The change in tangential velocity, caused by the fact that the man is moving radially outward, is

$$\Delta v_\theta = \omega(r + \Delta r) - \omega r = \omega \Delta r.$$

Dividing by Δt and letting Δt approach zero yields

$$a'' = \frac{dv_\theta}{dt} = \omega \frac{dr}{dt} = \omega v_r.$$

Now both a' and a'' are magnitudes of vectors that point in the same direction,

---

* See "The Coriolis Effect," James E. McDonald, *Scientific American*, May 1952 and also "The Case of the Coriolis Force," Malcolm Correll, *The Physics Teacher*, January 1976.

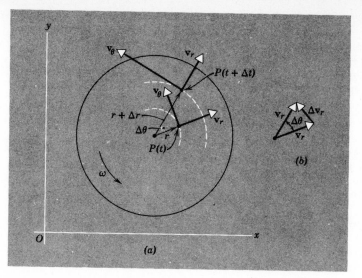

(a)

figure 1-1

(a) A merry-go-round, rotating about a fixed axis, is observed by an observer in inertial reference frame *x, y*. A man walks along a radial line at constant speed *v*. In a time interval Δt this line, as seen by the ground observer, sweeps through an angle $\Delta \theta$ and the man moves between the positions shown. His *r*- and θ-directed velocities are shown for each position. *(b)* Showing the change Δv_r in the walking man's *r*-directed velocity. Note that, as $\Delta t \to 0$, Δv_r points in the θ-direction at *P*.

namely the direction of increasing θ at point $P(t)$. The total acceleration in this direction is then

$$a' + a'' = v_r \omega + \omega v_r = 2\omega v_r,$$

which is just what we set out to prove.

If there is indeed an acceleration in the θ-direction in Fig. I-1, there must be a force in this direction. For a man walking outward along a radial line on a rotating merry-go-round this force can only be provided by the friction between his feet and the floor.

We remember that we can interpret classical mechanics most simply if we always view events from an inertial frame. If we do so, we can always associate accelerations with forces exerted by bodies that we can point to in the environment. We can still apply classical mechanics, however, if we select a noninertial reference frame, such as a rotating frame. The small penalty that we must pay is that we must introduce inertial forces, that is, forces that we cannot associate with objects in the environment and which cannot be detected by an observer in an inertial frame. In Section 6-4 we saw that centrifugal force is such an inertial force.

Consider an observer on the rotating merry-go-round watching a man walk along a radial line at a constant speed $v_r = dr/dt$. He would say that the man is in equilibrium because he has no acceleration. Yet the floor is exerting a (very real) frictional force on the soles of the man's feet. This force has one component $(-\mathbf{u}_r F_r)$ that points radially inward and one $(\mathbf{u}_\theta F_\theta)$ that points in the θ-direction, that is, in the direction of rotation.

From the point of view of the ground observer these forces are understandable and, indeed, quite necessary. F_r is associated with the centripetal acceleration $\omega^2 r$ and F_θ with the Coriolis acceleration $2\omega v_r$. The observer on the merry-go-round does not see either of these accelerations however; to him the walking man is in equilibrium. How can this be, in view of the frictional forces that act on the soles of the walking man's shoes? The man himself is well aware of these forces; if he did not lean to compensate for their turning effect, they would knock him off his feet!

The observer on the merry-go-round saves the situation by declaring that two inertial forces act on the walking man, just canceling the (real) frictional forces. One of these inertial forces, called the *centrifugal force*, has magnitude F_r and acts radially *outward*. The other, called the *Coriolis force*, has magnitude F_θ and acts in the negative θ-direction, that is, *opposite* to the direction of rotation. By introducing these forces, which seem quite "real" to him although he cannot point to any body in the environment that is causing them, the observer in the rotating (noninertial) reference frame can apply classical mechanics in the usual way. The ground observer, who is in an inertial frame, cannot detect these inertial forces. Indeed there is no need for them—and no room for them—in his applications of classical mechanics.

Equations I-1 and I-2 are general kinematical descriptions for the motion of a particle in two dimensions. An obvious extension, which we will not attempt here, is to derive corresponding descriptions for motion in three dimensions; this will require us to introduce a third unit vector to define the third dimension.*

SUPPLEMENTARY TOPIC II
POLAR VECTORS AND AXIAL VECTORS

Some vectors called *axial vectors*, such as ω, α, τ, and l, differ in a rather important way from other vectors called *polar vectors*, of which r, v, a, F, and p are examples. Although we shall not need to take this difference into account in this book, it may prove to be instructive and interesting to examine briefly what the difference is.

Consider a typical polar vector such as r. If a student leaves his dormitory and goes to a classroom, his displacement vector r points *from* the dormitory *to* the classroom; there is no question as to our choice of direction. This direction is both "physical" and "natural." Similar remarks apply to the other typical polar vectors listed, namely, v, a, F, and p.

If a student sees a wheel rotating about a fixed axis, he can assign an angular velocity ω to the wheel and can give direction to ω by the right-hand rule (see Section 11-4). This direction, however, *is a convention only*, based on this arbitrary rule. A left-hand rule would have given the opposite direction. The things that are "physical" and "natural" about the wheel are the axis of rotation and the sense of rotation, that is, is it going clockwise or counterclockwise as the student looks at it from a particular end of the axis? Whether ω is chosen to point in one way or the other along the axis does not really matter as long as we are consistent. The same remarks apply to the angular acceleration α and to the other axial vectors listed, namely, $\tau (= r \times F)$ and $l (= r \times p)$. It is for this reason that we sometimes find it more comfortable to say "torque *around* an axis" than "torque *along* an axis" although they mean the same thing. All vectors defined as the vector product of two *polar* vectors are axial vectors because they all depend for their direction assignment on the (arbitrary) right-hand rule.

We have stressed that the laws of physics remain the same no matter how we change the inertial reference frame in which they are expressed. In Section 2-5 we discussed this for translations and rotations of the reference frame and noted that laws expressed in vector form remained unchanged (that is, *invariant*) under such transformations. We also noted that something special may occur when we change the reference frame in another way, namely, by substituting a left-handed frame for a right-handed one. There is an easy way to make such a transformation: Build a right-handed frame and look at its image in a mirror; it will be converted to a left-handed frame (see Fig. II-1) because of the well-known property of a mirror to reverse right and left.

Figure II-1a shows the vector displacement of a student from his dormitory to each of three classrooms. In the mirror each displacement is *still* from the dormitory D to a classroom C. In Fig. II-1b, however, we show a rotating wheel in three orientations. If we establish the directions of ω for both the wheels and their mirror images by the right-hand rule, we see that the image vectors are reversed in comparison to the corresponding image vectors in Fig. II-1a (toward the origin rather than away from the origin). Polar vectors and axial vectors behave differently when we transform reference frames by mirror reflection! This behavior of axial vectors under mirror reflection is not hard to understand. If we imagine ourselves physically applying the right-hand rule to a real rotating wheel, in the mirror, we shall *seem* to be applying a left-hand rule because the image of our right hand is our left hand. A left-hand rule, of course, will give us the opposite direction for ω.

Hence an axial vector is a vector whose sense of direction depends on the

---

* See, for example, *Mechanics*, Section 3-5, by Keith R. Symon, Addison-Wesley Publishing Co., 3d ed., 1971.

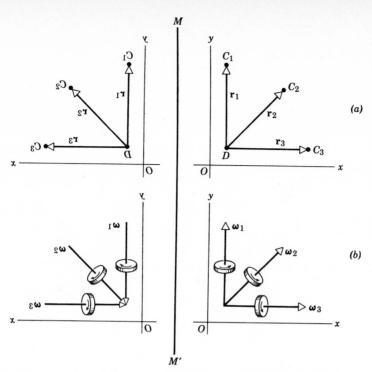

figure II-1

(a) Polar vectors, showing, on the right, the displacements $\mathbf{r}_1$, $\mathbf{r}_2$, and $\mathbf{r}_3$ between a dormitory D and three classrooms C_1, C_2, and C_3. On the left we have the mirror images of D, C_1, C_2, and C_3, along with the corresponding displacements. *(b)* Axial vectors, showing, on the right, the angular velocities $\boldsymbol{\omega}_1$, $\boldsymbol{\omega}_2$, and $\boldsymbol{\omega}_3$ of three wheels rotating as shown. On the left we have the mirror images of these wheels, along with the angular velocities assigned using the usual right hand rule.

handedness of the reference frame. It is sometimes called a *pseudovector*. A polar vector is a vector that has a direction independent of the reference frame. We mention these facts (1) to stress the arbitrary character of the direction assigned to axial vectors and (2) to stress the importance of testing experiments and physical laws for invariance under translation, rotation, and mirror reflection of the inertial reference frame. In Section 2-5 we referred briefly to some experiments that were *not* invariant under a reflection transformation. This fact, which constituted a violation under certain circumstances of a law of physics previously thought to be well founded (the law of *conservation of parity*), has posed some challenging problems and is leading us to an understanding of the physical world at a deeper level.*

Figure III-1 shows a section of a long string which is under tension F. The string has been pulled transversely in the y-direction so that a displacement wave travels along the string in the x-direction. We consider a differential element of the string dx and apply Newton's second law of motion to it in order to find how the wave moves along the string.

Let μ be the mass per unit length of the string, so that the mass of element dx is $\mu\, dx$. The net force in the y-direction acting on this element is

$$F \sin \theta_{x+dx} - F \sin \theta_x.$$

We consider only small transverse displacements of the string, so that the restoring force will vary linearly with displacement and the principle of superposition will hold (see Section 19-4). This means that θ in Fig. III-1 will be small, so that we may replace $\sin \theta$ by $\tan \theta$. Now $\tan \theta$ is simply the slope of the string, that is, it equals $\partial y/\partial x$. We must use partial derivatives because the transverse displacement y depends not only on x but also on t. The net force in the y-direction is then

$$F\left(\frac{\partial y}{\partial x}\right)_{x+dx} - F\left(\frac{\partial y}{\partial x}\right)_x,$$

* See "The Overthrow of Parity," by Philip Morrison, *Scientific American*, April 1957.

SUPPLEMENTARY TOPIC III
THE WAVE EQUATION FOR A STRETCHED STRING

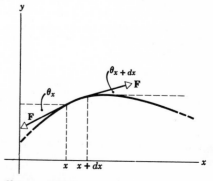

figure III-1

which we may write as

$$F\frac{\partial}{\partial x}\left(\frac{\partial y}{\partial x}\right)dx$$

or

$$F\frac{\partial^2 y}{\partial x^2}\,dx.$$

The mass of the element of the string is $\mu\,dx$ and its transverse acceleration is simply $\partial^2 y/\partial t^2$. Hence, Newton's second law, applied to the transverse motion of the string, is

$$F\frac{\partial^2 y}{\partial x^2}\,dx = (\mu\,dx)\frac{\partial^2 y}{\partial t^2}$$

or

$$\frac{\partial^2 y}{\partial x^2} = \frac{\mu}{F}\frac{\partial^2 y}{\partial t^2}. \tag{III-1}$$

Equation III-1, called the *wave equation*, is the differential equation that describes wave propagation in a string of mass per unit length μ and tension F.

To prove this we show that Eqs. 19-2 and 19-3

$$y = f(x \pm vt), \tag{III-2}$$

which is the general equation representing a wave of any shape traveling along x, is a solution of Eq. III-1. Recall that v in Eq. III-2 is the speed of the wave disturbance and f is any reasonable function of $(x \pm vt)$.

Let us see whether Eq. III-2 is indeed a solution of Eq. III-1 by substituting the former equation into the latter. To do so we note that the two second partial derivatives of y are

$$\frac{\partial^2 y}{\partial x^2} = f'' \quad\text{and}\quad \frac{\partial^2 y}{\partial t^2} = v^2 f''$$

in which f'' is the second derivative of the function f of Eq. III-2 with respect to $(x \pm vt)$. Substitution of these derivatives into Eq. III-1 yields

$$f'' = \frac{\mu}{F}v^2 f'',$$

which we may write as (see Eq. 19-12)

$$v = \sqrt{\frac{F}{\mu}}. \tag{III-3}$$

Thus we conclude that Eq. III-2 is indeed a solution of the partial differential equation Eq. III-1 if the speed of the wave disturbance described by this equation is given by Eq. III-3.

In particular, let us check that Eq. 19-10

$$y = y_m \sin(kx \pm \omega t) \tag{19-10}$$

is a solution of Eq. III-1. We know that it must be because Eq. 19-10 is simply a special case of the general relation Eq. III-2, which we have just shown to be a solution. Even so it is instructive to test this important specific function of $(x \pm vt)$ by substitution into Eq. III-1.

The second derivatives of Eq. 19-10 are

$$\frac{\partial^2 y}{\partial x^2} = -k^2 y_m \sin(kx \pm \omega t)$$

and

$$\frac{\partial^2 y}{\partial t^2} = -\omega^2 y_m \sin(kx \pm \omega t).$$

Substitution into Eq. III-1 yields

$$-k^2 y_m \sin(kx \pm \omega t) = \left(\frac{\mu}{F}\right)[-\omega^2 y_m \sin(kx \pm \omega t)]$$

or

$$\frac{\omega}{k} = \sqrt{\frac{F}{\mu}}.$$

Since $\omega/k = v$ (see Eq. 19-11), this relation is identical with Eq. III-3, and Eq. 19-10, as we expect, is indeed a solution of Eq. III-1.

Boltzmann, in 1876, derived the Maxwell speed distribution law (see Eq. 24-2) from this line of argument: Let a uniform gravitational field $\mathbf{g}$ act on an ideal gas maintained at a fixed temperature T. The number of molecules per unit volume n_v will then decrease with altitude z according to the law of atmospheres (see Example 1, Chapter 17). From what we know about the statistical-mechanical interpretation of temperature, however, the speed distribution law —whose form we assume that we do not yet know—must remain the same at all altitudes because it depends only on the temperature. However, this law determines the rate at which molecules move vertically in the atmosphere at any altitude and must thus be intimately related to the decrease of n_v with z. By exploring this relationship in detail we can, in fact, deduce the speed distribution law.

The weight of gas per unit area between the levels z and $z + dz$ in Fig. IV-1 is $n_v mg\, dz$ in which m is the mass of a single molecule. For equilibrium, this weight per unit area must equal the difference in pressure between z and $z + dz$, or

$$n_v mg\, dz = -dp \tag{IV-1}$$

in which we have inserted a minus sign because p decreases as z increases.

We can write the equation of state of an ideal gas, $pV = nRT$, as

$$p = n_v kT \tag{IV-2}$$

because $n = n_v V/N_0$, where N_0 $(= R/k)$ is Avogadro's number, the number of molecules per mole, and k is Boltzmann's constant. Combining Eqs. IV-1 and IV-2 yields

$$\frac{dp}{p} = \frac{dn_v}{n_v} = -\frac{mg}{kT}\, dz.$$

For a constant temperature, we can integrate this relation to yield

$$n_v = \text{constant } e^{-mgz/kT} \tag{IV-3}$$

which, in view of Eq. IV-2, agrees with the result of Example 1, Chapter 17.

We can find the change in n_v as we go from z to $z + dz$ by differentiating Eq. IV-3, or

$$dn_v = -\text{constant } e^{-mgz/kT}\, dz. \tag{IV-4}$$

We associate this decrease in n_v over the interval dz with the fact that, at $z = 0$ (which can be any level we choose) there are some upward-directed molecules —we call them "special molecules" temporarily for convenience—whose vertical velocity components lie in a particular range v_z to $v_z + dv_z$ such that (neglecting collisions; see below) they can rise as high as z but not as high as $z + dz$. Such molecules pass upward through the level z, reverse their direction and pass downward again, as Fig. IV-1 shows. At this point we see more clearly the relationship between Eq. IV-3 and the speed distribution law. Molecules that pass through the interval dz (from above or below) or molecules that never reach the interval cannot contribute to the decrease dn_v of Eq. IV-4.

The rate per unit area at which "special molecules" leave level $z = 0$ (and arrive at level z) is $v_z n_v(v_z)\, dv_z$. Here $n_v(v_z)\, dv_z$ is the number of molecules per unit volume whose vertical velocity components lie between v_z and $v_z + dv_z$.

Now the rate per unit area at which the "special molecules" arrive at level z, but not as high as level $z + dz$, is proportional to the magnitude of the density difference dn_v between z and $z + dz$, or, from Eq. IV-4,

SUPPLEMENTARY TOPIC IV
DERIVATION OF MAXWELL'S SPEED DISTRIBUTION LAW

figure IV-1

$$v_z n_r(v_z)\, dv_z = \text{constant } e^{-mgz/kT}\, dz, \qquad\qquad \text{(IV-5)}$$

in which the constant is independent of z. Equation IV-5, which requires that the change dn_v be accounted for by the "special molecules" is, in fact, the defining equation for $n_v(v_z)$.

From conservation of energy the special molecules have the property that*

$$\tfrac{1}{2}mv_z{}^2 = mgz$$

or

$$mv_z\, dv_z = mg\, dz.$$

We use these two relations to eliminate z and dz from Eq. IV-5, obtaining, as you should verify,

$$n_v(v_z)\, dv_z = \text{constant } e^{-mv_z{}^2/2kT}\, dv_z \qquad\qquad \text{(IV-6a)}$$

in which $n_v(v_z)\, dv_z$ is the number of molecules per unit volume whose vertical velocity components lie between v_z and $v_z + dv_z$. Note that Eq. IV-6a does not contain g or z. The gravitational field of Fig. IV-1, introduced to allow us to calculate the speed distribution, has served its purpose. We may apply Eq. IV-6a to a gas for which $g = 0$ or in which gravitational effects are negligible. In such a case the vertical direction, which we have identified as the z-direction, no longer has any special meaning. That is, the speed distribution for one component of velocity should be the same for another component of velocity since there is no special or preferred direction in a gas in equilibrium free of external forces. Thus we can write

$$n_v(v_x)\, dv_x = \text{constant } e^{-mv_x{}^2/2kT}\, dv_x \qquad\qquad \text{(IV-6b)}$$

and

$$n_v(v_y)\, dv_y = \text{constant } e^{-mv_y{}^2/2kT}\, dv_y, \qquad\qquad \text{(IV-6c)}$$

for the other two velocity components.

We now seek to find Maxwell's speed distribution (Eq. 24-2); it is expressed in terms of the speed v, rather than in terms of the separate components v_x, v_y, and v_z. We are not concerned here with the direction of $\mathbf{v}$, because we assume it to be completely random. We can represent any velocity $\mathbf{v}$ as a vector extending from the origin in Fig. IV-2; the projections of the vector in the x-y- and z-directions are v_x, v_y, and v_z, respectively. We commonly say that the axes of Fig. IV-2 define a "velocity space," which has many formal similarities to ordinary (or coordinate) space, in which the axes are x, y, and z.

We also show in Fig. IV-2 a small "volume" element, whose sides are dv_x, dv_y, and dv_z; we say that this element has a volume $dv_x\, dv_y\, dv_z$ in velocity space. A point in this element corresponds to a particle whose velocity components lie between v_x and $v_x + dv_x$; v_y and $v_y + dv_y$; and v_z and $v_z + dv_z$. We can regard $n_v(v_z)$ in Eq. IV-6a as giving the *probability* that a given molecule will have a velocity component in the specified range v_z to $v_z + dv_z$, with similar interpretations for $n_v(v_x)$ and $n_v(v_y)$. The probability that a given molecule will have *all three* of its velocity components in the specified ranges, that is, the probability that the tip of the velocity vector $\mathbf{v}$ will lie inside the volume element of Fig. IV-2, is the *product* of the three (independent) probabilities given in Eq. IV-6, or

$$\text{constant } e^{-mv_x{}^2/2kT} e^{-mv_y{}^2/2kT} e^{-mv_z{}^2/2kT}\, dv_x\, dv_y\, dv_z$$

which, since

$$v^2 = v_x{}^2 + v_y{}^2 + v_z{}^2,$$

we may write as

$$\text{constant } e^{-mv^2/2kT}(dv_x\, dv_y\, dv_z). \qquad\qquad \text{(IV-7)}$$

figure IV-2

---

*If we consider collisions, this result is still true *on the average* for the many molecules that start at $z = 0$ with a given value of v_z and move to the interval z to $z + dz$, having $v_z = 0$ there, even though such molecules would follow very erratic paths because of the collisions.

The quantity in parentheses above is a volume element in velocity space. Since in Maxwell's speed distribution law we are not concerned with the direction of molecular velocities but only with their speeds, it is more convenient to substitute a different volume element for the above, namely, one corresponding to all molecules whose speeds lie between v and $v + dv$, *regardless of direction*. This volume element is not a "cube" but is the space between two concentric spheres, one of radius v and one of radius $v + dv$. The volume of this element in velocity space is $(4\pi v^2)(dv)$. Substituting this for the quantity enclosed in parentheses in Eq. IV-7 yields for the number of molecules per unit volume whose speeds lie between v and $v + dv$,

$$n_v(v)\ dv = \text{constant } e^{-mv^2/2kT}(4\pi v^2\ dv)$$

or

$$n_v(v) = Cv^2 e^{-mv^2/2kT}$$

in which C is a constant. If we sum up over all possible speeds, we simply obtain the total number of molecules per unit volume, regardless of speed. Hence, we can find C by requiring that

$$\int_0^\infty n_v(v)\ dv = n_v,$$

where n_v is the total number of particles per unit volume, regardless of speed. Guided by the methods of Example 3 (Chapter 24), you should show that

$$C = 4\pi n_v(m/2\pi kT)^{3/2}$$

so that

$$n_v(v) = 4\pi n_v(m/2\pi kT)^{3/2}v^2 e^{-mv^2/2kT}. \qquad \text{(IV-8)}$$

Let us consider a finite number N of molecules contained in a box of volume V. If we multiply each side of the above equation by V, we can replace $n_v V$ on the right by N and $n_v(v)V$ on the left by $N(v)$, which gives us Eq. 24-2.

SUPPLEMENTARY TOPIC V
SPECIAL RELATIVITY— A SUMMARY OF CONCLUSIONS*

V-1
Introduction

Here we simply display in one place some conclusions drawn from the special theory of relativity (hereafter, SR) proposed by Einstein in 1905. We omit all proofs and make only a modest attempt to make the conclusions acceptable in terms of "common sense."

V-2
The Postulates (RR, Section 1.9)

Einstein based his theory on two postulates and *all* of the conclusions of SR derive from them.

a. The First Postulate. From the time of Galileo it was known that the laws of mechanics were the same in all inertial frames (see Fig. V-1 and p. 66). This means that all inertial observers having relative motion, even though they may measure different values for the velocities, momenta, etc., of the particles involved in a given experiment (a game of pool, perhaps), would nevertheless agree on the laws of mechanics involved (conservation of linear momentum, etc.) and on the outcome of the experiment (who won).

Einstein took the bold step of extending this invariance principle to *all* of physics, not only mechanics, including especially electromagnetism. Einstein's first postulate is:

---

* For a fuller treatment, geared to the level of this book, see *Introduction to Special Relativity*, Robert Resnick, John Wiley and Sons, Inc., New York, 1968. References to this work will be in the style RR, p. 187; RR, Section 1.9, etc.

The laws of physics are the same in all inertial frames. No preferred inertial frame exists.

b. The Second Postulate. In pre-SR days a vexing question was this: Given that the speed of light c is 2.988×10^8 m/s, with respect to what is this speed measured? For sound waves in air the answer is simple—it is with respect to the medium (air) through which the sound wave travels. Light, however, travels through a vacuum. Even so, is there a tenuous medium (the luminiferous, or light carrying, ether) that plays the same role for light that air does for sound? If such an ether exists, can we detect it? Alternatively, should c be measured with respect to the source that emits the light?

All attempts to make experimental verifications along these lines failed completely (see Section 45-8* and RR, Sections 1.5 through 1.8). Einstein made a second bold postulate.

The speed of light is the same in all inertial frames.

Note that no ether is needed or involved. This second postulate means, for example, that if you consider three light sources (a) one at rest with respect to you, (b) one moving toward you at speed 0.9 c, say, and (c) one moving away from you at speed 0.9 c, you would measure the *same* speed of light from all three sources.

This second postulate has been tested directly (see RR, p. 34) using as a moving "light" source π^0 mesons generated in a proton synchrotron at speeds of 0.99975 c. These mesons disintegrate by emitting γ-rays which, like light, are electromagnetic in character and travel with the same speed. The speed measured for the radiation emitted by these fast moving sources was, within experimental error, just c, as Einstein's second postulate predicts.

Many of the conclusions of SR simply don't seem reasonable on the basis of everyday experience. Even Einstein's second postulate seems to violate common sense. If you catch a pitched baseball thrown by a pitcher (a) at rest with respect to you, (b) moving toward you (in an automobile, say) at 30 mi/h and, (c) moving away from you at this same speed, you expect a different baseball speed in each case with respect to you. But if you extend this experience to a source (the pitcher) emitting light (photons), you would contradict Einstein's second postulate. And yet experiment shows that light does have the same speed in each case, in support of Einstein's postulate.

The solution to this dilemma comes about when we realize that the basis of our "common sense" experience is very limited indeed. It is restricted to speeds v such that $v \ll c$, where c is the speed of light. For example, the speed of a satellite in earth orbit may be about 8000 m/s, which seems fast to us, but in terms of the speed of light (3.0×10^8 m/s) it is only 0.000027 c. We simply have no personal experience in regions of high relative velocity.

As an example, to accelerate an average person (to say nothing of a spaceship) to 0.90 c would require no less than 13 percent of this country's 1971 total energy consumption. However, the particles of physics (electrons, mesons, protons, etc.) can readily be accelerated to high speeds. Electrons emerging from the two-mile long linear accelerator at Stanford University have speeds of 0.999 c, for example. In the arena of particle physics SR is absolutely necessary for the solution of mechanical problems.

It turns out that in nature there is a certain finite speed that cannot be exceeded and which we call the limiting speed. This limiting speed is the speed

V-3

Special Relativity and Newtonian Mechanics (RR, Section 2.8)

---

*This book is published in a combined volume (Chapters 1–50) and separate volumes (Part I, Chapters 1–25; Part II, Chapters 26–50). Whether cited references are accessible depends on which of these three volumes you are reading.

of light, c, the greatest speed with which signals can be transmitted. Classical physics assumes that signals can be transmitted with infinite speed, but nature contradicts that assumption, and it really does seem fanciful that such a signal could exist. Experiment confirms c as the limiting speed, so that in a sense the speed of light plays the role in relativity that infinity does in classical physics. It is then not difficult to understand—in fact, it becomes very plausible —that the finite speed of the source of light cannot affect the measured value of the speed of an emitted signal already having the limiting value.

The world in which we live and develop our sense perceptions is a world of Newtonian mechanics, in which $v \ll c$. Newtonian mechanics is revealed as a special case of SR for the limit of low speeds. Indeed, a test of SR is to allow $c \to \infty$ (in which case $v \ll c$ always holds true) and see that the corresponding formulas of Newtonian mechanics emerge.

Newtonian mechanics, although a special case, is an all-important one. It describes the essential motions of our solar system, the tides, our space ventures, the behavior of baseballs and pinball machines, etc. It works beautifully in the vastly important realm $v \ll c$. But it breaks down at speeds approaching that of light.

Few theories have been subject to more rigorous experimental tests than SR. Not the least among them is the fact that particle accelerators work. They are designed using SR at the level of engineering and technology. An accelerator designed on the basis of Newtonian mechanics simply would not work. Nuclear reactors and, alas, nuclear bombs, are further proof that SR really works.

Einstein once said that no number of experiments could prove him right but a single experiment could prove him wrong. To date this single experiment has not been found.

The basic observation made in SR (or in Newtonian mechanics for that matter) is this. Consider observers to be in different inertial frames, S and S' (Fig. V-1). The corresponding axes of S and S' are parallel, the x-x' axes being common, and S' moves to the right with speed v as seen by S; the two origins coincide at $t = t' = 0$. Each observer, S and S', records the same event, which might be the detonation of a tiny flashbulb, and assigns space and time coordinates to the event, namely, x, y, z, t and x', y', z', t'. What are the relations between these two sets of numbers written down in the observers' notebooks?

Before SR the accepted relations were

$$x' = x - vt \qquad y' = y$$
$$t' = t \qquad z' = z, \tag{V-1}$$

called the *Galilean transformation equations* (RR, Section 1.2). Though impressively correct in the important region $v \ll c$ they nevertheless fail as $v \to c$.

The corresponding equations used in SR, called the *Lorentz transformation equations*, are (RR, Table 2-1)

$$x' = \frac{x - vt}{\sqrt{1 - (v/c)^2}} \qquad y' = y$$
$$t' = \frac{t - (v/c^2)x}{\sqrt{1 - (v/c)^2}} \qquad z' = z \tag{V-2}$$

Note certain things about these equations. (a) Space and time coordinates are thoroughly intertwined. In particular, time is not the same for each observer; t' depends on x as well as on t. (b) If we let $c \to \infty$, the Lorentz equations reduce to the Galilean equations, as promised! Finally, (c) We must have $v < c$ or else the quantities x' and t' become indeterminate ($v = c$) or imaginary ($v > c$). The speed of light is an upper limit for the speeds of material objects.

The Lorentz equations, like everything else in SR, can be derived from Einstein's two postulates (RR, Section 2.2).

V-4

The Transformation Equations (RR, Section 2.2)

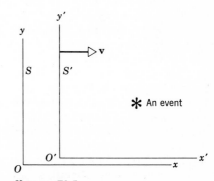

figure V-1
Two inertial frames with parallel axes, the $x - x'$ axis being common. S' moves to the right with speed v as seen by S. At $t = t' = O$ the two origins, O and O', coincide.

Let S' observe two events that occur at the same place in his reference frame. They might be two successive positions of the hand of a clock located at a fixed position, x'. Let S' measure the time interval $\Delta t'$ between these events. S, for whom the clock appears to be moving, observes the same two events and measures a different time interval Δt, which is given by

$$\Delta t = \frac{\Delta t'}{\sqrt{1 - (v/c)^2}} . \tag{V-3}$$

This fact, that $\Delta t > \Delta t'$, is called *time dilation*, and we often verbalize it as "moving clocks run slow." Observer S records a longer time interval than shown to have transpired on the moving clock.

Equation V-3 has been tested experimentally and found to be correct. In one test the "moving clocks" were fast particles called pions $(\pi \pm)$. Pions are radioactive, and their rate of radioactive decay is a measure of their time-keeping ability. See RR, Example 3, p. 75.

Now let us consider a rod, parallel to the $x - x'$ axes, to be at rest in the S' frame. S' will measure a length $\Delta x'$ for it. S, however, measuring the same rod, which is moving with respect to him, would find a length Δx, which is given by

$$\Delta x = \sqrt{1 - (v/c)^2} \, \Delta x' \tag{V-4}$$

This fact, that $\Delta x < \Delta x'$, is called *length contraction*.

The length contraction has been verified in the design of, say, the linear electron accelerator at Stanford University. At an exit speed of $v = 0.999975c$, each meter of the accelerating tube seems like 7.1 mm to an observer moving with the electron. If these length contraction considerations had not been taken into account the machine simply would not work.

The simplest way to understand these results—the time dilation and the length contraction—is to note that one observer, S', is at rest with respect to what he is measuring (clock or rod) whereas for the other observer, S, the objects are in motion. Relativity therefore asserts that *motion affects measurement*. If we had interchanged frames, letting the clock and rod be at rest in S, for example, we would have found the observers again disagreeing on the measured values, but now we would have $\Delta x' < \Delta x$ and $\Delta t' > \Delta t$. So the results are reciprocal, neither observer being "absolutely" right or wrong.

What both observers *will* agree on however, is the *rest length* of a given rod (they will both measure the rod to have the same length when the rod is at rest with respect to their measuring instruments) and the *proper time interval* of a given clock (they will both measure the successive positions of the hand of the clock to have taken the same elapsed time when the clock is at rest with respect to their measuring instruments).

That motion should affect measurement is not a strange idea, even in classical physics. For example, the measured frequency of sound or of light depends on the motion of the source with respect to the observer; we call this the Doppler effect and everyone is familiar with it. And, in mechanics the measured values of the speed, the momentum, or kinetic energy of moving particles are different for observers on the ground than those on a moving train. However, in classical physics measurements of space intervals and time intervals *are* absolute whereas in SR such measurements are relative to the observer. Not only does experiment contradict classical physics but only by adopting the relativity of space and time do we arrive at the invariance (the absoluteness) of all of the laws of physics for all observers. Surely, giving up the absoluteness of the laws of physics (would they then be laws?), as classical notions of time and length require, would leave us with an arbitrary and complex world. By comparison, relativity is absolute and simple.

Let S observe a particle moving with speed u' parallel to the x'-axis. What speed u would S measure? From the Galilean transformation equations (Eq. V-1) we can easily show that

$$u = u' + v \tag{V-5}$$

This relation, which seems to most of us to be intuitively obvious, is, alas, not

V-5
Time Dilation and Length Contraction (RR, Sections 2.3 and 2.4)

V-6
Relativistic Addition of Velocities and the Doppler Effect (Sections 4.6, 6.5, 42.4, and 42.5; RR, Sections 2.6 and 2.7)

true (except for the very important special case of $v \ll c$). The Lorentz transformation equations lead us to

$$u = \frac{u' + v}{1 + (u'v/c^2)}.$$

(V-6)

As we expect, for $c \to \infty$, Eq. V-6 reduces to Eq. V-5. Prove that if $u' < c$ and $v < c$, then it must always be true that $u < c$. There is no way to generate speeds $\geqslant c$ by compounding velocities.

Using the relativistic velocity addition result (Eq. V-6), we can deduce the Doppler effect for light. In relativity theory there is no difference between the two cases, which are different in classical theory (namely, source at rest—observer moving and observer at rest—source moving); only the relative motion v of source and observer counts. This fact and the result

$$\nu = \nu' \sqrt{\frac{c \pm v}{c \mp v}}$$

(V-7)

are in agreement with experiment. Here, ν' is the frequency of the source at rest in S' and ν is the frequency observed in frame S with respect to which the source moves at speed v; the upper signs refer to source and observer moving *toward* one another and the lower signs refer to source and observer moving *away* from one another. Equation V-7 is called the *longitudinal* Doppler effect, and v refers to the relative velocity of source and observer along the line connecting them.

There is in relativity, however, an effect not predicted by classical physics, namely a *transverse* Doppler effect; that is, if the relative velocity v is at right angles to the line connecting source and observer, we find

$$\nu = \nu' \sqrt{1 - v^2/c^2}.$$

(V-8)

This result, confirmed by experiment, can be interpreted simply as a time dilation, moving clocks appearing to run slow.

We have seen that time and length measurements are functions of velocity v. Should mass be any different? SR tells us that the *relativistic mass* m of a particle moving at speed v with respect to the observer is

$$m = \frac{m_0}{\sqrt{1 - (v/c)^2}}$$

(V-9)

in which m_0 is the rest mass, that is, the mass measured when the particle is at rest $(v = 0)$ with respect to the observer.

It is m and not m_0 that must be taken into account when designing magnets to bend charged particles in arcs of circles. By these techniques, Eq. V-9 has been thoroughly tested. Incidentally, the ratio m/m_0 for electrons emerging from the Stanford University linear accelerator at $K = 30$ GeV is the order of 60,000.

To preserve the law of conservation of linear momentum in SR, we redefine the momentum of a particle of rest mass m_0 and speed v as,

$$p = mv = \frac{m_0 v}{\sqrt{1 - (v/c)^2}}.$$

As a result of the considerations above, in SR the kinetic energy of a particle is no longer given by $\frac{1}{2} m_0 v^2$ but by

$$K = mc^2 - m_0 c^2$$

$$= m_0 c^2 \left(\frac{1}{\sqrt{1 - (v/c)^2}} - 1 \right).$$

(V-10)

Can you show that $K \to \frac{1}{2} m_0 v^2$ as $c \to \infty$?

V-7

Mass, Momentum, and Kinetic Energy (Sections 8.9 and 9.3; RR, Sections 3.3 and 3.5)

The best known result of SR is the so-called mass-energy equivalence. That is, the conservation of total energy is equivalent to the conservation of relativistic mass. Mass and energy are equivalent; they form a single invariant that we can call mass-energy. The relation

$$E = mc^2 \qquad \text{(V-11)}$$

expresses the fact that mass-energy can be expressed in energy (E) units or equivalently in mass ($m = E/c^2$) units. In fact, it has become common practice to refer to masses in terms of electron volts, such as saying that the rest mass of an electron is 0.51 MeV, for convenience in energy calculations. Likewise entities of zero rest mass, such as photons, may be assigned an effective mass equivalent to their energy. We associate mass with each of the various forms of energy.

In classical physics we had two separate conservation principles: (1) the conservation of (classical) mass, as in chemical reactions, and (2) the conservation of energy. In relativity, these merge into one conservation principle, that of the conservation of mass-energy. The two classical laws may be viewed as special cases that would be expected to agree with experiment only if energy transfers into or out of the system are so small compared with the system's rest mass that the corresponding fractional change in rest mass of the system is too small to be measured.

For example, the rest mass of a hydrogen atom is 1.00797 u (= 938.8 MeV). If enough energy (13.58 eV) is added to ionize hydrogen, that is, to break it up into its constituent parts, a proton and an electron, the fractional change in rest mass of the system is

$$\frac{13.58 \text{ eV}}{938.8 \times 10^6 \text{ eV}} = 1.45 \times 10^{-8}$$

or 1.45×10^{-6} percent, too small to measure. However, for a nucleus such as the deuteron, whose rest mass is 2.01360 u (= 1876.4 MeV), one must add an energy of 2.22 MeV to break it up into its constituent parts, a proton and a neutron. The fractional change in rest mass of the system is

$$\frac{2.22 \text{ MeV}}{1876.4 \text{ MeV}} = 1.18 \times 10^{-3}$$

or 0.12 percent, which is readily measureable. This is characteristic of the fractional rest-mass changes in nuclear reactions, so that one must use the relativistic law of conservation of mass-energy to get agreement between theory and experiment in nuclear reactions. The classical (rest) mass is *not* conserved, but total energy (mass-energy) is.

SUPPLEMENTARY TOPIC VI*
THE DIFFERENTIAL FORM OF MAXWELL'S EQUATIONS AND THE ELECTROMAGNETIC WAVE EQUATION

VI-1
Introduction

In Chapter 41 we sought to make the existence of electromagnetic waves plausible by showing that such waves are consistent with Maxwell's equations as expressed in Table 40-2. Here we seek to start from Maxwell's equations and derive from them a differential equation whose solutions will describe electromagnetic waves. We will show directly that the speed c of such waves is given by Eq. 40-1, or $c = 1/\sqrt{\epsilon_0 \mu_0}$.

We followed a similar program in Supplementary Topic III for mechanical waves on a stretched string. Starting from Newton's laws of motion we derived a differential equation (Eq. III-1) whose solutions (Eq. III-2) described such waves. We showed further that the speed v of these waves is given by Eq. III-3, or $v = \sqrt{F/\mu}$.

In Table 40-2 we wrote Maxwell's equations as

$$\epsilon_0 \oint \mathbf{E} \cdot d\mathbf{S} = q, \qquad \text{(VI-1)}$$

* Supplementary Topics I to V appear in Part I.

$$\oint \mathbf{B} \cdot d\mathbf{S} = 0, \tag{VI-2}$$

$$\oint \mathbf{B} \cdot d\mathbf{l} = \mu_0(i + \epsilon_0 \, d\Phi_E/dt), \tag{VI-3}$$

and

$$\oint \mathbf{E} \cdot d\mathbf{l} = -d\Phi_B/dt. \tag{VI-4}$$

These equations are said to be written in integral form. The field variables **E** and **B**, which are usually the unknown quantities, appear in the integrands. Only in a few symmetric cases (see Sections 28-8 and 34-2, for example) can we "factor them out." In more general problems we cannot do so.

The situation is somewhat analogous to computing the density ρ of a body if we know its mass m and volume τ. In general these are related by the integral equation

$$m = \int \rho \, d\tau.$$

Only if ρ is a constant over all parts of the volume can we factor it out and write $\rho = m/\tau$.

To carry out our program it is desirable to recast Maxwell's equations in the form of equalities that apply at each *point* in space rather than as integrals that apply to various *regions* of space. In other words, we wish to convert Maxwell's equations from the integral form of Eqs. VI-1 to 4 into *differential form*. We will then be able to relate **E** and **B** at a point to the charge density and current density at that point.

VI-2
The Operator ∇

To transform Maxwell's equations into differential form we must deepen our understanding of vector methods and, in particular, become familiar with the vector operator ∇.

In Section 29-7 we saw how to obtain the components of the (vector) electrostatic field **E** at any point from the (scalar) potential function $V(x,y,z)$ by partial differentiation. Thus,

$$E_x = -\frac{\partial V}{\partial x}, \quad E_y = -\frac{\partial V}{\partial y}, \quad \text{and} \quad E_z = -\frac{\partial V}{\partial z}$$

so that the electrostatic field

$$\mathbf{E} = \mathbf{i}E_x + \mathbf{j}E_y + \mathbf{k}E_z$$

can be written as

$$\mathbf{E} = -\left(\mathbf{i}\frac{\partial V}{\partial x} + \mathbf{j}\frac{\partial V}{\partial y} + \mathbf{k}\frac{\partial V}{\partial z}\right). \tag{VI-5}$$

We can write Eq. VI-5 in compact vector notation as

$$\mathbf{E} = -\nabla V,$$

where ∇ ("del") is a vector operator defined as

$$\nabla = \mathbf{i}\frac{\partial}{\partial x} + \mathbf{j}\frac{\partial}{\partial y} + \mathbf{k}\frac{\partial}{\partial z}. \tag{VI-6}$$

This operator is useful in dealing with scalar and vector fields (see Sections 16-8 and 18-7 for examples of such fields). Given any scalar field ψ we may form a vector field, called the gradient of ψ and written as grad ψ or $\nabla\psi$, simply by applying the operator ∇ to ψ. Given a vector field $\mathbf{U} = U_x\mathbf{i} + U_y\mathbf{j} + U_z\mathbf{k}$ we may apply the operator ∇ to it in two different ways. One way is to take the dot product of ∇ and **U**, yielding the scalar field called the divergence of **U** and written as div **U** or $\nabla \cdot \mathbf{U}$. The other way is to take the cross product of ∇ and **U**, yielding the vector field called the curl of **U** and written curl **U** or $\nabla \times \mathbf{U}$. These operations may be summarized as

$$\text{grad } \psi \equiv \nabla \psi = \mathbf{i}\frac{\partial \psi}{\partial x} + \mathbf{j}\frac{\partial \psi}{\partial y} + \mathbf{k}\frac{\partial \psi}{\partial z}, \tag{VI-7}$$

$$\text{div } \mathbf{U} \equiv \nabla \cdot \mathbf{U} = \frac{\partial U_x}{\partial x} + \frac{\partial U_y}{\partial y} + \frac{\partial U_z}{\partial z}, \tag{VI-8}$$

$$\text{curl } \mathbf{U} \equiv \nabla \times \mathbf{U} = \mathbf{i}\left(\frac{\partial U_z}{\partial y} - \frac{\partial U_y}{\partial z}\right)$$

$$+ \mathbf{j}\left(\frac{\partial U_x}{\partial z} - \frac{\partial U_z}{\partial x}\right)$$

$$+ \mathbf{k}\left(\frac{\partial U_y}{\partial x} - \frac{\partial U_x}{\partial y}\right). \tag{VI-9}$$

Note that grad ψ and curl $\mathbf{U}$ are vectors, whereas div $\mathbf{U}$ is a scalar. You can gain some familiarity with these operations by the following exercises: (1) prove that curl (grad ψ) = 0 and (2) prove that div (curl $\mathbf{U}$) = 0.

Another frequently occurring operator is ∇^2 ("del squared"). It is simply $\nabla \cdot \nabla$, or, as the student can show from Eq. VI-6,

$$\nabla^2 = \nabla \cdot \nabla = \frac{\partial^2}{\partial x^2} + \frac{\partial^2}{\partial y^2} + \frac{\partial^2}{\partial z^2}.$$

When we apply ∇^2 to a scalar field ψ, we obtain

$$\nabla^2 \psi = \frac{\partial^2 \psi}{\partial x^2} + \frac{\partial^2 \psi}{\partial y^2} + \frac{\partial^2 \psi}{\partial z^2}. \tag{VI-10}$$

For a vector field $\mathbf{U}$, the operation $\nabla^2 \mathbf{U}$ is

$$\nabla^2 \mathbf{U} = \mathbf{i}\left(\frac{\partial^2}{\partial x^2} + \frac{\partial^2}{\partial y^2} + \frac{\partial^2}{\partial z^2}\right)U_x + \mathbf{j}\left(\frac{\partial^2}{\partial x^2} + \frac{\partial^2}{\partial y^2} + \frac{\partial^2}{\partial z^2}\right)U_y + \mathbf{k}\left(\frac{\partial^2}{\partial x^2} + \frac{\partial^2}{\partial y^2} + \frac{\partial^2}{\partial z^2}\right)U_z \tag{VI-11}$$

As an exercise, show that curl (curl $\mathbf{U}$) = $-\nabla^2\mathbf{U}$ + grad (div $\mathbf{U}$).

In this section we show how to cast the first two of Maxwell's equations (Eqs. VI-1, 2) into differential form. Let us apply Eq. VI-1 to a differential volume element shaped like a rectangular parallelepiped and containing a point P at (and near) which an electric field exists (see Fig. VI-1a). Point P is located at x, y, z in the reference frame of Fig. VI-1b and the edges of the parallelepiped have lengths dx, dy, and dz.

We can write the surface area vector for the rear face of the parallelepiped as $d\mathbf{S} = -\mathbf{i}\,dy\,dz$. The minus sign enters because $d\mathbf{S}$ is defined to point in the direction of the *outward* normal, which is defined by $-\mathbf{i}$. For the front face we have $d\mathbf{S} = +\mathbf{i}\,dy\,dz$.

If the electric field at the rear face is $\mathbf{E}$, that at the front face, which is a distance dx away from the rear face, is $\mathbf{E} + (\partial\mathbf{E}/\partial x)\,dx$, the latter term representing the change in $\mathbf{E}$ associated with the change dx in x.

The flux through the entire surface of the parallelepiped is $\oint \mathbf{E} \cdot d\mathbf{S}$ and the contribution to this flux due to these two faces alone is

$$(\mathbf{E}) \cdot (-\mathbf{i}\,dy\,dz) + \left(\mathbf{E} + \frac{\partial\mathbf{E}}{\partial x}\,dx\right) \cdot (+\mathbf{i}\,dy\,dz)$$

$$= dx\,dy\,dz\left(\frac{\partial\mathbf{E}}{\partial x} \cdot \mathbf{i}\right) = dx\,dy\,dz\,\frac{\partial}{\partial x}(\mathbf{E} \cdot \mathbf{i})$$

$$= dx\,dy\,dz\,\frac{\partial E_x}{\partial x}.$$

With similar contributions from the other four faces the total electric flux becomes

$$\oint \mathbf{E} \cdot d\mathbf{S} = dx\,dy\,dz\left(\frac{\partial E_x}{\partial x} + \frac{\partial E_y}{\partial y} + \frac{\partial E_z}{\partial z}\right).$$

VI-3
Maxwell's Equations in Differential Form—I

(a)   *(b)*

From Eq. VI-8 we may write this as

$$\oint \mathbf{E} \cdot d\mathbf{S} = dx \, dy \, dz \, \text{div } \mathbf{E}. \qquad \text{(VI-12)}$$

Now the right-hand side of Eq. VI-1, which gives the charge enclosed by the surface, may be written in general as $q = \int \rho \, d\tau$ and, in particular, for the differential volume element at P, as

$$q = \rho \, dx \, dy \, dz, \qquad \text{(VI-13)}$$

where ρ is the charge per unit volume at P. Substituting Eqs. VI-12 and 13 into Eq. VI-1 and canceling the common factor $dx \, dy \, dz$, we have finally

$$\epsilon_0 \, \text{div } \mathbf{E} = \rho, \qquad \text{(VI-14)}$$

which is Maxwell's first equation (Eq. VI-1) in differential form.

Using the same technique, we can express Maxwell's second equation (Eq. VI-2) in differential form as

$$\text{div } \mathbf{B} = 0. \qquad \text{(VI-15)}$$

We now seek to transform Maxwell's third and fourth equations (Eqs. VI-3, 4) into differential form. We start by applying Eq. VI-3 to a differential surface element of rectangular shape at a point P in some region of a magnetic field, as shown in Fig. VI-2a. The point P is located at x, y, z in the reference frame of Fig. VI-2b and the sides of the rectangle, which is parallel to the x-y plane, have lengths dx and dy. Going around the path, as shown by the arrows, we have

$$\oint \mathbf{B} \cdot d\mathbf{l} = \mathbf{B} \cdot (-\mathbf{j} \, dy) \qquad \text{(rear side)}$$

$$+ \mathbf{B} \cdot (+\mathbf{i} \, dx) \qquad \text{(left side)}$$

$$+ \left(\mathbf{B} + \frac{\partial \mathbf{B}}{\partial x} dx \right) \cdot (+\mathbf{j} \, dy) \qquad \text{(front side)}$$

$$+ \left(\mathbf{B} + \frac{\partial \mathbf{B}}{\partial y} dy \right) \cdot (-\mathbf{i} \, dx), \qquad \text{(right side)}$$

where $\mathbf{B}$ is the magnetic field at P.

Collecting terms we obtain

$$\oint \mathbf{B} \cdot d\mathbf{l} = dx \, dy \left(\frac{\partial \mathbf{B}}{\partial x} \cdot \mathbf{j} - \frac{\partial \mathbf{B}}{\partial y} \cdot \mathbf{i} \right)$$

$$= dx \, dy \left[\frac{\partial}{\partial x} (\mathbf{B} \cdot \mathbf{j}) - \frac{\partial}{\partial y} (\mathbf{B} \cdot \mathbf{i}) \right]$$

$$= dx \, dy \left(\frac{\partial B_y}{\partial x} - \frac{\partial B_x}{\partial y} \right). \qquad \text{(VI-16)}$$

Now in the right-hand side of Eq. VI-3 i is the current enclosed by the path and $d\Phi_E/dt$ is the change in electric flux through the enclosed surface. Hence, if $\mathbf{J}$ is taken to represent the current density and $d\mathbf{S}(= \mathbf{k} \, dx \, dy)$ the surface area vector, we can write

$$i = \mathbf{J} \cdot d\mathbf{S} = \mathbf{J} \cdot (\mathbf{k} \, dx \, dy) = dx \, dy \, J_z \qquad \text{(VI-17)}$$

and

$$\frac{d\Phi_E}{dt} = \frac{\partial \mathbf{E}}{\partial t} \cdot d\mathbf{S} = \frac{\partial \mathbf{E}}{\partial t} \cdot (\mathbf{k} \, dx \, dy)$$

or

$$\frac{d\Phi_E}{dt} = \frac{\partial E_z}{\partial t} dx \, dy. \qquad \text{(VI-18)}$$

Substituting Eqs. VI-16, 17, and 18 into Eq. VI-3 and canceling the common factor $dx \, dy$, we get

$$\frac{1}{\mu_0} \left(\frac{\partial B_y}{\partial x} - \frac{\partial B_x}{\partial y} \right) = J_z + \epsilon_0 \frac{\partial E_z}{\partial t}. \qquad \text{(VI-19)}$$

VI-4
Maxwell's Equations in Differential Form—II

(a)   (b)

We could have proceeded exactly as above if we had started with a rectangle parallel to the y-z plane or one parallel to the z-x plane. Each rectangle would have given us a different component of an arbitrarily oriented differential surface at P. Equation VI-19 is obviously the z-component equation corresponding to Eq. VI-3. If we multiply it by $\mathbf{k}$ and add to it the two similar vector equations, which may be obtained by cyclically permuting x, y, z and $\mathbf{i}$, $\mathbf{j}$, $\mathbf{k}$, corresponding to the x-component and y-component equations, we obtain

$$\text{curl } \mathbf{B} = \mu_0(\mathbf{J} + \epsilon_0 \, \partial\mathbf{E}/\partial t), \qquad \text{(VI-20)}$$

which is the third Maxwell equation in differential form.

Similarly, starting with Eq. VI-4, we may show that

$$\text{curl } \mathbf{E} = -\partial\mathbf{B}/\partial t, \qquad \text{(VI-21)}$$

which is the fourth Maxwell equation in differential form.

We have derived four differential equations (see VI-22 to VI-25 below) from the four integral equations (VI-1 to VI-4). It can be shown that the integral equations can be derived from the differential equations, that is, the two sets of equations are *equivalent*.

VI-5
The Wave Equation

We have now obtained from their integral form the four basic equations of electromagnetism, Maxwell's equations, in differential form. Corresponding to the integral equations, Eqs. VI-1, 2, 3, and 4 respectively, we have

$$\epsilon_0 \text{ div } \mathbf{E} = \rho, \qquad \text{(VI-22)}$$

$$\text{div } \mathbf{B} = 0, \qquad \text{(VI-23)}$$

$$\text{curl } \mathbf{B} = \mu_0(\mathbf{J} + \epsilon_0 \, \partial\mathbf{E}/\partial t), \qquad \text{(VI-24)}$$

$$\text{curl } \mathbf{E} = -\partial\mathbf{B}/\partial t, \qquad \text{(VI-25)}$$

which are four coupled partial differential equations. They apply at each point of space in an electromagnetic field.

We will now derive the wave equation for electromagnetic waves in free space. In free space, the charge density ρ and the current density $\mathbf{J}$ are zero, so that the Maxwell equations there become

$$\text{div } \mathbf{E} = 0,$$

$$\text{div } \mathbf{B} = 0,$$

$$\text{curl } \mathbf{B} = \mu_0\epsilon_0 \, \partial\mathbf{E}/\partial t,$$

and

$$\text{curl } \mathbf{E} = -\partial\mathbf{B}/\partial t.$$

Let us take the curl of the equation for curl $\mathbf{E}$; we obtain

$$\text{curl curl } \mathbf{E} = -\text{curl } \frac{\partial\mathbf{B}}{\partial t} = -\frac{\partial}{\partial t} \text{ curl } \mathbf{B}.$$

But, from above, curl $\mathbf{B} = \mu_0\epsilon_0(\partial\mathbf{E}/\partial t)$, so that

$$\text{curl curl } \mathbf{E} = -\mu_0\epsilon_0 \frac{\partial^2\mathbf{E}}{\partial t^2}. \qquad \text{(VI-26)}$$

From the exercise in VI-2, we know that curl curl $\mathbf{E} = -\nabla^2 E + \text{grad div } \mathbf{E}$ and from above that div $\mathbf{E} = 0$. Thus,

$$\text{curl curl } \mathbf{E} = -\nabla^2\mathbf{E}. \qquad \text{(VI-27)}$$

Combining Eqs. VI-26 and VI-27, we obtain finally

$$\nabla^2\mathbf{E} = \mu_0\epsilon_0 \frac{\partial^2\mathbf{E}}{\partial t^2}. \qquad \text{(VI-28}a\text{)}$$

Proceeding as above, show that $\mathbf{B}$ satisfies the same equation, or

$$\nabla^2\mathbf{B} = \mu_0\epsilon_0 \frac{\partial^2\mathbf{B}}{\partial t^2}. \qquad \text{(VI-28}b\text{)}$$

Equations VI-28 are the equations of electromagnetic wave motion. Being vector equations, they are equivalent to six scalar equations, one for each component of **E** and of **B**.

There are many solutions of Eqs. VI-28, corresponding to different kinds of electromagnetic waves—plane, spherical, and cylindrical waves being three examples. Let us consider a solution in which two components of **E** and two of **B** vanish, that is, in which

$$E_x = E_z = 0 \quad \text{and} \quad B_x = B_y = 0.$$

Equations VI-28 are satisfied for these assumptions. For the nonvanishing components, E_y and B_z, Eqs. VI-28 reduce to (see Eq. VI-11)

$$\frac{\partial^2 E_y}{\partial x^2} + \frac{\partial^2 E_y}{\partial y^2} + \frac{\partial^2 E_y}{\partial z^2} = \mu_0 \epsilon_0 \frac{\partial^2 E_y}{\partial t^2} \tag{VI-29a}$$

and

$$\frac{\partial^2 B_z}{\partial x^2} + \frac{\partial^2 B_z}{\partial y^2} + \frac{\partial^2 B_z}{\partial z^2} = \mu_0 \epsilon_0 \frac{\partial^2 B_z}{\partial t^2}. \tag{VI-29b}$$

If we make the additional assumption that E_y and B_z are functions of x and t only, the simplified wave equation that results, namely

$$\frac{\partial^2 E_y}{\partial x^2} = \mu_0 \epsilon_0 \frac{\partial^2 E_y}{\partial t^2} \quad \text{and} \quad \frac{\partial^2 B_z}{\partial x^2} = \mu_0 \epsilon_0 \frac{\partial^2 B_z}{\partial t^2}, \tag{VI-29c}$$

is similar to Eq. III-1 for the vibrating string.

A solution to these equations, as you may verify by substitution, is

$$E_y = E_m \sin(kx - \omega t) \tag{VI-30a}$$

and

$$B_z = B_m \sin(kx - \omega t). \tag{VI-30b}$$

We interpret Eqs. VI-30 as an electromagnetic wave traveling in the positive x-direction, as in Fig. 41-13, with a speed $c = \omega/k$. Show that substituting Eq. VI-30a into Eq. VI-29a (or Eq. VI-30b into Eq. VI-29b) yields

$$c = \omega/k = 1/\sqrt{\epsilon_0 \mu_0},$$

which (see Eq. 40-1) gives the speed of electromagnetic waves in free space.

appendices

SI base units[a]

Quantity	Name	Symbol	Definition
length	meter	m	" . . . the length equal to 1,650,763.73 wavelengths in vacuum of the radiation corresponding to the transition between the levels $2p_{10}$ and $5d_5$ of the krypton-86 atom." (1960)
mass	kilogram	kg	" . . . this prototype [a certain platinum-iridium cylinder] shall henceforth be considered to be the unit of mass." (1889)
time	second	s	" . . . the duration of 9,192,-631,770 periods of the radiation corresponding to the transition between the two hyperfine levels of the ground state of the cesium-133 atom." (1967)

* Adapted from "The International System of Units (SI)," National Bureau of Standards Special Publication 330, 1972 edition.

[a] The definitions of these base units were adopted by the General Conference of Weights and Measures, an international body, on the dates shown. In this book we will not use the candela.

SI base units (Continued)

Quantity	Name	Symbol	Definition
electric current	ampere	A	" . . . that constant current which, if maintained in two straight parallel conductors of infinite length, of negligible circular cross section, and placed 1 meter apart in vacuum, would produce between these conductors a force equal to 2×10^{-7} newton per meter of length." (1946)
thermodynamic temperature	kelvin	K	" . . . the fraction 1/273.16 of the thermodynamic temperature of the triple point of water." (1967)
amount of substance	mole	mol	". . . the amount of substance of a system which contains as many elementary entities as there are atoms in 0.012 kilogram of carbon-12." (1971)
luminous intensity	candela	cd	" . . . the luminous intensity, in the perpendicular direction, of a surface of 1/600,000 square meter of a blackbody at the temperature of freezing platinum under a pressure of 101,325 newton per square meter." (1967)

Some SI derived units with special names

Quantity	SI unit			
	Name	Symbol	Expression in terms of other units	Expression in terms of SI base units
frequency	hertz	Hz		s^{-1}
force	newton	N		$m \cdot kg/s^2$
pressure	pascal	Pa	N/m^2	$kg/m \cdot s^2$
energy, work, quantity of heat	joule	J	$N \cdot m$	$kg \cdot m^2/s^2$
power, radiant flux	watt	W	J/s	$kg \cdot m^2/s^3$
quantity of electricity, electric charge	coulomb	C		$A \cdot s$
electric potential, potential difference, electromotive force	volt	V	W/A	$kg \cdot m^2/A \cdot s^3$
capacitance	farad	F	C/V	$A^2 \cdot s^4/kg \cdot m^2$
electric resistance	ohm	Ω	V/A	$kg \cdot m^2/A^2 \cdot s^3$
conductance	siemens	S	A/V	$A^2 \cdot s^3/kg \cdot m^2$
magnetic flux	weber	Wb	$V \cdot s$	$kg \cdot m^2/A \cdot s^2$
magnetic field	tesla	T	Wb/m^2	$kg/A \cdot s^2$
inductance	henry	H	Wb/A	$kg \cdot m^2/A^2 \cdot s^2$

Some symbols for units of physical quantities

SI Symbols		Symbols other than SI that are Commonly Used	
Name	Abbreviation	Name	Abbreviation
ampere	A	angstrom	Å
candela	cd	British thermal unit	Btu
coulomb	C	calorie	cal
farad	F	day	d
henry	H	degree	°
hertz	Hz	dyne	dyn
joule	J	electron volt	eV
kelvin	K	foot	ft
kilogram	kg	gauss	G
meter	m	gram	g
mole	mol	horsepower	hp
newton	N	hour	h
ohm	Ω	inch	in.
pascal	Pa	mile	mi
radian	rad	minute (of arc)	'
second	s	minute (of time)	min
siemens	S	pound	lb
steradian	sr	revolution	rev
tesla	T	second (of arc)	"
volt	V	standard atmosphere	atm
watt	W	unified atomic mass unit	u
weber	Wb	year	yr

Over the years many hundreds of measurements of fundamental physical quantities, alone and in combination, have been made by hundreds of scientists in many countries. These measurements have different precisions and, most important, they are interdependent. For example, the direct measurements of e, e/m, h/e, etc., are obviously interrelated. Sorting out the best values of e, m, h, etc., from a large mass of overlapping data is not simple.†

For most problems in this book three significant figures will do, and the computational (rounded) values may be used.

APPENDIX B
SOME FUNDAMENTAL CONSTANTS OF PHYSICS*

Constant	Symbol	Computational value	Best (1973) Value Value[a]	Uncertainty[b]
Speed of light in a vacuum	c	3.00×10^8 m/s	2.99792458	0.004
Elementary charge	e	1.60×10^{-19} C	1.6021892	2.9
Electron rest mass	m_e	9.11×10^{-31} kg	9.109534	5.1
Permittivity constant	ϵ_0	8.85×10^{-12} F/m	8.854187818	0.008
Permeability constant	μ_0	1.26×10^{-6} H/m	4π (exactly)	—
Electron charge to mass ratio	e/m_e	1.76×10^{11} C/kg	1.7588047	2.8
Proton rest mass	m_p	1.67×10^{-27} kg	1.6726485	5.1
Ratio of proton mass to electron mass	m_p/m_e	1840	1836.15152	0.38
Neutron rest mass	m_n	1.68×10^{-27} kg	1.6749543	5.1
Muon rest mass	m_μ	1.88×10^{-28} kg	1.883566	5.6
Planck constant	h	6.63×10^{-34} J·s	6.626176	5.4
Electron Compton wavelength	λ_C	2.43×10^{-12} m	2.4263089	1.6
Molar gas constant	R	8.31 J/mol·K	8.31441	31
Avogadro constant	N_A	6.02×10^{23}/mol	6.022045	5.1
Boltzmann constant	k	1.38×10^{-23} J/K	1.380662	32
Molar volume of ideal gas at STP[c]	V_m	2.24×10^{-2} m³/mol	2.241383	31
Faraday constant	F	9.65×10^4 C/mol	9.648456	2.8
Stefan-Boltzmann constant	σ	5.67×10^{-8} W/m²·K⁴	5.67032	125
Rydberg constant	R	1.10×10^7/m	1.097373177	0.075
Gravitational constant	G	6.67×10^{-11} m³/s²·kg	6.6726	75
Bohr radius	a_0	5.29×10^{-11} m	5.2917706	0.82
Electron magnetic moment	μ_e	9.28×10^{-24} J/T	9.284832	3.9
Proton magnetic moment	μ_p	1.41×10^{-26} J/T	1.4106171	3.9
Bohr magneton	μ_B	9.27×10^{-24} J/T	9.274078	3.9
Nuclear magneton	μ_N	5.05×10^{-27} J/T	5.050824	3.9

[a] Same unit and power of ten as the computational value.
[b] Parts per million.
[c] STP-standard temperature and pressure = 0° C and 1.0 atm.

* The values in this table were selected from a longer list developed by E. Richard Cohen and B. N. Taylor, *Journal of Physical and Chemical Reference Data*, vol. 2, no. 4 (1973).
† See "A Pilgrim's Progress in Search of the Fundamental Constants," by J. W. M. Du Mond, *Physics Today*, October 1965, and "The Fundamental Physical Constants" by Taylor, Langenberg, and Parker, *Scientific American*, October, 1970.

The sun

Mass	1.99×10^{30} kg
Radius	6.96×10^{5} km
Mean density	1,410 kg/m³
Surface gravity	274 m/s²
Surface temperature	6000 K
Total radiation rate	3.92×10^{26} W

The earth

Mass	5.98×10^{24} kg
Equatorial radius	6.378×10^{6} m
	3963 mi
Polar radius	6.357×10^{6} m
	3950 mi
Radius of a sphere of the same volume	6.371×10^{6} m
	3959 mi
Mean density	5522 kg/m³
Acceleration of gravity[a]	9.80665 m/s²
	32.1740 ft/s²
Mean orbital speed	29,770 m/s
	18.50 mi/s
Angular speed	7.29×10^{-5} rad/s
Solar constant[b]	1340 W/m²
Magnetic field (at Washington, D.C.)	5.7×10^{-5} T
Magnetic dipole moment	8.1×10^{22} A·m²
Standard atmosphere	1.013×10^{5} Pa
	14.70 lb/in.²
	760.0 mm-Hg
Density of dry air at STP[c]	1.29 kg/m³
Speed of sound in dry air at STP	331.4 m/s
	1089 ft/s
	742.5 mi/h

[a] This value, adopted by the General Committee on Weights and Measures in 1901, approximates the value at 45° latitude at sea level.
[b] This is the rate per unit area at which solar energy falls, at normal incidence, just outside the earth's atmosphere.
[c] STP = standard temperature and pressure = 0° C and 1 atm.

The moon

Mass	7.36×10^{22} kg
Radius	1738 km
Mean density	3340 kg/m³
Surface gravity	1.67 m/s²
Mean earth-moon distance	3.80×10^{5} km

APPENDIX D
*THE SOLAR SYSTEM**

	MERCURY	VENUS	EARTH	MARS	JUPITER	SATURN	URANUS	NEPTUNE	PLUTO
Maximum distance from sun (10^6 km)	69.7	109	152.1	249.1	815.7	1,507	3,004	4,537	7,375
Minimum distance from sun (10^6 km)	45.9	107.4	147.1	206.7	740.9	1,347	2,735	4,456	4,425
Mean distance from sun (10^6 km)	57.9	108.2	149.6	227.9	778.3	1,427	2,869.6	4,496.6	5,900
Mean distance from sun (astronomical units)	.387	.723	1	1.524	5.203	9.539	19.18	30.06	39.44
Period of revolution	88 d	224.7 d	365.26 d	687 d	11.86 y	29.46 y	84.01 y	164.8 y	247.7 y
Rotation period	59 d	−243 d retrograde	23 h 56 min 4 s	24 h 37 min 23 s	9 h 50 min 30 s	10 h 14 min	−11 h retrograde	16 h	6 d 9 h
Orbital velocity (km/s)	47.9	35'	29.8	24.1	13.1	9.6	6.8	5.4	4.7
Inclination of axis	<28°	3°	23°27'	23°59'	3°05'	26°44'	82°5'	28°48'	?
Inclination of orbit to ecliptic	7°	3.4°	0°	1.9°	1.3°	2.5°	.8°	1.8°	17.2°
Eccentricity of orbit	.206	.007	.017	.093	.048	.056	.047	.009	.25
Equatorial diameter (km)	4,880	12,104	12,756	6,787	142,800	120,000	51,800	49,500	6,000 (?)
Mass (earth = 1)	.055	.815	1	.108	317.9	95.2	14.6	17.2	.1 (?)
Volume (earth = 1)	.06	.88	1	.15	1,316	755	67	57	.1 (?)
Density (water = 1)	5.4	5.2	5.5	4.0	1.3	.7	1.2	1.7	?
Oblateness	0	0	.003	.009	.06	.1	.06	.02	?
Atmosphere (main components)	none	carbon dioxide	nitrogen, oxygen	carbon dioxide, argon	hydrogen, helium	hydrogen, helium	hydrogen, helium, methane	hydrogen, helium, methane	none detected
Mean temperature at visible surface (degrees Celsius) S = solid, C = clouds	350(S) d −170(S) night	−33 (C) 480 (S)	22 (S)	−23 (S)	−150 (C)	−180 (C)	−210 (C)	−220 (C)	−230(?)
Atmospheric pressure at surface (millibars)	10^{-9}	90,000	1,000	6	?	?	?	?	?
Surface gravity (earth = 1)	.37	.88	1	.38	2.64	1.15	1.17	1.18	?
Mean apparent diameter of sun as seen from planet	1°22'40"	44'15"	31'59"	21'	6'09"	3'22"	1'41"	1'04"	49"
Known satellites	0	0	1	2	13	10	5	2	0

* Reprinted by permission from "The Solar System," Carl Sagan, *Scientific American*, September 1975.

APPENDIX E
PERIODIC TABLE OF THE ELEMENTS

atomic number →
atomic mass →

1
H
1.0079

NOBLE GASES

IA	IIA												IIIA	IVA	VA	VIA	VIIA	O
1 **H** 1.00797																		2 **He** 4.00260
3 **Li** 6.941	4 **Be** 9.01218												5 **B** 10.81	6 **C** 12.01115	7 **N** 14.0067	8 **O** 15.9994	9 **F** 18.99840	10 **Ne** 20.179
11 **Na** 22.98977	12 **Mg** 24.305	IIIB	IVB	VB	VIB	VIIB		VIII		IB	IIB		13 **Al** 26.98154	14 **Si** 28.086†	15 **P** 30.97376	16 **S** 32.06	17 **Cl** 35.453	18 **Ar** 39.948
19 **K** 39.098	20 **Ca** 40.08	21 **Sc** 44.9559	22 **Ti** 47.90	23 **V** 50.9414	24 **Cr** 51.996	25 **Mn** 54.9380	26 **Fe** 55.847	27 **Co** 58.9332	28 **Ni** 58.71	29 **Cu** 63.546	30 **Zn** 65.38		31 **Ga** 69.72	32 **Ge** 72.59	33 **As** 74.9216	34 **Se** 78.96	35 **Br** 79.904	36 **Kr** 83.80
37 **Rb** 85.4678	38 **Sr** 87.62	39 **Y** 88.9059	40 **Zr** 91.22	41 **Nb** 92.9064	42 **Mo** 95.94	43 **Tc** 98.9062	44 **Ru** 101.07	45 **Rh** 102.9055	46 **Pd** 106.4	47 **Ag** 107.868	48 **Cd** 112.40		49 **In** 114.82	50 **Sn** 118.69	51 **Sb** 121.75	52 **Te** 127.60	53 **I** 126.9045	54 **Xe** 131.30
55 **Cs** 132.9054	56 **Ba** 137.34	57 ***La** 138.9055	72 **Hf** 178.49	73 **Ta** 180.9479	74 **W** 183.85	75 **Re** 186.2	76 **Os** 190.2	77 **Ir** 192.22	78 **Pt** 195.09	79 **Au** 196.9665	80 **Hg** 200.59		81 **Tl** 204.37	82 **Pb** 207.19	83 **Bi** 208.9804	84 **Po** (210)	85 **At** (210)	86 **Rn** (222)
87 **Fr** (223)	88 **Ra** 226.0254	89 **†Ac** (227)	104 **Rf** (261)	105 **Ha** (260)	106 ****** (263)	107 ******		109 ******										

PERIODS

*
58 **Ce** 140.12	59 **Pr** 140.9077	60 **Nd** 144.24	61 **Pm** (147)	62 **Sm** 150.4	63 **Eu** 151.96	64 **Gd** 157.25	65 **Tb** 158.9254	66 **Dy** 162.50	67 **Ho** 164.9304	68 **Er** 167.26	69 **Tm** 168.9342	70 **Yb** 173.04	71 **Lu** 174.97

†
90 **Th** 232.0381	91 **Pa** 231.0359	92 **U** 238.029	93 **Np** 237.0482	94 **Pu** (244)	95 **Am** (243)	96 **Cm** (247)	97 **Bk** (247)	98 **Cf** (251)	99 **Es** (254)	100 **Fm** (257)	101 **Md** (258)	102 **No** (255)	103 **Lr** (256)

**Discovery of these three elements has been reported but names for them have not yet been proposed.

Family name	Particle name	Symbol		Spin	Charge, e	Strangeness	Rest mass, MeV	Mean life, seconds	Typical decay mode
		Particle	Antiparticle						
—	Photon	γ	γ	1	0	0	0	Stable	—
L E P T O N S	Electron	e^-	$\overline{e}^-$	$\frac{1}{2}$	∓ 1	0	0.5110	Stable	—
	Muon	μ^+	$\overline{\mu}^+$	$\frac{1}{2}$	± 1	0	105.7	2.197×10^{-6}	$e + \nu + \overline{\nu}$
	Electron's neutrino	ν_e	$\overline{\nu}_e$	$\frac{1}{2}$	0	0	0	Stable	—
	Muon's neutrino	ν_μ	$\overline{\nu}_\mu$	$\frac{1}{2}$	0	0	0	Stable	—
H A D R O N S — M E S O N S	Pion	π^+ π^0	$\overline{\pi}^+$ π^0	0 0	± 1 0	0 0	139.6 135.0	2.603×10^{-8} 8.28×10^{-17}	$\mu + \nu$ $\gamma + \gamma$
	K-meson	K^+	$\overline{K}^+$	0	± 1	± 1	493.7	1.237×10^{-8}	$\mu + \nu$
		K^0	$\overline{K}^0$	0	0	± 1	497.7	$\begin{cases} 8.930 \times 10^{-11} \\ 5.181 \times 10^{-8} \end{cases}$	$\pi^+ + \pi^-$ $\pi^0 + \pi^0 + \pi^0$
	Eta-meson	η^0	η^0	0	0	0	548.8	?	$\gamma + \gamma$
(BARYONS) NUCLEON	Proton	p	$\overline{p}$	$\frac{1}{2}$	± 1	0	938.3	Stable	—
	Neutron	n	$\overline{n}$	$\frac{1}{2}$	0	0	939.6	918	$p + e^- + \nu$
B A R Y O N S	Lambda particle	Λ^0	$\overline{\Lambda}^0$	$\frac{1}{2}$	0	∓ 1	1116	2.578×10^{-10}	$p + \pi^-$
	Sigma particle	Σ^+	$\overline{\Sigma}^+$	$\frac{1}{2}$	$+1$	∓ 1	1189	8.00×10^{-11}	$p + \pi^0$
		Σ^0	$\overline{\Sigma}^0$	$\frac{1}{2}$	0	∓ 1	1192	$< 1.0 \times 10^{-14}$	$\Lambda^0 + \gamma$
		Σ^-	$\overline{\Sigma}^-$	$\frac{1}{2}$	-1	∓ 1	1197	1.482×10^{-10}	$n + \pi^-$
	Xi particle	Ξ^0	$\overline{\Xi}^0$	$\frac{1}{2}$	0	∓ 2	1315	2.96×10^{-10}	$\Lambda^0 + \pi^0$
		Ξ^-	$\overline{\Xi}^-$	$\frac{1}{2}$	∓ 1	∓ 2	1321	1.652×10^{-10}	$\Lambda^0 + \pi^-$
	Omega particle	Ω^-	$\overline{\Omega}^-$	$\frac{3}{2}$	∓ 1	∓ 3	1672	1.3×10^{-10}	$\Xi^0 + \pi^-$

* See (1) "Review of Particle Properties," *Reviews of Modern Physics*, vol. 48, no. 2, Part II, April (1976).
(2) "Quarks with Color and Flavor," by Sheldon Lee Glashow, *Scientific American*, October (1975).
(3) "The New Elementary Particles and Charm," by Lewis Ryder, *Physics Education*, January (1976) for fuller information. Particle physics is one of the sharp cutting edges of contemporary physics.

Conversion factors may be read off directly from the tables. For example, 1 degree $= 2.778 \times 10^{-3}$ revolutions, so $16.7° = 16.7 \times 2.778 \times 10^{-3}$ rev. The SI quantities are capitalized. The prefix "ab" refers to electromagnetic units (emu); "stat" refers to electrostatic units (esu). Adapted in part from G. Shortley and D. Williams, *Elements of Physics*, Prentice-Hall, Englewood Cliffs, N.J., 1965.

Plane angle

	°	'	"	RADIAN	rev
1 degree =	1	60	3600	1.745×10^{-2}	2.778×10^{-3}
1 minute =	1.667×10^{-2}	1	60	2.909×10^{-4}	4.630×10^{-5}
1 second =	2.778×10^{-4}	1.667×10^{-2}	1	4.848×10^{-6}	7.716×10^{-7}
1 RADIAN =	57.30	3438	2.063×10^5	1	0.1592
1 revolution =	360	2.16×10^4	1.296×10^6	6.283	1

Solid angle

1 sphere $= 4\pi$ steradians $= 12.57$ steradians

Length

	cm	METER	km	in.	ft	mi
1 centimeter =	1	10^{-2}	10^{-5}	0.3937	3.281×10^{-2}	6.214×10^{-6}
1 METER =	100	1	10^{-3}	39.3	3.281	6.214×10^{-4}
1 kilometer =	10^5	1000	1	3.937×10^4	3281	0.6214
1 inch =	2.540	2.540×10^{-2}	2.540×10^{-5}	1	8.333×10^{-2}	1.578×10^{-5}
1 foot =	30.48	0.3048	3.048×10^{-4}	12	1	1.894×10^{-4}
1 mile =	1.609×10^5	1609	1.609	6.336×10^4	5280	1

1 angstrom $= 10^{-10}$ m   1 light-year $= 9.4600 \times 10^{12}$ km   1 yard $= 3$ ft
1 nautical mile $= 1852$ m   1 parsec $= 3.084 \times 10^{13}$ km   1 rod $= 16.5$ ft
  $= 1.151$ miles $= 6076$ ft   1 fathom $= 6$ ft   1 mil $= 10^{-3}$ in.

Area

	METER²	cm²	ft²	in.²	circ mil
1 SQUARE METER =	1	10^4	10.76	1550	1.974×10^9
1 square centimeter =	10^{-4}	1	1.076×10^{-3}	0.1550	1.974×10^5
1 square foot =	9.290×10^{-2}	929.0	1	144	1.833×10^8
1 square inch =	6.452×10^{-4}	6.452	6.944×10^{-3}	1	1.273×10^6
1 circular mil =	5.067×10^{-10}	5.067×10^{-6}	5.454×10^{-9}	7.854×10^{-7}	1

1 square mile $= 2.788 \times 10^7$ ft² $= 640$ acres   1 acre $= 43,560$ ft²
1 barn $= 10^{-28}$ m²

Volume

	METER³	cm³	li	ft³	in.³
1 CUBIC METER =	1	10^6	1000	35.31	6.102×10^4
1 cubic centimeter =	10^{-6}	1	1.000×10^{-3}	3.531×10^{-5}	6.102×10^{-2}
1 liter =	1.000×10^{-3}	1000	1	3.531×10^{-2}	61.02
1 cubic foot =	2.832×10^{-2}	2.832×10^4	28.32	1	1728
1 cubic inch =	1.639×10^{-5}	16.39	1.639×10^{-2}	5.787×10^{-4}	1

1 U. S. fluid gallon = 4 U. S. fluid quarts = 8 U. S. pints = 128 U. S. fluid ounces = 231 in.³
1 British imperial gallon = 277.4 in.³   1 liter = 10^{-3} m³.

Mass

Quantities in the shaded areas are not mass units but are often used as such. When we write, for example, 1 kg "=" 2.205 lb this means that a kilogram is a *mass* that *weighs* 2.205 pounds under standard condition of gravity (g = 9.80665 m/s²).

	gm	KG	slug	u	oz	lb	ton
1 gram =	1	0.001	6.852×10^{-5}	6.024×10^{23}	3.527×10^{-2}	2.205×10^{-3}	1.102×10^{-6}
1 KILOGRAM =	1000	1	6.852×10^{-2}	6.024×10^{26}	35.27	2.205	1.102×10^{-3}
1 slug =	1.459×10^4	14.59	1	8.789×10^{27}	514.8	32.17	1.609×10^{-2}
1 u =	1.660×10^{-24}	1.660×10^{-27}	1.137×10^{-28}	1	5.855×10^{-26}	3.660×10^{-27}	1.829×10^{-30}
1 ounce =	28.35	2.835×10^{-2}	1.943×10^{-3}	1.708×10^{25}	1	6.250×10^{-2}	3.125×10^{-5}
1 pound =	453.6	0.4536	3.108×10^{-2}	2.732×10^{26}	16	1	0.0005
1 ton =	9.072×10^5	907.2	62.16	5.465×10^{29}	3.2×10^4	2000	1

Density

Quantities in the shaded areas are weight densities and, as such, are dimensionally different from mass densities. See note for mass table.

	slug/ft³	KG/METER³	g/cm³	lb/ft³	lb/in.³
1 slug per ft³ =	1	515.4	0.5154	32.17	1.862×10^{-2}
1 KILOGRAM per METER³ =	1.940×10^{-3}	1	0.001	6.243×10^{-2}	3.613×10^{-5}
1 gram per cm³ =	1.940	1000	1	62.43	3.613×10^{-2}
1 pound per ft³ =	3.108×10^{-2}	16.02	1.602×10^{-2}	1	5.787×10^{-4}
1 pound per in.³ =	53.71	2.768×10^4	27.68	1728	1

Time

	yr	d	h	min	SECOND
1 year =	1	365.2	8.766×10^3	5.259×10^5	3.156×10^7
1 day =	2.738×10^{-3}	1	24	1440	8.640×10^4
1 hour =	1.141×10^{-4}	4.167×10^{-2}	1	60	3600
1 minute =	1.901×10^{-6}	6.944×10^{-4}	1.667×10^{-2}	1	60
1 SECOND =	3.169×10^{-8}	1.157×10^{-5}	2.778×10^{-4}	1.667×10^{-2}	1

Speed

	ft/s	km/h	METER/SECOND	mi/h	cm/s	knot
1 foot per second =	1	1.097	0.3048	0.6818	30.48	0.5925
1 kilometer per hour =	0.9113	1	0.2778	0.6214	27.78	0.5400
1 METER per SECOND =	3.281	3.6	1	2.237	100	1.944
1 mile per hour =	1.467	1.609	0.4470	1	44.70	0.8689
1 centimeter per second =	3.281×10^{-2}	3.6×10^{-2}	0.01	2.237×10^{-2}	1	1.944×10^{-2}
1 knot =	1.688	1.852	0.5144	1.151	51.44	1

1 knot = 1 nautical mi/h   1 mi/min = 88.00 ft/s = 60.00 mi/h

Force

Quantities in the shaded areas are not force units but are often used as such. For instance, if we write 1 gram-force "=" 980.7 dynes, we mean that a gram-*mass* experiences a *force* of 980.7 dynes under standard conditions of gravity ($g = 9.80665$ m/s²).

	dyne	NEWTON	lb	pdl	gf	kgf
1 dyne =	1	10^{-5}	2.248×10^{-6}	7.233×10^{-5}	1.020×10^{-3}	1.020×10^{-6}
1 NEWTON =	10^5	1	0.2248	7.233	102.0	0.1020
1 pound =	4.448×10^5	4.448	1	32.17	453.6	0.4536
1 poundal =	1.383×10^4	0.1383	3.108×10^{-2}	1	14.10	1.410×10^{-2}
1 gram-force =	980.7	9.807×10^{-3}	2.205×10^{-3}	7.093×10^{-2}	1	0.001
1 kilogram-force =	9.807×10^5	9.807	2.205	70.93	1000	1

Pressure

	atm	dyne/cm²	inch of water	cm-Hg	PASCAL	lb/in.²	lb/ft²
1 atmosphere =	1	1.013×10^6	406.8	76	1.013×10^5	14.70	2116
1 dyne per cm² =	9.869×10^{-7}	1	4.015×10^{-4}	7.501×10^{-5}	0.1	1.450×10^{-5}	2.089×10^{-3}
1 inch of water[a] at 4° C =	2.458×10^{-3}	2491	1	0.1868	249.1	3.613×10^{-2}	5.202
1 centimeter of mercury[a] at 0° C =	1.316×10^{-2}	1.333×10^4	5.353	1	1333	0.1934	27.85
1 PASCAL =	9.869×10^{-6}	10	4.015×10^{-3}	7.501×10^{-4}	1	1.450×10^{-4}	2.089×10^{-2}
1 pound per in.² =	6.805×10^{-2}	6.895×10^4	27.68	5.171	6.895×10^3	1	144
1 pound per ft² =	4.725×10^{-4}	478.8	0.1922	3.591×10^{-2}	47.88	6.944×10^{-3}	1

[a] Where the acceleration of gravity has the standard value 9.80665 m/s².

1 bar = 10^6 dyne/cm² = 0.1 MPa   1 millibar = 10^3 dyne/cm² = 10^2 Pa

Energy, work, heat

Quantities in the shaded areas are not properly energy units but are included for convenience. They arise from the relativistic mass-energy equivalence formula $E = mc^2$ and represent the energy released if a kilogram or unified atomic mass unit (u) is completely converted to energy.

	Btu	erg	ft·lb	hp·h	JOULE	cal	kW·h	eV	MeV	kg	u
1 British thermal unit =	1	1.055×10^{10}	777.9	3.929×10^{-4}	1055	252.0	2.930×10^{-4}	6.585×10^{21}	6.585×10^{15}	1.174×10^{-14}	7.074×10^{12}
1 erg =	9.481×10^{-11}	1	7.376×10^{-8}	3.725×10^{-14}	10^{-7}	2.389×10^{-8}	2.778×10^{-14}	6.242×10^{11}	6.242×10^{5}	1.113×10^{-24}	670.5
1 foot-pound =	1.285×10^{-3}	1.356×10^{7}	1	5.051×10^{-7}	1.356	0.3239	3.766×10^{-7}	8.464×10^{18}	8.464×10^{12}	1.509×10^{-17}	9.092×10^{9}
1 horsepower-hour =	2545	2.685×10^{13}	1.980×10^{6}	1	2.685×10^{6}	6.414×10^{5}	0.7457	1.676×10^{25}	1.676×10^{19}	2.988×10^{-11}	1.800×10^{16}
1 JOULE =	9.481×10^{-4}	10^{7}	0.7376	3.725×10^{-7}	1	0.2389	2.778×10^{-7}	6.242×10^{18}	6.242×10^{12}	1.113×10^{-17}	6.705×10^{9}
1 calorie =	3.968×10^{-3}	4.186×10^{7}	3.087	1.559×10^{-6}	4.186	1	1.163×10^{-6}	2.613×10^{19}	2.613×10^{13}	4.659×10^{-17}	2.807×10^{10}
1 kilowatt-hour =	3413	3.6×10^{13}	2.655×10^{6}	1.341	3.6×10^{6}	8.601×10^{5}	1	2.247×10^{25}	2.247×10^{19}	4.007×10^{-11}	2.414×10^{16}
1 electron volt =	1.519×10^{-22}	1.602×10^{-12}	1.182×10^{-19}	5.967×10^{-26}	1.602×10^{-19}	3.827×10^{-20}	4.450×10^{-26}	1	10^{-6}	1.783×10^{-36}	1.074×10^{-9}
1 million electron volts =	1.519×10^{-16}	1.602×10^{-6}	1.182×10^{-13}	5.967×10^{-20}	1.602×10^{-13}	3.827×10^{-14}	4.450×10^{-20}	10^{6}	1	1.783×10^{-30}	1.074×10^{-3}
1 kilogram =	8.521×10^{13}	8.987×10^{23}	6.629×10^{16}	3.348×10^{10}	8.987×10^{16}	2.147×10^{16}	2.497×10^{10}	5.610×10^{35}	5.610×10^{29}	1	6.025×10^{26}
1 unified atomic mass unit =	1.415×10^{-13}	1.492×10^{-3}	1.100×10^{-10}	5.558×10^{-17}	1.492×10^{-10}	3.564×10^{-11}	4.145×10^{-17}	9.31×10^{8}	931.0	1.660×10^{-27}	1

Power

	Btu/h	ft·lb/s	hp	cal/s	kW	WATT
1 British thermal unit per hour =	1	0.2161	3.929×10^{-4}	7.000×10^{-2}	2.930×10^{-4}	0.2930
1 foot-pound per second =	4.628	1	1.818×10^{-3}	0.3239	1.356×10^{-3}	1.356
1 horsepower =	2545	550	1	178.2	0.7457	745.7
1 calorie per second =	14.29	3.087	5.613×10^{-3}	1	4.186×10^{-3}	4.186
1 kilowatt =	3413	737.6	1.341	238.9	1	1000
1 WATT =	3.413	0.7376	1.341×10^{-3}	0.2389	0.001	1

Charge

	abcoul	A·h	COULOMB	statcoul
1 abcoulomb =	1	2.778×10^{-3}	10	2.998×10^{10}
1 ampere-hour =	360	1	3600	1.079×10^{13}
1 COULOMB =	0.1	2.778×10^{-4}	1	2.998×10^{9}
1 statcoulomb =	3.336×10^{-11}	9.266×10^{-14}	3.336×10^{-10}	1

1 electronic charge = 1.602×10^{-19} coulomb

Current

	abamp	AMPERE	statamp
1 abampere =	1	10	2.998×10^{10}
1 AMPERE =	0.1	1	2.998×10^{9}
1 statampere =	3.336×10^{-11}	3.336×10^{-10}	1

Potential, electromotive force

	abvolt	VOLT	statvolt
1 abvolt =	1	10^{-8}	3.336×10^{-11}
1 VOLT =	10^{8}	1	3.336×10^{-3}
1 statvolt =	2.998×10^{10}	299.8	1

Resistance

	abohm	OHM	statohm
1 abohm =	1	10^{-9}	1.113×10^{-21}
1 OHM =	10^{9}	1	1.113×10^{-12}
1 statohm =	8.987×10^{20}	8.987×10^{11}	1

Capacitance

	abf	FARAD	μF	statf
1 abfarad =	1	10^{9}	10^{15}	8.987×10^{20}
1 FARAD =	10^{-9}	1	10^{6}	8.987×10^{11}
1 microfarad =	10^{-15}	10^{-6}	1	8.987×10^{5}
1 statfarad =	1.113×10^{-21}	1.113×10^{-12}	1.113×10^{-6}	1

Inductance

	abhenry	HENRY	μH	mH	stathenry
1 abhenry =	1	10^{-9}	0.001	10^{-6}	1.113×10^{-21}
1 HENRY =	10^{9}	1	10^{6}	1000	1.113×10^{-12}
1 microhenry =	1000	10^{-6}	1	0.001	1.113×10^{-18}
1 millihenry =	10^{6}	0.001	1000	1	1.113×10^{-15}
1 stathenry =	8.987×10^{20}	8.987×10^{11}	8.987×10^{17}	8.987×10^{14}	1

Magnetic flux

	maxwell	WEBER
1 maxwell =	1	10^{-8}
1 WEBER =	10^{8}	1

Magnetic field

	gauss	TESLA	milligauss
1 gauss =	1	10^{-4}	1000
1 TESLA =	10^{4}	1	10^{7}
1 milligauss =	0.001	10^{-7}	1

1 tesla = 1 weber/meter^2

Mathematical Signs and Symbols

$=$ equals
$\cong$ equals approximately
$\neq$ is not equal to
$\equiv$ is identical to, is defined as
$>$ is greater than ($\gg$ is much greater than)
$<$ is less than ($\ll$ is much less than)
$\geq$ is more than or equal to (or, is no less than)
$\leq$ is less than or equal to (or, is no more than)
$\pm$ plus or minus ($\sqrt{4} = \pm 2$)
$\propto$ is proportional to (Hooke's law: $F \propto x$, or $F = -kx$)
Σ the sum of
$\bar{x}$ the average value of x

The Greek Alphabet

Alpha	A	α	Nu	N	ν
Beta	B	β	Xi	Ξ	ξ
Gamma	Γ	γ	Omicron	O	o
Delta	Δ	δ	Pi	Π	π
Epsilon	E	ϵ	Rho	P	ρ
Zeta	Z	ζ	Sigma	Σ	σ
Eta	H	η	Tau	T	τ
Theta	Θ	θ	Upsilon	Y	υ
Iota	I	ι	Phi	Φ	ϕ, φ
Kappa	K	κ	Chi	X	χ
Lambda	Λ	λ	Psi	Ψ	ψ
Mu	M	μ	Omega	Ω	ω

Geometry

Circle of radius r: circumference $= 2\pi r$; area $= \pi r^2$.
Sphere of radius r: area $= 4\pi r^2$; volume $= \frac{4}{3}\pi r^3$.
Right circular cylinder of radius r and height h: area $= 2\pi r^2 + 2\pi rh$;
  volume $= \pi r^2 h$.

Quadratic Formula

If $ax^2 + bx + c = 0$, then $x = \dfrac{-b \pm \sqrt{b^2 - 4ac}}{2a}$.

Trigonometric Functions of Angle θ

$$\sin \theta = \frac{y}{r} \qquad \cos \theta = \frac{x}{r}$$

$$\tan \theta = \frac{y}{x} \qquad \cot \theta = \frac{x}{y}$$

$$\sec \theta = \frac{r}{x} \qquad \csc \theta = \frac{r}{y}$$

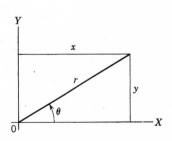

Pythagorean Theorem

$$x^2 + y^2 = r^2$$

APPENDIX H
MATHEMATICAL SYMBOLS AND THE GREEK ALPHABET

APPENDIX I
MATHEMATICAL FORMULAS

Trigonometric Identities

$$\sin^2 \theta + \cos^2 \theta = 1 \qquad \sec^2 \theta - \tan^2 \theta = 1 \qquad \csc^2 \theta - \cot^2 \theta = 1$$

$$\sin 2\theta = 2 \sin \theta \cos \theta$$

$$\cos 2\theta = \cos^2 \theta - \sin^2 \theta = 2 \cos^2 \theta - 1 = 1 - 2 \sin^2 \theta$$

$$\sin \theta = \frac{e^{i\theta} - e^{-i\theta}}{2i} \qquad \cos \theta = \frac{e^{i\theta} + e^{-i\theta}}{2}$$

$$e^{\pm i\theta} = \cos \theta \pm i \sin \theta$$

$$\sin (\alpha \pm \beta) = \sin \alpha \cos \beta \pm \cos \alpha \sin \beta$$

$$\cos (\alpha \pm \beta) = \cos \alpha \cos \beta \mp \sin \alpha \sin \beta$$

$$\tan (\alpha \pm \beta) = \frac{\tan \alpha \pm \tan \beta}{1 \mp \tan \alpha \tan \beta}$$

$$\sin \alpha \pm \sin \beta = 2 \sin \tfrac{1}{2}(\alpha \pm \beta) \cos \tfrac{1}{2}(\alpha \mp \beta)$$

Taylor's Series

$$f(x_0 + x) = f(x_0) + f'(x_0)x + f''(x_0) \frac{x^2}{2!} + f'''(x_0) \frac{x^3}{3!} + \cdots$$

Binomial Expansion

$$(1 + x)^n = 1 + \frac{nx}{1!} + \frac{n(n-1)}{2!} x^2 + \cdots$$

Exponential Expansion

$$e^x = 1 + x + \frac{x^2}{2!} + \frac{x^3}{3!} + \cdots$$

Logarithmic Expansion

$$\ln(1 + x) = x - \tfrac{1}{2}x^2 + \tfrac{1}{3}x^3 - \cdots$$

Trigonometric Expansions (θ in radians)

$$\sin \theta = \theta - \frac{\theta^3}{3!} + \frac{\theta^5}{5!} - \cdots$$

$$\cos \theta = 1 - \frac{\theta^2}{2!} + \frac{\theta^4}{4!} - \cdots$$

Derivatives and Indefinite Integrals

In what follows, the letters u and v stand for any functions of x, and a and m are constants. To each of the integrals should be added an arbitrary constant of integration. The *Handbook of Chemistry and Physics* (Chemical Rubber Publishing Co.) gives a more extensive tabulation.

1. $\dfrac{dx}{dx} = 1$

1. $\int dx = x$

2. $\dfrac{d}{dx}(au) = a \dfrac{du}{dx}$

2. $\int au\ dx = a \int u\ dx$

3. $\dfrac{d}{dx}(u + v) = \dfrac{du}{dx} + \dfrac{dv}{dx}$

3. $\int (u + v)\ dx = \int u\ dx + \int v\ dx$

4. $\dfrac{d}{dx} x^m = m x^{m-1}$

4. $\int x^m\ dx = \dfrac{x^{m+1}}{m + 1} \qquad (m \neq -1)$

5. $\dfrac{d}{dx} \ln x = \dfrac{1}{x}$

5. $\displaystyle\int \dfrac{dx}{x} = \ln |x|$

6. $\dfrac{d}{dx} (uv) = u\dfrac{dv}{dx} + v\dfrac{du}{dx}$

6. $\displaystyle\int u\dfrac{dv}{dx}\,dx = uv - \int v\dfrac{du}{dx}\,dx$

7. $\dfrac{d}{dx} e^x = e^x$

7. $\displaystyle\int e^x\,dx = e^x$

8. $\dfrac{d}{dx} \sin x = \cos x$

8. $\displaystyle\int \sin x\,dx = -\cos x$

9. $\dfrac{d}{dx} \cos x = -\sin x$

9. $\displaystyle\int \cos x\,dx = \sin x$

10. $\dfrac{d}{dx} \tan x = \sec^2 x$

10. $\displaystyle\int \tan x\,dx = \ln |\sec x|$

11. $\dfrac{d}{dx} \cot x = -\csc^2 x$

11. $\displaystyle\int \cot x\,dx = \ln |\sin x|$

12. $\dfrac{d}{dx} \sec x = \tan x \sec x$

12. $\displaystyle\int \sec x\,dx = \ln |\sec x + \tan x|$

13. $\dfrac{d}{dx} \csc x = -\cot x \csc x$

13. $\displaystyle\int \csc x\,dx = \ln |\csc x - \cot x|$

14. $\dfrac{d}{dx} \arctan x = \dfrac{1}{1 + x^2}$

14. $\displaystyle\int \dfrac{dx}{1 + x^2} = \arctan x$

15. $\dfrac{d}{dx} \arcsin x = \dfrac{1}{\sqrt{1 - x^2}}$

15. $\displaystyle\int \dfrac{dx}{\sqrt{1 - x^2}} = \arcsin x$

16. $\dfrac{d}{dx} \operatorname{arcsec} x = \dfrac{1}{x\sqrt{x^2 - 1}}$

16. $\displaystyle\int \dfrac{dx}{x\sqrt{x^2 - 1}} = \operatorname{arcsec} x$

Vector Products

Let $\mathbf{i}, \mathbf{j}, \mathbf{k}$ be unit vectors in the x, y, z directions. Then

$$\mathbf{i} \cdot \mathbf{i} = \mathbf{j} \cdot \mathbf{j} = \mathbf{k} \cdot \mathbf{k} = 1, \qquad \mathbf{i} \cdot \mathbf{j} = \mathbf{j} \cdot \mathbf{k} = \mathbf{k} \cdot \mathbf{i} = 0,$$

$$\mathbf{i} \times \mathbf{i} = \mathbf{j} \times \mathbf{j} = \mathbf{k} \times \mathbf{k} = 0,$$

$$\mathbf{i} \times \mathbf{j} = \mathbf{k}, \qquad \mathbf{j} \times \mathbf{k} = \mathbf{i}, \qquad \mathbf{k} \times \mathbf{i} = \mathbf{j}.$$

Any vector $\mathbf{a}$ with components a_x, a_y, a_z along the x, y, z axes can be written

$$\mathbf{a} = a_x\mathbf{i} + a_y\mathbf{j} + a_z\mathbf{k}.$$

Let $\mathbf{a}, \mathbf{b}, \mathbf{c}$ be arbitrary vectors with magnitudes a, b, c. Then

$$\mathbf{a} \times (\mathbf{b} + \mathbf{c}) = \mathbf{a} \times \mathbf{b} + \mathbf{a} \times \mathbf{c}$$

$$(s\mathbf{a}) \times \mathbf{b} = \mathbf{a} \times (s\mathbf{b}) = s(\mathbf{a} \times \mathbf{b}) \qquad (s = \text{a scalar}).$$

Let θ be the smaller of the two angles between $\mathbf{a}$ and $\mathbf{b}$. Then

$$\mathbf{a} \cdot \mathbf{b} = \mathbf{b} \cdot \mathbf{a} = a_xb_x + a_yb_y + a_zb_z = ab \cos \theta$$

$$\mathbf{a} \times \mathbf{b} = -\mathbf{b} \times \mathbf{a} = \begin{vmatrix} \mathbf{i} & \mathbf{j} & \mathbf{k} \\ a_x & a_y & a_z \\ b_x & b_y & b_z \end{vmatrix} = (a_yb_z - b_ya_z)\mathbf{i} + (a_zb_x - b_za_x)\mathbf{j} + (a_xb_y - b_xa_y)\mathbf{k}$$

$$|\mathbf{a} \times \mathbf{b}| = ab \sin \theta$$

$$\mathbf{a} \cdot (\mathbf{b} \times \mathbf{c}) = \mathbf{b} \cdot (\mathbf{c} \times \mathbf{a}) = \mathbf{c} \cdot (\mathbf{a} \times \mathbf{b})$$

$$\mathbf{a} \times (\mathbf{b} \times \mathbf{c}) = (\mathbf{a} \cdot \mathbf{c})\mathbf{b} - (\mathbf{a} \cdot \mathbf{b})\mathbf{c}$$

Degrees	Radians	Sine	Tangent	Cotangent	Cosine		
0	0	0	0	∞	1.0000	1.5708	**90**
1	.0175	.0175	.0175	57.290	.9998	1.5533	89
2	.0349	.0349	.0349	28.636	.9994	1.5359	88
3	.0524	.0523	.0524	19.081	.9986	1.5184	87
4	.0698	.0698	.0699	14.301	.9976	1.5010	86
5	.0873	.0872	.0875	11.430	.9962	1.4835	**85**
6	.1047	.1045	.1051	9.5144	.9945	1.4661	84
7	.1222	.1219	.1228	8.1443	.9925	1.4486	83
8	.1396	.1392	.1405	7.1154	.9877	1.4312	82
9	.1571	.1564	.1584	6.3138	.9877	1.4137	81
10	.1745	.1736	.1763	5.6713	.9848	1.3963	**80**
11	.1920	.1908	.1944	5.1446	.9816	1.3788	79
12	.2094	.2079	.2126	4.7046	.9781	1.3614	78
13	.2269	.2250	.2309	4.3315	.9744	1.3439	77
14	.2443	.2419	.2493	4.0108	.9703	1.3265	76
15	.2618	.2588	.2679	3.7321	.9659	1.3090	**75**
16	.2793	.2756	.2867	3.4874	.9613	1.2915	74
17	.2967	.2924	.3057	3.2709	.9563	1.2741	73
18	.3142	.3090	.3249	3.0777	.9511	1.2566	72
19	.3316	.3256	.3443	2.9042	.9455	1.2392	71
20	.3491	.3420	.3640	2.7475	.9397	1.2217	**70**
21	.3665	.3584	.3839	2.6051	.9336	1.2043	69
22	.3840	.3746	.4040	2.4751	.9272	1.1868	68
23	.4014	.3907	.4245	2.3559	.9205	1.1694	67
24	.4189	.4067	.4452	2.2460	.9135	1.1519	66
25	.4363	.4226	.4663	2.1445	.9063	1.1345	**65**
26	.4538	.4384	.4877	2.0503	.8988	1.1170	64
27	.4712	.4540	.5095	1.9626	.8910	1.0996	63
28	.4887	.4695	.5317	1.8807	.8829	1.0821	62
29	.5061	.4848	.5543	1.8040	.8746	1.0647	61
30	.5236	.5000	.5774	1.7321	.8660	1.0472	**60**
31	.5411	.5150	.6009	1.6643	.8572	1.0297	59
32	.5585	.5299	.6249	1.6003	.8480	1.0123	58
33	.5760	.5446	.6494	1.5399	.8387	.9948	57
34	.5934	.5592	.6745	1.4826	.8290	.9774	56
35	.6109	.5736	.7002	1.4281	.8192	.9599	**55**
36	.6283	.5878	.7265	1.3764	.8090	.9425	54
37	.6458	.6018	.7536	1.3270	.7986	.9250	53
38	.6632	.6157	.7813	1.2799	.7880	.9076	52
39	.6807	.6293	.8098	1.2349	.7771	.8901	51
40	.6981	.6428	.8391	1.1918	.7660	.8727	**50**
41	.7156	.6561	.8693	1.1504	.7547	.8552	49
42	.7330	.6691	.9004	1.1106	.7431	.8378	48
43	.7505	.6820	.9325	1.0724	.7314	.8203	47
44	.7679	.6947	.9657	1.0355	.7193	.8029	46
45	.7854	.7071	1.0000	1.0000	.7071	.7854	**45**
		Cosine	Cotangent	Tangent	Sine	Radians	Degrees

1901	Wilhelm Konrad Röntgen	1845–1923	for the discovery of the remarkable rays subsequently named after him
1902	Hendrik Antoon Lorentz Pieter Zeeman	1853–1928 1865–1943	for their researches into the influence of magnetism upon radiation phenomena
1903	Antoine Henri Becquerel	1852–1908	for his discovery of spontaneous radioactivity

* See *Nobel Lectures, Physics,* 1901–1970, Elsevier Publishing Company, for the Nobel presentations, lectures and biographies. The attributions are, almost without exception, quotations from the Nobel Prize citations.

	Pierre Curie Marie Sklowdowska-Curie	1859–1906 1867–1934	for their joint researches on the radiation phenomena discovered by Professor Henri Becquerel
1904	Lord Rayleigh (John William Strutt)	1842–1919	for his investigations of the densities of the most important gases and for his discovery of argon
1905	Philipp Eduard Anton von Lenard	1862–1947	for his work on cathode rays
1906	Joseph John Thomson	1856–1940	for his theoretical and experimental investigations on the conduction of electricity by gases
1907	Albert Abraham Michelson	1852–1931	for his optical precision instruments and metrological investigations carried out with their aid
1908	Gabriel Lippmann	1845–1921	for his method of reproducing colors photographically based on the phenomena of interference
1909	Guglielmo Marconi Carl Ferdinand Braun	1874–1937 1850–1918	for their contributions to the development of wireless telegraphy
1910	Johannes Diderik van der Waals	1837–1923	for his work on the equation of state for gases and liquids
1911	Wilhelm Wien	1864–1928	for his discoveries regarding the laws governing the radiation of heat
1912	Nils Gustaf Dalén	1869–1937	for his invention of automatic regulators for use in conjunction with gas accumulators for illuminating lighthouses and buoys
1913	Heike Kamerlingh Onnes	1853–1926	for his investigations of the properties of matter at low temperatures which led, *inter alia*, to the production of liquid helium
1914	Max von Laue	1879–1960	for his discovery of the diffraction of Röntgen rays by crystals
1915	William Henry Bragg William Lawrence Bragg	1862–1942 1890–1971	for their services in the analysis of crystal structure by means of Röntgen rays
1917	Charles Glover Barkla	1877–1944	for his discovery of the characteristic Röntgen radiation of the elements
1918	Max Planck	1858–1947	for his discovery of energy quanta
1919	Johannes Stark	1874–1957	for his discovery of the Doppler effect in canal rays and the splitting of spectral lines in electric fields
1920	Charles-Edouard Guillaume	1861–1938	for the service he has rendered to precision measurements in Physics by his discovery of anomalies in nickel steel alloys
1921	Albert Einstein	1879–1955	for his services to Theoretical Physics, and especially for his discovery of the law of the photoelectric effect
1922	Niels Bohr	1885–1962	for the investigation of the structure of atoms, and of the radiation emanating from them
1923	Robert Andrews Millikan	1868–1953	for his work on the elementary charge of electricity and on the photoelectric effect
1924	Karl Manne Georg Siegbahn	1886–1954	for his discoveries and research in the field of x-ray spectroscopy
1925	James Franck Gustav Hertz	1882–1964 1887–1975	for their discovery of the laws governing the impact of an electron upon an atom

1926	Jean Baptiste Perrin	1870–1942	for his work on the discontinuous structure of matter, and especially for his discovery of sedimentation equillibrium
1927	Arthur Holly Compton	1892–1962	for his discovery of the effect named after him
	Charles Thomson Rees Wilson	1869–1959	for his method of making the paths of electrically charged particles visible by condensation of vapor
1928	Owen Willans Richardson	1879–1959	for his work on the thermionic phenomenon and especially for the discovery of the law named after him
1929	Prince Louis-Victor de Broglie	1892–	for his discovery of the wave nature of electrons
1930	Sir Chandrasekhara Venkata Raman	1888–1970	for his work on the scattering of light and for the discovery of the effect named after him
1932	Werner Heisenberg	1901–1976	for the creation of quantum mechanics, the application of which has, among other things, led to the discovery of the allotropic forms of hydrogen
1933	Erwin Schrödinger Paul Adrien Maurice Dirac	1887–1961 1902–1984	for the discovery of new productive forms of atomic theory
1935	James Chadwick	1891–1974	for his discovery of the neutron
1936	Victor Franz Hess	1883–1964	for his discovery of cosmic radiation
	Carl David Anderson	1905–1984	for his discovery of the positron
1937	Clinton Joseph Davisson George Paget Thomson	1881–1958 1892–1975	for their experimental discovery of the diffraction of electrons by crystals
1938	Enrico Fermi	1901–1954	for his demonstrations of the existence of new radioactive elements produced by neutron irradiation, and for his related discovery of nuclear reactions brought about by slow neutrons
1939	Ernest Orlando Lawrence	1901–1958	for the invention and development of the cyclotron and for results obtained with it, especially with regard to artificial radioactive elements
1943	Otto Stern	1888–1969	for his contribution to the development of the molecular ray method and his discovery of the magnetic moment of the proton
1944	Isidor Isaac Rabi	1898–	for his resonance method for recording the magnetic properties of atomic nuclei
1945	Wolfgang Pauli	1900–1958	for the discovery of the Exclusion Principle, also called the Pauli Principle
1946	Percy Williams Bridgman	1882–1961	for the invention of an apparatus to produce extremely high pressures, and for the discoveries he made therewith in the field of high-pressure physics
1947	Sir Edward Victor Appleton	1892–1965	for his investigations of the physics of the upper atmosphere, especially for the discovery of the so-called Appleton layer
1948	Patrick Maynard Stuart Blackett	1897–1974	for his development of the Wilson cloud chamber method, and his discoveries therewith in the fields of nuclear physics and cosmic radiation

1949	Hideki Yukawa	1907–1981	for his prediction of the existence of mesons on the basis of theoretical work on nuclear forces
1950	Cecil Frank Powell	1903–1969	for his development of the photographic method of studying nuclear processes and his discoveries regarding mesons made with this method
1951	Sir John Douglas Cockcroft Ernest Thomas Sinton Walton	1897–1967 1903–	for their pioneer work on the transmutation of atomic nuclei by artificially accelerated atomic particles
1952	Felix Bloch Edward Mills Purcell	1905–1983 1912–	for their development of new methods for nuclear magnetic precision methods and discoveries in connection therewith
1953	Frits Zernike	1888–1966	for his demonstration of the phase-contrast method, especially for his invention of the phase-contrast microscope
1954	Max Born	1882–1970	for his fundamental research in quantum mechanics, especially for his statistical interpretation of the wave function
	Walther Bothe	1891–1957	for the coincidence method and his discoveries made therewith
1955	Willis Eugene Lamb	1913–	for his discoveries concerning the fine structure of the hydrogen spectrum
	Polykarp Kusch	1911–	for his precision determination of the magnetic moment of the electron
1956	William Shockley John Bardeen Walter Houser Brattain	1910– 1908– 1902–	for their researches on semiconductors and their discovery of the transistor effect
1957	Chen Ning Yang Tsung Dao Lee	1922– 1926–	for their penetrating investigation of the so-called parity laws which has led important discoveries regarding the elementary particles
1958	Pavel Aleksejevič Čerenkov Il' ja Michajlovič Frank Igor' Evgen'evič Tamm	1904– 1908– 1895–1971	for the discovery and the interpretation of the Čerenkov effect
1959	Emilio Gino Segrè Owen Chamberlain	1905– 1920–	for their discovery of the antiproton
1960	Donald Arthur Glaser	1926–	for the invention of the bubble chamber
1961	Robert Hofstadter	1915–	for his pioneering studies of electron scattering in atomic nuclei and for his thereby achieved discoveries concerning the structure of the nucleons
	Rudolf Ludwig Mössbauer	1929–	for his researches concerning the resonances absorption of γ-radiation and his discovery in this connection of the effect which bears his name
1962	Lev Davidovič Landau	1908–1968	for his pioneering theories of condensed matter, especially liquid helium
1963	Eugene P. Wigner	1902–	for his contributions to the theory of the atomic nucleus and the elementary particles, particularly through the discovery and application of fundamental symmetry principles
	Maria Goeppert Mayer J. Hans D. Jensen	1906–1972 1907–1973	for their discoveries concerning nuclear shell structure

1964	Charles H. Townes Nikolai G. Basov Alexander M. Prochorov	1915– 1922– 1916–	for fundamental work in the field of quantum electronics which has led to the construction of oscillators and amplifiers based on the maser-laser principle
1965	Sin-Itiro Tomonaga Julian Schwinger Richard P. Feynman	1906–1979 1918– 1918–	for their fundamental work in quantum electrodynamics, with deep-ploughing consequences for the physics of elementary particles
1966	Alfred Kastler	1902–1984	for the discovery and development of optical methods for studying Hertzian resonance in atoms
1967	Hans Albrecht Bethe	1906–	for his contributions to the theory of nuclear reactions, especially his discoveries concerning the energy production in stars
1968	Luis W. Alvarez	1911–	for his decisive contribution to elementary particle physics, in particular the discovery of a large number of resonance states, made possible through his development of the technique of using hydrogen bubble chamber and data analysis
1969	Murray Gell-Mann	1929–	for his contributions and discoveries concerning the classification of elementary particles and their interactions
1970	Hannes Alfvén	1908–	for fundamental work and discoveries in magneto-hydro-dynamics with fruitful applications in different parts of plasma physics
	Louis Néel	1904–	for fundamental work and discoveries concerning antiferromagnetism and ferrimagnetism which have led to important applications in solid state physics
1971	Dennis Gabor	1900–	for his discovery of the principles of holography
1972	John Bardeen Leon N. Cooper J. Robert Schrieffer	1908– 1930– 1931–	for their development of a theory of superconductivity
1973	Leo Esaki	1925–	for his discovery of tunneling in semiconductors
	Ivar Giaever	1929–	for his discovery of tunneling in superconductors
	Brian D. Josephson	1940–	for his theoretical prediction of the properties of a super-current through a tunnel barrier
1974	Antony Hewish	1924–	for the discovery of pulsars
	Sir Martin Ryle	1918–1984	for his pioneering work in radioastronomy
1975	Aage Bohr Ben Mottelson James Rainwater	1922– 1926– 1917–1986	for the discovery of the connection between collective motion and particle motion and the development of the theory of the structure of the atomic nucleus based on this connection
1976	Burton Richter Samuel Chao Chung Ting	1931– 1936–	for their (independent) discovery of an important fundamental particle.
1977	Philip Warren Anderson Nevill Francis Mott John Hasbrouck Van Vleck	1923– 1905– 1899–1980	for their fundamental theoretical investigations of the electronic structure of magnetic and disordered solids

1978	Peter L. Kapitza	1894–1984	for his basic inventions and discoveries in the area of low-temperature physics
1979	Arno A. Penzias	1926-	for their discovery of cosmic microwave background radiation
	Robert Woodrow Wilson	1936-	
	Sheldon Lee Glashow	1932-	for their unified model of the action of the weak and electromagnetic forces and for their prediction of the existence of neutral currents for the discovery of violations of fundamental symmetry principles in the decay of neutral K mesons.
	Abdus Salam	1926-	
	Steven Weinberg	1933-	
1980	James W. Cronin	1931-	
	Val L. Fitch	1923-	
1981	Nicolaas Bloembergen	1920-	for their contribution to the development of laser spectroscopy
	Arthur Leonard Schawlow	1921-	
	Kai M. Siegbahn	1918-	for his contribution to the development of high-resolution electron spectroscopy
1982	Kenneth Geddes Wilson	1936-	for his method of analyzing the critical phenomena inherent in the changes of matter under the influence of pressure and temperature.
1983	Subrahmanyan Chandrasekhar	1910–	for his theoretical studies of the structure and evolution of stars
	William A. Fowler	1911–	for his studies of the formation of the chemical elements in the universe
1984	Carlo Rubbia	1934–	for their decisive contributions to the large project, which led to the discovery of the field particles W and Z, communicators of the weak interaction
	Simon van der Meer	1934–	
1985	Klaus von Klitzing	1943–	for his discovery of the quantized Hall resistance
1986	Ernst Ruska	1906–	for his invention of the electron microscope
	Gerd Binnig	1947–	for their invention of the scanning tunneling microscope
	Heinrich Rohrer	1933–	

APPENDIX L
THE GAUSSIAN SYSTEM OF UNITS

Much of the literature of physics is written, and continues to be written, in the Gaussian system of units. In electromagnetism many equations have slightly different forms depending on whether it is intended, as in this book, that mks variables be used or that Gaussian variables be used. Equations in this book can be cast in Gaussian form by replacing the symbols listed below under "SI" by those listed under "Gaussian." For example, Eq. 37-26,

$$\mathbf{B} = \mu_0(\mathbf{H} + \mathbf{M})$$

becomes

$$\frac{\mathbf{B}}{c} = \left(\frac{4\pi}{c^2}\right)\left(\frac{c}{4\pi}\mathbf{H} + c\mathbf{M}\right)$$

or

$$\mathbf{B} = \mathbf{H} + 4\pi\mathbf{M}$$

in Gaussian form. Symbols used in this book that are not listed below remain unchanged. The quantity c is the speed of light.

Quantity	SI	Gaussian
Permittivity constant	ϵ_0	$1/4\pi$
Permeability constant	μ_0	$4\pi/c^2$
Electric displacement	**D**	$\mathbf{D}/4\pi$
Magnetic induction	**B**	$\mathbf{B}/c$
Magnetic flux	Φ_B	Φ_B/c
Magnetic field strength	**H**	$c\mathbf{H}/4\pi$
Magnetization	**M**	$c\mathbf{M}$
Magnetic dipole moment	$\boldsymbol{\mu}$	$c\boldsymbol{\mu}$

In addition to casting the equations in the proper form it is of course necessary to use a consistent set of units in those equations. Below we list some equivalent quantities in SI and Gaussian units. This table can be used to transform units from one system to the other.

Quantity	Symbol	SI	Gaussian system
Length	l	1 m	10^2 cm
Mass	m	1 kg	10^3 g
Time	t	1 s	1 s
Force	**F**	1 N	10^5 dynes
Work or Energy	W, E	1 J	10^7 ergs
Power	P	1 W	10^7 ergs/s
Charge	q	1 C	3×10^9 statcoul
Current	i	1 A	3×10^9 statamp
Electric field strength	**E**	1 V/m	$\frac{1}{3} \times 10^{-4}$ statvolt/cm
Electric potential	V	1 V	$\frac{1}{300}$ statvolt
Electric polarization	**P**	1 C/m^2	3×10^5 statcoul/cm^2
Electric displacement	**D**	1 C/m^2	$12\pi \times 10^5$ statvolt/cm
Resistance	R	1 Ω	$\frac{1}{9} \times 10^{-11}$ s$\cdot$cm^{-1}
Capacitance	C	1 F	9×10^{11} cm
Magnetic flux	Φ_B	1 Wb	10^8 maxwells
Magnetic induction	**B**	1 T	10^4 gauss
Magnetic field strength	**H**	1 A-turn/m	$4\pi \times 10^{-3}$ oersted
Magnetization	**M**	1 Wb/m^2	$1/4\pi \times 10^4$ gauss
Inductance	I	1 H	$\frac{1}{9} \times 10^{-11}$

All factors of 3 in the above table, apart from exponents, should be replaced by 2.99792458 for accurate work; this arises from the numerical value of the speed of light. For example the SI unit of capacitance (= 1 farad) is actually $8.987551787 \times 10^{11}$ cm rather than 9 (= 3^2) $\times 10^{11}$ cm as listed above. This example also shows that not only units but also the dimensions of physical quantities may differ between the two systems.

Consult *Classical Electromagnetism*, second edition, by J. D. Jackson (John Wiley and Sons, 1975) for a fuller treatment of units and dimensions.

index to part two

Polarization charge, 665
Polarized atom, 631
Polarized light, applications of,
  1073
  circularly, 1081, 1084, 1090
    on polarizing sheet,
    1082-1083
  partially, 1088
  plane, 1069, 1070, 1071
    on quarter-wave plate, 1083
  production, 1071
  reflection and, 1074
  scattering and, 1085
  unpolarized, 1071
Polarized material, 661
Polarizing angle, 1074
Polarizing sheet, 1071
  as analyzer, 1072
  law of Malus and, 1072
Polar material, 660
Population inversion, 1220
Positron decay, 1296
Positronium, 1131
Potential gradient, 636
Potentiometer, 704
Power, 688
  in ac circuits, 864
  average, 864, 865
  induction and, 775
Power factor, 866
  of transformer, 874
Poynting, J., 900
Poynting vector, 909-910, 919
  average, 912
  capacitor and, 917
  for plane wave, 911
  resistor and, 916
  resonant cavity and, 910-911
Priestley, J., 608
Principal index of refraction,
  1077
Principle of complementarity,
  1162
Prism, constant deviation, 955
  formation of spectrum by, 940
  minimum deviation of, 943,
    954
  thin, 954
Prism spectrograph, 1053
Probability density, 1155

Project Seafarer, 912
Proton, 571
  magnetic moment of, 830,
    831, 1272
Proton-proton cycle, 1318
Ptolemy, C., 953
p-type semiconductor, 1246
Pulsars, 897
Purcell, E. M., 830

"Q" of circuit, 856, 878
Quadropole moment, 598
Quantization, 571, 1097
Quantum electrodynamics, 1186
Quantum number, 1097
Quantum number principle,
  1211
Quark, 1266
Quarter-wave plate, 1081
  circularly polarized light on,
    1083
  plane polarized light on, 1083
Quartz, index of refraction of,
  939
Quasars, 896

Rad, 1280
Radial probability density, 1155
Radiancy, 1093
Radiation field, from dipole, 904
Radiation pressure, 919
  measurement of, 920
Radioactive decay, 1273
Radiometer effect, 921
Radius of curvature, plane
    mirror, 957
  spherical mirror, 965
  thin lens, 975
Rare earths, 1214
Ray, 957
  paraxial, 963
Rayleigh-Jeans law, 1099
Rayleigh's criterion, 1035, 1036
RC circuit, 705, 798
  charge in, 706
  current, 706
  discharging, 708
  potential in, 708-709
  time constant of, 706

RCL circuit, 847, 856, 858
  with alternating emf, 862-863
  charge in, 877
  current, 863, 865
  power of, 866
  resonance of, 867
  charge oscillations of, 848, 855
  current, 855
  "Q" of, 856
Real image, 960
  refracting surface, 970
  spherical mirror, 963
  thin lens, 975
Rectification, 868
Rectifier, 868
  full and half-wave, 870
Red giant, 1320
Red shift, 937
Reference position of potential,
  623
Reflection, diffuse, 941
  Fermat's principle and, 950
  Huygen's principle and, 944
  intensity of, 949-950
  law of, 939
  phase change in, 1006-1007
  polarization and, 1074
  specular, 941
  test of law, 940
Reflection grating, 1051
Refracting surface, image
    distance, 970
  plane, 971-972
  real image in, 970
  R and V-sides, 970
  spherical, 969
  virtual image in, 971
Refraction, double, 1077
  Fermat's principle and, 951
  Huygen's principle and, 944
  index of, 939
  intensity in, 949-950
  law of, 939
  polarization and, 1074
  spectrum by, 939-940
Rem, 1281
Resistance, 678, 684
  equivalent, 699, 703
    parallel, 703
    series, 699

SOME USEFUL NUMBERS

$\sqrt{2} = 1.414$ $\sqrt{3} = 1.732$ $\sqrt{10} = 3.162$ $\pi = 3.142$

$\pi^2 = 9.870$ $\sqrt{\pi} = 1.773$ $\log \pi \doteq 0.4971$ $4\pi = 12.57$

$e = 2.718$ $1/e = 0.3679$ $\log e = 0.4343$ $\ln 2 = 0.6932$

$\sin 30° = \cos 60° = 0.5000$ $\cot 30° = \tan 60° = 1.7321$

$\cos 30° = \sin 60° = 0.8660$ $\sin 45° = \cos 45° = 0.7071$

$\tan 30° = \cot 60° = 0.5774$ $\tan 45° = \cot 45° = 1.0000$

Change of base

$\log x = \ln x/\ln 10 = 0.4343 \ln x$

$\ln x = \log x/\log e = 2.303 \log x$